COMPUTER VISION

COMPUTER VISION

Linda G. Shapiro

Department of Computer Science and Engineering
Department of Electrical Engineering
University of Washington
Seattle, Washington
shapiro@cs.washington.edu

George C. Stockman

Department of Computer Science and Engineering
Michigan State University
East Lansing, Michigan
stockman@cse.msu.edu

PRENTICE HALL, Upper Saddle River, New Jersey 07458

Library of Congress Cataloging-in-Publication Data

Shapiro, Linda G.
Computer Vision / Linda G. Shapiro, George C. Stockman.
p. cm.
Includes bibliographical references and index.
1. Computer Vision. I. Stockman, George C., 1943– II. Title.
ISBN 0-13-030796-3

TA1634.S48 2001
006.3′7—dc21 00-066556

Vice president and editorial director, ECS: *Marcia J. Horton*
Publisher: *Tom Robbins*
Associate editor: *Alice Dworkin*
Editorial assistant: *Jessica Power*
Vice president and director of production and manufacturing, ESM: *David W. Riccardi*
Production editor: *Chanda Wakefield*
Director of creative services: *Paul Belfanti*
Creative director: *Carole Anson*
Executive managing editor: *Vince O'Brien*
Managing editor: *David A. George*
Art director: *Jayne Conte*
Cover designer: *Bruce Kenselaar*
Art editor: *Adam Velthaus*
Manufacturing manager: *Trudy Pisciotti*
Manufacturing buyer: *Dawn Murrin*
Marketing manager: *Holly Stark*
Cover credit: *Cover photo taken on safari in Kenya by Asya Ollis, who discovered that the world is only black-and-white when we make it so.*

Printed in the United States of America
10 9 8 7 6 5 4 3

ISBN 0-13-030796-3

Prentice-Hall International (UK) Limited, *London*
Prentice-Hall of Australia Pty. Limited, *Sydney*
Prentice-Hall Canada Inc., *Toronto*
Prentice-Hall Hispanoamericana, S.A., *Mexico*
Prentice-Hall of India Private Limited, *New Delhi*
Prentice-Hall of Japan, Inc., *Tokyo*
Pearson Education Asia Pte. Ltd., *Singapore*
Editora Prentice-Hall do Brasil, Ltda., *Rio de Janeiro*

Contents

PREFACE **xvii**

1 INTRODUCTION **1**

1.1 Machines that See? 2

1.2 Application Problems 3

1.2.1 A Preview of the Digital Image, 3
1.2.2 Image Database Query, 3
1.2.3 Inspecting Crossbars for Holes, 4
1.2.4 Examining the Inside of a Human Head, 6
1.2.5 Processing Scanned Text Pages, 8
1.2.6 Accounting for Snow Cover Using a Satellite Image, 8
1.2.7 Understanding a Scene of Parts, 9

1.3 Operations on Images 10

1.3.1 Changing Pixels in Small Neighborhoods, 10
1.3.2 Enhancing an Entire Image, 11
1.3.3 Combining Multiple Images, 12
1.3.4 Computing Features from an Image, 13
1.3.5 Extracting Non-iconic Representations, 14

1.4 The Good, the Bad, and the Ugly 14

1.5 Use of Computers and Software 15

1.6 Related Areas 15

1.7 The Rest of the Book 16

1.8 References 17

1.9 Additional Exercises 18

2 IMAGING AND IMAGE REPRESENTATION 21

2.1 Sensing Light 21

2.2 Imaging Devices 22
- *2.2.1 CCD Cameras, 22*
- *2.2.2 Image Formation, 24*
- *2.2.3 Video Cameras, 26*
- *2.2.4 The Human Eye, 26*

2.3 Problems in Digital Images* 27
- *2.3.1 Geometric Distortion, 27*
- *2.3.2 Scattering, 27*
- *2.3.3 Blooming, 28*
- *2.3.4 CCD Variations, 28*
- *2.3.5 Clipping or Wrap-Around, 28*
- *2.3.6 Chromatic Distortion, 29*
- *2.3.7 Quantization Effects, 29*

2.4 Picture Functions and Digital Images 29
- *2.4.1 Types of Images, 29*
- *2.4.2 Image Quantization and Spatial Measurement, 31*

2.5 Digital Image Formats* 35
- *2.5.1 Image File Header, 36*
- *2.5.2 Image Data, 36*
- *2.5.3 Data Compression, 36*
- *2.5.4 Commonly Used Formats, 36*
- *2.5.5 Run-Coded Binary Images, 37*
- *2.5.6 PGM: Portable Gray Map, 37*
- *2.5.7 GIF Image File Format, 38*
- *2.5.8 TIFF Image File Format, 38*
- *2.5.9 JPEG Format for Still Photos, 38*
- *2.5.10 PostScript, 39*
- *2.5.11 MPEG Format for Video, 39*
- *2.5.12 Comparison of Formats, 40*

2.6 Richness and Problems of Real Imagery 41

2.7 3D Structure from 2D Images 42

2.8 Five Frames of Reference 42
- *2.8.1 Pixel Coordinate Frame* **I**, *43*
- *2.8.2 Object Coordinate Frame* **O**, *44*

2.8.3 *Camera Coordinate Frame* **C**, *44*
2.8.4 *Real Image Coordinate Frame* **F**, *44*
2.8.5 *World Coordinate Frame* **W**, *44*

2.9 Other Types of Sensors* 45
2.9.1 *Microdensitometer*, 45*
2.9.2 *Color and Multispectral Images*, 45*
2.9.3 *X-ray*, 46*
2.9.4 *Magnetic Resonance Imaging (MRI)*, 47*
2.9.5 *Range Scanners and Range Images*, 47*

2.10 References 49

3 BINARY IMAGE ANALYSIS **51**

3.1 Pixels and Neighborhoods 51

3.2 Applying Masks to Images 53

3.3 Counting the Objects in an Image 54

3.4 Connected Components Labeling 56

3.5 Binary Image Morphology 63
3.5.1 *Structuring Elements, 63*
3.5.2 *Basic Operations, 65*
3.5.3 *Some Applications of Binary Morphology, 68*
3.5.4 *Conditional Dilation, 71*

3.6 Region Properties 73

3.7 Region Adjacency Graphs 81

3.8 Thresholding Gray-Scale Images 83
3.8.1 *The Use of Histograms for Threshold Selection, 83*
3.8.2 *Automatic Thresholding: The Otsu Method*, 85*

3.9 References 89

4 PATTERN RECOGNITION CONCEPTS **92**

4.1 Pattern Recognition Problems 92

4.2 Common Model for Classification 94
4.2.1 *Classes, 94*
4.2.2 *Sensor/Transducer, 94*
4.2.3 *Feature Extractor, 94*
4.2.4 *Classifier, 95*
4.2.5 *Building the Classification System, 95*
4.2.6 *Evaluation of System Error, 96*
4.2.7 *False Alarms and False Dismissals, 96*

4.3 Precision Versus Recall 97

4.4 Features Used for Representation 98

4.5 Feature Vector Representation 100

4.6 Implementing the Classifier 101
4.6.1 Classification Using the Nearest Class Mean, 101
4.6.2 Classification Using the Nearest Neighbors, 103

4.7 Structural Techniques 104

4.8 The Confusion Matrix 106

4.9 Decision Trees 107

4.10 Bayesian Decision-Making 114
4.10.1 Parametric Models for Distributions, 116

4.11 Decisions Using Multidimensional Data 117

4.12 Machines that Learn 119

4.13 Artificial Neural Nets* 119
4.13.1 The Perceptron Model, 120
4.13.2 The Multilayer Feedforward Network, 123

4.14 References 126

5 FILTERING AND ENHANCING IMAGES **128**

5.1 What Needs Fixing? 129
5.1.1 An Image Needs Improvement, 129
5.1.2 Low-Level Features Must Be Detected, 129

5.2 Gray-Level Mapping 130
5.2.1 Histogram Equalization, 132

5.3 Removal of Small Image Regions 134
5.3.1 Removal of Salt-and-Pepper Noise, 134
5.3.2 Removal of Small Components, 135

5.4 Image Smoothing 135

5.5 Median Filtering 137
5.5.1 Computing an Output Image from an Input Image, 139

5.6 Detecting Edges Using Differencing Masks 141
5.6.1 Differencing 1D Signals, 141
5.6.2 Difference Operators for 2D Images, 144

5.7 Gaussian Filtering and LOG Edge Detection 149
5.7.1 Detecting Edges with the LOG Filter, 151

5.7.2 On Human Edge Detection, 153
5.7.3 Marr-Hildreth Theory, 155

5.8 The Canny Edge Detector 157

5.9 Masks as Matched Filters* 158
5.9.1 The Vector Space of All Signals of n Samples, 158
5.9.2 Using an Orthogonal Basis, 160
5.9.3 Cauchy-Schwartz Inequality, 162
5.9.4 The Vector Space of $m \times n$ Images, 162
5.9.5 A Roberts Basis for 2×2 Neighborhoods, 162
5.9.6 The Frei-Chen Basis for 3×3 Neighborhoods, 163

5.10 Convolution and Cross Correlation* 167
5.10.1 Defining Operations via Masks, 167
5.10.2 The Convolution Operation, 169
5.10.3 Possible Parallel Implementations, 172

5.11 Analysis of Spatial Frequency using Sinusoids* 172
5.11.1 A Fourier Basis, 174
5.11.2 2D Picture Functions, 175
5.11.3 Discrete Fourier Transform, 179
5.11.4 Bandpass Filtering, 181
5.11.5 Discussion of the Fourier Transform, 181
5.11.6 The Convolution Theorem, 182*

5.12 Summary and Discussion 184

5.13 References 185

6 COLOR AND SHADING 187

6.1 Some Physics of Color 188
6.1.1 Sensing Illuminated Objects, 189
6.1.2 Additional Factors, 190
6.1.3 Sensitivity of Receptors, 190

6.2 The RGB Basis for Color 191

6.3 Other Color Bases 193
6.3.1 The CMY Subtractive Color System, 193
6.3.2 HSI: Hue-Saturation-Intensity, 194
6.3.3 YIQ and YUV for TV Signals, 197
6.3.4 Using Color for Classification, 198

6.4 Color Histograms 199

6.5 Color Segmentation 201

6.6 Shading 203
6.6.1 Radiation from One Light Source, 203

6.6.2 *Diffuse Reflection, 204*
6.6.3 *Specular Reflection, 205*
6.6.4 *Darkening with Distance, 206*
6.6.5 *Complications, 207*
6.6.6 *Phong Model of Shading*, 208*
6.6.7 *Human Perception Using Shading, 208*

6.7 Related Topics* 209
6.7.1 *Applications, 209*
6.7.2 *Human Color Perception, 209*
6.7.3 *Multispectral Images, 210*
6.7.4 *Thematic Images, 210*

6.8 References 210

7 TEXTURE **212**

7.1 Texture, Texels, and Statistics 213

7.2 Texel-Based Texture Descriptions 214

7.3 Quantitative Texture Measures 215
7.3.1 *Edge Density and Direction, 215*
7.3.2 *Local Binary Partition, 217*
7.3.3 *Co-occurrence Matrices and Features, 217*
7.3.4 *Laws Texture Energy Measures, 220*
7.3.5 *Autocorrelation and Power Spectrum, 221*

7.4 Texture Segmentation 223

7.5 References 224

8 CONTENT-BASED IMAGE RETRIEVAL **226**

8.1 Image Database Examples 226

8.2 Image Database Queries 228

8.3 Query-by-Example 229

8.4 Image Distance Measures 230
8.4.1 *Color Similarity Measures, 231*
8.4.2 *Texture Similarity Measures, 233*
8.4.3 *Shape Similarity Measures, 235*
8.4.4 *Object Presence and Relational Similarity Measures, 240*

8.5 Database Organization 244
8.5.1 *Standard Indexes, 244*
8.5.2 *Spatial Indexing, 247*

8.5.3 Indexing for Content-Based Image Retrieval with Multiple Distance Measures, 248

8.6 References 248

9 MOTION FROM 2D IMAGE SEQUENCES 251

9.1 Motion Phenomena and Applications 251

9.2 Image Subtraction 253

9.3 Computing Motion Vectors 254
9.3.1 The Decathlete Game, 255
9.3.2 Using Point Correspondences, 256
9.3.3 MPEG Compression of Video, 261
9.3.4 Computing Image Flow, 262*
9.3.5 The Image Flow Equation, 263*
9.3.6 Solving for Image Flow by Propagating Constraints, 264*

9.4 Computing the Paths of Moving Points 265
9.4.1 Integrated Problem-Specific Tracking, 271

9.5 Detecting Significant Changes in Video 272
9.5.1 Segmenting Video Sequences, 273
9.5.2 Ignoring Certain Camera Effects, 274
9.5.3 Storing Video Subsequences, 277

9.6 References 277

10 IMAGE SEGMENTATION 279

10.1 Identifying Regions 280
10.1.1 Clustering Methods, 281
10.1.2 Region Growing, 289

10.2 Representing Regions 291
10.2.1 Overlays, 292
10.2.2 Labeled Images, 292
10.2.3 Boundary Coding, 292
10.2.4 Quadtrees, 294
10.2.5 Property Tables, 294

10.3 Identifying Contours 295
10.3.1 Tracking Existing Region Boundaries, 295
10.3.2 The Canny Edge Detector and Linker, 297
10.3.3 Aggregating Consistent Neighboring Edgels into Curves, 301
10.3.4 Hough Transform for Lines and Circular Arcs, 303

10.4 Fitting Models to Segments 312

10.5 Identifying Higher-level Structure 317
10.5.1 Ribbons, 317
10.5.2 Detecting Corners, 320

10.6 Segmentation Using Motion Coherence 321
10.6.1 Boundaries in Space-Time, 321
10.6.2 Aggregrating Motion Trajectories, 321

10.7 References 324

11 MATCHING IN 2D **326**

11.1 Registration of 2D Data 326

11.2 Representation of Points 328

11.3 Affine Mapping Functions 329

11.4 A Best 2D Affine Transformation* 339

11.5 2D Object Recognition via Affine Mapping 341

11.6 2D Object Recognition via Relational Matching 350

11.7 Nonlinear Warping 364

11.8 Summary 368

11.9 References 368

12 PERCEIVING 3D FROM 2D IMAGES **371**

12.1 Intrinsic Images 371

12.2 Labeling of Line Drawings from Blocks World 377

12.3 3D Cues Available in 2D Images 383

12.4 Other Phenomena 388
12.4.1 Shape from X, 388
12.4.2 Vanishing Points, 392
12.4.3 Depth from Focus, 393
12.4.4 Motion Phenomena, 393
12.4.5 Boundaries and Virtual Lines, 393
12.4.6 Alignments are Non-accidental, 394

12.5 The Perspective Imaging Model 395

12.6 Depth Perception from Stereo 397
12.6.1 Establishing Correspondences, 400

12.7 The Thin Lens Equation* 403

12.8 Concluding Discussion 406

12.9 References 407

13 3D SENSING AND OBJECT POSE COMPUTATION 410

13.1 General Stereo Configuration 411

13.2 3D Affine Transformations 413
13.2.1 Coordinate Frames, 413
13.2.2 Translation, 415
13.2.3 Scaling, 415
13.2.4 Rotation, 415
13.2.5 Arbitrary Rotation, 418
13.2.6 Alignment via Transformation Calculus, 419

13.3 Camera Model 422
13.3.1 Perspective Transformation Matrix, 423
13.3.2 Orthographic and Weak Perspective Projections, 426
13.3.3 Computing 3D Points Using Multiple Cameras, 428

13.4 Best Affine Calibration Matrix 431
13.4.1 Calibration Jig, 431
13.4.2 Defining the Least-Squares Problem, 431
13.4.3 Discussion of the Affine Method, 436

13.5 Using Structured Light 437

13.6 A Simple Pose Estimation Procedure 439

13.7 An Improved Camera Calibration Method* 444
13.7.1 Intrinsic Camera Parameters, 445
13.7.2 Extrinsic Camera Parameters, 445
13.7.3 Calibration Example, 449

13.8 Pose Estimation* 453
13.8.1 Pose from 2D-3D Point Correspondences, 455
13.8.2 Constrained Linear Optimization, 456
13.8.3 Computing the Transformation $\mathbf{Tr} = \{\mathbf{R}, \mathbf{T}\}$*, 458*
13.8.4 Verification and Optimization of Pose, 460

13.9 3D Object Reconstruction 460
13.9.1 Data Acquisition, 461
13.9.2 Registration of Views, 463
13.9.3 Surface Reconstruction, 464
13.9.4 Space-Carving, 464

13.10 Computing Shape from Shading 468
13.10.1 Photometric Stereo, 471
13.10.2 Integrating Spatial Constraints, 472

13.11 Structure from Motion 472

13.12 References 475

14 3D MODELS AND MATCHING **479**

14.1 Survey of Common Representation Methods 480
14.1.1 3D Mesh Models, 480
14.1.2 Surface-Edge-Vertex Models, 480
14.1.3 Generalized-Cylinder Models, 483
14.1.4 Octrees, 484
14.1.5 Superquadrics, 486

14.2 True 3D Models versus View-Class Models 488

14.3 Physics-Based and Deformable Models 489
14.3.1 Snakes: Active Contour Models, 489
14.3.2 Balloon Models for 3D, 493
14.3.3 Modeling Motion of the Human Heart, 494

14.4 3D Object Recognition Paradigms 495
14.4.1 Matching Geometric Models via Alignment, 496
14.4.2 Matching Relational Models, 504
14.4.3 Matching Functional Models, 513
14.4.4 Recognition by Appearance, 516

14.5 References 523

15 VIRTUAL REALITY **527**

15.1 Features of Virtual Reality Systems 528

15.2 Applications of VR 529

15.3 Augmented Reality (AR) 530

15.4 Teleoperation 533

15.5 Virtual Reality Devices 535

15.6 Summary of Sensing Devices for VR 539

15.7 Rendering Simple 3D Models 540

15.8 Composing Real and Synthetic Imagery 542

15.9 HCI and Psychological Issues 546

15.10 References 546

16 CASE STUDIES 548

16.1 Veggie Vision: A System for Checking Out Vegetables 548
16.1.1 Application Domain and Requirements, 549
16.1.2 System Design, 550
16.1.3 Identification Procedure, 551
16.1.4 More Details on the Process, 551
16.1.5 Performance, 554

16.2 Identifying Humans via the Iris of an Eye 554
16.2.1 Requirements for Identification Systems, 555
16.2.2 System Design, 557
16.2.3 Performance, 560

16.3 References 561

Preface

This book is intended as an introduction to computer vision for a broad audience. It provides necessary theory and examples for students and practitioners who will work in fields where significant information must be extracted automatically from images. The book should be a useful resource for professionals, a text for both undergraduate and beginning graduate courses, and a resource for enrichment of college or even high school projects. Our goals were to provide a basic set of fundamental concepts and algorithms and also discuss some of the exciting evolving application areas. This book is unique in that it contains chapters on image databases (Chapter 8) and on virtual and augmented reality (Chapter 15), two exciting evolving application areas. A final chapter (Chapter 16) gives a complete view of real-world systems that use computer vision.

Due to recent progress in the computer field, economical and flexible use of computer images is now pervasive. Computing with images is no longer just for the realm of the sciences, but also for the arts and social sciences and even for hobbyists. The book should serve an established and growing audience including those interested in multimedia; art and design; geographic information systems; and image databases, in addition to the traditional areas of automation, image science, medical imaging, remote sensing, and computer cartography.

A broad purpose at first seems impossible to achieve. However, there are other kinds of texts that already do this in other areas—calculus, physics, and general computing. We hope we have made at least a good beginning—we wanted a book that would be useful in the classroom and also to the independent reader. We find the chosen topics interesting and sometimes exciting, and hope that they are accessible to a large audience. It is assumed that use of the text in a graduate, or even senior level, computer vision course would be supplemented by papers from the archival literature. Coverage is not intended to be comprehensive; only a modest set of papers are cited at the end of each chapter.

The early chapters begin at an intuitive level and progress towards mathematical models with the goal of intuitive understanding before formal characterization. Sections marked by an asterisk (*) are more mathematical or more advanced and need not be covered in a less technical course. To strengthen the intuitive approach, we have stayed with the processing of iconic imagery for the first eleven chapters and have delayed 3D computer vision until the later chapters, but it should be easy for experienced instructors to resequence them to fit a particular course or teaching style. There are many viable applications that are entirely 2D, and many concepts and algorithms are more simply taught in their 2D form. We provide some basics of pattern recognition in Chapter 4, so that students can consider complete recognition systems before the full coverage of image features and matching. A reader should have a good idea of 2D image processing applications after Chapter 4; Chapters 5, 6, and 7 add in gray-tone, color, and texture features. Chapter 8 treats image databases, a popular recent topic. Although some colleagues advised us to place this material near the end of the book, our goal of positioning it early in the chapter sequence is to reinforce the concepts of the prior chapters and to provide material that can lead to an excellent half-term project. Segmentation and matching are treated in their 2D forms in Chapters 10 and 11, so that the basic concepts are presented in a simple form, without introducing the complexities of 3D transformations.

Characteristics of the 3D world are briefly introduced in Chapter 2 and then are studied in much more detail in Chapter 12. Chapter 12 surveys qualitatively many aspects of how a 3D world can be perceived from 2D images: It concludes with quantitative models of stereo and study of the thin lens equation for depth-from-focus and resolving power. The transition to 3D computer vision is made in Chapter 13: The authors have found from their own teaching that the difficulty increases abruptly for students at this point. The use of matrices to model homogeneous transformations are included within the chapter rather than in appendices; the 3D versions are extensions of the simpler 2D versions given in Chapter 11. Least-squares fitting, introduced in a simple 2D context in Chapter 11, is also extended in Chapter 13. Non-linear optimization is introduced in a simple P3P context and then used for camera calibration including the modeling of radial distortion in a lens. Chapter 14 treats 3D models and the matching of models to 3D sensed data: it is of mixed difficulty. Chapter 15 discusses applications in virtual and augmented (mixed) reality and the role of computer vision techniques.

Programming Language Issue

The book does not rely on any programming language, but uses a generic algorithmic notation. Commitment to a particular language is unnecessary and would be the wrong language for many readers. Students who are programmers should have little trouble implementing the algorithms, as our own students have shown. Examples will eventually be provided on the World Wide Web when appropriate and available, primarily so students can quickly experiment, secondarily so that they can study some sample code.

Several tools and libraries are available to instructors and students; for example, Khoros, NIH-Image, XView, gimp, MATLAB, etc. There are also packages that can be purchased from companies that make machine vision hardware. The authors have decided not to base the text on any specific software because, first, most readers would be using something else, and second, it would be counterproductive to bury the essence of the image operations within the complex framework of data structures and methods needed in an industrial strength system. Having first studied principles in an environment with few variables, the reader will then be better able to successfully choose and use an industrial system.

Ways to Use the Text

The book material can be selected, and sometimes sequenced, in different ways according to the goal of the course and interests of the instructor and students.

- Chapter 3, with brief summary of Chapter 2
  A minimum usage would be 1–3 lectures in a data structures and algorithms course. Chapter 3, with some background from Chapter 2 contains motivational applications and programming exercises on 2D arrays, depth-first search, and the union-find data structure for sets.
- Chapters 1, 2, and 3, and optionally some of Chapters 4, 5, and 6
  This could serve as an enrichment unit of 1 to 3 weeks for high school or lower division undergrads. The objective could be as simple as a term paper or as complex as group work on a program to, say, create a 2D parts recognition system based on connected components and prototype matching of feature vectors.
- Much of Chapters 1–11
  This would be a survey of 2D material for an elective course for students in geography, natural resources or microbiology, for example, provided that many of the optional sections are passed over. If most sections of Chapters 1–11 are covered, this would constitute a semester undergraduate course in image processing and analysis with an introduction to computer vision.
- Most of the text
  This would constitute a semester course in computer vision for the senior or first year graduate student level. There is more material in the book than can be covered well in one semester. Some sections will have to be ignored or surveyed and the reader should not be expected to be able to work homework problems in all sections. For the quarter system, Chapters 1–4, 6–12, and 14 make a good introduction to computer vision for undergraduates. For a one quarter graduate course, Chapters 1–4 can be minimally covered with the emphasis on Chapters 6–14 and a brief coverage of Chapter 15. For any graduate level course, it is expected that some papers from the current literature would also be covered.

We are grateful to our many colleagues, teachers, and students with whom we have shared our interests. They have contributed much to our growing field and shared their work and excitement. Many have generously supported this book with encouragement and with contributions of ideas, figures, and algorithms. Specific citations are given throughout the book. With regret we have left out some important contributions—a text can only be so large. The several reviewers and many colleagues who have given us feedback have significantly improved our work. In particular, for careful editing, we are indebted to Mohammad Ghavamzadeh, Nick Dutta, Kevin Bowyer, Adam Clark, Yu-Yu Chou, Habib Abi-Rached, and Valentin Razmov. We take responsibility for any errors remaining in the book and for providing corrections in the future.

This book was four years in the making. We are indebted to Paul Becker of Addison Wesley-Longman for much guidance in getting the project going and to Tom Robbins of Prentice Hall for finishing it off. We thank Cathy Davison and Lorraine Evans for their persistence in helping to resolve the many cases where permissions needed to be tracked down. We are grateful to Rose Rummel-Eury and Chanda Wakefield of ICC for meticulous editing of our notation and English, and for pushing the schedule. Creating the book was not light work and it certainly helped to have a team with both skill and humor.

Linda Shapiro
shapiro@cs.washington.edu

George Stockman
stockman@cse.msu.edu

COMPUTER VISION

1

Introduction

This book is an introduction to the broad field of computer vision. Without a doubt, machines can be built to see; for example, machines inspect millions of light bulb filaments and miles of fabric each day. Automatic teller machines (ATMs) have been built to scan the human eye for user identification and cars have been driven by a computer using camera input. This chapter introduces several important problem areas where computer vision provides solutions. After reading this chapter, you should have a broad view of some problems and methods of computer vision.[1]

1 Definition. The goal of **computer vision** is to make useful decisions about real physical objects and scenes based on sensed images.

In order to make decisions about real objects, it is almost always necessary to construct some description or model of them from the image. Because of this, many experts will say that *the goal of computer vision is the construction of scene descriptions from images.* Although our study of computer vision is problem-oriented, fundamental issues will be addressed. Critical issues raised in this chapter and studied in the remainder of the text include:

Sensing: How do sensors obtain images of the world? How do the images encode properties of the world, such as material, shape, illumination, and spatial relationships?

[1]In this book, we generally use the terms *machine vision* and *computer vision* to mean the same thing. However, we often use the term machine vision in the context of industrial applications and the term computer vision with the field in general.

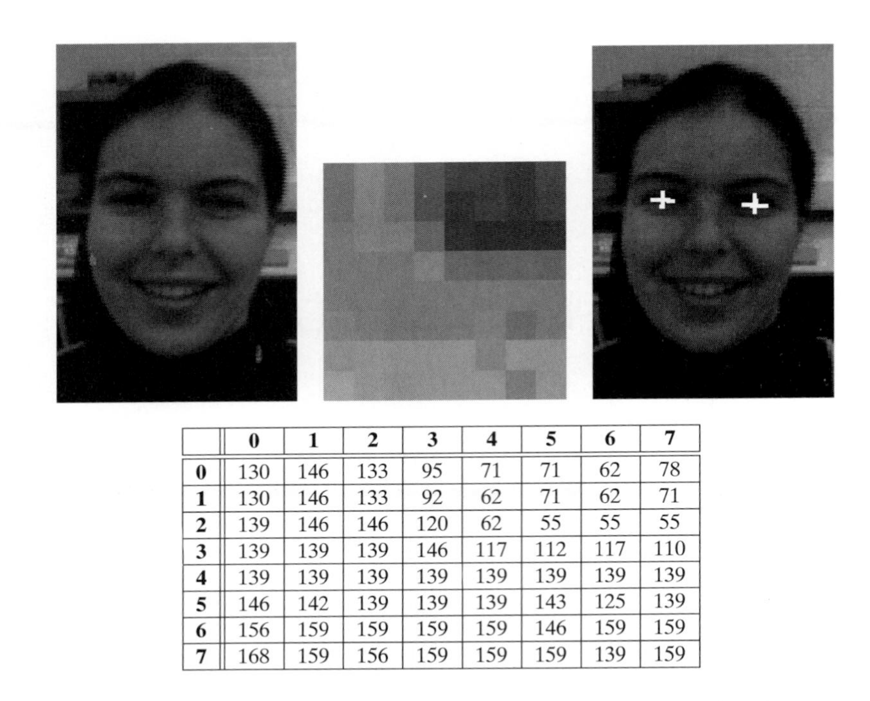

	0	1	2	3	4	5	6	7
0	130	146	133	95	71	71	62	78
1	130	146	133	92	62	71	62	71
2	139	146	146	120	62	55	55	55
3	139	139	139	146	117	112	117	110
4	139	139	139	139	139	139	139	139
5	146	142	139	139	139	143	125	139
6	156	159	159	159	159	146	159	159
7	168	159	156	159	159	159	139	159

Figure 1.1 (top left) Image of a face; (top center) subimage of 8 × 8 pixels from the right eye region; (top right) eye location detected by a computer program; and (bottom) intensity values from the 8 × 8 subimage. (Images courtesy of Vera Bakic.)

Encoded Information: How do images yield information for understanding the 3D world, including the geometry, texture, motion, and identity of objects in it?

Representations: What representations should be used for stored descriptions of objects, their parts, properties, and relationships?

Algorithms: What methods are there to process image information and construct descriptions of the world and its objects?

These issues and others will be studied in the following chapters. We now introduce various applications and some important issues that arise in their context.

1.1 MACHINES THAT SEE?

Scientists and science fiction writers have been fascinated by the possibility of building intelligent machines, and the capability of understanding the visual world is a prerequisite that some would require of such a machine. Much of the human brain is dedicated to vision. Humans solve many visual problems effortlessly, yet most have little analytical understanding of visual cognition as a process. Alan Turing, one of the fathers of both the modern digital computer and the field of artificial intelligence, believed that a digital computer would achieve intelligence and the ability to understand scenes. Such lofty goals have proved difficult to achieve and the richness of human imagination is not yet matched by our engineering. However, there has been surprising progress along some lines of research. While building practical systems is a primary theme of this text and artificial intelligence is not, we will sometimes ponder the deeper questions, and, where we can, make some assessment of progress. Consider, for example, the following scenario, which could be realized within the next few years. A TV camera at your door provides images to your home computer which you have trained to recognize some faces of people important to you. When you call in to your home message center, your computer not only reports the phone messages, but it also reports probable visits from your sister, Eleanor and Chad, the paper boy. We will discuss such current research ideas at various places in the book.

1.2 APPLICATION PROBLEMS

The applications of computers in image analysis are virtually limitless. Only a small sample of applications can be included here, but these will serve us well for both motivation and orientation to the field of study.

1.2.1 A Preview of the Digital Image

A digital image might represent a cartoon, a page of text, a person's face, a map of Katmandu, or a product for purchase from a catalog. A digital image contains a fixed number of rows and columns of *pixels,* short for *picture elements*. Pixels are like little tiles holding quantized values—small numbers, often between 0 and 255, that represent the brightness at the points of the image. Depending on the coding scheme, 0 could be the darkest and 255 the brightest, or vice-versa. At the top left in Figure 1.1 is a printed digital image of a face that is 257 rows high by 172 columns wide. At the top center is an 8×8 subimage extracted from the right eye of the left image. At the bottom of the figure are the 64 numbers representing the brightness of the pixels in that subimage. The numbers below 100 in the upper right of the subimage represent the lower reflection from the dark of the eye (iris), while the higher numbers represent the brighter white of the eye. A color image would have three numbers for each pixel, perhaps one value for red, one for blue, and one for green. Digital images are most commonly displayed on a monitor, which is basically a television screen with a digital image memory. A color image that has 500 rows and 500 columns is roughly equivalent to what you see at one instant of time on your TV. A pixel is displayed by energizing a small spot of luminescent material; displaying color requires energizing three neighboring spots of different materials. A high resolution computer display has roughly $1{,}200 \times 1{,}000$ pixels. Chapter 2 discusses digital images in more detail, while coding and interpretation of color in digital images is treated in Chapter 6.

1.2.2 Image Database Query

Huge digital memories, high bandwidth transmission, and multimedia personal computers have facilitated the development of image databases. Good use of the many existing images requires good retrieval methods. Standard database techniques apply to images that have

Figure 1.2 Image query by example: (left) query image; and two most similar images produced by an image database system. (Courtesy of Graphic-sha, Tokyo.)

been augmented with text keys; however, *content-based* retrieval is needed and is a topic of much current research. Suppose that a newly formed company wants to design and protect a new logo and that an artist has created several candidates for the company to consider. A logo cannot be used if it is too similar to one of an existing company, so a database of existing logos must be searched. This operation is analagous to patent search and is done by humans, but could be greatly aided by machine vision methods (see Figure 1.2). There are many similar problems. Suppose an architect or an art historian wants to search for buildings with a particular kind of entryway. It would be desirable to just provide a picture, perhaps fetched from the database itself, and request the system to produce other similar pictures. In Chapter 8, you will see how geometric, color, and texture features can be used to aid in answering such an image database query. Suppose that an advertising agency wants to search for existing images of young children enjoying eating. This semantic requirement, which is simple for humans to understand, presents a very high level of difficulty for machine vision. Characterizing *children, enjoyment,* and *eating* would require complex use of color, texture, and geometric features. We note in passing that a computer algorithm has been devised that decides whether or not a color image contains a naked person. This could be useful for parents who want to screen images that their children retrieve from the web. Image database retrieval methods are treated in Chapter 8.

1.2.3 Inspecting Crossbars for Holes

In the late 1970s an engineer in Milwaukee implemented a machine vision system that successfully counted the number of bolt holes in crossbars made for truck companies. The truck companies demanded that every crossbar be inspected before being shipped to them, because a missing bolt hole on a partly assembled truck was a very costly defect. Either the assembly line would have to be stopped while the needed hole was drilled, or worse, a worker might ignore placing a required bolt in order to keep the production line running. To create a digital image of the truck crossbar, lights were placed beneath the existing transfer line and a digital camera above it. When a crossbar came into the field of view, an image was taken. Dark pixels inside the shadow of the crossbar were represented as 1s indicating steel, and pixels in the bright holes were represented as 0s, indicating that the hole was drilled. The number of holes can be computed as the number of *external corners* minus the number of *internal corners* all divided by four. Figure 1.3 shows three bright holes (0s) in a background of 1s. An *external corner* is just a 2×2 set of neighboring pixels containing exactly 3 ones, while an *internal corner* is a 2×2 set of neighboring pixels containing exactly 3 zeroes. Example processing of an image with 7 rows and 16 columns is shown

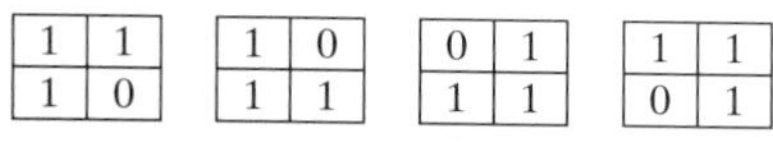

(*a*) 2 × 2 *external corner* patterns

0	0
0	1

0	1
0	0

1	0
0	0

0	0
1	0

(*b*) 2 × 2 *internal corner* patterns

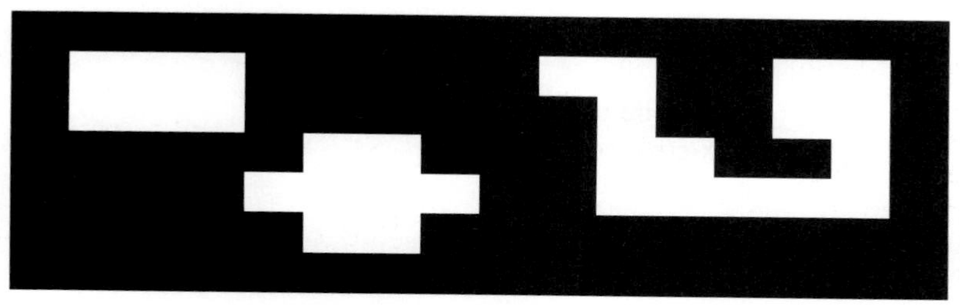

(*c*) Three bright holes in dark background

	0	1	2	3	4	5	6	7	8	9	0	1	2	3	4	5	e	i
0	1	1	1	1	1	1	1	1	1	1	1	1	1	1	1	1		
1	1	0	0	0	1	1	1	1	1	0	0	1	1	0	0	1		
2	1	0	0	0	1	1	1	1	1	1	0	1	1	0	0	1		
3	1	1	1	1	1	0	0	1	1	1	0	0	1	1	0	1		
4	1	1	1	1	0	0	0	0	1	1	0	0	0	0	0	1		
5	1	1	1	1	1	0	0	1	1	1	1	1	1	1	1	1		
6	1	1	1	1	1	1	1	1	1	1	1	1	1	1	1	1		

(*d*) Binary input image 7 rows high and 16 columns wide

	0	1	2	3	4	5	6	7	8	9	0	1	2	3	4	5	e	i
0	e			e					e		e		e		e		6	0
1									e	i							1	1
2	e			e	e		e				i	e	e	i			6	2
3				e	i		i	e				i		i			2	4
4				e	i		i	e		e					e		4	2
5					e		e										2	0
6																	0	0

(*e*) External corner patterns marked with e; internal corners marked i

Figure 1.3 Counting the number of holes in a binary image: 21 external corner patterns (e) minus 9 internal corner patterns (i) divided by 4 yields a count of 3 holes. Why?

Input a binary image and output the number of holes it contains.

M is a binary image of **R** rows of **C** columns.
1 represents material through which light has not passed;
0 represents absence of material indicated by light passing.
Each region of 0s must be 4-connected and all image border pixels must be 1s.
E is the count of *external corners* (3 ones and 1 zero)
I is the count of *internal corners* (3 zeros and 1 one)

```
integer procedure Count_Holes(M)
{
    examine entire image, 2 rows at a time;
    count external corners E;
    count internal corners I;
    return(number_of_holes = (E - I)/4);
}
```

Algorithm 1.1 Skeleton of algorithm for counting holes in a binary image.

in the figure and a skeleton algorithm is also shown. Holecounting is only one example of many simple, but powerful operations possible with digital images. (As the exercises below show, the holecounting algorithm is correct only if the holes are *4-connected* and *simply connected*—that is, they have no background pixels inside them. These concepts are discussed further in Chapter 3 and in more detail in the text by Rosenfeld.)

Exercise 1.1: Holecounting.

Consider the following three images, which are 4×5, 4×4, and 4×5 respectively.

1	1	1	1	1
1	0	1	0	1
1	0	1	0	1
1	1	1	1	1

1	1	1	1
1	1	0	1
1	0	1	1
1	1	1	1

1	1	1	1	1
1	0	1	0	1
1	0	0	0	1
1	1	1	1	1

In scanning for corner patterns, 12, 9, and 12 2×2 neighbors are checked by Algorithm 1.1 for the three images above. Each 2×2 neighborhood matches one of these patterns e, i, n, for external corner, internal corner, and neither. (a) For each of the three images, how many of each 2×2 pattern are there? (b) Does the holecounting formula work for all three images?

1.2.4 Examining the Inside of a Human Head

Magnetic resonance imaging (MRI) devices can sense materials in the interior of 3D objects. Figure 1.4 shows a section through a human head: brightness is related to movement of material; this is actually a picture of blood flow. One can see important blood vessels.

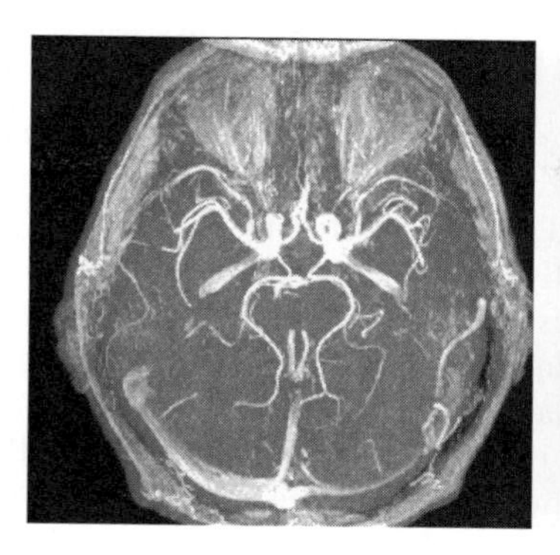

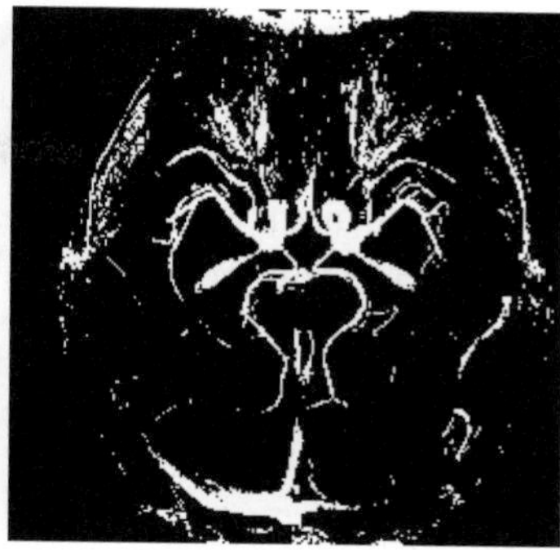

Figure 1.4 (left) Magnetic resonance image where brightness relates to blood flow; and (right) binary image resulting from changing all pixels with value 208 or above to 255 and those below 208 to 0. (Image courtesy of James Siebert, Michigan State Radiology.)

The whispy comet-like structures are associated with the eyes. MRI images are used by doctors to check for tumors or blood flow problems such as abnormal vessel constrictions or expansions. The image at the right in Figure 1.4 was made from a copy of the one on the left by making every pixel of value 208 or more bright (255) and those below 208 dark (0). Most pixels correctly show blood vessels versus background, but there are many incorrectly colored pixels of both types. Machine vision techniques are often used in medical image analysis, although usually to aid in data presentation and measurement rather than diagnosis itself. Wouldn't it be great if we could see thoughts occuring in the brain? Well, it turns out that MRI can sense organic activity related to thought processes. This is a very exciting current area of research.

Exercise 1.2: How many pixels per hole?

Consider the application of counting holes in truck crossbars at a more detailed level. Suppose that the area of the crossbar that is imaged is 50 inches long and 10 inches wide and suppose that this area almost fills up a digital image of 100 rows and 500 columns of pixels. Suppose a particular bolt hole in the crossbar is $1/2$ inch in diameter. What would you expect the radius and area of its image to be in terms of pixels?

Exercise 1.3: Imaging coins as pixels.

This problem is related to the one above. Obtain some graph paper (0.25 inch squares would be good) and a quarter. Randomly place the quarter on the graph paper and trace its circumference: do this five times. For each of the five placements, estimate the area of the image of the quarter in pixel units (a) by deciding whether a pixel is part of the quarter or not (no fractions) and (b) for each pixel cut by the circumference, estimate to the nearest tenth of a pixel how much of the pixel is part of the image of the quarter. After doing these measurements, compute the mean and standard deviation of the image area separately for methods (a) and (b).

1.2.5 Processing Scanned Text Pages

Converting information from paper documents into digital form for information systems is a common problem. For example, we might want to make an old book available on the Internet, or we might need to convert a blueprint of some object into a geometry file so that the part can be made by a numerically controlled machine tool.

Figure 1.5 shows the same message in both Chinese and English. The Chinese characters were written on paper and scanned into an image of 482 rows and 405 columns. The postscript file encoding the graphics and printed in the figure has a size of 68,464 bytes. The English version is stored in a file of 115 bytes, each holding one ASCII character. There is an entire range of important applications in processing documents. Recognizing individual characters from the dots of the scanner or FAX files is one such application that is done fairly well today, provided that the characters conform to standard patterns. Providing a semantic interpretation of the information, possibly to be used for indexing in a large database, is a harder problem.

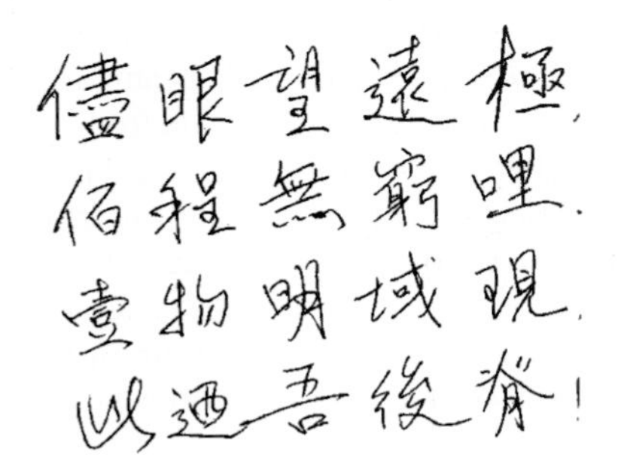

I looked as hard as I could see,
beyond 100 plus infinity
an object of bright intensity
—it was the back of me!

Figure 1.5 (left) Chinese characters; and (right) English equivalent. Is it possible that a machine could automatically translate one into the other? (English poem by George Stockman; translation into Chinese characters by John Weng.)

1.2.6 Accounting for Snow Cover Using a Satellite Image

Much of the earth's surface is scanned regularly from satellites, which transmit their images to earth in digital form. These images can then be processed to extract a wealth of information. For example, inventory of the amount of snow in the watershed of a river may be critical for regulating a dam for flood control, water supply, or wildlife habitat. Estimates of snow mass can be made by accounting for the number of pixels in the image that appear as snow. A pixel from a satellite image might result from sensing a $10m \times 10m$ spot of earth, but some satellites reportedly can see much smaller spots than that. Often, the satellite image must be compared to a map or other image to determine which pixels are in a particular area or watershed. This operation is usually manually aided by a human user interacting with the image-processing software and will be discussed more in Chapter 11 where image matching is covered. Figure 1.6 is a photograph taken on a space shuttle flight managed by the Johnson Space Center in Houston, Texas. It shows the town of Wenatchee, Washington, where the Wenatchee River flows into the Columbia River.

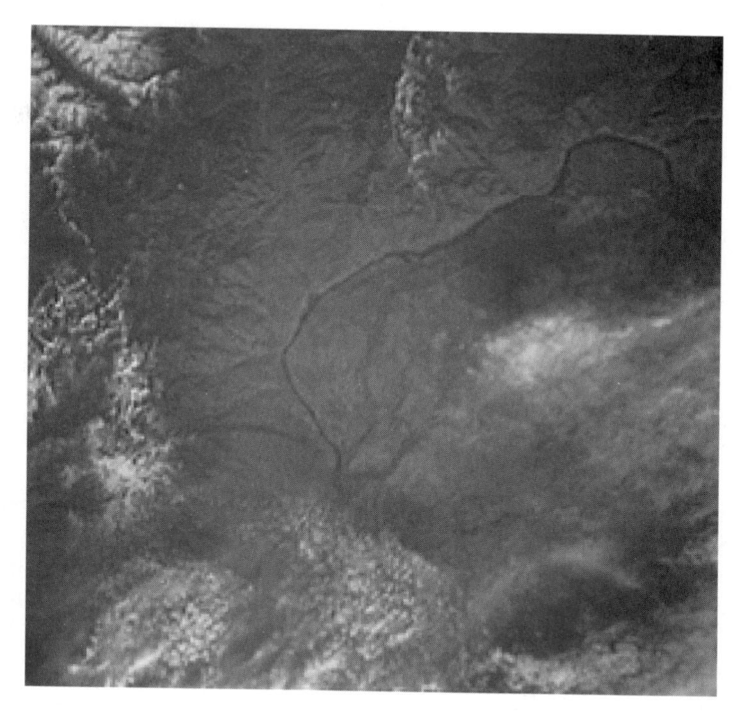

Figure 1.6 Photo of the Wenatchee and Columbia Rivers in Washington State. (Courtesy of Johnson Space Center; see http://www.earthrise.sdsc.edu.)

Computers are known for their ability to handle large amounts of data; certainly the earth-scanning satellites produce a tremendous amount of data useful for many purposes. For example, counts and locations of snow pixels might be input to a computer program that simulates the hydrology for that region. (Temperature information for the region must be input to the program as well.) Another related application takes inventory of crops and predicts harvests. Yet another takes inventory of buildings for tax purposes; this is usually done manually with pictures taken from airplanes.

1.2.7 Understanding a Scene of Parts

At many points of manufacturing processes, parts are transferred on conveyors or in boxes. Parts must be individually placed in machines, packed, inspected, etc. If the operation is dull or dangerous, a vision-guided robot might provide a solution. The underlying image of Figure 1.7 shows three workpieces in a robot's workspace. By recognizing edges and holes, the robot vision system is able to guess at both the identity of a part and its position in the workspace. Using a 3D model made by computer-aided-design (CAD) for each guessed part and its guessed position, the vision system then compares the sensed image data with a computer graphic generated from the model and its position in space. Bad matches are rejected, while good matches cause the guess to be refined. The bright lines in Figure 1.7 show three such refined matches between the image and models of the objects it contains.

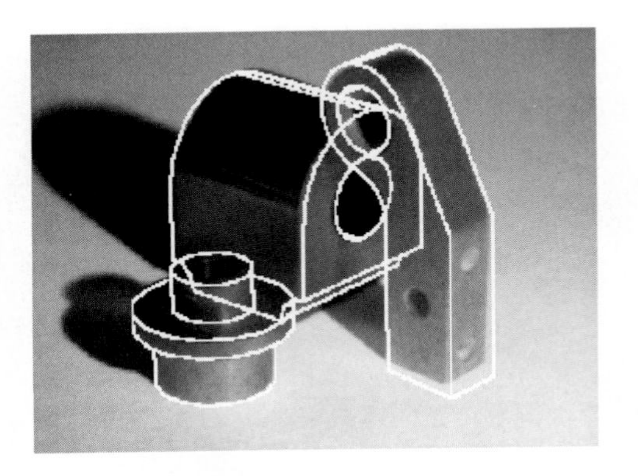

Figure 1.7 An inspection or assembly robot matches stored 3D models to a sensed 2D image. (Courtesy of Mauro Costa.)

Finally, the robot eye-brain can tell the robot arm how to pick up a part and where to put it. The problems and techniques of 3D vision are covered in Chapters 13 and 14 of this text.

Exercise 1.4: Other problem areas.

Describe a problem different from those already discussed for which machine vision might provide a solution. If you do not have a special application area in mind, choose one for the moment. What kind of scenes would be sensed? What would an image be like? What output would be produced?

Exercise 1.5: Examining problem context.

Problems can be solved in different ways and a problem solver should not get trapped early in a specific approach. Consider the problem of identifying cars in various situations: (a) entering a parking lot or secured area, (b) passing through a tollgate, (c) exceeding the speed limit. Several groups are developing or have developed machine vision methods to read a car's license plate. Suggest an alternative to machine vision. How do the economic and social costs compare to the machine vision approach?

1.3 OPERATIONS ON IMAGES

This book presents a large variety of image operations. Operations can be grouped into different categories depending on their structure, level, or purpose. Some operations are for the purpose of improving the image solely for human consumption, while others are for extracting information for downstream automatic processing. Some operations create new output images, while others output non-image descriptions. A few important categories of image operations follow.

1.3.1 Changing Pixels in Small Neighborhoods

Pixel values can be changed according to how they relate to a small number of neighboring pixels, for example, neighbors in adjacent rows or columns. Frequently, isolated 1s or 0s in a binary image will be reversed in order to make them the same as their neighbors. The

Figure 1.8 (left) Binary image of bacteria (in the original microscope image, the bacteria were blue due to their fluorescence); and (right) cleaner image resulting from changing black pixels that had one or more white neighbors to white. (Original image courtesy of Frank Dazzo.)

purpose of this operation could be to remove likely noise from the digitization process. Or, it could be just to simplify image content; for example, to ignore tiny islands in a lake or imperfections in a sheet of paper. Another common operation is to change *border pixels* to be *background pixels* as shown in Figure 1.8. The images of bacteria have fuzzy borders and often fuse together. By changing the black border pixels to white, the bacteria images, although smaller, have clearer borders and some formerly fusing pairs are separated. These operations are treated in Chapter 3.

Exercise 1.6

Identify some defects remaining in the right image of Figure 1.8 and describe simple neighborhood operations that will improve the image.

1.3.2 Enhancing an Entire Image

Some operations treat the entire image in a uniform manner. The image might be too dark—say its maximum brightness value is 120—so all brightness values can be scaled up by a factor of 2 to improve its displayed appearance. Noise or unnecessary detail can be removed by replacing the value of every input pixel with the average of all nine pixels in its immediate neighborhood. Alternatively, details can be enhanced by replacing each pixel value by the contrast between it and its neighbors. Figure 1.9 shows a simple contrast computation applied at all pixels of an input image. Note how the boundaries of most

Figure 1.9 Contrast in the left image is shown in the right image. The top 10 percent of the pixels in terms of contrast are made bright while the lower 90 percent are made dark. Contrast is computed from the 3×3 neighborhood of each pixel.

objects are well detected. The output image results from computations made only on the local 3×3 neighborhoods of the input image. Chapter 5 describes several of these kinds of operations. Perhaps an image is taken using a fish eye lens and we want to create an output image with less distortion: in this case, we have to move the pixel values to other locations in the image to move them closer to the image center. Such an operation is called *image warping* and is covered in Chapter 11.

1.3.3 Combining Multiple Images

An image can be created by adding or subtracting two input images. Image subtraction is commonly used to detect change over time. Figure 1.10 shows two images of a moving part and the difference image resulting from subtracting the corresponding pixel values of the second image from those of the first image. Image subtraction captures the boundary of the moving object, but not perfectly. (Since negative pixel values were not used, not all changes were saved in the output image.) In another application, urban development might be more easily seen by subtracting an aerial image of a city taken five years ago from a current image of the city. Image addition is also useful. Figure 1.11 shows an image of Thomas Jefferson added to an image of the great arch opening onto the lands of the Louisiana Purchase; more work is needed in this case to blend the images better.

Figure 1.10 (left and center) Images of a moving part; and (right) a difference image that captures the boundary of the part.

Figure 1.11 (left) Image of the great archway at St. Louis; (center) face of Jefferson; and (right) combination of the two.

1.3.4 Computing Features from an Image

We have already seen the example of counting holes. More generally, the regions of 0s corresponding to holes in the crossbar inspection problem could be images of objects, often called *blobs*—perhaps these are microbes in a water sample. Important features might be average object area, perimeter, direction, etc. We might want to output these important features separately for every detected object. Chapter 3 describes such processing. Chapters 6 and 7 discuss means of quantitatively summarizing the color or texture content of regions of an image. Chapter 4 shows how to classify objects according to these features; for example, is the extracted region the image of microbe A or B? Figure 1.12 shows output from a well-known algorithm applied to the bacteria image of Figure 1.8 giving features of separate regions identified in the image, including the region area and location. Regions with area of a few hundred pixels correspond to isolated bacterium while the large region is due to several touching bacteria.

Object	Area	Bounding Box	Centroid
1	247	[(20 , 26), (32 , 56)]	(26.1, 42.0)
2	6	[(25 , 22), (26 , 24)]	(25.5, 23.0)
3	116	[(35 , 72), (54 , 86)]	(44.1, 79.4)
4	4	[(37 , 69), (38 , 70)]	(37.5, 69.5)
5	15	[(46 , 86), (50 , 89)]	(47.6, 87.4)
6	586	[(49 , 122), (95 , 148)]	(71.7, 134.8)
7	300	[(54 , 91), (77 , 112)]	(65.6, 101.9)
8	592	[(57 , 138), (108 , 163)]	(83.6, 150.8)
9	562	[(57 , 158), (104 , 183)]	(81.2, 171.4)
10	5946	[(74 , 195), (221 , 313)]	(138.0, 256.5)
11	427	[(204 , 115), (229 , 151)]	(217.1, 132.3)
12	797	[(242 , 42), (286 , 97)]	(264.9, 71.8)
13	450	[(248 , 170), (278 , 204)]	(262.7, 188.1)
14	327	[(270 , 182), (291 , 216)]	(279.9, 200.3)
15	264	[(293 , 195), (311 , 221)]	(300.8, 206.7)
16	145	[(304 , 179), (316 , 193)]	(310.4, 186.4)

total object area = 10784 pixels

Figure 1.12 Components automatically identified in the bacteria image in Figure 1.8 (right). Single bacterium have an area of a few hundred pixels: the large component consists of several touching bacteria, and the tiny objects 2, 4, and 5 are due to noise.

1.3.5 Extracting Non-iconic Representations

Higher-level operations usually extract representations of the image that are non-iconic, that is, data structures that are not like an image. (Recall that extraction of such descriptions is often defined to be the goal of computer vision.) Figure 1.12 shows a non-iconic description derived from the bacteria image. In addition to examples already mentioned, consider a report of the count of microbes of type A and B in a slide from a microscope or the volume of traffic flow between two intersections of a city computed from a video taken from a utility pole. In another important application, the (iconic) input might be a scanned magazine article and the output a hypertext structure containing sections of recognized ASCII text and sections of raw images for the figures. As a final example, in the application illustrated in Figure 1.7, the machine vision system would output a set of three detections, each encoding a part number, three parameters of part position and three parameters of the orientation of the part. This scene description could then be turned over to the motion-planning system, which would decide on how to manipulate the three parts.

1.4 THE GOOD, THE BAD, AND THE UGLY

Having cited many applications of machine vision, we cannot proceed without saying that success usually is hard won. Often, implementors have to accept environmental constraints that compromise system flexibility. For example, scene lighting might have to be carefully controlled, or objects might have to be mechanically separated or positioned before imaging. This is because the real world yields exorbitant variations in the input image, challenging the best computer algorithms in their task of extracting the essence, or *invariant features* of objects. Appearance of an object can vary significantly due to changes in illumination or presence of other objects, which might be unexpected. Consider, for example, the shadows in Figure 1.7 and Figure 1.9. Moreover, decisions about object structure must often be made by integrating a variety of information from many pixels of the image. For example, the brightness of the tops of the glasses on the counter in Figure 1.9 is the same as that of the wall, so no glass-wall boundary is evident at the pixel level. In order to recognize each glass as a separate object, pixels from a wider area must be grouped and organized. Humans are quite good at this, but developing flexible grouping processes for machine vision has proved difficult. Problems of occlusion hamper recognition of 3D objects. Can a vision system recognize the person or the chair in Figure 1.9, even though neither appears to have legs? At a higher level yet, what model of a *dog* could empower a machine to recognize the diverse individuals that could be imaged? These difficulties, and others, will be discussed throughout this book.

Exercise 1.7

Which invariant features of the following objects enable you to recognize them in rain or sunshine, alone or alongside other objects, from the front or side: (a) your tennis shoes; (b) your front door; (c) your mother; and (d) your favorite make of automobile?

1.5 USE OF COMPUTERS AND SOFTWARE

Computers are legendary for accurate accounting of quantitative information. Computing with images has gone on for over thirty years—initially mostly in research labs with mainframe computers or in production shops with special-purpose computers. Recently, large inexpensive memories and high-speed general-purpose processors have brought image computing potential to every multimedia personal computer user, including the hobbyist working in her dining room.

One can compute with images in different ways. The easiest is to acquire an existing program that can perform many of the needed image operations. Some programs are free to the public; others must be purchased. Many free images are available from the World Wide Web. To control your own image input, you can buy a flatbed scanner or a digital camera, each available for a few hundred dollars. Software libraries are available which contain many subroutines for processing images: The user writes an application program which calls the library routines to perform the required operations on the user's image data. Most companies selling input devices for machine vision also provide libraries for image operations and even driver programs with nice graphical user interfaces (GUI). Special hardware is available for speeding up image operations that can take many seconds, or even minutes, on a general purpose processor. Many of the early parallel computers costing millions of dollars were designed with image processing as a primary task; however, today most of the critical operations can be provided by sets of boards costing a few thousand dollars. Usually, special hardware is only needed for high production rates or real time response. Special programming languages with images and image operations as language primitives have been defined; sometimes, these have been combined with operations for controlling an industrial robot. Today, it is apparent that much good image processing can and will be done using a general purpose language, such as C, and a general purpose computer available via mail order or the local computer store. This bodes exceedingly well for the machine vision field, since challenging problems will now be attacked from all directions possible! The reader is invited to join in.

1.6 RELATED AREAS

Computer vision is related to many other disciplines; we are not able to pursue all of these relations in depth in this text. First, it is important to distinguish between *image processing* and *image understanding*. Image processing is primarily concerned with the transformation of images into images, whereas, image understanding is concerned with making decisions based on images and explicitly constructing the scene descriptions needed to do so. Image processing is quite often used in support of image understanding and thus will be treated to some extent in this book. Books concerned with image processing typically are based on the model of an image as a continuous function $f(x, y)$ of two spatial parameters x and y, whereas this text will concentrate on the model of an image as a discrete 2D array $\mathbf{I[r, c]}$ of integer brightness samples. In this book, we use the terms *computer vision, machine vision,* and *image understanding* interchangeably; however, experts would certainly debate their nuances.

The psychology of human perception is very important for two reasons; first, the creator of images for human consumption must be aware of the characteristics of the client, and secondly, study of the tremendous human capability in image understanding can guide our development of algorithms. While this text includes some discussion of human perception and cognition, its approach is primarily hands-on problem solving. The physics of light, including optics and color science, is important to our study. We will present the basic material necessary; however, readers who want to be experts on illumination, sensing, or lenses will need to access the related literature. A variety of mathematical models are used throughout the text. For mastery, the reader must be comfortable with the notions of functions, probability, calculus, and analytical geometry. The intuitive concepts of image processing often strengthen the mathematical concepts. Finally, any book about computer vision must be strongly related to computer graphics. Both fields are concerned with how objects are viewed and how objects are modeled; the prime distinction is one of direction—computer vision is concerned with description and recognition of objects from images, while computer graphics is concerned with generation of images from object descriptions. Recently, there has been a great deal of integration of these two areas: Computer graphics is needed to display computer vision results and computer vision is needed to make object models. Digital images are commonly used as input for computer graphics products.

1.7 THE REST OF THE BOOK

The previous sections informally introduced many of the concepts in the book and indicated the chapters in which they are treated. The reader should now appreciate the range of problems attacked by machine vision and a few of its methods. The chapters that immediately follow describe 2D machine vision. In those chapters, the image is analyzed in self-referencing terms of pixels, rows, intersections, colors, textures, etc. To be sure, knowledge about how the image was taken from the real 3D world is present, but the relationship between image pixels and real world elements is obvious—only the scale is different. For example, a radiologist can readily tell from an image if a blood vessel is constricted without knowing much about the physics of the sensor or about what portion of the body a pixel represents. So can a machine vision program. Similarly, the essence of a character recognition algorithm has nothing to do with the real font size being scanned. Consequently, the material in Chapters 2 to 11 has a 2D character and is more generic and simpler than material in Chapters 12 to 16. In Chapters 13 to 15, the 3D nature of objects and the viewpoints used to image them are crucial. The analysis cannot be done with the coordinates of a single image because we need to relate multiple images or images and models; or, we need to relate a sensor's view to a robot's view. We are analyzing 3D scenes in Chapters 13 to 15, not 2D images, and the most important tool for the analysis is 3D analytical geometry. As in computer graphics, the step from 2D to 3D is a large one in terms of both modeling abstraction and computational effort.

1.8 REFERENCES

The machine vision literature is highly specialized according to application area. For example, the paper by Fleck and others (1996) addresses how pornographic images might be detected so they could be screened from a child's computer. Our discussion of the hole-counting algorithm was derived from the work of Kopydlowski (1983), which described its use in the inspection of truck crossbars. The design of sensors for satellites is very different from the design of medical instruments, for example. Manufacturing systems are different still. Some references to the special areas are Nagy (1972) and Hord (1982) for remote sensing, Glasby and Horgan (1995) for biological sciences, Ollus (1987) and QCAV (1999) for industrial applications, and ASAE (1983) for agricultural applications. One of several articles on the early development of color CCD cameras is by Dillon and others (1978). Problems, methods, and theory shared among several application areas are, of course, topics for textbooks, this one included. Perhaps the first textbook on picture processing using a computer by Rosenfeld (1969) contains material on image processing without use of higher-level models. The book by Ballard and Brown (1982), perhaps the first *Computer Vision* text, concentrated on image analysis using higher-level models. Levine's text (1985) is noteworthy due to its inclusion of significant material on the human visual system. The two-volume set by Haralick and Shapiro (1992) is a modern resource for algorithms and their mathematical justification. The text by Jain, Kasturi, and Schunk (1995) is a modern introduction to machine vision with primarily an engineering viewpoint.

1. ASAE. 1983. Robotics and intelligent machines in agriculture. *Proc. 1st Int. Conf. Robotics and Intelligent Machines in Agriculture* (2–4 Oct. 1983), American Society of Agricultural Engineers, Tampa: FL, St. Joseph, MI.
2. Ballard, D. H., and C. M. Brown. 1982. *Comput. Vision.* Prentice-Hall, Englewood Cliffs, NJ.
3. Dillon, P., D. Lewis, and F. Kaspar. 1978. Color imaging system using a single CCD area array. *IEEE Trans. Electron Devices,* ED-25(2):102–107.
4. Fleck, M., D. Forsyth, and C. Pregler. 1996. Finding naked people [in images]. *Proc. European Conf. Comput. Vision.* Springer-Verlag, New York, 593–602.
5. Glasby, C. A., and G. W. Horgan. 1995. *Image Analysis for the Biological Sciences.* John Wiley & Sons, Chichester, England.
6. Haralick, R., and L. Shapiro. 1992/3. *Computer and Robot Vision, Volumes I and II.* Addison-Wesley, New York.
7. Hord, R. M. 1982. *Digital Image Processing of Remotely Sensed Data.* Academic Press, New York.
8. Igarashi, T., ed. 1983. *World Trademarks and Logotypes.* Graphic-sha, Tokyo.
9. ——— ed. 1987. *World Trademarks and Logotypes II: A Collection of International Symbols and Their Applications.* Graphic-sha, Tokyo.
10. Jain, R., R. Kasturi, and B. Schunk. 1995. *Machine Vision.* McGraw-Hill, New York.

11. Kopydlowski, D. 1983. 100% inspection of crossbars using machine vision. Publication MS83-210, Society of Manufacturing Engineers, Dearborn, MI.
12. Levine, M. D. 1985. *Vision in Man and Machine.* McGraw-Hill, New York.
13. Nagy, G., 1972, Digital image processing activites in remote sensing for Earth resources. *Proc. IEEE,* v. 60(10):1177–1200.
14. Ollus, M., ed. 1987. Digital image processing in industrial applications. *Proc. 1st IFAC Workshop,* Espoo, Finland (10–12 June 1986), Pergamon Press, Oxford.
15. Pratt, W. 1991. *Digital Image Processing,* 2nd ed. John Wiley, New York.
16. QCAV. 1999. Quality control by artificial vision. *Proc. 5th Int. Conf. Quality Control by Artificial Vision* (18–21 May 1999), Trois-Rivieres, Canada.
17. Rosenfeld, A. 1969. *Picture Processing by Computer.* Academic Press, New York.

1.9 ADDITIONAL EXERCISES

Next are several questions requiring essay-type answers. A few questions require quantitative thinking. Most of these questions will be revisited in more detail later in the text. Throughout this text, questions marked with an asterisk (*) present more difficulty because they require some research, very careful reasoning, or careful programming.

Exercise 1.8: The problem of selling produce.

Consider the problem a grocery store checker has in charging you for your purchases. Barcode technology makes some items easy; a soup can need only be dragged over the barcode reader until an acknowledging *beep* is heard. This system does not work for produce picked by the shopper in variable unpackaged quantities and the checker has to stop to do special actions. What actions? Do you think that a camera could be placed above or inside the combined scale and barcode-reader to tell the cash register which kind of produce it is? Suppose using methods of this textbook, a machine-vision system could tell the difference between spinach greens and collard greens, Fuji apples versus McIntosh, etc. Describe how this machine-vision system would integrate with the other technology already in place to help the checker compute your bill.

Exercise 1.9: Counting bacteria.

Examine the image of bacteria in Figure 1.8 and the sample of automatically computed features in Figure 1.12. Is there potential for obtaining a count of bacteria, say within 5 percent accuracy, shown in this example? Explain.

Exercise 1.10: 3D model from video?

Suppose you are given a video of Notre Dame Cathedral in Paris. The video was taken by a person walking around the outside and inside of the cathedral; so many viewpoints are available. Do you think that you could make a reasonable 3D model of that cathedral using only the video? (If you lack confidence, assume you're an architect.) If not, why not? If so, how could you construct a 3D model when all you have are 2D images?

Exercise 1.11: On computing contrast.

Think of a method that computes the contrast of a 3×3 image neighborhood similar to what is shown in Figure 1.9. Assume that the 9 input pixel values are brightness values between 0 and 255 and that the output pixel value is a single value between 0 and 255 measuring the amount of contrast. (The picture at the right in Figure 1.9 actually uses only the two pixel values 0 and 255; however, you may use the entire range.)

Exercise 1.12: On face interpretation.

(a) Is it easy for you to decide the gender and approximate age of persons pictured in magazine ads? (b) Psychologists might tell us that humans have the ability to see a face and immediately decide on the age, sex, and degree of hostility of the person. Assume that this ability exists for humans. If you think it would be based on image features, then what are they? If you think that image features are not used, then explain how humans might make such decisions?

Exercise 1.13: Is a picture worth a thousand words?

Consider the following passage from *The Sound and the Fury* by William Faulkner (Vintage Books Edition, 1987, Copyright 1984 by Jill Faulkner Summers, p. 195). Do you think that a machine could extract such a description from a video of the scenes discussed?

> *I could smell the curves of the river beyond the dusk and I saw the last light supine and tranquil upon tideflats like pieces of broken mirror, then beyond them lights began in the pale clear air, trembling a little like butterflies hovering a long way off.*

Exercise 1.14: Toward the Correctness of Holecounting.*

This question requires some thought and reading that extends beyond this chapter and should be regarded as enrichment. (a) How many possible 2×2 neighborhood patterns are there in a binary image? List them all. (b) Which of the patterns of part (a) cannot occur in a binary image that is 4-connected? Define *border point* to be the center grid point of a 2×2 pixel neighborhood that contains both 0 and 1 pixels. (c) Argue that a single hole can be accounted for by just counting the number of e and i patterns along its border and that the formula $n = (e - i)/4$ is correct when one hole is present. (d) Argue that no two holes can have a common border point. (e) Argue that the formula is correct when an arbitrary number of holes is present.

Exercise 1.15: Are binary images suitable?

Consider imaging some scene and then processing it to produce a binary image as output, such as the red blood cell image, where target objects image as regions of 0s and the background, or non-objects, image as 1s. Think about whether or not this can be achieved for the following scenes. In your opinion, why can or cannot we create such a binary image?

1. The input image is of a paper containing typing and it is scanned using a page scanner. Our overall objective is to recognize most of the typed characters and make an ASCII file so that we can edit the text using a word processor.
2. The input image is an X-ray of someone's head. We would like to find bullets or tumors as regions of 0s in a background of 1s.
3. The input image is a satellite image of Richmond, VA taken in spring. By tuning the sensor or by some simple computer algorithm, we would like to create an image where a 0 indicates the presence of an azalea bush and a 1 indicates no bush.
4. We want to check the width of the stem of an auto engine valve by just counting the number of pixels across its shadow. Since we will be making hundreds of thousands of valves per day, detailed control of the environment and costly equipment can be justified.

2

Imaging and Image Representation

Humans derive a great deal of information about the world through their visual sense. Light reflects off objects and sometimes passes through objects to create an image on the retina of each eye. From this pair of images much of the structure of the 3D environment is derived. The important components are thus (a) a scene of objects; (b) illumination of the objects; and, (c) sensing the illumination reflecting off the objects (or passing through them).

The major purpose of this chapter is to describe how sensors produce digital images of 2D or 3D scenes. Different kinds of radiation that reflect from or penetrate objects in the physical world can be sensed by different imaging devices. The 2D digital image is an array of intensity samples reflected from or transmitted through objects: This image is processed by a machine or computer program in order to make decisions about the scene. Often, a 2D image represents a projection of a 3D scene; this is the most common representation used in machine vision and in this book. At the end of the chapter, we discuss some relationships between structures in the 3D world and structures in the 2D image.

Various sections of this chapter are marked by an asterisk (*) to indicate that they provide technical details that can be skipped by a reader who is not particularly interested in them at this point.

2.1 SENSING LIGHT

Much of the history of science can be told in terms of the progress of devices created to sense and produce different types of electromagnetic radiation, such as radio waves, X-rays, microwaves, etc. The chemicals in the receptors of the human eye are sensitive to radiation (light) with wavelengths ranging from roughly 400 nanometers (violet) to 800 nanometers (red). Snakes and CCD sensors (see Figure 2.2) can sense wavelengths

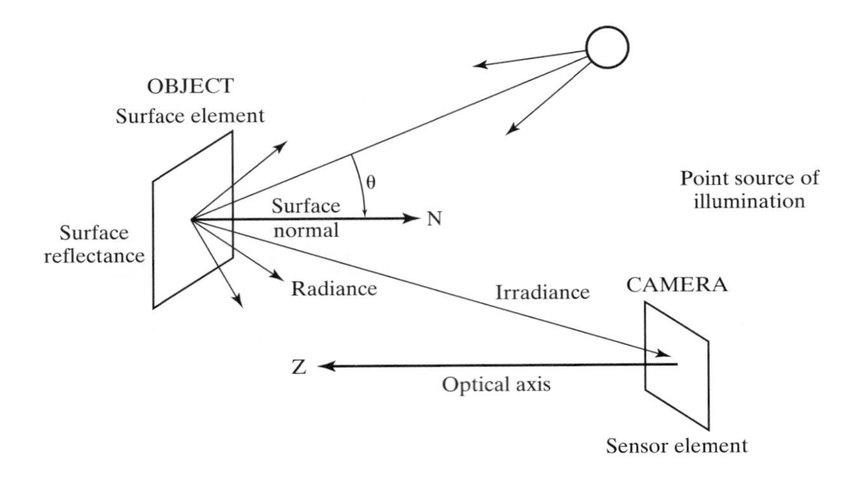

Figure 2.1 Reflection of radiation received from a single source of illumination.

longer than 800 nanometers (infrared). There are devices to detect very short length X-rays and those which detect long radio waves. Different wavelengths of radiation have different properties; for example, X-rays can penetrate human bone while longer wavelength infrared might not penetrate even clouds.

Figure 2.1 shows a simple model of common photography: A surface element, illuminated by a single source (the sun or a flash bulb) reflects radiation toward the camera, which senses it via chemicals on film. More details of this situation are covered in Chapter 6. Wavelengths in the light range result from generating or reflecting mechanisms very near the surface of objects. We are concerned with many properties of electromagnetic radiation in this book; however, we will usually give a qualitative description of phenomena and leave the quantitative details to books about physics or optics. Application engineering requires some knowledge of the material being sensed and the radiation and sensor used.

2.2 IMAGING DEVICES

There are many different devices that produce digital images. They differ in the phenomena sensed as well as in their electromechanical design. Several different sensors are described in this chapter; the most common ones are discussed in this section, others are left to optional Section 2.9. Our intent is to disclose the important functional and conceptual aspects of each sensor, leaving most technical information to outside reading.

2.2.1 CCD Cameras

Figure 2.2 shows a camera built using charge-coupled device (CCD) technology, the most flexible and common input device for machine-vision systems. The CCD camera is very much like a 35*mm* film camera commonly used for family photos, except on the image

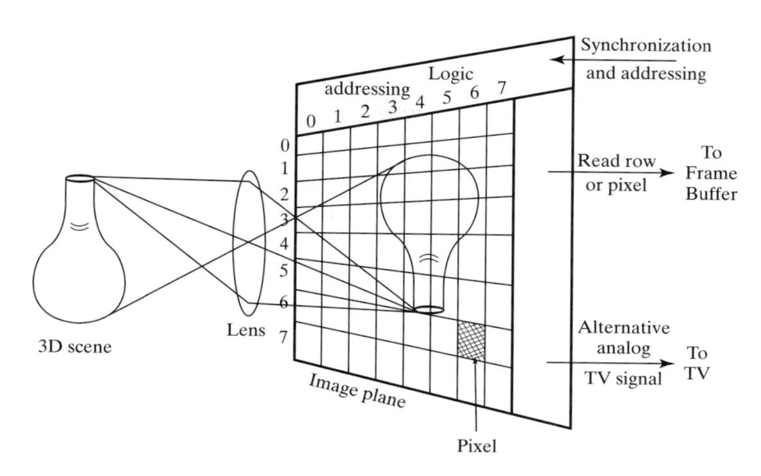

Figure 2.2 A CCD (charge-coupled device) camera imaging a vase; discrete cells convert light energy into electrical charges, which are represented as small numbers when input to a computer.

plane, instead of chemical film reacting to light, tiny solid state cells convert light energy into electrical charge. Each cell converts the light energy it receives into an electrical charge. All cells are first cleared to 0, and then they begin to integrate their response to the light energy falling on them. A shutter may or may not be needed to control the sensing time. The image plane acts as a digital memory that can be read row by row by a computer input process. The figure shows a simple monochrome camera.

If the digital image has 500 rows and 500 columns of byte-sized gray values, a memory array of a quarter of a million bytes is obtained. A CCD camera sometimes plugs into a computer board, called a *frame grabber* which contains memory for the image and perhaps control of the camera. New designs now allow for direct digital communication (such as using the IEEE 1394 standard). Today major camera manufacturers offer digital cameras that can store a few dozen images in memory within the camera body itself; some contain a floppy disk for this purpose. These images can be input for computer processing at any time. Figure 2.3 sketches an entire computer system with both camera input and graphics output. This is a typical system for an industrial-vision task or medical-imaging task. It is also typical for *multimedia* computers, which may have an inexpensive camera available to take images for teleconferencing purposes. The role of a *frame buffer* as a high speed image store is central here: the camera provides an input image which is stored in digital form in the frame buffer after analog to digital conversion where it is available for display to the user and for processing by various computer algorithms. The frame buffer actually may store several images or their derivatives.

A computer program processing a digital image might refer to pixel values as **I[r, c]** or **I[r][c]** where **I** is an array name and **r** and **c** are row and column numbers, respectively. This book uses such notation in the algorithms presented. Some cameras can be set so that

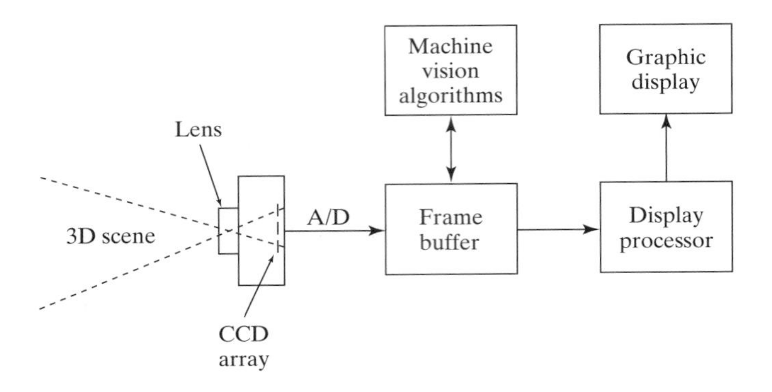

Figure 2.3 Central role of the *frame buffer* in image processing.

they produce a *binary image*—pixels are either 0 or 1 representing dark versus bright, or the reverse. A simple algorithm can produce the same effect by changing all pixels below some *threshold* value t to 0 and all pixels at or above it to 1. An example was given in Chapter 1 where a magnetic resonance image was thresholded to contrast high blood flow versus low blood flow.

2.2.2 Image Formation

The geometry of image formation can be conceptualized as the projection of each point of the 3D scene through the *center of projection* or *lens center* onto the image plane. The intensity at the image point is related to the intensity radiating from the 3D surface point; the actual relationship is complex as we will later learn. This projection model can be physically justified since a *pin-hole* camera can actually be made by using a camera box with a small hole and no lens at all. A CCD camera usually will employ the same kind of lens as $35mm$ film cameras used for family photos. A single lens with two convex surfaces is shown in Figure 2.2, but most actual lenses are compound with more than two refracting surfaces. There are two very important points to be made. First, the lens is a light collector: light reaches the image point via an entire cone of rays reaching the lens from the 3D point. Three rays are shown projecting from the top of the vase in Figure 2.2; these determine the extremes of the cone of rays collected by the lens for only the top of the vase. A similar cone of rays exists for all other scene points. Because of geometric imperfections in the lens, different bending of different colors of light, and other phenomena, the cone of rays actually results in a finite or blurred spot on the image plane called the *circle of confusion*. Second, the CCD sensor array is constructed of physically discrete units and not infinitesimal points; thus, each sensor cell integrates the rays received from many neighboring points of a 3D surface. These two effects cause blurring of the image and limit its sharpness and the size of the smallest scene details that can be sensed.

CCD arrays are manufactured on chips typically measuring about $1cm \times 1cm$. If the array has 640×480 pixels or 512×512 pixels, then each pixel has a real width of roughly 0.001 inch. There are other useful ways of placing CCD sensor cells on the image plane (or

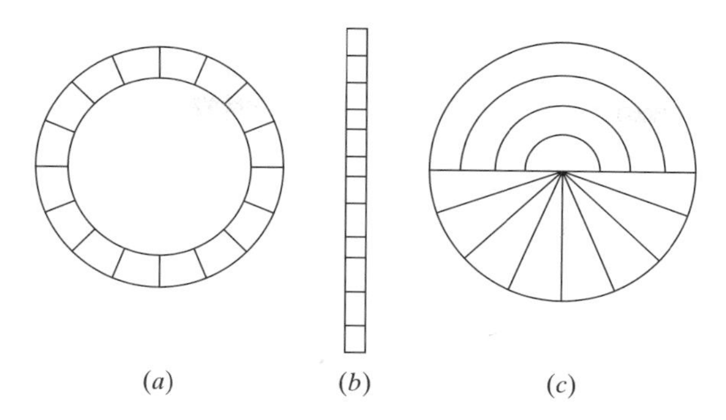

Figure 2.4 Other useful array geometries: (*a*) circular; (*b*) linear; and (*c*) *ROSA*.

image line) as shown in Figure 2.4. A linear array can be used in cases where we only need to measure the width of objects or where we may be imaging and inspecting a continuous web of material flowing by the camera. With a linear array, 1,000 to 5,000 pixels are available in a single row. Such an array can be used in a push broom fashion where the linear sensor is moved across the material being scanned as done with a hand-held scanner or in highly accurate mechanical scanners, such as flatbed scanners. Currently, many flatbed scanners are available for a few hundred dollars and are used to acquire digital images from color photos or print media. Cylindrical lenses are commonly used to focus a line in the real world onto the linear CCD array. The circular array would be handy for inspecting analog dials such as on watches or speedometers; the object is positioned carefully relative to the camera and the circular array is scanned for the image of the needle. The interesting *ROSA* partition shown in Figure 2.4(*c*) provides a hardware solution to integrating all the light energy falling into either sectors or bands of the circle. It was designed for quantizing the power spectrum of an image, but might have other simple uses as well. Chip manufacturing technology presents opportunities for implementing other custom designs.

Exercise 2.1: Examination of a CCD camera.

If you have access to a CCD camera, obtain permission to explore its construction. Remove the lens and note its construction; does it have a shutter to close off all light, does it have an aperature to change the size of the cone of rays passing through? Is there a means of changing the focal length—which is the distance between the lens and CCD? Inspect the CCD array. How large is the active sensing area? Can you see the individual cells—do you need a magnifying glass?

Exercise 2.2

Suppose that an analog clock is to be read using a CCD camera that stares directly at it. The center of the clock images at the center of a 256×256 digital image and the hour hand is twice the width of the minute hand but 0.7 times its length. To locate the images of the hands of the clock we need to scan the pixels of the digital image in a circular fashion.

(a) Give a formula for computing $\mathbf{r}(t)$ and $\mathbf{c}(t)$ for pixels **I[r, c]** on a circle of radius R centered at the image center **I[128, 128]**, where t is the angle made between the ray to **I[r, c]** and the horizontal axis. (b) Is there a problem in controlling t so that a unique sequence of pixels of a *digital circle* is generated? (*c) Do some outside reading in a text on computer graphics and report on a practical method for generating such a digital circle.

2.2.3 Video Cameras

Video cameras creating imagery for human consumption record sequences of images at a rate of 30-per-second, enabling a representation of object motion over time in addition to the spatial features represented in the single images or frames. To provide for smooth human perception, 60 half frames-per-second are used; these half frames are all odd image rows followed by all even image rows in alternate succession. An audio signal is also encoded. Video cameras creating imagery for machine consumption can record images at whatever rate is practical and need not use the half frame technique.

Frames of a video sequence are separated by markers and some image compression scheme is usually used to reduce the amount of data. The analog TV standards have been carefully designed to satisfy multiple requirements: the most interesting features allow for the same signal to be used for either color or monochrome TVs and to carry sound or text signals as well. For details, the interested reader can consult standards documents for TV and MPEG encoding given in Section 2.5. We continue here with the notion of digital video being just a sequence of 2D digital images.

CCD camera technology for machine vision has sometimes suffered from display standards designed for human consumption. First, the interlacing of odd and even frames in a video sequence, needed to give a smooth picture to a human makes unnecessary complexity for machine vision. Second, many CCD arrays have had pixels with a 4:3 ratio of width to height because most displays for humans have a 4:3 size ratio. Square pixels and a single scale parameter would benefit machine vision. The huge consumer market has driven device construction toward human standards and machine vision developers have had to either adapt or pay more for devices made in limited quantities.

2.2.4 The Human Eye

Crudely speaking, the human eye is a spherical camera with a 20*mm* focal length lens at the outside focusing the image on the *retina* which is opposite the lens and fixed on the inside of the surface of the sphere (see Figure 2.5). The *iris* controls the amount of light

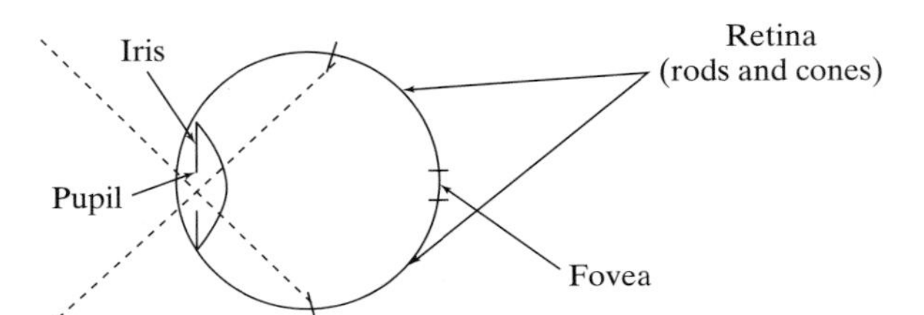

Figure 2.5 Crude sketch of the human eye as camera. (Much more detail can be obtained in the 1985 book by Levine.)

passing through the lens by controlling the size of the *pupil*. Each eye has one hundred million receptor cells—quite a lot compared to a typical CCD array. Moreover, the retina is unevenly populated with sensor cells. An area near the center of the retina, called the *fovea,* has a very dense concentration of color receptors, called *cones*. Away from the center, the density of cones decreases while the density of black-white receptors, the *rods*, increases. The human eye senses three separate intensities for three constituent colors of a single surface spot imaging on the fovea, because the light received from that spot falls on three different types of cones. Each type of cone has a special pigment that is sensitive to wavelengths of light in a certain range. One of the most intriguing properties of the human eye-brain is its ability to smoothly perceive a seamless and stable 3D world even though the eyes are constantly moving. These *saccades* of the eye are necessary for proper human visual perception. A significant part of the human brain is engaged in processing visual input. Other characteristics of the human visual system will be discussed at various points in the book: in particular, more details of color perception are given in Chapter 6.

Exercise 2.3

Assume that a human eyeball is 1 inch in diameter and that 10^8 rods and cones populate a fraction of $1/\pi$ of its inner surface area. What is the average size of area covered by a single receptor? (Remember, however, that foveal receptors are packed much more densely than this average, while peripheral receptors are more sparse.)

2.3 PROBLEMS IN DIGITAL IMAGES*

Several problems affect the sensing process, some of the most important of which are listed next. Usually, our idealized view given previously is only an approximation to the real physics. The overall effect of the combination of these problems is an image that has some distortion in both its geometry and intensities. Methods for correcting some of these problems are given in Chapter 11; methods for making decisions despite such imperfections are more common, however.

2.3.1 Geometric Distortion

Geometric distortion is present in several ways in the imaging process. The lens may be imperfect so that the beams of light being collected from a scene surface element are not bent exactly as intended. Barrel distortion is commonly observed for small focal length lenses; straight lines at the periphery of the scene appear to bow away from the center of the image as shown at the right in Figure 2.6.

2.3.2 Scattering

Beams of radiation can be bent or dispersed by the medium through which they pass. Aerial and satellite images are particularly susceptible to such effects, which are caused by water vapor or temperature gradients that give lens-like characteristics to the atmosphere.

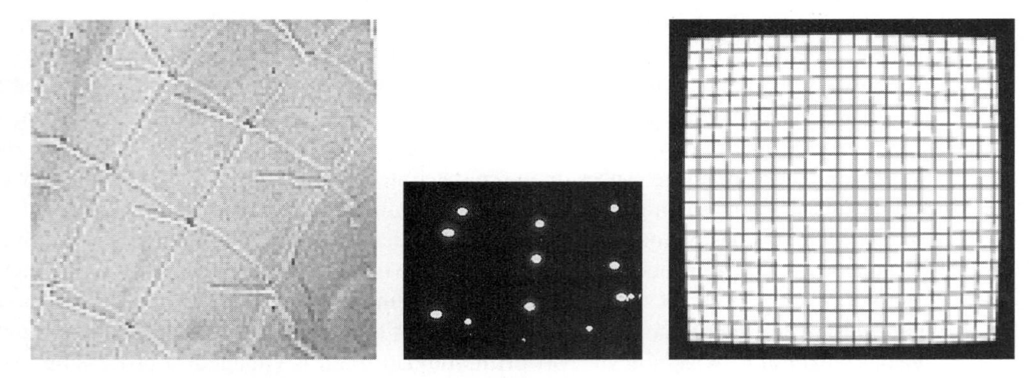

Figure 2.6 Images showing various distortions. (left) Gray-level clipping during A/D conversion occurs at the intersection of some bright stripes; (center) blooming increases the intensity at the neighbors of bright pixels; and (right) barrel distortion is often observed when short focal length lenses are used.

2.3.3 Blooming

Because discrete detectors, such as CCD cells, are not perfectly insulated from each other, charge collected at one cell can leak into a neighboring cell. The term *blooming* arises from the phenomena where such leakage spreads out from a very bright region on the image plane, resulting in a bright "flower" in the image that is larger than it actually should be as shown in Figure 2.6 (center).

2.3.4 CCD Variations

Due to imperfections in manufacturing, there may be variations in the responses of the different cells to identical light intensity. For precise interpretation of intensity, it may be necessary to determine a full array of scale factors $\mathbf{s[r, c]}$ and shifts $\mathbf{t[r, c]}$, one for each pixel, by calibration with uniform illumination so that intensity can be restored as $\mathbf{I_2[r, c] = s[r, c]I_1[r, c] + t[r, c]}$. In an extreme case, the CCD array may have some *dead cells* which give no response at all. Such defects can be detected by inspection: one software remedy is to assign the response of a dead cell to be the average response of the neighbors.

2.3.5 Clipping or Wrap-Around

In the analog to digital conversion, a very high intensity may be clipped off to a maximum value, or, its high order bits may be lost, causing the value to be wrapped-around into some encoding for a lower intensity. The result of wrap-around is seen in a gray-scale image as a bright region with a darker core; in a color image it can result in a noticeable change in color. The image at the left in Figure 2.6 shows wrap-around: some intersections of bright lines result in pixels darker than those for either line.

2.3.6 Chromatic Distortion

Different wavelengths of light are bent differently by a lens (the *index of refraction* of the lens varies with wavelength). As a result, energy in different wavelengths of light from the same scene spot may actually image a few pixels apart on the detector. For example, the image of a very sharp black-white boundary in the periphery of the scene may result in a *ramp* of intensity change spread over several pixels in the image.

2.3.7 Quantization Effects

The digitization process collects a sample of intensity from a discrete area of the scene and maps it to one of a discrete set of gray values and thus is susceptible to both mixing and rounding problems. These are addressed in more detail in the next section.

2.4 PICTURE FUNCTIONS AND DIGITAL IMAGES

We now discuss some concepts and notation important for both the theory and programming of image-processing operations.

2.4.1 Types of Images

In computing with images, it is convenient to work with both the concepts of *analog image* and *digital image*. The picture function is a mathematical model that is often used in analysis where it is fruitful to consider the image as a function of two variables. All of functional analysis is then available for analyzing images. The digital image is merely a 2D rectangular array of discrete values. Both image space and intensity range are quantized into a discrete set of values, permitting the image to be stored in a 2D computer memory structure. It is common to record intensity as an 8-bit (1-byte) number which allows values of 0 to 255. 256 different levels is usually all the precision available from the sensor and also is usually enough to satisfy the consumer. And, bytes are convenient for computers. For example, an image might be declared in a C program as **char I[512][512].** Each pixel of a color image would require 3 such values. In some medical applications, 10-bit encoding is used, allowing 1,024 different intensity values, which approaches the limit of humans in discerning them.

The following definitions are intended to clarify important concepts and also to establish notation used throughout this book. We begin with an ideal notion of an analog image created by an ideal optical system, which we assume to have infinite precision. Digital images are formed by *sampling* this analog image at discrete locations and representing the intensity at a location as a discrete value. All real images are affected by physical processes that limit precision in both position and intensity.

2 Definition. An **analog image** is a 2D image **F(x, y)** which has infinite precision in spatial parameters x and y and infinite precision in intensity at each spatial point **(x, y)**.

3 Definition. A **digital image** is a 2D image **I[r, c]** represented by a discrete 2D array of intensity samples, each of which is represented using a limited precision.

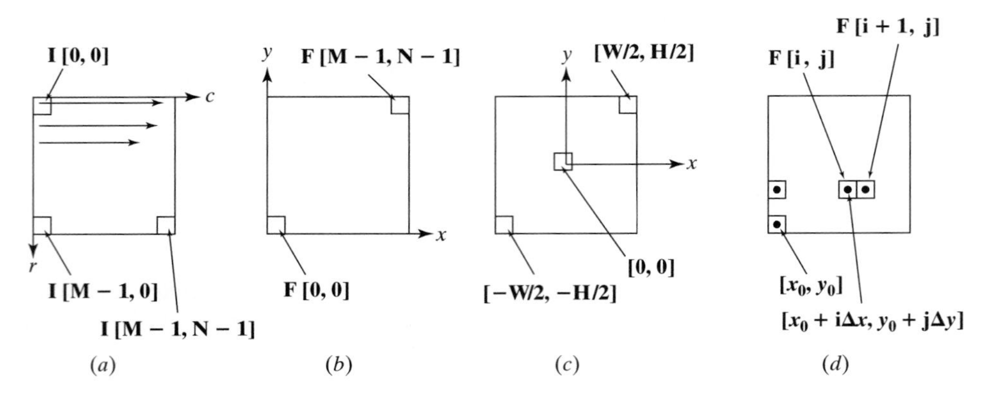

Figure 2.7 Different coordinate systems used for images: (*a*) *raster oriented* uses row and column coordinates starting at [0, 0] from the top left; (*b*) Cartesian coordinate frame with [0, 0] at the lower left; (*c*) Cartesian coordinate frame with [0, 0] at the image center; and (*d*) Relationship of pixel center point [x, y] to area element sampled in array element **I[i, j]**.

The mathematical model of an image as a function of two real spatial parameters is enormously useful in both describing images and defining operations on them. Figure 2.7(*d*) shows how the pixels of an image are samples of a continuous image taken at various points **[x, y]** of the image plane. If there are M samples in the X-direction across a distance of w, then the x-spacing Δx between pixels is w/M. The formula relating the center point of a pixel to the array cell containing the intensity sample is given in Figure 2.7.

4 Definition. A **picture function** is a mathematical representation **f(x, y)** of a picture as a function of two spatial variables **x** and **y**. **x** and **y** are real values defining points of the picture and **f(x, y)** is usually also a real value defining the intensity of the picture at point **(x, y)**.

5 Definition. A **gray-scale image** is a monochrome digital image **I[r, c]** with one intensity value per pixel.

6 Definition. A **multispectral image** is a 2D image **M[x, y]** which has a vector of values at each spatial point or pixel. If the image is actually a color image, then the vector has 3 elements.

7 Definition. A **binary image** is a digital image with all pixel values 0 or 1.

8 Definition. A **labeled image** is a digital image **L[r, c]** whose pixel values are *symbols* from a finite alphabet. The symbol value of a pixel denotes the outcome of some decision made for that pixel. Related concepts are **thematic image** and **pseudo-colored image.**

A coordinate system must be used to address individual pixels of an image; to operate on it in a computer program, to refer to it in a mathematical formula, or to address it relative to device coordinates. Different systems used in this book and elsewhere are shown in Figure 2.7. Unfortunately, different computer tools often use different systems and the user will need to get accustomed to them. Fortunately, concepts are not tied to a coordinate system. In this book, concepts are usually discussed using a Cartesian coordinate system consistent with mathematics texts while image-processing algorithms usually use raster coordinates.

2.4.2 Image Quantization and Spatial Measurement

Each pixel of a digital image represents a sample of some elemental region of the real image as is shown in Figure 2.2. If the pixel is projected from the image plane back out to the source material in the scene, then the size of that scene element is the *nominal resolution* of the sensor. For example, if a 10 inch square sheet of paper is imaged to form a 500 × 500 digital image, then the nominal resolution of the sensor is 0.02 inches. This concept may not make sense if the scene has a lot of depth variation, since the nominal resolution will vary with depth and surface orientation. The *field of view* of an imaging sensor is a measure of how much of the scene it can see. The *resolution* of a sensor is related to its precision in making spatial measurements or in detecting fine features. (With careful use, and some model information, a 500 × 500 pixel image can be used to make measurements to an accuracy of 1 part in 5,000, which is called *subpixel resolution*.)

> **9 Definition.** The **nominal resolution** of a CCD sensor is the size of the scene element that images to a single pixel on the image plane.

> **10 Definition.** The term **resolution** refers to the precision of the sensor in making measurements, but is formally defined in different ways. If defined in real world terms, it may just be the nominal resolution, as in "the resolution of this scanner is one meter on the ground" or it may be in the number of line pairs per millimeter that can be *resolved* or distinguished in the sensed image. A totally different concept is the number of pixels available—"the camera has a resolution of 640 by 480 pixels." This later definition has an advantage in that it states into how many parts the field of view can be divided, which relates to both the capability to make precise measurements and to cover a certain region of a scene. If precision of measurement is a fraction of the nominal resolution, this is called **subpixel resolution.**

Figure 2.8 shows four images of the same face to emphasize resolution effects: humans can recognize a familiar face using 64 × 64 resolution, and maybe using 32 × 32 resolution, but 16 × 16 is insufficient. In solving a problem using computer vision, the implementor should use an appropriate resolution; too little resolution will produce poor recognition or imprecise measurements while too much will unnecessarily slow down algorithms and waste memory.

> **11 Definition.** The **field of view** of a sensor **(FOV)** is the size of the scene that it can sense, for example 10 inches by 10 inches. Since this may vary with depth, it may be more meaningful to use **angular field of view,** such as 55 degrees by 40 degrees.

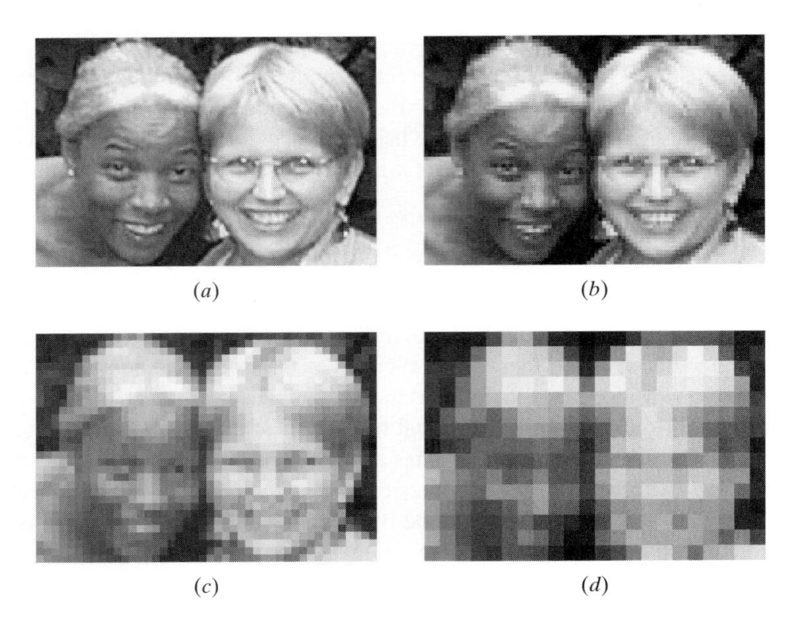

Figure 2.8 Four digital images of two faces; (*a*) 127 rows of 176 columns; (*b*) (126×176) created by averaging each 2×2 neighborhood of (*a*) and replicating the average four times to produce a 2×2 average block; (*c*) (124×176) created in same manner from (*b*); and (*d*) (120×176) created in same manner from (*c*). Effective nominal resolutions are (127×176), (63×88), (31×44), and (15×22) respectively. (Try looking at the blocky images by squinting; it usually helps by blurring the annoying sharp boundaries of the squares.) (Photo courtesy of Frank Biocca.)

Since a pixel in an image measures an area in the real world and not a point, its value often is determined by a mixture of different materials. For example, consider a satellite image where each pixel samples from a spot of the earth $10m \times 10m$. Clearly, that pixel value may be a sample of water, soil, and vegetation combined. The problem appears in a severe form when binary images are formed. Reconsider the above example of imaging a sheet of paper with 10 characters per inch. Many image pixels will overlap a character boundary and hence receive a mixture of higher intensity from the background and lower intensity from the character; the net result being a value in between background and character that could be set to either 0 or 1. Whichever value it is, it is partly incorrect!

Figure 2.9 gives details of quantization problems. Assume that the 2D scene is a 10×10 array of black (brightness 0) and white (brightness 8) tiles as shown at the left in the figure. The tiles form patterns that are two bright spots and two bright lines of different widths. If the image of the scene falls on a 5×5 CCD array such that each 2×2 set of adjacent tiles falls precisely on one CCD element the result is the digital image shown in Figure 2.9(*b*). The top left CCD element senses intensity $2 = (0 + 0 + 0 + 8)/4$ which is the average intensity from four tiles. The set of four bright tiles at the top right falls on two CCD elements, each of which integrates the intensity from two bright and two dark tiles. The single row of bright tiles of intensity 8 images as a row of CCD elements

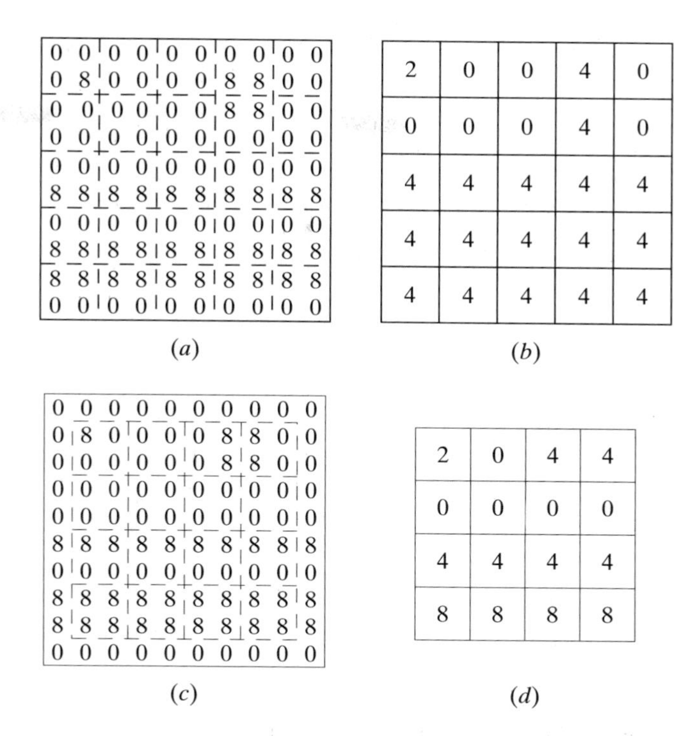

Figure 2.9 (*a*) 10 × 10 field of tiles of brightness 0 or 8; (*b*) Intensities recorded in a 5 × 5 image of precisely the brightness field at the left where each pixel senses the average brightness of a 2 × 2 neighborhood of tiles; (*c*) Image sensed by shifted camera one tile down and one tile to the right. Note that the quantized brightness values depend on both the actual pixel size and position relative to the brightness field; and (*d*) Intensities recorded from the shifted camera in the same manner as in (*b*). Interpretation of the actual scene features will be problematic with either image (*b*) or (*d*).

sensing intensity 4, while the double row images as two rows of intensity 4; however, the two lines in the scene are blended together in the image. If the image is thresholded at $t = 3$, then a bright pattern consisting of one tile will be lost from the image and the three other features will all fuse into one region! If the camera is displaced by an amount equivalent to one tile in both the horizontal and vertical direction, then the image shown in Figure 2.9(*d*) results. The shape of the 4-tile bright spot is distorted in (*d*) in a different manner than in (*b*) and the two bright lines in the scene result in a *ramp* in (*d*) as opposed to the constant gray region of (*b*); moreover, (*d*) shows three object regions whereas (*b*) shows two. Figure 2.9 shows that the images of scene features that are nearly the size of one pixel are unstable.

Figure 2.9, shows how *spatial quantization effects* impose limits on measurement accuracy and detectability. Small features can be missed or fused and even when larger features are detected, their spatial extent might be poorly represented. Note how the bright set of four tiles images as either a vertical or horizontal pair of CCD elements of intensity 4.

Perhaps we should expect an error as bad as 0.5 pixels in the placement of a boundary due to rounding of a *mixed pixel* when a binary image is created by thresholding; this implies a one pixel expected error in a measurement made across two boundaries. Moreover, *if we expect to detect certain features in a binary image, then we must make sure that their image size is at least two pixels in diameter: this includes gaps between objects*. Consider a *period* ending a sentence in a FAX whose image is one pixel in diameter but falls exactly centered at the point where 4 CCD cells meet: Each of the 4 pixels will be mixed with more background than character and it is likely that the character will be lost when a binary image is formed!

12 Definition. A **mixed pixel** is an image pixel whose intensity represents a sample from a mixture of material types in the real world.

Exercise 2.4: On variation of area.

Consider a dark rectangle on a white paper that is imaged such that the rectangle measures exactly 5.9×8.1 pixels on the real image. A binary image is to be produced such that pixel values are 0 or 1 depending upon whether the pixel sees more object or more background. Allow the rectangle to translate with its sides parallel to the CCD rows and columns. What is the smallest area in pixels in the binary image output? What is the largest area?

Exercise 2.5: On loss of thin features.

Consider two bright parallel lines of conductor on a printed circuit board. The width of each line is 0.8 pixels on the image plane. Is there a situation where one line and not the other would disappear in a binary image created as in the exercise above? Explain.

In Chapter 13, the thin lens equation from optics is reviewed and is studied with respect to how it relates camera resolution, image blur, and depth of field; the interested reader will be able to understand that section at this point. Having taken some care to discuss the characteristics of sensing and the notions of resolution and mixed pixels, we now have enough background to begin working on certain 2D machine vision applications. We might want to find certain objects using a microscope, inspect a PC board, or recognize the shadow of a backlit 3D object. The imaging environment must be engineered so that the features that must be seen are of the proper size in the image. Assuming that there is no significant 3D character remaining in the image after the scaling from world to image is considered, the images can then be analyzed using the 2D methods described in Chapters 3 to 10.

Exercise 2.6: Sensing paper money denominations.

Consider the design in a sensor for a vending machine that takes U.S. paper money in denominations of \$1, \$5, \$10, and \$20. You only need to create a representation for the recognizer to use; you need not design a recognition algorithm, nor should you be concerned about detecting counterfeit bills. (Be sure to obtain some samples before answering.) Assume that a linear CCD array must be used to digitize a bill as it enters the machine. (a) What kind of lens and what kind of illumination should be used? (b) How many pixels are needed in the linear array? Explain.

2.5 DIGITAL IMAGE FORMATS*

Use of digital images is widespread in communication, databases, and machine vision and standard formats have been developed so that different hardware and software can share data. Figure 2.10 sketches this situation. Unfortunately, there are dozens of different formats still in use. A few of the most important ones are briefly discussed in this section. A *raw image* may be just a stream of bytes encoding the image pixels in row-by-row order, called *raster order,* perhaps with line-feeds separating rows. Information such as image type, size, time taken, and creation method is not part of a raw image. Such information might be handwritten on a tape label or in someone's research notebook—this is inadequate. (One project, in which one author took part, videotaped a barcode before videotaping images from the experiment. The computer program would then process the barcode to obtain overall non-image information about the experimental treatment.) Most recently developed standard formats contain a header with non-image information necessary to label the data and to decode it.

Several formats originated with companies creating image processing or graphics tools; in some cases but not in others, public documentation and conversion software are available. The details provided next should provide the reader with practical information for handling computer images. Although the details are rapidly changing with technology, there are several general concepts contained in this section that should endure.

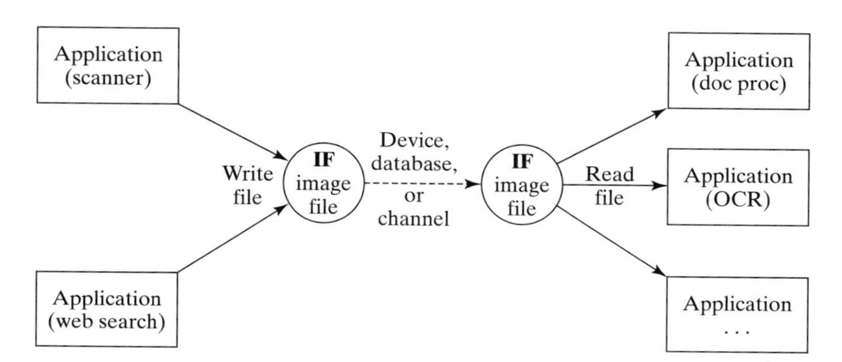

Figure 2.10 Many devices or application programs create, consume, or convert image data. Standard format image files (**IF**s) are needed to do this productively for a family of devices and programs.

2.5.1 Image File Header

A *file header* is needed to make an image file self-describing so that image-processing tools can work with them. The header should contain the image dimensions, type, date of creation, and some kind of title. It may also contain a color table or coding table to be used to interpret pixel values. A nice feature not often available is a *history section* containing notes on how the image was created and processed.

2.5.2 Image Data

Some formats can handle only limited types of images, such as binary and monochrome; however, those surviving today have continued to grow to include more image types and features. Pixel size and image size limits typically differ between different file formats. Several formats can handle a sequence of frames. *Multimedia* formats are evolving and include image data along with text, graphics, music, etc.

2.5.3 Data Compression

Many formats provide for *compression* of the image data so that all pixel values are not directly encoded. Image compression can reduce the size of an image to 30 percent or even 3 percent of its raw size depending on the quality required and method used. Compression can be *lossless* or *lossy*. With lossless compression, the original image can be recovered exactly. With lossy compression, the pixel representations cannot be recovered exactly: sometimes a loss of quality is perceived, but not always. To implement compression, the image file must include some overhead information about the compression method and parameters. Most digital images are very different from symbolic digital information—loss or change of a few bits of digital image data will have little or no effect on its consumer, regardless of whether it is a human or machine. The situation is quite different for most other computer files; for example, changing a single bit in an employee record could change the salary field by $8,192 or the apartment address from A to B. Image compression is an exciting area that spans the gamut from signal processing to object recognition. Image compression is discussed at several points of this textbook, but is not systematically treated.

> **13 Definition.** An image compression method is **lossless** if a decompression method exists to precisely recover (every bit of) the original image representation. Otherwise, the compression method is **lossy.**

2.5.4 Commonly Used Formats

Many of the images in this book passed through multiple formats. Some images were received from a colleague or retrieved from an image database in GIF, JPG or even PS format. Some were scanned from photos and their original digital format was GIF or TIFF. Simple image processing might have been done using the image tool xv and more complex operations were done with hips tools or with special C or C++ programs. Some of the most commonly used formats are briefly described next. **Image/Graphics file formats are still evolving** with a trend for each to be more inclusive. The reader should be aware that some of the details given below will have to be updated by consulting the latest reference document.

```
                  0000000000111111111122222222223333333333444444444
Column c       :  0123456789012345678901234567890123456789012345678

Image Row r    :  0000000011111000000000000111000000011111111100000

Run-code A     :  8(0)5(1)12(0)3(1)7(0)9(1)5(0)

Run-code B     :  (8,12) (25,27) (35,43)
```

Figure 2.11 Run-coding encodes the runs of consecutive 0 or 1 values, and for some domains, yields an efficiently compressed image.

2.5.5 Run-Coded Binary Images

Run-coding is an efficient coding scheme for binary or labeled images: not only does it reduce memory space, but it can also speed up image operations, such as set operations. Run-coding works well when there is a lot of redundancy in pixels along the image rows. Assume a binary image; for each image row, we could record the *number* of 0s followed by the number of 1s alternating across the entire row. Figure 2.11A gives an example. Run-code B of the figure shows a more compact encoding of just the 1-runs from which we can still recover the original row. We will use such encodings for some algorithms in this book. Run-coding is often used for compression within standard file formats.

2.5.6 PGM: Portable Gray Map

One of the simplest file formats for storing and exchanging image data is the **PBM** or **P**ortable **B**it **M**ap family of formats (PBM/PGM,PPM). The image header and pixel information are encoded in ASCII. The image file representing an image of 8 rows of 16 columns with maximum gray value of 192 is shown in Figure 2.12. Two graphic renderings are also shown, each is the output of image conversion tools applied to the original text input. The image at the lower left was made by replicating the pixels to make a larger image of 32 rows of 64 columns each; the image at the lower right was made by first converting to JPG format with lossy compression. The first entry of the PGM file is the *Magic Value,* "P2" in our example, indicating how the image information is coded (ASCII grey levels in our example). Binary, rather than ASCII pixel coding is available for large pictures. (The magic number for binary is "P4").

Exercise 2.7: Creating a PPM picture.

A color image can be coded in PBM format by using the magic value "P3" and three (R,G,B) intensity values for each pixel, similar to the coded monochrome "P2" file shown in Figure 2.12. Using your editor, create a file bullseye.ppm encoding 3 concentric circular regions of different colors. For each pixel, the three color values follow in immediate succession, rather than encoding three separate monochrome images as is done in some other formats. Display your picture using an image tool or Web browser.

```
P2
# sample small picture 8 rows of 16 columns, max gray value of 192
# making an image of the word "Hi".
  16 8   192

 64 64  64  64  64  64  64  64  64 64 64  64  64 64 64 64
 64 64 128 128  64  64  64 128 128 64 64 192 192 64 64 64
 64 64 128 128  64  64  64 128 128 64 64 192 192 64 64 64
 64 64 128 128 128 128 128 128 128 64 64  64  64 64 64 64
 64 64 128 128 128 128 128 128 128 64 64 128 128 64 64 64
 64 64 128 128  64  64  64 128 128 64 64 128 128 64 64 64
 64 64 128 128  64  64  64 128 128 64 64 128 128 64 64 64
 64 64  64  64  64  64  64  64  64 64 64  64  64 64 64 64
```

Figure 2.12 Text (ASCII) file representing an image of the word "Hi"; 64 is the background level, 128 is the level of "H" and the lower part of "i", and 192 is the level of the dot of the "i". At the lower left is a printed picture made from the above text file using image format conversion tools. At the bottom right is an image made using a lossy compression algorithm.

2.5.7 GIF Image File Format

The **G**raphics **I**nterchange **F**ormat (GIF) originated from CompuServe, Inc. and has been used to encode a huge number of images on the World Wide Web or in current databases. GIF files are relatively easy to work with, but cannot be used for high-precision color, since only 8-bits are used to encode color. The 256 color values available are typically sufficient for computer displayed images; a more compact 16-color option can also be used. Lempel-Ziv-Welch (LZW) nonlossy compression is available.

2.5.8 TIFF Image File Format

Originated by Aldus Corp., TIFF or TIF is very general and very complex. It is used on all popular platforms and is often the format used by scanners. **T**ag **I**mage **F**ile **F**ormat supports multiple images with 1 to 24 bits of color per pixel. Options are available for either lossy or lossless compression.

2.5.9 JPEG Format for Still Photos

JPEG (JFIF/JFI/JPG) is a more recent standard from the Joint Photographic Experts Group; the major purpose was to provide for practical compression of high-quality color still images. The JPEG coding scheme is stream-oriented and allows for real-time hardware for coding and decoding. An image can have up to 64K $\times$ 64K pixels of 24 bits each, although there is only one image per file. The header can contain a thumbnail image of up to 64k uncompressed bytes. JPEG is independent of the color coding system, a major advantage. More details on color systems are given in Chapter 6. To achieve high compression, a flexible, but complex lossy coding scheme is used which often can compress a high quality image 20:1 without noticeable degradation. The compression works well when the image has large regions of nearly constant color and when high frequency variation in regions of detail is not important to the consumer. (JPEG has a little used lossless compression option, which might achieve 2:1 compression using *predictive coding*.) The compression scheme uses the *discrete cosine transformation,* which is discussed in Chapter 5, followed by *Huffman coding*, which is not treated in this book. JPEG is *not* designed for video.

Exercise 2.8

Locate an image viewing toolset on your computer system. (These might be available just by clicking on your image file icon.) Use one image of a face and one of a landscape; both, should originally be of high quality, say 800 $\times$ 600 color pixels, from a flatbed scanner or digital camera. Convert the image among different formats—GIF, TIFF, JPEG, etc. Record the size of the encoded image files in bytes and note the quality of the image; consider the overall scene plus small details.

Exercise 2.9: JPEG study.*

(a) Research the JPEG compression scheme for 8 $\times$ 8 image blocks. (b) Implement and test the DCT scheme in a lossless manner (except for possible roundoff error). (c) Create a lossy compression using some existing image tool. (d) Using the 64 coefficients from lossy compression, regenerate an 8 $\times$ 8 image and compare its pixel values with those of the original 8 $\times$ 8 image.

2.5.10 PostScript

The family of formats BDF/PDL/EPS store image data using printable ASCII characters and are often used with X11 graphics displays and printers. PDL is a page description language and EPS is encapsulated postscript (originally from Adobe), which is commonly used to contain graphics or images to be inserted into a larger document. Pixel values are encoded via 7-bit ASCII codes, so these files can be examined and changed by a text editor. 75 to 3,000 dots per inch of gray scale or color can be represented and newer versions include JPEG compression. A PDL header contains the bounding box of the image on the page where it is to appear. Most of the images in this book have been included as EPS files.

2.5.11 MPEG Format for Video

MPEG (MPG/MPEG-1/MPEG-2) is a stream-oriented encoding scheme for video, audio, text, and graphics. MPEG stands for **M**otion **P**icture **E**xperts **G**roup, an international group of representatives from industry and governments. The MPEG family of standards is currently evolving rapidly along with the technology of computers and communication. MPEG-1 is primarily designed for multimedia systems and provides for a data rate of 0.25 Mbits per second of compressed audio and 1.25 Mbits of compressed video. These rates are suitable for multimedia for popular personal computers, but are too low for high-quality TV. The MPEG-2 standard provides for up to 15 Mbits per second data rates to handle high definition TV rates. The compression scheme takes advantage of both spatial redundancy, as used in JPEG, and temporal redundancy and generally provides a useful compression ratio of 25 to 1, with 200 to 1 ratios possible. *Temporal redundancy* essentially means that many regions do not change much from one frame to the next and an encoding scheme can just encode changes and even predict frames from frames before and after in the video sequence. (Future versions of MPEG will have codes for recognized objects and program code to generate their images.) Media quality is determined at the time of encoding. Motion JPEG is a hybrid scheme which just applies JPEG compression to video single frames and does not take advantage of temporal redundancy. While encoding and decoding is simplified using Motion JPEG, compression is not as good, so memory usage and transmission will be poorer than with MPEG. Use of motion vectors by MPEG for compression of video is described in Chapter 9.

2.5.12 Comparison of Formats

Table 2.1 compares some popular image formats in terms of storage size. The left columns of the table apply to the tiny 8×16 gray-scale picture "Hi" whereas the right column applies to a 347×489 color image. It is possible to obtain different size pictures for the *same* image by using different sequences of format conversions. For example, the "Cars" TIF file output from the scanner was 509,253 bytes, whereas a conversion to a GIF file with only 256 colors required 138,267 bytes, and a "TIF" file derived from that required 171,430 bytes. This final TIF file had fewer bits in the color codes, but appeared to be qualitatively the same viewed on the CRT. The JPEG file one third its size also displayed the same. While the lossy JPEG is a clear winner in terms of space, this is at a cost of decoding complexity which may require hardware for real-time performance.

TABLE 2.1 FILE SIZES (IN BYTES) FOR THE SAME IMAGE ENCODED IN DIFFERENT FORMATS: 8×16 GRAY-LEVEL "HI" IMAGE SHOWN IN FIGURE 2.12 AND 347×489 COLOR "CARS" IMAGE SHOWN IN FIGURE 2.13.

Image File Format	No. Bytes "Hi"	No. Bytes "Cars"
PGM	595	509,123
GIF	192	138,267
TIF	918	171,430
PS	1,591	345,387
HIPS	700	160,783
JPG (lossless)	684	49,160
JPG (lossy)	619	29,500

2.6 RICHNESS AND PROBLEMS OF REAL IMAGERY

A brief walk with open eyes and mind will confirm what the artist already knows about the richness of the natural visual world. This richness enhances human experience but causes problems for machine vision. (See Figure 2.13, for example.) The intensity or color at an image point depends in complex ways on material, geometry, and lighting; not only is the type of material important, but so is its orientation relative to the sensor, light sources, and other objects. There are specularities on shiny surfaces, shadows, mutual reflection, and transparent materials, for example. For recognition of many surfaces or objects, color may be of little importance relative to shape or texture, characteristics that depend upon many pixels, not just one. Cases where we have little control over the environment, such as monitoring traffic patterns, can be interesting and difficult.

Figure 2.13 A complex scene with many kinds of depth cues for human perception.

Problems still remain even in well-engineered industrial environments or TV studios. As we shall see in Chapter 6, reflection from a shiny metal cylinder illuminated by a point source can vary in intensity over a range of 100,000 to 1 and most sensors cannot handle such a dynamic range. Sunlight or artificial light can heat surfaces, causing them to radiate differently over time, brightening CCD images with increased infrared or leaving shadows of airplanes on a runway after their departure. Controlled monochrome laser light can greatly assist some imaging operations, but it can also be totally absorbed by certain surfaces or be dominated by secondary reflections on others.

In many applications of automation, problems can be solved by engineering. Irrelevant bandwidths of light can be filtered out; for example, bruises in dark red cherries can be seen more clearly if a filter is used to allow only infrared light to pass. Moving objects that would cause a blurred image under steady illumination can be illuminated by a *strobe light* for a very short period of time, so that they appear still in an image formed by a highly sensitive detector. Use of *structured light* can make surface measurement and inspection much easier: for example, turbine blades can be illuminated by precise alternating stripes of red and green light, so that many surface defects appear as obvious breaks in the smoothness of the stripes in the 2D image. We will return to these methods at various points of the text.

2.7 3D STRUCTURE FROM 2D IMAGES

The human vision system perceives the structure of the 3D world by integrating several different cues. Here we give only a qualitative description. Cognitive psychologist J. J. Gibson outlined quantitative models for many of these cues. Implementation and demonstration of these models was intensely pursued by computer vision researchers in the 1980s, and several of the quantitative models will be discussed at several points in the book.

The imaging process records complex relationships between the 3D structure of the world and the 2D structure of the image. Assume the perspective projection model that was described with Figure 2.2 and refer to Figure 2.13. *Interposition* is perhaps the most important depth cue: objects that are closer occlude parts of objects that are farther away; recognition of occlusions gives relative depth. A person seen within the region of a wall is clearly closer to the sensor than the wall. A person recognized behind a car is farther away than the car. Relative size is also an important cue. The image of a car $20m$ away will be much smaller than the image of a car $10m$ away, even though the far car might be a larger car. Cars appear to us to be both tiny and moving slowly in the distance; our experience has taught us how to relate the size and speed to the distance. As we walk down the railroad track, the rails appear to meet at a point in the distance (the *vanishing point*), although we know that they must maintain the same separation in 3D. A door that is open into our room images on our retina as a trapezoid and not the rectangle we know it to be. The far edge of the door appears shorter than the near edge; this is the *foreshortening* effect of perspective projection and conveys information about the 3D orientation of the door. A related cue is *texture gradient*. The texture of surfaces changes with both the distance from the viewer and the surface orientation. In the park, we can see individual blades of grass or maple leaves up close, while far away we see only green color. The change of image texture due to perspective viewing of a surface receding in the distance is called the *texture gradient*. Chapter 12 gives much more discussion of the issues just mentioned.

Exercise 2.10: Observe as an artist.

Consciously make observations in two different environments and sketch some of the cues discussed above. For example, try a busy cafeteria, or a city street corner observed from a height of a few floors, or a spot in a woods.

2.8 FIVE FRAMES OF REFERENCE

Reference frames are needed in order to do either qualitative or quantitative analysis of 3D scenes. Five frames of reference are needed for general problems in 3D scene analysis, such as controlling operations in a work cell with robots and sensors or providing a virtual 3D environment for human interaction. Several of these frames are not only important for robotics, but are also important to psychologists and the understanding of human spatial perception. The five types of frames are illustrated in Figure 2.14; actually, six reference frames are shown since there are two different objects in the scene, a block and a pyramid, each with its own reference frame. In all of these coordinate frames, coordinates are real numbers along continuous axes, except for image coordinates, which are integer subscripts of the pixel array. For the examples in the discussion you should also imagine an analogous situation where the camera is a TV camera covering a baseball game and objects in the scene are the players, bases, balls, bats, etc.

2.8.1 Pixel Coordinate Frame I

In the pixel array, each point has integer pixel coordinates. In Figure 2.14, the image of the tip of the pyramid **A** falls within pixel $\mathbf{a} = [a_r, a_c]$ where a_r and a_c are integer row and column, respectively. Many things about a scene can be determined by analysis of the image in terms of only pixel rows and columns. For example, if a pick-and-place robot or other transfer mechanism always delivered a block (or box of laundry detergent) roughly frontal to the camera, then the markings on its frontal surface could be inspected using only the image as a matrix of rows and columns of pixels. In the baseball game analogy, using only the image, one could determine if a batter was using a black bat. Using only image **I**,

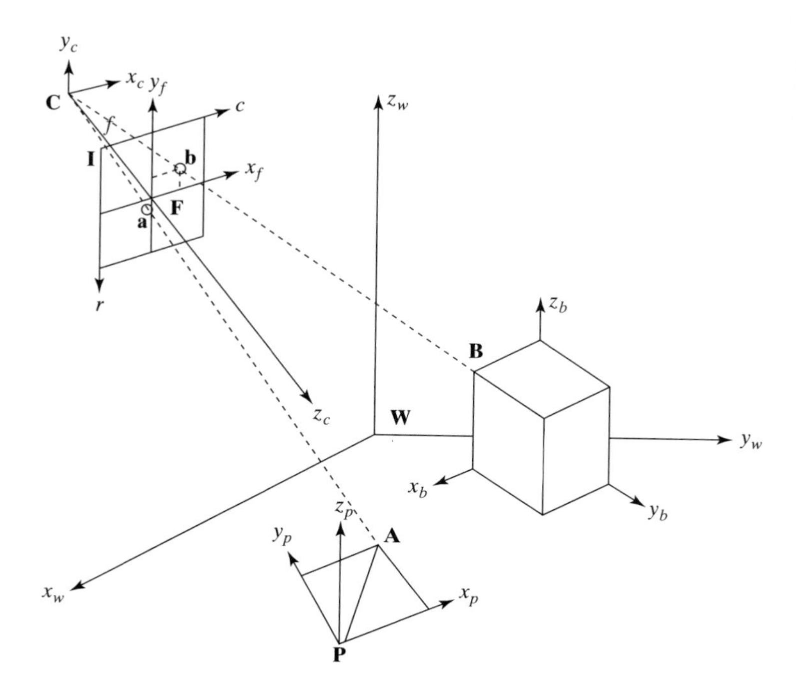

Figure 2.14 Five coordinate frames needed for 3D scene analysis: world **W**, object **O** (for pyramid $\mathbf{O_p}$ or block $\mathbf{O_b}$), camera **C**, real image **F** and pixel image **I**.

however, and no other information, we cannot determine which object is actually larger in 3D or whether or not objects are on a collision course.

2.8.2 Object Coordinate Frame O

An object coordinate frame is used to model ideal objects in both computer graphics and computer vision. For example, Figure 2.14 shows two object coordinate frames, one for a block $\mathbf{O_b}$ and one for a pyramid $\mathbf{O_p}$. The coordinates of 3D corner point **B** relative to the object coordinate frame are $[x_b, 0, z_b]$. These coordinates remain the same, regardless of how the block is posed relative to the world or workspace coordinate frame **W**. The object coordinate frame is needed to inspect an object; for example, to check if a particular hole is in proper position relative to other holes or corners.

2.8.3 Camera Coordinate Frame C

The camera coordinate frame C is often needed for an *egocentric* (cameracentric) view; for example, to represent whether or not an object is just in front of the sensor, moving away, etc. A ball whose image continues to enlarge in the center of your retina is likely to hit you. A seeing robot or human are both object and sensor[s] so their object and sensor coordinate systems may be almost, but not exactly, the same. (Did you ever run into a doorway even though it looked as if you'd pass through without contact?) Computer graphics systems allow the user to select different camera views of the 3D scene being viewed. A play at first base might be better viewed by a camera pointing there.

2.8.4 Real Image Coordinate Frame F

Camera coordinates are real numbers, usually in the same units as the world coordinates—say, inches or mm—*including the depth coordinate* z_c. 3D points project to the real image plane at coordinates $[x_f, y_f, f]$ where f is the focal length. x_f and y_f are not subscripts of pixels in the image array, but are related to the pixel size and pixel position of the optical axis in the image. In Figure 2.14, both coordinates of real image point **a** relative to frame **F** are negative. Frame **F** *contains* the picture function that is digitized to form the digital image in the pixel array **I**.

2.8.5 World Coordinate Frame W

The coordinate frame **W** is needed to relate objects in 3D; for example, to determine whether or not a runner is far off a base or if the runner and second baseman will collide. In a robotics cell or virtual environment, actuators and sensors often communicate via world coordinates; for example, the image sensor tells the robot where to pick up a bolt and in which hole to insert it.

Geometrical and mathematical relationships among these coordinate systems will be very important later in the book. For the next several chapters, however, we process only the information contained in the pixel array under the assumption that there is a straightforward correspondence to the real world. The reader who must work with the algebra of the perspective transformation or its scaling effect will be able at this point to go forward to Chapter 12 to study the perspective imaging model.

2.9 OTHER TYPES OF SENSORS*

This section includes the description of several more sensors. A reader might bypass this section on first reading, unless a particular sensor is very important to an application of current interest. Sensor technology is advancing rapidly; we should expect not only new sensors in the future, but also better performance of current devices.

2.9.1 Microdensitometer*

Slides or film can be scanned by passing a single beam of light *through* the material: a single sensor on the opposite side from the light records the optical density of the material at location **[r, c]**. The material is moved very precisely by mechanical stages until an entire rectangular area is scanned. Having a single sensor gives one advantage over the CCD array, there should be less variation in intensity values due to manufacturing differences. Another advantage is that many more rows and columns can be obtained. Such an instrument is slow however and cannot be used in automation environments.

The reader might find some interest in the following history of a related scanning technique. In the 1970s in the lab of Azriel Rosenfeld, many pictures were input for computer processing in the following manner. Black and white pictures were taken and wrapped around a steel cylinder. Usually 9×9 inch pictures or collages were scanned at once. The cylinder was placed in a standard lathe which spun all spots of the picture area in front of a small LED and sensor that measured light reflecting off each spot. Each revolution of the cylinder produced a row of 3,600 pixels that were stored as a block on magnetic tape whose recording speed was in sync with the lathe! The final tape file had $3{,}600 \times 3{,}600$ pixels, usually containing many experimental data sets that were then separated by software.

2.9.2 Color and Multispectral Images*

Because the human eye has separate receptors for sensing light in separate wavelength bands, it may be called a *multispectral* sensor. Some color CCD cameras are built by placing a thin refracting film just in front of the CCD array. The refracting film disperses a single beam of white light into 4 beams falling on 4 neighboring cells of the CCD array. The digital image it produces can be thought of as a set of four interleaved color images, one for each of the 4 differently refracted wavelength components. The gain in spectral information is traded for a loss in spatial resolution. In a different design, a color wheel is synchronously rotated in the optical pathway so that during one period of time only red light is passed; then blue, then green. (A color wheel is a disk of transparent film with equal size sectors of each color.) The CCD array is read 3 times during one rotation of the color wheel to obtain 3 separate images. In this design, sensing speed is traded for color sensitivity; a point on a rapidly moving object may actually image to different pixels on the image plane during acquisition of the 3 separate images.

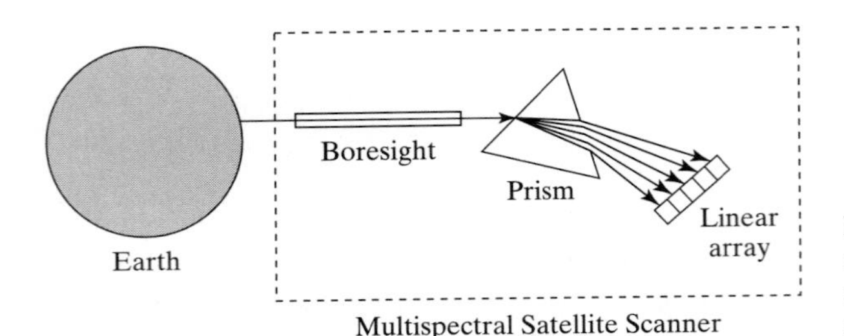

Figure 2.15 Boresighted multispectral scanner aboard a satellite. Radiation from a single surface element is refracted into separate components based on wavelength.

Some satellites use the concept of *sensing through a straw or boresight*: each spot of the earth is viewed through a *boresight* so that all radiation from that spot is collected at the same instant of time, while radiation from other spots is masked. See Figure 2.15. The beam of radiation is passed through a prism which disperses the different wavelengths onto a linear CCD array which then simultaneously samples and digitizes the intensity in the several bands used. (Recall that light of shorter wavelength is bent more by the prism than light of longer wavelength.) Figure 2.15 shows a spectrum of 5 different bands resulting in a pixel that is a vector $[\ b_1, b_2, b_3, b_4, b_5\ ]$ of 5 intensity values. A 2D image is created by moving the boresight or using a scanning mirror to get columns of a given row. The motion of the satellite in its orbit around the earth yields the different rows of the image. As you might expect, the resulting image suffers from motion distortion—the set of all scanned spots form a trapezoidal region on the earth whose form can be obtained from the *rectangular* digital image file using the warping methods of Chapter 11. By having a *spectrum of intensity values* rather than just a single intensity for a single spot of earth, it is often possible to classify the ground type as water or forest or asphalt, etc.

2.9.3 X-ray*

X-ray devices transmit X-ray radiation through material, often parts of the human body, but also welded pipes and jars of applesauce. Sensors record transmitted energy at image points on the far side of the emitter in much the same manner as with the microdensitometer. Low energy at one sensed image point indicates an accumulation of material density along the entire ray projected from the emitter. It is easy to imagine a 2D X-ray film being exposed to X-rays passing through a body. 3D sensing can be accomplished using a CT scanner (cat scanner), which mathematically constructs a 3D volume of density values from data collected by projecting X-rays along many different rays through the body. At the right in Figure 2.16 is a 2D computer graphic rendering of high density 3D voxels from a CT scan of a dog: these voxels are rendered as if they were nontransparent reflecting surfaces seen from a particular viewpoint. A diagnostician can examine the sensed bone structure from any viewpoint.

Exercise 2.11

Think about some of your own dental X-rays: what was bright and what was dark – the dense tooth or the softer cavity? Why?

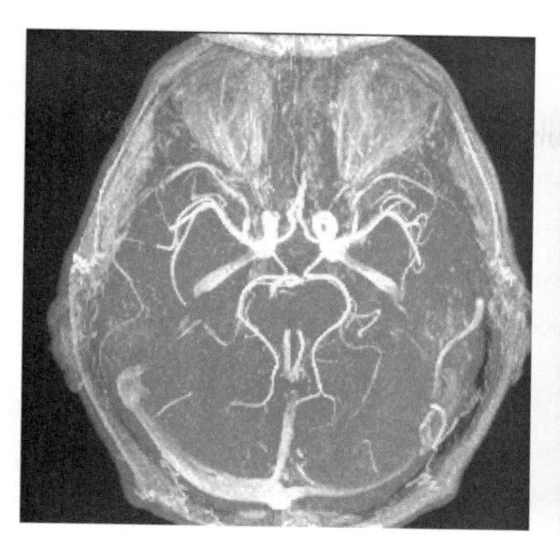

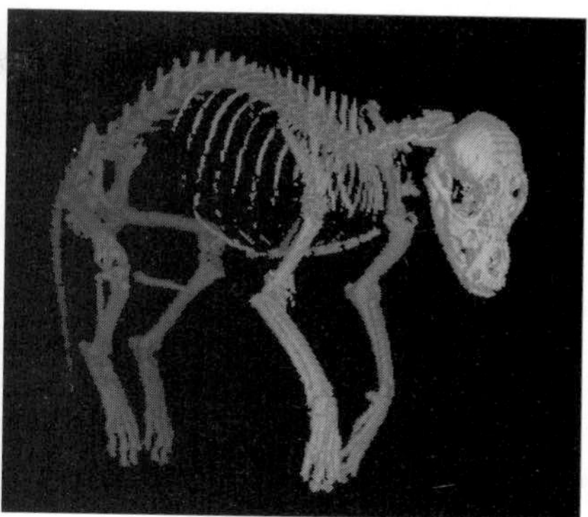

Figure 2.16 (left) A maximum intensity projection (MIP) made by projecting the brightest pixels from a set of MRA slices from a human head (provided by MSU Radiology); and (right) a computer generated image displaying high density voxels of a set of CT scans as a set of illuminated surface elements. (Data courtesy of Theresa Bernardo.)

2.9.4 Magnetic Resonance Imaging (MRI)*

Magnetic resonance imaging (MRI) produces 3D images of materials, usually parts of the human body. The data produced is a 3D array **I[s, r, c]**, where **s** indicates a slice through the body and **r** and **c** are as before. Each small volume element or *voxel* represents a sample perhaps 2 *mm* in diameter and the intensity measured there is related to the chemistry of the material. Magnetic resonance angiography (MRA) produces intensities related to the speed (of blood flow) of material at the voxel. Such scanners can cost a million dollars and a single scanning can cost a thousand dollars, but their value is well established for diagnosis. MRI scanning can detect internal defects in fruits and vegetables and may be used for such in the future if the cost of the device drops. Figure 2.16(left) shows a digital image extracted from 3D MRA data. This *maximum intensity projection,* or **MIP[r, c]**, could be produced by choosing the brightest voxel **I[s, r, c]** over all slices s. A computer algorithm can actually generate a MIP image by projecting in any view direction. Diagnosis is typically done using a wall full of such printed 2D images, but true 3D displays are now available and radiologists are learning to use them.

2.9.5 Range Scanners and Range Images*

Devices are available that sense depth or range to a 3D surface element, rather than just the intensity of radiation received from it. Samples of the surface shape of objects are directly available in a range image, whereas in an intensity image, surface shape can only be derived from difficult and error prone analysis. A LIDAR device, shown in Figure 2.17, transmits an amplitude modulated laser beam to a spot of the 3D surface and receives the reflected signal back. By comparing the change of phase (delay) between the transmitted and received signal, the LIDAR can measure the distance in terms of the period of the

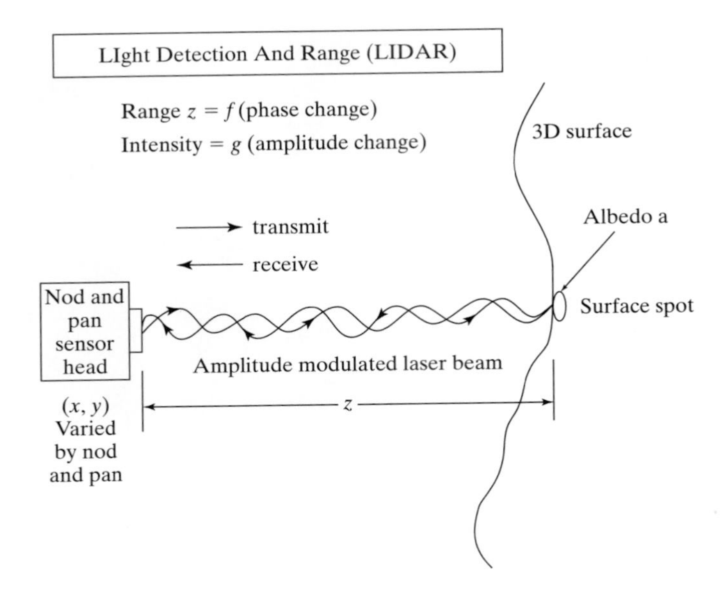

Figure 2.17 A LIDAR sensor can produce pixels containing both range and intensity.

modulation of the laser beam. This works only for distances covered by one period because of the ambiguity—a spot at distance of $d + n\lambda/2$ produces the same response as a spot at distance d, where λ is the period of modulation. Moreover, by comparing the received intensity to the transmitted intensity, the LIDAR also estimates the reflectivity of the surface spot for this wavelength of laser light. Thus, the LIDAR produces two registered images—a range image and an intensity image. The LIDAR is slower than a CCD camera because of the *dwell time* needed to compute the phase change for each spot: it is also much more expensive because of the mechanical parts needed to steer the laser beam. This expense has been justified in mining robots and in robots that explore other bodies in our solar system.

A variation of the 5,000-year-old surveying method of triangulation can be used to obtain 3D surface measurements as shown in Figure 2.18. A plane of light is projected onto the surface of the object and the bright line it produces is observed by a camera. Each bright image point $[x_c, y_c]$ is the image of some corresponding illuminated 3D point $[x_w, y_w, z_w]$.

So, the sensing device "knows" the plane of light and the ray of light from the camera center through the image point out into 3D space. From intuitive geometry, we know that the imaging ray will pierce the plane of light in a unique point. The coordinates x_w, y_w, z_w can be determined by analytical geometry: we have one equation in those 3 unknowns from the light sheet and 2 equations in those 3 unknowns from the imaging ray; solving these 3 simultaneous linear equations yields the location of the 3D surface point. In Chapter 13, *calibration* methods are given that enable us to derive the necessary equations from several measurements made on the workbench.

The above argument is even simpler if a single beam of light is projected rather than an entire plane of light. Many variations exist and a sensor is usually chosen according to

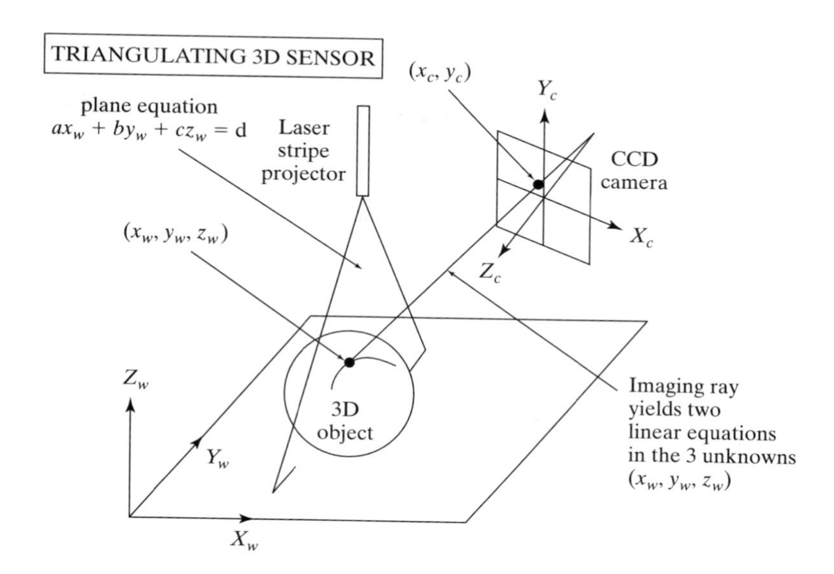

Figure 2.18 A light striping sensor produces 3D points by triangulation.

the particular application. To scan an entire scene, a light sheet or beam must be swept across the scene. Scanning mirrors can be used to do this, or objects can be translated past the sheet over time using a conveyor system. Many creative designs can be found in the literature. Machines using multiple light sheets are used to do automobile wheel alignment and to check the fit of car doors during manufacture. When looking at specific objects in very specific poses, image analysis may just need to verify if a particular image stripe is close enough to some ideal image position. The stream of observations from the sensor is used to adjust online manufacturing operations for quality control and for reporting offline.

2.10 REFERENCES

More specific information about the design of imaging devices can be found in the text by Schalkoff (1989). Tutorials and technical specifications of charge-coupled devices are readily found on the web using a search engine: one example is a tutorial provided by the University of Wisconsin at www.mrsec.wisc.edu/edetc/ccd.html. One of several early articles on the early development of color CCD cameras is by Dillon and others (1978). Discussion and modeling of many optical phenomena can be found in the book by Hecht and Zajac (1976).

Many fundamental observations leading to computer vision techniques can be found in the book by pyschologist J. J. Gibson (1950). Properties of animal vision systems and the human visual system from the perspective of an engineer are given in the text by Levine (1985). The book by Nalwa (1993) begins with a discussion of the capabilities and faults of the human visual system and gives a good intuitive description of imaging and the perspective transformation. Margaret Livingstone (1988) gives a popular treatment of human perception with an orientation to art appreciation. Many known mathematical properties of the perspective transformation are contained in Haralick and Shapiro, Volume II (1992). Practical details and an integrating overview on image file formats is provided in the encyclopedia by Murray and VanRyper (1994); a CD of common software utilities collected from several different sources is included.

1. Dillon, P., D. Lewis, and F. Kaspar. 1978. Color imaging system using a single CCD area array. *IEEE Trans. Electron Devices,* ED-25(2):102–107.
2. Gibson, J. J. 1950. *The Perception of the Visual World.* Houghton-Mifflin, Boston.
3. Hecht, E., and A. Zajac. 1974. *Optics.* Addison-Wesley, Reading, MA.
4. Haralick, R., and L. Shapiro. 1992. *Computer and Robot Vision, Volumes I and II.* Addison-Wesley, Reading, MA.
5. Levine, M. D. 1985. *Vision in Man and Machine.* McGraw-Hill, New York.
6. Livingstone, M. 1988. Art, illusion and the visual system. *Sci. Am.* (Jan. 1988) 78–85.
7. Murray, J., and W. VanRyper. 1994. *Encyclopedia of Graphics File Formats.* O'Reilly and Associates, Inc., 103 Morris St., Suite A, Sebastopol, CA 95472.
8. Nalwa, V. 1993. *A Guided Tour of Computer Vision.* Addison-Wesley, Reading, MA.
9. Schalkoff, R. J. 1989. *Digital Image Processing and Computer Vision.* John Wiley & Sons, New York.

3

Binary Image Analysis

In a number of applications, such as document analysis and some industrial machine vision tasks, binary images can be used as the input to algorithms that perform useful tasks. These algorithms can handle tasks ranging from very simple counting tasks to much more complex recognition, localization, and inspection tasks. Thus by studying binary image analysis before going on to gray-tone and color images, one can gain insight into the entire image analysis process.

In this chapter, the basic operations of binary machine vision are described. First, a simple object-counting algorithm is used to show the reader how a very simple algorithm can be used to accomplish a useful task. Next we discuss the connected components labeling operator, which gives each separate connected group of pixels a unique label and is a predecessor to most later steps of processing. Then a set of thinning and thickening operators is introduced. The operators of mathematical morphology can be used to join and separate components, close up holes, and find features of interest in an image. Once a set of components has been isolated, a number of important properties of each component can be computed for use in higher-level tasks such as recognition and tracking. A set of basic properties is defined and the accuracy of the algorithms that compute it discussed. Finally, the problem of automatically thresholding a gray-scale or color image to produce a useful binary image is studied.

3.1 PIXELS AND NEIGHBORHOODS

A binary image **B** can be obtained from a gray-scale or color image **I** through an operation that selects a subset of the image pixels as *foreground* pixels, the pixels of interest in an image analysis task, leaving the rest as *background pixels* to be ignored. The selection operation can be as simple as the thresholding operator that chooses pixels in a certain range of gray tones or subspace of color space, or it may be a complex classification algorithm. Thresholding will be discussed at the end of this chapter, while more advanced selection operators will appear in various parts of the text. For the beginning of this chapter, we will assume that the binary image **B** is the initial input to our tasks. Figure 3.1 illustrates the concept with four binary images of hand-printed characters.

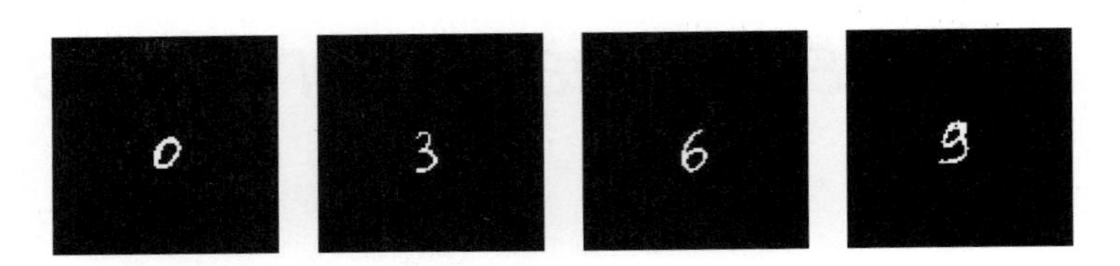

Figure 3.1 Binary images of hand-printed characters.

The pixels of a binary image **B** are 0s and 1s; the 1s will be used to denote foreground pixels and the 0s background pixels. The term **B[r, c]** denotes the value of the pixel located at row **r**, column **c** of the image array. An $\mathbf{M} \times \mathbf{N}$ image has **M** rows numbered from **0** to $\mathbf{M-1}$ and **N** columns numbered from **0** to $\mathbf{N-1}$. Thus **B[0, 0]** refers to the value of the upper leftmost pixel of the image and $\mathbf{B[M-1, N-1]}$ refers to the value of the lower rightmost pixel.

In many algorithms, not only the value of a particular pixel, but also the values of its neighbors are used when processing that pixel. The two most common definitions for neighbors are the *4-neighbors* and the *8-neighbors* of a pixel. The 4-neighborhood $\mathbf{N_4[r, c]}$ of pixel **[r, c]** includes pixels $\mathbf{[r-1, c]}$, $\mathbf{[r+1, c]}$, $\mathbf{[r, c-1]}$, and $\mathbf{[r, c+1]}$, which are often referred to as its north, south, west, and east neighbors, respectively. The 8-neighborhood $\mathbf{N_8[r, c]}$ of pixel **[r, c]** includes each pixel of the 4-neighborhood plus the diagonal neighbor pixels $\mathbf{[r-1, c-1]}$, $\mathbf{[r-1, c+1]}$, $\mathbf{[r+1, c-1]}$, and $\mathbf{[r+1, c+1]}$, which can be referred to as its northwest, northeast, southwest, and southeast neighbors, respectively. Figure 3.2 illustrates these concepts.

	N	
W	*	E
	S	

(*a*) 4-neighborhood N_4

NW	N	NE
W	*	E
SW	S	SE

(*b*) 8-neighborhood N_8

Figure 3.2 The two most common neighborhoods of a pixel.

Either the 4-neighborhood or the 8-neighborhood (or some alternate definition) can be used as the *neighborhood* of a pixel in various algorithms. To be general, we will say that a pixel **[r′, c′]** *neighbors* a pixel **[r, c]** if **[r′, c′]** lies in the selected type of neighborhood of **[r, c]**.

3.2 APPLYING MASKS TO IMAGES

A basic concept in image processing is that of applying a *mask* to an image. The concept comes from the image processing operation of convolution, but is used in a general sense in image analysis as a whole. A mask is a set of pixel positions and corresponding values called *weights*. Figure 3.3 shows three different masks. The first two (*a*) and (*b*) are square masks, one with equal weights, all of value one and one with unequal weights. The third mask (*c*) is rectangular and has equal weights.

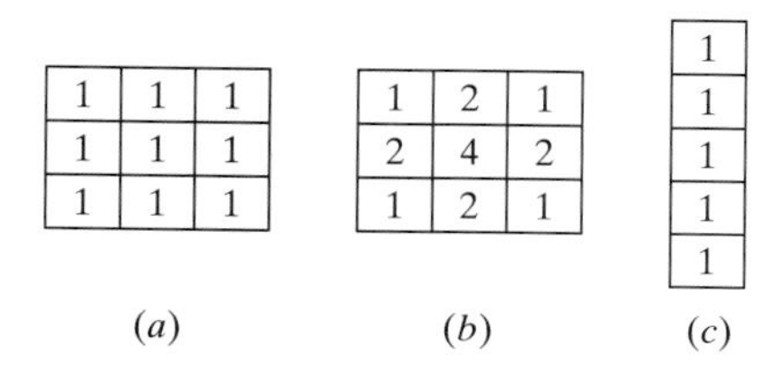

Figure 3.3 Three masks that can be applied to an image.

Each mask has an *origin,* which is usually one of its positions. Usually the origins of symmetric masks, such as (*a*) and (*b*) of Figure 3.3, are their center pixels. For nonsymmetric masks, any pixel may be chosen as the origin, depending on the intended use. The top pixel of mask (*c*) might be chosen as its origin.

The application of a mask to an input image yields an output image of the same size as the input. For each pixel in the input image, the mask is conceptually placed on top of the image with its origin lying on that pixel. The value of each input image pixel under the mask is multiplied by the weight of the corresponding mask pixel. The results are summed together to yield a single output value that is placed in the output image at the location of the pixel being processed on the input. Figure 3.4 illustrates the application of mask (*b*) of Figure 3.3 to a gray-tone image.

The original gray-tone image is shown in Figure 3.4(*a*). Notice that when the center of the mask lies on top of one of the perimeter pixels of the image, some of the pixels of the mask lie outside of the image. In order to make the output image come out the same size as the input image, we must add some virtual rows and columns to the input image around the edges. In the next example, we have added two virtual rows (one above the image and one below) and two virtual columns (one to the left and one to the right). The values in these virtual rows and columns can be set arbitrarily to zero or some other constant or, as has been done here, they can merely duplicate the closest row (or column) to them. Thus the virtual row added to the top of the input image would duplicate the values 40, 40, 80, 80, 80; the virtual column on the left would be all 40s; the virtual column on the right would be all 80s; and the virtual row added to the bottom would again have values 40, 40, 80, 80, 80. The output image (*c*) produced by the application of the mask (*b*) is a smoothed version of the input (*a*); however, the values are all much bigger than in the original. To normalize, we can divide the value obtained for each pixel by the sum of the weights in the mask, in this case 16, obtaining the final image shown in (*d*). The original and final images are shown in gray-tone in (*e*) expanded to 120×120 for visibility. Because of the expansion, each pixel

40	40	80	80	80
40	40	80	80	80
40	40	80	80	80
40	40	80	80	80
40	40	80	80	80

(*a*) Original gray-tone image

1	2	1
2	4	2
1	2	1

(*b*) 3 × 3 mask

640	800	1120	1280	1280
640	800	1120	1280	1280
640	800	1120	1280	1280
640	800	1120	1280	1280
640	800	1120	1280	1280

(*c*) Result of applying the mask to the image

40	50	70	80	80
40	50	70	80	80
40	50	70	80	80
40	50	70	80	80
40	50	70	80	80

(*d*) Normalized result after division by the sum of the weights in the mask (16)

(*e*) Original image and result, expanded to 120 × 120 for viewing

Figure 3.4 Application of a mask with weights to a gray-scale image.

in the final image is shown as a 24-pixel strip; thus the smoothness is at a strip level, instead of at a pixel level.

3.3 COUNTING THE OBJECTS IN AN IMAGE

Chapter 1 presented an application where it was important to count the number of holes in an object. Counting the number of foreground objects is an equivalent problem that can be performed with the same algorithm by merely swapping the roles of the two sets: **E** and **I**. For counting foreground objects, the external corner patterns are 2 × 2 masks that have three 0s and one 1-pixel. The internal corner patterns are 2 × 2 masks that have three 1s and one 0-pixel. Figure 3.5 illustrates the two sets of masks. Note that the algorithm expects each object to be a 4-connected set of 1-pixels with no interior holes.

0	0		0	0		1	0		0	1
0	1		1	0		0	0		0	0

(*a*) **E:** external corners

1	1		1	1		1	0		0	1
1	0		0	1		1	1		1	1

(*b*) **I:** internal corners

Figure 3.5 The 2 × 2 masks for counting the foreground objects in a binary image. The 1s represent foreground pixels, and the 0s represent background pixels.

The application of one of these masks to a binary image can be visualized as placing the mask on the image so that the top left pixel of the mask lines up with the particular pixel being considered on the image. In this case the mask is defining a neighborhood of the image pixel consisting of the pixel, its neighbor to the right, and the two pixels below them. If all four image pixels that fall under the mask have exactly the same value as the corresponding mask pixel, then the type of corner defined by that mask is identified with that image pixel. Suppose that the function *external_match*(**L, P**) sequences through the four external masks and returns *true* if the subimage with top left pixel **[L, P]** matches one of them, false otherwise. Similarly, the function *internal_match*(**L, P**) returns *true* if the subimage with top left pixel **[L, P]** matches one of the internal masks and false otherwise. The object-counting function *count_objects*(**B**) takes in a binary image **B**, loops through each pixel of the image, excluding pixels of the last row and the last column, where the 2×2 mask cannot be placed, and returns the number of objects in the image.

Conventions for defining algorithms Pseudo-code for the object-counting procedure is given next. We will use this syntax for all procedures given in the text. Note that all routines are called *procedures,* but those that are functions include a *return* statement (as in C) to return a value. To keep the procedures short and simple, we will often use utility procedures within them such as *external_match* and *internal_match*. The code for very straightforward utility procedures such as these is usually omitted. We also omit type declarations, which are language-dependent, but we specify the required types in the text

Compute the number of foreground objects of binary image B.

Objects are 4-connected and simply connected.
E is the number of external corners.
I is the number of internal corners.

```
procedure count_objects(B);
{
E := 0;
I := 0;
for L := 0 to MaxRow - 1
   for P := 0 to MaxCol - 1
      {
      if external_match(L, P) then E := E + 1;
      if internal_match(L, P) then I := I + 1;
      };
return((E - I) / 4);
}
```

Algorithm 3.1 Counting foreground objects.

and explain important variables in comments. Finally, we use global constants for various sizes rather than clouding the procedure calls with extra arguments.

In the object-counting procedure, the constant **MaxRow** is the row number of the last row in the image, while **MaxCol** is the column number of the last column. The first row and the first column are assumed to be row and column zero, the default for C arrays.

Exercise 3.1: Efficiency of counting objects.

What is the maximum number of times that procedure *count_objects* examines each pixel of the image? How can procedures *external_match* and *internal_match* be coded to be as efficient as possible?

Exercise 3.2: Driving around corners.

Obtain some graph paper to represent a pixel array and blacken some region of connected squares (keep it small at first). The blackened squares correspond to the foreground pixels and the empty squares correspond to the background. Imagine that the pixels are all city blocks and you are driving around the blackened region in a clockwise direction. Do your right turns correspond to **E** corners or **I** corners? What about left turns? Is there a relationship between the number of left turns and the number of right turns made in driving the complete perimeter? If so, what is it? In driving the entire perimeter, did you ever cross over or touch a previously visited intersection? Is that ever possible? Why or why not? Before answering, consider the case of only two blackened blocks touching diagonally across a single shared intersection. Do your left-right counting rules still hold? Does the object-counting formula still hold?

3.4 CONNECTED COMPONENTS LABELING

Suppose that **B** is a binary image and that $\mathbf{B[r,c]} = \mathbf{B[r',c']} = v$ where either $v = 0$ or $v = 1$. The pixel $\mathbf{[r,c]}$ is *connected* to the pixel $\mathbf{[r',c']}$ with respect to value v if there is a sequence of pixels $\mathbf{[r,c]} = \mathbf{[r_0,c_0]}, \mathbf{[r_1,c_1]}, \ldots, \mathbf{[r_n,c_n]} = \mathbf{[r',c']}$ in which $\mathbf{B[r_i,c_i]} = v$, $i = 0, \ldots, \mathbf{n}$, and $\mathbf{[r_i,c_i]}$ neighbors $\mathbf{[r_{i-1},c_{i-1}]}$ for each $i = 1, \ldots, \mathbf{n}$. The sequence of pixels $\mathbf{[r_0,c_0]}, \ldots, \mathbf{[r_n,c_n]}$ forms a connected *path* from $\mathbf{[r,c]}$ to $\mathbf{[r',c']}$. A *connected component* of value v is a set of pixels C, each having value v, and such that every pair of pixels in the set are connected with respect to v. Figure 3.6(a) shows a binary image with five such connected components of 1s; these components are actually connected with respect to either the 8-neighborhood or the 4-neighborhood definition.

14 Definition. A **connected components labeling** of a binary image **B** is a labeled image **LB** in which the value of each pixel is the label of its connected component.

A label is a symbol that uniquely names an entity. While character labels are possible, positive integers are more convenient and are most often used to label the connected components. Figure 3.6(b) shows the connected components labeling of the binary image of Figure 3.6(a).

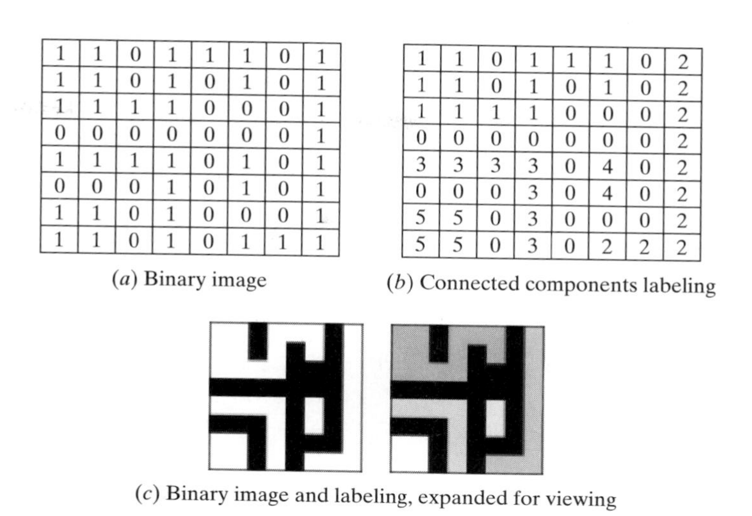

1	1	0	1	1	1	0	1
1	1	0	1	0	1	0	1
1	1	1	1	0	0	0	1
0	0	0	0	0	0	0	1
1	1	1	1	0	1	0	1
0	0	0	1	0	1	0	1
1	1	0	1	0	0	0	1
1	1	0	1	0	1	1	1

(*a*) Binary image

1	1	0	1	1	1	0	2
1	1	0	1	0	1	0	2
1	1	1	1	0	0	0	2
0	0	0	0	0	0	0	2
3	3	3	3	0	4	0	2
0	0	0	3	0	4	0	2
5	5	0	3	0	0	0	2
5	5	0	3	0	2	2	2

(*b*) Connected components labeling

(*c*) Binary image and labeling, expanded for viewing

Figure 3.6 A binary image with five connected components of the value 1.

There are a number of different algorithms for the connected components labeling operation. Some algorithms assume that the entire image can fit in memory and employ a simple, recursive algorithm that works on one component at a time, but can move all over the image while doing so. Other algorithms were designed for larger images that may not fit in memory and work on only two rows of the image at a time. Still other algorithms were designed for massively parallel machines and use a parallel propagation strategy. We will look at two different algorithms in this chapter: The recursive search algorithm and a row-by-row algorithm that uses a special union-find data structure to keep track of components.

A Recursive Labeling Algorithm Suppose that **B** is a binary image with **MaxRow** + 1 rows and **MaxCol** + 1 columns. We wish to find the connected components of the 1-pixels and produce a labeled output image **LB** in which every pixel is assigned the label of its connected component. The strategy, adapted from the Tanimoto Artificial Intelligence (AI) text, is to first negate the binary image, so that all the 1-pixels become -1s. This is needed to distinguish unprocessed pixels (-1) from those of component label 1. We will accomplish this with a function called *negate* that inputs the binary image **B** and outputs the negated image **LB**, which will become the labeled image. Then the process of finding the connected components becomes one of finding a pixel whose value is -1 in **LB**, assigning it a new label, and calling procedure *search* to find its neighbors that have value -1 and recursively repeat the process for these neighbors. The utility function *neighbors*(**L, P**) is given a pixel position defined by **L** and **P**. It returns the set of pixel positions of all of its neighbors, using either the 4-neighborhood or 8-neighborhood definition. Only neighbors that represent legal positions on the binary image are returned. The neighbors are returned

Compute the connected components of a binary image.

B is the original binary image.
LB will be the labeled connected component image.

```
procedure recursive_connected_components(B, LB);
{
LB := negate(B);
label := 0;
find_components(LB, label);
print(LB);
}

procedure find_components(LB, label);
{
for L := 0 to MaxRow
   for P := 0 to MaxCol
      if LB[L,P] == -1 then
         {
         label := label + 1;
         search(LB, label, L, P);
         }
}

procedure search(LB, label, L, P);
{
LB[L,P] := label;
Nset := neighbors(L, P);
for each L[', P'] in Nset
   {
   if LB[L', P'] == -1
   then search(LB, label, L', P');
   }
}
```

Algorithm 3.2 Recursive connected components.

in scan-line order as shown in Figure 3.7. The recursive connected components labeling algorithm is a set of six procedures, including *negate, print,* and *neighbors,* which are left for the reader to code.

Figure 3.8 illustrates the application of the recursive connected components algorithm to the first (top leftmost) component of the binary image of Figure 3.6.

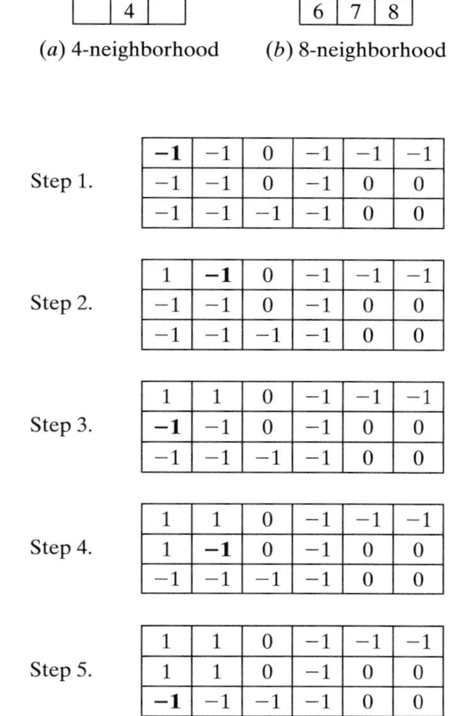

Figure 3.7 Scan-line order for returning the neighbors of a pixel.

Figure 3.8 The first five steps of the recursive labeling algorithm applied to the first component of the binary image of Figure 3.6. The image shown is the (partially) labeled image **LB**. The boldface pixel of the image is the one being processed by the search procedure. Using the neighborhood orderings shown in Figure 3.7, the first unprocessed neighbor of the boldface pixel whose value is -1 is selected at each step as the next pixel to be processed.

A Row-by-Row Labeling Algorithm The classical algorithm, deemed so because it is based on the classical connected components algorithm for graphs, was described in Rosenfeld and Pfaltz (1966). The algorithm makes two passes over the image: One pass to record equivalences and assign temporary labels and the second pass to replace each temporary label by the label of its equivalence class. In between the two passes, the recorded set of equivalences, stored as a binary relation, is processed to determine the equivalence classes of the relation. Since that time, the *union-find* algorithm, which dynamically constructs the equivalence classes as the equivalences are found, has been widely used in computer science applications. The union-find data structure allows efficient construction and manipulation of equivalence classes represented by tree structures. The addition of this data structure is a useful improvement to the classical algorithm.

Union-Find Structure The purpose of the union-find data structure is to store a collection of disjoint sets and to efficiently implement the operations of *union* (merging two sets into one) and *find* (determining which set a particular element is in). Each set is stored as a tree structure in which a node of the tree represents a label and points to its one parent node. This is accomplished with only a vector array **PARENT** whose subscripts are the set of possible labels and whose values are the labels of the parent nodes. A parent value of zero

PARENT

1	**2**	**3**	**4**	**5**	**6**	**7**	**8**
2	3	0	3	7	7	0	3

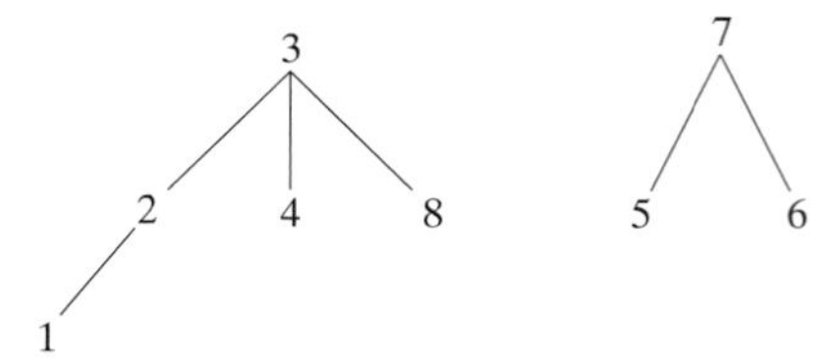

Figure 3.9 The union-find data structure for two sets of labels. The first set contains the labels $\{1,2,3,4,8\}$, and the second set contains labels $\{5,6,7\}$. For each integer label **i**, the value of **PARENT**[**i**] is the label of the parent of **i** or zero if **i** is a root node and has no parent.

means that this node is the root of the tree. Figure 3.9 illustrates the tree structure for two sets of labels $\{1,2,3,4,8\}$ and $\{5,6,7\}$. Label 3 is the parent node and set label for the first set; label 7 is the parent node and set label for the second set. The values in array **PARENT** tell us that nodes 3 and 7 have no parents, label 2 is the parent of label 1, label 3 is the parent of labels 2, 4, and 8, and so on. Note that element 0 of the array is not used, since 0 represents the background label, and a value of 0 in the array means that a node has no parent.

The *find* procedure is given a label **X** and the parent array **PARENT.** It merely follows the parent pointers up the tree to find the label of the root node of the tree that **X** is in. The *union* procedure is given two labels **X** and **Y** and the parent array **PARENT.** It modifies the structure (if necessary) to merge the set containing **X** with the set containing **Y**. It starts at labels **X** and **Y** and follows the parent pointers up the tree until it reaches the roots of the two sets. If the roots are not the same, one label is made the parent of the other. The procedure for *union* given here arbitrarily makes **X** the parent of **Y**. It is also possible to keep track of the set sizes and to attach the smaller set to the root of the larger set; this has the effect of keeping the tree depths down.

Find the parent label of a set.

X is a label of the set.
PARENT is the array containing the union-find data structure.

```
procedure find(X, PARENT);
{
j := X;
while PARENT[j] <> 0
  j := PARENT[j];
return(j);
}
```

Algorithm 3.3 Find.

Construct the union of two sets.

X is the label of the first set.
Y is the label of the second set.
PARENT is the array containing the union-find data structure.

```
procedure union(X, Y, PARENT);
{
j := X;
k := Y;
while PARENT[j] <> 0
  j := PARENT[j];
while PARENT[k] <> 0
  k := PARENT[k];
if j <> k then PARENT[k] := j;
}
```

Algorithm 3.4 Union.

The Classical Connected Components Algorithm using Union-Find The union-find data structure makes the classical connected components labeling algorithm more efficient. The first pass of the algorithm performs label propagation to propagate a pixel's label to its neighbors to the right and below it. Whenever a situation arises in which two different labels can propagate to the same pixel, the smaller label propagates and each such equivalence found is entered in the union-find structure. At the end of the first pass, each equivalence class has been completely determined and has a unique label, which is the root of its tree in the union-find structure. A second pass through the image then performs a translation, assigning to each pixel the label of its equivalence class.

The procedure uses two additional utility functions: *prior_neighbors* and *labels*. The *prior_neighbors* function returns the set of neighboring 1-pixels above and to the left of a given one and can be coded for a 4-neighborhood (in which case the north and west neighbors are returned) or for an 8-neighborhood (in which case the northwest, north, northeast, and west neighbors are returned). The *labels* function returns the set of labels currently assigned to a given set of pixels.

Figure 3.10 illustrates the application of the classical algorithm with union-find to the binary image of Figure 3.6. Figure 3.10(a) shows the labels for each pixel after the first pass. Figure 3.10(b) shows the union-find data structure indicating that the equivalence classes determined in the first pass are $\{\{1, 2\}, \{3, 7\}, 4, 5, 6\}$. Figure 3.10($c$) shows the final labeling of the image after the second pass. The connected components represent regions of the image for which both shape and intensity properties can be computed. We will discuss some of these properties in Section 3.5.

1	1	0	2	2	2	0	3
1	1	0	2	0	2	0	3
1	1	1	1	0	0	0	3
0	0	0	0	0	0	0	3
4	4	4	4	0	5	0	3
0	0	0	4	0	5	0	3
6	6	0	4	0	0	0	3
6	6	0	4	0	7	7	3

(*a*) After Pass 1

1	1	0	1	1	1	0	3
1	1	0	1	0	1	0	3
1	1	1	1	0	0	0	3
0	0	0	0	0	0	0	3
4	4	4	4	0	5	0	3
0	0	0	4	0	5	0	3
6	6	0	4	0	0	0	3
6	6	0	4	0	3	3	3

(*c*) After Pass 2

PARENT

1	**2**	**3**	**4**	**5**	**6**	**7**
0	1	0	0	0	0	3

(*b*) Union-find structure showing equivalence classes

Figure 3.10 The application of the classical algorithm with the union-find data structure to the binary image of Figure 3.6.

Using Run-Length Encoding for Connected Components Labeling As introduced in Chapter 2, a *run-length encoding* of a binary image is a list of contiguous horizontal runs of 1s. For each run, the location of the starting pixel of the run and either its length or the location of its ending pixel must be recorded. Figure 3.11 shows a sample run-length data structure. Each run in the image is encoded by its starting- and ending-pixel locations. (**ROW, START_COL**) is the location of the starting pixel and (**ROW, END_COL**) is the location of the ending pixel, **LABEL** is the field in which the label of the connected component to which this run belongs will be stored. It is initialized to zero and assigned temporary values in pass 1 of the algorithm. At the end of pass 2, the **LABEL** field contains the final, permanent label of the run. This structure can then be used to output the labels back to the corresponding pixels of the output image.

Initialize the data structures for classical connected components.

```
procedure initialize();
"Initialize global variable label and array PARENT."
{
"Initialize label."
label := 0;
"Initialize the union-find structure."
for i := 1 to MaxLab
  PARENT[i] := 0;
}
```

Algorithm 3.5 Initialization for classical connected components.

	0	1	2	3	4
0	1	1	0	1	1
1	1	1	0	0	1
2	1	1	1	0	1
3	0	0	0	0	0
4	0	1	1	1	1

(*a*)

	ROW_START	ROW_END
0	1	2
1	3	4
2	5	6
3	0	0
4	7	7

(*b*)

	ROW	START_COL	END_COL	LABEL
1	0	0	1	0
2	0	3	4	0
3	1	0	1	0
4	1	4	4	0
5	2	0	2	0
6	2	4	4	0
7	4	1	4	0

(*c*)

Figure 3.11 Binary image (*a*) and its run-length encoding (*b*) and (*c*). Each run of 1s is encoded by its row (**ROW**) and the columns of its starting and ending points (**START_COL** and **END_COL**). In addition, for each row of the image, **ROW_START** points to the first run of the row and **ROW_END** points to the last run of the row. The **LABEL** field will hold the component label of the run; it is initialized to zero.

3.5 BINARY IMAGE MORPHOLOGY

The word *morphology* refers to form and structure; in computer vision it can be used to refer to the shape of a region. The operations of *mathematical morphology* were originally defined as set operations and shown to be useful for processing sets of 2D points. In this section, we define the operations of binary morphology and show how they can be useful in processing the regions derived from the connected components labeling operation.

3.5.1 Structuring Elements

The operations of binary morphology input a binary image **B** and a *structuring element* **S**, which is another, usually much smaller, binary image. The structuring element represents a shape; it can be of any size and have arbitrary structure that can be represented by a binary image. However, there are a number of common structuring elements such as a rectangle of specified dimensions [BOX(l,w)] or a circular region of specified diameter [DISK(d)]. Some image processing packages offer a library of these primitive structuring elements. Figure 3.12 illustrates some common structuring elements and several nonstandard ones.

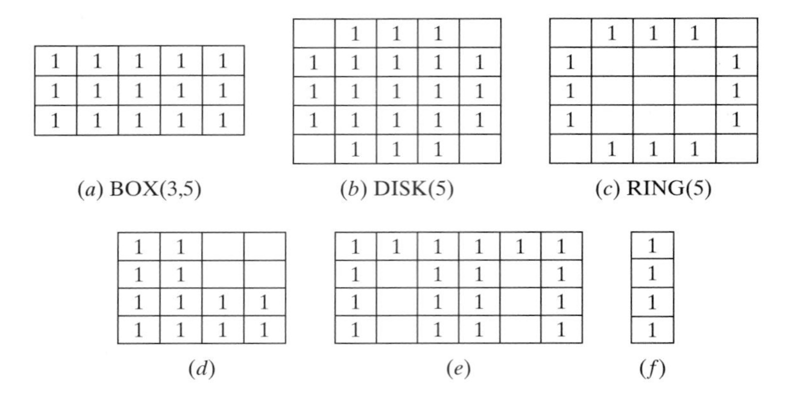

Figure 3.12 Examples of structuring elements (blanks represent 0s).

Exercise 3.3: Labeling Algorithm Comparison.

Suppose a binary image has one foreground region, a rectangle of size $1{,}000 \times 1{,}000$. How many times does the recursive algorithm look at (read or write) each pixel? How many times does the classical procedure look at each pixel?

Exercise 3.4: Relabeling.

Because equivalent labels are merged into one equivalence class, some of the initial labels from Pass 1 are lost in Pass 2, producing a final labeling whose numeric sequence of labels often has many gaps. Write a relabeling procedure that converts the labeling to one that has a contiguous sequence of numbers from 1 to the number of components in the image.

Exercise 3.5: Run-Length Encoding.

Design and implement a row-by-row labeling algorithm that uses the run-length encoding of a binary image instead of the image itself and uses the **LABEL** field of the structure to store the labels of the runs.

The purpose of the structuring elements is to act as probes of the binary image. One pixel of the structuring element is denoted as its *origin;* this is often the central pixel of a symmetric structuring element, but may in principle be any chosen pixel. Using the origin as a reference point, translations of the structuring element can be placed anywhere on the image and can be used to either enlarge a region by that shape or to check whether or not the shape fits inside a region. For example, we might want to check the size of holes by seeing if a smaller disk fits entirely within a region, while a larger disk does not.

Compute the connected components of a binary image.

B is the original binary image.
LB will be the labeled connected component image.

```
procedure classical_with_union-find(B, LB);
{
"Initialize structures."
initialize();
"Pass 1 assigns initial labels to each row L of the image."
for L := 0 to MaxRow
  {
  "Initialize all labels on line L to zero"
  for P := 0 to MaxCol
    LB[L,P] := 0;
  "Process line L."
  for P := 0 to MaxCol
    if B[L,P] == 1 then
      {
      A := prior_neighbors(L,P);
      if isempty(A)
      then {M := label; label := label + 1;};
      else M := min(labels(A));
      LB[L,P] := M;
      for X in labels(A) and X <> M
        union(M, X, PARENT);
      }
  }
"Pass 2 replaces Pass 1 labels with equivalence class labels."
for L := 0 to MaxRow
  for P := 0 to MaxCol
    if B[L, P] == 1
    then LB[L,P] := find(LB[L,P], PARENT);
};
```

Algorithm 3.6 Classical connected components with union-find.

3.5.2 Basic Operations

The basic operations of binary morphology are *dilation, erosion, closing,* and *opening*. As the names indicate, a dilation operation enlarges a region, while an erosion makes it smaller. A closing operation can close up internal holes in a region and eliminate *bays* along the boundary. An opening operation can get rid of small portions of the region that jut out from the boundary into the background region. The mathematical definitions are as follows:

15 Definition. The **translation** X_t of a set of pixels X by a position vector t is defined by

$$X_t = \{x + t \mid x \in X\} \tag{3.1}$$

Thus the translation of a set of 1s in a binary image moves the entire set of 1s by the specified amount. The translation t would be specified as an ordered pair $(\Delta r, \Delta c)$ where Δr is the amount to move in rows and Δc is the amount to move in columns.

16 Definition. The **dilation** of binary image **B** by structuring element **S** is denoted by $\mathbf{B} \oplus \mathbf{S}$ and is defined by

$$B \oplus S = \bigcup_{b \in B} S_b \tag{3.2}$$

This union can be thought of as a neighborhood operator. The structuring element **S** is swept over the image. Each time the origin of the structuring element touches a binary 1-pixel, the entire translated structuring element shape is ORed to the output image, which has been initialized to all zeros. Figure 3.13(*a*) shows a binary image, and Figure 3.13(*c*) illustrates its dilation by the 3×3 rectangular structuring element shown in Figure 3.13(*b*).

To follow the mathematical definition, consider the first 1-pixel of the binary image **B**. Its coordinates are [1,0] meaning row 1, column 0 of the image. The translation $\mathbf{S}_{(1,0)}$ means that the structuring element S is ORed into the output image so that its origin (which is its center) coincides with position (1,0). As a result of this OR, the output image (initially all 0) has 1-pixels at positions [0,0], [0,1], [1,0], [1,1], [2,0], and [2,1], which are real positions, and at positions [0,−1], [1,−1], and [2,−1], which are virtual positions and are ignored. For the next pixel [1,1] of **B**, the translation $\mathbf{S}_{(1,1)}$ is added to the output by ORing in 1-pixels at positions [0,0], [0,1], [0,2], [1,0], [1,1], [1,2], [2,0], [2,1], [2,2]. This continues until a copy of the structuring element has been ORed into the output image for every pixel of the input image, producing the final result of Figure 3.13(*c*).

17 Definition. The **erosion** of binary image **B** by structuring element **S** is denoted by $\mathbf{B} \ominus \mathbf{S}$ and is defined by

$$B \ominus S = \{b \mid b + s \in B \ \forall s \in S\} \tag{3.3}$$

The erosion operation also sweeps the structuring element over the entire image. At each position where every 1-pixel of the structuring element covers a 1-pixel of the binary image, the binary image pixel corresponding to the origin of the structuring element is ORed to the output image. Figure 3.13(*d*) illustrates an erosion of the binary image of Figure 3.13(*a*) by the 3×3 rectangular structuring element.

Dilation and erosion are the most primitive operations of mathematical morphology. There are two more common operations that are composed of these two: closing and opening.

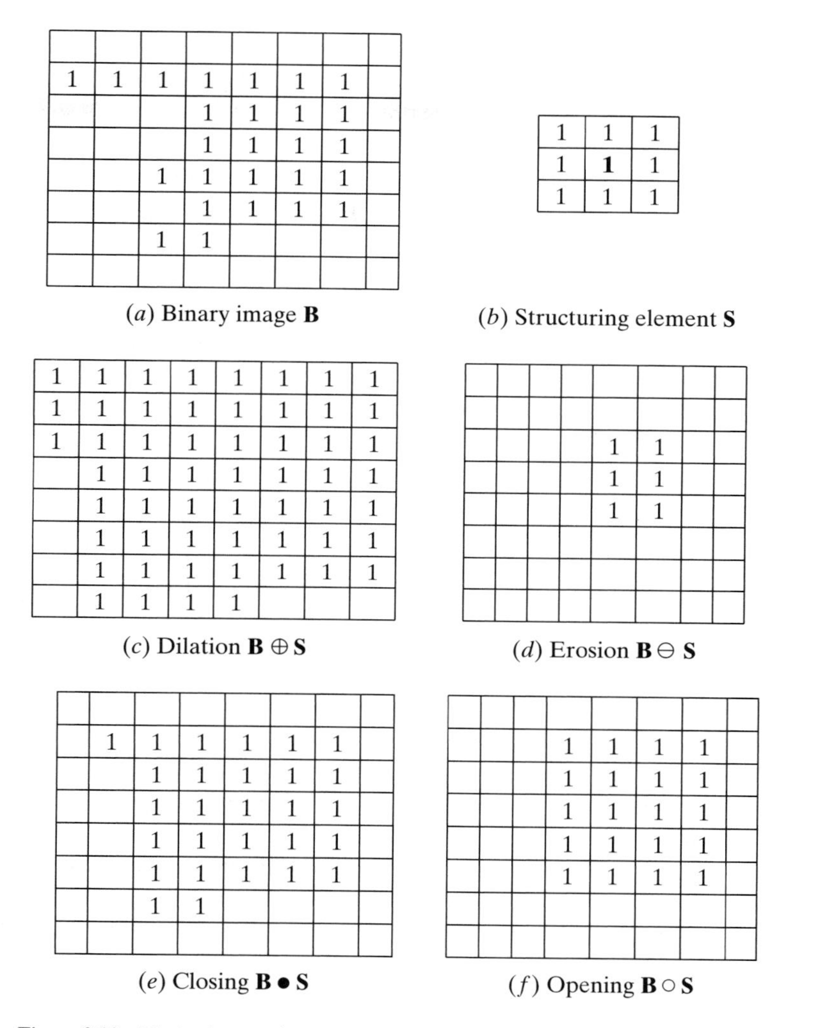

Figure 3.13 The basic operations of binary morphology. Foreground pixels are shown as 1s. Background pixels, whose value is 0, are shown as blanks.

18 Definition. The **closing** of binary image **B** by structuring element **S** is denoted by **B • S** and is defined by

$$B \bullet S = (B \oplus S) \ominus S \tag{3.4}$$

19 Definition. The **opening** of binary image **B** by structuring element **S** is denoted **B ∘ S** and is defined by

$$B \circ S = (B \ominus S) \oplus S \tag{3.5}$$

Figure 3.13(*e*) illustrates the closing of the binary image of Figure 3.13(*a*) by the 3×3 rectangular structuring element; Figure 3.13(*f*) illustrates the opening of the binary image by the same structuring element.

Exercise 3.6: Using elementary operations of binary morphology.

A camera takes an image I of a penny, a dime, and a quarter lying on a white background and not touching one another. Thresholding is used successfully to create a binary image B with 1 bits for the coin regions and 0 bits for the background. You are given the known diameters of the coins D_P, D_D, and D_Q. Using the operations of mathematical morphology (dilation, erosion, opening, closing) and the logical operators AND, OR, NOT, and MINUS (set difference), show how to produce three binary output images: P, D, and Q. P should contain just the penny (as 1 bits), D should contain just the dime, and Q should contain just the quarter.

3.5.3 Some Applications of Binary Morphology

Closings and openings are useful in imaging applications where thresholding—or some other initial process—produces a binary image with tiny holes in the connected components or with a pair of components that should be separate joined by a thin region of foreground pixels. Figure 3.14(*a*) is a 512×512 16-bit gray-scale medical image, Figure 3.14(*b*) is the result of thresholding to select pixels with gray tones above 1,070, and Figure 3.14(*c*) is the result of performing an opening operation to separate the organs and a closing to get rid of small holes. The structuring element used in the opening was DISK(13), and the structuring element used in the closing was DISK(2).

Binary morphology can also be used to perform very specific inspection tasks in industrial machine vision. Sternberg (1985) showed how a watch gear could be inspected to check whether it had any missing or broken teeth. Figure 3.15(*a*) shows a binary image of a watch gear. The watch gear has four holes inside of the main object and is surrounded by a number of teeth, which are individually visible in the image. In order to process the watch gear images, Sternberg defined several special purpose structuring elements whose shapes and sizes were derived from the physical properties of the watch gear. The following structuring elements are used in the watch-gear inspection algorithm:

- **hole_ring:** a ring of pixels whose diameter is slightly larger than the diameters of the four holes in the watch gears. It fits just around these holes and can be used to mark a few pixels at their centers.
- **hole_mask:** an octagon that is slightly larger than the holes in the watch gears.
- **gear_body:** a disk structuring element that is as big as the gear minus its teeth.
- **sampling_ring_spacer:** a disk structuring element that is used to move slightly outward from the gear body.
- **sampling_ring_width:** a disk structuring element that is used to dilate outward to the tips of the teeth.
- **tip_spacing:** a disk structuring element whose diameter spans the tip-to-tip space between teeth.

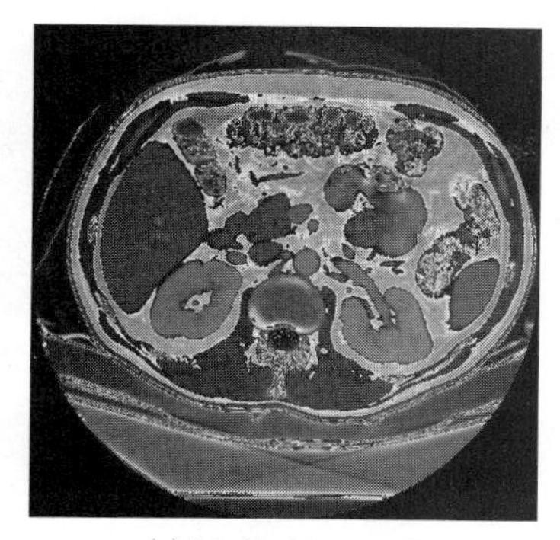

(*a*) Medical image G

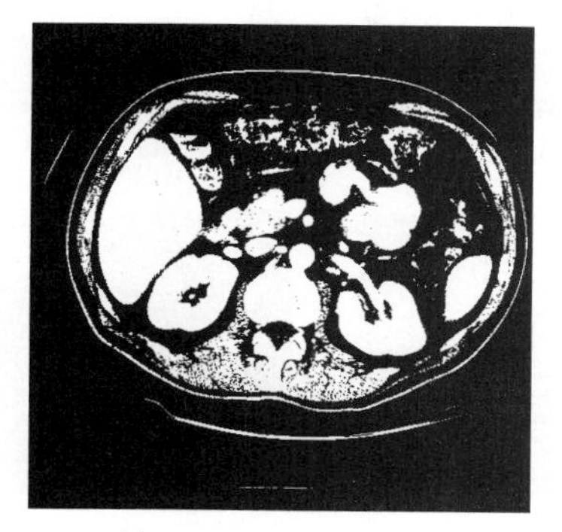

(*b*) Thresholded image B

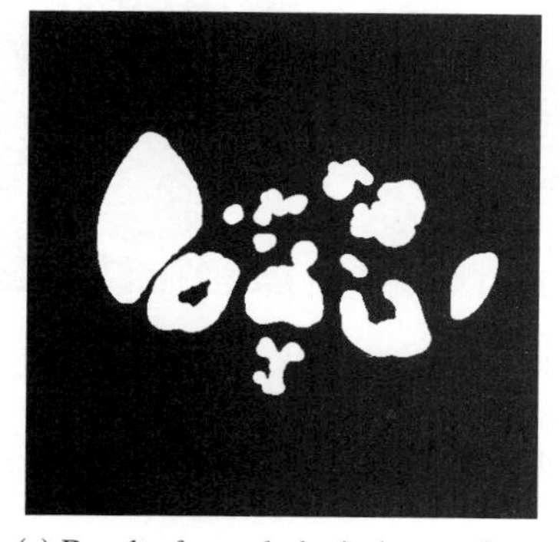

(*c*) Result of morphological operations

Figure 3.14 Use of binary morphology in medical imaging. The 512×512 16-bit medical image shown in (*a*) is thresholded (at 1,070) to produce the binary image shown in (*b*). Opening with a DISK(13) structuring element and closing with a DISK(2) gives the results shown in (*c*).

- **defect_cue:** a disk structuring element whose purpose is to dilate defects in order to show them to the user.

Figure 3.15 illustrates the gear-tooth inspection procedure. Figure 3.15(*a*) shows the original binary image to be inspected. Figure 3.15(*b*) shows the result of eroding the original image with the **hole_ring** structuring element. The result image has 1 pixels in a tiny cluster

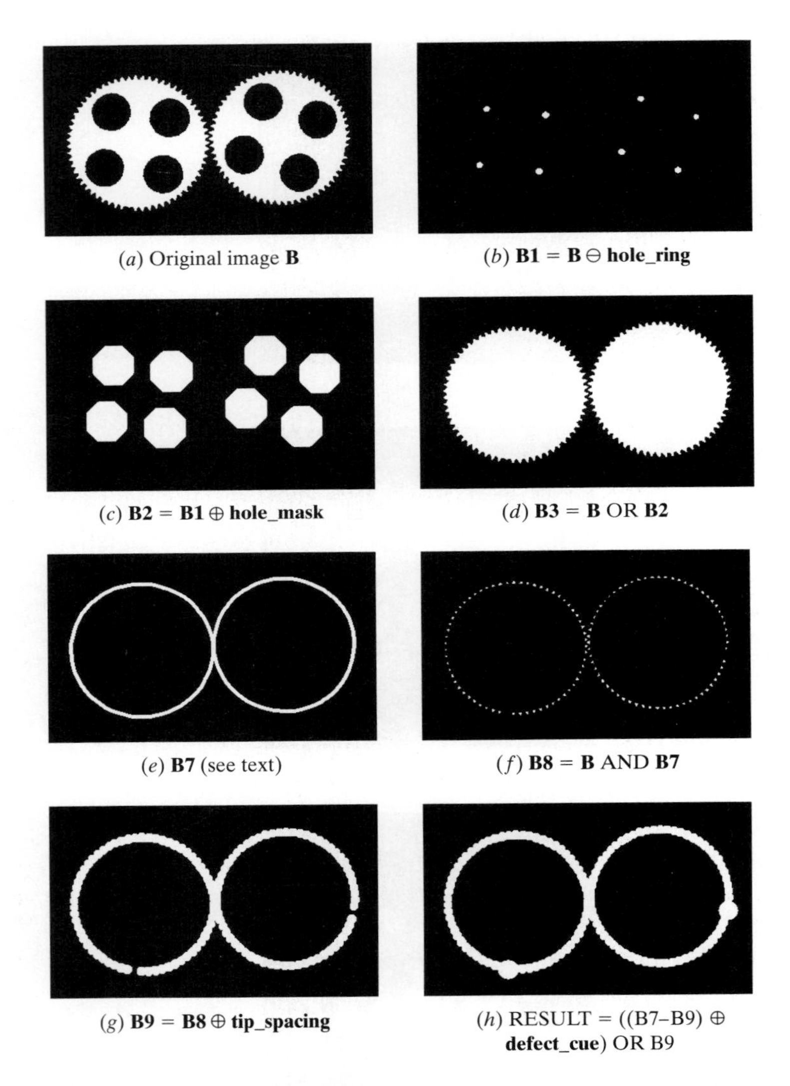

(*a*) Original image **B**

(*b*) **B1** = **B** ⊖ **hole_ring**

(*c*) **B2** = **B1** ⊕ **hole_mask**

(*d*) **B3** = **B** OR **B2**

(*e*) **B7** (see text)

(*f*) **B8** = **B** AND **B7**

(*g*) **B9** = **B8** ⊕ **tip_spacing**

(*h*) RESULT = ((B7–B9) ⊕ **defect_cue**) OR B9

Figure 3.15 The gear-tooth inspection procedure. (Courtesy of Stanley R. Sternberg with permission of Academic Press.)

in the center of each hole. These are the only pixel locations where the **hole_ring** structuring element completely overlapped the object region. Figure 3.15(*c*) shows the result of dilating the previous image with structuring element **hole_mask.** The result here is four octagons covering the original four holes. Figure 3.15(*d*) shows the result of ORing the four octagons into the original binary image. The result is the gear with the four holes filled in.

The next step is to produce a sampling ring that can be used to check the teeth. It is produced by taking the image of Figure 3.15(*d*), opening it with structuring element

gear_body to get rid of the teeth, dilating that with structuring element **sampling_ring_spacer** to bring it out to the base of the teeth, dilating that with the structuring element **sampling_ring_width** to bring the next image out to the tip of the teeth, and subtracting the second to the last result from the last result to get a ring that just fits over the teeth. The sampling ring is shown in Figure 3.15(*e*).

Once we have the sampling ring, it is ANDed with the original image to produce an image of just the teeth, as shown in Figure 3.15(*f*). The gaps are already visible, but not marked. Dilating the teeth image with the structuring element **tip_spacing** produces the solid ring image shown in Figure 3.15(*g*) which has spaces in the solid ring wherever there are defects in the teeth. Subtracting this result from the sampling ring leaves only the defects, which are dilated by structuring element **defect_cue** and shown to the user as large blobs on the screen.

Exercise 3.7: Structuring element choices.

Sternberg used a ring structuring element to detect the centers of the holes in the gear-tooth inspection task. If your system only supports disk and box structuring elements, what can you do to detect the centers of the holes?

Exercise 3.8: Morphological processing application.

Suppose a satellite image of a region can be thresholded so that the water pixels are 1s. However, bridges across rivers produce thin lines of 0s cutting across the river regions. (a) Describe how to restore the bridge pixels to the water region. (b) Describe how to detect the thin bridges as separate objects.

Binary morphology can also be used to extract primitive features of an object that can be used to recognize the object. For instance, the corners of flat two-dimensional objects can be good primitives in shape recognition. If an object with sharp corners is opened with a disk structuring element, the corners are chopped off as shown in Figure 3.16. If the resultant opening is subtracted from the original binary image of the shape, only the corners remain and can be used in a structural recognition algorithm. A shape matching system can use morphological feature detection to rapidly detect primitives that are useful in object recognition.

3.5.4 Conditional Dilation

One use of binary morphology is to identify components of a binary image that satisfy certain shape and size constraints. It is often possible to derive a structuring element that when applied to a binary image removes the components that do not satisfy the constraints

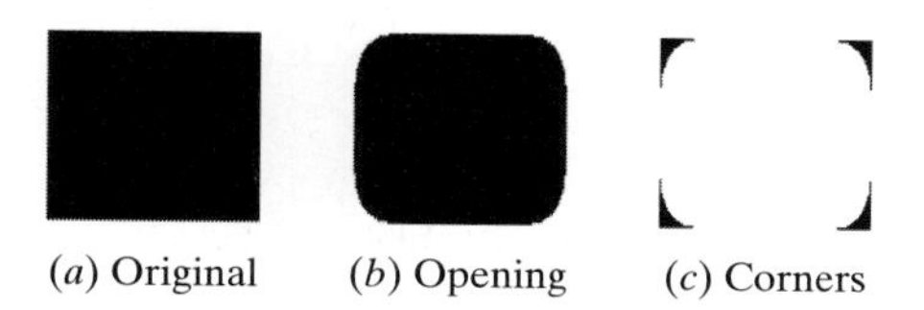

Figure 3.16 The use of binary morphology to extract shape primitives.

and leaves a few 1-pixels of those components that do satisfy the constraints. But we want the entire components, not just what remains of them after the erosion. The *conditional dilation* operation was defined to solve this problem.

20 Definition. Given an original binary image **B**, a processed binary image **C**, and a structuring element **S**, let $\mathbf{C_0} = \mathbf{C}$ and $\mathbf{C_n} = (\mathbf{C_{n-1}} \oplus \mathbf{S}) \cap \mathbf{B}$. The **conditional dilation** of **C** by **S** with respect to **B** is defined by

$$C \oplus |_B \, S = C_m \tag{3.6}$$

where the index m is the smallest index satisfying $C_m = C_{m-1}$.

This definition is intended for discrete sets of points arising from finite digital images. It says that the set $\mathbf{C} = \mathbf{C_0}$ is repeatedly dilated by structuring element **S**, and each time the result is reduced to only the subset of pixels that were 1s in the original binary image **B**. Figure 3.17 illustrates the operation of conditional dilation. In the figure, the binary image

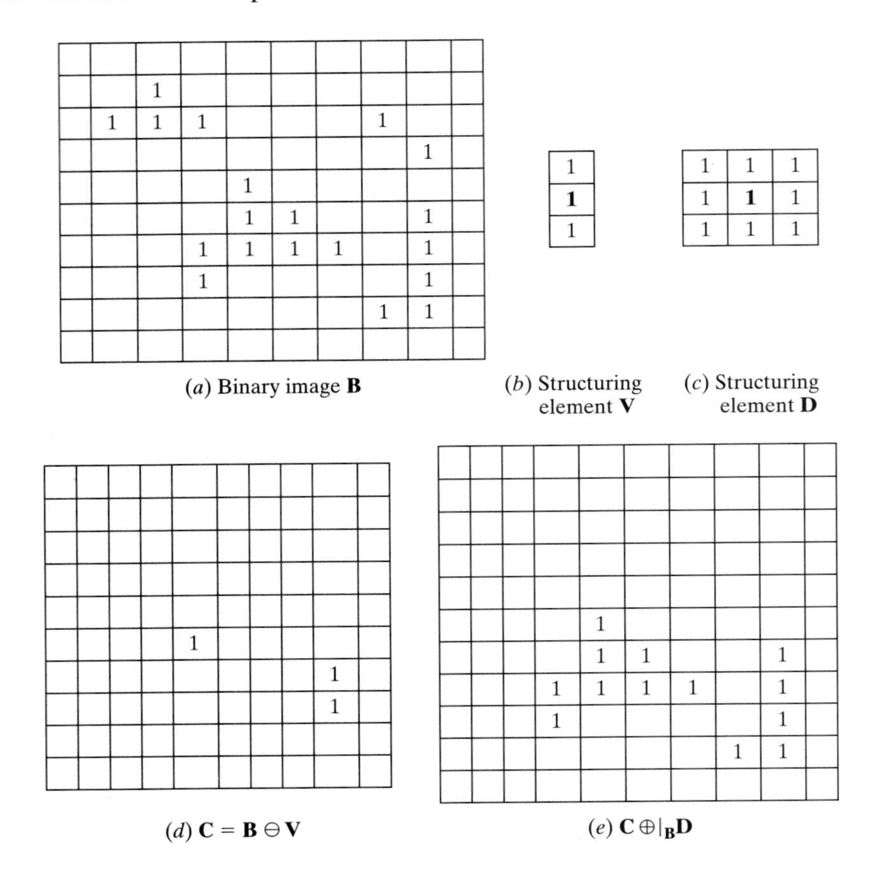

Figure 3.17 The operation of conditional dilation.

B was eroded by structuring element **V** to select components in which 3-pixel-long vertical edges could be found. Two of the components were selected, as shown in the result image **C**. In order to see these entire components, **C** is conditionally dilated by **D** with respect to the original image **B** to produce the results.

3.6 REGION PROPERTIES

Once a set of regions has been identified, the properties of the regions become the input to higher-level procedures that perform decision-making tasks such as recognition or inspection. Most image processing packages have operators that can produce a set of properties for each region. Common properties include geometric properties such as the area of the region, the centroid, and the extremal points; shape properties such as measures of the circularity and elongation; and intensity properties such as mean gray tone and various texture statistics. In this section we give the definitions of some of the most useful geometric and shape properties and explain how they may be used in decision-making tasks. Gray-level properties are covered in Chapter 7 on Image Texture.

In the discussion that follows, we denote the set of pixels in a region by R. The simplest geometric properties are the region's area A and centroid $(\bar{r}, \bar{c})$. Assuming square pixels, we define these properties by

area:

$$A = \sum_{(r,c)\in R} 1 \tag{3.7}$$

which means that the area is just a count of the pixels in the region R.
centroid:

$$\bar{r} = \frac{1}{A} \sum_{(r,c)\in R} r \tag{3.8}$$

$$\bar{c} = \frac{1}{A} \sum_{(r,c)\in R} c \tag{3.9}$$

The centroid $(\bar{r}, \bar{c})$ is thus the *average* location of the pixels in the set R. Note that even though each **[r, c]** $\in R$ is a pair of integers, $(\bar{r}, \bar{c})$ is generally not a pair of integers; often a precision of tenths of a pixel is justifiable for the centroid.

Exercise 3.9: Using the area property.

The gear-tooth example was designed to use only morphological and logical operations that could be rapidly executed on a specially designed machine. Given that we are looking for larger-than-normal gaps between the teeth, how could the detection be performed in a way that minimizes the morphological operations for general purpose machines on which they do not run rapidly?

The length of the *perimeter* P of a region is another global property. A simple definition of the perimeter of a region without holes is the set of its interior border pixels. A pixel of a region is a border pixel if it has some neighboring pixel that is outside the region. When

8-connectivity is used to determine whether a pixel inside the region is connected to a pixel outside the region, the resulting set of perimeter pixels is 4-connected. When 4-connectivity is used to determine whether a pixel inside the region is connected to a pixel outside the region, the resulting set of perimeter pixels is 8-connected. This motivates the following definition for the 4-connected perimeter P_4 and the 8-connected perimeter P_8 of a region R.

perimeter:

$$P_4 = \{(r, c) \in R | N_8(r, c) - R \neq \emptyset\}$$

$$P_8 = \{(r, c) \in R | N_4(r, c) - R \neq \emptyset\}$$

Exercise 3.10: Region from perimeter.

Describe an algorithm to generate a binary image of a region without holes, given only its perimeter.

Exercise 3.11: Area from perimeter.

Design an algorithm to compute the area of a region without holes, given only its perimeter. Is it possible to perform the task without regenerating the binary image?

To compute length $|P|$ of perimeter P, the pixels in P must be ordered in a sequence $P = < (r_o, c_o), \ldots, (r_{K-1}, c_{K-1}) >$, each pair of successive pixels in the sequence being neighbors, including the first and last pixels. Then the *perimeter length* $|P|$ is defined by

perimeter length:

$$\begin{aligned} |P| = & |\{k | (r_{k+1}, c_{k+1}) \in N_4(r_k, c_k)\}| \\ & + \sqrt{2} |\{k | (r_{k+1}, c_{k+1}) \in N_8(r_k, c_k) - N_4(r_k, c_k)\}| \end{aligned} \tag{3.10}$$

where $k + 1$ is computed modulo K, the length of the pixel sequence. Thus two vertically or horizontally adjacent pixels in the perimeter cause value 1 to be added to the total, while two diagonally adjacent pixels cause about 1.4 to be added.

With the area A and perimeter P defined, a common measure of the circularity of the region is the length of the perimeter squared divided by the area.

circularity(1):

$$C_1 = \frac{|P|^2}{A} \tag{3.11}$$

However, for digital shapes, $|P|^2/A$ assumes its smallest value not for digital circles, as it would for continuous planar shapes, but for digital octagons or diamonds depending on whether the perimeter is computed as the number of its 4-neighboring border pixels or as the length of the border, counting 1 for vertical or horizontal moves and $\sqrt{2}$ for diagonal moves. To solve this problem, Haralick (1974) proposed a second circularity measure

circularity(2):

$$C_2 = \frac{\mu_R}{\sigma_R} \tag{3.12}$$

where μ_R and σ_R are the mean and standard deviation of the distance from the centroid of the shape to the shape boundary and can be computed according to the following formulas.

mean radial distance:

$$\mu_R = \frac{1}{K} \sum_{k=0}^{K-1} \|(r_k, c_k) - (\bar{r}, \bar{c})\| \tag{3.13}$$

standard deviation of radial distance:

$$\sigma_R = \left(\frac{1}{K} \sum_{k=0}^{K-1} [\|(r_k, c_k) - (\bar{r}, \bar{c})\| - \mu_R]^2 \right)^{1/2} \tag{3.14}$$

where the set of pixels (r_k, c_k), $k = 0, \ldots, K - 1$ lie on the perimeter P of the region. The circularity measure C_2 increases monotonically as the digital shape becomes more circular and is similar for digital and continuous shapes.

Figure 3.18 illustrates some of these basic properties on a simple labeled image having three regions: an ellipse, a rectangle, and a 3×3 square.

```
0 0 0 0 0 0 0 0 0 0 0 0 0 0 0 0
0 0 0 0 0 0 0 0 0 0 0 0 0 0 0 0
0 0 0 0 0 0 0 0 0 0 0 0 0 0 0 0
0 0 0 0 0 0 0 0 0 0 1 1 1 1 0 0
2 2 2 2 0 0 0 0 0 1 1 1 1 1 1 0
2 2 2 2 0 0 0 0 1 1 1 1 1 1 1 1
2 2 2 2 0 0 0 0 1 1 1 1 1 1 1 1
2 2 2 2 0 0 0 0 1 1 1 1 1 1 1 1
2 2 2 2 0 0 0 0 0 1 1 1 1 1 1 0
2 2 2 2 0 0 0 0 0 0 1 1 1 1 0 0
2 2 2 2 0 0 0 0 0 0 0 0 0 0 0 0
2 2 2 2 0 0 0 0 0 0 0 0 0 0 0 0
2 2 2 2 0 0 3 3 3 0 0 0 0 0 0 0
2 2 2 2 0 0 3 3 3 0 0 0 0 0 0 0
2 2 2 2 0 0 3 3 3 0 0 0 0 0 0 0
2 2 2 2 0 0 0 0 0 0 0 0 0 0 0 0
```

(*a*) Labeled connected-components image

region num.	**region area**	**row of center**	**col of center**	**perim. length**	**circu- larity$_1$**	**circu- larity$_2$**	**radius mean**	**radius var.**
1	44	6	11.5	21.2	10.2	15.4	3.33	.05
2	48	9	1.5	28	16.3	2.5	3.80	2.28
3	9	13	7	8	7.1	5.8	1.2	0.04

(*b*) Properties of the three regions

Figure 3.18 Basic properties of image regions.

Exercise 3.12: Using properties.

Suppose you have a collection of two-dimensional shapes. Some of them are triangles, some are rectangles, some are octagons, some are circles, and some are ellipses or ovals. Devise a recognition strategy for these shapes. You may use the operations of mathematical morphology and the properties defined so far.

bounding box and extremal points:

It is often useful to have a rough idea of where a region is in an image. One useful concept is its *bounding box,* which is a rectangle with horizontal and vertical sides that encloses the region and touches its topmost, bottommost, leftmost, and rightmost points. As shown in Fig. 3.19, there can be as many as eight distinct extremal pixels to a region: topmost right, rightmost top, rightmost bottom, bottommost right, bottommost left, leftmost bottom, leftmost top, and topmost left. Each extremal point has an extremal coordinate value in either its row or column coordinate position. Each extremal point lies on the bounding box of the region.

Extremal points occur in opposite pairs: topmost left with bottommost right; topmost right with bottommost left; rightmost top with leftmost bottom; and rightmost bottom with leftmost top. Each pair of opposite extremal points defines an axis. Useful properties of the axis include its axis length and orientation. Because the extremal points come from a spatial digitization or quantization, the standard Euclidean distance formula will provide distances that are biased slightly low. (Consider, for example, the length covered by two pixels horizontally adjacent. From the left edge of the left pixel to the right edge of the right pixel is a length of 2 but the distance between the pixel centers is only 1.) The appropriate calculation for distance adds a small increment to the Euclidean distance to account for this. The increment depends on the orientation angle θ of the axis and is

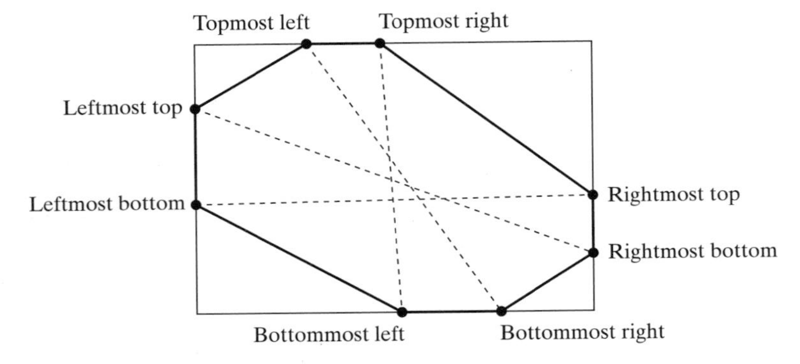

Figure 3.19 The eight extremal points of a region and the normally oriented bounding box that encloses the region. The dotted lines pair together opposite extremal points and form the extremal point axes of the shape.

given by

$$Q(\theta) = \begin{cases} \frac{1}{|\cos\theta|} & : \quad |\theta| < 45^\circ \\ \frac{1}{|\sin\theta|} & : \quad |\theta| > 45^\circ \end{cases} \tag{3.15}$$

With this increment, the length of the extremal axis from extremal point $[\mathbf{r_1}, \mathbf{c_1}]$ to extremal point $[\mathbf{r_2}, \mathbf{c_2}]$ is

extremal axis length:

$$D = \sqrt{(r_2 - r_1)^2 + (c_2 - c_1)^2} + Q(\theta) \tag{3.16}$$

Spatial moments are often used to describe the shape of a region. There are three second order *spatial moments* of a region. They are denoted by μ_{rr}, μ_{rc}, and μ_{cc} and are defined as follows:

second-order row moment:

$$\mu_{rr} = \frac{1}{A} \sum_{(r,c) \in R} (r - \bar{r})^2 \tag{3.17}$$

second-order mixed moment:

$$\mu_{rc} = \frac{1}{A} \sum_{(r,c) \in R} (r - \bar{r})(c - \bar{c}) \tag{3.18}$$

second-order column moment:

$$\mu_{cc} = \frac{1}{A} \sum_{(r,c) \in R} (c - \bar{c})^2 \tag{3.19}$$

Thus μ_{rr} measures row variation from the row mean, μ_{cc} measures column variation from the column mean, and μ_{rc} measures row and column variation from the centroid. These quantities are often used as simple shape descriptors, as they are invariant to translation and scale change of a 2D shape.

The second spatial moments have value and meaning for a region of any shape, the same way that the covariance matrix has value and meaning for any two-dimensional probability distribution. If the region is an ellipse, there is an algebraic meaning that can be given to the second spatial moments.

If a region R is an ellipse whose center is the origin, then R can be expressed as

$$R = \{(r,\ c) \mid dr^2 + 2erc + fc^2 \leq 1\} \tag{3.20}$$

A relationship exists between the coefficients d, e, and f of the equation of the ellipse and the second moments μ_{rr}, μ_{rc}, and μ_{cc}. It is given by

$$\begin{pmatrix} d & e \\ e & f \end{pmatrix} = \frac{1}{4(\mu_{rr}\mu_{cc} - \mu_{rc}^2)} \begin{pmatrix} \mu_{cc} & -\mu_{rc} \\ -\mu_{rc} & \mu_{rr} \end{pmatrix} \tag{3.21}$$

Since the coefficients d, e, and f determine the lengths of the major and minor axes and the orientation of the ellipse, this relationship means that the second moments μ_{rr}, μ_{rc}, and μ_{cc} also determine the lengths of the major and minor axes and the orientation of the ellipse. Ellipses are frequently the result of imaging circular objects. Ellipses also provide a rough approximation to other elongated objects.

Lengths and Orientations of Ellipse Axes* To determine the lengths of the major and minor axes and their orientations from the second-order moments, we must consider the following four cases.

1. $\mu_{rc} = 0$ and $\mu_{rr} > \mu_{cc}$
   The major axis is oriented at an angle of $-90°$ counterclockwise from the column axis and has a length of $4\mu_{rr}^{1/2}$. The minor axis is oriented at an angle of $0°$ counterclockwise from the column axis and has a length of $4\mu_{cc}^{1/2}$.
2. $\mu_{rc} = 0$ and $\mu_{rr} \le \mu_{cc}$
   The major axis is oriented at an angle of $0°$ counterclockwise from the column axis and has a length of $4\mu_{cc}^{1/2}$. The minor axis is oriented at an angle of $-90°$ counterclockwise from the column axis and has a length of $4\mu_{rr}^{1/2}$.
3. $\mu_{rc} \neq 0$ and $\mu_{rr} \le \mu_{cc}$
   The major axis is oriented at an angle of

   $$\tan^{-1}\left\{\frac{-2\mu_{rc}}{\mu_{rr} - \mu_{cc} + \left[(\mu_{rr} - \mu_{cc})^2 + 4\mu_{rc}^2\right]^{1/2}}\right\}$$

   counterclockwise with respect to the column axis and has a length of

   $$\left\{8\left(\mu_{rr} + \mu_{cc} + \left[(\mu_{rr} - \mu_{cc})^2 + 4\mu_{rc}^2\right)^{1/2}\right]\right\}^{1/2}$$

   The minor axis is oriented at an angle $90°$ counterclockwise from the major axis and has a length of

   $$\left[8\left\{\mu_{rr} + \mu_{cc} - \left[(\mu_{rr} - \mu_{cc})^2 + 4\mu_{rc}^2\right]^{1/2}\right\}\right]^{1/2}$$

4. $\mu_{rc} \neq 0$ and $\mu_{rr} > \mu_{cc}$
   The major axis is oriented at an angle of

   $$\tan^{-1}\frac{\left[\left\{\mu_{cc} + \mu_{rr} + \left[(\mu_{cc} - \mu_{rr})^2 + 4\mu_{rc}^2\right]^{1/2}\right\}\right]^{1/2}}{-2\mu_{rc}}$$

counterclockwise with respect to the column axis and has a length of

$$\left[8\left\{\mu_{rr} + \mu_{cc} + \left[(\mu_{rr} - \mu_{cc})^2 + 4\mu_{rc}^2\right]^{1/2}\right\}\right]^{1/2}$$

The minor axis is oriented at an angle of 90° counterclockwise from the major axis and has a length of

$$\left[8\left\{\mu_{rr} + \mu_{cc} - \left[(\mu_{rr} - \mu_{cc})^2 + 4\mu_{rc}^2\right]^{1/2}\right\}\right]^{1/2}$$

Best Axis* Some image regions (objects) have a natural axis; for example, a pencil or hammer, or the characters 'I', '/' and '-'. A *best axis* for an object can be computed as that axis about which the region pixels have least second moment. Using an analogy from mechanics, this is an axis of least inertia—an axis about which we could spin the pixels with least energy input. Note that for a circular disk, all axes have equal minimum (and maximum) inertia. It is known that an axis of least inertia must pass through the centroid $(\bar{r}, \bar{c})$ of our set of pixels (unit masses), and we will assume this here. First, we compute the second moment of a point set about an arbitrary axis; then we'll find the axis of least second moment. A set of moments about a selected set of axes might provide a good set of features for recognizing objects, as we shall see in the next chapter. For example, the second moment of character 'I' about a vertical axis through its centroid is very small, whereas that of the character '/' or '-' is not small.

Figure 3.20 shows a set of pixels and an axis making angle α with the row axis. The angle $\beta = \alpha + 90$ is the angle that a perpendicular to the axis makes with the row axis. To compute the second moment of the point set about the axis, we need to sum the squares of the distances d for all pixels: we normalize by the number of pixels to obtain a feature that

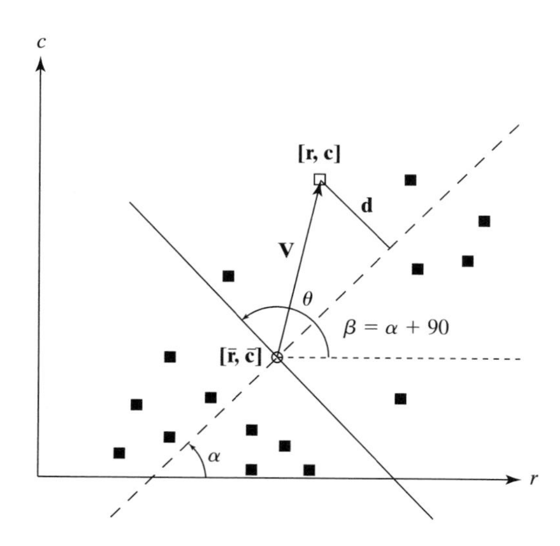

Figure 3.20 Moment about an axis is computed by summing the squared distance of each pixel from the axis.

does not change significantly with the number of pixels making up the shape. Note that, since we are summing d^2, the angles α and β can be changed $+/-\ \pi$ with no change to the second moment. Equation 3.22 gives the formula for computing the second moment: $\circ$ is the vector scalar product that is used to project the vector $\bar{V}$ onto the unit vector in direction β, giving length d. Any axis can be specified by the three parameters $\bar{r}$, $\bar{c}$ and α.

second moment about axis:*

$$\begin{aligned}
\mu_{\bar{r},\bar{c},\alpha} &= \frac{1}{A} \sum_{(r,c)\in R} d^2 \\
&= \frac{1}{A} \sum_{(r,c)\in R} (\bar{V} \circ (\cos\beta, \sin\beta))^2 \\
&= \frac{1}{A} \sum_{(r,c)\in R} ((r-\bar{r})\cos\beta + (c-\bar{c})\sin\beta)^2
\end{aligned} \tag{3.22}$$

where $\beta = \alpha + \pi/2$.

Exercise 3.13: Program to compute point set features.

Write a program module, or C++ class, that manages a *bag* of 2D points and provides the following functionality. A *bag* is different from a *set* in that duplicate points are allowed.

- construct an initially empty bag of 2D points **[r, c]**
- add point **[r, c]** to the bag
- compute the centroid of the current bag of points
- compute the row and column moments of the current bag of points
- compute the bounding box
- compute the best and worst axes and the second moments about them

---

Exercise 3.14: Program to compute features from images.

After creating the feature extraction module of the previous exercise, enhance it to compute the second moments about horizontal, vertical and diagonal axes through the centroid of points. Thus, five different second moments will be available for any bag of points. Create a set of 20×20 binary images of digits from 0 to 9 for test data, or access some existing data. Write a program that scans an image of a digit and computes the five moments. Study whether or not the five moments have potential for recognizing the input digit.

The above formula can be used to compute several moments to capture some information about the shape of the point set; for example, moments about the vertical, horizontal, and diagonal axes are useful for classifying alphabetic characters in standard orientation. The least (and most) inertia is an invariant property of the point set and translates and rotates with the point set. The axis of least inertia can be obtained by minimizing $\mu_{\bar{r},\bar{c},\alpha}$. Assuming now that the best axis must pass through the centroid, we need only differentiate the formula with respect to α to determine the best $\hat{\alpha}$.

axis with least second moment:*

$$\begin{aligned}
\tan 2\hat{\alpha} &= \frac{2\sum(r-\bar{r})(c-\bar{c})}{\sum(r-\bar{r})(r-\bar{r})-\sum(c-\bar{c})(c-\bar{c})} \\
&= \frac{\frac{1}{A}2\sum(r-\bar{r})(c-\bar{c})}{\frac{1}{A}\sum(r-\bar{r})(r-\bar{r})-\frac{1}{A}\sum(c-\bar{c})(c-\bar{c})} \\
&= \frac{2\,\mu_{rc}}{\mu_{rr}-\mu_{cc}} \qquad (3.23)
\end{aligned}$$

There are two extreme values for α, a minimum and a maximum, which are 90 degrees apart. We have already seen the method to distinguish the two in the above discussion about the major and minor axes of an ellipse. In fact, the above formula allows us to compute an ellipse that approximates the point set in the sense of these moments. Note that highly symmetrical objects, such as squares and circles, will cause a zero-divide in the above formula; hence the case analysis used with the elliptical data must also be done here.

Exercise 3.15: Compute the extremes of inertia.

Differentiate the formula in Equation 3.22 and show how the best (and worst) axes are obtained in Equation 3.23.

Exercise 3.16: Verify that the best axis passes through the centroid.

Verify that the axis of least inertia must pass though the centroid. Consult the references at the chapter's end or other references on statistical regression or mechanics; or, prove it yourself.

3.7 REGION ADJACENCY GRAPHS

In addition to properties of single regions, relationships among groups of regions are also useful in image analysis. One of the simplest, but most useful relationships is *region adjacency*. Two regions are adjacent if a pixel of one region is a neighbor of a pixel of the second region. In binary images, there are only two kinds of regions: foreground regions and background regions. All of the foreground regions are adjacent to the background and not to one another. If the background is one single, connected region, then there is nothing further to compute. Suppose instead that the foreground regions can have holes in them, each hole belonging to the background. Applying the connected components labeling operation to the foreground pixels yields a labeled image in which the foreground regions each have a numeric label and the background regions all have label zero. But it is also possible to apply the connected components operator to the background. In this case, all the background regions can be assigned labels, too. One of these regions will be large and will start at the top left of the image. This one can be given a special label, such as 0. The rest of the background regions are the holes in the foreground regions. Given the image of foreground labels and the image of background labels, it is useful to determine which background regions are adjacent to each foreground region or vice versa. The structure for keeping track of adjacencies between pairs of regions is called a *region adjacency graph*. It can be used for keeping track of adjacencies between foreground and background regions in the binary case and for keeping track of all adjacencies in the general image segmentation case.

21 Definition. A **region adjacency graph (RAG)** is a graph in which each node represents a region of the image, and an edge connects two nodes if the two regions are adjacent.

Figure 3.21 gives an example of a region adjacency graph for a binary image of foreground and background regions. The foreground regions have been labeled as usual with positive integers. The background regions have been labeled with zero for the large region that starts at the upper left pixel of the image and with negative integers for the hole regions.

The algorithm for constructing a region adjacency graph is straightforward. It processes the image, looking at the current row and the one above it. It detects horizontal and vertical adjacencies, and if 8-adjacency is specified, diagonal adjacencies between points with different labels. As new adjacencies are detected, new edges are added to the region adjacency graph data structure being constructed. There are two issues related to the efficiency of this algorithm. The first is with respect to space. It is possible for an image to have tens of thousands of labels. In this case, it may not be feasible, or at least not suitable in a paging environment, to keep the entire structure in internal memory at once. The second issue relates to execution time. When moving along an image, point by point, the same adjacency (that is, the same two region labels) will be detected over and over again. It is desirable to enter the adjacency into the data structure as infrequently as possible. These issues are addressed in Exercise 17.

Exercise 3.17: Efficient RAG construction.

Design a data structure for keeping track of adjacencies while constructing a region adjacency graph. Give algorithms that construct the graph from an arbitrary labeled image and that attempt to minimize references to the data structure. Discuss how you would store the final RAG in permanent storage (on disk) and how you would handle the case where the RAG is too large to keep in internal memory during its construction.

0	0	0	0	0	0	0	0	0	0
0	1	1	1	1	1	0	2	2	0
0	1	−1	−1	−1	1	0	2	2	0
0	1	1	1	1	1	0	2	2	0
0	0	0	0	0	0	0	2	2	0
0	3	3	3	0	2	2	2	2	0
0	3	−2	3	0	2	−3	−3	2	0
0	3	−2	3	0	2	−3	−3	2	0
0	3	3	3	0	2	2	2	2	0
0	0	0	0	0	0	0	0	0	0

(*a*) Labeled image of foreground and background regions

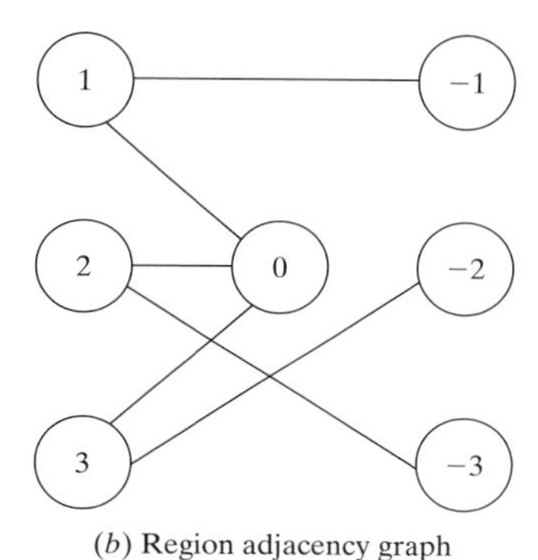

(*b*) Region adjacency graph

Figure 3.21 A labeled image and its region adjacency graph.

3.8 THRESHOLDING GRAY-SCALE IMAGES

Binary images can be obtained from gray-scale images by thresholding operations. A thresholding operation chooses some of the pixels as the foreground pixels that make up the objects of interest and the rest as background pixels. Given the distribution of gray tones in a given image, certain gray-tone values can be chosen as threshold values that separate the pixels into groups. In the simplest case, a single threshold value t is chosen. All pixels whose gray-tone values are greater than or equal to t become foreground pixels and all the rest become background. This threshold operation is called *threshold above*. There are many variants including *threshold below,* which makes the pixels with values less than or equal to t the foreground; *threshold inside,* which is given a lower threshold and an upper threshold and selects pixels whose values are between the two as foreground; and *threshold outside,* which is the opposite of threshold inside. The main question associated with these simple forms of thresholding is how to choose the thresholds.

3.8.1 The Use of Histograms for Threshold Selection

Thresholds can be selected interactively by a user of an interactive package, but for image analysis processes that must run automatically, we would like to be able to compute the thresholds automatically. The basis for choosing a threshold is the *histogram* of the gray-tone image.

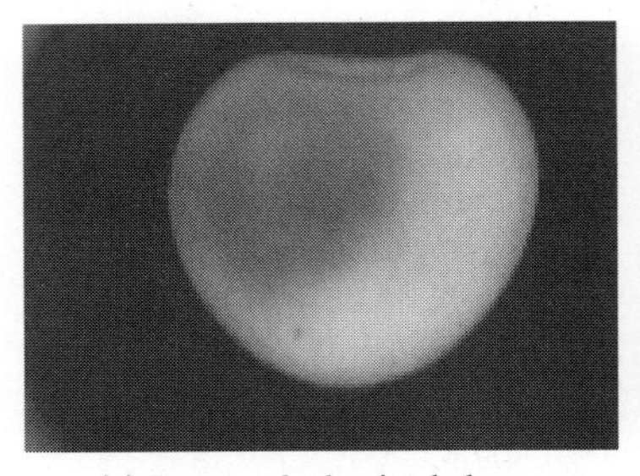

(*a*) Image of a bruised cherry

(*b*) Histogram of the cherry image

Figure 3.22 Histogram of the image of a bruised cherry displaying two modes: (*a*) one representing the bruised portion and (*b*) the nonbruised portion. (Courtesy of Patchrawat Uthaisombut.)

22 Definition. The **histogram h** of gray-tone image **I** is defined by

$$h(m) = |\{(r, c) \mid I(r, c) = m\}|,$$

where m spans the gray-level values.

Figure 3.22 shows the image of a bruised cherry and its histogram. The histogram has two distinct modes representing the bruised portion and nonbruised portion of the cherry.

A histogram can be computed by using an array data structure and a very simple procedure. Let **H** be a vector array dimensioned from 0 to **MaxVal,** where 0 is the value of the smallest possible gray-level value and **MaxVal** is the value of the largest. Let **I** be the two-dimensional image array with row values from 0 to **MaxRow** and column values from 0 to **MaxCol** as in the previous sections. The histogram procedure is given by the following code:

Compute the histogram H of gray-tone image I.

```
procedure histogram(I,H);
{
"Initialize the bins of the histogram to zero."
for i := 0 to MaxVal
  H[i] := 0;
"Compute values by accumulation."
for L := 0 to MaxRow
  for P := 0 to MaxCol
    {
    grayval := I[r,c];
    H[grayval] := H[grayval] + 1;
    };
}
```

Algorithm 3.7 Image histogram.

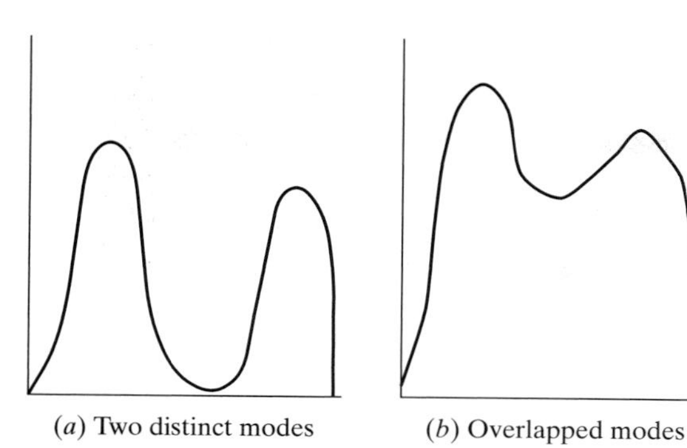

Figure 3.23 Two image histograms. (*a*) The histogram has two easily separable modes; (*b*) the histogram has overlapped modes that make it more difficult to find a suitable threshold.

This histogram procedure assumes that each possible gray tone of the image corresponds to a single bin of the histogram. Sometimes we instead want to group several gray tones into a single bin, usually for purposes of displaying the histogram when there are many possible gray tones. In this case the procedures can easily be modified to calculate the bin number as a function of the gray-tone. If *binsize* is the number of gray tones per bin, then *grayval* / *binsize* truncated to its integer value gives the correct bin subscript.

Given the histogram, automatic procedures can be written to detect peaks and valleys of the histogram function. The simplest case is when we are looking for a single threshold that separates the image into dark pixels and light pixels. If the distributions of dark pixels and bright pixels are widely separated, then the image histogram will be bimodal, one mode corresponding to the dark pixels and one mode corresponding to the bright pixels. With little distribution overlap, the threshold value can easily be chosen as any value in the valley between the two dominant histogram modes as shown in Figure 3.23(*a*). However, as the distributions for the bright and dark pixels become more and more overlapped, the choice of threshold value becomes more difficult, because the valley begins to disappear as the two distributions begin to merge together as shown in Figure 3.23(*b*).

3.8.2 Automatic Thresholding: the Otsu Method*

Several different methods have been proposed for automatic threshold determination. We discuss here the Otsu method, which selects the threshold based on the minimization of the within-group variance of the two groups of pixels separated by the thresholding operator. For this discussion, we will specify the histogram function as a probability function P where $P(0), \ldots, P(I)$ represent the histogram probabilities of the observed gray values $0, \ldots, I$; $P(i) = |\{(r, c) \mid Image(r, c) = i\}| / |R \times C|$, where $R \times C$ is the spatial domain of the image. If the histogram is bimodal, the histogram thresholding problem is to determine a best threshold t separating the two modes of the histogram from each other. Each threshold t determines a variance for the group of values that are less than or equal to t and a variance for the group of values greater than t. The definition for best threshold suggested by Otsu is that threshold for which the weighted sum of *within-group variances* is minimized. The weights are the probabilities of the respective groups.

We motivate the within-group variance criterion by considering the situation that sometimes happens at a ski school. A preliminary test of capabilities is given and the histogram of the resulting scores is bimodal. There are advanced skiers and novices. Lessons that are aimed at the advanced skiers go too fast for the others, and lessons that are aimed at the level of the novices are boring to the advanced skiers. To fix this situation, the teacher decides to divide the class into two mutually exclusive and homogeneous groups based on the test score. The question is to determine which test score to use as the dividing criterion. Ideally, each group should have test scores that have a unimodal bell-shaped histogram, one around a lower mean and one around a higher mean. This would indicate that each group is homogeneous within itself and different from the other.

A measure of group homogeneity is variance. A group with high homogeneity will have low variance. A group with low homogeneity will have high variance. One possible way to choose the dividing criterion is to choose a dividing score such that the resulting weighted sum of the within-group variances is minimized. This criterion emphasizes high group homogeneity. A second way to choose the dividing criterion is to choose a dividing score that maximizes the resulting squared difference between the group means. This difference is related to the between-group variance. Both dividing criteria lead to the same dividing score because the sum of the within-group variances and the between-group variances is a constant.

Let σ_W^2 be the weighted sum of group variances, that is, the *within-group variance*. Let $\sigma_1^2(t)$ be the variance for the group with values less than or equal to t and $\sigma_2^2(t)$ be the variance for the group with values greater than t. Let $q_1(t)$ be the probability for the group with values less than or equal to t and $q_2(t)$ be the probability for the group with values greater than t. Let $\mu_1(t)$ be the mean for the first group and $\mu_2(t)$ the mean for the second group. Then the within-group variance σ_W^2 is defined by

$$\sigma_W^2(t) = q_1(t)\,\sigma_1^2(t) + q_2(t)\,\sigma_2^2(t) \tag{3.24}$$

where

$$q_1(t) = \sum_{i=1}^{t} P(i)$$

$$q_2(t) = \sum_{i=t+1}^{I} P(i) \tag{3.25}$$

$$\mu_1(t) = \sum_{i=1}^{t} i\;P(i)/q_1(t)$$

$$\mu_2(t) = \sum_{i=t+1}^{I} i\;P(i)/q_2(t) \tag{3.26}$$

$$\sigma_1^2(t) = \sum_{i=1}^{t} [i - \mu_1(t)]^2 \, P(i)/q_1(t)$$

$$\sigma_2^2(t) = \sum_{i=t+1}^{I} [i - \mu_2(t)]^2 \, P(i)/q_2(t) \tag{3.27}$$

The best threshold t can then be determined by a simple sequential search through all possible values of t to locate the threshold t that minimizes $\sigma_W^2(t)$. In many situations this can be reduced to a search between the two modes. However, identification of the modes is really equivalent to the identification of separating values between the modes.

There is a relationship between the within-group variance $\sigma_W^2(t)$ and the total variance σ^2 that does not depend on the threshold. The total variance is defined by

$$\sigma^2 = \sum_{i=1}^{I} (i - \mu)^2 P(i)$$

where

$$\mu = \sum_{i=1}^{I} i \; P(i)$$

The relationship between the total variance and the within-group variance can make the calculation of the best threshold less computationally complex. By rewriting σ^2, we have

$$\begin{aligned}
\sigma^2 &= \sum_{i=1}^{t} [i - \mu_1(t) + \mu_1(t) - \mu]^2 \; P(i) + \sum_{i=t+1}^{I} [i - \mu_2(t) + \mu_2(t) - \mu]^2 \, P(i) \\
&= \sum_{i=1}^{t} \{[i - \mu_1(t)]^2 + 2[i - \mu_1(t)][\mu_1(t) - \mu] + [\mu_1(t) - \mu]^2\} P(i) \\
&\quad + \sum_{i=t+1}^{I} \{[i - \mu_2(t)]^2 + 2[i - \mu_2(t)][\mu_2(t) - \mu] + [\mu_2(t) - \mu]^2\} P(i)
\end{aligned}$$

But

$$\sum_{i=1}^{t} [i - \mu_1(t)][\mu_1(t) - \mu] P(i) = 0 \quad \text{and}$$

$$\sum_{i=t+1}^{I} [i - \mu_2(t)][\mu_2(t) - \mu)] P(i) = 0$$

Since

$$
\begin{aligned}
q_1(t) &= \sum_{i=1}^{t} P(i) \text{ and } q_2(t) = \sum_{i=t+1}^{I} P(i) \\
\sigma^2 &= \sum_{i=1}^{t} [i - \mu_1(t)]^2 P(i) + [\mu_1(t) - \mu]^2 q_1(t) \\
&\quad + \sum_{i=t+1}^{I} [i - \mu_2(t)]^2 P(i) + [\mu_2(t) - \mu]^2 q_2(t) \\
&= \left[q_1(t)\, \sigma_1^2(t) + q_2(t)\, \sigma_2^2(t)\right] \\
&\quad + \{q_1(t)\, [\mu_1(t) - \mu]^2 + q_2(t)\, [\mu_2(t) - \mu]^2\} \qquad (3.28)
\end{aligned}
$$

The first bracketed term is the within-group variance σ_W^2. It is just the sum of the weighted variances of each of the two groups. The second bracketed term is called the between-group variance σ_B^2. It is just the sum of the weighted squared distances between the means of each group and the grand mean. The between-group variance can be further simplified. Note that the grand mean μ can be written as

$$\mu = q_1(t)\, \mu_1(t) + q_2(t)\, \mu_2(t) \qquad (3.29)$$

Using Eq. (3.29) to eliminate μ in Eq. (3.28), substituting $1 - q_1(t)$ for $q_2(t)$, and simplifying, we obtain

$$\sigma^2 = \sigma_W^2(t) + q_1(t)[1 - q_1(t)]\,[\mu_1(t) - \mu_2(t)]^2$$

Since the total variance σ^2 does not depend on t, the t minimizing $\sigma_W^2(t)$ will be the t maximizing the between group variance $\sigma_B^2(t)$,

$$\sigma_B^2(t) = q_1(t)\,[1 - q_1(t)]\,[\mu_1(t) - \mu_2(t)]^2 \qquad (3.30)$$

To determine the maximizing t for $\sigma_B^2(t)$, the quantities determined by Eqs. (3.25) to (3.27) all have to be determined. However, this need not be done independently for each t. There is a relationship between the value computed for t and that computed for the next $t : t + 1$. We have directly from Eq. (3.25) the recursive relationship

$$q_1(t + 1) = q_1(t) + P(t + 1) \qquad (3.31)$$

with initial value $q_1(1) = P(1)$.

From Eq. (3.26) we obtain the recursive relation

$$\mu_1(t + 1) = \frac{q_1(t)\, \mu_1(t) + (t + 1)P(t + 1)}{q_1(t + 1)} \qquad (3.32)$$

with the initial value $\mu_1(0) = 0$. Finally, from Eq. (3.29) we have

$$\mu_2(t + 1) = \frac{\mu - q_1(t + 1)\, \mu_1(t + 1)}{1 - q_1(t + 1)} \qquad (3.33)$$

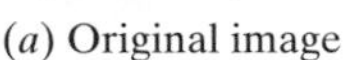

(*a*) Original image

(*b*) Pixels below 93

(*c*) Pixels above 93

Figure 3.24 A gray-tone image and the pixels below and above the threshold of 93 (shown in white) found by the Otsu automatic thresholding operator. (Original image courtesy of John Illingworth and Ata Etamadi.)

Automatic threshold-finding algorithms only work well when the images to be thresholded satisfy their assumptions about the distribution of the gray-tone values over the image. The Otsu automatic threshold finder assumes a bimodal distribution of gray-tone values. If the image approximately fits this constraint, it will do a good job. If the image is not at all bimodal, the results are not likely to be useful. Figure 3.24 illustrates the application of the Otsu operator to the gray-tone image of some toy blocks shown in (*a*). The operator returned a threshold of 93 from the possible range of 0 to 255. The pixels below and above the threshold are shown in (*b*) and (*c*), respectively. Only the very dark regions of the image have been isolated.

If the gray-tone values of an image are strongly dependent on the location within the image, for example lighter in the upper left corner and darker in the lower right, then it may be more appropriate to use local instead of global thresholds. This idea is sometimes called *dynamic* thresholding. In some applications, the approximate shapes and sizes of the objects to be found are known in advance. In this case a technique called *knowledge-based* thresholding, which evaluates the resultant regions and chooses the threshold that provides the best results, can be employed. Finally, some images are just not thresholdable, and alternate techniques must be used to find the objects in them.

Exercise 3.18: Automatic threshold determination.

Write a program to implement the Otsu automatic threshold finder. Try the program on several different types of scanned images.

3.9 REFERENCES

There are a number of different algorithms for the connected components labeling operation, each designed to address a certain task. Tanimoto (1990) assumes that the entire image can fit in memory and employs a simple, recursive algorithm that works on one component at a time, but can move all over the image. Other algorithms were designed for larger images that may not fit in memory and work on only two rows of the image at a time. Rosenfeld and Pfaltz (1966) developed the two-pass algorithm that uses a global equivalence table and is sometimes called the *classical* connected components algorithm. Lumia, Shapiro, and Zuniga (1983) developed another two-pass algorithm that uses a local equivalence table to avoid paging problems. Danielsson and Tanimoto (1983) designed an algorithm for massively parallel machines that uses a parallel propagation strategy. Any algorithms that keep track of equivalences can use the union-find data structure (Tarjan, 1975) to efficiently perform set-union operations.

Serra (1982) produced the first systematic theoretical treatment of mathematical morphology. Sternberg (1985) designed a parallel pipeline architecture for rapidly performing the operations and applied it to problems in medical imaging and industrial machine vision. He also extended the binary morphology operations to gray-scale morphology operations (1986), which have become standard image filtering operation. Haralick, Sternberg, and Zhuang (1987) published a tutorial paper on both binary and gray-scale morphology that has helped to show their value to the computer vision community. Shapiro, MacDonald, and Sternberg (1987) showed that morphological feature detection can be used for object recognition.

Automatic thresholding has been addressed in a number of papers. The method described in this text is due to Otsu (1979). Other methods have been proposed by Kittler and Illingworth (1986) and by Cho, Haralick, and Yi (1989). Sahoo and others (1988) give a general survey of thresholding techniques.

1. Tanimoto, S. L. 1990. *The Elements of Artificial Intelligence Using Common LISP.* W. H. Freeman and Company, New York.
2. Rosenfeld, A., and J. L. Pfaltz. 1966. Sequential operations in digital picture processing. *J. Assoc. Comput. Machinery,* v. 13:471–494.
3. Lumia, R., G. Shapiro, and O. Zuniga. 1983. A new connected components algorithm for virtual memory computers. *Comput. Vision, Graphics, and Image Proc.,* v. 22: 287–300.
4. Danielsson, P.-E., and S. L. Tanimoto. 1983. Time complexity for serial and parallel propagation in images. In *Architecture and Algorithms for Digital Image Processing,* A. Oosterlinck and P.-E. Danielsson, eds. *Proc. SPIE,* v. 435:60–67.
5. Tarjan, R. E. 1975. Efficiency of a good but not linear set union algorithm. *J. Assoc. Comput. Machinery,* v. 22:215–225.
6. Serra, J. 1982. *Image Analysis and Mathematical Morphology.* Academic Press, New York.
7. Sternberg, S. R. 1985. An overview of image algebra and related architectures. *Integrated Technology for Parallel Image Processing.* Academic Press, London, 79–100.
8. Sternberg, S. R. 1986. Grayscale morphology. *Comput. Vision, Graphics, and Image Proc.,* v. 35:333–355.
9. Haralick, R. M., S. R. Sternberg, and X. Zhuang. 1987. Image analysis using mathematical morphology. *IEEE Trans. Pattern Analysis and Machine Intelligence,* v. PMI-9:523–550.
10. Shapiro, L. G., R. S. MacDonald, and S. R. Sternberg. 1987. Ordered structural shape matching with primitive extraction by mathematical morphology. *Pattern Recog.,* v. 20(1)75–90.
11. Haralick, R. M. 1974. A measure of circularity of digital figures. *IEEE Trans. Syst., Man, and Cybern.,* v. SMC-4:394–396.

12. Otsu, N. 1979. A threshold selection method from gray-level histograms. *IEEE Trans. Syst., Man and Cybern.,* v. SMC-9:62–66.
13. Kittler, J., and J. Illingworth. 1986. Minimum error thresholding. *Pattern Recog.,* v. 19:41–47.
14. Cho, S., R. M. Haralick, and S. Yi. 1989. Improvement of Kittler and Illingworth's minimum error thresholding. *Pattern Recog.,* v. 22:609–617.
15. Sahoo, P. K., and others. 1988. A survey of thresholding techniques. *Comput. Vision, Graphics, and Image Proc.,* v. 41:233–260.

4

Pattern Recognition Concepts

This chapter gives a brief survey of methods used to recognize objects. These methods apply to the recognition of objects in images, and are applicable to any other kind of data as well. The basic approach views an instance to be recognized as a vector of measurements. Several examples are discussed; the central one being the recognition of characters. The reader is also introduced to some simple methods whereby a machine can learn to recognize objects by being taught from samples. After studying Chapters 1 through 4, the reader should understand the design of some complete machine vision systems and should be able to experiment with building a complete set of algorithms for some simple, yet real, problem.

4.1 PATTERN RECOGNITION PROBLEMS

In many practical problems, there is a need to make some decision about the content of an image or about the classification of an object that it contains. For example, the user of a notebook computer may be able to give input using handprinted characters. In this case, there would be $m = 128$ ASCII characters and each handprinted object would be classified into one of the m classes. See Figure 4.1. The classification of an object—whether it is an 'A' or an '8', etc.—would be based on the features of its optical image or perhaps of a pressure footprint, which is also an image-like representation. The classification process might actually fail, either because the character is badly made, or because the person invented a new character. Usually, a reject class is included in a system design in order to cover such cases. Image data put into the reject class might be examined again later at some higher level, might result in the formation of a new class, or might just be saved in raw form for viewing.

```
00000000000000000000   00000000000000000000
00000000010000000000   00000000011110000000
00000000110000000000   00000001100001100000
00000000101000000000   00000011000000110000
00000001100110000000   00000100000000010000
00000001000010000000   00001100000000011000
00000010000010000000   00001000000000001000
00000110000001000000   00001000000000011000
00000100000001000000   00001100000000010000
00000100000001100000   00000100000000110000
00001000000000100000   00000111000000100000
00001100111111110000   00000011100111100000
00001111110000010000   00000000111100000000
00011000000000011000   00000011000111000000
00010000000000001000   00000110000001100000
00010000000000001100   00001100000000110000
00110000000000000100   00011000000000011000
00110000000000000110   00110000000000001000
00100000000000000010   00100000000000001100
00100000000000000010   00010000000000011000
01100000000000000010   00011000000000010000
01000000000000000000   00001000000000110000
00000000000000000000   00001110000011100000
00000000000000000000   00000011111110000000
00000000000000000000   00000000000000000000
```

Figure 4.1 Binary images of 'A' and '8'.

Imagine an automatic bank teller machine (ATM) using a camera to verify that a current user is indeed authentic. Here, the image of the current person's face is to be matched to a stored image, or images, attached to the current account and stored either on a computer network or in the bank card itself.

> **23 Definition.** The process of matching an object instance to a single object prototype or class definition is called **verification.**

In another application, introduced in the exercises of Chapter 1, a food market recognition system would classify fruits and vegetables placed on the checker's scale. The classes would be the set of all identifiable produce items, such as Ida apples, Fuji apples, collard greens, spinach greens, mushrooms, etc., each with a separate name and per-pound charge.[1]

One definition for recognition is to know again. A recognition system must contain some memory of the objects that it is to recognize. This memory representation might be built in, perhaps as is the frog's model of a fly; or might be taught by provision of a large number of samples, as a schoolteacher teaches the alphabet; or it might be programmed in terms of specific image features, perhaps as a mother would teach a child to recognize fire trucks versus buses. Recognition and learning of patterns are subjects of considerable depth and interest to cognitive pyschology, pattern recognition, and computer vision. Chapter 4 takes a practical approach and describes methods that have had success in applications, leaving some pointers to the large theoretical literature in the references at the end of the chapter.

[1]Such a system, called *Veggie Vision,* has already been developed by IBM. See Chapter 16.

4.2 COMMON MODEL FOR CLASSIFICATION

We summarize the elements of the common model of **classification:** this breakdown is practical rather than theoretical and done so that pattern recognition systems can be designed and built using separately developed hardware and software modules.

4.2.1 Classes

There is a set of m known classes of objects. These are known either by some description or by having a set of examples for each of the classes. For example, for character classification, we have either a description of the appearance of each character or we have a set of samples of each. In the general case, there will be a special reject class for objects that cannot be placed in one of the known classes.

24 Definition. An ideal **class** is a set of objects having some important common properties: in practice, a class to which an object belongs is denoted by some **class label. Classification** is a process that assigns a label to an object according to some representation of the object's properties. A **classifier** is a device or algorithm that inputs an object representation and outputs a class label.

25 Definition. A **reject class** is a generic class for objects that cannot be placed in any of the designated known classes.

4.2.2 Sensor/Transducer

There must be some device to sense the actual physical object and output a (usually digital) representation of it for processing by machine. Most often, the sensor is selected from existing sensors (off-the-shelf) built for a larger class of problems. For example, to classify vegetables in the supermarket, we could first try using a general color camera that would provide an image representation from which color, shape, and texture features could be obtained. To recognize characters made by an impression using a stylus, we would use a pressure sensitive array.

Since this is a book about machine vision, sensors that produce 2D arrays of sensed data are of most interest. However, pattern recognition itself is more general and just as applicable to recognizing spoken phone numbers, for example, as phone numbers written on paper.

4.2.3 Feature Extractor

The feature extractor extracts information relevant to classification from the data input by the sensor. Usually, feature extraction is done in software. Software can be adapted to the sensor hardware on the input side and can evolve through research and development to output results highly relevant to classification. Many image features were defined in Chapter 3.

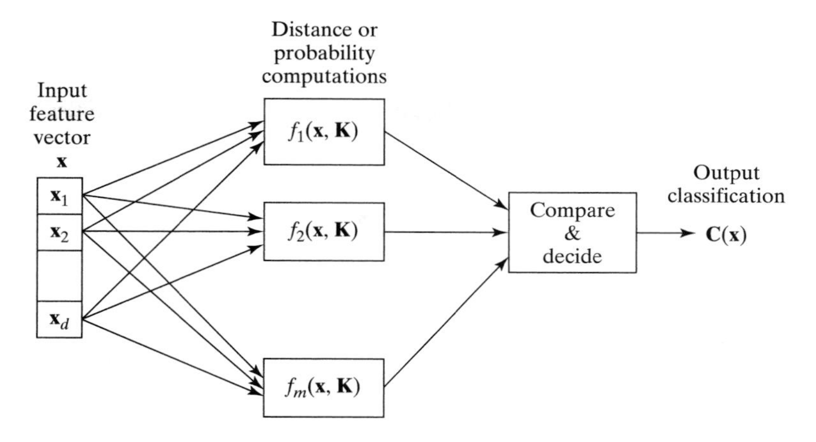

Figure 4.2 Classification system diagram: discriminant functions $f(\mathbf{x}, \mathbf{K})$ perform some computation on input feature vector **x** using some knowledge **K** from training and pass results to a final stage that determines the class.

4.2.4 Classifier

The classifier uses the features extracted from the sensed object data to assign the object to one of the m designated classes $C_1, C_2, \ldots, C_{m-1}, C_m = C_r$, where C_r denotes the reject class.

A block diagram of a classification system is given in Figure 4.2. A d-dimensional feature vector **x** is the input representing the object to be classified. The system has one block for each possible class, which contains some knowledge **K** about the class and some processing capability. Results from the m computations are passed to the final classification stage, which decides the class of the object. The diagram is general enough to model the three different types of classification discussed next: (a) classification using the nearest mean; (b) classification by maximum *a posteriori* probability, and, (c) classification using a feedforward artificial neural network.

4.2.5 Building the Classification System

Each of the system parts has many alternative implementations. Image sensors were treated in Chapter 2. Chapter 3 discussed how to compute a number of different features from binary images of objects. Computing color and texture features is treated in Chapters 6 and 7. The character recognition example is again instructive. Characters written in a 30×20 window would result in 600 pixels. Feature extraction might process these 600 pixels and output 10 to 30 features on which to base the classification decisions. This example is developed next.

Another common name for the feature extractor is the *preprocessor*. Between the sensor and the classifier, some filtering or noise cleaning must also be performed that should be part of the preprocessing. Some noise cleaning operations were seen in Chapter 3 and more will be studied in Chapter 5. The division of processing between feature extraction and classification is usually somewhat arbitrary and more due to engineering concerns than to some inherent properties of an application. Indeed, we will look at neural nets, which can do one step classification directly from the input image.

4.2.6 Evaluation of System Error

The *error rate* of a classification system is one measure of how well the system solves the problem for which it was designed. Other measures are speed, in terms of how many objects can be processed per unit time, and expense, in terms of hardware, software, and development cost. Performance is determined by both the errors and rejections made; classifying all inputs into the reject class means that the system makes no errors but is useless.

26 Definition. The classifier makes a **classification error** whenever it classifies the input object as class C_i when the true class is class C_j; $i \neq j$ and $C_i \neq C_r$, the reject class.

27 Definition. The **empirical error rate** of a classification system is the number of errors made on independent test data divided by the number of classifications attempted.

28 Definition. The **empirical reject rate** of a classification system is the number of rejects made on independent test data divided by the number of classifications attempted.

29 Definition. **Independent test data** are sample objects with true class known, including objects from the *reject class,* that were not used in designing the feature extraction and classification algorithms.

The previous definitions can be used in practice to test the performance of a classification system. We must be very careful to insure that the samples used to design and those used to test the system are representative of the samples that the system will have to process in the future; and, samples used to test the system must be independent of those used to design it. Sometimes, we assume that our data follows some theoretical distribution. With this assumption, we are able to compute a theoretical probability of error for future performance, rather than just an empirical error rate from testing. This concept will be discussed below.

Suppose a handprinted character recognition module for a hand-held computer correctly recognizes 95 percent of a user's input characters. Given that the user may have to edit an input document anyway, a 5 percent error rate may be acceptable. Interestingly, such a system might actually train the user, as well as the user training the system, so that performance gradually improves. For example, perhaps the user learns to more carefully close up 8s so that they are not confused with 6s. For a banking system that must read hand-printed digits on deposit slips, a 5 percent error rate might be intolerable.

4.2.7 False Alarms and False Dismissals

Some problems are special *two-class problems* where the meaning of the classes might be (a) good object versus bad object; (b) object present in the image versus object absent; or,

(c) person has disease D versus person does not have disease D. Here, the errors take on special meaning and are not symmetric. Case (c) is most instructive: if the system incorrectly says that the person does have disease D then the error is called a *false alarm* or *false positive;* whereas, if the system incorrectly says that the person does not have disease D, then the error is called a *false dismissal* or *false negative*. In the case of the false alarm, it probably means that the person will undergo the cost of more tests, or of taking medicine that is not needed. In the case of false dismissal, the diagnosis is missed and the person will not be treated, possibly leading to grave circumstances. Because the cost of the errors differ greatly, it may make sense to bias the decision in order to minimize false dismissals at the cost of increasing the number of false alarms. Case (a) is less dramatic when the problem is to cull out bruised cherries; a false alarm may mean that the cherry goes into a pie rather than in the produce bin where it would have had more value. False alarms in case (b) may mean we waste energy by turning on a light when there really was no motion in the scene or that we counted an auto on the highway when one really did not pass by; false dismissals in case (b) also have interesting consequences. Figure 4.3 shows a typical *receiver operating curve* (*ROC*), which relates false alarm rate to detection rate. In order to increase the percentage of objects correctly recognized, one usually has to pay a cost of incorrectly passing along objects that should be rejected.

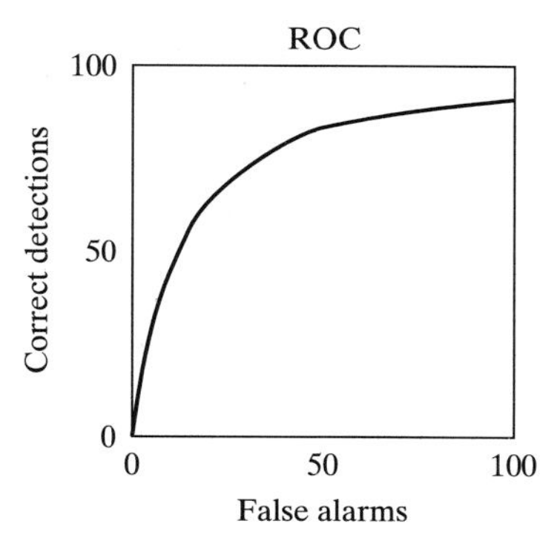

Figure 4.3 The *receiver operating curve* or *ROC* plots correct detection rate versus false alarm rate. Generally, the number of false alarms will go up as the system attempts to detect higher percentages of known objects. Modest detection performance can often be achieved at a low cost in false alarms, but correct detection of nearly all objects will cause a large percentage of unknown objects to be incorrectly classified into some known class.

4.3 PRECISION VERSUS RECALL

In the application of document retrieval (DR) or image retrieval, the objective is to retrieve interesting objects of class C_1 and not too many uninteresting objects of class C_2 according to features supplied in a user's query. For example, the user might be interested in retrieving images of sunsets, or perhaps horses. The performance of such a system is characterized by its *precision* and *recall.*

30 Definition. The **precision** of a DR system is the number of relevant documents (true C_1) retrieved divided by the total number of documents retrieved (true C_1 plus false alarms actually from C_2).

31 Definition. The **recall** of a DR system is the number of relevant documents retrieved by the system divided by the total number of relevant documents in the database. Equivalently, this is the number of true C_1 documents retrieved divided by the total of the true C_1 documents retrieved and the false dismissals.

For example, suppose an image database contains 200 sunset images that would be of interest to the user and that the user hopes will match the query. Suppose the system retrieves 150 of those 200 relevant images and 100 other images of no interest to the user. The precision of this retrieval (classification) operation is $150/250 = 60$ percent while the recall is $150/200 = 75$ percent. The system could obtain 100 percent recall if it returned all images in the database, but then its precision would be terrible. Alternatively, if the classification is tightly set for a low false alarm rate, then the precision would be high, but the recall would be low. Image database retrieval will be examined in some detail in Chapter 8.

4.4 FEATURES USED FOR REPRESENTATION

A crucial issue for both theory and practice is *what representation or encoding of the object is used in the recognition process?* Alternatively, what features are important for recognition? Let us return to the handprinted character recognition application. Suppose that individual characters can be isolated by a connected components algorithm or by requiring the writer to write them in designated boxes and that we have the following features of each computed by the methods of Chapter 3.

- the *area* of the character in units of black pixels.
- the *height* and *width* of the bounding box of its pixels.
- the *number of holes* inside the character.
- the *number of strokes* forming the character.
- the *center (centroid)* of the set of pixels.
- the *best axis direction* through the pixels as the axis of least inertia.
- the *second moments* of the pixels about the axis of least inertia and most inertia.

Using common-sense reasoning, we can make a table of the properties of characters in terms of these features. The table can be refined by studying the properties of many samples of each character. After doing this, we might have a short decision procedure to classify characters, or at least a set of prototypes to be used for comparison.

Table 4.1 shows 8 features for 10 different characters. For now, assume that there is no error in computing the features. A sequential decision procedure can be used to classify instances of these 10 classes as given in Algorithm 4.1. This structure for classification is called a *decision tree*. Decisions shown in the table are easily implemented in a computer program that has access to the feature values. At each point in the decision process, a small set of features is used to branch to some other points of the decision process; in the current example only one feature is used at each decision point. The branching process models the reduction in the set of possibilities as more features are successively considered.

Classify a character based on 3 features.

input: feature vector with [#holes, #strokes, moment of inertia]
output: class of character

```
case of #holes

   0: character is 1, W, X, *, -, or /

        case of moment about axis of least inertia

           low: character is  1, -, or /

                  case of best axis direction
                     0: character is  -
                    60: character is  /
                    90: character is  1

         large: character is W or X or *

                  case of #strokes
                     0: character is *
                     2: character is X
                     4: character is W

   1: character is A or O

        case of #strokes
           0: character is o
           3: character is A

   2: character is B or 8

        case of #strokes
           0: character is 8
           1: character is B
```

Algorithm 4.1 Simple decision procedure to classify characters from a set of eight possible characters.

TABLE 4.1 EXAMPLE FEATURES FOR A SAMPLE CHARACTER SET.

(Class) Character	Area	Height	Width	Number Holes	Number Strokes	(cx,cy) Center	Best Axis	Least Inertia
'A'	medium	high	3/4	1	3	1/2, 2/3	90	medium
'B'	medium	high	3/4	2	1	1/3, 1/2	90	large
'8'	medium	high	2/3	2	0	1/2, 1/2	90	medium
'0'	medium	high	2/3	1	0	1/2, 1/2	90	large
'1'	low	high	1/4	0	1	1/2, 1/2	90	low
'W'	high	high	1	0	4	1/2, 2/3	90	large
'X'	high	high	3/4	0	2	1/2, 1/2	?	large
'*'	medium	low	1/2	0	0	1/2, 1/2	?	large
'-'	low	low	2/3	0	1	1/2, 1/2	0	low
'/'	low	high	2/3	0	1	1/2, 1/2	60	low

The current example is used because of its intuitive value. It is naïve to suppose that the decision procedure sketched so far is close to a decision procedure that would perform well in a real handprinted character recognition system. Reliably defining and computing the number of strokes, for instance, is very difficult, although we will see a method that might work in Chapter 10. Moreover, methods from Chapters 3 and 5 are needed to remove some of the variations in the data before features are extracted. In some controlled industrial environments, one can set up such simple classification procedures and then adjust the quantitative parameters according to measurements made from sample images. We should expect variations of features within the same class and overlap of some features across classes. Some methods to handle such variation and overlap are studied next.

4.5 FEATURE VECTOR REPRESENTATION

Objects may be compared for similarity based on their representation as a vector of measurements. Suppose that each object is represented by exactly d measurements. The *ith* coordinate of such a feature vector has the same meaning for each object A; for example, the first coordinate might be object area, the second coordinate the row moment μ_{rr} defined in Chapter 3, the third coordinate the elongation, and so on. It is convenient for each measurement to be a real or floating point number. The similarity, or closeness, between the feature vector representations of two objects can then be described using the Euclidean distance between the vectors defined in Equation 4.1. As is shown in Figure 4.4 and discussed in the next section, sometimes the Euclidean distance between an observed vector and a stored class prototype can provide a useful classification function.

32 Definition. The **Euclidean distance** between two d-dimensional feature vectors $\mathbf{x}_1$ and $\mathbf{x}_2$ is

$$\|\mathbf{x}_1 - \mathbf{x}_2\| = \sqrt{\sum_{i=1,d} (\mathbf{x}_1[i] - \mathbf{x}_2[i])^2} \tag{4.1}$$

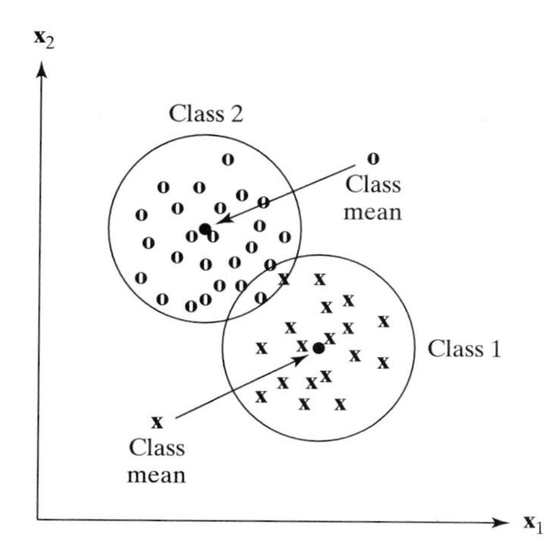

Figure 4.4 Two compact classes: Classification using nearest mean will yield a low error rate.

4.6 IMPLEMENTING THE CLASSIFIER

We now return to the classical paradigm, which represents an unknown object to be classified as a vector of atomic features. A recognition system can be designed in different ways based on feature vectors learned from samples or predicted from models. We examine two alternate methods of using a database of training samples. Assume that there are m classes of objects, not including a reject class, and that we have n_i sample vectors for class i. In our character recognition example from Algorithm 4.1, we had $m = 10$ classes of characters; perhaps we would have $n_i = 100$ samples from each. The feature vectors have dimension $d = 8$ in this case.

4.6.1 Classification Using the Nearest Class Mean

A simple classification algorithm is to summarize the sample data from each class using the class mean vector, or *centroid*, $\bar{\mathbf{x}}_\mathbf{i} = 1/n_i \sum_{j=1,n_i} \mathbf{x}_{i,j}$ where $\mathbf{x}_{i,j}$ is the *jth* sample feature vector from class i. An unknown object with feature vector $\mathbf{x}$ is classified as class i if it is [much] closer to the mean vector of class i than to any other class mean vector. We have the opportunity to put $\mathbf{x}$ into the reject class if it is not close enough to any of the sample means. This classification method is simple and fast and will work in some problems where the sample vectors from each class are compact and far from those of the other classes. A simple two class example with feature vectors of dimension $d = 2$ is shown in Figure 4.4: Sample vectors of class one are denoted by **x** and those of class two are denoted by **o**. Since there are samples of each class that are equidistant from both class centroids, the error rate will not be zero, although we expect it to be very low if the structure of the samples well represents the structure of future sensed objects. We now have one concrete interpretation for the function boxes of Figure 4.2: The *ith* function box computes the distance between unknown input **x** and the mean vector of training samples from that class. The training samples constitute the knowledge **K** about the class.

Exercise 4.1: Classifying coins.

Obtain ten samples of each coin used in the U.S. (penny, nickel, dime, quarter, half-dollar, dollar). Using a micrometer, measure the diameter and thickness of each of the 60 samples to the nearest 0.01 in. and then create a scatter plot of the six classes as done in Figure 4.4. (Always measure the thickness either at the center or across the edge.) Estimate the error rate of a classifier based on the nearest mean computation.

Difficulties can arise when the structure of class samples is complex. Figure 4.5 shows a case where class samples are well separated, but the structure is such that nearest mean classification will yield poor results for multiple reasons. First of all, class two (**o**) is *multimodal:* its samples lie in two separate compact clusters that are not represented well by the overall mean that lies midway between the two modes. Several of the samples from class one (**x**) are closer to the mean of class two than to the mean of class one. By studying the samples, we might discover the two modes of class two and be able to separate class two into two subclasses represented by two separate means. While this is simple using a 2D scatter plot as in Figure 4.5, it may be very difficult to understand the structure of samples when dimension d is much higher than 2. A second problem is due to the elongation of classes one and three. Clearly, samples of class three with large coordinate x_2 are closer to the mean of class two than they are to the mean of class three. Similarly, samples of class one (**x**) with small coordinate x_1 will still be closer to the mean of one of the subclasses of class two, even if class two is split into two modes. This problem can be reduced by modifying the distance computation to take into consideration the different spread of the samples along the different dimensions.

We can compute a modified distance from unknown feature vector $\mathbf{x}$ to class mean vector $\mathbf{x_c}$ by scaling by the spread, or *standard deviation,* σ_i of class c along each dimension i. The standard deviation is the square root of the variance.

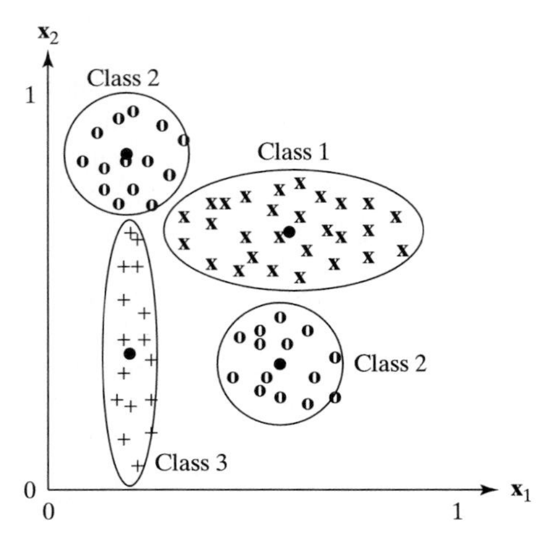

Figure 4.5 Three classes with complex structure: Classification using nearest mean will yield poor results.

33 Definition. Scaled Euclidean distance from x to class mean $\mathbf{x_c}$:

$$\|\mathbf{x} - \mathbf{x_c}\| = \sqrt{\sum_{i=1,d} ((\mathbf{x}[i] - \mathbf{x_c}[i])/\sigma_i)^2} \quad (4.2)$$

Scaling is almost always required due to the different units along the different dimensions. For example, suppose we were classifying vehicles using features $\mathbf{x}[\mathbf{1}]$ = length in feet and $\mathbf{x}[\mathbf{2}]$ = weight in pounds: without scaling, the Euclidean distance would be dominated by the large numbers of pounds, which would overshadow any discrimination due to vehicle length.

In the case shown in Figure 4.5, with separate mean vectors for the two modes of class two, such separate class-dependent scaling of the features x_1 and x_2 would yield good classification results. Most cases are not so easy, however. If the ellipses representing the distribution of class samples are not aligned with the coordinate axes, as they are in Figure 4.5, then coordinate transformations are needed in order to properly compute the distance of an unknown sample to a class mean. This is discussed below under the Bayesian classification scheme. A harder problem results if the set of samples has a curved structure in d-dimensional space.

4.6.2 Classification Using the Nearest Neighbors

A more flexible but more expensive method of classification is to classify unknown feature vector $\mathbf{x}$ into the class of the individual sample closest to it. This is the *nearest-neighbor rule*. Nearest-neighbor classification can be effective even when classes have complex structure in d-space and when classes overlap. No assumptions need to be made about models for the distribution of feature vectors in space; the algorithm uses only the existing training samples. A brute force approach (Algorithm 4.2) computes the distance from $\mathbf{x}$ to all samples in the database of samples and remembers the minimum distance. One advantage of this approach is that new labeled samples can be added to the database at any time. There are data structures that can be used to eliminate many unnecessary distance computations. Tree-structured or gridded data sets are two examples that are described in the chapter references.

A better classification decision can be made by examining the nearest k feature vectors in the database. $k > 1$ allows a better sampling of the distribution of vectors in d-space: this is especially helpful in regions where classes overlap. It has been shown that in the limit as the number of samples grows to infinity, the error rate for even $k = 1$ is no worse than twice the optimal error rate. In theory, we should do better with $k > 1$; but, effectively using a larger k depends on having a larger number of samples in each neighborhood of the space to prevent us from having to search too far from $\mathbf{x}$ for samples. In a two-class problem using $k = 3$, we would classify $\mathbf{x}$ into the class that has 2 of the 3 samples nearest $\mathbf{x}$. If there are more than two classes then there are more combinations possible and the decision is more complex. Algorithm 4.2 given below classifies the input vector into the reject class if there is no majority of the nearest k samples from any one class. This algorithm assumes no structure on the set of training samples; without such structure, the algorithm

Compute the k nearest neighbors of x and return the majority class.

S is a set of n labeled class samples s_i where s_i.**x** is a feature vector and $s_i.c$ is its integer class label.
x is the unknown input feature vector to be classified.
A is an array capable of holding up to k samples in sorted order by distance d.
The value returned is a class label in the range **[1, m]**

```
procedure K_Nearest_Neighbors(x, S)
{
make A empty;
for all samples s_i in S
{
  d = Euclidean distance between s_i and x;
  if A has less than k elements then insert (d, s_i) into A;
  else if d is less than max A
    then {
          remove the max from A;
          insert (d, s_i) in A;
         }
};
assert A has k samples from S closest to x;
if a majority of the labels s_i.c from A are class c_0
  then classify x into class c_o;
  else classify x into the reject class;
return(class_of_x);
}
```

Algorithm 4.2 Nearest-neighbor classification.

becomes slower with increasing number of samples n and k. Algorithms that use efficient data structures for the samples can be found via the references at the end of this chapter.

4.7 STRUCTURAL TECHNIQUES

Simple numeric or symbolic features of an object may not be sufficient for recognition. For example, consider the two characters shown in Figure 4.6. They have identical bounding boxes, the same numbers of holes and strokes, the same centroid, and the same second moments in the row and column directions, and their major axis directions are within 0.1 radian of being the same. Each of these characters has two *bays,* which are intrusions of the background into the character. Each bay has a *lid,* a virtual line segment that closes up the bay. The main differentiating feature of these two characters is a relationship, the spatial relationship between these two bays. In the character on the left, the lid of the upper bay is to the right of the lid of the lower bay. In the character on the right, the lid of the upper bay

0	0	0	0	0	0	0	0	0	0
0	0	**1**	**1**	**1**	**1**	**1**	**1**	**1**	0
0	**1**	0	0	0	0	0	0	**1**	0
0	**1**	0	0	0	0	0	0	0	0
0	**1**	0	0	0	0	0	0	0	0
0	**1**	**1**	**1**	**1**	**1**	**1**	0	0	0
0	0	0	0	0	0	0	**1**	0	0
0	0	0	0	0	0	0	0	**1**	0
0	0	0	0	0	0	0	**1**	**1**	0
0	**1**	0	0	0	**1**	**1**	**1**	0	0
0	0	**1**	**1**	**1**	**1**	0	0	0	0
0	0	0	0	0	0	0	0	0	0

0	0	0	0	0	0	0	0	0	0
0	**1**	**1**	**1**	**1**	**1**	**1**	**1**	0	0
0	**1**	0	0	0	0	0	0	**1**	0
0	0	0	0	0	0	0	0	**1**	0
0	0	0	0	0	0	0	0	**1**	0
0	0	0	**1**	**1**	**1**	**1**	**1**	**1**	0
0	0	**1**	0	0	0	0	0	0	0
0	**1**	0	0	0	0	0	0	0	0
0	**1**	**1**	0	0	0	0	0	0	0
0	0	**1**	**1**	**1**	0	0	0	**1**	0
0	0	0	0	**1**	**1**	**1**	**1**	0	0
0	0	0	0	0	0	0	0	0	0

Figure 4.6 Two characters that have the same global features but a different structure.

is to the left of the lid of the lower bay. This suggests that relationships among primitive features can be used as higher-level and potentially more powerful features for recognition. The field of *structural pattern recognition* has developed from this premise.

Statistical pattern recognition traditionally represents entities by feature vectors, which are typically vectors of atomic values, such as numbers and Boolean values (T or F). These values measure some global aspect of the entities, such as area or spatial moments. Our character example goes one step further in that it measures the number of holes and the number of strokes in each character. This implies the existance of a hole-finding procedure to find and count the holes and of some type of segmentation algorithm that can partition the character into strokes.

In structural pattern recognition, an entity is represented by its primitive parts, their attributes, and their relationships, as well as by its global features. Figure 4.7 illustrates three separate letter As that have approximately the same structure. Each can be broken up into 4 major strokes: two horizontal and two vertical or slanted. Each has a hole or *lake* near the top of the character with a bay below it; the lake and the bay are separated by a horizontal stroke.

When the relationships among primitives are binary relations, a structural description of an entity can be viewed as a graph structure. Suppose the following relationships over

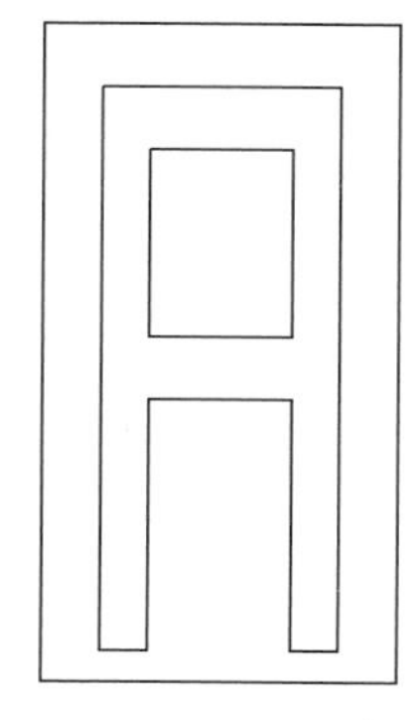

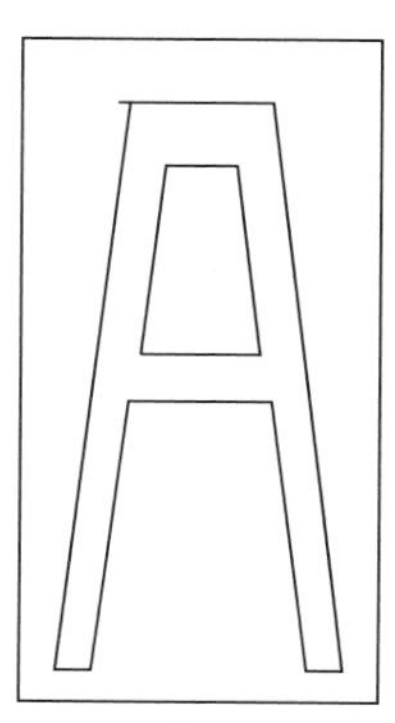

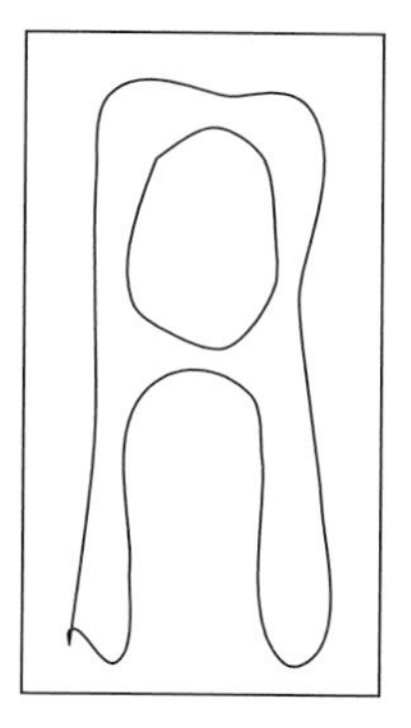

Figure 4.7 Three As with similar structure.

strokes, bays, and lakes are useful in recognizing characters:

- **CON:** specifies the connection of two strokes
- **ADJ:** specifies that a stroke region is immediately adjacent to a lake or bay region
- **ABOVE:** specifies that one hole (lake or bay) lies above another

Figure 4.8 shows a graph representation of the structural description of the character 'A' using these three binary relations. Higher-level relations, that is, ternary or even quaternary, can be used if they can be defined to provide even stronger constraints. For example, a ternary relationship exists among the lake, the horizontal stroke below it, and the bay below the stroke.

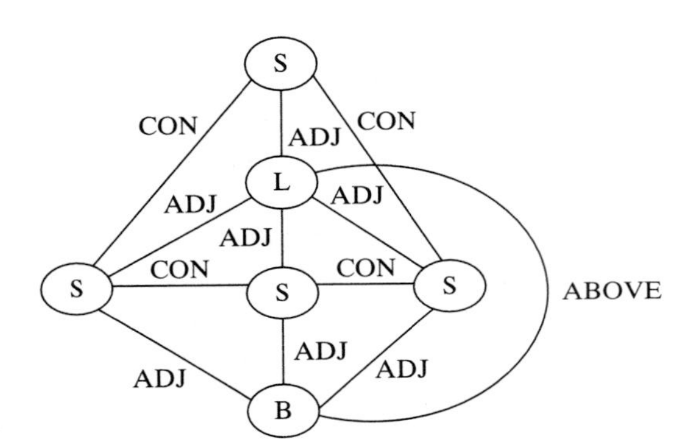

Figure 4.8 A graph structure representing the letter 'A'. 'S', 'L', 'B' denote *side, lake, bay* respectively.

Structural pattern recognition is often achieved through *graph-matching* algorithms, which will be covered in Chapter 11. However, the relationship between two primitives can itself be considered an atomic feature and thus can be used in a feature vector and incorporated into a statistical decision procedure. One simple way to do this is to merely count the number of times a particular relationship between two particular feature types (such as, a bay beneath a horizontal stroke) appears in a pattern. The integer value of the count becomes a feature for recognition of the overall pattern.

Structural methods are useful for recognition of complex patterns involving many sub-patterns. They also offer advantages in higher-level understanding of a scene, especially when multiple objects are present. From one general viewpoint, structural pattern recognition together with the other methods of this chapter emcompasses all of computer vision: within this view, the remaining chapters of the book can be taken to provide more methods of extracting features and parts from 2D or 3D objects and scenes.

4.8 THE CONFUSION MATRIX

34 Definition. The **confusion matrix** is commonly used to report results of classification experiments. Figure 4.9 gives an example. The entry in row i, column j records the number of times that an object labeled to be truly of class i was classified as class j.

		Class j output by the pattern recognition system										
		'0'	'1'	'2'	'3'	'4'	'5'	'6'	'7'	'8'	'9'	'R'
	'0'	97	0	0	0	0	0	1	0	0	1	1
	'1'	0	98	0	0	1	0	0	1	0	0	0
True	'2'	0	0	96	1	0	1	0	1	0	0	1
object	'3'	0	0	2	95	0	1	0	0	1	0	1
class	'4'	0	0	0	0	98	0	0	0	0	2	0
	'5'	0	0	0	1	0	97	0	0	0	0	2
	'6'	1	0	0	0	0	1	98	0	0	0	0
i	'7'	0	0	1	0	0	0	0	98	0	0	1
	'8'	0	0	0	1	0	0	1	0	96	1	1
	'9'	1	0	0	0	3	0	0	0	1	95	0

Figure 4.9 Hypothetical confusion matrix for digit recognition. 'R' is the reject class.

The confusion matrix diagonal, where $i = j$, indicates the successes: with perfect classification results, all off-diagonal elements are zero. High off-diagonal numbers indicate confusion between classes and force us to reconsider our feature extraction procedures or our classification procedure. If we have done thorough testing, the matrix indicates the kinds and rates of errors we expect in a working system. In the example shown in Figure 4.9, 7 of 1,000 vectors input to the system were rejected. Three inputs labeled as 9 were incorrectly classified as 4, while two inputs labeled as 4 were incorrectly classified as 9. Altogether, 25 of the input vectors were misclassified. Assuming that the test data was independent of that used to train the classification system, we would have an empirical reject rate of $7/1{,}000 = 0.007$ and an (overall) error rate of $25/1{,}000 = 0.025$. The error rate for just 9s, however, is $5/100 = 0.05$.

4.9 DECISION TREES

When a pattern recognition task is complex and involves many different potential features, comparing an entire unknown feature vector to many different pattern feature vectors may be too time consuming. It may even be impossible as in the case of medical diagnosis, where measurement of a feature usually means a costly and perhaps painful lab test. Use of a decision tree allows feature extraction and classification steps to be interleaved. The decision tree is a compact structure that uses one feature (or, perhaps a few features) at a time to split the search space of all possible patterns. The simple decision procedure of Algorithm 4.1 implements the flow of control shown in the decision tree of Figure 4.10. This tree has nodes that represent different features of the feature vector. Each branching node has one child per possible value for its feature. The decision procedure selects a child node based on the value of the specified feature in the unknown feature vector. A child node may specify another feature to be tested, or it may be a leaf node containing the classification associated with that path through the tree.

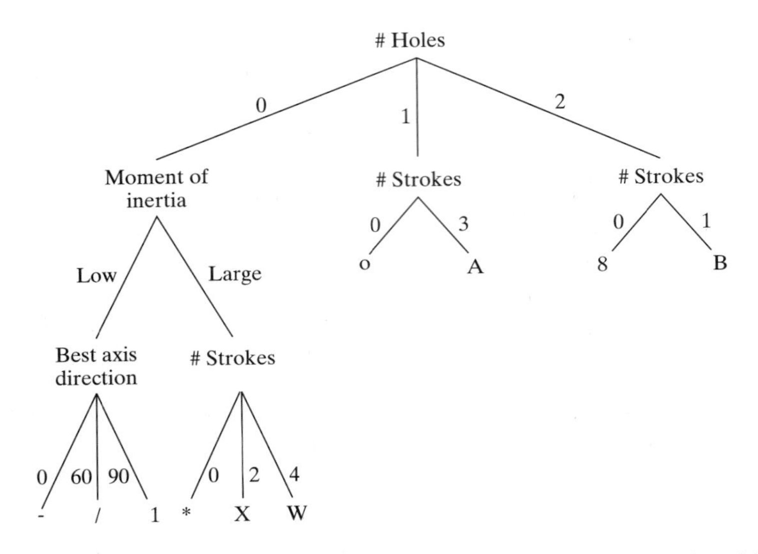

Figure 4.10 Decision tree that implements the classification procedure of Algorithm 4.1.

35 Definition. A **binary decision tree** is a binary tree structure that has a decision function associated with each node. The decision function is applied to the unknown feature vector and determines whether the next node to be visited is the left child or the right child of the current node.

In the simplest case with numeric feature values, the decision function at a node merely compares the value of a particular feature of the unknown feature vector to a threshold. The decision function selects the left child if the value of the feature is less than the threshold and the right child otherwise. In this case, only the feature to be used and the threshold value need to be stored in each branch node of the tree. Each leaf node stores the name of a pattern class; if the decision tree procedure reaches a leaf node, the unknown feature vector is classified as belonging to that pattern class. Figure 4.11 illustrates this type of decision tree, which was constructed to correctly classify the training data shown.

The tree of Figure 4.11 was constructed manually by looking at the data and picking suitable features and thresholds. The training data here is just a toy example; real data is likely to have many more features and many more samples. It is not uncommon to have several hundred features and thousands of training samples for real applications such as medical diagnosis. In this case, an automated procedure for producing decision trees is needed. Furthermore, for any given set of training samples, there may be more than one decision tree that can classify them. So it is important to select features that give the best decision tree by some criterion. Usually a tree that is simpler or one that has fewer levels and therefore fewer tests is preferred.

Consider the training data and two possible decision trees shown in Figure 4.12. Both trees can discriminate between the two classes, class I and class II, shown in the training

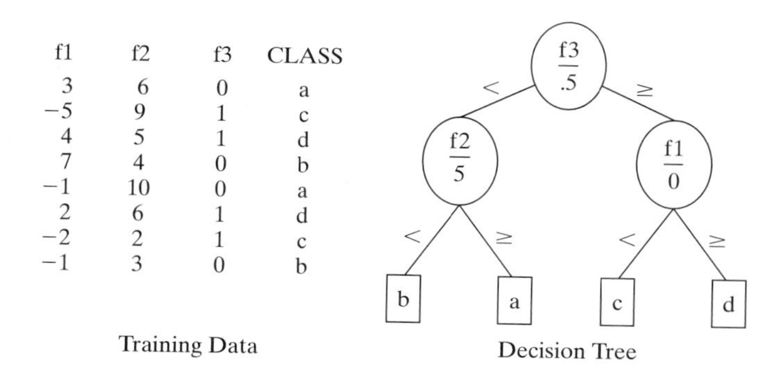

f1	f2	f3	CLASS
3	6	0	a
−5	9	1	c
4	5	1	d
7	4	0	b
−1	10	0	a
2	6	1	d
−2	2	1	c
−1	3	0	b

Figure 4.11 Binary decision tree based on a feature and threshold value at each node.

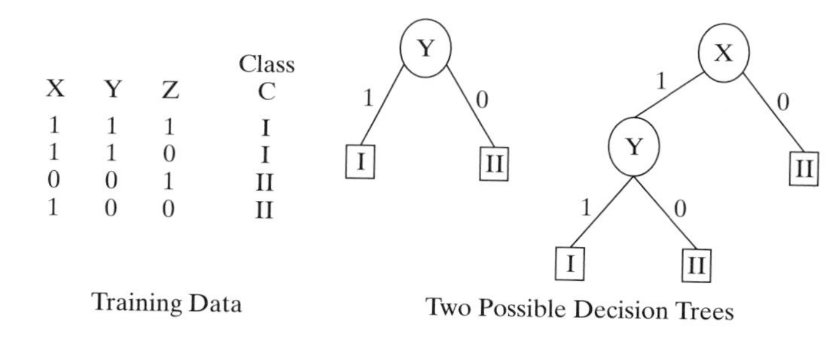

X	Y	Z	Class C
1	1	1	I
1	1	0	I
0	0	1	II
1	0	0	II

Figure 4.12 Two different decision trees that can each classify the given training samples.

data. The tree on the left is very simple; it is able to classify a feature vector with only a single comparison. The tree on the right is larger and requires more comparisons.

Automatic Construction of a Decision Tree There are a number of different methods for constructing optimal decision trees from training data, each with its own definition of optimality. (See Haralick and Shapiro, vol. I, Chapter 4, for an in-depth coverage.) One simple but effective method is grounded in information theory. The most basic concept in information theory is called *entropy*.

36 Definition. The **entropy of a set of events** $x = \{x_1, x_2, \ldots, x_n\}$ where each x_i is an event is

$$H(x) = -\sum_{i=1}^{n} p_i log_2 p_i \tag{4.3}$$

where p_i is the probability of event x_i.

Entropy can be interpreted as the average uncertainty of the information source. Quinlan (1986) used an entropy-based measure called information gain to evaluate features and produce optimal decision trees.

Examples of entropy computations for sets of possible events

Consider the set of three possible events with their associated probabilities.

$$X = \{(x_1, 3/4), (x_2, 1/8), (x_3, 1/8)\}$$

The computation of entropy is as follows:

$$\begin{aligned} H(x) &= -[(3/4)log_2(3/4) + (1/8)log_2(1/8) + (1/8)log_2(1/8)] \\ &= -[(3/4)(-0.415) + (1/8)(-3) + (1/8)(-3)] \\ &= 1.06 \end{aligned}$$

Similarly, a set of four equally likely events has entropy 2.0.

$$X = \{(x_1, 1/4), (x_2, 1/4), (x_3, 1/4)(x_4, 1/4)\}$$
$$H(x) = -[4((1/4)(-2))] = 2$$

Exercise 4.2

(a) Compute the entropy of a set of two equally likely events. (b) Compute the entropy of a set of four possible outcomes with probabilities $\{1/8, 3/4, 1/16, 1/16\}$.

Information theory allows us to measure the information content of an event. In particular, the information content of a class event with respect to each of the feature events is useful for our problem. The information content $I(C; F)$ of the class variable C with possible values $\{c_1, c_2, \ldots, c_m\}$ with respect to the feature variable F with possible values $\{f_1, f_2, \ldots, f_d\}$ is defined by

$$I(C; F) = \sum_{i=1}^{m}\sum_{j=1}^{d} P(C = c_i, F = f_j)log_2\frac{P(C = c_i, F = f_j)}{P(C = c_i)P(F = f_j)} \tag{4.4}$$

where $P(C = c_i)$ is the probability of class C having value c_i, $P(F = f_j)$ is the probability of feature F having value f_j, and $P(C = c_i, F = f_j)$ is the joint probability of class $C = c_i$ and variable $F = f_j$. These prior probabilities can be estimated from the frequency of the associated events in the training data. For example, since class I occurs in two out of the four training samples (see Figure 4.12), $P(C = I) = 2/4 = 0.5$. Since three of the four training samples have value 1 for feature X, $P(X = 1) = 3/4 = 0.75$.

We can use this information content measure to decide which feature is the best one to select at the root of the tree. We calculate $I(C, F)$ for each of the three features: X, Y, and Z.

$$
\begin{aligned}
I(C,X) = {} & P(C=I, X=1)log_2\frac{P(C=I,X=1)}{P(C=I)P(X=1)} \\
& + P(C=I, X=0)log_2\frac{P(C=I,X=0)}{P(C=I)P(X=0)} \\
& + P(C=II, X=1)log_2\frac{P(C=II,X=1)}{P(C=II)P(X=1)} \\
& + P(C=II, X=0)log_2\frac{P(C=II,X=0)}{P(C=II)P(X=0)} \\
= {} & .5log_2\frac{.5}{.5\times .75} + 0 + .25log_2\frac{.25}{.5\times .25} + .25log_2\frac{.25}{.5\times .75} \\
= {} & 0.311 \\
I(C,Y) = {} & .5log_2\frac{.5}{.5\times .5} + 0 + .5log_2\frac{.5}{.5\times .5} + 0 \\
= {} & 1.0 \\
I(C,Z) = {} & .25log_2\frac{.25}{.5\times .5} + .25log_2\frac{.25}{.5\times .5} + .25log_2\frac{.25}{.5\times .5} + .25log_2\frac{.25}{.5\times .5} \\
= {} & 0.0
\end{aligned}
$$

Feature Y, which has an information content of 1.0, gives the most information in determining class and so should be selected as the first feature to be tested at the root node of the decision tree. In the case of this simple example, the two classes are completely discriminated and the tree is complete with a single branch node. In the more general case, at any branch node of the tree when the selected feature does not completely separate a set of training samples into the proper classes, the set of samples is partitioned according to the decision procedure at that node and the tree construction algorithm is invoked recursively for the subsets of training samples at each of the child nodes for which more than one class is still present.

The algorithm described here was meant to operate on a decision tree like the one of Figure 4.10, a general tree that has branches for each possible value of the feature being tested at a node. In order to adapt it to the binary threshold type tree like the one of Figure 4.11, the information content of each feature-threshold pair would have to be considered for each possible threshold. Although this sounds like an infinite set of possibilities, only a finite number of values for each feature appear in the training data, and this finite set is all that need be considered.

The above example is very simple, but it is possible to automatically construct real, useful decision trees on tens or even hundreds of features. Consider again the character recognition problem, but this time for more difficult, hand-printed characters. Some useful

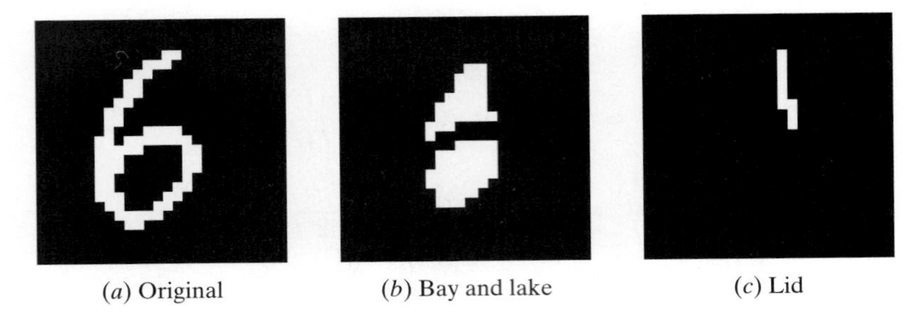

(*a*) Original (*b*) Bay and lake (*c*) Lid

Figure 4.13 (*a*) Image of a hand-printed character six; (*b*) the bay (above) and lake (below) extracted by morphological image processing; and (*c*) the lid of the bay extracted by further morphological processing.

features for this type of characters are *lakes, bays,* and *lids,* as discussed in Section 4.6. Lakes are holes in the character (regions of label 0 completely surrounded by character pixels of label 1), bays are intrusions into the character (regions of label 0 that are only partially surrounded by character pixels of label 1), and lids are segments that can be used to close up the bays. Figure 4.13 shows a hand-printed character six (*a*), its bay and lake features (*b*), and its lid feature (*c*). The operations of mathematical morphology described in Chapter 3 can be used to extract these useful primitive features. From them, the following numeric features can be computed:

- **lake_num:** the number of lakes extracted
- **bay_num:** the number of bays extracted
- **lid_num:** the number of lids extracted
- **bay_above_bay:** Boolean feature that is true if any bay is completely above another
- **lid_rightof_bay:** Boolean feature that is true if any lid is completely to the right of a bay
- **bay_above_lake:** Boolean feature that is true if any bay is completely above a lake
- **lid_bottom_of_image:** Boolean feature that is true if the lowest point of any lid is within a few pixels of the lowest point of the whole character

With sufficient training data, these features can be used to construct a decision tree that can classify the hand-printed numeric digits. Figure 4.14 illustrates a sample set of training data for the digits zero through nine.

Exercise 4.3: Decision tree construction.

Given the training data shown in Figure 4.14 write a program that uses information content to construct a decision tree to discriminate among the ten digit classes. How well does the tree constructed on all 40 samples work on the training data? What happens if you construct the tree from the last 20 samples and then test it on the first 20 samples?

lake_ num	bay_ num	lid_ num	bay_ above_ bay	lid_ rightof_ bay	bay_ above_ lake	lid_ bottomof_ image	class
1	0	0	F	F	F	F	0
1	0	0	F	F	F	F	0
0	0	0	F	F	F	F	1
0	2	2	F	T	F	T	2
0	2	2	T	F	F	F	3
1	1	1	F	F	F	T	4
1	1	1	F	T	T	F	6
0	2	2	T	T	F	F	2
0	2	2	T	T	F	T	4
0	1	1	F	F	F	F	7
0	0	0	F	F	F	F	1
1	0	0	F	F	F	F	0
1	1	1	F	F	F	F	9
0	2	2	T	T	F	F	2
1	1	1	F	F	F	F	9
0	1	1	F	F	F	F	1
0	2	2	T	F	F	T	4
0	2	2	T	T	F	F	5
1	1	1	F	T	T	F	6
0	2	2	T	F	F	F	3
0	1	1	F	F	F	F	1
1	0	0	F	F	F	F	0
0	2	2	T	T	F	F	5
1	1	1	F	T	T	F	6
0	1	1	F	F	F	T	7
2	0	0	F	F	F	F	8
1	1	1	F	F	F	F	9
1	0	0	F	F	F	F	0
2	0	0	F	F	F	F	8
1	1	1	F	T	T	F	6
0	2	2	F	T	F	F	7
1	1	1	F	F	F	F	9
1	0	0	F	F	F	F	0
0	2	2	T	F	F	F	3
0	2	2	T	T	F	F	5
0	2	2	T	T	F	F	2
0	2	2	T	T	F	F	2
0	1	1	F	F	F	F	1
0	1	1	F	F	F	F	7
0	2	2	T	T	F	F	5
0	2	2	T	F	F	T	4
0	2	2	T	F	F	F	3

Figure 4.14 Training data for hand-printed characters.

Exercise 4.4

(a) Describe how to extract a lake using morphological image processing from Chapter 3.
(b) Describe how to extract a lid, given that a bay has already been identified.

4.10 BAYESIAN DECISION-MAKING

We examine how the knowledge of probability distributions can be used to make classification decisions with least expected error rate. Suppose we take a single measurement x from an infrared image of a dark red cherry and use it to determine whether the cherry is bruised or not. An unbruised cherry is in class ω_1, while a bruised cherry is in class ω_2. Also, suppose that we have studied a very large number of surface elements from a large number of bruised and unbruised cherries, so that we have the knowledge captured in the distribution functions shown in Figure 4.15. The curve at the right, $p(x \mid \omega_1)$, shows the distribution of measurement x over a large number of surface samples of unbruised cherries. The curve on the left, $p(x \mid \omega_2)$, shows the distribution of measurement x over a large number of bruised surface samples. The data has been normalized so that the area under each curve is 1.0, making each a probability distribution. (Bruised tissue contains water, which absorbs infrared radiation more than unbruised tissue and thus low reflectance is much more likely for those cherries. The water content varies and so does the darkness of the skin color, causing the distributions to overlap—some dark unbruised cherries will reflect similar to some bright bruised cherries.)

If bruised cherries are as likely to occur as unbruised ones, and if all classification errors cost us the same, then we can make the decision ω_1 whenever $x > t$ and ω_2 otherwise. For such a decision policy, the hatched area to the right of t represents (twice) the false

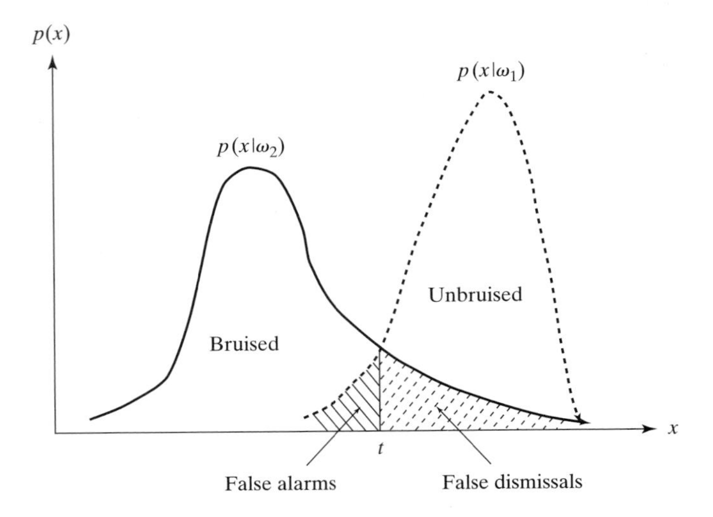

Figure 4.15 Distributions for intensity measurement x conditioned on whether x is taken from an unbruised or bruised cherry.

dismissal rate: This is the probability of getting the acceptably high measurement x from a bruised cherry. The area is twice the false dismissal rate because of the assumption that the *a priori* probability for each density is 0.5, and thus each density should be scaled down so that the total area under the curve is 0.5. The hatched area to the left of t represents (twice) the false alarm probability: this is the probability that a good cherry will be classified as bruised because $x < t$. Because bruised and unbruised cherries are assumed to be equally likely to occur as input to our system, each curve would actually represent only 0.5 of the overall probability so all areas are shown to be twice their actual size. The total error is the total hatched area under the curves. It is important to observe that moving decision threshold t either to the left or to the right will result in a *greater* hatched area and hence a greater error.

The example above considered only the special case of two equally likely classes with errors of equal cost. Now we extend the method to cover the case of m classes all with possibly different *a priori* probabilities. We keep the simplifying assumption that all errors are equally costly. Using Bayesian decision-making, we classify an object into the class to which it is most likely to belong.

37 Definition. A **Bayesian classifier** classifies an object into the class to which it is most likely to belong based on the observed features.

In order to compute the likelihoods given the measurement x, the following distributions are needed.

$$\textbf{class conditional distribution:} \quad p(x \mid \omega_i) \textit{ for each class } \omega_i \tag{4.5}$$

$$\textbf{a priori probability:} \quad P(\omega_i) \textit{ for each class } \omega_i \tag{4.6}$$

$$\textbf{unconditional distribution:} \quad p(x) \tag{4.7}$$

If all of the classes ω_i are disjoint possibilities covering all possible cases, we can apply Bayes Rule to compute *a posteriori* probabilities for each of the classes given the *a priori* probabilities of the classes and the distributions of x for each class.

$$P(\omega_i \mid x) = \frac{p(x \mid \omega_i)P(\omega_i)}{p(x)} = \frac{p(x \mid \omega_i)P(\omega_i)}{\sum_{j=1,m} p(x \mid \omega_j)P(\omega_j)} \tag{4.8}$$

Returning to the classifier sketched in Figure 4.2, inside each of the class computation boxes we make $f_i(x, K) = P(\omega_i \mid x)$, which by Bayes Rule in Equation 4.8 can be computed as $p(x \mid \omega_i)P(\omega_i)/p(x)$. Since $p(x)$ is the same for all the class computation boxes, we can ignore it and just make the classification decision ω_i for the maximum $p(x \mid \omega_i)P(\omega_i)$. To design our Bayes classifier, we must have as knowledge K the prior probability of each class $P(\omega_i)$ and the class conditional distribution of $p(x \mid \omega_i)$. Having such knowledge allows us to design for optimal future decisions. It is often difficult to establish these prior probabilities. For example, how would we know the probability that an arbitrary cherry entering our sorting machine is a bruised cherry? If this varies with the weather and picking crews, it may take too much sampling work to obtain the needed information whenever conditions change.

4.10.1 Parametric Models for Distributions

In practice, we must implement the computation of $p(x \mid \omega_i)$ in some manner. An empirical method is to quantize the range of x and record the frequency of occurences of x in the samples for each interval and store the result in an array or histogram. One could fit a smooth spline function to this data to produce a probability function valid for all real x. Note that we need to scale our results so that the sum over all possible values of x is 1.0. If we observe that the distribution of x follows some known parametric model, then we can represent the distribution by the small number of parameters that characterize it. Poisson, exponential, and normal (or Gaussian) distributions are commonly used. A normal distribution is the well-known *bell-shaped curve* sometimes used to assign grades in college courses.

38 Definition. A **normal distribution** characterized by mean μ and standard deviation σ is defined as follows.

$$p(x) = N(\mu, \sigma)(x) = \frac{1}{\sqrt{2\pi}\sigma} exp\left[-\frac{1}{2}\left(\frac{x-\mu}{\sigma}\right)^2\right] \tag{4.9}$$

A reference in statistics can be consulted for use of the χ^2 test to decide whether or not the sample data can actually be modeled well using the normal (or other) distribution. (For example, see the text by Hogg and Craig in the references.) It is easy to compute the mean and standard deviation from sample data and hence derive a normal distribution model. Many implementations will use the normal model because of its simplicity and because of other convenient mathematical properties, even when it is known to be a coarse approximation to the actual data.

Exercise 4.5

How could we estimate the following *a priori* probabilties? (a) that a customer in a market will buy spinach; (b) that a person at an ATM machine is an imposter; (c) that a just picked dark red cherry is bruised; and (d) that a person over forty years-old has stomach cancer?

By using a *parametric model,* such as the normal distribution, to model a distribution of class samples, simple formulas are available for comparing probabilities for Bayesian decision-making as implemented in Figure 4.2. Once the distributions $p(x \mid \omega_i)$ are known for each class i, Figure 4.16 can be used to set the thresholds on values of x to separate the classes. Moreover, the probability model can be directly used to estimate the probability of error because there is now a formula for the error regions shown in Figure 4.15.

Exercise 4.6: Classifying coins B.

Refer to the Exercise 4.1, which required measuring diameter and thickness of U.S. coins. Use the data for only the pennies, nickels, and dimes. (a) Let feature x be the thickness of the coin. Compute the mean and standard deviation for each of the three classes. Are there thresholds t_1 and t_2 that would separate the classes and give an overall error rate of less than 5 percent? Explain. (b) Repeat (a) using x as the diameter of the coin.

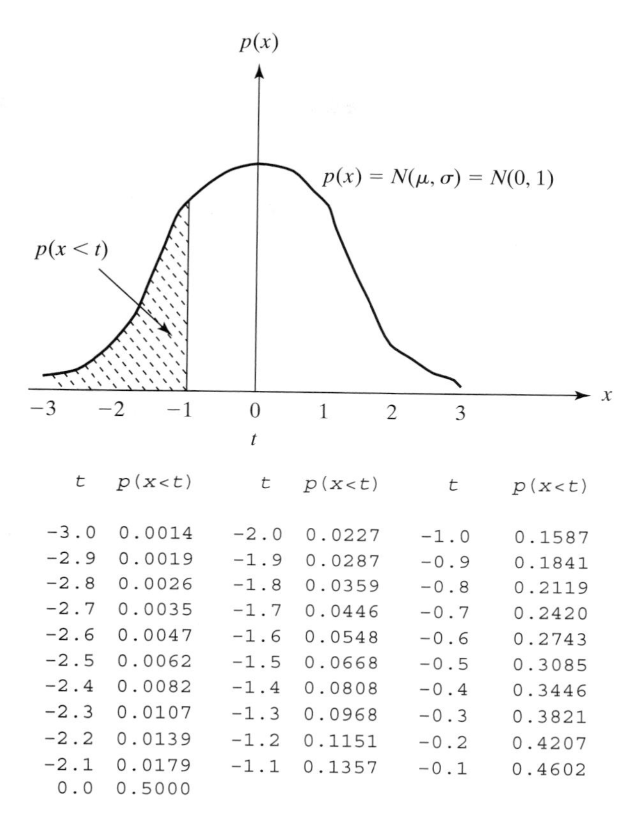

t	p(x<t)	t	p(x<t)	t	p(x<t)
-3.0	0.0014	-2.0	0.0227	-1.0	0.1587
-2.9	0.0019	-1.9	0.0287	-0.9	0.1841
-2.8	0.0026	-1.8	0.0359	-0.8	0.2119
-2.7	0.0035	-1.7	0.0446	-0.7	0.2420
-2.6	0.0047	-1.6	0.0548	-0.6	0.2743
-2.5	0.0062	-1.5	0.0668	-0.5	0.3085
-2.4	0.0082	-1.4	0.0808	-0.4	0.3446
-2.3	0.0107	-1.3	0.0968	-0.3	0.3821
-2.2	0.0139	-1.2	0.1151	-0.2	0.4207
-2.1	0.0179	-1.1	0.1357	-0.1	0.4602
0.0	0.5000				

```
Use symmetry to extend the table values from 0.0 to 3.0. For example,
  p ( -2.0 < x < 1.0 ) = p ( -2.0 < x < 0.0 ) + p ( 0.0 < x < 1.0 )
   = [ p ( x < 0.0 ) - p ( x < -2.0 ) ] + p ( -1.0 < x < 0.0 )
     [ 0.5000 - 0.0227 ] + 0.1587 = 0.6360.
```

Figure 4.16 Normal distribution with mean $\mu = 0$ and standard deviation $\sigma = 1$.

4.11 DECISIONS USING MULTIDIMENSIONAL DATA

In many real-world problems being tackled today, a dimension of $d = 10$ or more is common. As we saw previously, the nearest-neighbor classification procedure is defined for feature vectors of any dimension d. There are parametric probability models for multidimensional feature vectors $\mathbf{x}$; the reader should consult the references for their mathematical treatment. Here, we abstractly discuss the concept of multidimensional structure. A good intuitive grasp will reach a long way into the exciting current research cited in the references.

Consider two classes of samples in three dimensions—each shaped like a tree—with the two trees growing together. Class one data is shaped like a maple tree and is approximately a large sphere. Class two data is shaped like a pine, taller and much narrower than the maple, it is approximately an ellipsoid with a major axis much larger than its two minor axes. Class one samples correspond to the leaves of the maple, while class two samples correspond to the needles of the pine. Moreover, suppose the pine tree grows through the crown of the maple and well above it. The problem of classifying an unknown 3D feature vector **x** requires relating it to the known sample structure in the 3D space. If **x** is within the crown of the maple, but not near the trunk of the pine, then **x** is more likely to be maple (class one). On the other hand, if **x** is outside the crown of the maple or close to the trunk of the pine, then **x** is more likely to be pine (class two). There are positions in the space that are ambiguous because the samples of the two classes overlap. The most important point is that an understanding of the structure of the samples in the d-dimensional space allows us not only to make informed decisions but also to understand our errors. The structure of the space can be represented by a large database of samples, data structures summarizing subsets of samples, or by parameterized geometric models of subsets of samples.

Exercise 4.7: On error rates due to thresholding.*

This exercise deals with the potential error in area computations due to the thresholding of an image in an attempt to separate an object from background. Make the following set of assumptions.

- The image of some object covers *precisely* 3,932 pixels of the 512×512 pixel image. (There are no mixed pixels; object boundaries correspond to pixel boundaries exactly. There is no blurring of neighboring pixels due to the lens.)
- Due to variations in its surface, the intensity of pixels of the image of the object is distributed as N(80,5) (this means normally distributed with mean 80 and standard deviation 5).
- Similarly, background intensity is distributed as N(50,10).
- The gray level of any single pixel is determined without regard to the gray level of neighboring pixels.

1. If the image is thresholded at intensity 70 so that **LABEL[r, c]** = 1 *if* **I[r, c]** >= 70 and **LABEL[r, c]** = 0 otherwise, what is the expected number of pixels labeled as OBJECT?
2. Where in the image are the pixels labeled BACKGROUND that truly should be OBJECT? (These are *false dismissals.*)
3. Where in the image are the pixels labeled OBJECT that truly should be labeled BACKGROUND? (These are *false alarms.*)
4. What is the percentage error expected in the computation of object area by merely counting the number of 1 pixels in the labeled image?
5. *Now suppose that *salt-and-pepper noise* is removed from the labeled image by creating a new image where any pixel is replaced by the values of the neighbors if all 4-neighbors have the other value. What is the percentage error expected in the computation of object area by merely counting the number of 1 pixels in this new labeled image?

A second 3D example is also illuminating. Assume that the samples of class one are structured as a coil spring, or helix, and that the samples of class two are structured as a pencil, or rod, positioned as the axis of the helix. (Or, imagine two coil springs intertwined as they might be in a bin in a hardware store.) These highly structured classes are one-dimensional and can be easily separated *once their structure is known*. A nearest-mean classifer is useless because the means are the same. Rescaling any of the dimensions is not going to work either because the samples will still be intertwined. Nearest-neighbor classification will work, but we'll need to store a lot of samples. A practical alternative is to approximate the helical data by a union of many rods. A rod can be represented simply as a cylindrical section. Classification can be done by simple geometrical computations that check to see if unknown **x** is within any of the cylinders. A better alternative is to use a formula for the helix parameterized by its axis, radius, and rate of climb.

We note some important points in leaving these thought experiments. First, it is important to capture the *intrinsic structure and dimensionality* of the sample data. Structure can be represented by geometrical or statistical models: having models allows simple computations for our decisions as opposed to searching a large unstructured data base of samples. Second, the natural structure of the data may not be aligned with the axes of our measurement space. For example, the axis of the pine tree or the helix need not be along any of the axes **x[1]**, **x[2]** or **x[3]**. Methods for discovering structure or transforming coordinates are given in the references.

4.12 MACHINES THAT LEARN

We pause to summarize the important point that the methods discussed in this chapter provide a basic type of machine learning called *supervised learning*. The produce classification application in Chapter 16 is a clear example. We have assumed that labeled samples were available for all of the classes that were to be distinguished; in other words, the teacher knew the structure of the data and the desired outcomes. *Unsupervised learning* or *clustering* can also be done; in unsupervised learning, the machine must also determine the class structure; that is, what the classes are and how many there are. The reader can consult the references to study this topic.

When nearest-neighbor classification is done, all the data samples are merely input into the memory and then accessed in order to recognize an unknown object. The recognition behavior of the machine is completely determined by the training data. When parametric models are used, the parameters of the class models are learned from the training data and then are used to model the entire space of possible objects. In the optional section below, supervised learning is achieved using discriminant functions that are designed to model the neurons of living organisms. Machine learning is currently an area of intense research and development, which the reader is encouraged to explore with additional reading.

4.13 ARTIFICIAL NEURAL NETS*

Because of their learning capability, neurons of living organisms have been studied for application to machine learning. A simple model of a neuron is shown in Figure 4.17.

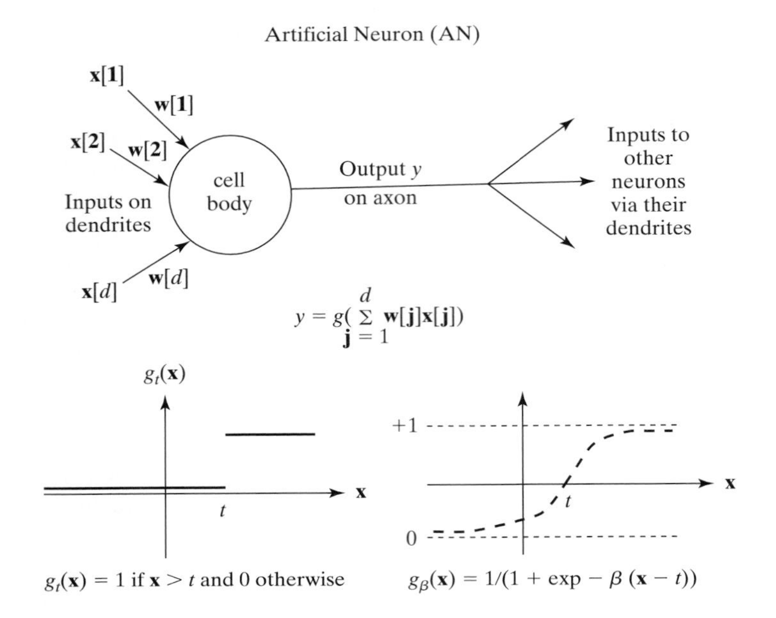

Figure 4.17 Simple model of a neuron and two possible output conditioning functions.

Although the model is only an approximation to what is known from biology, it has become very important in its own right as a model for computation. Networks of such model neurons, called *artificial neural networks or ANNs*, have proved to be successful in many machine vision problems, especially because of their learning capability. ANNs can learn the complex structure of samples in multidimensional space using less memory than what is needed for nearest-neighbor classification and can be implemented for massively parallel computation. Only a limited introduction to ANNs is given here; for more information about this large and still rapidly growing area, consult the references.

4.13.1 The Perceptron Model

As Figure 4.17 shows, the neuron (AN) receives its d inputs $\mathbf{x[j]}$ via dendritic connections from other neurons, perhaps sensor cells. The cell body sums the inputs after multiplying each by a gain factor $\mathbf{w[j]}$. The neuron output y is sent along the single axon, which eventually branches into many dendritic connections, providing input to other neurons in the neural net. One model for determining y from the summed scaled inputs to the cell is to compare the sum with a threshold t and output $y = 1$ if the sum exceeds the threshold and output $y = 0$ otherwise. This binary output behavior is shown in the lower left of Figure 4.17. To get a smooth output behavior between 0 and 1, the sigmoid function, shown in the lower right of the figure, can be used. The parameter β is the slope or gain at $x = t$, which determines how the input value is scaled to compute the output value near $x = t$. For notational and programming convenience, the negative of the threshold t for the neuron is stored as **w[0]** and its corresponding input **x[0]** is set to 1.0 giving Equation 4.10. The neuron learns by adapting the weights **w[j]** to the input vectors **x** that it experiences.

$$y = g\left(\sum_{j=0,d} w[j]x[j]\right) \tag{4.10}$$

Exercise 4.8: Simulating an AN.

(a) Study the behavior of an AN with two inputs **x[1]** and **x[2]**, with weights **w[1]** $= 0.8$ and **w[2]** $= 0.3$, threshold $t = 1.0$, and step function $G(\mathbf{x})$ at the output. Make a plot of the results by plotting a 1 when the output is 1 and a 0 when the output is 0 for the 16 possible input combinations where both **x[1]** and **x[2]** take values 0,1,2,3. (b) Create another plot, this time using the smooth sigmoid function with $\beta = 4$, keeping all other elements of the problem the same. Note that the outputs will now be real numbers rather than just 0 or 1.

Exercise 4.9: AND, OR, and NOT gates using an AN.

1. Design a single AN that has the same behavior as an OR gate. Let **x[1]** and **x[2]** be the two inputs that can have Boolean values of 0 or 1 only. The output of the AN should be 1 when either or both of the inputs have value 1, and should be 0 when both inputs are 0. Recall that **x[0]** $= 1$ and that the threshold is $-$**w[0]**. Complete the set of weights to define the AN. Plot the four input combinations using 2D axes and show the decision boundary implemented by the AN.
2. Repeat the above question for an AND gate. The output of an AND gate is 1 only when both inputs are 1.
3. Show how a single input AN can behave as a NOT gate. If the input to the NOT gate is 0 then the output is 1, and if the input is 1 then the output is 0.

The computational capability of the simple artificial neuron is of great interest both in theory and practice. From Exercise 4.9 we learn that an AN can model AND, OR, and NOT gates. The significance of this is that any Boolean function can be implemented by cascading several ANs. From Exercise 4.10 we learn that a single AN cannot even implement the simple exclusive or function. Many other important functions cannot be implemented by a single AN: results published by Minsky and Papert (1987) discouraged research for a period. After a few years there was a spate of successful work with multilayer ANNs, which are more complex and not as limited in computational capability. We leave the theory of the computational capabilities of ANNs to further reading and return to look at a simple version of a training algorithm for a single AN.

Assume that two classes of 2D samples can be separated by a line: this would be the case for Figure 4.4 if we removed the single 'X' and 'O' from the overlap area. Clearly, one could take the parameters of the separating line and construct a neuron to make the classification decision. For 3D samples, we would use a separating plane; in d dimensions, we would need a hyperplane, but the concept and construction would be the same. Rather surprisingly, if a separating hyperplane exists for a two-class problem, then a simple learning algorithm exists to find the hyperplane formula from training samples from the two classes (see Algorithm 4.3). Proof that the algorithm converges to a separating hyperplane is beyond the scope of this text but can be found in Duda and Hart (1973).

The perceptron learning algorithm begins with a random set of weights (including the threshold). It cycles through the labeled samples x and whenever the weight vector (perceptron) gives a positive output for a sample from Class 1, it subtracts gain * x from the weight vector. Similarly, if there is a negative output for a sample from Class 2, then gain * x is added to the weight vector. This policy moves the current separating line in the appropriate direction according to what was learned from the current sample. The gain controls the size of the change. The procedure training_pass carries out these adjustments. After all training samples are processed in one pass, the procedure check_samples is called to count how many samples are misclassified by an AN with these weights. If none are misclassified, then the algorithm exits with a solution. Otherwise, if the maximum allowed passes have not been done, then another training pass is made, this time with a halved gain factor. There are other implementations of the detailed control for the general algorithm.

Figure 4.18 shows the output from a program that implements the perceptron learning algorithm. By construction, all the Class 1 samples are below the line $y = 1 - x$, while all Class 2 samples are above that line. There is a corridor of space between these samples and the algorithm very quickly finds the line $-1 + 5/4x_1 + 5/4x_2 = 0$ to separate the classes. As the output shows, each sample from Class 1 produces a negative response while each sample from Class 2 produces a positive response.

Although the basic learning algorithm is simple, there are some nontrivial aspects. (1) What sequence of samples should be used to learn fast? Theory states that to guarantee convergence, each sample may have to be presented an arbitrary number of times. Some algorithms repeat training on a given sample until the sample is correctly classified. (2) Convergence will be affected by the gain factor used. The program used for the example output halved the gain factor with each pass through all the training samples. (3) For better performance on future samples, it may pay for the algorithm to search for a *best line* between the classes rather than just any line. (4) When training is taking a long time, how can we know whether or not it is due to the samples being inseparable? (5) How can we modify the learning algorithm so that in the case of samples that are linearly inseparable, we can find a line that yields the least misclassifications? These issues are left for possible outside research and experiments by the reader.

Exercise 4.10: Perceptron to implement exclusive or (XOR).

Show that a single AN cannot make the exclusive OR decision by plotting the following input data and trying to find a separating line. The inputs giving a positive response are **(0, 1)**, **(1, 0)** and the inputs giving a negative response are **(0, 0)**, **(1, 1)**.

```
Class 1 = { ( 0 , 0.5 ), ( 0.5 , 0 ), ( 0 , 0 ), ( 0.25 , 0.25 ) }
Class 2 = { ( 0 , 1.5 ), ( 1.5 , 0 ), ( 0.5 , 1 ), ( 1 , 0.5 ) }
Initial gain= 0.5
Limit to number of passes= 5
Number of samples in Class1= 4; Number of samples in Class2= 4

Training phase begins with weights:        -1     0.5     0.5

=====Adjust weights=====: gain= 0.5
 pattern vector x =        1       0       1.5
Input Weights:       -1    0.5     0.5
Output Weights:      -1    0.5     1.25

=====Adjust weights=====: gain= 0.5
 pattern vector x =        1       1.5     0
Input Weights:       -1    0.5     1.25
Output Weights:      -1    1.25    1.25

Weight Vector is:    -1    1.25        1.25     Classification for Class = 1

          Input Vector x / Response   / Error?

       1        0       0.5     -0.375      N
       1      0.5         0     -0.375      N
       1        0         0         -1      N
       1     0.25      0.25     -0.375      N

Weight Vector is:          -1    1.25     1.25   Classification for Class = 2

          Input Vector x / Response   / Error?

       1        0       1.5      0.875      N
       1      1.5         0      0.875      N
       1      0.5         1      0.875      N
       1        1       0.5      0.875      N

Errors for Class1: 0 Errors for Class2: 0
Final weights are:        -1     1.25     1.25
```

Figure 4.18 Output of computer perceptron learning program that learns the linear discriminant between two linearly separable classes.

Exercise 4.11: Program the perceptron learning algorithm.

Write a program to implement the perceptron learning algorithm for arbitrary d-dimensional feature vectors $\mathbf{x}$. Test it using 2D vectors and show that it can learn to discriminate as an OR gate and an AND gate. Show that learning does not converge for an XOR gate. Test on the following two classes of synthetic 3D samples: Class 1 is some set of random points in the first octant (x_1, x_2, x_3 all positive) while Class 2 is some set of points in any other octant.

4.13.2 The Multilayer Feedforward Network

A *feedforward network* is a special type of ANN where each neuron in the network is located at some level l. A neuron at level l receives input from all neurons of level $l-1$ and feeds

Compute weight vector w to discriminate Class 1 from 2.

S1 and **S2** are sets of *n* samples each.
gain is a scale factor used to change **w** when **x** is misclassified.
max_passes is maximum number of passes through all training samples.

```
procedure Perceptron_Learning(gain, max_passes, S1, S2)
{
input sample sets S1 and S2;
choose weight vector w randomly;
"let NE be the total number of samples misclassified"
NE = check_samples (S1, S2, w);
while (NE > 0 and passes < max_passes)
  {
    training_pass (S1, S2, w, gain);
    NE = check_samples (S1, S2, w);
    gain = 0.5 * gain;
    passes = passes + 1;
  }
report number of errors NE and weight vector w;
}
  procedure training_pass (S1, S2, w, gain);
{
  for i from 1 to size of Sk
  {
    "scalar, or dot, product ∘ implements AN computation"
    take next x from S1;
    if (w ∘ x > 0) w = w − gain ∗ x;
    take next x from S2;
    if (w ∘ x < 0) w = w + gain ∗ x;
  }
}
```

Algorithm 4.3 A perceptron learning algorithm for two linearly separable classes.

its output to all neurons at level $l+1$. Refer to Figure 4.19. We can model the inputs to lowest-level 1 neurons as sensor inputs and the outputs from highest level L neurons as classification results. The classification can be taken to be that c where output $\mathbf{y}[\mathbf{c}]$ is highest; or, all outputs can be considered to be a *fuzzy classification*. ANs between levels 1 and L are called *hidden units*. There is no feedback from any level to a lower level, thus the term *feedforward*. Because of this, the ANN works like a combinational circuit in the sense that its outputs are computed from its inputs without the use of memory about the prior sequence of inputs.

Previous exercises showed that single artificial neurons could have behavior equivalent to AND, OR, and NOT logic gates. This immediately implies that feedforward layers of ANs can implement any combinational logic function. Thus, such networks are surprisingly powerful and can simulate the behavior of many different computer programs. Moreover, since ANs are not limited to Boolean values, they can represent very complex geometrical partitions of d-dimensional space and can adaptively learn such structure from training samples. Figure 4.20 shows how a feedforward ANN can compute the exclusive OR function, something that was impossible with a single AN. The first layer uses ANs to implement AND and OR as in the exercises. There is only one AN at the last level: it has weight vector $\mathbf{w} = [\mathbf{0}, -\mathbf{1}, \mathbf{1}]$ and outputs a 1 if and only if $-1x_1 + 1x_2$ is positive. In order to see how a multilayer ANN can capture the geometric structure of a complex set of samples, the reader should do the following exercises.

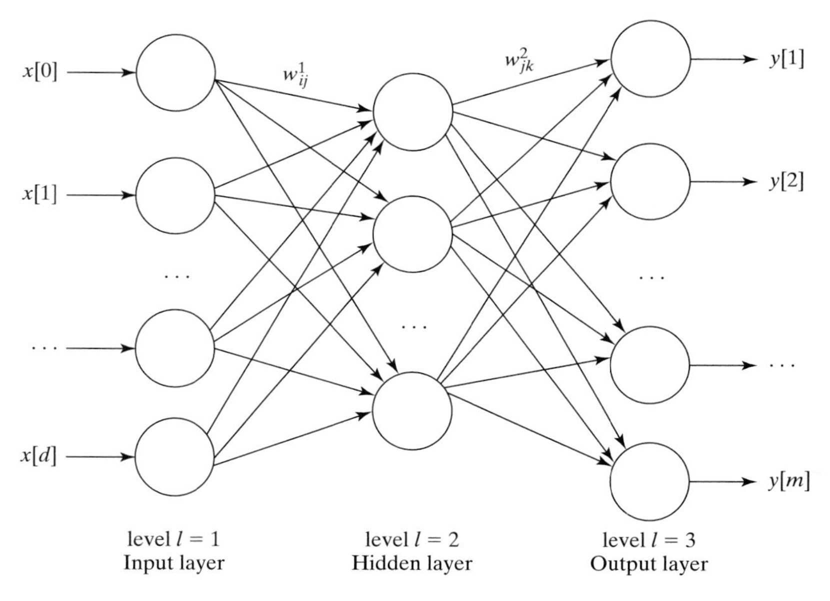

Figure 4.19 A multilayered feedforward artificial neural network: all neurons in level l receive input from all neurons of level $l-1$ and feed output to all neurons of level $l+1$.

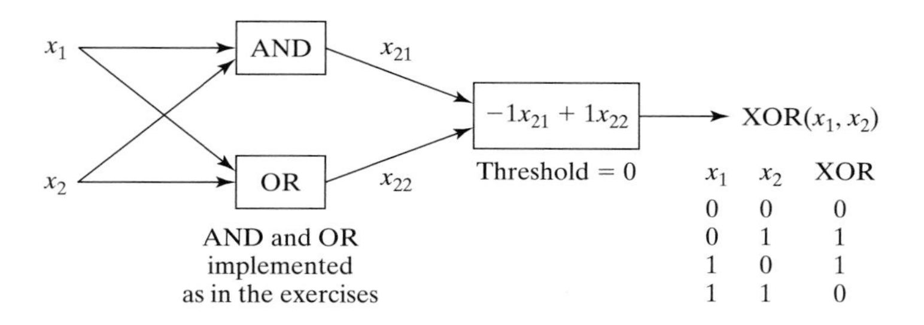

x_1	x_2	XOR
0	0	0
0	1	1
1	0	1
1	1	0

Figure 4.20 Implementation of XOR using a feedforward ANN.

A feedforward network can learn by adapting its weights to a sequence of training samples during a learning phase. A learning algorithm, called the *back-propagation algorithm* propagates classification errors from the output layers back toward the input layers. The sigmoid function is used to condition the output rather than thresholding in order to provide smooth control of the input/output relationship. Implementation and use of the back-propagation algorithm can be found in the references. Recently, the variety of successful applications of back propagation and other learning algorithms has provided renewed excitement to the fields of both pattern recognition and machine learning. Consult the references to learn of other types of networks and their many applications.

Exercise 4.12: An ANN for a triangular class structure in 2D.

Construct a feedforward ANN that yields output of 1 for all 2D points **x** inside the triangle with vertices (3,3), (6,6) and (9,1) and output of 0 for all 2D points outside of the triangle. Use the step function version of $G(x)$. Hint: use three ANs at the first level to establish the class boundaries that are the sides of the triangle and use one second level AN to integrate the three outputs from the first level.

Exercise 4.13: An ANN for a 3-class problem.

Show how a 2 layer feedforward network can recognize 2D input vectors from the following *disjoint* classes. Class 1 vectors are inside some triangle; class 2 vectors are inside some square, and class 3 vectors are inside some pentagon. Using the result of the above exercise, argue that an ANN exists to recognize each individual class versus its complement (you do not have to work with specific lines or equations, just call the subnetworks *triangle, square* and *pentagon*). The second layer output indicates the class of the input to the first layer.

4.14 REFERENCES

The early text by Duda and Hart (1973) is still a valuable reference for the problems and methods of classical statistical pattern recognition (there is a new edition with coauthor, David Stork, in 2000). Another classical text is that of Fukunaga (1972, 1990). These texts can be consulted for development of the Bayes classifier theory for d-dimensional feature vectors and they show the importance of the *covariance matrix* for modeling multidimensional structure. There are several good introductory texts on probability and statistics—one is by Hogg and Craig (1970) and another is by Feller (1957)—where one can study the theory of distributions such as the normal and chi-squared distributions. Jain and others (2000) is a survey of statistical pattern recognition containing many recent citations of contributions.

The *eigenvectors* of the covariance matrix give the natural directions of ellipsoidal clusters in space, while the corresponding *eigenvalues* give the spread of the samples. In addition, several ways of characterizing probability density are given. The recent texts by Schalkoff (1992) and Schurmann (1996) cover syntactic and structural pattern recognition in addition to statistical pattern recognition. A broad but brief treatment of ANNs can be found in the tutorial by Jain and others (1996); more extensive development is available in the texts by Haykin (1994), Hertz and others (1991) and by Schurmann (1996). The text by Tanimoto (1995) gives a good treatment of neural networks within the context of other learning schemes and also shows how to use symbolic features for input: Implementations for both perceptron learning and back-propagation are given in LISP. For an exciting theoretical treatment on what perceptrons can and cannot compute, consult Minsky and Papert (1989) or the original edition (1969).

1. Duda, R. O., and P. E. Hart. 1973. *Pattern Classification and Scene Analysis.* John Wiley & Sons, New York.
2. Duda, R. O., D. Stork, and P. Hart. 2000. *Pattern Classification.* John Wiley & Sons, New York.
3. Feller, W. 1957. *An Introduction to Probability Theory and Its Applications, vols. I and II.* John Wiley & Sons, New York.
4. Haykin, S. 1994. *Neural Networks: A Comprehensive Foundation.* Macmillan College Publishing, New York.
5. Hertz, J., A. Krogh, and R. Palmer. 1991. *Introduction to the Theory of Neural Computation.* Addison-Wesley, Reading, MA.
6. Hogg, R., and A. Craig. 1970. *Introduction to Mathematical Statistics.*
7. Jain, A. K., J. Mao, and K. M. Mohiuddin. 1996. Artificial neural networks: A tutorial. *IEEE Comput.* 29(3).
8. Jain, R. Duin, and J. Mao. 2000. Statistical pattern recognition, a review. *IEEE-TPAMI.* 22(1):4–37.
9. Fukunaga, K. 1990. *Introduction to Statistical Pattern Recognition,* 2nd ed. Academic Press, New York.
10. Kulkarni, A. 1994. *Artificial Neural Networks for Image Understanding.* Van Nostrand-Reinhold, New York.
11. Minsky, M., and S. Papert. 1989. *Perceptrons,* 2nd ed. MIT Press, Cambridge, MA.
12. Proakis, J. G. 1989. *Digital Communications.* McGraw-Hill, New York.
13. Quinlan, J. R. 1986. Induction of decision trees. *Machine Learning,* 1(1):81–106.
14. Schalkoff, R. 1992. *Pattern Recognition: Statistical, Structural, and Neural Approaches.* John Wiley & Sons, New York.
15. Schurmann, J. 1996. *Pattern Classification: A Unified View of Statistical and Neural Approaches.* John Wiley & Sons, New York.
16. Tanimoto, S. 1995. *The Elements of Artificial Intelligence with Common LISP,* 2nd ed. Computer Science Press, New York.

5

Filtering and Enhancing Images

This chapter describes methods to enhance images for either human consumption or for further automatic operations. Perhaps we need to reduce noise in the image; or, certain image details need to be emphasized or suppressed. Chapter 1 already introduced two methods of image filtering: first, in the bacteria image, isolated black or white pixels were removed from larger uniform regions; second, it was shown how a contrast operator could enhance the boundaries between different objects in the image, for example, improve the constrast between the pictures and the wall.

This chapter is about *image processing,* since the methods take an input image and create another image as output. Other appropriate terms often used are *filtering, enhancement,* or *conditioning*. The major notion is that the image contains some signal or structure, which we want to extract, along with uninteresting or unwanted variation, which we want to suppress. If decisions are made about the image, they are made at the level of a single pixel or its local neighborhood. We have already seen how we might label an image pixel as object versus background or boundary versus not boundary.

Image processing has both theory and methods that can fill several books. Only a few classical image processing concepts are treated here in detail. Most methods presented use the important notion that each pixel of the output image is computed from a local neighborhood of the corresponding pixel in the input image. However, a few of the enhancement methods are global in that all of the input image pixels are used in some way in creating the output image. The two most important concepts presented are those of (1) matching an image neighborhood with a pattern or mask (*correlation*), and (2) *convolution,* a single method that can implement many useful filtering operations.

5.1 WHAT NEEDS FIXING?

Before launching into the methods of this chapter, it is useful to review some of the problems that need them. Two general categories of problems follow.

5.1.1 An Image Needs Improvement

- On safari in Africa you got a shot of a lion chasing an antelope. Unfortunately, the sun was behind your main actor and much of your picture is too dark. The picture can be enhanced by boosting the lower intensities and not the higher ones.
- An old photo has a long bright scratch, but is otherwise fine. The photo can be digitized and the scratch removed. (See Figure 5.1.)
- A paper document needs to be scanned and converted into a text file. Before applying character recognition, noise pixels need to be cleaned from the background and dropouts in the characters need to be filled.

5.1.2 Low-Level Features Must Be Detected

- 0.12 inch diameter wire is made by a process that requires (closed loop) feedback from a vision sensor that constantly measures the wire diameter. The two sides of the

Figure 5.1 (left) Scratches from original photo of San Juan are removed; (center) intensity of photo of Alaskan Pipeline rescaled to show much better detail; and (right) image of airplane part has edges enhanced to support automatic recognition and measurement.

Figure 5.2 (left) Original sensed fingerprint; (center) image enhanced by detection and thinning of ridges; and (right) identification of special features called *minutia,* which can be used for matching to millions of fingerprint representations in a database. (Contributed by Shaoyun Chen and Anil Jain.)

wire can be located using an edge operator which accurately identifies the boundary between the wire and its background.

- An automatic pilot system for a car can steer by constantly monitoring the white lines on the highway. The white lines can be found in frames of a forward-looking video camera by finding two edges with opposite contrast and similar direction.
- A paper blueprint needs to be converted into a CAD (computer-aided-design) model. Part of the process involves finding the blueprint lines as dark streaks in the image about 1 pixel wide.

This chapter deals mostly with traditional methods of *image enhancement* and somewhat with *image restoration* (See Figure 5.2.). Before moving on, it is important to define and distinguish these terms.

39 Definition. Image enhancement operators improve the detectability of important image details or objects by man or machine. Example operations include noise reduction, smoothing, contrast stretching, and edge enhancement.

40 Definition. Image restoration attempts to retore a degraded image to an ideal condition. This is possible only to the extent that the physical processes of ideal image formation and image degradation are understood and can be modeled. The process of restoration involves inverting the degradation process to return to an ideal image.

5.2 GRAY-LEVEL MAPPING

It is common to enhance images by changing the intensity values of pixels. Most software tools for image processing have several options for changing the appearance of an image by transforming the pixels via a single function that maps an input gray value into a new output value. It is easy to extend this so that a user can indicate several different image regions and apply a separate mapping function to each. Remapping the gray values is often called *stretching* because it is common to stretch the gray values of an image that is too dark onto the full

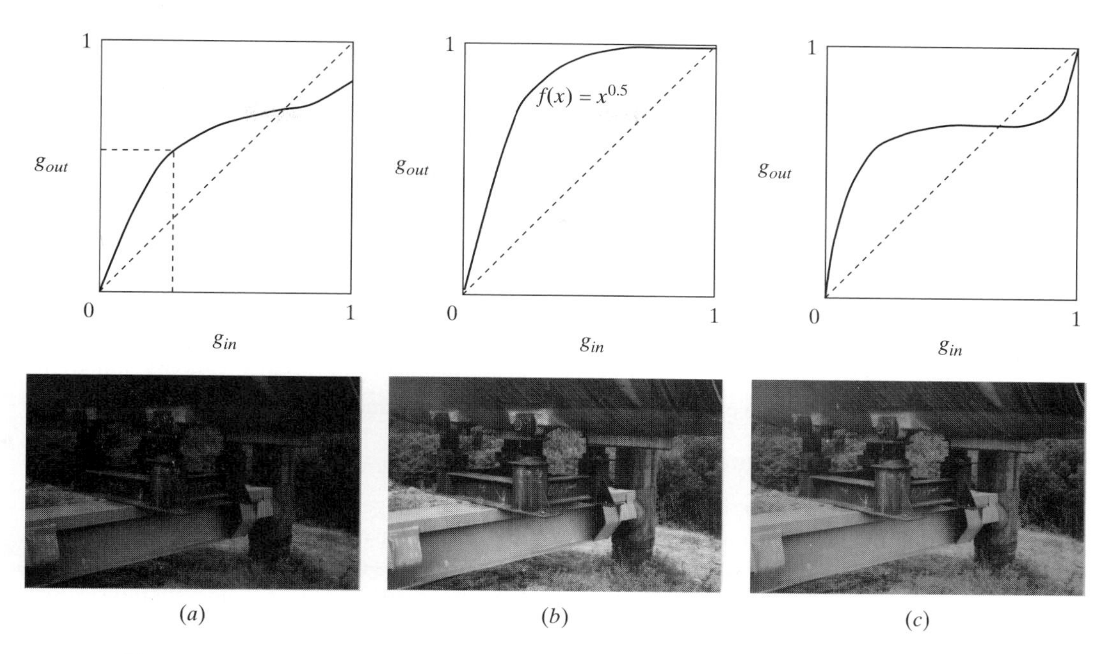

Figure 5.3 (top row) Intensity mapping functions f and (bottom row) output images resulting from applying f to the original image shown at the bottom left. (*a*) Original, fairly dark, Alaskan Pipeline image and example general intensity mapping function; (*b*) Gamma correction with $f(x) = x^{0.5}$ boosts dark pixels much more than bright ones; and (*c*) mapping using a tool that allows the user to interactively create the mapping curve shown, which boosts dark pixels while reducing bright ones. Note that different scene objects appear with different clarity in the different images.

set of gray values available. Figure 5.3 shows a picture whose intensity values are stretched according to two different mapping functions. Figure 5.3(*a*) shows the original image along with the general form of a mapping function. Figure 5.3(*b*) shows intensity mapping using the function $f(x) = x^{0.5}$, which nonlinearly boosts all intensities, but boosts lower intensities more than the higher ones. Using the mapping function $f(x) = x^{1/\gamma}$ is called *Gamma correction* and it might be the proper theoretical model to restore an image to an original form after undergoing known physical distortion. In this case $\gamma = 2.0$, which is a boosting value. Using reducing values, such as $\gamma = 0.3$, is impractical in this case since the scene was in the shade of the forest and of the pipeline itself. Figure 5.3(*c*) shows a more complex interactively defined mapping. The user defined the gray-level mapping function $g_{out} = f(g_{in})$ with an image processing tool controlled by the computer mouse: the image tool fit a smooth spline through points chosen by the user. The functions in Figure 5.3 stretch or extend at least some range of intensities to show more variation in the output. Image variation will be increased in the intensity ranges where the slope of function $f(x)$ is greater than 1.

41 Definition. A **point operator** applied to an image is an operator in which the output pixel is determined only by the input pixel, **Out[x, y]** $= f($**In[x, y]**$)$: possibly function f depends upon some global parameters.

42 Definition. A **contrast stretching** operator is a point operator that uses a piecewise smooth function $f(\mathbf{In[x, y]})$ of the input gray level to enhance important details of the image.

Because point operators map one input pixel to one output pixel, they can be applied to an entire pixel array in any sequential order or can be applied in parallel. Ad hoc mappings of intensities, including nonmonotonic ones as in Figure 5.3, can be very useful in enhancing images for general human consumption in graphics and journalism. However, in certain domains, such as radiology, one must be careful not to alter a meaningful intensity scale to which human experts and intricate sensors are carefully calibrated. Finally, we note that the performance of some algorithms for machine vision might not be changed by monotonic gray scale stretching ($f(g_2) > f(g_1)$ whenever gray level $g_2 > g_1$), although humans monitoring the process might appreciate the enhanced images.

5.2.1 Histogram Equalization

Histogram equalization is often tried to enhance an image. The two requirements on the operator are that (a) the output image should use all available gray levels and (b) the output image has approximately the same number of pixels of each gray level. Requirement (a) makes good sense but (b) is ad hoc and its effectiveness must be judged empirically. Figure 5.4 shows images transformed by histogram equalization. One can see that the remapping of gray levels does indeed change the appearance of some regions; for example, the welds on the arch are seen more easily seen. (A mapping similar to the right most example in Figure 5.3 might be better. Why?) The face image was cropped from a larger image and the arch did not have many lower intensity pixels. Requirement (b) will cause large homogeneous regions, such as sky, to be remapped into more gray levels and hence to show more texture. This may or may not help in image interpretation.

Requirements (a) and (b) mean that the target output image uses all gray values $z = z_1, z = z_2, \ldots, z = z_n$ and that each gray level z_k is used approximately $q = (R \times C)/n$ times, where R, C are the number of rows and columns of the image. The input image

Figure 5.4 Histogram equalization maps the gray scale such that the output image uses the entire range available and such that there are approximately the same number of pixels of each gray value in the output image: images after histogram equalization are at the right. (Portrait courtesy of F. Biocca.)

histogram $H_{in}[i]$ is all that is needed in order to define the stretching function f. $H_{in}[i]$ is the number of pixels of the input image having gray level z_i. The first gray level threshold t_1 is found by advancing i in the input image histogram until approximately q_1 pixels are accounted for: all input image pixels with gray level $z_k < t_1 - 1$ will be mapped to gray level z_1 in the output image. Threshold t_1 is formally defined by the following computational formula:

$$\sum_{i=1}^{t_1-1} H_{in}[i] \leq q_1 < \sum_{i=1}^{t_1} H_{in}[i].$$

This means that t_1 is the smallest gray level such that the original histogram contains no more than q pixels with lower gray values. The *kth* threshold t_k is defined by continuing the iteration:

$$\sum_{i=1}^{t_k-1} H_{in}[i] \leq (q_1 + q_2 + \cdots + q_k) < \sum_{i=1}^{t_k} H_{in}[i].$$

A practical implementation of the mapping f is a *lookup table* that is easily obtained from the above process. As the above formula is computed, the threshold values t_k are placed (possibly repetitively) in array $T[i]$ as long as the inequality holds: thus we'll have function $z_{out} = f(z_{in}) = T[z_{in}]$.

Exercise 5.1

An input image of 200 pixels has the following histogram: $H_{in} = [0, 0, 20, 30, 5, 5, 40, 40, 30, 20, 10, 0, 0, 0, 0, 0]$. (a) Using the formula for equalizing the histogram (of 15 gray levels), what would be the output image value $f(8)$? (b) Repeat question (a) for $f(11)$. (c) Give the lookup table $T[i]$ implementing f for transforming all input image values.

Exercise 5.2: An algorithm for histogram equalization.

Use pseudocode to define an algorithm for histogram equalization. Be sure to define all the data items and data structures being used.

Exercise 5.3: A histogram equalization program.

(a) Using the pseudocode from the previous exercise, implement and test a program for histogram equalization. (b) Describe how well it works on different images.

It is often the case that the range of output image gray levels is larger than the range of input image gray levels. It is thus impossible for any function f to remap gray levels onto the entire output range. If an approximately uniform output histogram is really wanted, a random number generator can be used to map an input value z_{in} to some neighborhood of $T[z_{in}]$. Suppose that the procedure described above calls for mapping $2q$ pixels of level g to output level g_1 and 0 pixels to level $g_1 + 1$. We can then simulate a coin flip so that an input pixel of level g has a 50-50 chance of mapping to either g_1 or $g_1 + 1$.

5.3 REMOVAL OF SMALL IMAGE REGIONS

Often, it is useful to remove small regions from an image. A small region might just be the result of noise, or it might represent low level detail that should be suppressed from the image description being constructed. Small regions can be removed by changing single pixels or by removing components after connected components are extracted.

Exercise 5.4

Give some arguments for and against *randomizing* function f in order to make the output histogram more uniform.

5.3.1 Removal of Salt-and-Pepper Noise

The introductory chapter briefly discussed how single anomalous pixels can be removed from otherwise homogeneous regions and methods were extended in Chapter 3. The presence of single dark pixels in bright regions, or single bright pixels in dark regions, is called *salt-and-pepper noise*. The analogy to real life is obvious. Often, salt-and-pepper noise is the natural result of creating a binary image via thresholding. Salt corresponds to pixels in a dark region that somehow passed the threshold for bright, and pepper corresponds to pixels in a bright region that were below threshold. Salt and pepper might be classification errors resulting from variation in the surface material or illumination, or perhaps noise in the analog/digital conversion process in the frame grabber. In some cases, these isolated pixels are not classification errors at all, but are tiny details contrasting with the larger neighborhood, such as a button on a shirt or a clearing in a forest, etc., and it may be that the description needed for the problem at hand prefers to ignore such detail.

Figure 5.5 shows the removal of salt-and-pepper noise from a binary image of bacteria. The operations on the input image are expressed in terms of *masks* given at the bottom of the figure. If the input image neighborhood matches the mask at the left, then it is changed into the neighborhood given by the mask at the right. Only two such masks are needed for this approach. In case the input image is a labeled image created by use of a set of thresholds or some other classification procedure, a more general mask can be used. As shown in the bottom row of Figure 5.5, any pixel label L that is isolated in an otherwise homogenous 8-neighborhood is coerced to the majority label X. L may be any of the k labels used in the image. The figure also shows that either the 8- or 4-neighborhood can be used for making the decision; in the case of the 4-neighborhood, the four corner pixels are not used in the decision. As shown in Chapter 3, use of different neighborhoods can result in different

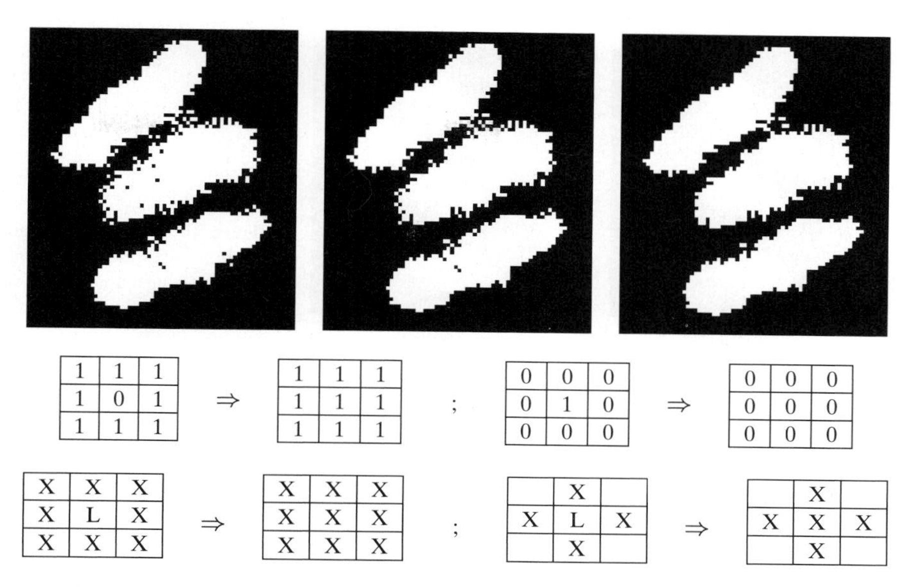

Figure 5.5 (top left) Binary image of bacteria, (top middle) salt-and-pepper noise removed using 8-neighbors; (top right), salt-and-pepper noise removed using 4-neighbors. Middle row: templates showing how binary pixel neighborhoods can be cleaned. Bottom row: templates defining isolated pixel removal for a general labeled input image; (bottom left) 8-neighborhood decision and (bottom right) 4-neighborhood decision. (Image cropped from bacteria image courtesy of Frank Dazzo.)

output images, as in the case with the bacteria image. Results in Figure 5.5 show some differences in the results depending upon whether the 8-neighborhood or 4-neighborhood is used.

5.3.2 Removal of Small Components

Chapter 3 discussed how to extract the connected components of a binary image and defined a large number of features that could be computed from the set of pixels comprising a single component. The description computed from the image is the set of components, each representing a region extracted from the background, and the features computed from that region. An algorithm can remove any of the components from this description based on computed features; for example, components with a small number of pixels or components that are very thin can be removed. This processing could remove some noise regions near the boundary of the bacteria. Once small regions are culled from the description it may not be necessary, or possible, to generate the corresponding output image. If an output image is necessary, information must be saved in order to return to the input image and correctly recode the pixels from the changed regions. Figure 5.6 shows the bacteria image after removal of salt-and-pepper noise and components of area 12 pixels or less.

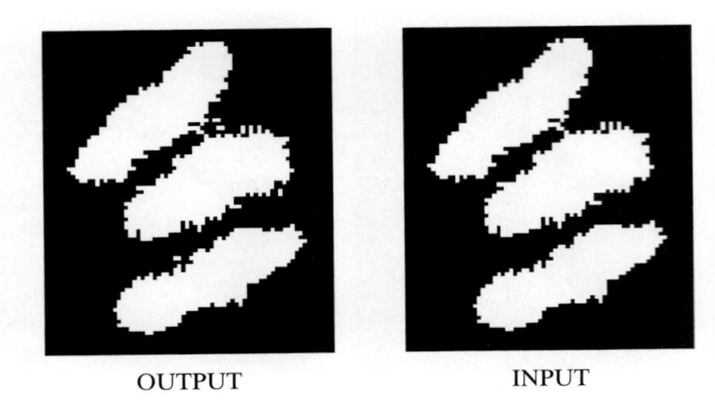

Figure 5.6 (left) Bacteria image of Figure 5.5 after salt-and-pepper noise was removed using a 4-neighborhood; and (right) image with small components removed. (Original image courtesy of Frank Dazzo.)

5.4 IMAGE SMOOTHING

Often, an image is composed of some underlying ideal structure, which we want to detect and describe, together with some random noise or artifact, which we would like to remove. For example, a simple model is that pixels from the image region of a uniform object have value $g_r + N(0, \sigma)$, where g_r is some expected gray level for ideal imaging conditions and $N(0, \sigma)$ is Gaussian noise of mean 0 and standard deviation σ. Figure 5.7(top left) shows an ideal checkerboard with uniform regions. Gaussian noise has been added to the ideal image to create the noisy image in the center: note that the noisy values have been clipped to remain within the interval [0, 255]. At the top right is a plot of the pixel values across a single (horizontal) row of the image.

Noise which varies randomly above and below a nominal brightness value for a region can be reduced by averaging a neighborhood of values.

$$OutputImage[r, c] = \textit{average of some neighborhood of } InputImage[r, c] \tag{5.1}$$

$$Out[r, c] = \left(\sum_{i=-2}^{+2} \sum_{j=-2}^{+2} In[r + i, c + j] \right) \Big/ 25 \tag{5.2}$$

Equation 5.2 defines a smoothing filter that averages 25 pixel values in a 5×5 neighborhood of the input image pixel in order to create a smoothed output image. Figure 5.7(bottom center) illustrates its use on the checkerboard image: the image row shown at the bottom right of the figure is smoother than the input image row shown at the top right. This row is not actually formed by averaging just within that row, but by using pixel values from five rows of the image. Also note that while the smooth image is cleaner than the original, it is not as sharp.

43 Definition. Smoothing an image by equally weighting a rectangular neighborhood of pixels is called using a **box filter.**

Rather than weight all input pixels equally, it is better to reduce the weight of the input pixels with increasing distance from the center pixel $\mathbf{I}[\mathbf{x_c}, \mathbf{y_c}]$. The Gaussian filter does this

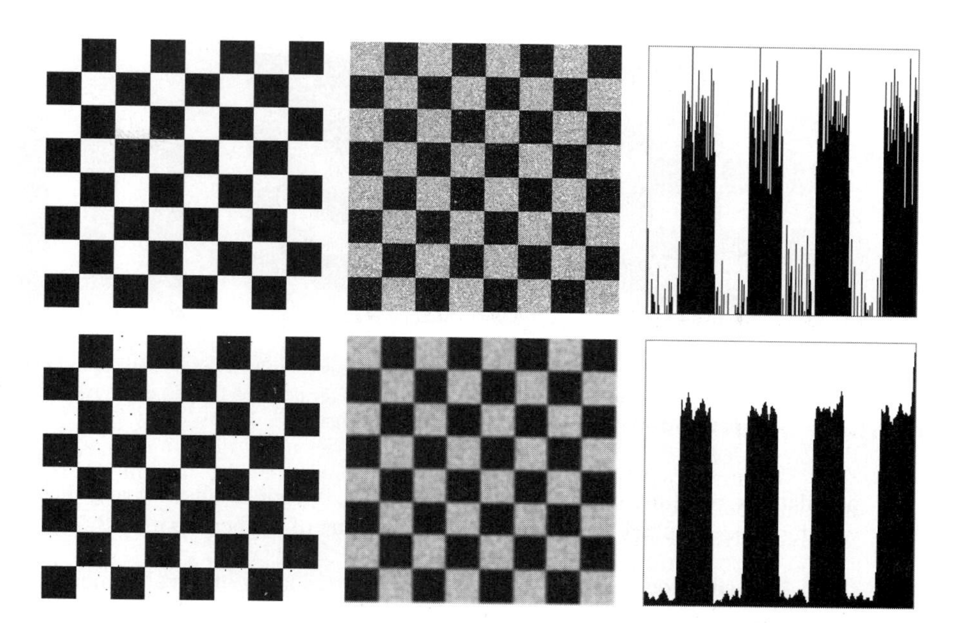

Figure 5.7 (top left) Ideal image of checkerboard with pixel values of 0 in the black squares and 255 in the white squares; (top center) image with added Gaussian noise of standard deviation 30; (top right) pixel values in a horizontal row 100 from the top of the noisy image; (bottom left) top center image thresholded in the valley of the image histogram shows some pepper noise; (bottom center) noise averaged using a 5 × 5 neighborhood centered at each pixel; and (bottom right) pixels across image row 100 from the top.

and is perhaps the most commonly used of all filters. Its desirable properties are discussed more fully in Section 5.7.

44 Definition. When a **Gaussian filter** is used, pixel **[x, y]** is weighted according to

$$g(x, y) = \frac{1}{\sqrt{2\pi}\,\sigma} e^{-\frac{d^2}{2\sigma^2}}$$

where $d = \sqrt{(x - x_c)^2 + (y - y_c)^2}$ is the distance of the neighborhood pixel **[x, y]** from the center pixel $[\mathbf{x_c}, \mathbf{y_c}]$ of the output image where the filter is being applied.

Later in this chapter, we will develop in more detail both the theory and methods of smoothing and will express edge detection within the same framework. Before proceeding in that direction, we introduce the useful and intuitive *median filter*.

5.5 MEDIAN FILTERING

Averaging is sure to produce a better estimate of **I[x, y]** when the average is taken over a homogeneous neighborhood with zero-mean noise. When the neighborhood straddles the boundary between two such regions, the estimate uses samples from two intensity

Figure 5.8 (left) Noisy checkerboard image; (center) result of setting output pixel to the median value of a 5×5 neighborhood centered at the pixel; and (right) display of pixels across image row 100 from the top; compare to Figure 5.7.

populations, resulting in blurring of the boundary. A popular alternative is the *median filter,* which replaces a pixel value with the *median* value of the neighborhood.

45 Definition. Let $\mathbf{A}[i]_{i=0\ldots(n-1)}$ be a sorted array of n real numbers. The **median** of the set of numbers in $\mathbf{A}$ is $\mathbf{A}[(n-1)/2]$

Sometimes, the cases of n being odd versus even are differentiated. For odd n, the array has a unique middle as defined above. If n is even, then we can define that there are two medians, at $\mathbf{A}[n/2]$ and $\mathbf{A}[n/2-1]$, or one median, which is the average of these two values. Although sorted order was used to define the median, the n values do not have to be fully sorted in practice to determine the median. The well-known quicksort algorithm can easily be modified so that it only recurses on the subarray of $\mathbf{A}$ that contains the $(n+1)/2th$ element; as soon as a sort pivot is placed in that array position the median of the entire set is known.

Figure 5.8 shows that the median filter can smooth noisy regions yet better preserve the structure of boundaries between them. When a pixel is actually chosen from one of the white squares, but near the edge, it is likely that the majority of the values in its neighborhood are white pixels with noise: if this is true, then neighboring pixels from a black square will not be used to determine the output value. Similarly, when computing the output pixel on the black side of an edge, it is highly likely that a majority of those pixels are black with noise, meaning that any neighborhood samples from the adjacent white region will not be used any further in computing the output value. Thus, unlike smoothing using averaging, the median filter tends to preserve edge structure while at the same time smoothing uniform regions. Median filtering can also remove salt-and-pepper noise and most other small artifacts that effectively replace a few ideal image values with noise values *of any kind*. Figure 5.9 shows how structured artifacts can be removed while at the same time reducing variation in uniform regions and preserving boundaries between regions.

Computing the median requires more computation time than computing a neighborhood average, since the neighborhood values must be partially sorted. Moreover, median filtering is not so easily implemented in special hardware that might be necessary for

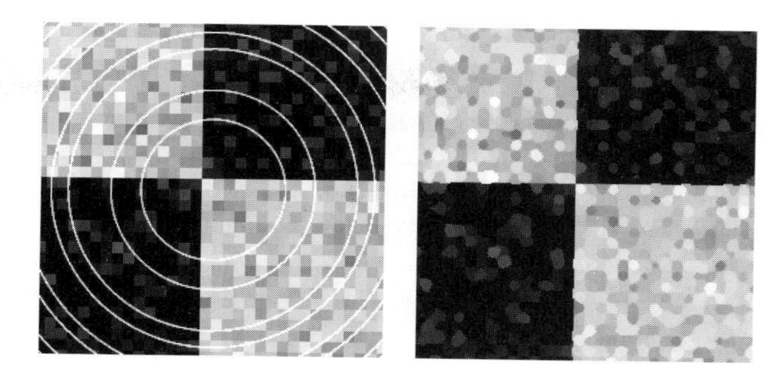

Figure 5.9 (left) Input image contains both Gaussian noise and bright ring artifacts added to four previously uniform regions; and (right) result of applying a 7×7 median filter.

real-time processing, such as in a video pipeline. However, in many image analysis tasks, its value in image enhancement is worth the time spent.

Exercise 5.5: Modifying Quicksort 1.

(a) Find pseudocode for the traditional quicksort algorithm from one of the many data structures and algorithms texts. Modify the algorithm to return only the median as soon as it can be decided. (b) What is the computational effort of finding the median relative to performing the entire sort? (c) Implement your algorithm in some programming language and test it on some example images.

Exercise 5.6: Modifying Quicksort 2.

Consider using the quicksort algorithm from above to detect steps in the picture function, such as the steps from black to white squares in the checkerboard images. Suppose the median of a neighborhood about **I[r, c]** has just been found by placing a pivot value in array position **A**$[n/2]$. Describe how the rest of the array can be processed to decide whether or not the pixel at **[r, c]** is or is not on the boundary between regions of different brightness.

5.5.1 Computing an Output Image from an Input Image

Now that examples have been presented showing what kinds of image enhancements will be done, it's important to consider how these operations on images can be performed. Different options for controlling the filtering of an input image to produce an enhanced output image are represented by the generic algorithm below.

Algorithm 5.1 shows simple sequential program control which considers each pixel of the output image **G** in raster scan order and computes the value of **G[r, c]** using a neighborhood of pixels around **F[r, c]**. It should be clear that the pixels of output image **G** could be computed in a random order rather than in row and column order, and, in fact,

Compute output image pixel G[r, c] from neighbors of input image pixel F[r, c].

F[r, c] is an input image of **MaxRow** rows and **MaxCol** columns;
F is unchanged by the algorithm.
G[r, c] is the output image of **MaxRow** rows and **MaxCol** columns.
The border of **G** are all those pixels whose neighborhoods
are not wholly contained in **G**.
w and h are the width and height, in pixels, defining a neighborhood.

```
procedure enhance_image(F,G,w,h);
{
for r := 0 to MaxRow - 1
   for c := 0 to MaxCol - 1
      {
      if [r, c] is a border pixel then G[r, c] := F[r, c];
      else G[r, c] := compute_using_neighbors (F, r, c, w, h);
      } ;
}
procedure compute_using_neighbors (IN, r, c, w, h)
{
using all pixels within w/2 and h/2 of pixel IN[r, c],
compute a value to return to represent IN[r, c]
}
```

Algorithm 5.1 Compute output image pixel **G[r, c]** from neighbors of input image pixel **F[r, c]**.

Exercise 5.7

Implement Algorithm 5.1 in some programming language. Code both the boxcar and median filtering operations and test them on some images such as in Figure 5.9.

could all be computed in parallel. This holds because the input image is unchanged by any of the neighborhood computations. Secondly, the procedure compute_using_neighbors could be implemented to perform either boxcar or median filtering. For boxcar filtering, the procedure need only add up the $w \times h$ pixels of the neighborhood of **F[r, c]** and then divide by the number of pixels $w \times h$. To implement a median filter, the procedure could copy those $w \times h$ pixel values into a local array **A** and partially sort it to obtain their median value.

Control could also be arranged so that only h rows of the image are in main memory at any one time. Outputs **G[r, c]** would be computed only for the middle row r. Then, a new row would be input to replace the oldest row in memory and the next output row of **G[r, c]** would be computed. This process is repeated until all possible output rows are computed.

Years ago, when computers had small main memories, the primary storage for images was on disk and many algorithms had to process images a few rows at a time. Today, such control is still of interest because it is used in image processing boards implementing a pipelined architecture.

5.6 DETECTING EDGES USING DIFFERENCING MASKS

Image points of high contrast can be detected by computing intensity differences in local image regions. Typically, such points form the border between different objects or scene parts. In this section, we show how to do this using neighborhood templates or *masks*. We start by using one-dimensional signals: this helps develop both the intuition and formalism, and is also very important in its own right. The 1D signals could just be rows or columns of a 2D image. The section ends by studying more general 2D situations.

5.6.1 Differencing 1D Signals

Figure 5.10 shows how masks can be used to compute representations of the derivatives of a signal. Given that the signal $\mathbf{S}$ is a sequence of samples from some function f, then $f'(x_i) \approx (f(x_i) - f(x_{i-1})/(x_i - x_{i-1})$. Assuming that the sample spacing is $\Delta x = 1$, the derivative of $f(x)$ can be approximated by applying the mask $\mathbf{M'} = [\mathbf{-1, 1}]$ to the samples in $\mathbf{S}$ as shown in Figure 5.10 to obtain an output signal $\mathbf{S'}$. As the figure shows,

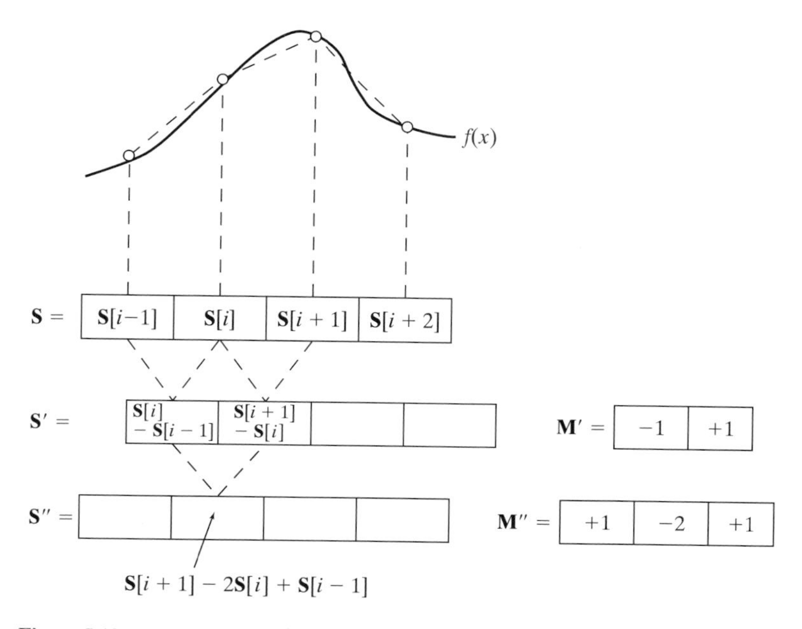

Figure 5.10 (left) The first ($\mathbf{S'}$) and second ($\mathbf{S''}$) difference signals are scaled approximations to the first and second derivatives of the signal $\mathbf{S}$; and (right) Masks $\mathbf{M'}$ and $\mathbf{M''}$ represent the derivative operations.

it's convenient to think of the values of $\mathbf{S}'$ as occurring in between the samples of $\mathbf{S}$. A high absolute value of $\mathbf{S}'[i]$ indicates where the signal is undergoing rapid change or *high contrast*. Signal $\mathbf{S}'$ itself can be differentiated a second time using mask $\mathbf{M}'$ to produce output $\mathbf{S}''$ which corresponds to the second derivative of the original function f. The important result illustrated in Figure 5.10 and derived in the equations below is that the approximate second derivative can be computed by applying the mask $\mathbf{M}''$ to the original sequence of samples $\mathbf{S}$.

$$\mathbf{S}'[i] = -S[i-1] + S[i] \tag{5.3}$$

$$\text{mask } \mathbf{M}' = [-1, +1] \tag{5.4}$$

$$\mathbf{S}''[i] = -S'[i] + S'[i+1] \tag{5.5}$$

$$= -(S[i] - S[i-1]) + (S[i+1] - S[i]) \tag{5.6}$$

$$= S[i-1] - 2S[i] + S[i+1] \tag{5.7}$$

$$\text{mask } \mathbf{M}'' = [1, -2, 1] \tag{5.8}$$

If only points of high contrast are to be detected, it is very common to use the absolute value after applying the mask at signal position $\mathbf{S}[i]$. If this is done, then the first derivative mask can be either $\mathbf{M}' = [-1, +1]$ or $[+1, -1]$ and the second derivative mask can be either

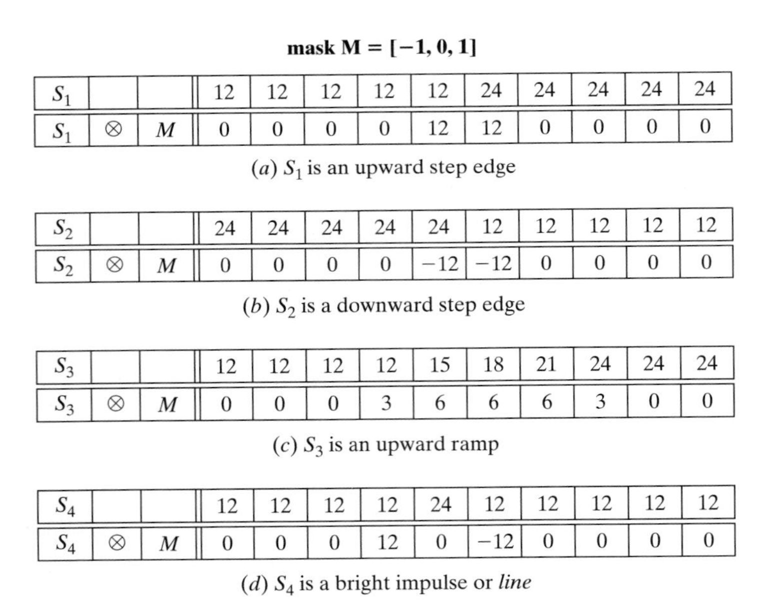

mask M = [−1, 0, 1]

S_1			12	12	12	12	12	24	24	24	24	24
S_1	⊗	M	0	0	0	0	12	12	0	0	0	0

(*a*) S_1 is an upward step edge

S_2			24	24	24	24	24	12	12	12	12	12
S_2	⊗	M	0	0	0	0	−12	−12	0	0	0	0

(*b*) S_2 is a downward step edge

S_3			12	12	12	12	15	18	21	24	24	24
S_3	⊗	M	0	0	0	3	6	6	6	3	0	0

(*c*) S_3 is an upward ramp

S_4			12	12	12	12	24	12	12	12	12	12
S_4	⊗	M	0	0	0	12	0	−12	0	0	0	0

(*d*) S_4 is a bright impulse or *line*

Figure 5.11 Cross correlation of four special signals with first derivative edge detecting mask $[-1, 0, 1]$; (*a*) upward step edge, (*b*) downward step edge, (*c*) upward ramp, and (*d*) bright impulse. Note that, since the coordinates of **M** sum to zero, output must be zero on a constant region.

mask M = [−1, 2, −1]

S_1			12	12	12	12	12	24	24	24	24	24
S_1	$\otimes$	M	0	0	0	0	−12	12	0	0	0	0

(a) S_1 is an upward step edge

S_2			24	24	24	24	24	12	12	12	12	12
S_2	$\otimes$	M	0	0	0	0	12	−12	0	0	0	0

(b) S_2 is a downward step edge

S_3			12	12	12	12	15	18	21	24	24	24
S_3	$\otimes$	M	0	0	0	−3	0	0	0	3	0	0

(c) S_3 is an upward ramp

S_4			12	12	12	12	24	12	12	12	12	12
S_4	$\otimes$	M	0	0	0	−12	24	−12	0	0	0	0

(d) S_4 is a bright impulse or *line*

Figure 5.12 Cross correlation of four special signals with second derivative edge detecting mask $\mathbf{M} = [\mathbf{-1, 2, -1}]$; ($a$) upward step edge, ($b$) downward step edge, ($c$) upward ramp, and ($d$) bright impulse. Since the coordinates of **M** sum to zero, response on constant regions is zero. Note how a *zero-crossing* appears at an output position where different trends in the input signal join.

$\mathbf{M}'' = [+1, -2, +1]$ or $[-1, +2, -1]$. A similar situation exists for 2D images as we shall soon see. Whenever only magnitudes are of concern, we will consider these patterns to be the same, and whenever the sign of the change is important, we will consider them to be different.

Use of another common first derivative mask is shown in Figure 5.11. This mask has 3 coordinates and is centered at signal point $\mathbf{S}[i]$ so that it computes the signal difference across the adjacent values. Because $\Delta x = 2$, it will give a high estimate of the actual derivative unless the result is divided by 2. Moreover, this mask is known to give a response on perfect step edges that is 2 samples wide, as is shown in Figure 5.11(a)–(b). Figure 5.12 shows the response of the second derivative mask on the sample signals. As Figure 5.12 shows, signal contrast is detected by a zero-crossing, which localizes and amplifies the change between two successive signal values. Taken together, the first and second derivative signals reveal much of the local signal structure. Figure 5.13 shows how smoothing of signals can be put into the same framework as differencing: the boxed table below compares the general properties of smoothing versus differencing masks.

Some properties of derivative masks follow:

- Coordinates of derivative masks have opposite signs in order to obtain a high response in signal regions of high contrast.

box smoothing mask M = [1/3, 1/3, 1/3]

S_1			12	12	12	12	12	24	24	24	24	24
S_1	$\otimes$	M	12	12	12	12	16	20	24	24	24	24

(*a*) S_1 is an upward step edge

S_4			12	12	12	12	24	12	12	12	12	12
S_4	$\otimes$	M	12	12	12	16	16	16	12	12	12	12

(*d*) S_4 is a bright impulse or *line*

Gaussian smoothing mask M = [1/4, 1/2, 1/4]

S_1			12	12	12	12	12	24	24	24	24	24
S_1	$\otimes$	M	12	12	12	12	15	21	24	24	24	24

(*a*) S_1 is an upward step edge

S_4			12	12	12	12	24	12	12	12	12	12
S_4	$\otimes$	M	12	12	12	15	18	15	12	12	12	12

(*d*) S_4 is a bright impulse or *line*

Figure 5.13 (top two rows) Smoothing of step and impulse with box mask [1/3, 1/3, 1/3]; and (bottom two rows) smoothing of step and impulse with Gaussian mask [1/4, 1/2, 1/4].

- The sum of coordinates of derivative masks is zero so that a zero response is obtained on constant regions.
- First derivative masks produce high absolute values at points of high contrast.
- Second derivative masks produce zero-crossings at points of high contrast.

For comparison, smoothing masks have these properties:

- Coordinates of smoothing masks are positive and sum to one so that output on constant regions is the same as the input.
- The amount of smoothing and noise reduction is proportional to the mask size.
- Step edges are blurred in proportion to the mask size.

5.6.2 Difference Operators for 2D Images

Contrast in the 2D picture function $f(x, y)$ can occur in any direction. From calculus, we know that the maximum change occurs along the direction of the gradient of the function, which is in the direction $[\frac{\partial f}{\partial x}, \frac{\partial f}{\partial y}]$ in the picture plane. Figure 5.14 shows that this is quite intuitive when one considers discrete approximations in digital images. We can estimate

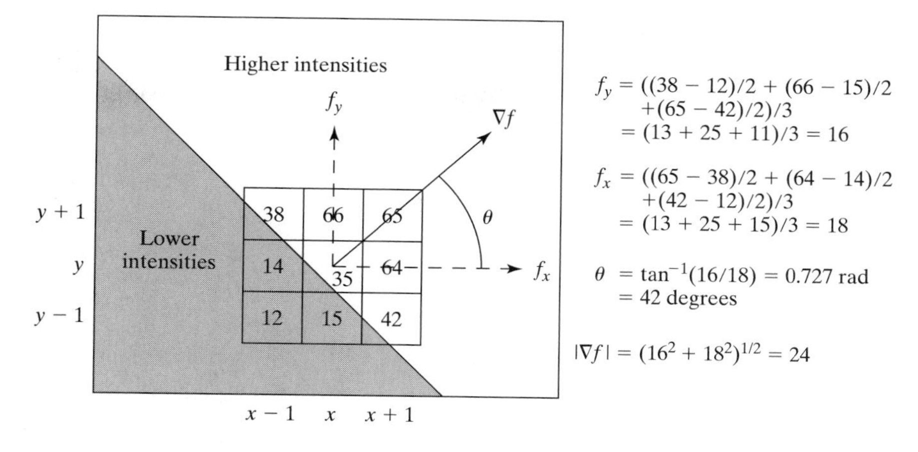

Figure 5.14 Estimating the magnitude and direction of contrast at **I[x, y]** by estimating the gradient magnitude and direction of picture function $f(x, y)$ using the discrete samples in the image array.

the contrast at image location **I[x, y]** along the x-direction by computing $(I[x + 1, y] - I[x - 1, y])/2$, which is the change in intensity across the left and right neighbors of pixel **[x, y]** divided by $\Delta x = 2$ pixel units. For the neighborhood shown in Figure 5.14, the constrast in the x direction would be estimated as $(64 - 14)/2 = 25$. Since our pixel values are noisy and since the edge might actually cut across the pixel array at any angle, it should help to average 3 different estimates of the contrast in the neighborhood of **[x, y]:**

$$\begin{aligned} \partial f/\partial x \equiv f_x \approx \tfrac{1}{3}[&(I[x + 1, y] - I[x - 1, y])/2 \\ &+ (I[x + 1, y - 1] - I[x - 1, y - 1])/2 \\ &+ (I[x + 1, y + 1] - I[x - 1, y + 1])/2)] \end{aligned} \tag{5.9}$$

This estimates the contrast in the x-direction by equally weighting the contrast across the row y with the row below and above it. The contrast in the y-direction can be estimated similarly:

$$\begin{aligned} \partial f/\partial y \equiv f_y \approx \tfrac{1}{3}[&(I[x, y + 1] - I[x, y - 1])/2 \\ &+ (I[x - 1, y + 1] - I[x - 1, y - 1])/2 \\ &+ (I[x + 1, y + 1] - I[x + 1, y - 1])/2)] \end{aligned} \tag{5.10}$$

Often, the division by 6 is ignored to save computation time, resulting in scaled estimates. Masks for these two contrast operators are denoted by M_x and M_y at the top of Figure 5.15. The gradient of the picture function can be estimated by applying the masks to the 8-neighborhood $N_8[x, y]$ of pixel **[x, y]** as given in Equations 5.11 to 5.14. These

Prewitt: $M_x = \begin{array}{|c|c|c|}\hline -1 & 0 & 1 \\ \hline -1 & 0 & 1 \\ \hline -1 & 0 & 1 \\ \hline \end{array}$; $M_y = \begin{array}{|c|c|c|}\hline 1 & 1 & 1 \\ \hline 0 & 0 & 0 \\ \hline -1 & -1 & -1 \\ \hline \end{array}$

Sobel: $M_x = \begin{array}{|c|c|c|}\hline -1 & 0 & 1 \\ \hline -2 & 0 & 2 \\ \hline -1 & 0 & 1 \\ \hline \end{array}$; $M_y = \begin{array}{|c|c|c|}\hline 1 & 2 & 1 \\ \hline 0 & 0 & 0 \\ \hline -1 & -2 & -1 \\ \hline \end{array}$

Roberts: $M_x = \begin{array}{|c|c|}\hline 0 & 1 \\ \hline -1 & 0 \\ \hline \end{array}$; $M_y = \begin{array}{|c|c|}\hline 1 & 0 \\ \hline 0 & -1 \\ \hline \end{array}$

Figure 5.15 Masks used to estimate the gradient of picture function $f(x, y)$: (top row) Prewitt; (middle row) Sobel; and (bottom row) Roberts.

masks define the *Prewitt operator* credited to Dr. Judith Prewitt who used them in detecting boundaries in biomedical images.

$$\frac{\partial f}{\partial x} \approx (1/6)(M_x \circ N_8[x, y]) \tag{5.11}$$

$$\frac{\partial f}{\partial y} \approx (1/6)(M_y \circ N_8[x, y]) \tag{5.12}$$

$$|\nabla f| \approx \sqrt{\frac{\partial f}{\partial x}^2 + \frac{\partial f}{\partial y}^2} \tag{5.13}$$

$$\theta \approx tan^{-1}\left(\frac{\partial f}{\partial y} / \frac{\partial f}{\partial x}\right) \tag{5.14}$$

The operation $M \circ N$ is defined formally in the next section: operationally, mask M is overlaid on image neighborhood N so that each intensity N_{ij} can be multiplied by weight M_{ij}; finally all these products are summed. The middle row of Figure 5.15 shows the two analogous *Sobel masks;* their derivation and interpretation is the same as for the Prewitt masks except that the assumption is that the center estimate should be weighted twice as much as the estimate to either side.

The *Roberts masks* are only 2×2; hence they are both more efficient to use and more locally applied. Often referred to as *the Roberts cross operator,* these masks actually compute a gradient estimate at the center of a 4-neighborhood and not at a center pixel. Moreover, the actual coordinate system in which the operator is defined is rotated 45° off the standard row direction. Application of the Roberts cross operator is shown in Figure 5.16: the original input image is shown in the upper left in part (a), while the output from slightly different Roberts operators is shown in (b) and (c). Parts (d) and (e) show the results of using just the intensity differences along the image columns and rows respectively, and (f) shows the row and column detections ORed together. Qualitatively, these results are typical of the several small-neighborhood operators—many pixels on many edges are detected but many are missed. There are responses on the textured grass region and the top of the garage is missed because its intensity matches the sky. The Roberts results should be compared to the results obtained by combining the simple 1D row and column masks

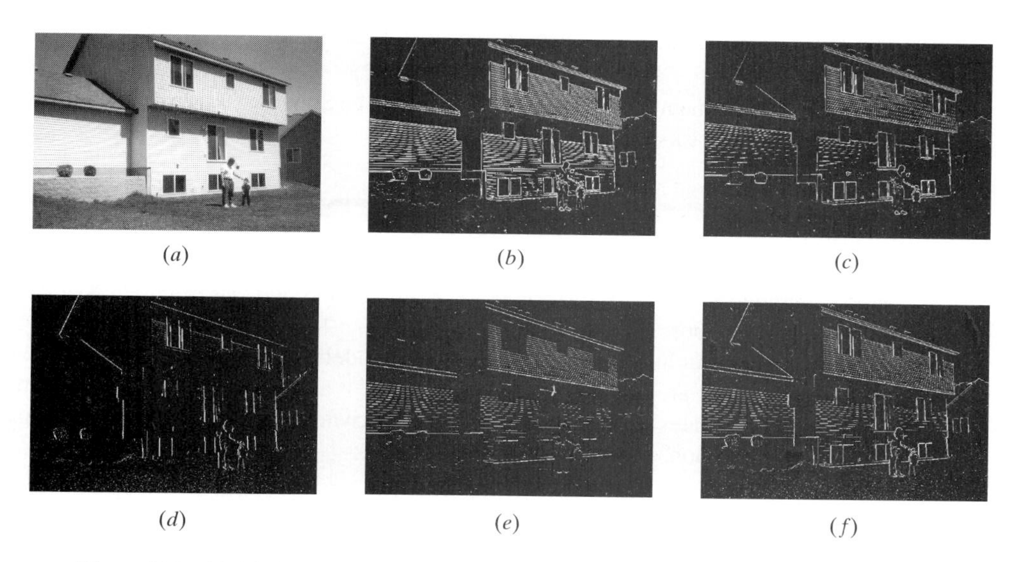

(*a*) (*b*) (*c*)

(*d*) (*e*) (*f*)

Figure 5.16 (*a*) Original image; (*b*) top 5 percent of the sum of the absolute values of the two Roberts mask response; (*c*) top 5 percent of the mean-squared value of the two Roberts masks; (*d*) top 2 percent by absolute value of the responses from y-direction edge mask $[-1, +1]$; (*e*) top 3 percent of the responses from x-direction edge mask $[-1, +1]^t$; and (*f*) images of (*d*) and (*e*) ORed together. There are minor differences among (*b*), (*c*), and (*f*). (Image courtesy of Ida Stockman.)

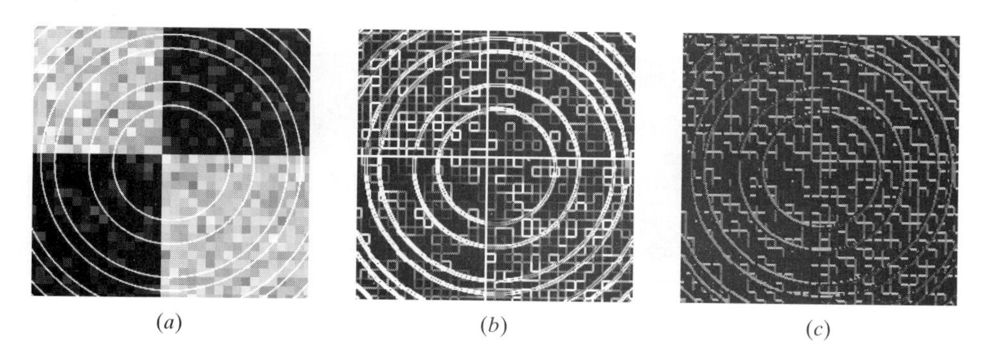

(*a*) (*b*) (*c*)

Figure 5.17 (*a*) Image of noisy squares and rings; (*b*) mean square response of 3×3 Sobel operator; and (*c*) coding of gradient direction computed by 3×3 Sobel operator.

shown in Figure 5.16(*d*-*f*). It is common to avoid computing a square root in order to compute gradient magnitude; alternatives are $max(|\frac{\partial f}{\partial x}|, |\frac{\partial f}{\partial y}|)$, $|\frac{\partial f}{\partial x}| + |\frac{\partial f}{\partial y}|$, or $(\frac{\partial f^2}{\partial x} + \frac{\partial f^2}{\partial y})/2$. Comparing Figure 5.16(*b*) and (*c*, *f*) seems to justify avoiding the square root operation. With these estimates, one must be careful when trying to interpret the actual gradient or gradient direction. Figure 5.17(*b*) shows the results of using the Sobel 3×3 operator to compute mean square gradient magnitude and Figure 5.17(*c*) shows an encoding of the gradient direction. The small squares in the original image are 8×8 pixels: the Sobel operator represents many, but not all, of the image edges.

Exercise 5.8

If a true gradient magnitude is needed, why might the Sobel masks provide a faster solution than the Prewitt masks?

Exercise 5.9: Optimality of Prewitt masks.*

The problem is to prove that Prewitt masks give the weights for the best-fitting plane approximating the intensity surface in a 3×3 neighborhood, assuming all 9 samples have equal weight. Suppose that the 9 intensities $I[r+i, c+j]$; $i, j = -1, 0, 1$ of a 3×3 image neighborhood are fit by the least squares planar model $I[r, c] = z = pr + qc + z_0$. (Recall that the 9 samples are equally spaced in terms of r and c.) Show that the Prewitt masks compute the estimates of p and q as the partial derivatives of the least squares planar fit of the intensity function.

Figure 5.18(b, c) show plots of the intensities of two rows of the room scene in (a). The lower row (b) cuts across four dark regions as seen in the image and the plot; (1) the coat on the chair at the left (columns 20 to 80), (2) Dr. Prewitt's chair and dress in the center (columns 170 to 240), (3) the shadow of the rightmost chair (columns 360 to 370) and (4) the electric wires (column 430). Note that the transitions between dark and bright are sharp except for the boundary between the chair and its shadow, which ramps down from brightness 220 to 20 over about 10 pixels. The upper row profile, shown in (c), shows sharp transitions where

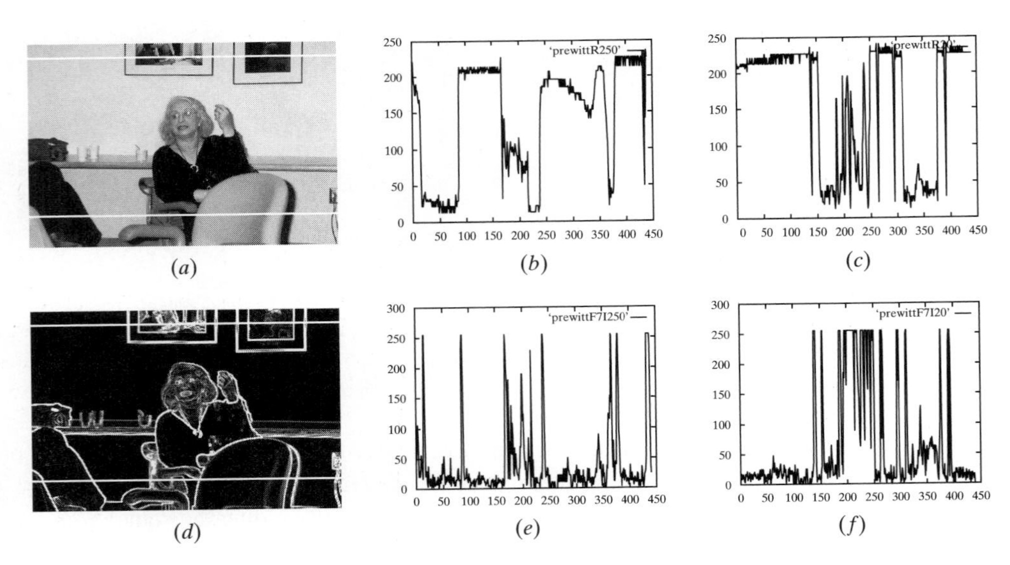

Figure 5.18 (a) Image of Judith Prewitt with two rows selected; (b) plot of intensities along selected lower row; (c) plot of intensities along selected upper row; (d) *gradient image* showing result of $|f_x| + |f_y|$ using the Prewitt 3×3 operator; (e) plot of selected lower row of gradient image; and (f) plot of intensities along gradient of selected upper row.

it cuts across the picture frames, mat, and pictures. The picture at the left shows much more intensity variation than the one at the right. Figure 5.18(d)–(f) shows the results of applying the 3×3 Prewitt gradient operator to the original image. The sum of the absolute values of the column and row gradients f_x and f_y is plotted for the same two image rows shown in parts (a)–(c) of the figure. The highest values of the Prewitt operator correspond well with the major boundaries crossed; however, the several medium level spikes from Dr. Prewitt's chair in (d) between columns 170 and 210 are harder to interpret. The contrasts in the upper row, graphed in (f), are interpreted similarly—major object boundaries correspond well to the object boundaries of the picture frame and mat; however, there is a lot of intensity variation in the leftmost picture on the wall. Generally, gradient operators work well for detecting the boundary of isolated objects, although some problems are common. Boundaries sometimes drop out due to object curvature or soft shadows: on the other hand, good contrast often produces boundaries that are several pixels wide, necessitating a thinning step afterward. Gradient operators will also respond to textured regions, as we shall study in more detail in Chapter 7.

5.7 GAUSSIAN FILTERING AND LOG EDGE DETECTION

The Gaussian function has important applications in many areas of mathematics, including image filtering. In this section, we highlight the characteristics that make it useful for smoothing images or detecting edges after smoothing.

46 Definition. A **Gaussian function** of one variable with spread σ is of the following form, where c is some scale factor.

$$g(x) = ce^{-\frac{x^2}{2\sigma^2}} \tag{5.15}$$

A Gaussian function of two variables is

$$g(x, y) = ce^{-\frac{(x^2+y^2)}{2\sigma^2}} \tag{5.16}$$

These forms have the same structure as the normal distribution defined in Chapter 4, where the constant c was set so that the area under the curve would be 1. To create a mask for filtering, we usually make c a large number so that all mask elements are integers. The Gaussian is defined centered on the origin and thus needs no location parameter μ as does the normal distribution: an image processing algorithm will translate it to wherever it will be applied in a signal or image. Figure 5.19 plots the Gaussian of one variable along with its first and second derivatives, which are also important in filtering operations. The derivation of these functions is given in Equations 5.17 to 5.22. The area under the function $g(x)$ is 1, meaning that it is immediately suitable as a smoothing filter that does not affect constant regions, $g(x)$ is a positive even function; $g'(x)$ is just $g(x)$ multiplied by odd function $-x$ and scaled down by σ^2. More structure is revealed in $g''(x)$. Equation 5.21 shows that $g''(x)$ is the difference of two even functions and that the central lobe will be negative with $x \approx 0$. Using Equation 5.22, it is clear that the zero crossings of the second derivative occur at

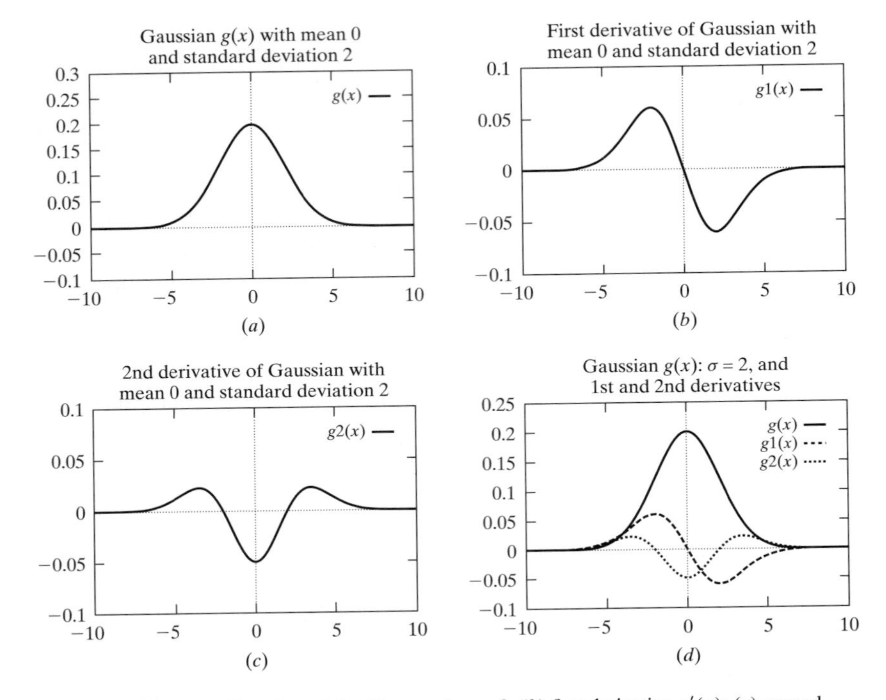

Figure 5.19 (a) Gaussian $g(x)$ with spread $\sigma = 2$; (b) first derivative $g'(x)$; (c) second derivative $g''(x)$, which looks like the cross section of a sombrero upside down from how it would be worn; (d) all three plots superimposed to show how the extreme slopes of $g(x)$ align with the extremas of $g'(x)$ and the zero crossings of $g''(x)$.

$x = \pm\sigma$, in agreement with the plots in Figure 5.19.

$$g(x) = \frac{1}{\sqrt{2\pi}\,\sigma} e^{-\frac{x^2}{2\sigma^2}} \tag{5.17}$$

$$g'(x) = \frac{-1}{\sqrt{2\pi}\,\sigma^3} x e^{-\frac{x^2}{2\sigma^2}} \tag{5.18}$$

$$= \frac{-x}{\sigma^2}\, g(x) \tag{5.19}$$

$$g''(x) = \left(\frac{x^2}{\sqrt{2\pi}\,\sigma^5} - \frac{1}{\sqrt{2\pi}\,\sigma^3} \right) e^{-\frac{x^2}{2\sigma^2}} \tag{5.20}$$

$$= \frac{x^2}{\sigma^4}\, g(x) - \frac{1}{\sigma^2}\, g(x) \tag{5.21}$$

$$= \left(\frac{x^2}{\sigma^4} - \frac{1}{\sigma^2} \right) g(x) \tag{5.22}$$

Understanding the properties of the 1D Gaussian, we can now intuitively create the corresponding 2D function $g(x, y)$ and its derivatives by making the substitution

$r = \sqrt{x^2 + y^2}$. This creates the 2D forms by just spinning the 1D form about the vertical axis yielding *isotropic* functions which have the same 1D Gaussian cross section in any cut through the origin. The second derivative form is well known as a sombrero or Mexican hat. From the mathematical derivations, the cavity for the head will be pointed upward along the $z = g(x, y)$ axis; however, it is usually displayed and used in filtering with the cavity pointed downward, or equivalently, with the center lobe positive and the rim negative.

Some Useful Properties of Gaussians

1. Weight decreases smoothly to zero with distance from the origin, meaning that image values nearer the central location are more important than values that are more remote; moreover, the spread parameter σ determines how broad or focused the neighborhood will be. 95 percent of the total weight will be contained within 2σ of the center.
2. Symmetry about the abscissa; flipping the function for convolution produces the same kernel.
3. Fourier transformation into the frequency domain produces another Gaussian form, which means convolution with a Gaussian mask in the spatial domain reduces high frequency image trends smoothly as spatial frequency increases.
4. The second derivative of a 1D Gaussian $g''(x)$ has a smooth center lobe of negative area and two smooth side lobes of positive area: the zero crossings are located at $-\sigma$ and $+\sigma$, corresponding to the inflection points of $g(x)$ and the extreme points of $g'(x)$.
5. A second derivative filter based on the Laplacian of the Gaussian is called a **LOG filter.** A LOG filter can be approximated nicely by taking the difference of two Gaussians $g''(x) \approx c_1 e^{-\frac{x^2}{2\sigma_1{}^2}} - c_2 e^{-\frac{x^2}{2\sigma_2{}^2}}$, which is often called a **DOG** filter (for **D**ifference **O**f **G**aussians). For a positive center lobe, we must have $\sigma_1 < \sigma_2$; also, σ_2 must be carefully related to σ_1 to obtain the correct location of zero crossings and so that the total negative weight balances the total positive weight.
6. The LOG filter responds well to intensity differences of two kinds—small blobs coinciding with the center lobe, and large step edges very close to the center lobe.

Two different masks for Gaussian smoothing are shown in Figure 5.20. Masks for edge detection are given below.

5.7.1 Detecting Edges with the LOG Filter

Two different masks implementating the LOG filter are given in Figures 5.21 and 5.22. The first is a 3×3 mask: The smallest possible implementation detects image details nearly the

$\mathbf{G}_{3\times3} =$

1	2	1
2	4	2
1	2	1

;

$\mathbf{G}_{7\times7} =$

1	3	7	9	7	3	1
3	12	26	33	26	12	3
7	26	55	70	55	26	7
9	33	70	90	70	33	9
7	26	55	70	55	26	7
3	12	26	33	26	12	3
1	3	7	9	7	3	1

Figure 5.20 (left) A 3×3 mask approximating a Gaussian obtained by matrix multiplication $[1, 2, 1]^t \otimes [1, 2, 1]$; and (right) a 7×7 mask approximating a Gaussian with $\sigma^2 = 2$ obtained by using Equation 5.16 to generate function values for integers x and y and then setting $c = 90$ so that the smallest mask element is 1.

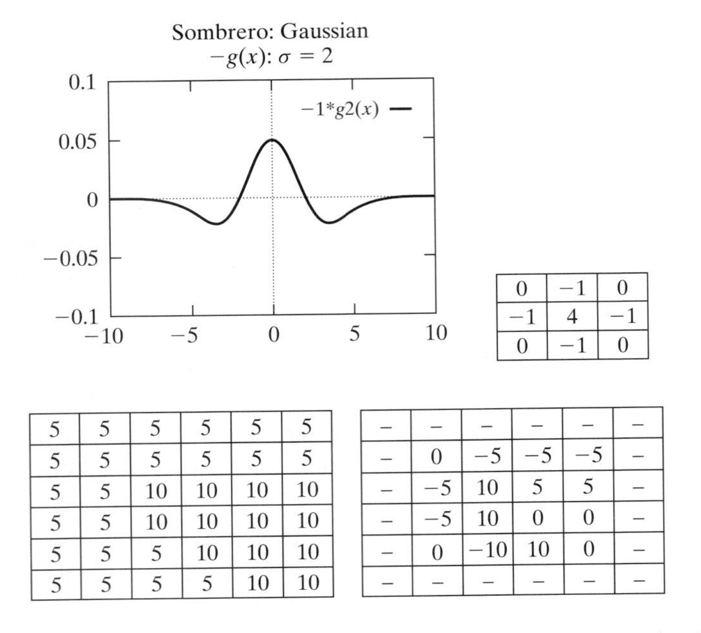

0	−1	0
−1	4	−1
0	−1	0

5	5	5	5	5	5
5	5	5	5	5	5
5	5	10	10	10	10
5	5	10	10	10	10
5	5	5	10	10	10
5	5	5	5	10	10

–	–	–	–	–	–
–	0	−5	−5	−5	–
–	−5	10	5	5	–
–	−5	10	0	0	–
–	0	−10	10	0	–
–	–	–	–	–	–

Figure 5.21 (top row) Cross section of the LOG filter and a 3×3 mask approximating it; and (bottom row) input image and result of applying the mask to it.

Exercise 5.10: Properties of the LOG filter.

Suppose that the 9 intensities of a 3×3 image neighborhood are perfectly fit by the planar model $I[r, c] = z = pr + qc + z_0$. (Recall that the 9 samples are equally spaced in terms of r and c.) Show that the simple LOG mask

0	−1	0
−1	4	−1
0	−1	0

has zero response on such a neighborhood. This means that the LOG filter has zero response on both constant regions and ramps.

0	0	0	−1	−1	−2	−1	−1	0	0	0
0	0	−2	−4	−8	−9	−8	−4	−2	0	0
0	−2	−7	−15	−22	−23	−22	15	−7	−2	0
−1	−4	−15	−24	−14	−1	−14	−24	−15	−4	−1
−1	−8	−22	−14	52	103	52	−14	−22	−8	−1
−2	−9	−23	−1	103	178	103	−1	−23	−9	−2
−1	−8	−22	−14	52	103	52	−14	−22	−8	−1
−1	−4	−15	−24	−14	−1	−14	−24	−15	−4	−1
0	−2	−7	−15	−22	−23	−22	15	−7	−2	0
0	0	−2	−4	−8	−9	−8	−4	−2	0	0
0	0	0	−1	−1	−2	−1	−1	0	0	0

Figure 5.22 An 11 × 11 mask approximating the Laplacian of a Gaussian with $\sigma^2 = 2$. (From Haralick and Shapiro, vol. I, p. 349.)

size of a pixel. The 11 × 11 mask computes a response by integrating the input of 121 pixels and thus responds to larger image features and not smaller ones. Integrating 121 pixels can take a lot more time than integrating 9 of them if done in hardware.

5.7.2 On Human Edge Detection

We now describe an artificial neural network (ANN) architecture which implements the LOG filtering operation in a highly parallel manner. The behavior of this network has been shown to simulate some of the known behaviors of the human visual system. Moreover, invasive techniques have also shown that the visual systems of cats and monkeys produce electrical signals consistent with the behavior of the neural network architecture. Figure 5.23 shows the processing of 1D signals. A step edge is sensed at various points by cells of the

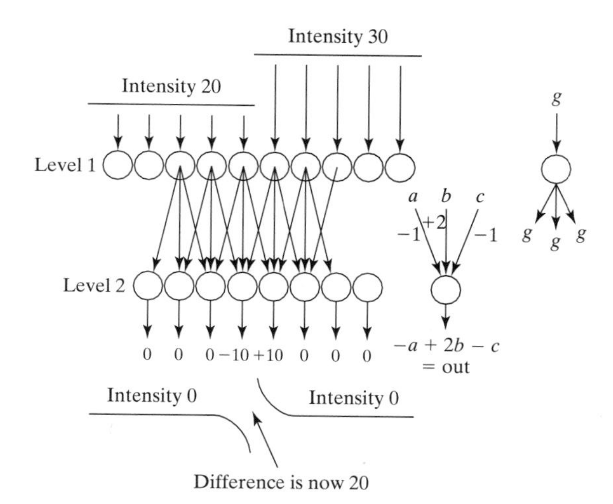

Figure 5.23 Producing the *Mach band effect* using an ANN architecture. Intensity is sensed by cells of the retina (Level 1), which then stimulate integrating cells in a higher level layer (Level 2).

retinal array. These level 1 cells stimulate the integrating cells at level 2. Each physical connection between level 1 cell i and level 2 cell j has an associated weight w_{ij} which is multiplied by the stimulus being communicated before it is summed in cell j. The output of cell j is $y_j = \sum_{i=1}^{N} w_{ij} x_i$ where x_i is the output of the *ith* first level cell and N is the total number of first level cells. (Actually, we need only account for those cells i directly connected to the second level cell j). By having the weights associated with each connection, it is possible, and common, to have the same cell i give positive input to cell j and negative input to cell $k \neq j$. Figure 5.23 shows that each cell j of the second level computes its output as $-a + 2b - c$, corresponding to the mask $[-1, 2, -1]$: the weight 2 is applied to the central input, while the (inhibitory) inputs a and b are each weighted by -1.

This kind of architecture can be defined for any kind of mask and permits a highly parallel implementation of cross correlation for filtering or feature detection. The psychologist Mach noted that humans perceive an edge between two regions as if it were pulled apart to exaggerate the intensity difference, as is done in Figure 5.23. Note that this architecture and mask create the *zero crossing* at the location of the edge between two cells, one of which produces positive output and the other negative output. The Mach band effect changes the perceived shape of joining surfaces and is evident in computer graphics systems that display polyhedral objects via shaded faces. Figure 5.24 shows seven constant regions, stepping from gray level 31 to 255 in steps of 32. Do you perceive concave panels in 3D such as on a Doric column from a Greek temple?

Figure 5.24 Seven constant stripes generated with gray levels $31 + 32k, k = 1, 7$. Due to the Mach band effect, humans perceive scalloped, or concave, panels.

Figure 5.25 extends Figure 5.23 to 2D images. Each set of retinal cells connected to integrating cell j comprise what is called the *receptive field* of that cell. To perform edge detection via a second derivative operation, each receptive field has a center set of cells which have positive weights w_{ij} with respect to cell j and a surrounding set of cells with negative weights. Retinal cells b and c are in the center of the receptive field of integrating cell A, whereas retinal cells a and d are in the surround and provide inhibitory input. Retinal cell d is in the center of the receptive field of integrating cell B, however, and cell c is in its surround. The sum of the weights from the center and the surround should be zero so that the integrating cell has neutral output on constant regions. Because the center and surround are circular, output will not be neutral whenever a straight region boundary just nips the center area, regardless of the angle. Thus, each integrating cell is an *isotropic* edge detector cell. Additionally, if a small region contrasting with the background images within the center of the receptive field the integrating cell will also respond, making it a spot

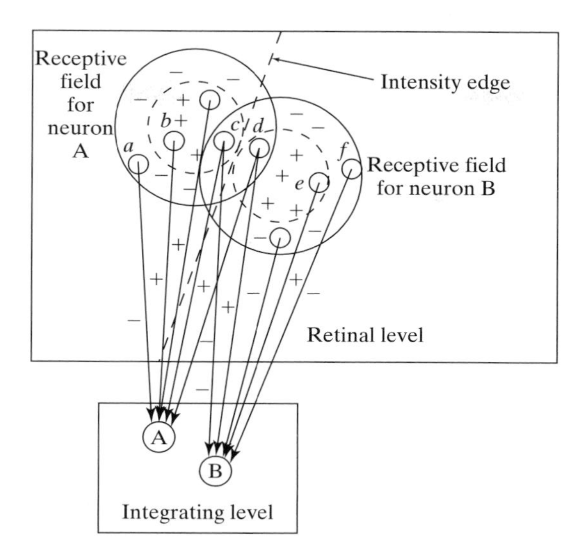

Figure 5.25 A 3D ANN architecture for LOG filtering.

detector as well. Figure 5.21 shows the result of convolving the smallest of the LOG masks with an image containing two regions. The result at the right of the figure shows how the operation determines the boundary between the regions via zero crossings. An 11×11 mask corresponding to the Laplacian of a Gaussian with $\sigma^2 = 2$ is shown in Figure 5.22. The tiny mask is capable of finding the boundary between tiny regions and is sensitive to high curvature bouundaries but will also respond to noise texture. The larger mask performs a lot of smoothing and will only respond to boundaries between larger regions with smoother perimeter.

Exercise 5.11

Give more detailed support to the above arguments that the integrating cells shown in Figure 5.25 (a) respond to contrasting spots or blobs that image within the center of the field, and (b) respond to boundaries between two large regions that just barely cross into the center of the field.

5.7.3 Marr-Hildreth Theory

David Marr and Ellen Hildreth proposed LOG filtering to explain much of the low level behavior of human vision. Marr proposed that the objective of low level human visual processing was the construction of a primal sketch which was a 2D description containing lines, edges, and blobs. (The primal sketches derived from the two eyes would then be processed further to derive 3D interpretations of the scene.) To derive a primal sketch, Marr and Hildreth proposed an organization based on LOG filtering with 4 or 5 different spreads σ. The mathematical properties outlined above explained the results of both perceptual experiments on humans and invasive experiments with animals. LOG filters with large σ would detect broad edges while those with small σ would focus on small detail. Coordination of the output from the different scales could be done at a higher level, perhaps with the large scale detections guiding those at the small scale. Subsequent work has produced different practical *scale space* methods of integrating output from detectors of different sizes.

Figure 5.26 shows an image processed at two different levels of Gaussian smoothing. The center image shows good representation of major objects and edges, whereas the rightmost image shows both more image detail and more noise. Note that the ship and part of the sand/water boundary is represented in the rightmost image but not in the center image. Marr's primal sketch also contained descriptions of virtual lines, which are formed by similar detected features organized along an image curve. These might be the images of a dotted line, a row of shrubs, etc. A synthetic image containing a virtual line is shown in Figure 5.27 along with the output of two different LOG filters. Both LOG filters give a

Figure 5.26 An input image (*a*) is smoothed using Gaussian filters of size (*b*) $\sigma = 4$ and (*c*) $\sigma = 1$ before performing edge detection. More detail and more noise is shown for the smaller filter. (Photo by David Shaffer 1998.)

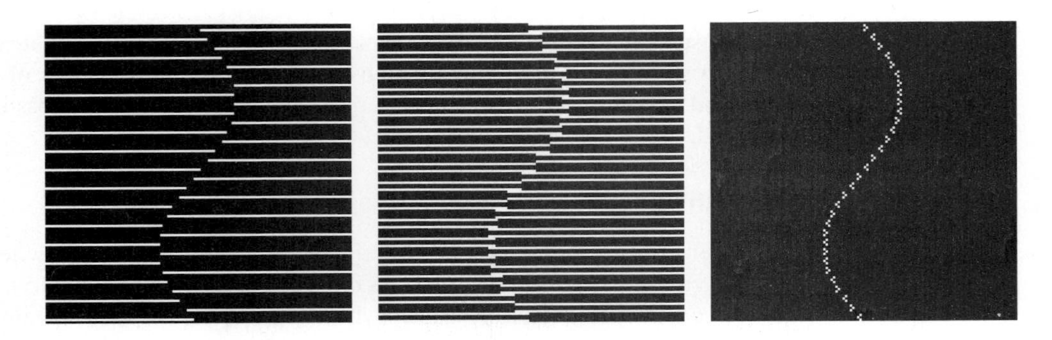

Figure 5.27 (left) A virtual line formed by the termination points of a line texture—perhaps this is two pieces of wrapping paper overlaid; (center) output from a specific 4×4 LOG filter responds to both lines and endpoints; and (right) a different 3×3 LOG filter responds only to the endpoints.

Figure 5.28 Picture obtained by thresholding: object boundaries are formed by virtual curves formed by the ends of stripes denoting cross sections of generalized cylinders. (Original photo by Eleanor Harding.)

response to the ends of the stripes; one is sensitive to the edges of the stripes as well, but the other is not. Figure 5.28 shows the same principle in a real image that was thresholded to obtain an artistic texture. Recent progress in researching the human visual system and brain has been rapid: results seem to complicate the interpretations of earlier work on which Marr and Hildreth based their mathematical theory. Nevertheless, use of variable-sized Gaussian and LOG filters is firmly established in computer vision.

5.8 THE CANNY EDGE DETECTOR

The Canny edge detector is a very popular and effective operator, and it is important to introduce it here, although details are placed in Chapter 10. The Canny operator first smoothes the intensity image and then produces extended contour segments by following high gradient magnitudes from one neighborhood to another. Figure 5.29 shows edge detection in fairly difficult outdoor images. The contours of the arch of St. Louis are detected quite well in Figure 5.29: using a parameter of $\sigma = 1$ detects some of the metal seams of the arch as well as some internal variations in the tree, whereas use of $\sigma = 4$ detects only the exterior boundaries of these objects. As shown in the bottom row of Figure 5.29, the operator isolates many of the texture elements of the checkered flags. For comparison, use of the Roberts operator with a low threshold on gradient magnitude is shown: more texture elements of the scene (grass and fence) are evident, although they have less structure than those in the Canny output. The algorithm for producing contour segments is treated in detail in Section 10.3.2.

Figure 5.29 (top left) Image of the great arch in St. Louis; (top center) results of Canny operator with $\sigma = 1$; (top right) results of Canny operator with $\sigma = 4$; (bottom left) image with textures; (bottom center) results of Canny operator with $\sigma = 1$; and (bottom right) results of Roberts operator thresholded to pass the top 20 percent of the pixels in gradient magnitude.

5.9 MASKS AS MATCHED FILTERS*

Here we give a theoretical basis for the concept that the response of a mask to a certain image neighborhood is proportional to how similar that neighborhood is to the mask. The important practical result of this is that we now know how to design masks to detect specific features—we just design a mask that is similar to the feature[s] we want to detect. This will serve us well in both edge and texture detection and also in detecting other special patterns such as holes or corners. We introduce the concepts using 1-dimensional signals, which are important in their own right, and which might correspond to rows or columns or any other cut through a 2D image. The concepts and mathematical theory immediately extend to the 2D case.

5.9.1 The Vector Space of all Signals of *n* Samples

For a given $n \geq 1$, the set of all vectors of n real coordinates forms a *vector space*. The practical and powerful vector space operations which we use are summarized next. The reader has probably already worked with vectors with $n = 2$ or $n = 3$ when studying analytical geometry or calculus. For $n = 2$ or $n = 3$, the notion of vector length is the same as that used in plane geometry and 3D analytical geometry since Euclid. In the domain of signals, length is related to energy of the signal, defined as the squared length of the signal, or equivalently, just the sum of the squares of all coordinates. Signal energy is an extremely useful concept as will be seen.

47 Definition. The **energy** of signal $S = [s_1, s_2, \ldots, s_n]$ is $\|S\|^2 = s_1^2 + s_2^2 + \cdots + s_n^2$.

Note that in many applications, the full range of real-valued signals do not arise because negative values are impossible for the coordinates. For example, a 12-dimensional vector recording the rainfall at a particular location for each of the 12 months should have no negative coordinates. Similarly, intensities along an image row are commonly kept in a non-negative integer range. Nevertheless, the vector space interpretation is still useful, as we shall see. Often, the mean signal value is subtracted from all coordinates in making an interpretation, and this may shift some of the coordinate values below zero. Moreover, it is quite common to have negative values in masks, which are templates or models of the shape of parts of signals.

Basic definitions for a vector space with a defined vector length

Let U and V be any two vectors; u_i and v_i be real numbers denoting the coordinates of these vectors; and let a, b, c, etc. be any real numbers denoting scalars.

48 Definition. For vectors $U = [u_1, u_2, \ldots, u_n]$ and $V = [v_1, v_2, \ldots, v_n]$ their **vector sum** is the vector $U \oplus V = [u_1 + v_1, u_2 + v_2, \ldots, u_n + v_n]$.

49 Definition. For vector $V = [v_1, v_2, \ldots, v_n]$ and real number (scalar) a the **product of the vector and scalar** is the vector $aV = [av_1, av_2, \ldots, av_n]$.

50 Definition. For vectors $U = [u_1, u_2, \ldots, u_n]$ and $V = [v_1, v_2, \ldots, v_n]$ their **dot product, or scalar product** is the real number $U \circ V = u_1v_1 + u_2v_2 + \cdots + u_nv_n$.

51 Definition. For vector $V = [v_1, v_2, \ldots, v_n]$ its **length, or norm**, is the non-negative real number $\|V\| = V \circ V = (v_1v_1 + v_2v_2 + \cdots + v_nv_n)^{1/2}$.

52 Definition. Vectors U and V are **orthogonal** if and only if $U \circ V = 0$.

53 Definition. The **distance between vectors** $U = [u_1, u_2, \ldots, u_n]$ and $V = [v_1, v_2, \ldots, v_n]$ is the length of their difference $d(U, V) = \|U - V\|$.

54 Definition. A **basis** for a vector space of dimension n is a set of n vectors $\{w_1, w_2, \ldots, w_n\}$ that are independent and that span the vector space. The spanning property means that any vector V can be expressed as a linear combination of basis vectors: $V = a_1w_1 \oplus a_2w_2 \oplus \cdots \oplus a_nw_n$. The independence property means that none of the basis vectors w_i can be represented as a linear combination of the others.

Properties of vector spaces that follow from the above definitions

1. $U \oplus V = V \oplus U$
2. $U \oplus (V \oplus W) = (U \oplus V) \oplus W$
3. There is a vector O such that for all vectors V, $O \oplus V = V$
4. For every vector V, there is a vector $(-1)V$ such that $V \oplus (-1)V = O$
5. For any scalars a, b and any vector V, $a(bV) = (ab)V$
6. For any scalars a, b and any vector V, $(a + b)V = aV \oplus bV$
7. For any scalar a and any vectors U, V, $a(U \oplus V) = aU \oplus aV$
8. For any vector V, $1V = V$
9. For any vector V, $(-1V) \circ V = -\|V\|^2$

Exercise 5.12

Choose any 5 of the 9 listed properties of vector spaces and show that each is true.

5.9.2 Using an Orthogonal Basis

Two of the most important results of the study of vector spaces are that (1) every vector can be expressed in only one way as a linear combination of the basis vectors, and (2) any set of basis vectors must have exactly n vectors. Having an *orthogonal basis* allows us a much stronger interpretation of the representation of any vector V as is shown in the example on page 161.

The example shows how to represent the signal [10, 15, 20] in terms of a basis of three given vectors $\{w_1, w_2, w_3\}$, which have special properties as we have seen. In general, let any signal $S = [a_1, a_2, a_3] = a_1 w_1 \oplus a_2 w_2 \oplus a_3 w_3$. Then $S \circ w_i = a_1(w_1 \circ w_i) \oplus a_2(w_2 \circ w_i) \oplus a_3(w_3 \circ w_i) = a_i(w_i \circ w_i) = a_i$, since $w_i \circ w_j$ is 0 when $i \neq j$ and 1 when $i = j$. So, it is very convenient to have an orthonormal basis; we can easily account for the energy in a signal separately in terms of its energy components associated with each basis vector. Suppose we repeat the above example using the signal $S_2 = [-5, 0, 5]$, which can be obtained by subtracting the mean signal value of S; $S_2 = S \oplus (-1[15, 15, 15])$. S_2 is the same as $S \circ w_1$ since the component along [1, 1, 1] has been removed. S_2 is just a scalar multiple of w_1: $S_2 = (10/\sqrt{2})w_1 = (10/\sqrt{2})((1/\sqrt{2}))[-1, 0, 1] = [-5, 0, 5]$ and we will say that S_2 has the same pattern as w_1. If w_1 is taken to be a filter, then it matches the signal S_2 very well; in some sense, it also matches the signal S very well. We will explore this idea further, but before moving on, we note that there are many different useful orthonormal bases for an n-dimensional space of signal vectors.

Example of representing a signal as a combination of basis signals.

Consider the vector space of all $n = 3$ sample signals $[v_1, v_2, v_3]$. Represented in terms of the *standard basis,* any vector $V = [v_1, v_2, v_3] = v_1[1, 0, 0] \oplus v_2[0, 1, 0] \oplus v_3[0, 0, 1]$. The standard basis vectors are orthogonal and have unit length; such a set is said to be *orthonormal*. We now study a different set of basis vectors $\{w_1, w_2, w_3\}$, where $w_1 = [-1, 0, 1]$, $w_2 = [1, 1, 1]$, and $w_3 = [-1, 2, -1]$. Any two of the basis vectors are orthogonal, since $w_i \circ w_j = 0$ for $i \neq j$ (show this). Scaling these to unit length, yields the new basis $\{\frac{1}{\sqrt{2}}[-1, 0, 1], \frac{1}{\sqrt{3}}[1, 1, 1], \frac{1}{\sqrt{6}}[-1, 2, -1]\}$.

Now represent the signal $S = [10, 15, 20]$ in terms of the orthogonal basis. $[10, 15, 20]$ is relative to the *standard basis*.

$$\begin{aligned}
S \circ w_1 &= \frac{1}{\sqrt{2}}(-10 + 0 + 20) \\
S \circ w_2 &= \frac{1}{\sqrt{3}}(10 + 15 + 20) \\
S \circ w_3 &= \frac{1}{\sqrt{6}}(-10 + 30 - 20) \\
S &= (S \circ w_1)w_1 \oplus (S \circ w_2)w_2 \oplus (S \circ w_3)w_3 \\
S &= (10/\sqrt{2})w_1 \oplus (45/\sqrt{3})w_2 \oplus 0w_3 \\
\|S\|^2 &= 100 + 225 + 400 = 725 \\
&= (10/\sqrt{2})^2 + (45/\sqrt{3})^2 + 0^2 = 725
\end{aligned} \tag{5.23}$$

The last two equations show that when an orthonormal basis is used, it is easy to account for the total energy by summing the energy associated with each basis vector.

Exercise 5.13

(*a*) Following the previous boxed example, represent the vector $[10, 14, 15]$ in terms of the basis $\{\frac{1}{\sqrt{2}}[-1, 0, 1], \frac{1}{\sqrt{3}}[1, 1, 1], \frac{1}{\sqrt{6}}[-1, 2, -1]\}$. (*b*) Now represent $[10, 19, 10]$: to which basis vector is it most similar? Why?

From the properties of vectors and the dot product, the *Cauchy-Schwartz Inequality* of Equation 5.24 is obtained. Its basic meaning is that the dot product of unit vectors must lie between -1 and 1. Thus we have a ready measure to determine the similarity between two vectors: note that if $U = V$, then $+1$ is obtained and if $U = -V$, then -1 is obtained. The *normalized dot product* is used to define the angle between two vectors. This angle is the same angle that one can compute in 2D or 3D space using trigonometry. For $n \geq 3$ the angle, or its cosine, is taken to be an abstract measure of similarity between two vectors; if their normalized dot product is 0, then they are dissimilar; if it is 1, then they are maximally similar scaled versions of each other; if it is -1, then one is a negative scaled version of the other, which may or may not be similar, depending on the problem domain.

5.9.3 Cauchy-Schwartz Inequality

$$\textit{For any two nonzero vectors } U \textit{ and } V, -1 \leq \frac{U \circ V}{\| U \| \| V \|} \leq +1 \tag{5.24}$$

55 Definition. Let U and V be any two nonzero vectors, then the **normalized dot product of** U and V is defined as $(\frac{U \circ V}{\|U\| \ \|V\|})$

56 Definition. Let U and V be any two nonzero vectors, then the **angle between** U and V is defined as $cos^{-1}(\frac{U \circ V}{\|U\| \ \|V\|})$

Exercise 5.14

Sketch the following five vectors and compute the normalized dot product, or *cos* of the angle between each pair of the vectors: $[5, 5]$, $[10, 10]$, $[-5, 5]$, $[-5, -5]$, $[-10, 10]$. Which pairs are perpendicular? Which pairs have the same direction? Which have opposite directions? Compare the relative directions to the value of the normalized dot product.

5.9.4 The Vector Space of $m \times n$ Images

The set of all $m \times n$ matrices with real-valued elements is a vector space of dimension $m \times n$. Here we interpret the vector space theory in terms of masks and image regions and show how it applies. In this section, our model of an image is that of an image function over a discrete domain of $m \times n$ sample points $I[x, y]$. We work mainly with 2×2 and 3×3 matrices, but everything easily generalizes to images or masks of any size.

5.9.5 A Roberts Basis for 2×2 Neighborhoods

The structure of a 2×2 neighborhood of an intensity image can be interpreted in terms of the basis shown in Figure 5.30, which we shall call the *Roberts basis*. Two of the four basis vectors were shown in Figure 5.15. As the exercise below shows, any 2×2 neighborhood of real intensity values can be expressed uniquely as a sum of these four basis vectors, each scaled as needed. The relative size of the scale factor directly indicates the amount of similarity between the image neighborhood and that basis vector and thus can be used to interpret the neighborhood structure. Several examples are given in Figure 5.30.

Exercise 5.15

Verify that the Roberts basis vectors shown in Figure 5.30 are orthonormal.

Roberts basis: $\mathbf{W}_1 = \frac{1}{2}\begin{array}{|c|c|}\hline 1 & 1 \\ \hline 1 & 1 \\ \hline\end{array}$ $\quad \mathbf{W}_2 = \frac{1}{\sqrt{2}}\begin{array}{|c|c|}\hline 0 & 1 \\ \hline -1 & 0 \\ \hline\end{array}$ $\quad \mathbf{W}_3 = \frac{1}{\sqrt{2}}\begin{array}{|c|c|}\hline 1 & 0 \\ \hline 0 & -1 \\ \hline\end{array}$ $\quad \mathbf{W}_4 = \frac{1}{2}\begin{array}{|c|c|}\hline -1 & 1 \\ \hline 1 & -1 \\ \hline\end{array}$

Constant region: $\begin{array}{|c|c|}\hline 5 & 5 \\ \hline 5 & 5 \\ \hline\end{array} = \frac{20}{2}\left(\frac{1}{2}\begin{array}{|c|c|}\hline 1 & 1 \\ \hline 1 & 1 \\ \hline\end{array}\right) = 10\mathbf{W}_1 \oplus 0\mathbf{W}_2 \oplus 0\mathbf{W}_3 \oplus 0\mathbf{W}_4$

Step Edge: $\begin{array}{|c|c|}\hline -1 & +1 \\ \hline -1 & +1 \\ \hline\end{array} = 0\mathbf{W}_1 \oplus \frac{2}{\sqrt{2}}\mathbf{W}_2 \oplus \frac{-2}{\sqrt{2}}\mathbf{W}_3 \oplus 0\mathbf{W}_4$

Step Edge: $\begin{array}{|c|c|}\hline +1 & +1 \\ \hline -3 & +1 \\ \hline\end{array} = 0\mathbf{W}_1 \oplus \frac{4}{\sqrt{2}}\mathbf{W}_2 \oplus 0\mathbf{W}_3 \oplus \frac{-4}{2}\mathbf{W}_4$

Line: $\begin{array}{|c|c|}\hline 0 & 8 \\ \hline 8 & 0 \\ \hline\end{array} = 8\mathbf{W}_1 \oplus 0\mathbf{W}_2 \oplus 0\mathbf{W}_3 \oplus 8\mathbf{W}_4$

Figure 5.30 (top row) A basis for all 2×2 images which contains the two Roberts gradient masks; (row 2) constant region is a multiple of the constant image; (row 3) vertical step edge has energy only along the gradient masks; (row 4) diagonal step edge has most energy along the matching gradient mask; and (row 5) line pattern has energy along constant mask $\mathbf{W_1}$ and line mask $\mathbf{W_4}$.

Exercise 5.16

Consider the vector space of all 2×2 images with real-valued pixels. (a) Determine the values of the a_j so that the image $\begin{array}{|c|c|}\hline 10 & 5 \\ \hline 5 & 0 \\ \hline\end{array}$ is represented as a linear combination of the four Roberts basis images $\mathbf{W}_j$. (b) Explain why we can always find unique a_j for any such 2×2 image.

Exercise 5.17

Suppose that the 2×2 image $\begin{array}{|c|c|}\hline a & b \\ \hline c & d \\ \hline\end{array}$ has energy e_1, e_2, e_3, e_4 along the respective four Roberts basis vectors $\mathbf{W_1}, \mathbf{W_2}, \mathbf{W_3}, \mathbf{W_4}$ respectively. What are the formulas for computing the four e_1 in terms of a, b, c, d?

5.9.6 The Frei-Chen Basis for 3 × 3 Neighborhoods

Usually, masks used for image processing are 3×3 or larger. A *standard basis* for 3×3 image neighborhoods is given in Figure 5.31. One advantage of the standard basis is that it is obvious how to expand any image neighborhood using this basis. However, the obvious expansion tells nothing about the 2D structure of the neighborhood. The Frei-Chen basis, shown in Figure 5.32, consists of a set of orthonormal masks which enable simple interpretation of the structure of the 3×3 neighborhood.

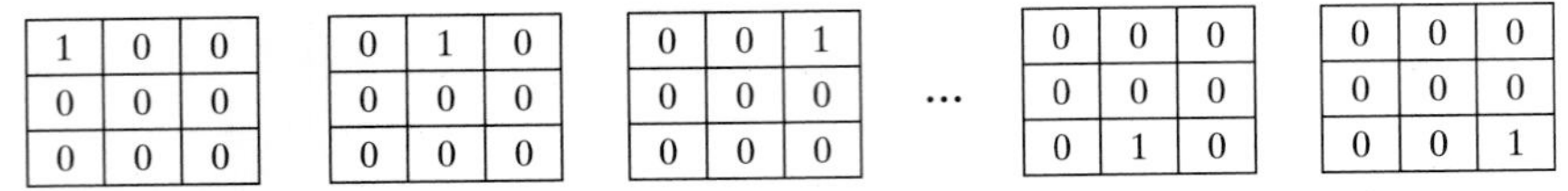

Nine *standard* basis vectors for the space of all 3 × 3 matrices.

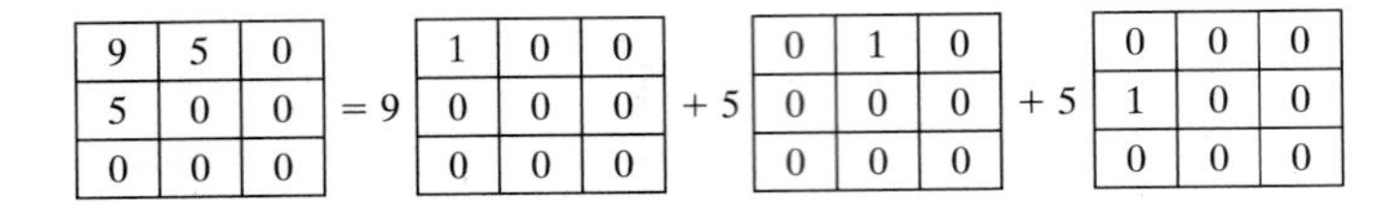

Figure 5.31 Any 3 × 3 matrix can be represented as a sum of no more than 9 scaled standard matrices: (top row) nine basis vectors and (bottom row) representation of an example matrix in terms of the basis.

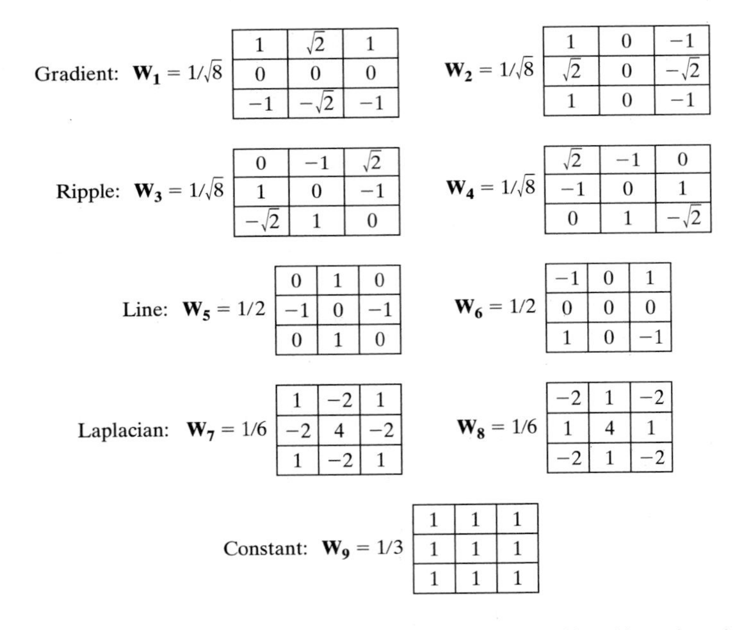

Figure 5.32 The Frei-Chen basis for the set of all 3 × 3 images with real intensity values.

Representation of an image neighborhood in terms of the Frei-Chen basis allows intepretation of the energy as gradient, ripple, and line, etc. Energy will be high when the intensity structure is similar to a basis vector, or mask. Each of the basis vectors has a specially designed structure. The basis vectors $\mathbf{W_1}$ and $\mathbf{W_2}$ are similar to the Prewitt and Sobel gradient masks, while the basis vectors $\mathbf{W_7}$ and $\mathbf{W_8}$ are similar to the common 3 × 3 Laplacian mask. The line masks will respond strongly to one-pixel wide lines passing through a 3 × 3 neighborhood, while the two ripple masks model two perpendicular waves with two peaks, two troughs, and three zero-crossings. The elements of the vectors differ slightly from the masks that we formerly designed in isolation because of the requirement that the set be orthogonal.

Algorithm 5.2 computes a binary image which detects intensity neighborhood structure with significant energy in a specified subspace. To detect edges, one could select pixels by their neighborhood energy along the basis vectors $\mathbf{W_1}$, $\mathbf{W_2}$, which would be indicated by setting $\mathbf{S} = \{1, 1, 0, 0, 0, 0, 0, 0, 0\}$. An example of the calculations projecting the intensity neighborhood on the Frei-Chen basis vectors is given also.

Detect neighborhoods with high energy in a selected subspace

F[r, c] is an input intensity image; **F** is unchanged by the algorithm.
S is a bit vector such that **S[j]** = **1** if and only if $\mathbf{W_j}$ is included in the subspace of interest.
thresh is a threshold on the fraction of energy required.
noise is the noise energy level.
G[r, c] is the output image, a binary image where **G[r, c]** = 1 indicates that **F[r, c]** has above threshold energy in the specified subspace **S**.

```
procedure detect_neighborhoods(F, G, S, thresh, noise);
{
        for r := 0 to MaxRow - 1
        for c := 0 to MaxCol - 1
          {
            if [r,c] is a border pixel then G[r, c] := 0;
            else G[r,c] := compute_using_basis (F, r, c, S, thresh, noise);
          };
}
procedure compute_using_basis(IN, r, c, thresh, noise)
{
        N[r, c] is the 3 × 3 neighborhood centered at pixel [r,c] of IN[i].
        average_energy := N[r, c] ∘ W9;
        subspace_energy := 0.0;
        for j := 1 to 8
        {
          if (S[ j]) subspace_energy := subspace_energy + (N[r, c] ∘ Wj)^2;
        }
        if subspace_energy < noise return 0;
        if subspace_energy/((N[r, c] ∘ N[r, c]) − average_energy) < thresh return 0;
        else return 1;
}
```

Algorithm 5.2 Compute detection image **G[r, c]** from input image **F[r, c]** and subspace **S**.

Example of representing an intensity neighborhood using the Frei-Chen basis

Consider the intensity neighborhood $\mathbf{N} =$

10	10	10
10	10	5
10	5	5

We find the component of this vector along each basis vector using the dot product as before. Since the basis is orthonormal, the total image energy is just the sum of the component energies and the structure of **N** can be interpreted in terms of the components.

$$N \circ \mathbf{W_1} = \frac{5+5\sqrt{2}}{\sqrt{8}} \approx 4.3;\ energy \approx 18$$

$$N \circ \mathbf{W_2} = \frac{5+5\sqrt{2}}{\sqrt{8}} \approx 4.3;\ energy \approx 18$$

$$N \circ \mathbf{W_3} = 0;\ energy = 0$$

$$N \circ \mathbf{W_4} = \frac{5\sqrt{2}-10}{\sqrt{8}} \approx -1;\ energy \approx 1$$

$$N \circ \mathbf{W_5} = 0;\ energy = 0$$

$$N \circ \mathbf{W_6} = 2.5;\ energy \approx 6$$

$$N \circ \mathbf{W_7} = 2.5;\ energy \approx 6$$

$$N \circ \mathbf{W_8} = 0;\ energy = 0$$

$$N \circ \mathbf{W_9} = 25;\ energy = 625$$

The total energy in **N** is $\mathbf{N} \circ \mathbf{N} = 675$, 625 of which is explained just by the average intensity level along $\mathbf{W_9}$. The energy along all other components is 50 of which 36, or 72 percent, is along the gradient basis vectors $\mathbf{W_1}$ and $\mathbf{W_2}$. Thus, the neighborhood center would be marked as a detected feature in case the gradient subspace is of interest.

Exercise 5.18

Verify that the set of nine vectors shown in Figure 5.32 is an orthonormal set.

Exercise 5.19

(*a*) Represent the intensity neighborhood

0	0	1
0	1	0
1	0	0

in terms of the basis vectors shown in Figure 5.32. Is all the energy distributed along the *line* basis vectors $\mathbf{W_5}$ and $\mathbf{W_6}$? (*b*) Repeat the question (*a*) for the intensity neighborhood

0	1	0
0	1	0
0	1	0

.

Exercise 5.20

(*a*) Represent the intensity neighborhood

10	10	10
10	20	10
10	10	10

in terms of the basis vectors shown in Figure 5.32. Is all the energy distributed along certain basis vectors as you expect? (*b*) Would the interpretation of the intensity neighborhood

0	0	0
0	1	0
0	0	0

be different: why or why not? (*c*) What kind of image neighborhoods give responses to only $\mathbf{W_7}$ and $\mathbf{W_8}$?

Exercise 5.21

Write a program to implement the detection of pixels using the Frei-Chen basis as outlined in the algorithm above. Allow the user of the program to input the subspace **S** of interest as a string of 9 bits. The user should also be able to input the noise energy level and the threshold determining the minimum energy required in the selected subspace. Test your program on some real images and also on some test patterns such as those in the exercises above.

5.10 CONVOLUTION AND CROSS CORRELATION*

The previous sections showed how many useful detections can be performed by matching a mask or pattern image to an image neighborhood. Also, we saw that image smoothing could be put in the same framework. In this section, we give definitions for the important operations of *cross correlation* and *convolution,* which formalize the notion of moving a mask around the image and recording the dot product of the mask with each image neighborhood.

5.10.1 Defining Operations via Masks

We begin by redefining simple image smoothing as the cross correlation of the image with a smoothing mask. A boxcar filter computes the ouptut image pixel as an equal weighting of a neighborhood of pixels of the corresponding input image pixel. This is equivalent to performing a dot product of an $m \times n$ image pattern of weights $\frac{1}{mn}$ as is shown in Figure 5.33 for a 3×3 mask. Assuming that m and n are both odd and the division by 2 ignores the remainder, Equation 5.25 defines the dot product used to compute the value of the output pixel $\mathbf{G[x, y]}$ from input image $\mathbf{F[x, y]}$ using mask $\mathbf{H[x, y]}$. In this formulation, mask $\mathbf{H}$ is centered at the origin so that $\mathbf{H[0, 0]}$ is the center pixel of the mask: it is obvious how $\mathbf{H}$ is used to weight the pixels from the neighborhood of $\mathbf{F[x, y]}$. An alternate formulation convenient to compute all output pixels of $\mathbf{G}$ results from a simple change of variables in Equation 5.25 yielding Equation 5.26 which can use masks $\mathbf{H[i, j]}$ with even dimensions.

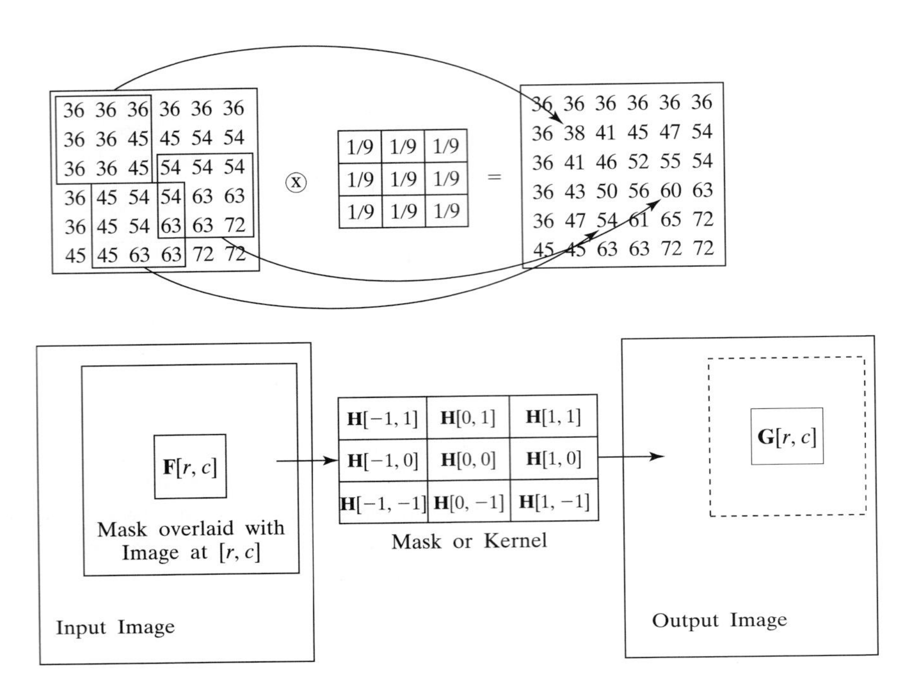

Figure 5.33 Smoothing an image using a 3 × 3 boxcar filter can be viewed as computing a dot product between every input image neighborhood and a mask (boxcar) that is a small image of equal values.

Exercise 5.22

Suppose that an image **F** has all 0 pixels except for a single 1 pixel at the image center. What output image **G** results from convolving **F** with the 3 × 3 boxcar shown in Figure 5.33?

Exercise 5.23

Design a single mask to detect edge elements making an angle of 30 degrees with the X-axis. The mask should not respond strongly to edge elements of other directions or to other patterns.

Exercise 5.24: Corner detection.

(a) Design a set of four 5 × 5 masks to detect the corners of any rectangle aligned with the image axes. The rectangle can be brighter or darker than the background. (b) Are your masks orthogonal? (c) Specify a decision procedure to detect a corner and justify why it would work.

57 Definition. The **cross correlation** of image $F[x, y]$ and mask $H[x, y]$ is defined as

$$G[x, y] = F[x, y] \otimes H[x, y]$$
$$= \sum_{i=-w/2}^{w/2} \sum_{j=-h/2}^{h/2} F[x + i, y + j]H[i, j] \quad (5.25)$$

See Figure 5.35.
For implementation of this computational formula:
mask $H[x, y]$ is assumed to be centered over the origin, so negative coordinates make sense;
image $F[x, y]$ need not be centered at the origin;
result $G[x, y]$ must be defined in an alternate manner when $H[i]$ does not completely overlap $F[i]$.

An alternate formulation does not require a mask with odd dimensions, but should be viewed as an entire image transformation and not just an operation centered on pixel $G[x, y]$.

$$G[x, y] = \sum_{i=0}^{w-1} \sum_{j=0}^{h-1} F[x + i, y + j]H[i, j] \quad (5.26)$$

5.10.2 The Convolution Operation

58 Definition. The **convolution** of functions $f(x, y)$ and $h(x, y)$ is defined as

$$g(x, y) = f(x, y) \star h(x, y)$$
$$\equiv \int_{x'=-\infty}^{+\infty} \int_{y'=-\infty}^{+\infty} f(x', y')h(x - x', y - y')\, dx'\, dy' \quad (5.27)$$

Convolution is closely related to cross correlation and is defined formally in terms of continuous picture functions in Equation 5.27. In order for the integrals to be defined and to have practical use, the 2D image functions $f(x, y)$ and $h(x, y)$ should have zero values outside of some finite rectangle in the *xy-plane* and have finite volume under their surface. For filtering, *kernel* function $h(x, y)$ will often be zero outside some rectangle that is much smaller than the rectangle that supports $f(x, y)$. For a global analysis of the spatial frequency of f, the rectangle supporting h will include all of the support of f. More details are given in the optional Section 5.11 on Fourier analysis. Figure 5.34 shows an interpretation of steps in computing the value of $g(x)$ from the two functions being convolved in the case of 1D signals. Kernel function $h(x)$ is first flipped about the origin and then translated to the point x at which $g(x)$ is being computed. $g(x)$ is then computed

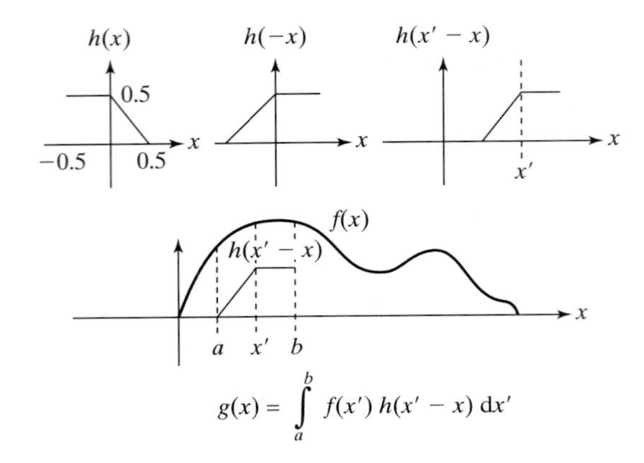

Figure 5.34 Computing convolution $g(x)$ of signal $f(x)$ with kernel $h(x)$ denoted $g(x) = f(x) \star h(x)$. For any domain point x, kernel h is flipped and then translated to x; then, $g(x)$ is computed by summing all the products of $f(x)$ and the values of the flipped and translated h.

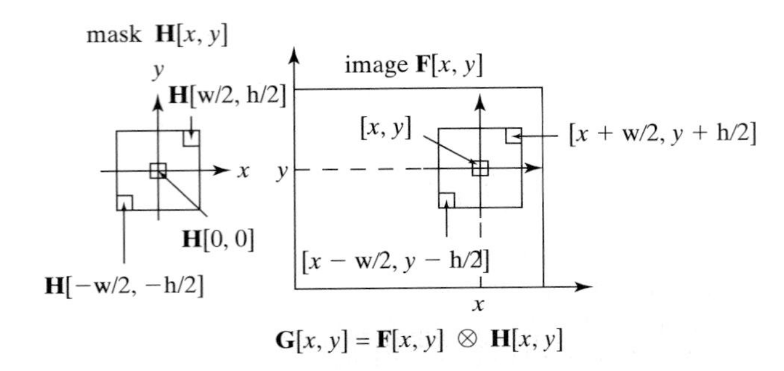

Figure 5.35 Computing the cross correlation $\mathbf{G}[x, y]$ of image $\mathbf{F}[x, y]$ with mask $\mathbf{H}[x, y]$: $\mathbf{G}[x, y] = \mathbf{F}[x, y] \otimes \mathbf{H}[x, y]$. To compute $\mathbf{G}[x, y]$, the mask $\mathbf{H}[x, y]$ is centered on input image point $\mathbf{F}[x, y]$ to sum all the products of image values of F and the corresponding overlaid weights of $\mathbf{H}$.

by integrating the products of the input function $f(x')$ times the relocated kernel $h(x' - x)$; since, that function is zero outside of the interval $[a, b]$, integration can be confined to this finite interval. Convolution is easily carried out with digital images by using discrete sums of products instead of the continuous integrals defined above.

Cross correlation translates the mask or kernel directly to the image point $[x, y]$ without flipping it as shown in Figure 5.35; otherwise it is the same operation used in convolution. Conceptually, it is easier not to have to think of flipping the kernel, but rather just placing it at some image location. If the kernel is symmetric so that the flipped kernel is the same as the original kernel, then the results of convolution and correlation will be identical. While this is the case for smoothing masks and other isotropic operators, many edge detection masks are asymmetric. Despite the formal difference between convolution and cross correlation, those who do image processing often loosely refer to either operation as convolution because of their similarity. The many masks used in this chapter, and others, have been presented assuming that they would not be flipped before being applied to an image. *Normalized cross correlation* will normalize by dividing $\mathbf{G}[x, y]$ by the magnitudes of both $\mathbf{F}[x, y]$ and $\mathbf{H}[x, y]$ so that the result can be interpreted as a result of matching the structure of $\mathbf{F}$ with the structure of $\mathbf{H}$ that is independent of scale factors, as has been discussed in previous sections.

5	5	5	5	5	5
5	5	5	5	5	5
5	5	10	10	10	10
5	5	10	10	10	10
5	5	5	10	10	10
5	5	5	5	10	10

$\otimes$

0	−1	0
−1	4	−1
0	−1	0

=

0	0	0	0	0	0
0	0	−5	−5	−5	0
0	−5	10	5	5	0
0	−5	10	0	0	0
0	0	−10	10	0	0
0	0	0	0	0	0

Cross correlation with a LOG mask produces zero-crossings at boundaries.

5	5	5	5	5	5
5	5	5	5	5	5
5	5	10	10	10	10
5	5	10	10	10	10
5	5	5	10	10	10
5	5	5	5	10	10

$\otimes$

−1	0	+1
−1	0	+1
−1	0	+1

=

0	0	0	0	0	0
0	5	5	0	0	0
0	10	10	0	0	0
0	10	15	5	0	0
0	5	10	10	5	0
0	0	0	0	0	0

Cross correlation with column derivative mask detects column boundaries.

5	5	5	5	5	5
5	5	5	5	5	5
5	5	10	10	10	10
5	5	10	10	10	10
5	5	5	10	10	10
5	5	5	5	10	10

$\otimes$

+1	+1	+1
0	0	0
−1	−1	−1

=

0	0	0	0	0	0
0	−5	−10	−15	−15	0
0	−5	−10	−15	−15	0
0	5	5	5	0	0
0	5	10	10	5	0
0	0	0	0	0	0

Cross correlation with row derivative mask detects column boundaries.

Figure 5.36 Cross correlation of an image with various masks to enhance region boundaries: (top) second derivative operator produces a zero-crossing at boundaries; (center) column (x) derivative operator detects changes across columns; and (right) row (y) derivative operator detects changes across rows.

Exercise 5.25: Rectangle detection.

Use the corner detection procedure of Exercise 5.24 to implement a program that detects rectangles in an image. (Rectangle sides are assumed to align with the sides of the image.) The first step should detect candidate rectangle corners. A second step should extract subsets of four candidate corners that form a proper rectangle according to geometric constraints. An optional third step might further check the four corners to make sure that the intensity within the candidate rectangle is uniform and contrasting with the backgorund. What is the expected behavior of your program if it is given a noisy checkerboard image? Test your program on a noisy checkerboard, such as in Figure 5.7 and an image of a building with rectangular windows, such as in Figure 5.42.

Exercise 5.26

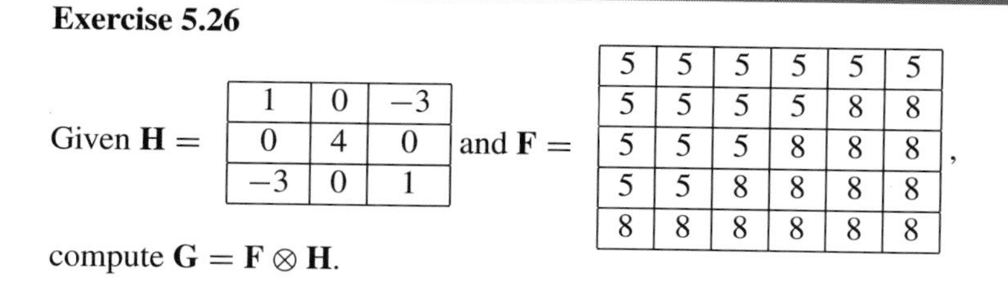

Given **H** =

1	0	−3
0	4	0
−3	0	1

and **F** =

5	5	5	5	5	5
5	5	5	5	8	8
5	5	5	8	8	8
5	5	8	8	8	8
8	8	8	8	8	8

,

compute $\mathbf{G} = \mathbf{F} \otimes \mathbf{H}$.

Exercise 5.27: Point spread.

Given kernel $\mathbf{H} = \begin{array}{|c|c|c|}\hline 1 & 2 & 1 \\ \hline 2 & 5 & 2 \\ \hline 1 & 2 & 1 \\ \hline\end{array}$, what is the result of convolving it with an image $\mathbf{F}[x, y]$ which has $\mathbf{F}[x_0, y_0] = 1$ and all other pixels 0?

Exercise 5.28

Suppose that function $h(x)$ takes value 1 for $-1/2 \leq x \leq 1/2$ and is zero elsewhere, and suppose that function $f(x)$ takes value 1 for $10 \leq x \leq 20$ and zero elsewhere. (a) Sketch the two functions f and h. (b) Compute and sketch the function $g(x) = f(x) \star h(x)$. (c) Compute and sketch the function $g(x) = h(x) \star h(x)$.

5.10.3 Possible Parallel Implementations

One can see from the definition of convolution, that $g(x_1, y_1)$ can be computed independently of $g(x_2, y_2)$; in fact, all the integrations can be done at the same time in parallel. Also, each integration computing an individual value $g(x, y)$ can form all of the products simultaneously, enabling a highly parallel system. Various computer architectures can be envisioned to perform all or some operations in parallel.

5.11 ANALYSIS OF SPATIAL FREQUENCY USING SINUSOIDS*

Fourier analysis is very important in signal processing; its theory and practice fills many books. We give only a brief snapshot here, building on the notions of a vector space already introduced.

The mathematician Fourier imagined the surface of the sea as a sum of sine waves. Large waves caused by the tide or ships had long wavelengths (low frequency), while smaller waves caused by the wind or dropped objects, etc., had short wavelengths (high frequency). The top row of Figure 5.37 shows three pure waves of 3, 16, and 30 cycles across a 1D space of $x \in [0, 512]$: in the bottom row are two functions, one which sums all three waves and one which sums only the first two. Similar sets of waves can be used to create 2D picture functions or even 3D density functions.

The Fourier theory in analysis shows how most real surfaces or real functions can be represented in terms of a basis of sinusoids. The energy along the basis vectors can be interpreted in terms of the structure of the represented surface (function). This is most useful when the surface has repetitive patterns across large regions of the surface; such as do city blocks of an aerial image of an urban area, waves on a large body of water, or the texture of a large forest or farm field. The idea is to expand the entire image—or various windows of it—using a Fourier basis and then filter the image or make decisions about it based on the image energy along various basis vectors. For example, high frequency noise can be removed by subtracting from the image all the components along high frequency sine/cosine waves. Equivalently, we can reconstruct our spatial image by adding up only the low frequency waves and ignoring the high frequency waves.

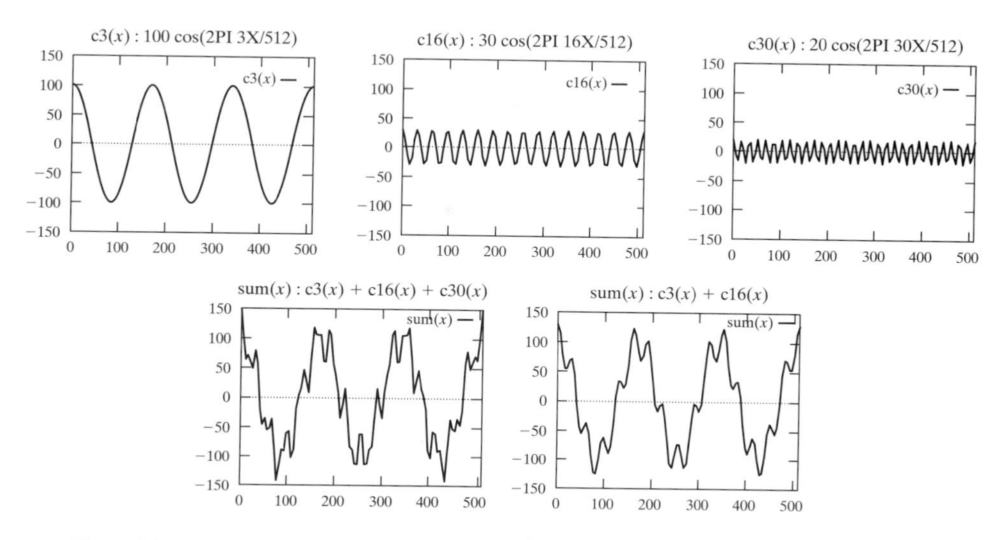

Figure 5.37 (top row) Three sinusoids, $100cos(2\pi \frac{3x}{512})$, $30cos(2\pi \frac{16x}{512})$ and $20cos(2\pi \frac{30x}{512})$; (bottom row left) sum of all three; and (bottom row right) sum of first two.

Exercise 5.29

The *pointilist* school of painters created their paintings by dabbing their paintbrush perpendicularly on the canvas and creating one dab/point of color at a time. Each dab is similar to a single pixel in a digital image. The human viewer of the artwork would stand back and a smooth picture would be perceived. Implement a program to do pointilist art. The program should present a palette of colors and some other options, such as choosing paintbrush size or whether 'OR' or 'XOR' would be used for dabbing, etc. When your program is working, create a painting of a starry night. Your program should work with an external file representing the painting in progress so that a user could save the session for later continuation.

Exercise 5.30

Implement a program to convolve a mask with an image. The program should read both the image and mask from input files in the same format. Perhaps you can test it on the artwork produced by the program from the previous exercise.

Exercise 5.31

Suppose in the search for extra terrestrial intelligence (SETI) deep space is scanned with the hope that interesting signals would be found. Suppose signal **S** is a concatenation of the first 100 prime numbers in binary form and that **R** is some received signal that is much longer than **S**. Assume that **R** has noise and has real valued coordinates. To find if **S** is embedded in **R**, would cross correlation or normalized cross correlation work? Why?

5.11.1 A Fourier Basis

For an intuitive introduction, let us assume an orthonormal set of sinusoidal basis images (or picture functions) $E_k \approx E_{u,v}(x, y)$. Presently, k and u, v are integers which define a finite set of basis vectors. It will soon become clear how parameters u, v determine the basis vectors, but right now, we'll use the single index k to keep our focus on the fundamental concept. The bottom row of Figure 5.37 shows two signals formed by summing three or two of the pure cosine waves shown in the top row of the figure. More complex functions can be created by using more pure cosine waves. Figure 5.39 shows the result of summing scaled versions of the three "pure" waves in Figure 5.38 to create a new picture function. By using the Fourier basis functions E_k, any picture function can be represented as $I[x, y] = \sum_{k=0}^{N-1} a_k E_k[x, y]$. As in previous sections, a_k is a measure of the similarity between $I[x, y]$ and $E_k[x, y]$ and the energy of image $I[x, y]$ along that particular component wave. Useful image processing operations can be done by operating on the values of a_k rather than on the individual intensities $I[x, y]$. Three major operations are described next.

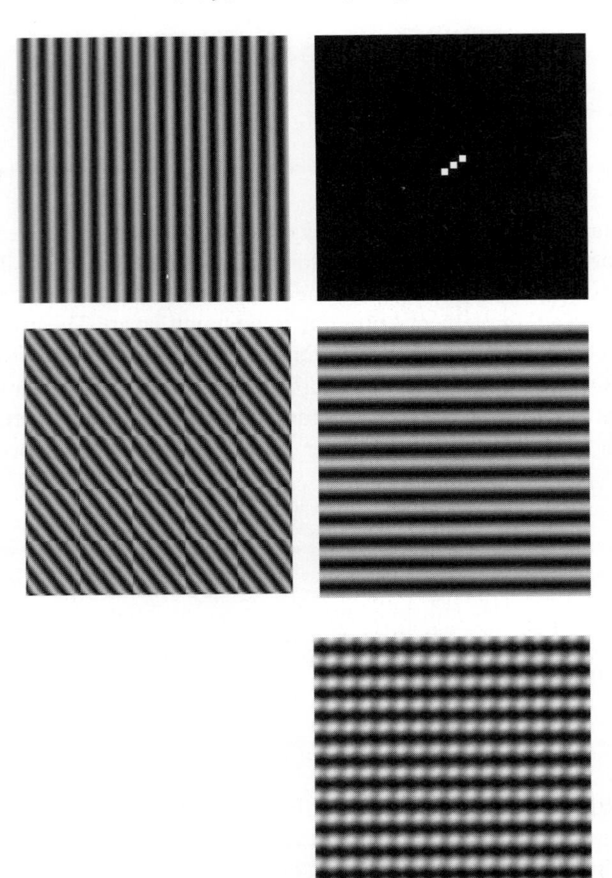

Figure 5.38 Different sinusoidal picture functions in the spatial domain $[x, y]$: The top left image was generated by the formula $100cos(2\pi(16x/512)) + 100$ and has 16 cycles along the x-axis. The bottom right image was generated by the formula $100cos(2\pi(12y/512)) + 100$ and has 12 cycles along the y-axis. The bottom left image was generated by the formula $100cos(2\pi(16x/512 + 12y/512)) + 100$: note how the waves of the bottom left image align with those of the top left and bottom right. The Fourier power spectrum is shown at the top right.

Figure 5.39 A picture function formed by the sum $I[x, y] = 100E_i + 30E_j + 10E_k$ where E_i is as in the bottom right of Figure 5.38, E_j is as in the top left, and E_k is as in the bottom left. (Perhaps it models an orchard or foam padding?)

Important Image Processing Operations using the Fourier Basis

1. The Fourier basis can be used to **remove high frequency noise** from the image or signal. The signal f is represented as $\sum_k a_k E_k$. The values of a_k for the high frequency sinusoids E_k are set to zero and a new signal $\hat{f}$ is computed by summing the remaining basis functions with $a_k \neq 0$.
2. The Fourier basis can be used to **extract texture features** that can be used to classify the type of object in an image region. After representing the image, or image region, in the Fourier basis, various a_k can be used to compute features to be used for the classification decision. Regions of waves on water or crops organized by rows are amenable to such a process. The a_k are useful for determining both the frequency and direction of textured regions.
3. The Fourier basis can also be used for **image compression.** A sender can send a subset of the a_k and a receiver can reconstruct the approximate picture by summing the known sinusoidal components. All of the a_k can be transmitted, if needed, in either order of energy, or in order of frequency: the receiver can terminate the transmission at any time based on the content received up to that point.

Our goal here is to produce a useful basis of picture functions and an understanding of how to use it in practice. Some important mathematical background using continuous functions is needed. For this development, lets assume that the origin of our coordinate system is at the center of our picture function, which is defined for a square region of the xy-plane. When we proceed to a digital version, the picture will be represented by digital image $I[x, y]$ with N^2 samples. First, we establish a set of sinusoids of different frequency as an orthogonal basis for continuous signals f. If m, n are any two different integers, then the two cosine waves with these frequency parameters are orthogonal over the interval $[-\pi, \pi]$. The reader should work through the following exercises to verify that the function set $\{\mathbf{1}, sin(mx), cos(nx), \ldots\}$ is an orthogonal set of functions on $[-\pi, \pi]$. The orthogonality of the cosine waves follows from Equation 5.28, since $sin(k\pi) = 0$ for all integers k.

$$\begin{aligned} \int_{-\pi}^{\pi} cos(m\theta)cos(n\theta)\, d\theta &= \frac{sin(m-n)\pi}{2(m-n)} + \frac{sin(m+n)(-\pi)}{2(m+n)} \\ &= 0 \ for \ m^2 \neq n^2 \end{aligned} \quad (5.28)$$

N symmetrically spaced samples of these cosine functions will create a set of vectors that will be orthogonal in the sense defined previously by the dot product.

5.11.2 2D Picture Functions

59 Definition. The complex valued picture function

$$\begin{aligned} \mathbf{E}_{u,v}(\mathbf{x}, \mathbf{y}) &\equiv e^{-j\ 2\pi(ux+vy)} \\ &= cos(2\pi(ux+vy)) - j sin(2\pi(ux+vy)) \end{aligned} \quad (5.29)$$

where u and v are spatial frequency parameters as shown in Figure 5.38 and $j = \sqrt{-1}$.

Exercise 5.32

Consider the set of all continuous functions f defined on the interval $x \in [x_1, x_2]$. Show that this set of functions $f, g, h, \ldots$, together with scalars $a, b, c, \ldots$ forms a vector space by arguing that the following properties hold.

$$f \oplus g = g \oplus f \qquad (f \oplus g) \oplus h = f \oplus (g \oplus h)$$
$$c(f \oplus g) = cf \oplus cg \qquad (a+b)f = af \oplus bf$$
$$(ab)f = a(bf) \qquad 1f = f$$
$$0f = \mathbf{0}$$

Exercise 5.33

For the space of continuous functions of $x \in [x_1, x_2]$, as in the exercise above, define a dot product and corresponding norm as follows.

$$f \circ g = \int_a^b f(x)g(x)\,dx; \quad \|f\| = \sqrt{f \circ f} \tag{5.30}$$

Argue why the following four properties of a dot product hold.

$$(f \oplus g) \circ h = (f \circ g) + (g \circ h)$$
$$f \circ f \geq 0$$
$$f \circ f = 0 \Longleftrightarrow f = \mathbf{0}$$
$$f \circ g = g \circ f$$
$$(cf) \circ g = c(f \circ g)$$

Exercise 5.34: Odd and even functions.

A function is an *even function* if $f(-x) = f(x)$ and a function is an *odd function* if $f(-x) = -f(x)$. (*a*) Show that $cos(mx)$ is an even function and that $sin(nx)$ is an odd function, where m, n are nonzero integers. (*b*) Let f and g be odd and even continuous functions on $[-L, L]$, respectively. Argue that $\int_{-L}^{L} f(x)g(x)dx = 0$.

Exercise 5.35

Using the definition of the dot product given in Exercise 5.33, show that the set of sinusoidal functions f_k defined below are orthogonal on the interval $[-\pi, \pi]$.
$f_0(x) = 1$; $f_1(x) = sin(x)$; $f_2(x) = cos(x)$; $f_3(x) = sin(2x)$; $f_4(x) = cos(2x)$; $f_5(x) = sin(3x)$; $f_6(x) = cos(3x)$; ...

Using a complex value is a convenience that allows us to separately account for both a cosine and sine wave of the same frequency: the sine wave has the same structure as the cosine wave but is 1/4 wavelength out of phase with the cosine wave. When one of these basis functions correlates highly with a picture function, it means that the picture function has high energy in frequency u, v. The *Fourier transform,* converts a picture function into an array of these correlations. We start with the integral form and then give the discrete summation for digital images below.

60 Definition. The 2D **Fourier Transform** transforms a spatial function $f(x, y)$ into the u, v frequency domain.

$$F(u, v) \equiv \int_{-\infty}^{\infty} \int_{-\infty}^{\infty} f(x, y) E_{u,v}(x, y)\, dx\, dy$$

$$= \int_{-\infty}^{\infty} \int_{-\infty}^{\infty} f(x, y) e^{-j\, 2\pi(ux+vy)}\, dx\, dy \tag{5.31}$$

The function f must be well-behaved; specifically, picture functions have the following properties which make the above formula well-defined. $\int_{-\infty}^{\infty} \int_{-\infty}^{\infty} | f(x, y) |\, dx\, dy$ is finite; moreover, $f(x, y) = 0$ outside of some rectangle R so the infinite limits in the formula can be replaced by the limits of R. Also, f has a finite number of extrema and no infinite discontinuities in R.

Exercise 5.36

What is the special meaning of $\mathbf{F}(0, 0)$, where $\mathbf{F}(u, v)$ is the Fourier transform of picture function $f(x, y)$?

Often, we want to work with the *power spectrum,* which combines energy from sine and cosine waves of the same frequency components u, v. A power spectrum is shown at the top right of Figure 5.38.

61 Definition. The **Fourier power spectrum** is computed as

$$P(u, v) \equiv (Real(F(u, v))^2 + Imaginary(F(u, v))^2\,)^{1/2} \tag{5.32}$$

Figure 5.40 shows how the actual wavelength of a 2D sinusoid relates to the wavelengths projected along each axis. u is the frequency along the X-axis in cycles per unit

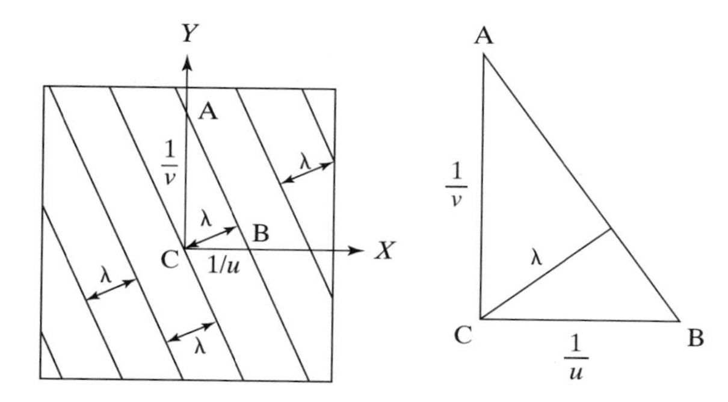

Figure 5.40 Relationships among the wavelengths $1/u_0$ and $1/v_0$ of a sinusoid intercepted by the X and Y axes and the wavelength λ of the 2D wave.

length and $1/u$ is the wavelength. v is the frequency along the Y-axis and $1/v$ is the wavelength. λ is the wavelength of the sinusoid along its natural axis, or direction of travel. By computing the area of the triangle shown at the right in Figure 5.40 in two different ways, we obtain the following formula which will help us interpret what the power spectrum tells us about the frequency content of the original image.

$$\begin{aligned} \lambda\sqrt{(1/u)^2 + (1/v)^2} &= (1/u)(1/v) \\ \lambda &= \frac{1}{\sqrt{u^2 + v^2}} \end{aligned} \tag{5.33}$$

Let us assume that the both the width and height of our picture have length 1. In Figure 5.38 there are $u = 16$ cycles across the width of the picture, so the wavelength is $1/u = 1/16$. Similarly, we have $1/v = 1/12$. Applying the formula in Equation 5.33, we obtain $\lambda = 1/20$. By counting the number of waves in Figure 5.38 at the bottom left, we see 27 waves across the 1.4 unit diagonal, yielding the expected frequency $27/1.4 \approx 20$ along the actual direction of the 2D wave.

Figure 5.41 shows the principal response in the power spectrum of the three sinusoidal picture functions originally shown in Figure 5.38. The power spectrum at the top right of Figure 5.38 actually shows heavy responses at three points, not just one. First of all, note that $\mathbf{F}(0, 0)$ is just the total energy in $f(x, y)$; since each sinusoid of Figure 5.38 has mean of 100 and not 0, they have significant average energy at *zero frequency*. Secondly, it is

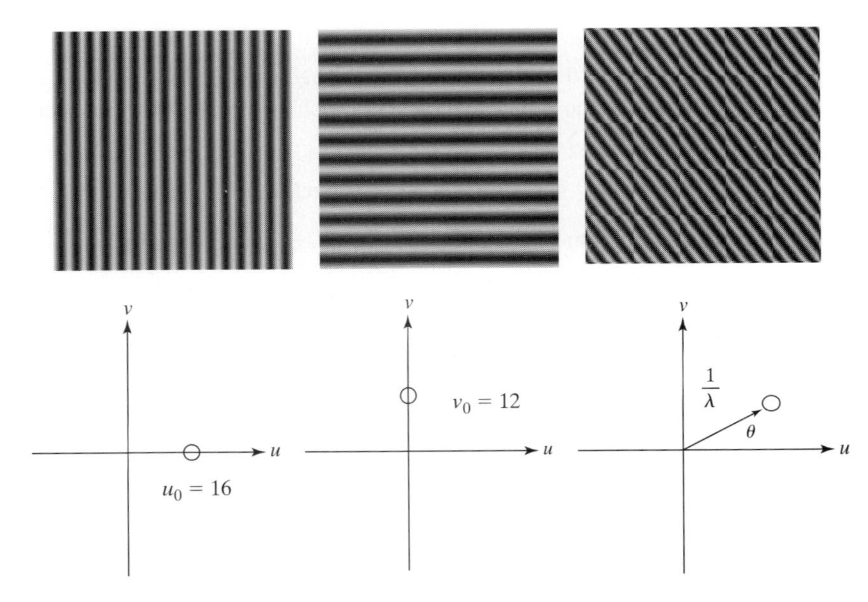

Figure 5.41 (top row) three sinusoidal waves and (bottom row) their principal response in the power spectrum.

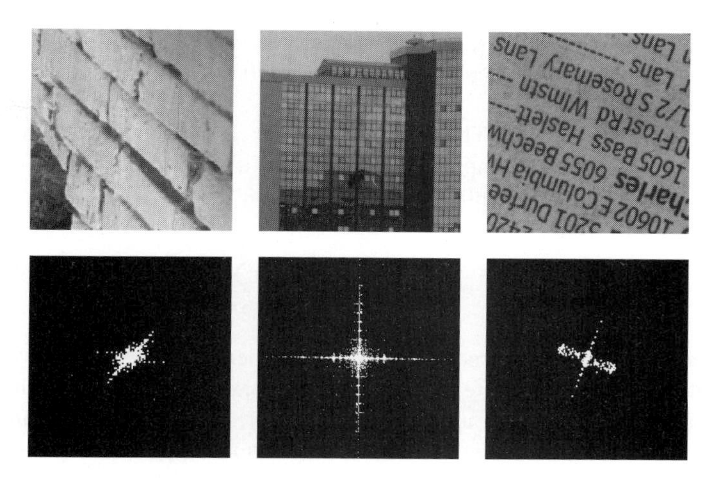

Figure 5.42 Three images (above) and their power spectrums (below). The power spectrum of the brick texture shows energy in many sinusoids of many frequencies, but the dominant direction is perpendicular to the 6 dark seams running about 45 degrees with the X-axis. There is noticeable energy at 0 degrees with the X-axis, due to the several short vertical seams. The power spectrum of the building shows high frequency energy in waves along the X-direction and the Y-direction. The image at the right, taken from a phone book, shows high frequency power at about 60 degrees with the X-axis, which represents the texture in the lines of text. Energy is spread more broadly in the perpendicular direction also in order to model the characters and their spacing. (Brick image from MIT Media Lab VisTex database. Image of Nairobi building courtesy of Ida Stockman.)

evident from the defintion that $P(-u, -v) = P(u, v)$, so the power spectrum is symmetric about the origin $u = 0$, $v = 0$. Figure 5.42 shows the power spectrums of four real images.

The power spectrum need not be interpreted as an image, but rather as a 2D display of the power in the original image versus the frequency parameters u and v. In fact, a transform can be computed via optics, and thus it can be realized as a physical image. In Chapter 2, there was brief mention of a sensor array partitioned into sectors and rings (refer back to the *ROSA* configuration in Figure 2.4(c) of Chapter 2). Since the power spectrum is symmetrical when rotated by π, as shown in Figure 5.42, the sectors can be used to sample directional power, while the rings sample frequency bands without regard to direction. Such sampling can also be done in software. In any case, if n_r rings and n_s sectors are sampled, one obtains $n_r + n_s$ features that may be useful for classification of the image neighborhood from which they were obtained.

5.11.3 Discrete Fourier Transform

The discrete Fourier transform, or DFT, which is immediately applicable to digital images is defined in Equation 5.34. We already know that a basis for the set of $N \times N$ images with real intensity values must have N^2 basis vectors. Each of these is determined by a pair of frequency parameters u, v, which range from 0 to $N - 1$ as used in the next formulas.

62 Definition. The **Discrete Fourier Transform (DFT)** transforms an image of $N \times N$ spatial samples $\mathbf{I}[x, y]$ into an $N \times N$ array $\mathbf{F}[u, v]$ of coefficients used in its frequency representation.

$$F[u, v] \equiv \frac{1}{N} \sum_{x=0}^{N-1} \sum_{y=0}^{N-1} I[x, y] e^{\frac{-2\pi j}{N}(xu+yv)} \tag{5.34}$$

To compute the single frequency domain element (*pixel*) $\mathbf{F}[u, v]$, we just compute the dot product between the entire image $\mathbf{I}[x, y]$ and a *mask* $E_{u,v}[x, y]$, which is usually not actually created, but computed implicitly in terms of u, v and the *cos* and *sin* functions as needed. We also define an inverse transform to transform a frequency domain representation $\mathbf{F}[u, v]$ into a spatial image $\mathbf{I}[x, y]$. Although it is useful to display the transform $\mathbf{F}$ as a 2D image, it might be less confusing to insist that it is NOT really an image. We have used the formal term *frequency representation* instead.

63 Definition. The **Inverse Discrete Fourier Transform(IDFT)** transforms an $N \times N$ frequency representation $\mathbf{F}[u, v]$ into an $N \times N$ image $\mathbf{I}[x, y]$ of spatial samples.

$$I[x, y] \equiv \frac{1}{N} \sum_{u=0}^{N-1} \sum_{v=0}^{N-1} F[u, v] e^{\frac{+2\pi j}{N}(ux+vy)} \tag{5.35}$$

If $\mathbf{F}[x, y]$ was computed by forward transforming $\mathbf{I}[x, y]$ in the first place, we want the inverse transform to return to that original image. This property does hold for the pair of definitions given above: proof is left to the guided exercises below. We first focus the discussion on the practical use of the **DFT & IDFT**. For storage or communication of an image, it might be useful to transform it into its frequency representation; the input image can be recovered by using the inverse transform. In image processing, it is common to perform some enhancement operation on the frequency representation before transforming back to recover the spatial image. For example, high frequencies can be reduced or even removed by reducing, or zeroing, the elements of $\mathbf{F}[u, v]$ that represent high frequency waves. The *convolution theorem* in Section 5.11.6 gives an elegant theoretical interpretation of this intuitive process.

Exercise 5.37: Some basics about complex numbers.

Use the definition $e^{j\omega} = cos\ \omega + j\ sin\ \omega$. (a) Show that $(e^{j\omega})^n = cos(n\omega) + j\ sin(n\omega)$. (b) Show that $x = e^{j\frac{2\pi k}{N}}$ is a solution to the equation $x^N - 1 = 0$ for $k = 0, 1, \ldots, N - 1$. (c) Let $x_0 = 1 = e^{j\frac{2\pi 0}{N}}, \ldots, x_k = e^{j\frac{2\pi k}{N}}$ be the N roots of the equation $x^N - 1 = 0$. Show that $x_1 + x_2 + x_3 + \cdots + x_{N-1} = 0$.

Exercise 5.38: Proof of invertability of DFT/IDFT transforms.

We want to prove that substituting the $\mathbf{F}[u, v]$ from Equation 5.34 into Equation 5.35 returns the exact original value $\mathbf{I}[x, y]$. Consider the following summation, where x, y, s, t are integer parameters in the range $[0, N-1]$ as used in the transform definitions:

$$G(x, y, s, t) = \sum_{u=0}^{N-1} \sum_{v=0}^{N-1} e^{\frac{j\,2\pi}{N}((x-s)u \,+\, (y-t)v)}.$$

(a) Show that if $s = x$ and $t = y$, then $G(x, y, s, t) = \sum_{u=0}^{N-1} \sum_{v=0}^{N-1} 1 = N^2$. (b) Show that if $s \neq x$ or $t \neq y$ then $G(x, y, s, t) = 0$. (c) Now show the major result that the inverse transform applied to the transform returns the original image.

5.11.4 Bandpass Filtering

Bandpass filtering is a common image processing operation performed in the frequency domain and is sketched in Figure 5.43. The **DFT** is used to transform the image into its frequency representation where the coefficients for some frequencies are reduced, perhaps to zero, while others are preserved. A sketch of the *low pass filter* is given at the left of Figure 5.43; intuitively, the high frequencies are erased and then the altered frequency representation is inverse transformed via Equation 5.35 to obtain a smoothed version of the original image. Instead of *erasing* elements of the frequency representation, we can take the dot product of $\mathbf{F}[u, v]$ with a 2D Gaussian, which will weight the low frequency components high and weight the high frequency components low. Figure 5.43 also shows how the frequency representation would be altered to accomplish high pass and bandpass filtering. The theory of convolution in Section 5.11.6 provides more insight to these operations.

5.11.5 Discussion of the Fourier Transform

The *Fast Fourier Transform* saves computational time by sharing common operations for different pairs of u, v and is usually used on square images of size $2^m \times 2^m$. Despite its

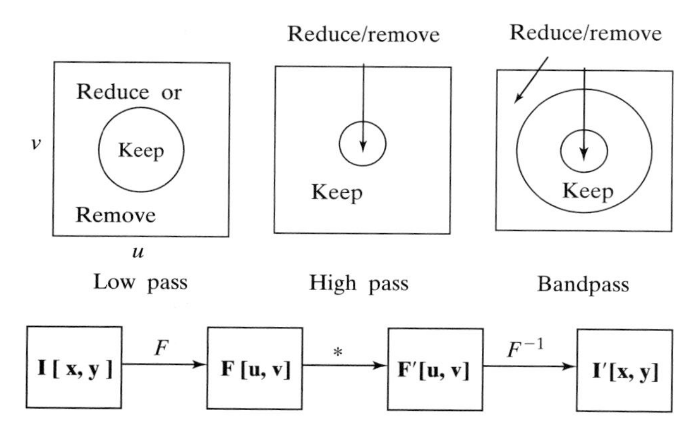

Figure 5.43 Bandpass filtering can be performed by a Fourier transformation into the frequency domain (**F[u, v]**) followed by multiplication (*) by a bandpass filter. The multiplication can set the *coefficients* u, v of various frequencies to zero, as shown in the top row. This modified frequency representation is then inverse transformed to obtain the modified spatial image **I′[x, y]**.

common use in image processing, the Fourier transform can cause unwanted degradation in local features of images because it is a *global transform* that uses all image pixels in every computation of $\mathbf{F}[u, v]$. For example, to represent hair or grass, or some other fine feature, high frequency waves must be used. First, such high frequencies might be filtered out as noise; however, even if not filtered out, a high frequency response $\mathbf{F}[u_h, v_h]$ would be computed by the dot product of a high frequency wave with the entire image. Since the region of hair or grass would only be a small part of the entire image, the value of the dot product could be high or low, depending on the rest of the image content. In the last ten years, there has been a strong movement toward using *wavelets* instead of image-size waves. Wavelets allow more sensitivity to local image variation, yet retain some major advantages of using global sinusoids. JPEG and other picture compression schemes use cosine wave representations on subimages to reduce the data size. Sometimes such compression must be suppressed in order to preserve needed local image detail.

5.11.6 The Convolution Theorem*

In this section, we sketch a proof of the important *convolution theorem,* which tells us that convolution of two functions in the spatial domain is equivalent to pointwise multiplication of their frequency representations. This equivalence has great practical utility as has already been seen.

Convolution Theorem:

If $f(x, y)$ and $h(x, y)$ are well-behaved functions of spatial parameters x, y, then $\mathbf{F}(f(x, y) \star h(x, y)) \equiv \mathbf{F}((f \star h)(x, y)) = \mathbf{F}(f(x, y))\mathbf{F}(h(x, y)) = \mathbf{F}(u, v)\mathbf{H}(u, v)$, where $\mathbf{F}$ is the Fourier transform operator and $\star$ is the convolution operator.

Before sketching a proof in the 1D case, we make the interpretation that is commonly used in signal processing: convolution can be carried out without applying a mask $h(x, y)$ to all points of an image $f(x, y)$.

For commonly used filters h, the transform H is likely to be available either in closed functional form or as a stored array in memory, thus short cutting step (2). Signal processing texts typically contain a table of *transform pairs* $< h, H >$ described both graphically and in functional form so that the user can select a filter with appropriate properties. We now sketch a proof of the convolution theorem for the 1D case; these same steps can be followed

Filter image $f(x, y)$ with mask $h(x, y)$

(1) Fourier transform the image $f(x, y)$ to obtain its frequency rep. $F(u, v)$.
(2) Fourier transform the mask $h(x, y)$ to obtain its frequency rep. $H(u, v)$
(3) multiply $F(u, v)$ and $H(u, v)$ pointwise to obtain $F'(u, v)$
(4) apply the inverse Fourier transform to $F'(u, v)$ to obtain the filtered image $f'(x, y)$.

Algorithm 5.3 Filtering image $f(x, y)$ with mask $h(x, y)$ using the Fourier transform for the 2D case. An intermediate step is to show what happens to the transform when the function is shifted.

Shift Theorem: $\mathbf{F}(f(x - x_0)) = e^{-j\,2\pi u x_0}\mathbf{F}(f(x))$

By definition $\mathbf{F}(f(x - x_0)) \equiv \int_{-\infty}^{+\infty} f(x - x_0)e^{-j\,2\pi u x}\,dx$. Making the change of variable $x' = x - x_0$, we obtain

$$\begin{aligned} \mathbf{F}(f(x - x_0)) &= \int_{-\infty}^{+\infty} f(x')e^{-j\,2\pi u(x'+x_0)}\,dx' \\ &= \int_{-\infty}^{+\infty} e^{-j\,2\pi u x_0} f(x')e^{-j\,2\pi u x'}\,dx' \\ &= e^{-j\,2\pi u x_0}\mathbf{F}(f(x)), \end{aligned} \tag{5.36}$$

since the first factor is constant with respect to the variable of integration x'. Note that

$$|e^{-j\,2\pi u x_0}|^2 = cos^2(2\pi u x_0) + sin^2(2\pi u x_0) = 1, \tag{5.37}$$

so shifting the function does not change the energy of $f(x)$ or $\mathbf{F}(u)$.

We can now use the above result to sketch the proof of the convolution theorem.

$$\mathbf{F}((f \star h)(x)) \equiv \int_{x=-\infty}^{x=+\infty} \left(\int_{t=-\infty}^{t=+\infty} f(t)h(x - t)\,dt \right) e^{-j\,2\pi u x}\,dx. \tag{5.38}$$

Using the good behavior assumed for functions f and h allows the order of integration to be interchanged.

$$\mathbf{F}((f \star h)(x)) = \int_{t=-\infty}^{t=+\infty} f(t) \left(\int_{x=-\infty}^{x=+\infty} h(x - t)e^{-j\,2\pi u x}\,dx \right) dt \tag{5.39}$$

Using the shift theorem,

$$\int_{x=-\infty}^{x=+\infty} h(x - t)e^{-j\,2\pi u x}\,dx = e^{-j\,2\pi u t}\mathbf{H}(u), \tag{5.40}$$

where $\mathbf{H}(u)$ is the Fourier transform of $h(x)$. We now have that

$$\begin{aligned} \mathbf{F}((f \star h)(x)) &= \int_{t=-\infty}^{t=+\infty} f(t)(e^{-j\,2\pi u t}\mathbf{H}(u))\,dt \\ &= \mathbf{H}(u) \int_{t=-\infty}^{t=+\infty} f(t)e^{-j\,2\pi u t}\,dt \\ &= \mathbf{H}(u)\mathbf{F}(u) = \mathbf{F}(u)\mathbf{H}(u) \end{aligned} \tag{5.41}$$

and the theorem is proved for the 1D case.

Exercise 5.39

Follow the steps given above for proving the Shift Theorem and Convolution Theorem in the 1D case and extend them to the 2D case.

5.12 SUMMARY AND DISCUSSION

This was a long chapter covering several methods and many examples: it is important to review major concepts. Methods were presented to enhance appearance of images for either human consumption or for automatic processing. A few methods were given for remapping intensity levels to enhance the appearance of scene objects; it was shown that often some image regions were improved at the expense of degrading others. The methods were presented only for gray-level images; however, most can be extended to color images, as you may already know from using some image enhancing tool. Edge enhancement was also presented as a method for human consumption. Hopefully, the artist's toolbox has been expanded.

The most important concept of the chapter is that of using a mask or kernel to define a local structure to be applied throughout the image. Convolution and cross correlation are two very powerful and related techniques which compute a result at $\mathbf{I}[x, y]$ by taking the sum of the pointwise products of the input image intensity and a corresponding mask value. These are linear operations that are pervasive in both theory and practice. Within this context, it was shown that the response to a particular mask at a particular image point (the *correlation*) is a measure of how similar the structure of the mask is to the structure of the image neighborhood. This notion provides a practical method for designing masks, or filters, for a variety of tasks, such as smoothing, edge detection, corner detection, or even texture detection.

There is a large literature on edge detection, and several different schemes were covered in this chapter. It should be clear that specifc edge detector output can be very useful for particular machine vision tasks. The dream of many developers has not been realized, however—to produce a single uniform solution to the low-level vision problem of representing significant object boundaries by some description based on detected edges. Perhaps it is unrealistic to expect this. After all, given an image of a car, how does the low-level system know whether or not a car is present, is of interest, is moving or still, or whether we are interested in inspecting for scratches in its paint or the fit of its doors versus just recognizing the make? Many of the edgemaps extracted in our examples seemed very promising for various applications. Indeed they are—many methods of subsequent chapters will be based on representations built using edge input. The reader should not be overly optimistic, however, because the reader has interpeted the images of this chapter using much higher level organization and knowledge of objects and the world. The higher level methods that we develop must be tolerant, because our edgemaps have gaps, noise, and multiple levels of structure, which will challenge the algorithms that build from them.

5.13 REFERENCES

Because few of the modern specialized journals existed when early work on image processing and computer vision was done, research papers are scattered across many different journals. Larry Roberts published his seminal thesis work in recognition of 3D blocks-world objects in 1965. Part of that work was the detection of edges using the operator that now bears his name. Roberts could not have anticipated the large amount of edge-detection work that would soon follow. Other early works were published by Prewitt in 1970 and Kirsch in 1971. The Kirsch masks are now regarded as corner detectors rather than edge detectors. Recent work by Shin and others (1998) supports the current popularity of the Canny step edge detector (1986) by showing that it is one of the best in both performance and efficiency, at least in a structure-from-motion task. The paper by Huertas and Medioni (1986) takes a deep practical look at implementing LOG filters and shows how to use them to locate edges to subpixel accuracy. The work on edge detection can easily be extended to 3D volume images as has been done by Zucker and Hummel (1981).

The book by Kreider and others (1966) gives required linear algebra background focused on its use in the analysis of functions, such as our picture functions: Chapters 1, 2, and 7 are most relevant. Chapters 9 and 10 give background on Fourier series approximation of 1D signals and are useful background for those wanting to explore the use of Fourier analysis of 2D picture functions. One can consult the optics text by Hecht and Zajac (1974) for Fourier interpretations of images as a superposition of waves. The book on algorithms by Cormen and others (1990) gives two algorithms for selecting the *ith* smallest of n numbers: the theoretical bound on one of them is $O(n)$ effort, which puts median filtering in the same theoretical complexity class as boxcar filtering.

1. Canny, J. 1986. A computational approach to edge detection. *IEEE Trans. Pattern Anal. and Machine Intelligence,* 8(6):679–698.
2. Cormen, T., C. Leiserson, and R. Rivest. 1990. *Introduction to Algorithms.* MIT Press, Cambridge, MA.
3. Duda, R., P. Hart, and D. Stark. 2000. *Pattern Classification,* 2nd ed. John Wiley & Sons, New York.
4. Frei, W., and C-C. Chen. 1977. Fast boundary detection: a generalization and new algorithm. *IEEE Trans. Comput.,* C-26(10):988–998.
5. Hecht, E., and A. Zajac. 1974. *Optics.* Addison-Wesley, New York.
6. Huertas, A., and G. Medioni. 1986. Detection of intensity changes with subpixel accuracy using Laplacian-Gaussian masks. *IEEE-T-PAMI,* v. 8(5):651–664.
7. Kirsch, R. 1971. Computer determination of the constituent structure of biological images. *Comput. Biomed. Res.,* v. 4(3):315–328.
8. Kreider, D., R. Kuller, D. Ostberg, and F. Perkins. 1966. *An Introduction to Linear Analysis.* Addison-Wesley, New York.

9. Prewitt, J. 1970. Object enhancement and extraction. In *Picture Processing and Psychopictorics,* B. Lipkin and A. Rosenfeld, eds. Academic Press, New York, 75–149.
10. Roberts, L. 1965. Machine perception of three-dimensional solids. In *Optical and Electro-Optical Information Processing,* J. Tippett and others, eds. MIT Press, Cambridge, MA, 159–197.
11. Shin, M., D. Goldgof, and K. Bowyer. 1998. An objective comparison methodology of edge detection algorithms using a structure from motion task. In *Empirical Evaluation Techniques in Computer Vision,* K. Bowyer and P. Philips, eds. IEEE Computer Society Press, Los Alamitos, CA.
12. Zucker, S., and R. Hummel. 1981. A three-dimensional edge operator. *IEEE-T-PAMI,* v. 3:324–331.

6

Color and Shading

The perception of color is very important for humans. Color perception depends upon both the physics of the light and complex processing by the eye-brain which integrates properties of the stimulus with experience. Humans use color information to distinguish objects, materials, food, places, and even the time of day. Figure 6.1 shows the same scene coded with different colors: Even though all shapes are the same, the right image is quite different from the left and the viewer might interpret it as an indoor scene of a housecat rather than a tiger in grass.

With recent innovation in economical devices, color processing by machine has become commonplace: We have color cameras, color displays and software that process color images. Color can also be used by machines for the same purposes as humans. Color is especially convenient because it provides multiple measurements at a single pixel of the image, often enabling classification to be done without complex spatial decision-making.

Careful study of the physics and perception of color would require many pages: Here we provide only a foundation that should be sufficient for beginning programming using color or as a guide to the literature. Some basic principles of the physics of color are given along with practical methods of coding color information in images. Then, we give some examples and methods for using color in recognizing objects and segmenting images.

We also study the shading of objects, which depends not just on the color of objects and the light illuminating them, but also on many other factors. These factors include the roughness of surfaces, the angles made between the surface and both the light sources and the viewer, and the distances of the surface from both the light sources and the viewer. Color and shading—important elements of art for centuries—are also important for interpreting a scene by computer vision algorithms.

Figure 6.1 (left) Naturally colored image of tiger in grass; (right) with transformed colors, recognition of a tiger is less secure—perhaps it is a cat on a rug? (Original image licensed from Corel Stock Photos.) See colorplate.

6.1 SOME PHYSICS OF COLOR

Electromagnetic radiation with wavelength λ in the range of between about 400 and 700 nanometers stimulates human neurosensors and produces the sensation of color (Figure 6.2). A nanometer is 10^{-9} meter: It is also referred to as a millimicron. For blue light, 400×10^{-9} *meters per wave* means 2.5×10^{6} *waves per meter* or 25,000 *waves per cm*. The speed of light in a vacuum is 3×10^{8} *m/sec,* which is equivalent to a frequency of 0.75×10^{15} blue light waves per second. This frequency is one thousandth of that for X-rays and one billion times that of broadcast radio waves.

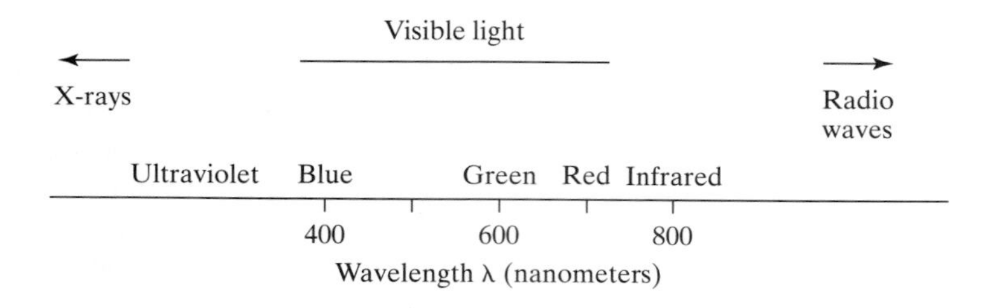

Figure 6.2 Visible part of the electromagnetic spectrum.

For the rest of this chapter, we refer to wavelength or frequency only in the context of the qualitative color it produces. Machines can detect radiation well beyond the range of human neurosensors; for example, short ultraviolet waves and extremely short X-rays, can be detected by special devices. Also, long infrared waves can be detected by many solid state cameras, and very long radio waves can be detected by a radio receiver. Science and engineering have developed many devices to sense and transduce pixel measurements into the visible spectrum: The X-ray machine and IR satellite weather scanner are two common examples.

Exercise 6.1

Suppose a piece of paper is 0.004 inches thick. What is its thickness in terms of the equivalent number of waves of blue light?

6.1.1 Sensing Illuminated Objects

Figure 6.3 shows light from a point source illuminating an object surface. As a result of the illuminating energy interacting with molecules of the object surface, light energy, or radiance, is emitted from the surface, some of which irradiates, or stimulates, a sensor element in a camera or organism's eye. The sensation, or perception, of an object's color depends upon three general factors:

- the spectrum of energy in various wavelengths illuminating the object surface,
- the spectral reflectance of the object surface, which determines how the surface changes the received spectrum into the radiated spectrum,
- the spectral sensitivity of the sensor irradiated by the light energy from the object's surface.

64 Definition. **White light** is composed of approximately equal energy in all wavelengths of the visible spectrum.

An object that is *blue* has a surface material that appears blue when illuminated with *white light*. This same object should appear violet if illuminated by only red light. A blue car under intense (white) sunlight will become hot to the touch and radiate energy in the IR range, which cannot be seen by the human eye but can be seen by an IR camera.

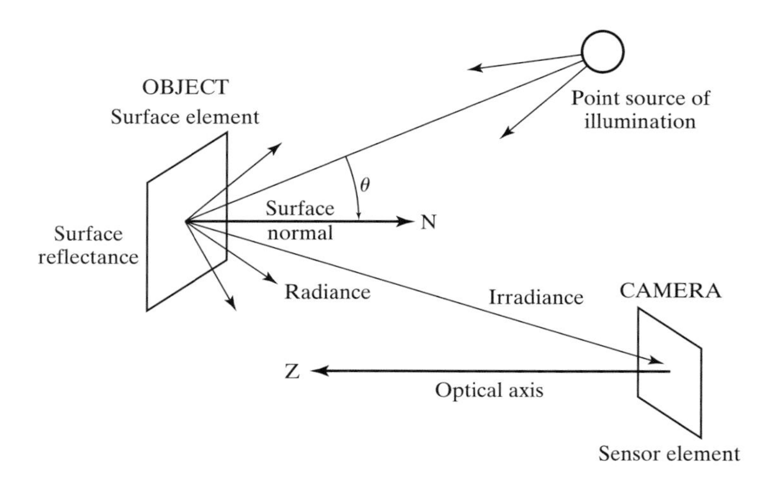

Figure 6.3 Light energy from a source reflects from an object surface and irradiates a sensor element.

6.1.2 Additional Factors

In addition to the three major factors given above, there are several complicating factors in both physics and human perception. Surfaces vary in specularity, that is, how much they act like a mirror. Matte surfaces reflect energy equally in all directions. The energy or intensity of radiation depends upon distance—surface elements farther from a point source of white light will receive less energy than closer surface elements. The effect is similar between the radiating object and the sensor elements. As a result, image intensities received from the same surface material might be nonuniform due to the nonuniform distances along the imaging rays. The orientation θ of the surface element relative to the source is even more important than distance in determining the energy reflected toward the sensor. These issues are discussed in more detail toward the end of this chapter.

Exercise 6.2: Variation of intensity with distance.

Point your computer's camera perpendicularly at a sheet of uniform white paper that is illuminated from an incandescent bulb off to one side. Record the image and study the image intensities. How much variation is there? Is there a systematic decrease of intensity as the distance from some brightest pixel increases?

Exercise 6.3: Variation of intensity with surface normal.

Repeat the above experiment using a spherical volleyball rather than a flat sheet of paper. Record the image and study the image intensities. Report on the variations and regularities.

6.1.3 Sensitivity of Receptors

Actual receptors react only to some wavelengths and are more sensitive to certain wavelengths than to others. Figure 6.4 shows sample sensitivity curves. Three of the curves correspond to three different kinds of cones in the human eye containing different chemical pigments sensitive to different wavelengths. The curve marked $human_1$ corresponds to a type of cone that is mildly sensitive to blue light between 400 and 500nm. The curve marked $human_2$ corresponds to cones that are very sensitive to green light and mildly sensitive to shorter wavelengths of blue and longer wavelengths of red. The brain fuses the responses from a local neighborhood of several cones to produce the perception of any visible color. It is somewhat remarkable that only three kinds of receptors are needed to do this, even though there are an infinite number of possible wavelengths of light. Many other seeing animals have only one or two types of receptors and perhaps perceive less rich color as a result. Solid state sensing elements usually have good sensitivity above the range for humans. It is important to remember this, since sometimes as the workday warms up, a machine vision system will see a scene differently from what a human operator sees. This is primarily due to the different sensitivity to IR radiation.

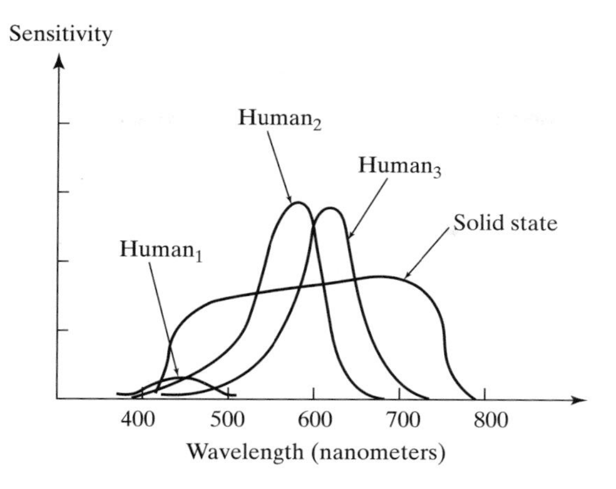

Figure 6.4 Comparison of relative sensitivities of three human pigments in cones and solid state sensor element.

Exercise 6.4: Favorite color.

Do you have a favorite color? If so, what is it? Why is it your favorite? Ask three other people what their favorite color is. Assuming you have multiple answers, how can you explain it given the known physics of color?

6.2 THE RGB BASIS FOR COLOR

Using only three types of receptors, humans can distinguish among thousands of colors; a more exact number is subject to argument. The *trichromatic* RGB (red-green-blue) encoding in graphics systems usually uses three bytes enabling $(2^8)^3$ or roughly 16 million distinct color codes. To be precise, we say 16 million codes and not 16 million colors because humans cannot actually perceive that many distinct colors. Machines can distinguish between any pair of different bit encodings, but the encodings may or may not represent differences that are significant in the real world. Each 3-byte or 24-bit RGB pixel includes one byte for each of red, green, and blue. The order in which each appears in memory can vary; order is irrelevant to theory but important for programming. Display devices whose color resolution matches the human eye are said to use true color. At least 16 bits are needed: A 15-bit encoding might use 5 bits for each of R, B, G, while a 16-bit encoding would better model the relatively larger green sensitivity using 6 bits.

The encoding of an arbitrary color in the visible spectrum can be made by combining the encoding of three primary colors (RGB) as shown in Figure 6.5. Red: (255, 0, 0) and green: (0, 255, 0) combined in equal amounts create yellow: (255, 255, 0). The amount of each primary color gives its intensity. If all components are of highest intensity, then the color white results. Equal proportions of less intensity create shades of grey: (c, c, c) for any constant $0 < c < 255$ down to black: (0, 0, 0). It is often more convenient to scale values in the range 0 to 1 rather than 0 to 255 when making decisions about color in our algorithms: use of such a range is device-independent.

	RGB	CMY	HSI
RED	(255, 0, 0)	(0,255,255)	(0.0 , 1.0, 255)
YELLOW	(255,255, 0)	(0, 0,255)	(1.05, 1.0, 255)
	(100,100, 50)	(155,155,205)	(1.05, 0.5, 100)
GREEN	(0,255, 0)	(255, 0,255)	(2.09, 1.0, 255)
BLUE	(0, 0,255)	(255,255, 0)	(4.19, 1.0, 255)
WHITE	(255,255,255)	(0, 0, 0)	(−1.0, 0.0, 255)
GREY	(192,192,192)	(63, 63, 63)	(−1.0, 0.0, 192)
	(127,127,127)	(128,128,128)	(−1.0, 0.0, 127)
	(63, 63, 63)	(192,192,192)	(−1.0, 0.0, 63)
	...		
BLACK	(0, 0, 0)	(255,255,255)	(−1.0, 0.0, 0)

Figure 6.5 Different digital trichromatic color encoding systems. It is often more convenient to scale values in the range 0 to 1 when making decisions in algorithms. HSI values are computed from RGB values using Algorithm 6.1: $H \in [0.0, 2\pi]$, $S \in [0.0, 1.0]$ and $I \in [0, 255]$. Byte codings exist for H and S.

The RGB system is an *additive color system* because colors are created by adding components to black: (0, 0, 0). This corresponds well to RGB displays (monitors) which have three types of phosphors to emit light. Three neighboring elements of phosphor corresponding to a pixel are struck by three electron beams of intensity c_1, c_2 and c_3 respectively: The human eye integrates their luminance to perceive color: (c_1, c_2, c_3). The light of three wavelengths from a small region of the CRT screen is thus physically added or mixed together.

Suppose that a color sensor encodes a pixel of a digital image as (R, G, B), where each coordinate is in the range [0, 255], for example. The computations shown in Equation 6.1 are one way to normalize image data for interpretation by both computer programs and people and for transformation to other color systems as discussed below. Imagine a color camera staring at a scene with variations in illumination; for example, object surface points are at varying distances from illumination sources and may even be in shadow relative to some of the light sources. An algorithm to aggregate green pixels corresponding to the image of a car would perform poorly unless the normalization for intensity were done first.

$$\begin{aligned} \textit{intensity}\ I &= (R + G + B)/3 \\ \textit{normalized red}\ r &= R/(R + G + B) \\ \textit{normalized green}\ g &= G/(R + G + B) \\ \textit{normalized blue}\ b &= B/(R + G + B) \end{aligned} \tag{6.1}$$

Using the normalization of Equation 6.1, the normalized values will always sum to 1. There are alternative normalizations; for instance, we could use $max(R, G, B)$ as the

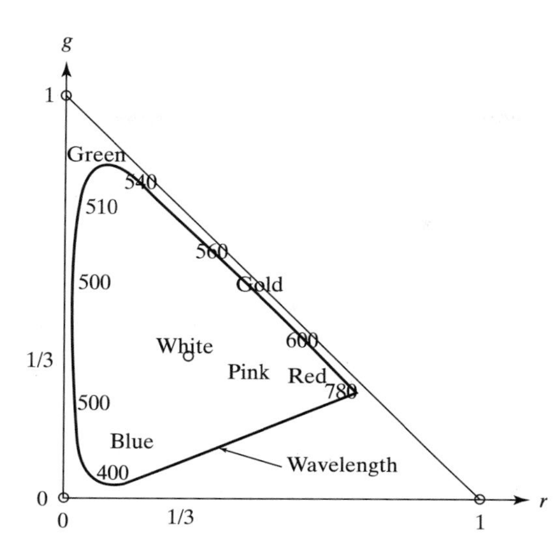

Figure 6.6 Color triangle for normalized RGB coordinates. The blue (b) axis is out of the page perpendicular to the r and g axes. Thus, the triangle is actually a slice through the points [1, 0, 0], [0, 1, 0] and [0, 0, 1] in 3D. The value for blue can be computed as $b = 1 - r - g$ for any pair of r-g values shown in the triangle.

divisor rather than the average RBG value. By using $r + g + b = 1$, the relationship of coordinate values to colors can be conveniently plotted via a 2D graph as in Figure 6.6. Pure colors are represented by points near the corners of the triangle. For example, a fire engine-red will be near the lower right corner with coordinates (1, 0) and a grass-green will be at the top with coordinates (0, 1) while white will be at the centroid $(1/3, 1/3)$. In Figure 6.6, the blue (b) axis is out of the page perpendicular to the r and g axes, and thus the triangle is actually a slice through the points [1, 0, 0], [0, 1, 0] and [0, 0, 1] in 3D. The value for blue can be computed as $b = 1 - r - g$ for any pair of r-g values shown inside the triangle.

Exercise 6.5: Experimenting with color codes.

Acquire an RGB color image and view it with some image tool. Exchange the green and blue bytes and report on the results. Double all and only the low blue values and report on the results.

6.3 OTHER COLOR BASES

Several other color bases exist which have special advantages relative to devices that produce color or relative to human perception. Some bases are merely linear transformations of others and some are not.

6.3.1 The CMY Subtractive Color System

The CMY color system models printing on white paper and subtracts from white rather than adds to black as the RGB system does. CMY coding is shown next to RGB in Figure 6.5. CMY is an abbreviation of *Cyan-Magenta-Yellow,* which are its three primary colors corresponding to three inks. Cyan absorbs red illumination, magenta absorbs green, and yellow absorbs blue, thus creating appropriate reflections when the printed image is illuminated with white light. The system is termed subtractive because of the encoding for absorption. Some trichromatic encodings are as follows; white: (0, 0, 0) because no white illumination should be absorbed, black: (255, 255, 255) because all components of white light should be absorbed and yellow: (0, 0, 255) because the blue component of incident white light should be absorbed by the inks, leaving the red and green components to create the perception of yellow.

6.3.2 HSI: Hue-Saturation-Intensity

The HSI system encodes color information by separating out an overall intensity value I from two values encoding *chromaticity*—hue H and saturation S. The color cube in Figure 6.7 is related to the RGB triangle shown in Figure 6.6. In the cube representation, each r, g, b value can range independently in [0.0, 1.0]. If we project the color cube along its major diagonal, we arrive at the hexagon at the left of Figure 6.8. In this representation, shades of gray that were formerly along the color cube diagonal now are all projected to the center white point while the red point [1, 0, 0] is now at the right corner and the green point [0, 1, 0] is at the top left corner of the hexagon. A related 3D representation, called a *hexacone,* is shown at the right in Figure 6.8: The 3D representation allows us to visualize the former cube diagonal as a vertical intensity axis **I**. Hue **H** is defined by an angle between 0 and 2π relative to the red-axis, with pure red at an angle of 0, pure green at $2\pi/3$ and pure blue at $4\pi/3$. Saturation **S** is the 3rd coordinate value needed in order to completely specify a point in this color space. Saturation models the purity of the color or hue, with 1 modeling a completely pure or saturated color and 0 modeling a completely unsaturated hue, that is, some shade of gray.

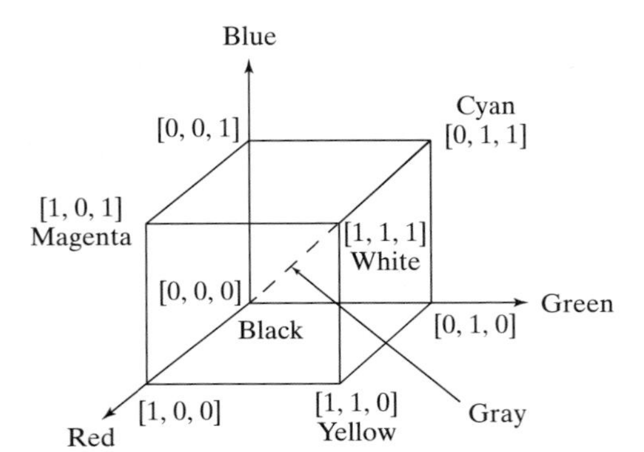

Figure 6.7 Color cube for normalized RGB coordinates: The triangle in Figure 6.6 is a projection of the plane through points [1, 0, 0], [0, 1, 0], and [0, 0, 1].

The HSI system is sometimes referred to as the *HSV* system using the term *value* instead of *intensity*. HSI is more convenient to some graphics designers because it provides direct control of brightness and hue. Pastels are centered near the **I** axis, while deep or rich colors are out at the periphery of the hexacone. HSI might also provide better support for computer vision algorithms because it can normalize for lighting and focus on the two chromaticity parameters that are more associated with the intrinsic character of a surface rather than the source that is lighting it.

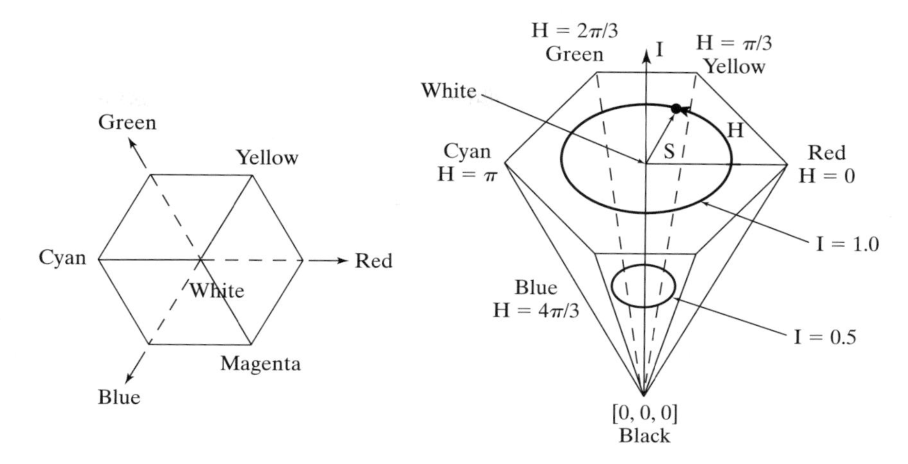

Figure 6.8 Color hexacone for HSI representation. At the left is a projection of the RGB cube perpendicular to the diagonal from [0, 0, 0] to [1, 1, 1]: Color names now appear at the vertices of a hexagon. At the right is a hexacone representing colors in HSI coordinates: Intensity (I) is the vertical axis; hue (H) is an angle from 0 to 2π with Red at 0.0; saturation (S) ranges from 0 to 1 according to how pure, or unlike white, the color is with S = 0.0 corresponding to the **I**-axis.

Derivation of HSI coordinates from RGB coordinates is given in Algorithm 6.1. The algorithm can convert input values (r, g, b) from the 3D color cube, or those normalized by Equation 6.1, or even byte-coded RGB values as in the left column of Figure 6.5. Intensity I is returned in the same range as the input values. Saturation S is not defined when intensity $I = 0$ and hue H is not defined when $S = 0$. H is in the range $[0, 2\pi]$. Whereas one might use a square root and inverse cosine to define mathematical conversion formulas, Algorithm 6.1 uses very simple computational operations so that it will run fast when converting an entire image of pixels from one encoding to another. Samples of the output of Algorithm 6.1 are given at the right in Figure 6.5.

Exercise 6.6

Using Algorithm 6.1, (a) convert the RGB code (100, 150, 200) into an HSI code and (b) convert the RGB code (0.0, 1.0, 0.0) to HSI.

Returning to Figure 6.6, we see how HSI values relate to the color triangle. Hue is related to the dominant wavelength of the light and corresponds approximately to a point on the sides of the triangle in Figure 6.6 with the lower values of λ near $400nm$ starting at the origin and increasing along the **g-axis** to about $520nm$ and further increasing toward $800nm$ down along the hypotenuse. Hue corresponds to the angle from the centroid corresponding to white toward some point (r, g) on a side of the triangle. The H and S values for 50 percent saturated gold is midway between the points marked white and gold in Figure 6.6. Figure 6.6 is an approximation to the painters color palette.

Conversion of RGB encoding to HSI encoding

R,G,B : input values of RGB all in range [0,1] or [0,255];
I : output value of intensity in same range as input;
S : output value of saturation in range [0,1];
H : output value of hue in range $[0,2\pi)$, -1 if S is 0;
R,G,B,H,S,I are all floating point numbers;

```
procedure RGB_to_HSI(in R,G,B; out H,S,I)
{
I := max (R, G, B);
min := min (R, G, B);
if (I ≥ 0.0) then S := (I – min )/I else S := 0.0;
if (S ≤ 0.0) then {H := −1.0; return;}
   "compute the hue based on the relative sizes of the RGB components"
diff := I – min;
"is the point within +/– 60 degrees of the red axis?"
if (R = I) then H := (π/3)*(G – B)/diff;
"is the point within +/– 60 degrees of the green axis?"
else if (G = I) then H := (2 ∗ π/3) + π/3 *(B – R)/diff;
"is the point within +/– 60 degrees of the blue axis?"
else if (B = I) then H := (4 ∗ π/3) + π/3 *(R – G)/diff;
if (H ≤ 0.0) H := H + 2π;
}
```

Algorithm 6.1 Conversion of RGB to HSI.

Figure 6.9 shows the transformation of an image by changing its saturation. The original input image is at the left. The center image is the result of increasing the saturation S of all individual pixels by 40 percent and the right image is the result of decreasing S by 20 percent. Relative to our experience, colors in the center image appear overdone while those in the right image appear washed out. It is important to note that hue H is unchanged in the three images and should thus be a reliable feature for color segmentation despite variations in intensity of white light under which a machine vision system might have to operate.

Exercise 6.7

Develop an algorithm to convert r,g,b color coordinates in [0, 1] to H,S,I using the following approach based on analytical geometry. Construct a perpendicular from point $[r, g, b]$ to the color cube diagonal through $[0, 0, 0]$ to $[1, 1, 1]$ and compute H,S,I accordingly.

Figure 6.9 (left) Input RGB image; (center) saturation S increased by 40 percent; and (right) saturation S decreased by 20 percent. (Photo by Frank Biocca.) See colorplate.

6.3.3 YIQ and YUV for TV Signals

The NTSC television standard is an encoding that uses one luminance value Y and two chromaticity values I and Q; only luminance is used by black and white TVs, while all three are used by color TVs. An approximate linear transformation from RGB to YIQ is given in Equation 6.2. In practice, the Y value is encoded using more bits than used for the values of I and Q because the human visual system is more sensitive to luminance (intensity) than to the chromaticity values.

$$\begin{aligned} \textit{luminance}\ Y &= 0.30R + 0.59G + 0.11B \\ \textit{R-cyan}\ I &= 0.60R - 0.28G - 0.32B \\ \textit{magenta-green}\ Q &= 0.21R - 0.52G + 0.31B \end{aligned} \tag{6.2}$$

YUV encoding is used in some digital video products and compression algorithms such as JPEG and MPEG. The conversion of RGB to YUV is as follows.

$$\begin{aligned} Y &= 0.30R + 0.59G + 0.11B \\ U &= 0.493 * (B - Y) \\ V &= 0.877 * (R - Y) \end{aligned} \tag{6.3}$$

YIQ and YUV have better potential for compression of digital images and video than do other color encoding schemes, because luminance and chrominance can be coded using different numbers of bits, which is not possible with RGB.

Exercise 6.8: Color code conversion.

Suppose a color camera encodes a given pixel in RGB as (200, 50, 100), where 255 is the highest (most energy) value. (a) What should be the equivalent triple in the HSI system? (b) What should be the equivalent triple in the YIQ system?

Exercise 6.9

Is the transformation from RGB to YIQ invertible? If so, compute the inverse.

Exercise 6.10: Recoding images.

Assuming that you have a display and software to view an RGB image, perform the following experiment. First, create an HSI image such that the upper right quarter is saturated red, the lower left corner is saturated yellow, the upper left quarter is 50 percent saturated blue and the lower right quarter is 50 percent saturated green. Invert the RGB to HSI conversion of Algorithm 6.1 and convert the HSI image to RGB. Display the image and study the colors in the four image quarters.

6.3.4 Using Color for Classification

The color of a pixel contains good information for classifying that pixel in many applications. In Section 6.5 a color model for human skin color is described that goes a long way toward finding a human face in a color image. Confusion is possible, however. For example, pixels from a brown cardboard box can pass the skin color test and region shape might be needed to distinguish a polyhedral box face from an elipsoidal human face. Figure 6.10 shows the result of extracting white regions of an image by passing pixels that are close to some sample pixel from training. Sample pixels were obtained from the symbols on the sign. Several unwanted regions are also formed by other white objects and by specular reflections. Character recognition algorithms could recognize many of the characters and discard most of the unwanted components.

In general, interpretation of the color of an individual pixel is error prone. The image at the left in Figure 6.9 was taken with a flash from the camera and some of the faces of the pineapple chunks appear white because of specular reflection (described below in Section 6.6.3). A classifier that broadens the definition of yellow to include these white pixels is also likely to include pixels from specular reflections off a blue cup, for instance.

Figure 6.10 White pixels are segmented from the color image at the left. Individual connected components of white pixels are arbitrarily labeled by a coloring algorithm as decribed in Chapter 3. (Analysis contributed by David Moore.) See colorplate.

Interpetation problems occur in particular regions of the color space: When saturation is close to zero computation and interpretation of hue is unreliable and when intensity is low interpretation of saturation is also unreliable.

Exercise 6.11

Show that conversion from RGB to HSI is unstable when either saturation or intensity is close to 0 by performing the following experiments. Implement Algorithm 6.1 as a program. (a) Convert the RGB codes $(L + \Delta L_R, L + \Delta L_G, L + \Delta L_B)$ to HSI for L large and $\Delta L_X \in \{-2, -1, 1, 2\}$. Are the values of H consistent? (b) Repeat this experiment for L small (about 10) and Δ as above. Are the values for S consistent?

6.4 COLOR HISTOGRAMS

A histogram of a color image can be a useful representation of that image for the purpose of image retrieval or object recognition. A histogram counts the number of pixels of each kind and can be rapidly created by reading each image pixel just once and incrementing the appropriate bin of the histogram. Retrieval of images from image databases using color histograms is treated in Chapter 8. Color histograms are relatively invariant to translation, rotation about the imaging axis, small off-axis rotations, scale changes and partial occlusion. Here, we sketch the method of color histogram matching originally proposed by Swain and Ballard (1991) for use in object recognition.

A simple and coarse method of creating a histogram to represent a color image is to concatenate the higher order two bits of each RGB color code. The histogram will have $2^6 = 64$ bins. It is also possible to compute three separate histograms, one for each color, and just concatenate them into one. For example, separate RGB histograms quantized into 16 levels would yield an overall $k = 48$ bin histogram as used by Jain and Vailaya (1996). Two color images and histograms derived from them are shown in Figure 6.11.

The intersection of image histogram $h(I)$ and model histogram $h(M)$ is defined as the sum of the minimum over all K corresponding bins as denoted in Equation 6.4. The intersection value is normalized by dividing by the number of pixels of the model to get a match value. This match value is a measure of how much color content of the model is present in the image and is not diminished due to background pixels of the image that are not in the model. Other similarity measures can be defined; for example, we could normalize the histograms into frequencies by dividing each bin count by the total number of pixels and then use the Euclidean distance to compare two images.

$$
\begin{aligned}
intersection(h(I), h(M)) &= \sum_{j=1}^{K} min\{h(I)[j], h(M)[j]\} \\
match(h(I), h(M)) &= \frac{\sum_{j=1}^{K} min\{h(I)[j], h(M)[j]\}}{\sum_{j=1}^{K} h(M)[j]}
\end{aligned}
\tag{6.4}
$$

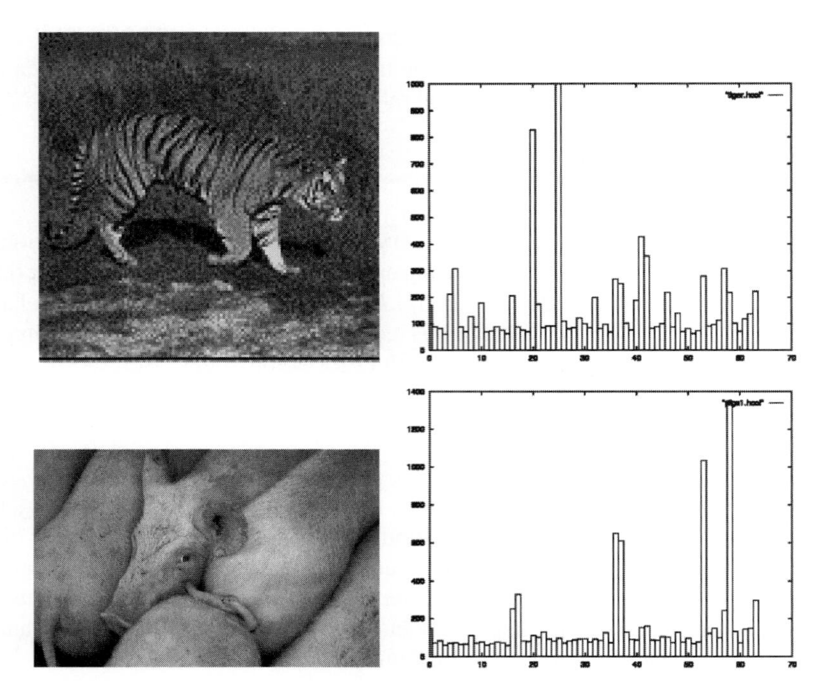

Figure 6.11 Color images and their 64-bin histograms. (Histograms courtesy of A. Vailaya. Tiger image licensed from Corel Stock Photos. Pig image licensed from Corbis. Creditline:\051 Clive Druett; Papilio/CORBIS.) See colorplate.

Experiments have shown that histogram match values can be good indicators of image similarity under variations mentioned above and also using different spatial quantizations of the image. Swain and Ballard also developed a *backprojection* algorithm which could locate a region in an image approximately the size of the model object that best matched the model histogram. Thus, they have developed two color-based algorithms, one to recognize if an image contains a known object and another to determine where that object is located. If images are taken under different lighting conditions, then intensity should be factored out first. One should also consider smoothing the histograms so that a good match will still be obtained with minor shifts of the reflectance spectrum. An alternate method is to match the cumulative distributions rather than the frequencies themselves.

Exercise 6.12: Matching images with permuted pixels.

Suppose that given image A, we create image B by randomly permuting the locations of the pixels in A. (This could be done as follows: First, copy A to B. Then, for each pixel $\mathbf{I}[r, c]$ in B, choose pixel $\mathbf{I}[x, y]$ at random and swap $\mathbf{I}[r, c]$ with $\mathbf{I}[x, y]$.) What would be the resulting match value between the histograms from A and B?

Exercise 6.13: Recognizing produce.

Obtain 3 bananas, 3 oranges, 3 red apples, 3 green apples, 3 green peppers, and 3 red tomatoes. For each of these six sets of produce, take three images, varying the arrangement of the three different produce items each time. This will result in 18 images. Construct a color histogram for each image. Use the first histogram of each set to be the model (6 models in all), then compute the histogram match value between each of the 6 models and the 12 other histograms. Report the results: Do your results support the possibility of a supermarket produce recognition system to recognize produce placed on the cashier's scale?

6.5 COLOR SEGMENTATION

We now describe work on finding a face in a color image taken from a workstation camera. The ultimate goal of the work is better man-machine communication. An algorithm is sketched which finds the main region of the image corresponding to the user's face. First, a training phase is undertaken to determine the nature of face pixels using samples from different people. Second, face pixels are identified in a new image according to where their (r, g) values fall within the training data. Figure 6.12 shows a plot of pixels (r, g) taken from

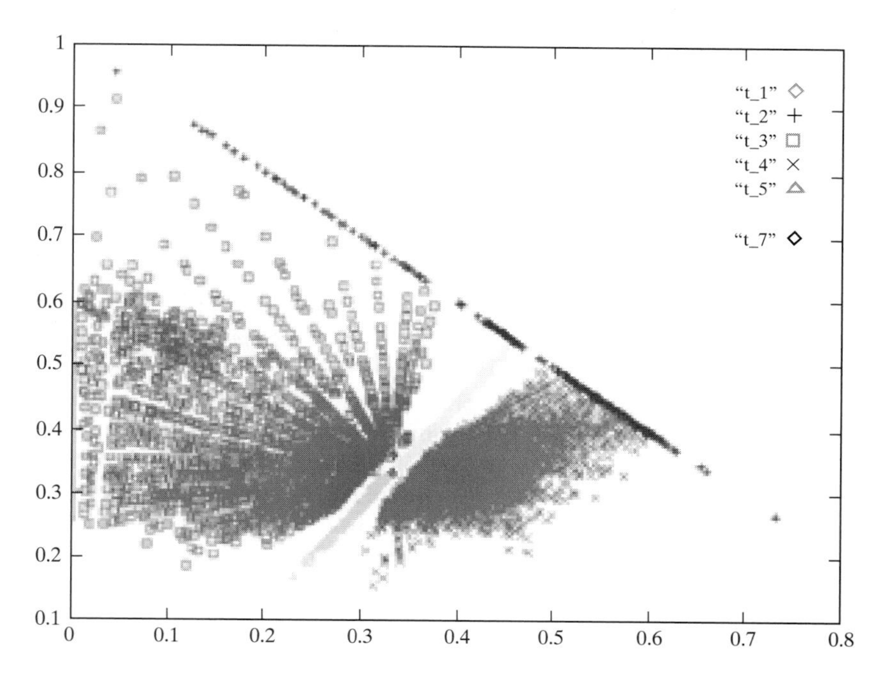

Figure 6.12 Skin color clusters obtained from training: The horizontal axis is R_{norm} and the vertical axis is G_{norm}. The cluster labeled as t_4 is the primary face color, clusters t_5 and t_6 are secondary face clusters associated with shadowed or bearded areas of a face. (Figure from V. Bakic.) See colorplate.

Figure 6.13 Face extraction examples: (left) input image; (middle) labeled image; and (right) boundaries of the extracted face region. (Images from V. Bakic.) See colorplate.

different images containing faces: normalized red and green values are used as computed from Equations 6.1. Six classes of pixels are easy to define by decision boundaries using the methods of Chapter 4; three of them contain face pixels, a primary class and two classes of pixels from shadows or beards.

Three major steps are used to identify the face region. The input to the first step is a labeled image using the labels 1, 2, . . . ,7 which result from classification according to the training data (label 7 is used for a pixel that is not in any of the other six classes). The middle images of Figure 6.13 show labeled images for two different faces: Most of the face pixels are correctly labeled as are most of the background pixels; however, there are many small areas of error. Components are aggregated and then merged or deleted according to their size and location relative to the major face region. First, connected component processing is done as described in Chapter 3 with only pixels labeled 4, 5, or 6 as foreground pixels. The second step selects as the face object the largest suitable component. This step also discards components that are too small or too big using heuristics learned from processing many examples. Typically, less than 100 components remain, the majority of which are in the shadow classes. The third step discards remaining components or merges them with the selected face object. Several heuristics using knowledge of the human face are applied; also, it is assumed that there is only one face in the scene. Example results are shown at the right in Figure 6.13. The program is fast enough to do these computations roughly 30 times per second (real-time), including computing the locations of the eyes and nose, which has not been described. This example generalizes to many other problems. A key stage is the clustering of the thousands of color codes in the original image to obtain the labeled image with only a few labels. In the face extraction example, clusters were decided by hand, but sometimes automatic clustering is done. Segmentation is covered thoroughly in Chapter 10.

▲ **Figure 6.1**

(left) Naturally colored image of tiger in grass; and (right) with transformed colors, recognition of a tiger is less secure–perhaps it's a cat on a rug? (Original image licensed from Corel Stock Photos.)

▲ **Figure 6.9**

(left) Input RGB image; (center) saturation S increased by 40 percent; and (right) saturation S decreased by 20 percent. (Photo by Frank Biocca.)

Figure 6.10 ▶

White pixels are segmented from the color image at the left. Individual connected components of white pixels are arbitrarily labeled by a coloring algorithm as described in Chapter 3. (Analysis contributed by David Moore.)

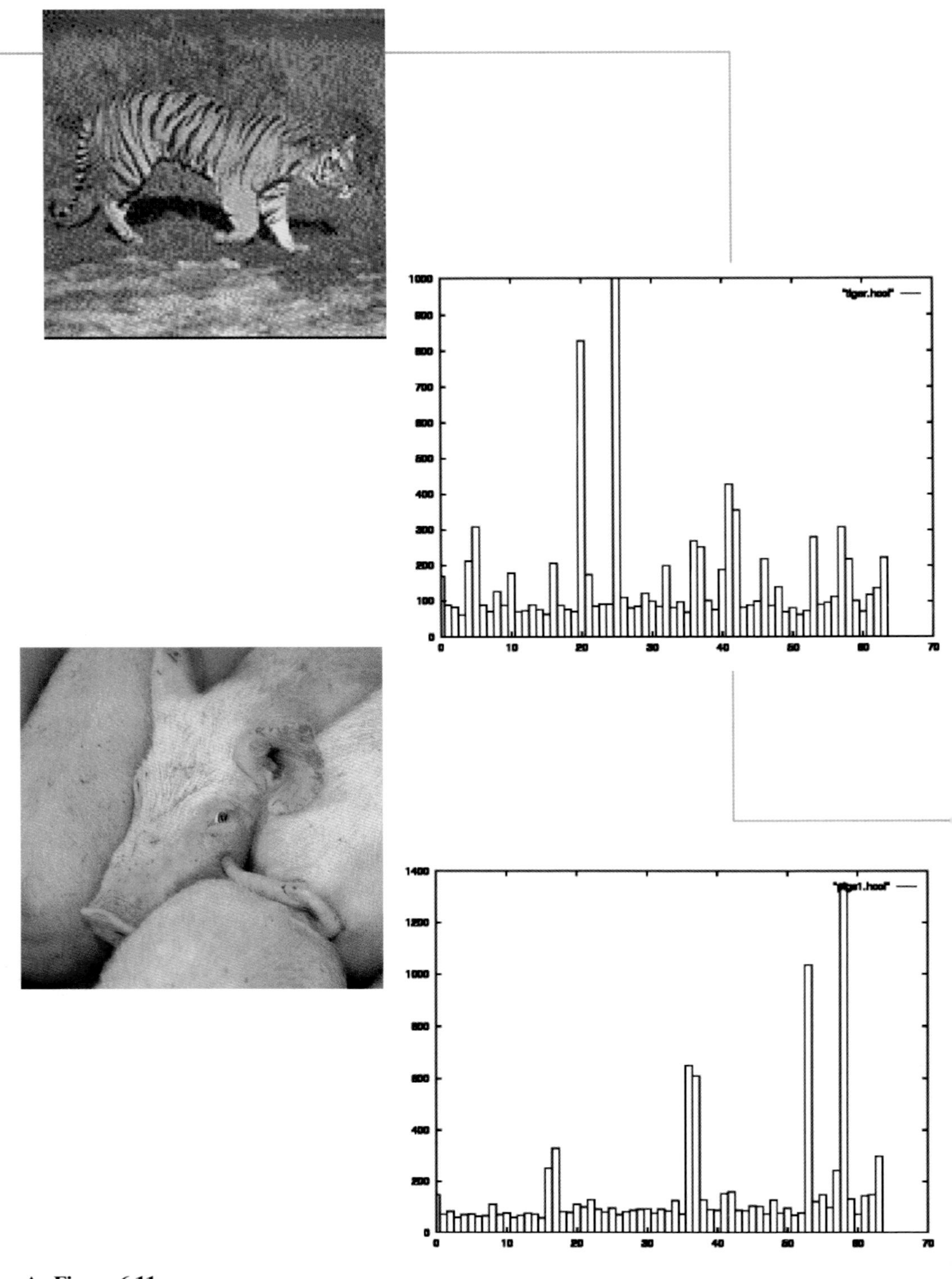

▲ **Figure 6.11**
Color images and their 64-bin histograms (Histograms courtesy of A. Vailaya. Images licensed from Corel Stock Photos.)

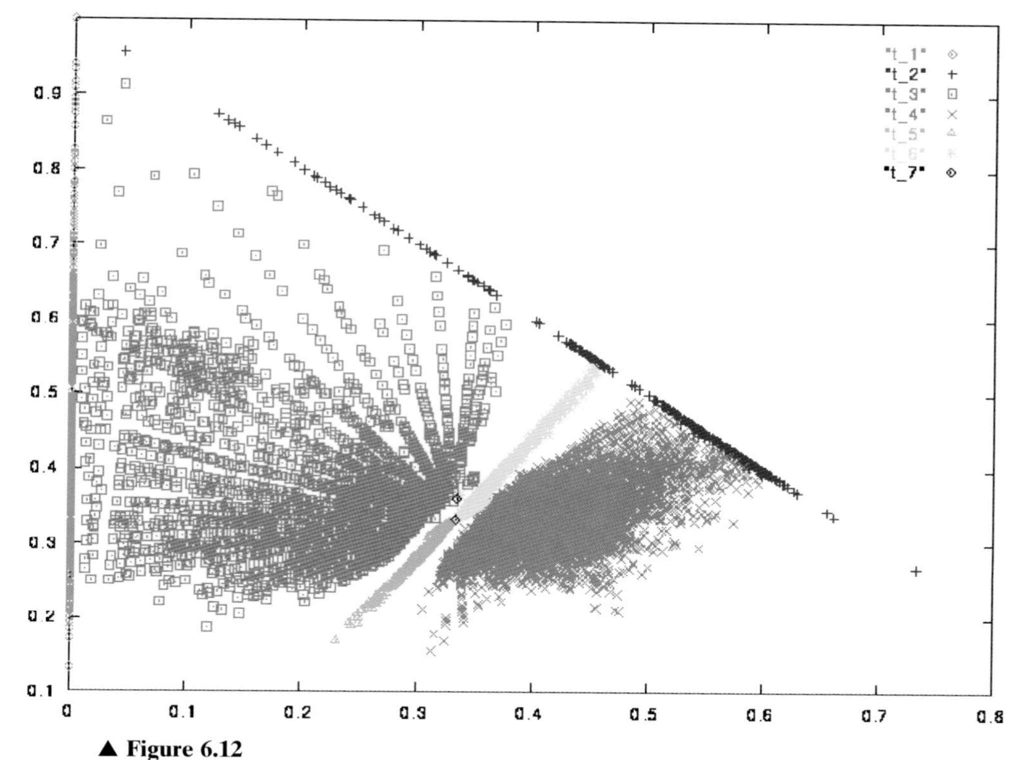

▲ **Figure 6.12**
Skin color clusters obtained from training: the horizontal axis is Rnorm and the vertical axis is G_{norm}. The cluster labeled as `t_4` is the primary face color, cluster `t_5` and `t_6` are secondary face clusters associated with shadowed or bearded areas of a face. (Figure from V. Bakic.)

▼ **Figure 6.13**
Face extraction examples: (left) input image; (middle) labeled image; and (right) boundaries of the extracted face region. (Images from V.Bakic.)

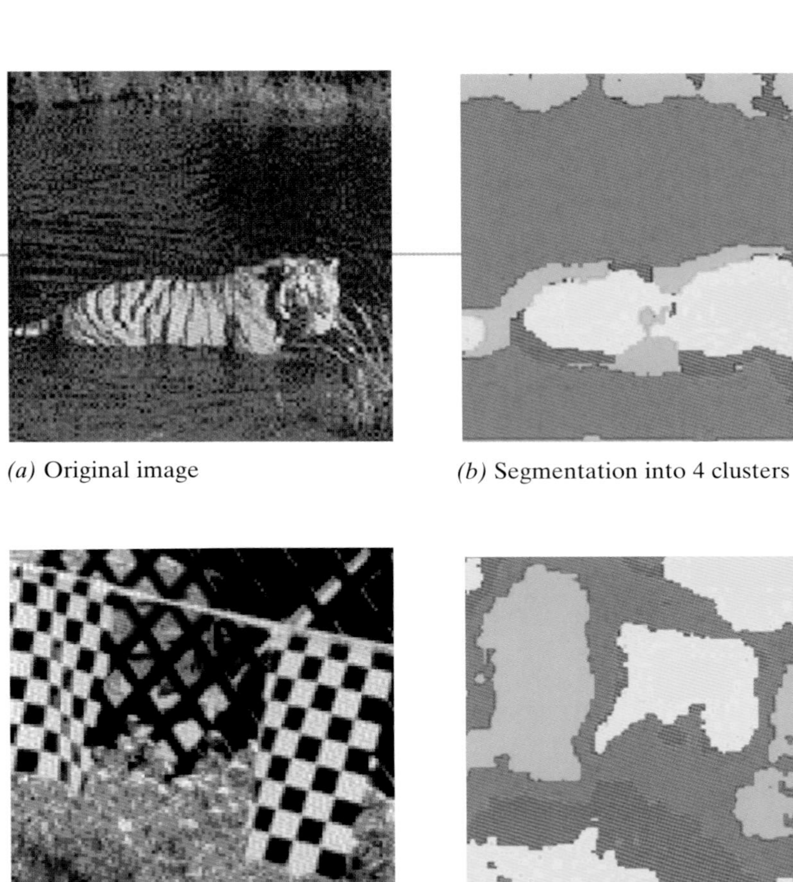

(a) Original image

(b) Segmentation into 4 clusters

(c) Original image

(d) Segmentation into 4 clusters

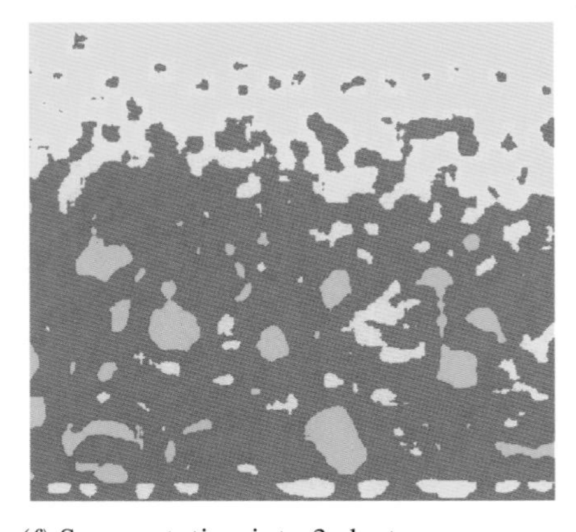

(e) Original image

(f) Segmentation into 3 clusters

▲ **Figure 7.8**
Examples of image segmentation using the Laws texture energy measures (original images are from Corel Stock Photos and from the MIT Media Lab VisTex database.)

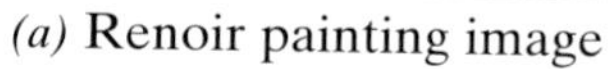

(a) Renoir painting image

(b) Amethyst image

▲ **Figure 8.1**
Images from digital collections. (Pierre-Auguste Renoir painting image, Landscape at Beaulieu, c. 1893, courtesy of the Fine Arts Museums of San Francisco, Mildred Anna Williams Collection, 1944.9; amethyst image courtesy of the Smithsonian Institution, 1992.)

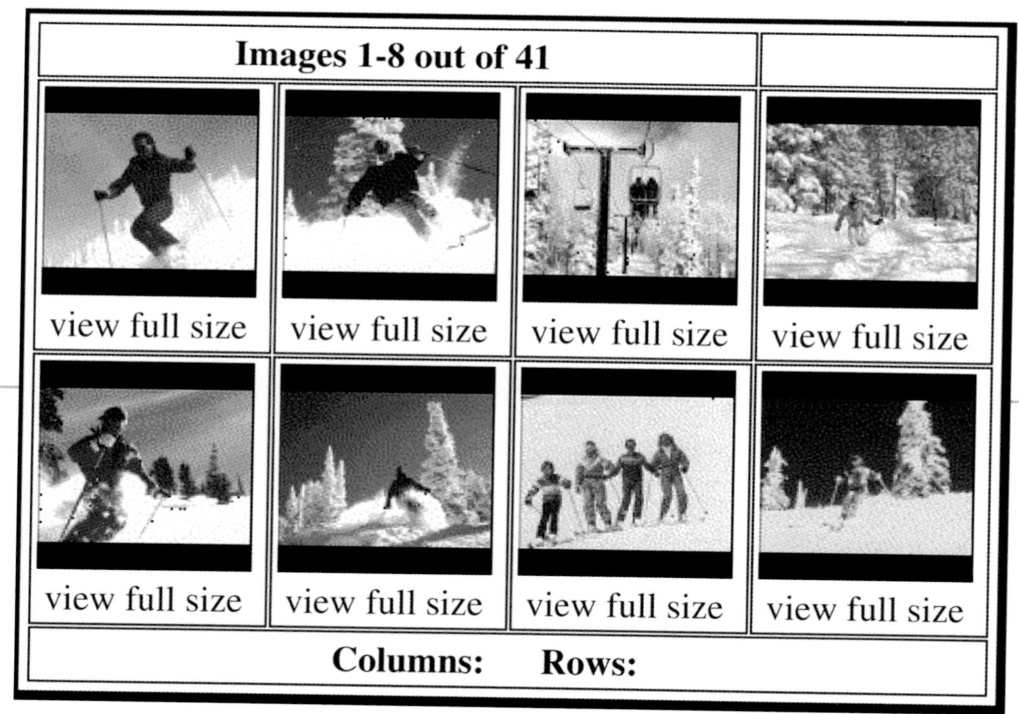

▲ **Figure 8.3**
Results of QBIC search based on color layout similarity; the query is the example image shown in the top left position. (Images courtesy of Egames.)

Figure 8.4 ▶
Results of a QBIC search based on color percentages; the query specified 40 percentage red, 30 percentage yellow, and 10 percentage black. (Images courtesy of Egames.)

▼ **Figure 8.5**
Results of an image database search in which the query is a painted grid. (Images from the MIT Media Lab VisTex database: http://vismod.www.media.mit.edu/vismod/imagery/VisionTexture/vistex.html.)

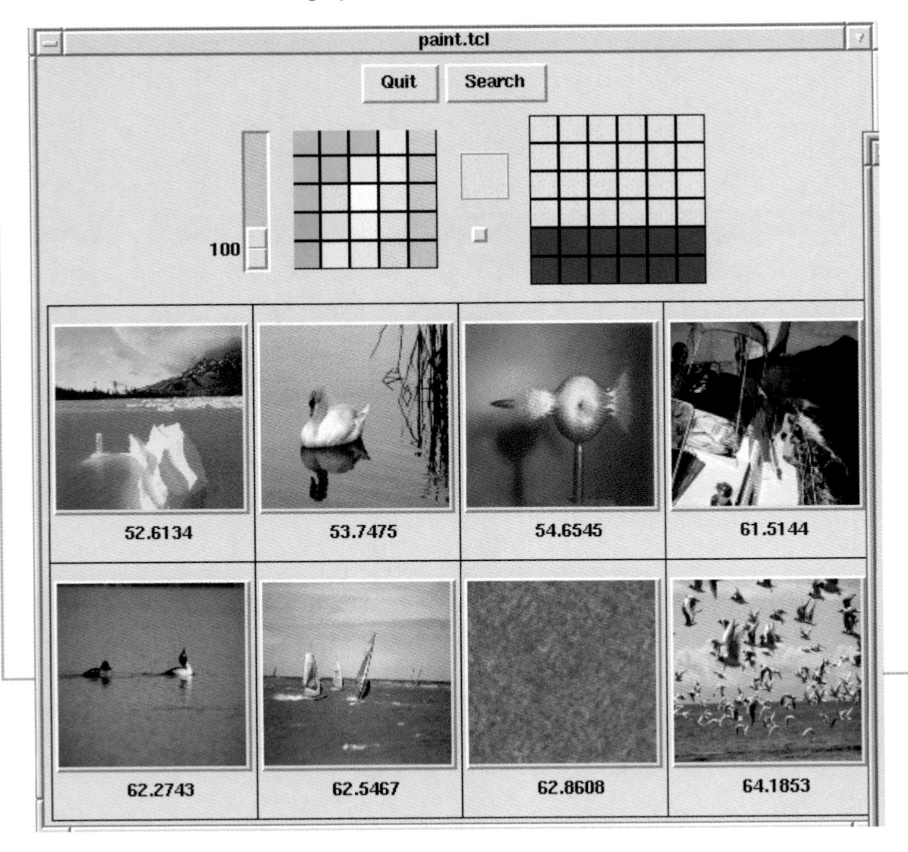

▼ **Figure 8.6**
Results of an image database search based on texture similarity. (Images from the MIT Media Lab VisTex database: http://vismod.www.mediamit.edu/vismod/imagery/VisionTexture/vistex.html.)

Figure 8.7 ▶
Image retrieval by elastic matching. (Courtesy of Alberto Del Bimbo.)

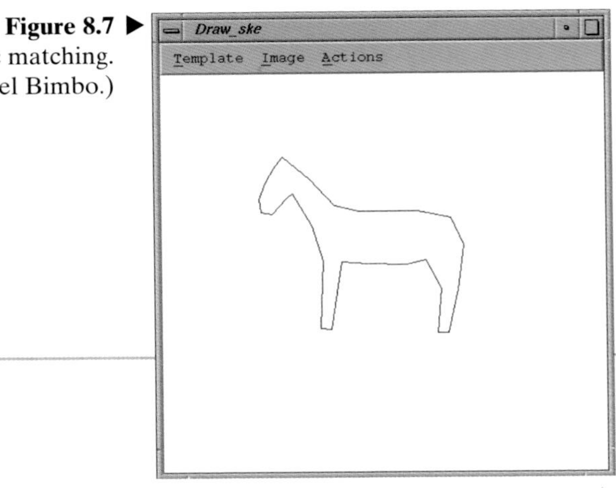

(a) The user's query shape.

(b) Two of the retrieved images.

(c) Another retrieved image in which two horses were found.

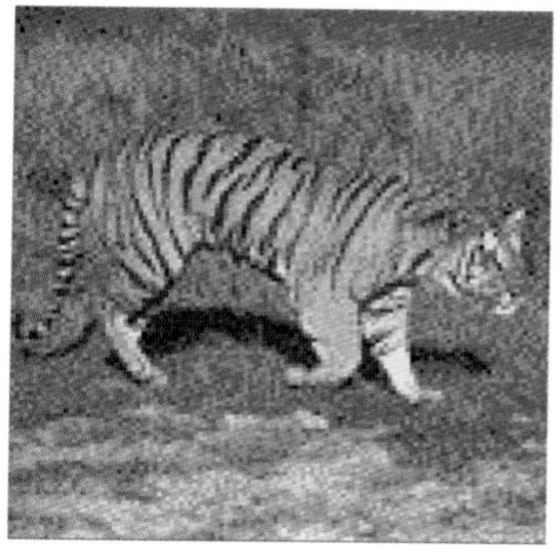

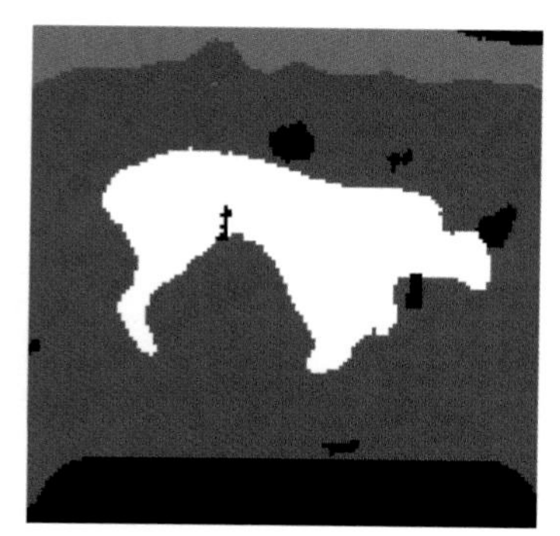

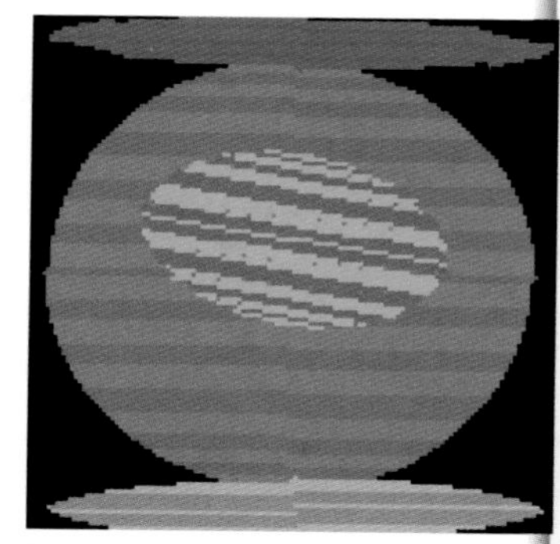

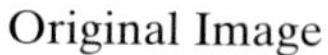

Original Image　　Segmentation　　Symbolic Representation

▲ **Figure 8.9**
Objects and spatial relationships that can be extracted from images and used for retrieval. (Original image licensed from Corel Stock Photos.)

▲ **Figure 9.6**
Results and applying Algorithm 9.3. At the right is the image at time t1. At the left is the image at time t2 with the motion analysis overlaid. Red squares indicate the location of the original neighborhoods detected by the interest operator in the image at the left. Blue squares indicate the best matches to these neighborhoods in the image at the right. There are three coherent sets of motion vectors (green lines) corresponding to three moving objects. The leftmost object moved down and slightly rigth. The lowest object moves right and slightly down; the rightmost object moves left and slightly up. (Analysis courtesy of Adam T. Clark.)

▲ **Figure 10.1**
(left) Football image and (right) segmentation into regions. Each region is a set of connected pixels that are similar in color.

▲ **Figure 10.4**
(left) football image and (right) $K = 6$ clusters resulting from a K-means clustering procedure shown as distinct gray tones. The six clusters correspond to the six main colors in the original image: dark green, medium green, dark blue, white, silver, and black.

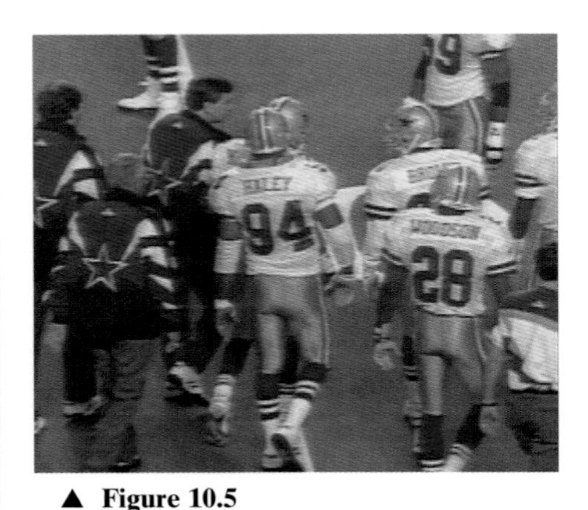

▲ **Figure 10.5**
(left) Football image and (right) $K = 5$ clusters resulting from an isodata clustering procedure shown as distinct gray tones. The five clusters correspond to five color groups: green, dark blue, white, silver, and black.

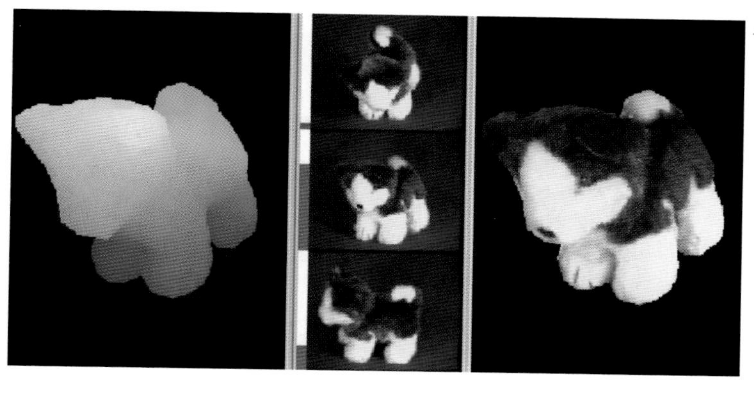

◀ **Figure 15.17**
(left) range data of a dog model (center); three real color images from nearby viewpoints; and (right) the rendered image using a weighted combination of pixels from these views. (Courtesy of Kari Pull:.)

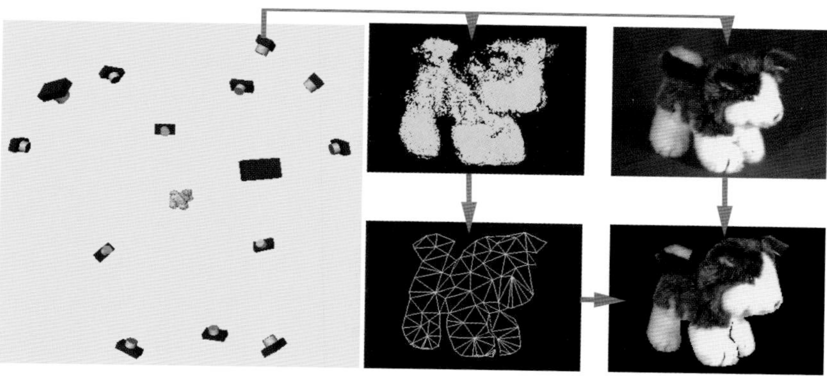

▲ **Figure 15.18**
Registered range and color images from a small number of views of an object can be used to produce a high-quality rendered image, without ever constructing a full model of the object. (left) Potential viewpoints; (top center) range data from one of the viewpoints; (top right) color data from the same viewpoint, (bottom center) mesh constructed from the range data; (bottom right) rendered image achieved by texture-mapping the color data onto the mesh. (Courtesy of Kari Pull:.)

▲ **Figure 15.19**
The same technique applied to an object for which construction of a full 3D model is nearly impossible, due to the thinness of parts of the object. (upper left) Three different color images of the object, (lower left) three images from a different, selected viewpoint constructed by mapping the pixels of the three original images onto the new viewpoint; and (right) and final rendered image, which is a weighted combination of the three constructed images. (Courtesy of Kari Pull:.)

6.6 SHADING

There are several complicating factors in both the physics of lighting and human perception. Surfaces vary in specularity, that is, how much they act like a mirror. Highly specular surfaces reflect a ray of incident energy in a restricted cone about a ray of reflection. Matte surfaces reflect energy equally in all directions. Thus, a surface not only has a wavelength dependent bias in reflecting incident radiation, but it also has directional bias. Moreover, the energy or intensity of radiation depends upon distance—surface elements farther from a point source of white light will receive less energy than closer surface elements. The effect is similar between the radiating object and the sensor elements. As a result, image intensities will be nonuniform due to the nonuniform distances along the imaging rays. The orientation θ of the surface element relative to the source is also very important.

6.6.1 Radiation from One Light Source

Consider radiation from a single distant light source reaching an object surface as shown in Figure 6.14. Currently, there is no view position from which we observe the surface; we are only considering how the surface is irradiated by the light source. We assume that the light source is far enough away so that the direction from all surface elements of the illuminated object to the light source can be represented by a single unit length direction vector **s**. The light energy per unit area (*intensity i*) that reaches each surface element A_j is proportional to the area of the surface element times the cosine of the angle that the surface element makes with the illumination direction **s**. The cosine of the angle is $\mathbf{n} \circ \mathbf{s}$, where **n** is the unit vector normal to the surface element A_j. Thus our mathematical model for the intensity of radiation received by a surface element is

$$\text{received } i \sim \mathbf{n} \circ \mathbf{s}. \tag{6.5}$$

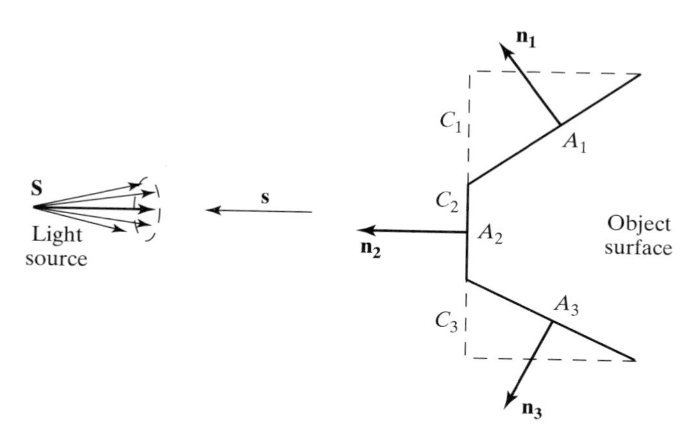

Figure 6.14 Object surface elements A_j irradiated by light source **S** receive energy proportional to the cross section $C_j = A_j \, cos\theta_j$ presented to the source. Intensity of received radiation is thus $i \sim \mathbf{n} \circ \mathbf{s}$ where **n** is the unit normal to the surface and **s** is the unit direction toward the source. θ_j is the angle between the surface normal $\mathbf{n_j}$ and **s**.

The radiation received is directly proportional to the power of the light source, which may or may not be known. The light source may radiate energy in all directions, or may be like a spotlight radiating light only in a small cone of directions. In either case, the power of the light source is expressed in watts per steradian, or energy per unit area of a conical sector of a unit sphere centered at the source. This simple model for irradiating surface elements extends readily to curved surfaces by considering that the rectangular surface elements become infinitesimally small in the limit. The fraction of the incident radiation that a surface element reflects is called its *albedo*.

65 Definition. The **albedo** of a surface element is the ratio of the total reflected illumination to the total received illumination.

We have assumed that albedo is an intrinsic property of the surface: For some surfaces this is not the case because the fraction of illumination reflected will vary with the direction of lighting relative to the surface normal.

6.6.2 Diffuse Reflection

We now extend our model to consider the reflection from object surfaces; moreover, we model how the surface element appears from some viewing position **V**. Figure 6.15 shows *diffuse* or *Lambertian* reflection. Light energy reaching a surface element is reflected evenly in all directions of the hemisphere centered at that surface element. Diffuse reflections occur from surfaces that are rough relative to the wavelength of the light. The intensity of the reflected illumination is proportional to the intensity of the received illumination: The constant factor is the albedo of the surface, which is low for dark surfaces and high for light surfaces.

$$\text{diffuse reflected } i \sim \mathbf{n_j} \circ \mathbf{s} \tag{6.6}$$

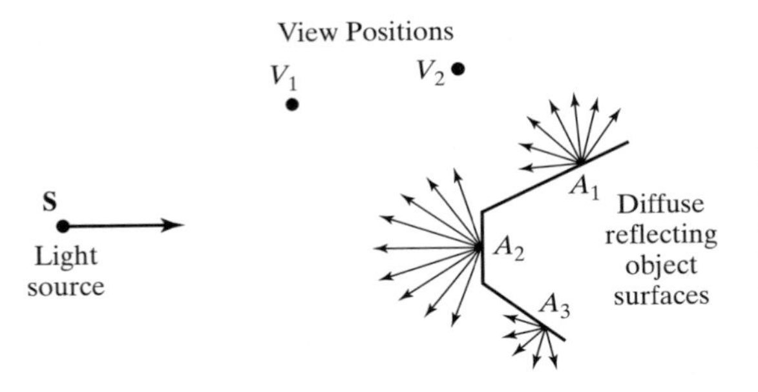

Figure 6.15 Diffuse, or Lambertian, reflection distributes energy uniformly in all directions of the hemisphere centered at a surface element. Thus an entire planar patch will appear to be of uniform brightness for all viewpoints from which that surface is visible.

66 Definition. A **diffuse** reflecting surface reflects light uniformly in all directions. As a result, it appears to have the same brightness from all viewpoints.

The critical characteristic is that the surface element will have the same brightness when viewed from the entire hemisphere of directions, because its brightness is independent of the viewer location. From Figure 6.15, surface element A_1 will have the same brightness when viewed from positions V_1 or V_2; similarly, surface A_2 will also appear to be of the same brightness when viewed from either V_1 or V_2. If all three surface elements are made of the same material, they have the same albedo and thus A_2 will appear brighter than A_1, which will appear brighter than A_3, due to the angles that these surfaces make with the direction of the illumination. Surface element A_3 will not be seen at all from either position V_1 or V_2. (A surface element will not be visible if $\mathbf{n} \circ \mathbf{v} < 0$ where $\mathbf{v}$ is the direction toward the viewer.)

A convincing example of diffuse reflection is given in Figure 6.16, which shows intensity of light reflected off an egg and a blank ceramic vase. Intensities from a row of the image are much like a cosine curve, which demonstrates that the shape of the object surface is closely related to the reflected light as predicted by Equation 6.6.

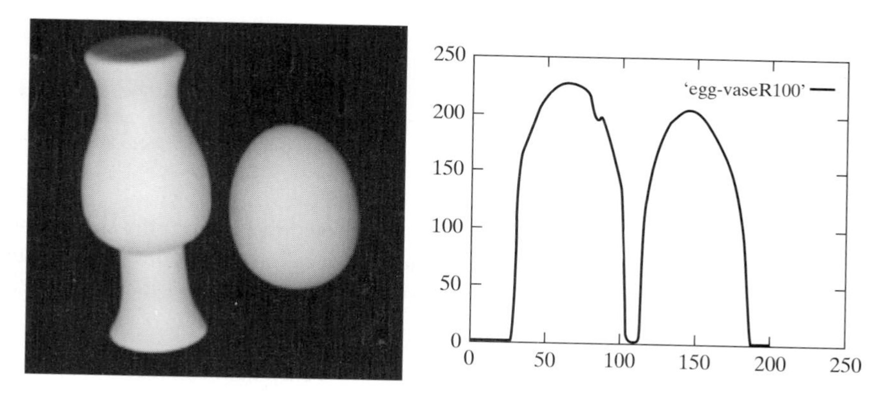

Figure 6.16 Diffuse reflection from Lambertian objects—a vase and an egg—and a plot of intensities across the highlighted row. The intensities are closely related to the object shape. (Image courtesy of Deborah Trytten.)

Exercise 6.14

Consider a polyhedral object of diffusely reflective material such that face F directly faces a distant light source S. Another face A adjacent to F appears to be half as bright as F. What is the angle made between the normals of faces A and F?

6.6.3 Specular Reflection

Many smooth surfaces behave much like a mirror, reflecting most of the received illumination along the ray of reflection as shown in Figure 6.17. The ray of reflection(**R**) is coplanar with the normal(**N**) to the surface and the ray of received illumination(**S**), and, makes equal angles with them. A perfect mirror will reflect all of the light energy received from source **S** along ray **R**. Moreover, the reflected energy will have the same wavelength composition as the received light regardless of the actual color of the object surface. Thus a red apple will have a white highlight, or twinkle, where it reflects a white light source. Equation 6.7 gives a mathematical model of specular reflection commonly used in computer graphics.

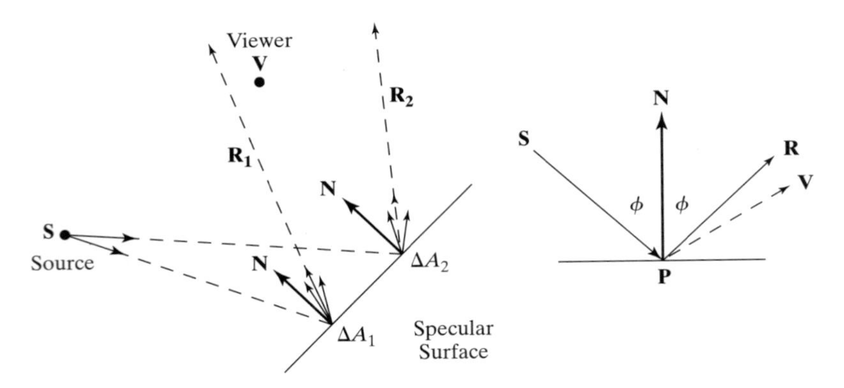

Figure 6.17 Specular, or mirrorlike, relection distributes energy in a narrow cone about the ray of reflection **R**. The viewpoint **V** receives some reflected energy from surface element ΔA_1 but very little from surface element ΔA_2. The intensity received at **V** is $e_i \sim (\mathbf{R} \circ \mathbf{V})^\alpha$, where **R** is the ray of reflection, **V** is the direction from the surface element toward the viewpoint and α is the *shininess* parameter.

Equation 6.8 defines how to compute the reflected ray **R** from the surface normal and the direction toward the source. The parameter α is called the shininess of the surface and has a value of 100 or more for very shiny surfaces. Note that as α increases, $cos\phi^\alpha$ decreases more sharply as ϕ moves away from 0.

$$\text{specular reflected } \boldsymbol{i} \sim (\mathbf{R} \circ \mathbf{V})^\alpha \tag{6.7}$$

$$\mathbf{R} = 2\mathbf{N}(\mathbf{N} \circ (-\mathbf{S})) \oplus \mathbf{S} \tag{6.8}$$

67 Definition. **Specular reflection** is mirrorlike reflection. Light reflected off the surface is radiated out in a tight cone about the ray of reflection. Moreover, the wavelength composition of the reflected light is similar to that of the source and independent of the surface color.

68 Definition. A **highlight** on an object is a bright spot caused by the specular reflection of a light source. Highlights indicate that the object is waxy, metallic, or glassy, etc.

6.6.4 Darkening with Distance

The intensity of light energy reaching a surface decreases with the distance of that surface from the light source. Certainly our Earth receives less intense radiation from the sun than does Mercury. A model of this phenomena is sketched in Figure 6.18. Assuming that a source radiates a constant energy flux per unit of time, any spherical surface enclosing that source must intercept the same amount of energy per unit time. Since the area of the spherical surface increases proportional to the square of its radius, the energy per unit area must decrease proportional to the inverse of the radius squared. Thus, the intensity of light received by any object surface will decrease with the square of its distance from the

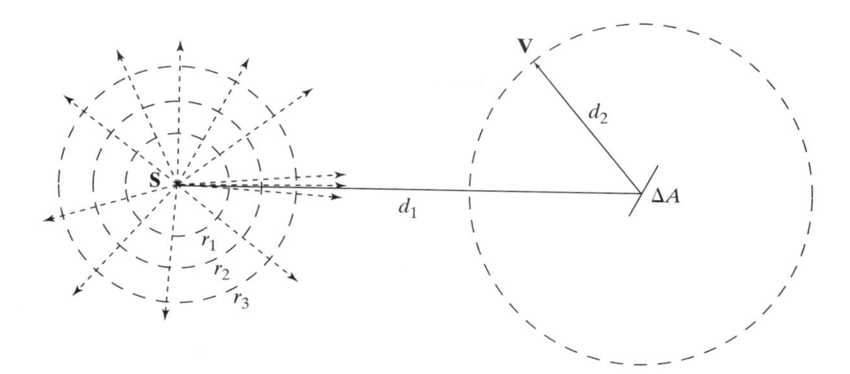

Figure 6.18 The total energy radiating from a point source through any enclosing spherical surface is the same: thus the energy per unit area of surface, or intensity, must decrease inversely with the square of the radius of the enclosing sphere (d_1). Similarly, light energy reflecting off a surface element must descrease in intensity with the distance (d_2) from which that surface is viewed.

source. Such a distance is labeled d_1 in Figure 6.18. The same model applies to light energy reflected from the object surface elements: Thus, a viewer at position V in space will observe a surface brightness (d_2) inversely proportional to the square of the distance from that surface element. This inverse square model is commonly used in computer graphics to compute the shading of rendered surfaces so that 3D distance, or depth, can be communicated to a user.

Exercise 6.15

An inventor wants to sell the traffic police the following device for detecting the speed of cars at night. The device emits a very short flash of light at times t_1 and t_2 and senses the reflection back from the car. From the intensities of the reflections, it computes the distances d_1 and d_2 for the two time instants using the principle in Figure 6.18. The speed of the car is simply computed as the change in distance over the change in time. Critique the design of this instrument. Will it work?

6.6.5 Complications

For most surfaces, a good reflection model must combine both diffuse and specular reflections. If we view an apple with a flashlight, we will actually see a reddish object with a whitish highlight on it: The reddish reflections are from diffuse reflection, while the highlight is from specular reflection. Were the apple entirely specular, then we wouldn't be able to observe most of its surface.

Often there are many light sources illuminating a scene and many more surface elements reflecting light from these sources. We might not be able to account for all the exchanges of energy, except by saying that there is *ambient light* in the scene. In computer graphics, it is common to use an ambient light factor when shading a surface.

69 Definition. **Ambient light** is steady state light energy everywhere in the scene resulting from multiple light sources and the interreflections off many surfaces.

Some surfaces actually emit light. These might be light bulbs or perhaps an object that absorbs one kind of energy and then emits it in the visible band. Such objects will reflect light as well as emit it. Finally, all of our emitting or reflecting phenomena are wavelength dependent. A source emits an entire spectrum of different wavelengths (unless it is a monchromatic laser) and a surface reflects or absorbs energy in some wavelengths more than others. Machines can be built to be sensitive to these wavelength phenomena; for example, multispectral scanners can produce 200 values for reflection from a single surface element. For humans, however, we can summarize a sample of visible light using a combination of only three values, such as RGB or HSI. Computer graphics commonly describes both illumination and surface reflection in terms of only RGB components.

Exercise 6.16

An amateur photographer took a photo of friends at the rim of the Grand Canyon just after sunset. Although a flash was used and the images of the friends were good, the beautiful canyon wall background was almost black. Why?

6.6.6 Phong Model of Shading*

A popular shading model used in computer graphics is the Phong shading model, which accounts for several phenomena; (a) ambient light, (b) diffuse reflection, (c) specular reflection, and (d) darkening with distance. Components (b), (c), and (d) are summed for each separate light source. We assume that we know the details of the surface element imaging at image point $\mathbf{I}[x, y]$ and the position and characteristics of all light sources. The reflective properties of this surface element are represented by $K_{d\lambda}$ for diffuse reflectivity and $K_{s\lambda}$ for specular reflectivity, where $K_{q\lambda}$ is a vector of coefficients of reflection for different wavelengths λ—usually three of them for RGB.

$$I_\lambda[x, y] = I_{a\lambda} K_{d\lambda} + \sum_{m=1}^{M} \left(\frac{1}{cd_m^2} I_{m\lambda} [K_{d\lambda}(\mathbf{n} \circ \mathbf{s}) + K_{s\lambda}(\mathbf{R}_m \circ \mathbf{V})^\alpha] \right) \tag{6.9}$$

Equation 6.9 uses ambient illumination $I_{a\lambda}$ and a set of M light sources $I_{m\lambda}$. The equation can be thought of as a vector equation treating each wavelength λ similarly. $I_{a\lambda}$ is the intensity of ambient light for wavelength λ, $I_{m\lambda}$ is the intensity of the light source m for wavelength λ. The *m-th* light source is a distance d_m from the surface element and makes reflection ray $\mathbf{R}_m$ off the surface element.

6.6.7 Human Perception Using Shading

There is no doubt that human perception of three-dimensional object shape is related to perceived surface shading. Moreover, the phenomena described account for shading that we perceive, although the above models of illumination and reflection are simplified. The simplified models are of central importance in computer graphics, and various approximations are used in order to speed up rendering of lit surfaces. In controlled environments, computer vision systems can even compute surface shape from shading using the above formulas: These methods are discussed in Chapter 13. We could, for example, compute surface normals for the surface points shown in Figure 6.16 by calibrating our formulas. In uncontrolled scenes, such as outdoor scenes, it is much more difficult to account for the different phenomena.

6.7 RELATED TOPICS*

6.7.1 Applications

Color features make some pattern recognition problems much simpler compared to when only intensity, texture, or shape information are available. Color measurements are local; aggregation methods and shape analysis may not be needed. For example, as indicated in Exercise 6.13, pixel level color information goes a long way in classification of fruits and vegetables for automatic charging at the grocery store or for quality sorting in a distribution center. A second example is the creation of a filter to remove pornographic images from the World Wide Web. The face detection algorithm as described in Section 6.5 first detects skin color according to the training data: Regions of skin pixels can then be aggregated and geometric relations between skin regions computed. If it is probable that bare body parts fill a significant part of the image, then that image could be blocked. Color is useful for access to image databases, as described in Chapter 8, and for understanding of biological images taken through a microscope.

6.7.2 Human Color Perception

Characteristics of human color perception are important for two reasons; first, the human visual system is often an efficient system to study and emulate, second, the main goal of graphics and image displays is to communicate with humans. The machine vision engineer often wants to learn how to duplicate or replace human capabilities, while the graphic artist must learn how to optimize communication with humans.

Humans in general have biased interpretations of colors. For example, wall colors are generally unsaturated pastels and not saturated colors; reds tend to stimulate, while blues tend to relax. Perhaps 8 percent of humans have some kind of color blindness, meaning that color combinations should be chosen carefully for communication. In the human retina, red and green sensitive receptors greatly outnumber blue receptors; this is accentuated in the high resolution fovea where blue receptors are rare. As a result, much color processing occurs in neurons that integrate input from the receptors. Various theories have been proposed to explain color processing in terms of processing by neurons. This higher-level processing is not fully understood and human visual processing is constantly under study. The color of single pixels of a display cannot be accurately perceived, but humans can make good judgments about the color of an extended surface even under variations of illumination, including illumination by only two principal wavelengths. Often, separate edge-based processing of intensity (Chapter 5) that is faster than color processing yields object recognition before color processing is complete. Theories usually address how human color processing might have evolved on top of more primitive intensity processing. The reader can pursue the vast area of human visual perception by referring to the references and following other references given there.

6.7.3 Multispectral Images

As discussed in Chapter 2, a sensor that obtains three color measurements per pixel is a multispectral sensor. However, sensing can be done in bands of the electromagnetic spectrum that are not perceived as color by humans; for example, in infrared bands of the spectrum. In IR bands of a satellite image, hot asphalt roads should appear bright and cold bodies of water should appear dark. Having multiple measurements at a single pixel is often useful for classifying the surface imaged there using simple procedures. The scanning system can be expensive, since it must be carefully designed in order to insure that the several frequency bands of radiation are indeed collected from the same surface element. The parameters of MRI scanning (refer to Chapter 2) can be changed to get multiple 3D images, effectively yielding m intensities for each voxel of the volume scanned. These n measurements can be used to determine whether the voxel material is fat, blood, or muscle tissue, etc. The reader might be alarmed to learn that it can take a full hour to obtain a 3D volume of MRI data, implying that some noise due to motion will be observed, particularly near the boundaries between different tissues where the material element sampled is most likely to change during the scanning process due to small motions caused by circulation or respiration.

6.7.4 Thematic Images

Thematic images use *pseudo color* to encode material properties or use of space represented in an image. For example, pixels of a map or satellite image might be labeled for human consumption so that rivers are blue, urban areas are purple, and roads are red. These are not the natural colors recorded by sensors but communicate image content well in our culture. Weather maps might show a temperature theme with red for hot and blue for cold. Similarly, thematic images can encode surface depth, local surface orientation or geometry, texture, density of some feature or any other scalar measurement or nominal classification. The two center images in Figure 6.13 are thematic images: The yellow, blue, and purple colors are just labels for three clusters in the real color space. It is important to remember that thematic images do not show actual physical sensor data but rather transduced or classified data for better visualization by a human.

6.8 REFERENCES

For a detailed treatment of light and optics, one can consult the text by Hecht and Zajac (1974). Some of the treatment of practical digital encoding of color was derived from Murray and VanRyper (1994): The reader can consult that book for many details on the many file formats used to store digital images. Details of the design of color display hardware, especially the shadow-mask technology for color displays, can be found in the graphics text by Foley and others (1996). The book by Levine (1985) contains the discussion of several different biological vision systems and their characteristics as devices. More detail is given in the book by Overington (1992), which takes a technical signal processing approach. Livingstone (1988) is a good start in the psychology literature. The discussion of matching color histograms was drawn from Swain and Ballard (1991) and Jain and Vailaya (1996). More details on the face extraction work can be found in a report by Bakic and Stockman (1999). Work on multispectral analysis of the brain using MRI can be found in the paper by Taxt and Lundervold (1994).

1. Bakic, V., and G. Stockman. 1999. Menu selection by facial aspect. *Proc. Vision Interface '99* (19–21 May 1999). Trois Rivieres, Quebec.
2. Fleck, M., D. Forsyth, and C. Pregler. 1966. Finding naked people. *Proc. Euro. Conf. Comput. Vision.* Springer-Verlag, New York, 593–602.
3. Foley, J., A. van Dam, S. Feiner, and J. Hughes. 1996. *Computer Graphics: Principles and Practice,* 2nd Ed in C. Addison-Wesley, New York.
4. Hecht, E., and A. Zajac. 1974. *Optics.* Addison-Wesley, New York.
5. Jain, A., and A. Vailaya. 1996. Image retrieval using color and shape. *Pattern Recog.,* v. 29(8):1233–1244.
6. Levine, M. 1985. *Vision in Man and Machine.* McGraw-Hill, New York.
7. Livingstone, M. 1988. Art, illusion and the visual system. *Sci. Am.* (Jan. 1988), 78–85.
8. Murray, J., and W. VanRyper. 1994. *Encyclopedia of Graphical File Formats.* O'Reilly and Associates, Sebastopol, CA.
9. Overington, I. 1992. *Computer Vision: A Unified, Biologically-Inspired Approach.* Elsevier, Amsterdam.
10. Swain, M., and D. Ballard. 1991. Color indexing. *Inter. J. Comput. Vision,* v. 7(1): 11–32.
11. Taxt, T., and A. Lundervold. 1994. Multispectral analysis of the brain in magnetic resonance imaging. *Proc. IEEE Workshop on Biomed. Image Anal.* (24–25 June 1994), Seattle, 33–42.

7

Texture

Texture is another feature that can help to segment images into regions of interest and to classify those regions. In some images, it can be the defining characteristic of regions and critical in obtaining a correct analysis. The image of Figure 7.1 has three very distinct textures: the texture of the tiger, the texture of the jungle, and the texture of the water. These textures can be quantified and used to identify the object classes they represent.

Texture gives us information about the spatial arrangement of the colors or intensities in an image. Suppose that the histogram of a region tells us that it has 50 percent white pixels and 50 percent black pixels. Figure 7.2 shows three different images of regions with this intensity distribution that would be considered three different textures. The leftmost image has two big blocks: one white and one black. The center image has eighteen small white blocks and eighteen small black blocks forming a checkerboard pattern. The rightmost image has six long blocks, three white and three black, in a striped pattern.

The images of Figure 7.2 were artificially created and contain geometric patterns constructed from black and white rectangles. Texture is commonly found in natural scenes, particularly in outdoor scenes containing both natural and man-made objects. Sand, stones, grass, leaves, bricks, and many more objects create a textured appearance in images. Figure 7.3 illustrates some of these natural textures. Note that the two different brick textures and two different leaf textures shown are quite different. Thus, textures must be described by more than just their object classifications. This chapter discusses what texture is, how it can be represented and computed, and its use in image analysis.

Figure 7.1 An image containing several different regions, each having a distinct texture. (Licensed from Corel Stock Photos.)

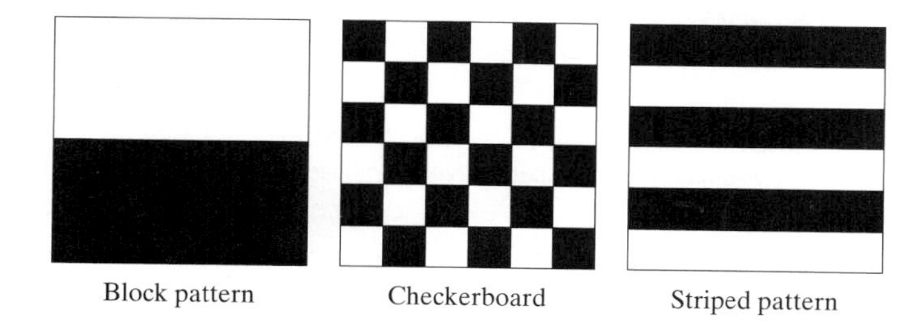

Figure 7.2 Three different textures with the same distribution of black and white.

7.1 TEXTURE, TEXELS, AND STATISTICS

The artificial textures of Figure 7.2 are made up of primitive rectangular regions in white or black. In the checkerboard, the regions are small squares arranged in a 2D grid of alternating colors. In the striped pattern, the regions are long stripes arranged in a vertical sequence of alternating colors. It is easy to segment out these single-color regions and to recognize these simple patterns.

Now, consider the two leaf textures of Figure 7.3. The first has a large number of small, round leaves, while the second has a smaller number of larger, pointed leaves. It is difficult to describe the spatial arrangements of the leaves in words; the arrangements are not regular, but there is some quality of the image that would make one argue that there is a noticeable arrangement in each image.

Part of the problem in texture analysis is defining exactly what texture is. There are two main approaches:

1. **structural approach:** Texture is a set of primitive *texels* in some regular or repeated relationship.

Figure 7.3 Natural textures. (From the MIT Media Lab VisTex database: http://vismod.www.media.mit.edu/vismod/imagery/VisionTexture/vistex.html.)

2. **statistical approach:** Texture is a quantitative measure of the arrangement of intensities in a region.

While the first approach is appealing and can work well for man-made, regular patterns, the second approach is more general and easier to compute and is used more often in practice.

7.2 TEXEL-BASED TEXTURE DESCRIPTIONS

A texture can be thought of as a set of primitive texels in a particular spatial relationship. A structural description of a texture would then include a description of the texels and a specification of the spatial relationship. Of course, the texels must be segmentable and the relationship must be efficiently computable. One very nice geometry-based description was proposed by Tuceryan and Jain (1990). The texels are image regions that can be extracted through some simple procedure such as thresholding. The characterization of their spatial relationships is obtained from a Voronoi tesselation of the texels as explained below.

Suppose that we have a set of already-extracted texels, and that we can represent each one by a meaningful point, such as its centroid. Let S be the set of these points. For any pair of points P and Q in S, we can construct the perpendicular bisector of the line joining them. This perpendicular bisector divides the plane into two half planes, one of which is the set of points that are closer to P and the other of which is the set of points that are closer to Q. Let $H^Q(P)$ be the half plane that is closer to P with respect to the perpendicular bisector of P and Q. We can repeat this process for each point Q in S. The *Voronoi polygon* of P

is the polygonal region consisting of all points that are closer to P than to any other point of S and is defined by

$$V(P) = \bigcap_{Q \in S, Q \neq P} H^Q(P)$$

Figure 7.4 illustrates the Voronoi polygons for a set of circular texels. This pattern produces hexagonal polygons for internal texels; texels that border the image boundary have various other shapes.

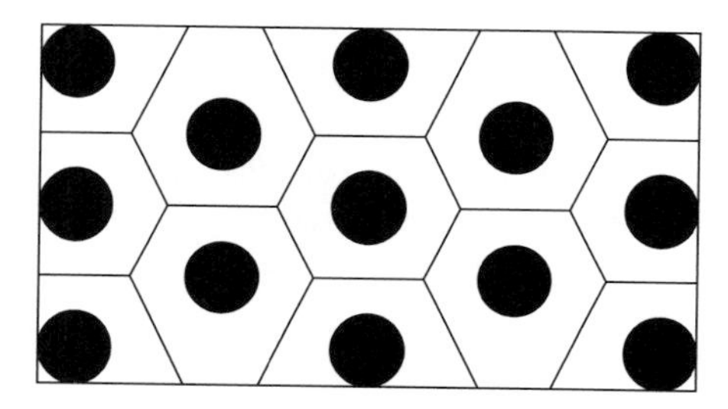

Figure 7.4 The Voronoi tesselation of a set of circular texels.

Once the texels have been extracted from an image and their Voronoi tesselation computed, shape features of the polygons are calculated and used to group the polygons into clusters that define uniformly textured regions. The type of pattern shown in Figure 7.4 extended to a large image would produce a single region of uniform texture characterized by the shape features of the regular hexagons.

Exercise 7.1: Texel-Based Descriptions.

Find or create a set of five images showing textures that have obvious texels that can be detected via a simple procedure such as thresholding based on gray-tone or color ranges. Try to find at least one texture that has more than one kind of texel. Draw the Voronoi tesselation for a small area on this image.

7.3 QUANTITATIVE TEXTURE MEASURES

For the most part, segmenting out the texels is more difficult in real images than in artificially generated patterns. Instead, numeric quantities or statistics that describe a texture can be computed from the gray tones (or colors) themselves. This approach is less intuitive, but is computationally efficient and can work well for both segmentation and classification of textures.

7.3.1 Edge Density and Direction

Since edge detection is a well-known and simple-to-apply feature detection scheme, it is natural to try to use an edge detector as the first step in texture analysis. The number of edge pixels in a given fixed-size region gives some indication of the busyness of that region. The directions of these edges, which are usually available as a biproduct of the edge-detection process, can also be useful in characterizing the texture pattern.

Consider a region of N pixels. Suppose that a gradient-based edge detector is applied to this region producing two outputs for each pixel p: 1) the gradient magnitude $Mag(p)$ and 2) the gradient direction $Dir(p)$, as defined in Chapter 5. One very simple texture feature is *edgeness per unit area* which is defined by

$$F_{edgeness} = \frac{|\{p \mid Mag(p) \geq T\}|}{N} \tag{7.1}$$

for some threshold T. Edgeness per unit area measures the busyness, but not the orientation of the texture.

This measure can be extended to include both busyness and orientation by employing histograms for both the gradient magnitude and the gradient direction. Let $H_{mag}(R)$ denote the normalized histogram of gradient magnitudes of region R, and let H_{dir} denote the normalized histogram of gradient orientations of region R. Both of these histograms are over a small, fixed number (such as 10) of bins representing groups of magnitudes and groups of orientations. Both are normalized according to the size N_R of region R. Then

$$F_{mag\,dir} = (H_{mag}(R), H_{dir}(R)) \tag{7.2}$$

is a quantitative texture description of region R.

Consider the two 5×5 images shown in Figure 7.5. The image on the left is busier than the image on the right. It has an edge in every one of its 25 pixels, so its edgeness per unit area is 1.0. The image on the right has 6 edges out of its 25 pixels, so its edgeness per unit area is only 0.24. For the gradient-magnitude histograms, we will assume there are two bins representing dark edges and light edges. For the gradient-direction histograms, we will use three bins for horizontal, vertical, and diagonal edges. The image on the left has 6 dark edges and 19 light edges, so its normalized gradient-magnitude histogram is (0.24,0.76), meaning that 24 percent of the edges are dark and 76 percent are light. It also has 12 horizontal edges, 13 vertical edges, and no diagonal edges, so its normalized gradient-direction histogram is (0.48,0.52,0.0), meaning that 48 percent of the edges are horizontal, 52 percent are vertical, and 0 percent are diagonal. The image on the right has no dark edges and 6 light edges, so its normalized gradient-magnitude histogram is (0.0,0.24). It also has no horizontal edges, no vertical edges, and 6 diagonal edges, so its normalized

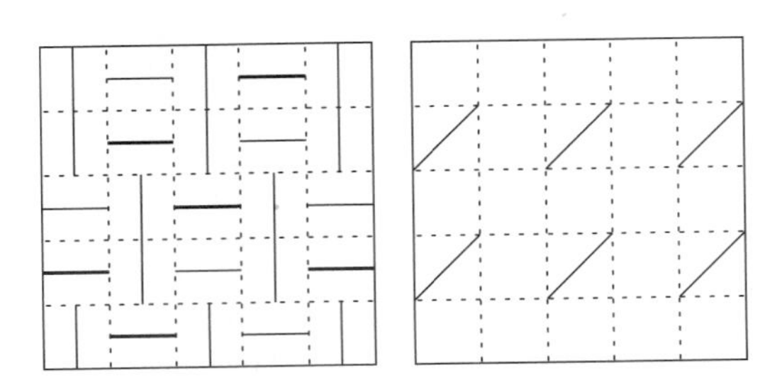

Figure 7.5 Two images with different edgeness and edge-direction statistics.

gradient-direction histogram is (0.0,0.0,0.24). In the case of these two images, the edgeness-per-unit-area measure is sufficient to distinguish them, but the histogram measure provides a more powerful descriptive mechanism in general. Two n-bin histograms H_1 and H_2 can be compared by computing their L_1 *distance*

$$L_1(H_1, H_2) = \sum_{i=1}^{n} | H_1[i] - H_2[i] | \tag{7.3}$$

Exercise 7.2: Edge-Based Texture Measures.

Obtain a set of images that have lots of man-made structures with clearly defined edges. Write a program to compute the texture measure $F_{mag\,dir}$ of Equation 7.2 for each of these images, and compare them using the L_1 distance of Equation 7.3.

7.3.2 Local Binary Partition

Another very simple, but useful texture measure is the local binary partition measure. For each pixel p in the image, the eight neighbors are examined to see if their intensity is greater than that of p. The results from the eight neighbors are used to construct an eight-digit binary number $b_1b_2b_3b_4b_5b_6b_7b_8$ where $b_i = 0$ if the intensity of the *ith* neighbor is less than or equal to that of p and 1 otherwise. A histogram of these numbers is used to represent the texture of the image. Two images or regions are compared by computing the L_1 distance between their histograms as defined above.

Exercise 7.3: LBP Texture Measures.

Using the images from the previous exercise, write another program to compute the histogram representing the LBP texture measure of each image. Again compute the L_1 distances between pairs of images using this measure. Compare to your previous results.

7.3.3 Co-occurrence Matrices and Features

A *co-occurrence* matrix is a two-dimensional array $\mathbf{C}$ in which both the rows and the columns represent a set of possible image values $\mathbf{V}$. For example, for gray-tone images $\mathbf{V}$ can be the set of possible gray tones and for color images $\mathbf{V}$ can be the set of possible colors. The value of $\mathbf{C(i,j)}$ indicates how many times value $\mathbf{i}$ co-occurs with value $\mathbf{j}$ in some designated spatial relationship. For example, the spatial relationship might be that value $\mathbf{i}$ occurs immediately to the right of value $\mathbf{j}$. To be more precise, we will look specifically at the case where the set $\mathbf{V}$ is a set of gray tones and the spatial relationship is given by a vector $\mathbf{d}$ that specifies the displacement between the pixel having value $\mathbf{i}$ and the pixel having value $\mathbf{j}$.

Let $\mathbf{d}$ be a displacement vector $\mathbf{(dr, dc)}$ where $\mathbf{dr}$ is a displacement in rows (downward) and $\mathbf{dc}$ is a displacement in columns (to the right). Let $\mathbf{V}$ be a set of gray tones. The gray-tone

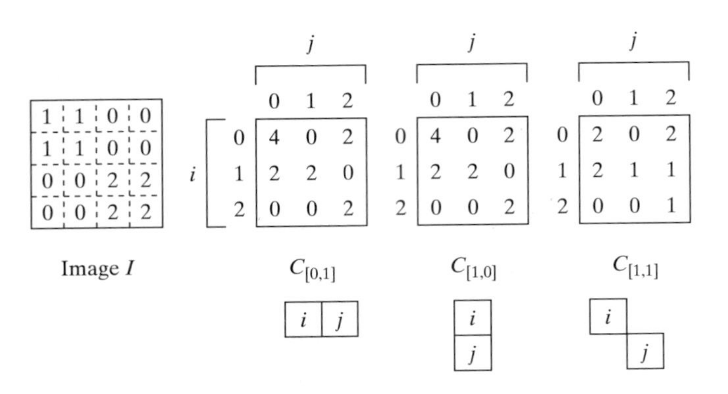

Figure 7.6 Three different co-occurrence matrices for a gray-tone image.

co-occurrence matrix $\mathbf{C_d}$ for image $\mathbf{I}$ is defined by

$$C_d[i, j] = |\{[r, c] \mid I[r, c] = i \textit{ and } I[r + dr, c + dc] = j\}| \tag{7.4}$$

Figure 7.6 illustrates this concept with a 4×4 image I and three different co-occurrence matrices for $\mathbf{I}$: $\mathbf{C}_{[0,1]}$, $\mathbf{C}_{[1,0]}$, and $\mathbf{C}_{[1,1]}$.

In $\mathbf{C}_{[0,1]}$ note that position [1,0] has a value of 2, indicating that $\mathbf{j} = 0$ appears directly to the right of $\mathbf{i} = 1$ two times in the image. However, position [0,1] has a value of 0, indicating that $\mathbf{j} = 1$ never appears directly to the right of $\mathbf{i} = 0$ in the image. The largest co-occurence value of 4 is in position [0,0], indicating that a 0 appears directly to the right of another 0 four times in the image.

Exercise 7.4: Co-occurence Matrices.

Construct the gray-tone co-occurrence matrices $\mathbf{C}_{[1,2]}$, $\mathbf{C}_{[2,2]}$, and $\mathbf{C}_{[2,3]}$ for the image of Figure 7.6.

There are two important variations of the standard gray-tone co-occurrence matrix. The first is the *normalized* gray-tone co-occurrence matrix $\mathbf{N_d}$ defined by

$$N_d[i, j] = \frac{C_d[i, j]}{\sum_i \sum_j C_d[i, j]} \tag{7.5}$$

which normalizes the co-occurrence values to lie between zero and one and allows them to be thought of as probabilities in a large matrix. The second is the *symmetric* gray-tone co-occurence matrix $\mathbf{S_d}$ defined by

$$S_d[i, j] = C_d[i, j] + C_{-d}[i, j] \tag{7.6}$$

which groups pairs of symmetric adjacencies.

Exercise 7.5: Normalized Co-occurrence.

Compute the normalized co-occurrence matrix $\mathbf{N}_{[1,1]}$ for the image of Figure 7.7 assuming that the black pixels have gray-tone 0, the gray pixels have gray-tone 1, and the white pixels have gray-tone 2. How does it represent the texture pattern of the image?

Figure 7.7 An image with a diagonal texture pattern.

Co-occurrence matrices capture properties of a texture, but they are not directly useful for further analysis, such as comparing two textures. Instead, numeric features are computed from the co-occurrence matrix that can be used to represent the texture more compactly. The following are standard features derivable from a normalized co-occurrence matrix.

$$Energy = \sum_i \sum_j N_d^2[i, j] \tag{7.7}$$

$$Entropy = -\sum_i \sum_j N_d[i, j] log_2 N_d[i, j] \tag{7.8}$$

$$Contrast = \sum_i \sum_j (i - j)^2 N_d[i, j] \tag{7.9}$$

$$Homogeneity = \sum_i \sum_j \frac{N_d[i, j]}{1 + |i - j|} \tag{7.10}$$

$$Correlation = \frac{\sum_i \sum_j (i - \mu_i)(j - \mu_j) N_d[i, j]}{\sigma_i \sigma_j} \tag{7.11}$$

where μ_i, μ_j are the means and σ_i, σ_j are the standard deviations of the row and column sums $\mathbf{N_d[i]}$ and $\mathbf{N_d[j]}$ defined by

$$N_d[i] = \sum_j N_d[i, j]$$

$$N_d[j] = \sum_i N_d[i, j]$$

One problem with deriving texture measures from co-occurrence matrices is how to choose the displacement vector **d**. A solution suggested by Zucker and Terzopoulos is to use a χ^2 statistical test to select the value(s) of **d** that have the most structure; that is, to maximize the value:

$$\chi^2(d) = \left(\sum_i \sum_j \frac{N_d^2[i, j]}{N_d[i] N_d[j]} - 1 \right)$$

7.3.4 Laws Texture Energy Measures

Another approach to generating texture features is to use local masks to detect various types of texture. Laws developed a texture-energy approach that measures the amount of variation within a fixed-size window. A set of nine 5×5 convolution masks is used to compute texture energy, which is then represented by a vector of nine numbers for each pixel of the image being analyzed. The masks are computed from the following vectors, which are similar to those studied in Chapter 5.

$$\begin{array}{lll} \text{L5} & \text{(Level)} & = [\ \ 1 \quad\ 4 \quad 6 \quad\ 4 \quad\ \ 1] \\ \text{E5} & \text{(Edge)} & = [-1 \quad -2 \quad 0 \quad\ 2 \quad\ \ 1] \\ \text{S5} & \text{(Spot)} & = [-1 \quad\ 0 \quad 2 \quad\ 0 \quad -1] \\ \text{R5} & \text{(Ripple)} & = [\ \ 1 \quad -4 \quad 6 \quad -4 \quad\ \ 1] \end{array}$$

The names of the vectors describe their purposes. The L5 vector gives a center-weighted local average. The E5 vector detects edges, the S5 vector detects spots, and the R5 vector detects ripples. The 2D convolution masks are obtained by computing outer products of pairs of vectors. For example, the mask E5L5 is computed as the product of E5 and L5 as:

$$\begin{bmatrix} -1 \\ -2 \\ 0 \\ 2 \\ 1 \end{bmatrix} \times \begin{bmatrix} 1 & 4 & 6 & 4 & 1 \end{bmatrix} = \begin{bmatrix} -1 & -4 & -6 & -4 & -1 \\ -2 & -8 & -12 & -8 & -2 \\ 0 & 0 & 0 & 0 & 0 \\ 2 & 8 & 12 & 8 & 2 \\ 1 & 4 & 6 & 4 & 1 \end{bmatrix}$$

The first step in the Laws procedure is to remove effects of illumination by moving a small window around the image, and subtracting the local average from each pixel, to produce a preprocessed image, in which the average intensity of each neighborhood is near to zero. The size of the window depends on the class of imagery; a 15×15 window was used for natural scenes. After the preprocessing, each of the sixteen 5×5 masks are applied to the preprocessed image, producing sixteen filtered images. Let $\mathbf{F_k[i,j]}$ be the result of filtering with the *kth* mask at pixel $\mathbf{[i,j]}$. Then the *texture energy map* $\mathbf{E_k}$ for filter $\mathbf{k}$ is defined by

$$E_k[r,c] = \sum_{j=c-7}^{c+7} \sum_{i=r-7}^{r+7} |F_k[i,j]| \tag{7.12}$$

Each texture energy map is a full image, representing the application of the *kth* mask to the input image.

Once the sixteen energy maps are produced, certain symmetric pairs are combined to produce the nine final maps, replacing each pair with its average. For example, E5L5 measures horizontal edge content, and L5E5 measures vertical edge content. The average of these two maps measures total edge content. The nine resultant energy maps are

L5E5/E5L5	L5S5/S5L5
L5R5/R5L5	E5E5
E5S5/S5E5	E5R5/R5E5
S5S5	S5R5/R5S5
R5R5	

The result of all the processing gives nine energy map images or, conceptually, a single image with a vector of nine texture attributes at each pixel. Table 7.1 shows the nine texture attributes for the main texture of each of the grass/stones/brick images of Figure 7.3. These texture attributes can be used to cluster an image into regions of uniform texture. Figure 7.8 illustrates the segmentation of several multi-texture images into clusters.

TABLE 7.1 LAWS TEXTURE ENERGY MEASURES FOR THE IMAGES OF FIGURE 7.3.

Image	E5E5	S5S5	R5R5	E5L5	S5L5	R5L5	S5E5	R5E5	R5S5
Leaves1	250.9	140.0	1309.2	703.6	512.2	1516.2	187.5	568.8	430.0
Leaves2	257.7	121.4	988.7	820.6	510.1	1186.4	172.9	439.6	328.0
Grass	197.8	107.2	1076.9	586.9	410.5	1208.5	144.0	444.8	338.1
Brick1	128.1	60.2	512.7	442.1	273.8	724.8	86.6	248.1	176.3
Brick2	72.4	28.6	214.2	263.6	130.9	271.5	43.2	93.3	68.5
Stone	224.6	103.2	766.8	812.8	506.4	1311.0	150.4	413.5	281.1

7.3.5 Autocorrelation and Power Spectrum

The autocorrelation function of an image can be used to detect repetitive patterns of texture elements and also describes the fineness/coarseness of the texture. The autocorrelation function $\rho(\mathbf{dr,dc})$ of an $(\mathbf{N}+1) \times (\mathbf{N}+1)$ image for displacement $\mathbf{d} = \mathbf{(dr,dc)}$ is given by

$$\rho(dr,dc) = \frac{\sum_{r=0}^{N}\sum_{c=0}^{N} I[r,c]I[r+dr,c+dc]}{\sum_{r=0}^{N}\sum_{c=0}^{N} I^2[r,c]} \tag{7.13}$$

$$= \frac{I[r,c] \circ I_d[r,c]}{I[r,c] \circ I[r,c]} \tag{7.14}$$

using the ideas of Chapter 5.

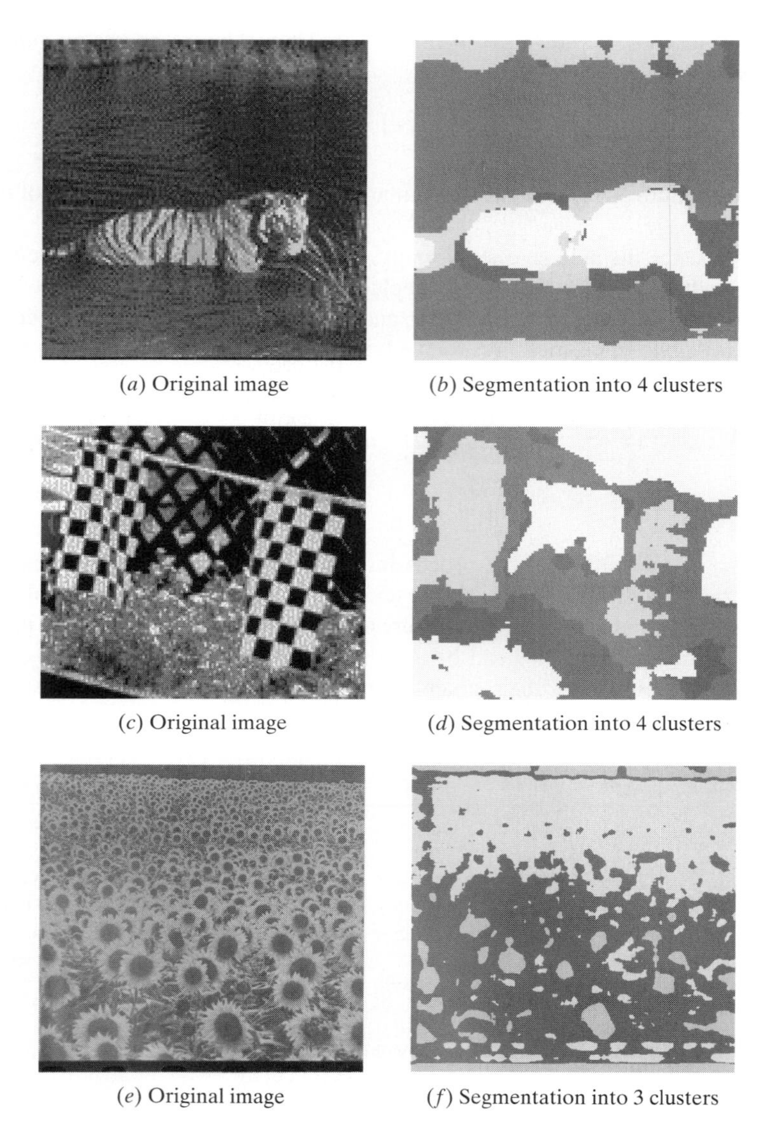

(*a*) Original image (*b*) Segmentation into 4 clusters

(*c*) Original image (*d*) Segmentation into 4 clusters

(*e*) Original image (*f*) Segmentation into 3 clusters

Figure 7.8 Examples of image segmentation using the Laws texture energy measures. (Original images are from Corel Stock Photos and from the MIT Media Lab VisTex database.) See colorplate.

Exercise 7.6: Laws Texture Energy Measures.

Write a program to compute the Laws texture energy measures that inputs a gray-scale image and outputs a set of nine images, one for each of the texture energy measures. Obtain a set of images of both man-made and natural textures and perform a sequence of tests on them. For each test, make one of the images the *test image* and call the others the *database images*. Write an interactive front end that allows a user to select a pixel of the test image. It then looks for those database images that have a texture similar to that of the selected pixel anywhere in that database image using the L_1 distance on the set of nine texture energy measures available for each pixel. The brute force way to do this is merely to compare the nine values for the test image pixel with the nine values for each pixel of each database image and select an image as soon as any of its pixels has similar enough texture energy measure values. How might you do this more efficiently?

If the texture is coarse, then the autocorrelation function drops off slowly; otherwise, it will drop off very rapidly. For regular textures, the autocorrelation function will have peaks and valleys. Since $\mathbf{I[r+dr,\ c+dc]}$ is undefined at the boundaries of the image, a method for computing these virtual image values must be defined.

The autocorrelation function is related to the power spectrum of the Fourier transform. If $\mathbf{I[r,c]}$ is the image function and $F(u, v)$ is its Fourier transform, the quantity $|F(u, v)|^2$ is defined as the power spectrum where $|\cdot|$ is the modulus of a complex number. The frequency domain can be divided into n_r regions bounded by circular rings (for frequency content) and n_d wedges (for orientation content) and the total energy in each region is computed to produce a set of texture features, as introduced in Chapter 5.

7.4 TEXTURE SEGMENTATION

Any texture measure that provides a value or vector of values at each pixel, describing the texture in a neighborhood of that pixel, can be used to segment the image into regions of similar textures. Like any other segmentation algorithm, texture segmentation algorithms are of two major types: region-based approaches and boundary-based approaches. Region-based approaches attempt to group or cluster pixels with similar texture properties. Boundary-based approaches attempt to find texture edges between pixels that come from different texture distributions. We leave the discussion of segmentation algorithms to Chapter 10 on Image Segmentation. Figure 7.8 shows the segmentation of several images using the Laws texture energy measures and a clustering algorithm to group pixels into regions.

In Figure 7.8(a) and (b), the tiger image has been segmented into regions representing tiger, water, and some other miscellaneous areas of the image. In Figure 7.8(c) and (d) a multi-object image has been segmented into regions that roughly correspond to the grass, the two flags, the black mesh fence, and some background. In Figure 7.8(e) and (f), the sunflower image has been segmented into three types of texture: dark borders found at the top and bottom of the image, small, far-away sunflowers at the back of the field and large, close-up sunflowers at the front of the field. Table 7.2 shows the mean Laws texture energy measures for the main regions of each of these images. Table 7.3 gives a comparison of the Laws measures for tiger regions in several different images.

In the sunflower image, the dark centers of some of the larger flowers are grouped with the dark border textures, because the mask used to compute the texture is smaller than these large flower centers. In general, these segmentation results are imperfect; they can do no better than the operators that define them. Segmentation based on both color *and* texture can do better, but segmentation of natural scenes is an unsolved problem. See Chapter 10 for a more comprehensive treatment of segmentation in general.

TABLE 7.2 LAWS TEXTURE ENERGY MEASURES FOR MAJOR REGIONS OF THE IMAGES OF FIGURE 7.8.

Region	E5E5	S5S5	R5R5	E5L5	S5L5	R5L5	S5E5	R5E5	R5S5
Tiger	168.1	84.0	807.7	553.7	354.4	910.6	116.3	339.2	257.4
Water	68.5	36.9	366.8	218.7	149.3	459.4	49.6	159.1	117.3
Flags	258.1	113.0	787.7	1057.6	702.2	2056.3	182.4	611.5	350.8
Fence	189.5	80.7	624.3	701.7	377.5	803.1	120.6	297.5	215.0
Grass	206.5	103.6	1031.7	625.2	428.3	1153.6	146.0	427.5	323.6
Small flowers	114.9	48.6	289.1	402.6	241.3	484.3	73.6	158.2	109.3
Big flowers	76.7	28.8	177.1	301.5	158.4	270.0	45.6	89.7	62.9
Borders	15.3	6.4	64.4	92.3	36.3	74.5	9.3	26.1	19.5

TABLE 7.3 LAWS TEXTURE ENERGY MEASURES FOR TIGER REGIONS OF SEVERAL DIFFERENT IMAGES.

Image	E5E5	S5S5	R5R5	E5L5	S5L5	R5L5	S5E5	R5E5	R5S5
Tiger1	171.2	96.8	1156.8	599.4	378.9	1162.6	124.5	423.8	332.3
Tiger2a	146.3	79.4	801.1	441.8	302.8	996.9	106.5	345.6	256.7
Tiger2b	177.8	96.8	1177.8	531.6	358.1	1080.3	128.2	421.3	334.2
Tiger3	168.8	92.2	966.3	527.2	354.1	1072.3	124.0	389.0	289.8
Tiger4	168.1	84.0	807.7	553.7	354.4	910.6	116.3	339.2	257.4
Tiger5	146.9	80.7	868.7	474.8	326.2	1011.3	108.2	355.5	266.7
Tiger6	170.1	86.8	913.4	551.1	351.3	1180.0	119.5	412.5	295.2
Tiger7	156.3	84.8	954.0	461.8	323.8	1017.7	114.0	372.3	278.6

Exercise 7.7: Texture Segmentation.

Use your program that computes the Laws texture energy measures for an image to investigate how well they would perform for texture segmentation. Write another interactive front end tnat allows the user to draw boxes around sections of the image that contain a single category of texture such as flowers or grass or sky. For each box, compute the average values for the nine texture features. Produce a table that lists each texture category by name and prints the nine averages for that category. Compare the results on several different categories.

7.5 REFERENCES

Texture analysis is one of the oldest areas of computer vision, dating back to early remote sensing applications in the late 1960s and early 1970s. Haralick, Shanmugam, and Dinstein (1973) developed the classic co-occurrence matrix features for analysis of remotely sensed images. Zucker and Terzopoulos (1980) developed a statistical test for determining the best displacements to use with co-occurrence techniques, and Trivedi (1984) used gray-tone co-occurrence features for object detection.

Julesz (1975) ran a set of now-famous experiments on human perception of texture. Tamura, Mori, and Yamawaki (1978) developed textural features that were meant to correspond to human visual perception. Tomita, Shirai, and Tsuji (1982) and Tuceryan and Jain

(1990) developed structural approaches to texture analysis. Wang and He (1990) proposed a spectral approach, while Cross and Jain (1983) developed a Markov random field method. The Laws texture energy masks are widely used in practice to produce numeric texture signatures. A survey article by Weszka, Dyer, and Rosenfeld (1976) describes the early work in the field. A newer survey by Tuceryan and Jain (1994) describes the full range of available methods up to 1995. Recent work by Leung and Malik (1999) gives a new method for learning texture primitives from training samples.

1. Cross, G. R., and A. K. Jain. 1983. Markov random field texture models. *IEEE Trans. Pattern Anal. Machine Intelligence,* v. PAMI-5:25–39.
2. Haralick, R. M., K. Shanmugam, and I. Dinstein. 1973. Textural features for image classification. *IEEE Trans. Systems, Man, and Cybernetics,* v. 3:610–621.
3. Julesz, B. 1975. Experiments in the visual perception of texture. *Sci. Am.,* 34–43.
4. Laws, K. 1980. Rapid texture identification. *SPIE Image Processing for Missile Guidance,* v. 238:376–380.
5. Leung, T., and J. Malik. 1999. Recognizing surfaces using three-dimensional textons. *Int. Conf. Comput. Vision* (Sept. 1999), 1010–1017.
6. Tamura, H., S. Mori, and T. Yamawaki. 1978. Textural features corresponding to visual perception. *IEEE Trans. Systems, Man, and Cybernetics,* v. 8(6):460–473.
7. Tomita, F., Y. Shirai, and S. Tsuji. 1982. Description of textures by a structural analysis. *IEEE Trans. Pattern Anal. Machine Intelligence,* v. PAMI-4:183–191.
8. Trivedi, M. M. 1984. Object detection based on gray level cooccurrence. *Comput. Vision, Graphics, and Image Proc.,* v. 28:199–219.
9. Tuceryan, M., and A. K. Jain. 1994. Texture analysis. In *Handbook of Pattern Recognition and Vision,* C. H. Chen, L. F. Pau, and P. S. P. Wang, Eds. World Scientific Publishing Co., Singapore, 235–276.
10. Tuceryan, M., and A. K. Jain. 1990. Texture segmentation using Voronoi polygons. *IEEE Trans. Pattern Anal. Machine Intelligence,* v. 12(2):211–216.
11. Wang, L., and D. C. He. 1990. Texture classification using texture spectrum. *Pattern Recog. Lett.,* v. 13:905–910.
12. Weszka, J., C. R. Dyer, and A. Rosenfeld. 1976. A comparative study of texture measures for terrain classification. *IEEE Trans. Systems, Man, and Cybernetics,* v. SMC-6: 269–285.
13. Zucker, S. W., and D. Terzopoulos. 1980. Finding structure in co-occurrence matrices for texture analysis. *Comput. Graphics and Image Proc.,* v. 2:286–308.

8

Content-Based Image Retrieval

As large amounts of both internal and external memory become increasingly less expensive and processors become increasingly more powerful, image databases have gone from an expectation to a firm reality. Image databases exist for storing art collections, satellite images, medical images, and general collections of photographs. Uses vary according to application. Art collection users may wish to find work by a certain artist or to find out who painted a particular image they have seen. Medical database users may be medical students studying anatomy or doctors looking for sample instances of a given disease. General collections might be accessed by illustrators looking for just the right picture for an article or book. The domain of this application is enormous; one user might want to find images of horses, another might want sunsets, and a third might be looking for an abstract concept, such as love.

Image databases can be huge, containing hundreds of thousands or millions of images. In most cases they are only indexed by keywords that have to be decided upon and entered into the database system by a human categorizer. However, images can be retrieved according to their content, where content might refer to color distributions, texture, region shapes, or object classification. While the state of segmentation and recognition algorithms is still primitive, commercial and research systems have been built and are already in use; demo systems are often available on the World Wide Web. This chapter explores the methods by which humans can retrieve images without resorting to a keyword search.

8.1 IMAGE DATABASE EXAMPLES

Some image databases have been constructed just to show how a particular retrieval system works. The IBM Query by Image Content (QBIC) Project Database is an example of this kind. QBIC is a research system that led to a commercial product developed and sold

(*a*) Renoir painting image

(*b*) Amethyst image

Figure 8.1 Images from digital collections. (Pierre-Auguste Renoir painting image, Landscape at Beaulieu, c. 1893, courtesy of the Fine Arts Museums of San Francisco, Mildred Anna Williams Collection, 1944.9; amethyst image courtesy of the Smithsonian Institution, 1992.) See colorplate.

by IBM. It retrieves images based on visual content, including such properties as color percentages, color layout, and texture. Virage, Inc. developed a competing product, the Virage search engine, which can retrieve images based on color, composition, texture, and structure. These and other image search engines can be used to search databases that are provided by other institutions. For example, the Fine Arts Museums of San Francisco have provided QBIC access to their Imagebase, a collection of digitized paintings. A Renoir painting from this collection is shown in Figure 8.1(*a*). Similar digital art collections are being created in many major cities of the world.

In addition to art collections, there are general image collections whose individual images are available for licensing by private customers who may want them for marketing products or illustrating articles. One of the biggest is the Corbis Archive, which contains more than 17 million images, with nearly one million in digital format and growing. This archive tries to capture the full range of human expression and perception; it contains such categories as history, art, entertainment, science and industry, and animals. Corbis provides retrieval of its images by keyword and by browsing. Another company, Getty Images, now provides several online image databases organized by categories and searchable through keywords.

In addition to artwork and photographs, there are scientific and medical image collections. The National Library of Medicine provides a database of X-rays, CT scans, MRI images, and color cross-sections, taken at very small intervals along the bodies of a male and a female cadaver. There are over 14,000 images available for people who want to use them for medical research. The National Aeronautics and Space Administration (NASA) collects huge databases of images from its satellites and makes them available for public acquisition (for a fee). The United States Geological Survey (USGS) provides a Web search capability for users who wish to find and order data sets including digital satellite and aerial images. Finally, the World Wide Web itself is a database that contains both text and large numbers of images. Search engines for finding images on the Web, based on keywords and, to a limited extent, on image content are being developed.

8.2 IMAGE DATABASE QUERIES

Given a database of images, we must have some way, other than searching through the whole set, to retrieve them. Companies that make databases of images available to their customers will generally have a selection process for determining which images should be added to the collection and a categorization process for assigning general categories and other keywords to the selected images. Images appearing on the World Wide Web will usually have a caption from which keywords can be obtained automatically.

In a relational database system, entities can be retrieved based on the values of their textual attributes. Attributes used to retrieve images might include general category, names of objects present, names of people present, date of creation, and source. The images can be indexed according to these attributes, so that they can be rapidly retrieved when a query is issued. This type of textual query can be expressed in the SQL relational database language, which is available for all standard relational systems. For example, the query

```
SELECT * FROM IMAGEDB
WHERE CATEGORY = 'GEMS' AND SOURCE = 'SMITHSONIAN'
```

would find and return all images from the set named IMAGEDB whose CATEGORY attribute was set to 'GEMS' and whose SOURCE attribute was set to 'SMITHSONIAN'. The object would be to retrieve images of the gem collection of the Smithsonian Institute. Figure 8.1(b) shows an amethyst image from this collection. The amethyst image would be retrieved along with many other gem images. In order to allow more selective retrievals, a descriptive set of keywords would have to be stored for each image. In a relational database, KEYWORD would be an attribute that could have multiple values for each image. So the amethyst image might have values of 'AMETHYST', 'CRYSTAL', and 'PURPLE' as its keywords, and could be retrieved according to all three or any of the three, depending on the desires of the user. For example, the SQL query

```
SELECT * FROM IMAGEDB
WHERE CATEGORY = 'GEMS' AND SOURCE = 'SMITHSONIAN'
      AND (KEYWORD = 'AMETHYST' OR KEYWORD = 'CRYSTAL'
           OR KEYWORD = 'PURPLE')
```

retrieves all images from the set named IMAGEDB, whose CATEGORY is 'GEMS', whose SOURCE is 'SMITHSONIAN', and which has a KEYWORD value of 'AMETHYST', 'CRYSTAL', or 'PURPLE'. This will retrieve more than just the amethyst image; the user will be able to browse through and select images from the set returned.

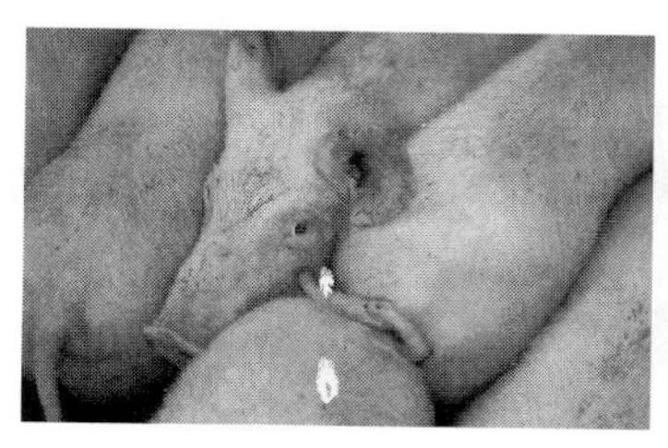

(*a*) Image with pigs

(*b*) Image with no pigs

Figure 8.2 Pig images returned by keyword search. (Image (*a*) was licensed from Corbis. Credit line: \051 Clive Druett; Papilio/CORBIS.)

There is a limit to what can be done with the keyword approach. Human coding of keywords is expensive and is bound to leave out some terms by which users will want to reference an image. In Web databases, using HTML captions can help to automate, but also provides only limited indexing capability. Furthermore, some of the returned images may turn out to look very different than the user expects from the automatically derived keywords. Figure 8.2 shows two images returned from a web search with the keyword 'pigs'.

Given that keywords alone are insufficient, we will explore other methods for retrieving images that can be used instead of or in addition to keywords.

Exercise 8.1: Keyword Queries.

Give an SQL query that will retrieve the image of Figure 8.2(*a*), but will not retrieve the image of Figure 8.2(*b*). Use whatever categories and keywords you think are appropriate.

8.3 QUERY-BY-EXAMPLE

Query-by-example is database terminology for a query that is formulated by filling in values and constraints in a table and can be converted by the system to SQL. The first QBE system was developed by IBM; today, Microsoft Access is a good example of this type of system. In standard relational databases, where the values of attributes are mainly text or numeric values, query-by-example merely provides a convenient interface for the user, without any additional power.

In image databases, the very idea of query-by-example is exciting. Instead of typing a query, the image database user should be able to show the system a sample image, or paint one interactively on the screen, or just sketch the outline of an object. The system should then be able to return similar images or images containing similar objects. This is the goal of all content-based image retrieval systems; each one has its own ways of specifying queries, determining similarity between a query and an image in the database, and of selecting the images to be returned.

To keep our discussion general, let us consider a query as an example image plus a set of constraints. The image may be a digital photograph; a user-painted rough example; a

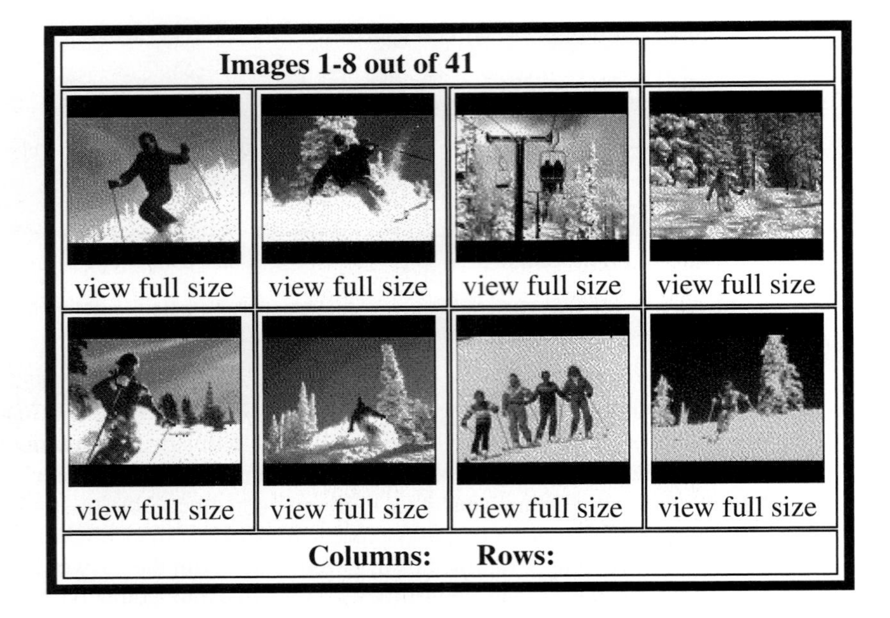

Figure 8.3 Results of a QBIC search based on color layout similarity; the query is the example image shown in the top left position. (Images courtesy of Egames.) See colorplate.

line-drawing sketch; or empty, in which case the retrieval set must only satisfy the constraints. The constraints may be keywords that should be present in some indexing system or may specify objects that should be in the image and even spatial relationships among them. In the most common case, the query is a digital image that is compared to images in the database according to an *image distance measure*. When the distance returned is zero, the image exactly matches the query. Values larger than zero indicate various degrees of similarity to the query. Image search engines usually return a set of images in order of their distance to the query. Figure 8.3 illustrates the results of a QBIC search based on its color layout distance measure. The images shown were the eight most similar images to the query image, which is shown in the upper left position as it is most similar to itself.

8.4 IMAGE DISTANCE MEASURES

The judgment of how similar a database image is to a query is dependent on which image distance measure or measures are used to judge similarity. There are four major classes of similarity measures:

1. color similarity,
2. texture similarity,
3. shape similarity, and
4. object and relationship similarity.

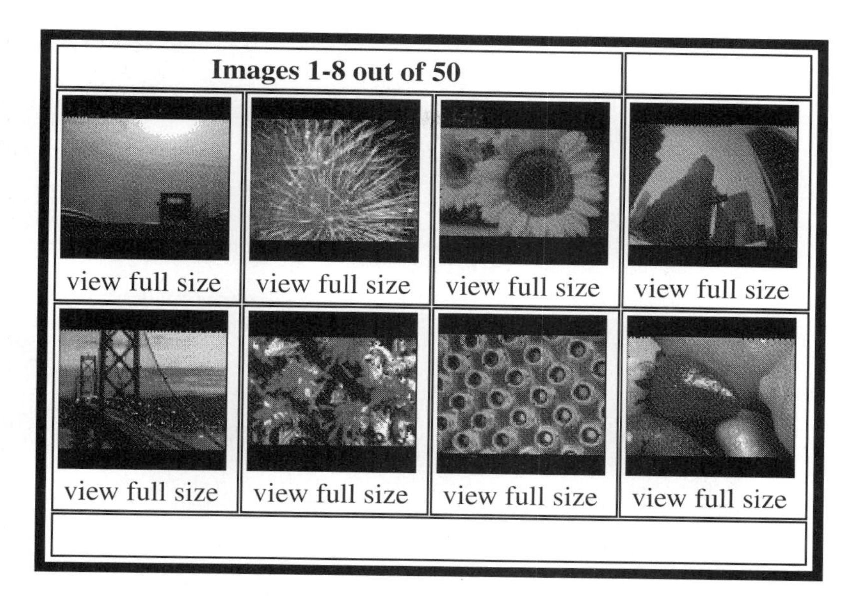

Figure 8.4 Results of a QBIC search based on color percentages; the query specified 40 percent red, 30 percent yellow, and 10 percent black. (Images courtesy of Egames.) See colorplate.

8.4.1 Color Similarity Measures

Color similarity measures are often very simple. They compare the color content of one image with the color content of a second image or of a query specification. For example, QBIC allows users to specify a query in terms of color percentages. The user chooses up to five colors from a color table and indicates the desired percentage of each color. QBIC looks for images that are closest to having these color percentages. The particular placement of the colors within the image is not a factor in the search. Figure 8.4 shows a set of images returned for a query that specified 40 percent red, 30 percent yellow, and 10 percent black. While the colors returned are very similar in each of the returned images, the images have very different compositions.

A related technique is color histogram matching, as discussed in Chapter 6 and used for vegetable recognition in Chapter 16. Users can provide a sample image and ask the system to return images whose color histogram distance to the sample is low. Color histogram distances should include some measure of how similar two different colors are. For example, QBIC defines its color histogram distance as

$$d_{hist}(I, Q) = (h(I) - h(Q))^T A \, (h(I) - h(Q)) \tag{8.1}$$

where $\mathbf{h(I)}$ and $\mathbf{h(Q)}$ are the K-bin histograms of images $\mathbf{I}$ and $\mathbf{Q}$, respectively, and $\mathbf{A}$ is a $K \times K$ similarity matrix. In this matrix colors that are very similar should have similarity values close to one, while colors that are very different should have similarity values close to or equal to zero.

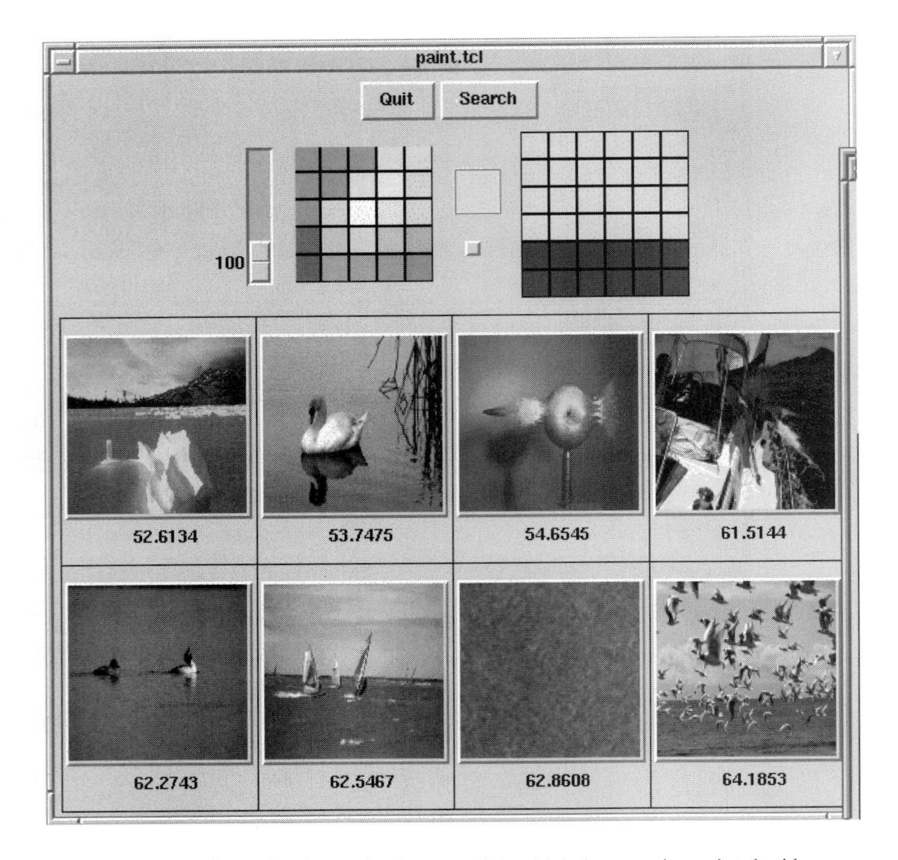

Figure 8.5 Results of an image database search in which the query is a painted grid. (Images from the MIT Media Lab VisTex database: http://vismod.www.media.mit.edu/vismod/imagery/VisionTexture/vistex.html.) See colorplate.

Color layout is another possible distance measure. It is common for the user to begin with an empty grid representing the query and to choose colors for each of the grid squares from a table. In Figure 8.5, the user has selected two colors from the color matrix shown at the upper left and has painted them onto the 6×6 spatial layout grid shown at the upper right. The images shown are those that were judged most similar to the query according to a simple color layout distance measure. As was shown in Figure 8.3, it is also possible to start with an example image and have the system return other images that have spatial color distributions that are similar to it.

Color layout measures that use a grid require a grid square color distance measure $\hat{d}_{color}$ that compares each grid square of the query to the corresponding grid square of a potential matching image and combines the results into a single image distance

$$d_{gridded\_color}(I, Q) = \sum_{g} \hat{d}_{color}(C^I(g), C^Q(g)) \tag{8.2}$$

where $\mathbf{C^I}(\mathbf{g})$ represents the color in grid square $\mathbf{g}$ of a database image $\mathbf{I}$ and $\mathbf{C^Q}(\mathbf{g})$ represents the color in the corresponding grid square $\mathbf{g}$ of the query image $\mathbf{Q}$. The representation of the color in a grid square itself can be very simple or more complicated. Some suitable representations are

1. the mean color in the grid square,
2. the mean and standard deviation of the color, and
3. a multi-bin histogram of the color.

The grid square distance $\hat{d}$ must be able to operate on the color representation to produce meaningful distances. For example, if mean color is represented as a triple (R, G, B), then the measure $\hat{d} = \|(R^Q, G^Q, B^Q) - (R^I, G^I, B^I)\|^2$ would be an obvious choice, but not necessarily the best one. Instead of comparing (R, G, B) values, some systems partition color space into a set of 3D bins and keep a table of the numerical similarities between pairs of bins. This is the same technique that was used in QBIC's histogram distance as discussed in Equation 8.1.

Exercise 8.2: Color Histogram Distances.

Implement a $4 \times 4 \times 4$ color histogram distance measure that can input two images and compare either the whole images or selected subimages of each. Use this basic measure to implement a gridded-color distance measure that allows the user to specify the dimensions of the grid and combines the distances between each pair of corresponding grid squares into a single distance as shown in Equation 8.2. Try your gridded histogram distance measure on several pairs of color images and with grid dimensions 1×1, 4×4, and 8×8.

8.4.2 Texture Similarity Measures

Texture similarity is more complex than color similarity. An image that has similar texture to a query should have the same spatial arrangements of colors (or gray tones), but not necessarily the same colors (or gray tones). The texture measures described in Chapter 7 can be used to judge the similarity between two textures. Figure 8.6 illustrates texture-based image retrieval, using a distance based on the Laws texture energy measures. As can be seen from the query results, this distance is independent of the colors in the image. However, it is also possible to develop distance measures that look for both color and texture similarity.

Texture distance measures have two aspects:

1. the representation of texture, and
2. the definition of similarity with respect to that representation.

The most commonly used texture representation is a *texture description vector,* which is a vector of numbers that summarizes the texture in a given image or image region. The vector of Haralick's five co-occurence-based texture features and that of Laws' nine texture energy features are examples of texture description vectors. While a texture description vector can be used to summarize the texture in an entire image, this is only a good method

Figure 8.6 Results of an image database search based on texture similarity. (Images from the MIT Media Lab VisTex database: http://vismod.www.media.mit.edu/vismod/imagery/VisionTexture/vistex.html.) See colorplate.

for describing single texture images. For more general images, texture description vectors are calculated at each pixel for a small (such as 15 × 15) neighborhood about that pixel. Then the pixels are grouped by a clustering algorithm that assigns a unique label to each different texture category it finds.

Given that pixels can be assigned a texture description vector and labeled as belonging to a texture class, various texture distances can be defined. The simplest texture distance is the *pick-and-click* distance. The user selects a texture by clicking on a pixel in a textured region of the query image or by selecting the texture from a predetermined set of choices. The selected texture is represented by its texture description vector, which is compared to the texture description vectors associated with the database. The distance measure is defined by

$$d_{pick\_and\_click}(I, Q) = min_{i \in I} \|T(i) - T(Q)\|^2 \tag{8.3}$$

where **T(i)** is the texture description vector at pixel **i** of image **I** and **T(Q)** is the texture description vector at the selected pixel or for the selected texture class of the query. Although this looks like it might be computationally intensive, most of the computation can be avoided by representing a database image by a list of its texture categories as determined by the clustering procedure. For each database image, the query texture description vector needs only to be compared to the texture description vectors in its list. Indexing can provide even faster retrieval.

The pick-and-click distance requires the user to select a given texture; it cannot operate automatically on a query image. A more general texture measure is a generalization of the gridded measures discussed above from color to texture. A grid is placed over the query image and a texture description vector is calculated for each grid square. The same process is applied to the database image. The gridded texture distance is given by

$$d_{gridded\_texture}(I, Q) = \sum_{g} \hat{d}_{texture}(T^{I}(g), T^{Q}(g)) \tag{8.4}$$

where $\hat{d}_{texture}$ can be Euclidean distance or some other distance metric. Texture histogram distances can be defined in a similar manner to color histograms. For each texture category, the histogram specifies the number of pixels whose texture description vector falls into that category. One interesting and easy to compute texture histogram measure has been developed in terms of pairs of touching line segments. A line finder (see Chapter 10) is applied to the image to detect line segments. Pairs of line segments that touch or almost touch are found, and the angle between each such pair of segments is computed. These angles are the variables used to produce the texture histogram that describes the image.

8.4.3 Shape Similarity Measures

Color and texture are both global attributes of an image. Distance measures based on these quantities try to determine if a given image has a specified color or texture and whether or not it occurs in the same approximate position as in the query image. Shape is not an image attribute; it does not make sense to ask what the shape of an image is. Instead, shape tends to refer to a specific region of an image. Shape goes one step further than color and texture in that it requires some kind of region identification process to precede the shape similarity measure. In many cases, this has to be done manually, but automated segmentation is possible in some domains. Segmentation is still a crucial problem to be solved, before shape-based retrieval can be made widely available. Segmentation will be discussed in Chapter 10; shape matching is covered here.

Exercise 8.3: Texture Distance Measures.

Pick several different texture measures from Chapter 7 and implement them as image distance measures that compare the texture in a subimage of a query image to that in a subimage of a database image. Then write a program that implements a gridded-texture distance measure and can call on any of these as the individual measures used in each of the grid squares. Compare results on a set of images using each of the individual measures and trying several different grid sizes. Test on a database of images that each have several regions of different textures.

Two-dimensional shape recognition is an important aspect of image analysis. In Chapter 3, a number of properties of image regions were defined; these are what we call *global shape properties,* since they refer to the shape as a whole. Two shapes can be compared according to global properties by any of the statistical pattern recognition methods described in Chapter 4. Shape matching can also use structural techniques whereby a shape is described by its primitive components and their spatial relationships. Since the representation is a relational graph, graph-matching methods can be used for matching. Graph matching is powerful, because it is based on spatial relationships that are invariant to most 2D transformations. However, graph matching can be a very slow process; the computation time is exponential in the number of components. In the context of content-based image retrieval, we need methods that can quickly decide how similar an image shape is to a query shape. Often, we require shape matching methods to be invariant to translation and to size. Sometimes we also want rotational invariance, so that an object can be identified whether it is right-side-up or in some other orientation. However, this property is not always required in image retrieval. There are many scenes in which objects usually appear in the correct orientation. Buildings, trees, and trucks in outdoor scenes are common examples.

Shape measures abound in the computer vision literature. They range from crude global measures that help with, but do not perform, object recognition to very detailed measures that look for objects with very specific shapes. Shape histograms are examples of simple measures that can rule out shapes that could not possibly match, but that will return many false positives, just as color histograms do. Boundary techniques are more specific; they work with a representation of the boundary of a shape and look for shapes with similar boundaries. Sketch matching can be even more specific, looking not just for a single object boundary, but for a set of image segments involving one or more objects that match a query drawn or supplied by a user. We now discuss each of these categories.

Shape Histograms Given that histogram distances are fast and easy to compute and that they have been used for both color and texture matching, it is natural to extend them to shape matching. The main problem is to define the variable on which the histogram is defined. Consider the shape as a region of 1s in a binary image whose other pixels are all 0s. One kind of histogram matching is *projection matching* using horizontal and vertical projections of the shape. Suppose the shape has n rows and m columns. Each row and each column become a bin in the histogram. The count that is stored in a bin is the number of 1-pixels that appear in that row or column. This leads to a histogram of $n + m$ bins, which is useful only if the shape is always the same size. To make projection matching size invariant, the number of row bins and the number of column bins can be fixed. By defining the bins from the top left to the bottom right of the shape, translational invariance is achieved. Projection matching is not rotationally invariant, but may work with small rotations or other small geometric distortions. One way to make it rotationlly invariant is to compute the axes of the best-fitting ellipse (as discussed in Chapter 3) and rotate the shape until the major axis is vertical. Because we don't know which is the top of the shape, two possible rotations must be tried. Furthermore, if the major and minor axes are about the same length, four possible rotations must be considered. Projection matching has been successfully used in logo retrieval.

Another possibility is to construct the histogram over the tangent angle at each pixel on the boundary of the shape. This measure is automatically size and translationally invariant, but it is not rotationally invariant, because the tangent angles are computed from a fixed orientation of the shape. There are several different ways to solve this problem. One way is to rotate the shape according to its major axis as described above. Another simpler way is to rotate the histogram instead. If the histogram has K bins, there are K possible rotations. Incorrect rotations can be ruled out rapidly as soon as the histogram distance being computed becomes too large. Or, the histograms can be normalized by always choosing the bin with the largest count to be the first bin. Because of possible noise and distortion, several largest bins should be tried.

Exercise 8.4: Shape Histograms.

Write a program that implements a shape-histogram distance measure using the tangent angle at each pixel on the boundary of the shape. Make it rotationally invariant by rotating the histogram of the query image so that each bin gets a turn as the first bin and the result is the minimum distance returned by each of these rotations. Use your distance measure to compare shapes that you extract from real images either by thresholding or interactively.

Boundary Matching Boundary matching algorithms require the extraction and representation of the boundaries of the query shape and the image shape. The boundary can be represented as a sequence of pixels or may be approximated by a polygon. For a sequence of pixels, one classical kind of matching uses Fourier descriptors to compare two shapes. In continuous mathematics, the Fourier descriptors are the coefficients of the Fourier series expansion of the function that defines the boundary of the shape. In the discrete case, the shape is represented by a sequence of m points $<V_0, V_1, \ldots, V_{m-1}>$. From this sequence of points, a sequence of unit vectors

$$v_k = \frac{V_{k+1} - V_k}{|V_{k+1} - V_k|} \tag{8.5}$$

and a sequence of cumulative differences

$$\begin{aligned} l_k &= \sum_{i=1}^{k} |V_i - V_{i-1}|, \quad k > 0 \\ l_0 &= 0 \end{aligned} \tag{8.6}$$

can be computed. The Fourier descriptors $\{a_{-M}, \ldots, a_0, \ldots, a_M\}$ are then approximated by

$$a_n = \frac{1}{L\left(\frac{n2\pi}{L}\right)^2} \sum_{k=1}^{m} (v_{k-1} - v_k) e^{-jn(2\pi/L)l_k} \tag{8.7}$$

These descriptors can be used to define a shape distance measure. Suppose $\mathbf{Q}$ is the query shape and $\mathbf{I}$ is the image shape to be compared to $\mathbf{Q}$. Let $\{a_n^Q\}$ be the sequence of FDs for the query and $\{a_n^I\}$ be the sequence of FDs for the image. Then the Fourier distance measure is given by

$$d_{Fourier}(I, Q) = \left[\sum_{n=-M}^{M} |a_n^I - a_n^Q|^2 \right]^{\frac{1}{2}} \quad (8.8)$$

As described, this distance is only invariant to translation. If the other invariants are required, it can be used in conjunction with a numeric procedure that solves for the scale, rotation, and starting point that minimize $d_{Fourier}(\mathbf{I}, \mathbf{Q})$.

If the boundary is represented by a polygon, the lengths of the sides and the angles between them can be computed and used to represent the shape. A shape can be represented by a sequence of junction points $\langle X_i, Y_i, \alpha_i \rangle$ where a pair of lines meet at coordinate location (X_i, Y_i) with angle magnitude α_i. Given a sequence $\mathbf{Q} = \mathbf{Q}_1, \mathbf{Q}_2, \ldots, \mathbf{Q}_n$ of junction points representing the boundary of a query object $\mathbf{Q}$ and a similar sequence $\mathbf{I} = \mathbf{I}_1, \mathbf{I}_2, \ldots \mathbf{I}_m$ representing the boundary of an image object $\mathbf{I}$, the goal is to find a mapping from $\mathbf{Q}$ to $\mathbf{I}$ that maps line segments of the query to similar-length line segments of the image and requires that a pair of adjacent query line segments that meet at a particular angle α should map to a pair of adjacent image line segments that meet at a similar angle α'.

Another boundary matching technique is *elastic matching* in which the query shape is deformed to become as similar as possible to the image shape. The distance between the query shape and the image shape depends on two components: (1) the energy required to deform the query shape into a shape that best matches the image shape, and (2) a measure of how well the deformed query actually matches the image. Figure 8.7 shows the retrieval of images of horses through elastic matching to a query in which a user drew a rough outline of the shape he wanted to retrieve.

Exercise 8.5: Boundary Matching.

While there are many known algorithms for boundary shape matching, they have not been heavily used so far in content-based retrieval. Can you explain why?

Sketch Matching Sketch matching systems allow the user to input a rough sketch of the major edges in an image and look for full color or gray-scale images that have matching edges. In the ART MUSEUM system, the database consists of full-color images of famous paintings. The color images are preprocessed as follows to obtain an intermediate form called an *abstract image*.

1. Apply an affine transform to reduce the image to a prespecified size, such as 64×64 pixels and a median filter to remove noise. The result is a normalized image.
2. Detect edges using a gradient-based edge-finding algorithm. The edge finding is performed in two steps: first global edges are found with a global threshold that is based on the mean and variance of the gradient; then, local edges are selected from the global according to locally computed thresholds. The result is called the refined edge image.

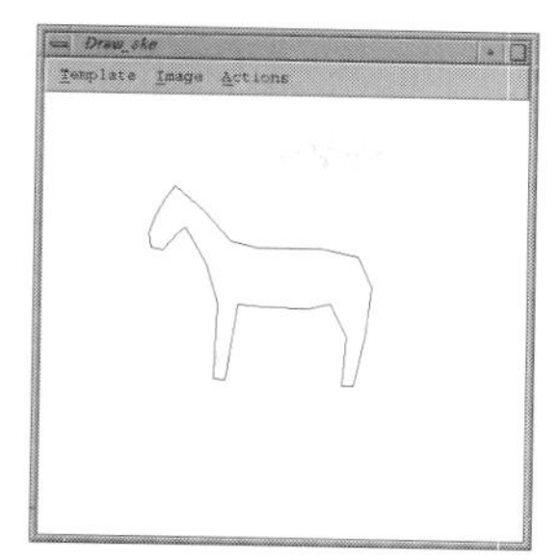

(*a*) The user's query shape.

(*b*) Two of the retrieved images.

(*c*) Another retrieved image in which two horses were found.

Figure 8.7 Image retrieval by elastic matching. (Courtesy of Alberto Del Bimbo.) See colorplate.

3. Perform thinning and shrinking on the refined edge image. The final result is called the abstract image. It is a relatively clean sketch of the edges of the original image.

When the user enters a rough sketch as a query, it is also converted to the normalized size, binarized, thinned, and shrunk. The result of this processing is called the linear sketch. Now the linear sketch must be matched to the abstract images. The matching algorithm is correlation-based. The two images are divided into grid squares. For each grid square of the query image, the local correlation with the corresponding grid square of the database image is computed. In order to be more robust, the local correlation is computed for several different shifts in the position of the grid square on the database image and the maximum correlation over all the shifts is the result for that query grid square. The final similarity measure is the sum of each of the local correlations. The distance measure is the inverse of this similarity measure. In terms of our previous notation, it can be expressed as

$$d_{sketch}(I, Q) = \frac{1}{\sum_g max_n[\hat{d}_{correlation}(shift_n(A^I(g)), L^Q(g))]} \tag{8.9}$$

where $\mathbf{A^I(g)}$ refers to grid square $\mathbf{g}$ of the abstract image computed from database image $\mathbf{I}$, *shift*($\mathbf{A^I(g)}$) refers to a shifted version of grid square $\mathbf{g}$ of the same abstract image, and $\mathbf{L^Q(g)}$ refers to grid square $\mathbf{g}$ of the linear sketch resulting from query image $\mathbf{Q}$.

Exercise 8.6: Sketch Matching.

Design and implement a sketch-matching distance measure along the lines of the ART MUSEUM system. Use it to retrieve a set of known images according to the user's sketch.

8.4.4 Object Presence and Relational Similarity Measures

Although most of the distance measures offered by the first image search engines involve color, texture, and shape, these are not the quantities that most end users want to see. End users tend to ask for images containing certain entities, which can be particular objects, such as people or dogs or can be abstract concepts, such as happiness or poverty. The first systems to offer object recognition have looked for such objects as human faces, human bodies, and horses. This is an area that will require further research in object recognition in order to be useful in image retrieval.

Face Finding Face finding is important because it allows us to search for images containing people. It is difficult because faces are found in all sizes and locations in an image, can be in a frontal or other view, and come in a variety of colors. A system developed at Carnegie-Mellon University employs a multiresolution approach to solve the size problem. It converts color images to gray scale to avoid color differences, normalizes for lighting, and expands the gray-tone range through histogram equalization. It then uses a neural net classifier that was trained on 16,000 images of faces and non-faces to perform the recognition. The neural net receives an image of $20 \times 20 = 400$ intensity values as its input and classifies it as a face or non-face. While it is difficult to extract an exact algorithm from a neural net, a sensitivity analysis showed that the network relies most heavily on the eyes,

Figure 8.8 Faces detected by a neural-net-based face finder. (Courtesy of Henry Rowley and Takeo Kanade.)

then on the nose, and then on the mouth area of the 20×20 image. The method works well, finding most, but not all frontal views of faces, as shown in Figure 8.8. It is not generally extensible to other objects unless they have a very particular pattern that shows up in the same way that the eyes, nose, and mouth do in gray-tone images.

Flesh Finding Another way of finding an object is to find regions in images that have the color and texture usually associated with that object. One of the first efforts in this area was focused on finding images of naked people, which has the useful application of filtering out pornography from query result sets. The method developed by Fleck, Forsyth, and Bregler (1996) has two main steps: (1) finding large regions of potential flesh and (2) grouping these regions to find potential human bodies.

The flesh filter operates at the pixel level. The initial RGB image is transformed into log-opponent space as

$$I = L(G) \tag{8.10}$$

$$R_g = L(R) - L(G) \tag{8.11}$$

$$B_y = L(B) - \frac{L(G) + L(R)}{2} \tag{8.12}$$

where $L(x)$ is defined by

$$L(x) = 105\ log_{10}(x + 1 + n) \tag{8.13}$$

and n is a random noise value from the range [0,1]. The I component is used to produce a texture amplitude map as follows:

$$texture = med_2(|I - med_1(I)|) \tag{8.14}$$

where med_1 and med_2 are two separate median filters of different sizes (med_2 is 1.5 times the size of med_1). The texture amplitude map is used to find regions of low texture, since skin in images tends to have a very smooth texture.

Hue and saturation are used to select the regions whose color matches that of skin. The R_g and B_y images are also median filtered before being used in the calculations. The conversion from log opponent space to hue and saturation is given by

$$hue = atan(R_g, B_y) \tag{8.15}$$

$$saturation = \sqrt{R_g^2 + B_y^2} \tag{8.16}$$

If a pixel falls into either of the following two ranges, it is marked as a skin pixel.

1. $texture < 5,\ 110 < hue < 150,\ 20 < saturation < 60$
2. $texture < 5,\ 130 < hue < 170,\ 30 < saturation < 130$

Note that all the constants used are from the original work and can be modified for different data sets or user preference.

The skin map is a binary array where 1-pixels are skin pixels and 0-pixels are non-skin pixels. This array can be processed by a morphological closing operation to produce a cleaner result. Once the images with skin regions have been found, they can be checked for (1) having enough flesh to be considered pornography (30 percent of the image was used) and (2) having regions in appropriate spatial relations to be considered human body parts.

Exercise 8.7: Flesh and Face Finding.

Implement a flesh finder to find regions of flesh color. Select regions of a specified size and larger and try to find evidence of facial features; in particular, the eyes, nose, and mouth, in that order of priority. Based on the features you find, assign to each region the probability of being a face.

Spatial Relationships Once objects can be recognized, their spatial relationships can also be determined, and queries can be formulated that require a certain set of named objects in predetermined spatial relationships. This is the final step in the image retrieval hierarchy. In recent work at Berkeley (Forsyth and others, 1996) and in similar work at Santa Barbara (Ma and Manjunath, 1997) researchers have successfully used both color and texture to segment images into regions that often correspond to objects or scene backgrounds. Such objects as tigers and zebras that stand out well and have a particular color and/or texture pattern can be found in this way. Backgrounds such as jungle or sky or beach can also be isolated. Figure 8.9 gives an example of this type of segmentation process. The original color image is shown at left, and its segmentation into regions is shown at center. A symbolic representation of the image in which the regions of interest are depicted as ellipses is shown at right. This representation can be used to construct a relational graph whose nodes are the classifications of the regions and whose edges represent spatial relationships. Now

Original Image

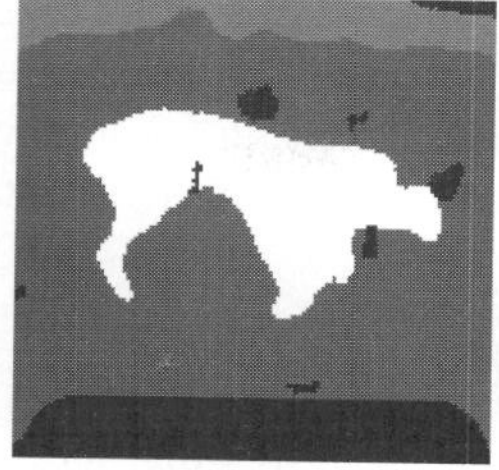

Segmentation

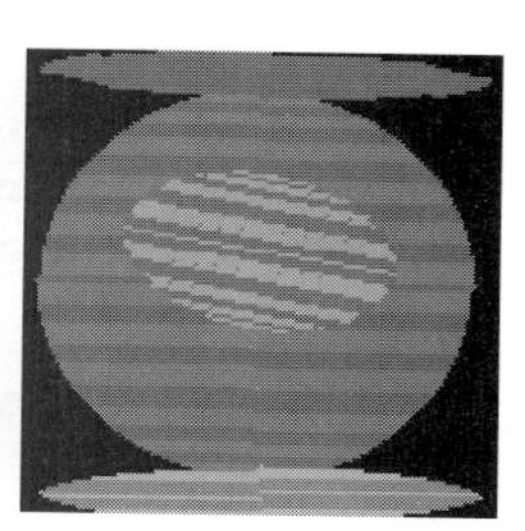

Symbolic Representation

Figure 8.9 Objects and spatial relationships that can be extracted from images and used for retrieval. (Original image licensed from Corel Stock Photos.) See colorplate.

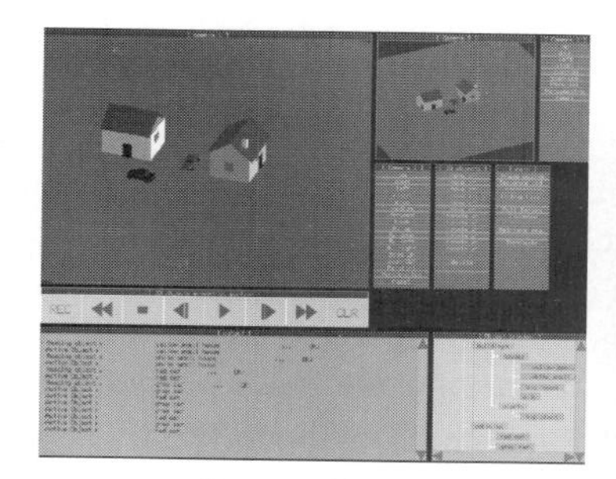

Query Window

Retrieved Image

Figure 8.10 Results of a spatial-relationship query. (Figure courtesy of Alberto Del Bimbo with permission of IEEE. Reprinted from "Symbolic Description and Visual Querying of Image Sequences using Spatio-Temporal Logic," by A. Del Bimbo, E. Vicario, D. Zingoni, *IEEE Transactions on Knowledge and Data Engineering*, vol. 7, no. 4, Aug. 1995. © 1995 IEEE.)

relational matching techniques can be used to create relational distance measures for image retrieval. Although the system shown here does not go that far, Del Bimbo has developed a retrieval system that could use this representation as its input. This system allows users to construct queries by placing selected icons in spatial relationships on the query screen and returns images having the corresponding objects in those relationships. Figure 8.10 shows an example of retrieval by this spatial query system.

Exercise 8.8: Retrieval by Objects and Relationships.

Obtain or write a program that segments a color image into regions based on color and, if possible, texture. Run the program on a set of training images in which each class of object, such as tiger, sky, jungle, is present in several images. Train a classification algorithm on these known regions according to their color and texture properties. Write a program that uses the segmenter and classifier to produce a set of labeled regions for an input image and then computes the spatial relationships *above, below, left-of, right-of,* and *adjacent-to* between pairs of regions. Then write an interactive front end that allows users to input a graph structure in which the nodes are objects from the training set and the edges are the required relationships. The program should return all database images that satisfy the user's query.

8.5 DATABASE ORGANIZATION

Large databases of images, like any other large databases, are too big to search the whole database for images that satisfy a given query. Instead, the images must be organized and indexed so that only a fraction of them are even considered for any one query. There are standard methods for indexing numeric and textual data that are used in most relational database systems. Methods for indexing spatial data also exist and have been used, in particular, for geographic information systems. Methods for indexing images for content-based image retrieval are being developed for current research systems.

8.5.1 Standard Indexes

In most relational databases, the user can specify an attribute on which an index is to be built. Usually this attribute is an important key associated with each data record. For example, if the database contains records of the employees who work for a certain company, then the social security number would be a good attribute to use to index the data. Since everyone has a unique social security number, this attribute is called a *primary key*. If the data is often accessed by some other attribute, such as last name of employee, a separate index can be built for it.

In a relational database, an *index* is a data structure with which the system can look up a given attribute value and rapidly find the set of all records in the database that have that value for that attribute. There are two common types of indexes used in relational database systems: hash indexes and B-tree indexes. Hash indexes allow rapid determination of the set of data records that have exactly the attribute value specified in the query. B-tree or B^+-tree indexes allow a speedy search for records whose attribute values lie in the range specified by a query.

Hash Indexes A hash index applies the theory of the *hash table* to access a large set of records in a database. The assumption is that there potentially is a large set of possible key values and only a fraction of these are ever present in the database at one time. Suppose that the database consists of a file of N records. Each record contains several different fields including one field for the key values. The access mechanism for the hash index is a *hash function* that maps any key value into an address in the file that contains (or sometimes points to) a database record containing the particular key value. If the key values are numeric, then one simple hash function is $f(x) = x \, mod \, N$, which reduces to dividing x by N and using the remainder of the division as the record number of the record to be accessed. Figure 8.11 shows a hash index for a database with numeric keys. The hash table has ten positions numbered 0 through 9 (a real hash table would be much bigger). The hash function being used is $f(x) = x \, mod \, 10$. The query shown is asking for all records whose key value is equal to 45, which hashes to position 5 in the hash table.

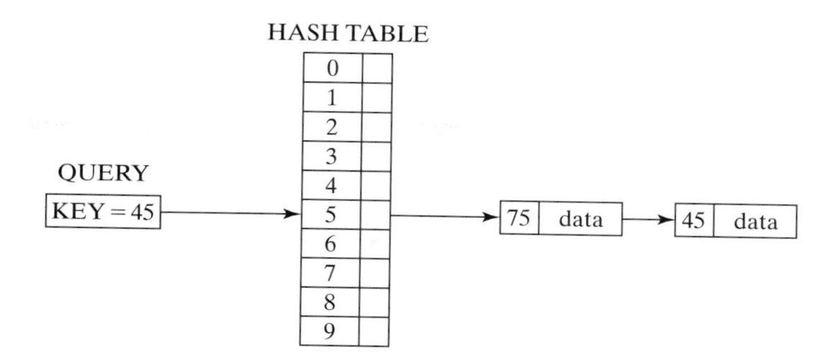

Figure 8.11 A sample hash index.

If each key value hashes to a different location in the table, then the access time for any given key is constant. Generally, this doesn't happen. Instead, several different keys may hash to the same location. This phenomenon is called a *collision* and solutions to it can be found in any data structures text. The solution shown in the figure is to keep a linked list of all records whose keys hash to the same location. The end result—in any case—is that the access process may involve a small amount of search, rather than being a simple direct access, but for good hash functions and tables that are not too full, the access time is still approximately constant. Because of the nature of the hashing process, it is most suitable for an exact match constraint of the form $KEY = VALUE$ and is less useful for range queries.

B$^+$-Tree Indexes B-trees and B$^+$-trees are balanced multi-way search trees that can be used for indexing and are suitable for range queries. B-trees have values of keys and data at both internal and leaf nodes, while B$^+$-trees have data only at leaf nodes. We will concentrate on B$^+$-trees, since the data in a database should be separate from the index.

A search tree of order p is a tree in which each node contains $\leq p - 1$ key values and p pointers. A B$^+$-tree is a search tree that has one format for internal nodes and a second format for leaf nodes. Each internal node of a B$^+$-tree has the following constraints:

1. It has the format $\langle P_1, K_1, P_2, K_2, \ldots, P_{q-1}, K_{q-1}, P_q \rangle$ where each P_i is a pointer to another node and each K_i is a key value. Intuitively, P_{i-1} points to a subtree whose nodes contain keys with values less than or equal to K_i, while P_i points to a subtree whose nodes contain keys with values greater than K_i.
2. If it is a nonroot node, it has at least $\lceil (p/2) \rceil$ subtree pointers.
3. If it is a root node, it has at least 2 subtree pointers.

Each leaf node of a B$^+$-tree satisfies the following:

1. It has the form $\langle K_1, Pr_1, K_2, Pr_2, \ldots, K_{q-1}, Pr_{q-1} >, P_{next} \rangle$ where K_i is a key, Pr_i is a data pointer, and P_{next} points to the next leaf node.
2. Pr_i points to a record whose search key is K_i or to a block of such records if the search field of the index is not a key of the file.

3. Each leaf node has $\lfloor(p/2)\rfloor$ values.
4. All leaf nodes are at the same level of the tree.

Internal nodes may have a different order p than leaf nodes; the goal is usually to fit each type of node into the physical block of storage that is the unit amount transferred from disk to internal memory.

To find a given key value or range of values in a B$^+$-tree, the retrieval system starts at the root of the tree. It reads that node into memory and performs a binary search of the keys in that node. If it finds two adjacent key values in the node between which the given key value lies, the pointer between them points to the subtree that will contain the given key value or the smallest value in the given key range. If the given key value is less than the first key value in the node, the pointer to the left of this key value references the appropriate subtree. Similarly, if the given key value is greater than the last key value in the node, the pointer to the right of this key value references the correct subtree. Once the appropriate subtree has been identified, it becomes the root of the tree to be searched. The retrieval system performs this operation recursively until it reaches a leaf node.

In the leaf node, a binary search is again performed to find the given key or starting key value K_i. The associated pointer P_i then points to the data record containing this key value. If only one key value is sought, the associated record can now be returned. If this is a range search, the P_{next} pointers in the data records can be used to find the remaining data records until the end of the specified key range is encountered.

Figure 8.12 gives an example B$^+$-tree that indexes database records with numeric keys and image data. The internal nodes are shown with solid rectangles, and the leaf nodes are shown with dashed rectangles. The root node points to three different subtrees: those with key values less than or equal to 100, those with key values between 100 and 200, and those with key values greater than 200. The subtree with key values between 100 and 200 is shown. Its root points to two leaf nodes: one with key values less than 110 and one with key values between 110 and 150. The leaf nodes contain some actual key values and associated image data files.

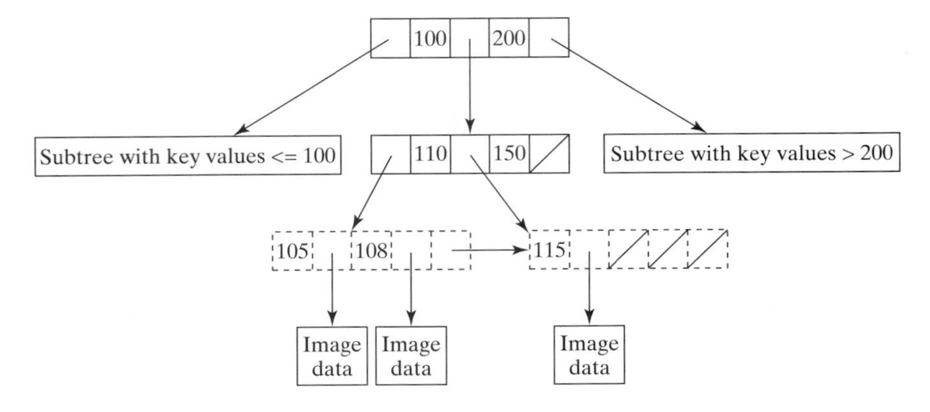

Figure 8.12 A sample B$^+$-tree index.

B^+-trees are flexible and efficient and are used heavily in relational database systems. They can be used in image database systems to index single numeric or text fields associated with an image. They were not intended to be used to index multi-dimensional data.

8.5.2 Spatial Indexing

Spatial information systems contain data that is multi-dimensional. A number of structures have been proposed for spatial indexing. Quadtrees are hierarchical structures of degree four that break up the search space for 2D data into four sub-quadrants at each level of the tree. Quadtrees can be used to represent regions in binary images. K-d trees are an extension of binary search trees, which allow search for k-dimensional data. R-trees are an extension of B-trees to higher dimensions and are suitable for a variety of spatial information system applications. In an R-tree, a data object is indexed by an n-dimensional minimum bounding rectangle (MBR), which bounds the space occupied by the object. Each actual data object is referenced by a unique identifier (ID). The leaf nodes of the R-tree contain the data object IDs. The internal nodes contain entries of the form (MBR,CHILD) where CHILD is a pointer to a lower node in the R-tree and MBR covers all the rectangles in the lower node's entries. Figure 8.13 shows a sample R-tree index for a collection of 2D objects. The

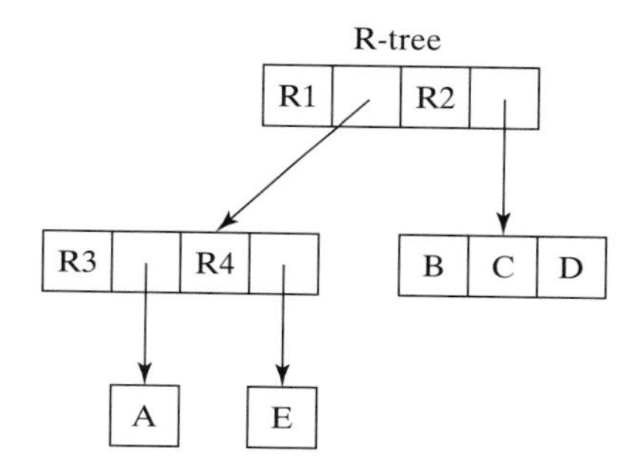

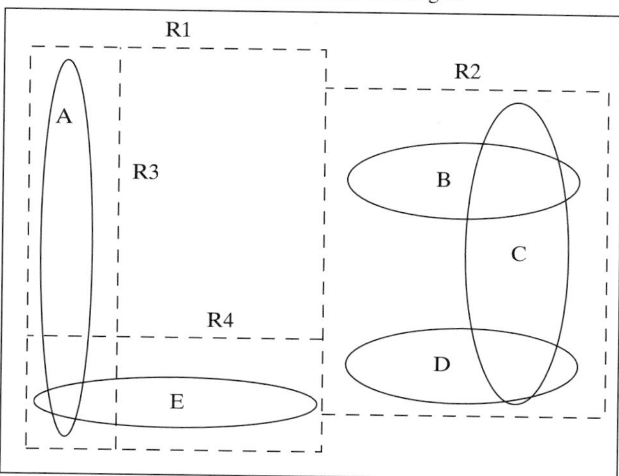

Figure 8.13 A sample R-tree index for 2D data. The ellipses represent the data objects and the indexing rectangles are shown with dashed lines. Rectangle R1 has been broken down further into rectangles R3 and R4, each of which contain a single data object. Rectangle R2 has not been broken down further and contains three data objects: B, C, and D.

distribution of the rectangles depends on the order in which the tree was constructed and the exact R-tree construction algorithm used. Variants such as R^+-trees and R^*-trees also exist.

8.5.3 Indexing for Content-Based Image Retrieval with Multiple Distance Measures

The above methods can be used to index images for retrieval via simple distance measures that are based on a single attribute or a small number of attributes. They are not suitable for a larger, general system that provides the user with the choice of a number of base distance measures and methods to combine them. Such a system requires a more flexible form of organization and indexing. If the base measures are metrics, then the triangle inequality property can be used to provide a nonstandard indexing method. The triangle inequality says that if **Q** is a query image, **I** is a database image, and **K** is a specially selected *key image*, then

$$d(I, Q) \geq |d(I, K) - d(Q, K)|$$

for any image distance measure **d**. Thus by comparing the database and query images to a third key image, a lower bound on the distance between the query image and the database image can be obtained.

Consider first the case of a single distance metric **d**. A set of key images can be selected from the database; intuitively, they should represent the different classes of scenes in the database. The query image **Q** is compared to each of the keys, $\mathbf{K_1}, \mathbf{K_2}, \ldots, \mathbf{K_M}$, obtaining a set of distances $\mathbf{d(Q, K_1)}, \mathbf{d(Q, K_2)}, \ldots, \mathbf{d(Q, K_M)}$. Suppose that the user has specified that all images with distance less than **T** from the query **Q** should be returned. Then for each key $\mathbf{K_i}$, all images **I** such that

$$|d(I, K_i) - d(Q, K_i)| > T$$

can be immediately ruled out, since **d(I,Q)** is already known to be too large. A data structure called the *triangle-tree* has been designed to take advantage of this approach and to rule out most of the images in a database from direct comparison to the query. The technique has also been extended to handle dynamically defined distance measures that are linear, or Boolean combinations of the base distance measures.

Exercise 8.9: Indexing.

Suppose a set of images is to be indexed according to Laws' texture energy measures. Explain how you could use R-trees as your indexing mechanism for this system.

8.6 REFERENCES

Content-based image retrieval is a relatively new application of computer vision techniques. The first general-purpose commercial system, QBIC, was developed by Niblack and his research group (1993) at IBM, Almaden; it used color, texture, and shape for retrieval.

Another very early system was the ART MUSEUM system of Kato and others (1992) which retrieved paintings from a collection based on user-drawn sketches. Minka and Picard (1996) showed how learning from user choices (relevance feedback) could be used to improve the performance of a retrieval system. Del Bimbo, Pala, and Santini (1994) created a system that could retrieve images containing objects that matched a user-specified shape through elastic matching.

Fleck, Forsyth, and Bregler (1996) developed an algorithm for finding unclothed people in images, using color and spatial relationships among regions. This work was later extended to other objects, such as horses. Rowley, Baluja, and Kanade (1996) developed a method for finding people's faces using a neural net trained on thousands of images. Carson, Belongie, Greenspan, and Malik (1997) at Berkeley and Ma and Manjunath (1997) at Santa Barbara developed region-based approaches to querying, using color- and texture-based segmentation algorithms to produce the regions of interest. Baeza-Yates and his group (1994) and Berman (1994) both developed tree structures for use in proximity matching. Berman and Shapiro (1999) developed an image retrieval system using extensions of these techniques. Samet's 1990 book describes general-purpose spatial data structures.

1. Baeza-Yates, R., W. Cunto, U. Manber, and S. Wu. 1994. Proximity matching using fixed queries trees. *Combinatorial Pattern Matching.* Springer-Verlag, New York, 198–212.
2. Berman, A. P. 1994. A new data structure for fast approximate matching, technical report 94-03-02. Department of Computer Science and Engineering, University of Washington.
3. Berman, A. P., and L. G. Shapiro. 1999. A flexible image database system for content-based retrieval. *Comput. Vision and Image Understanding,* v. 75(1–2):175–195.
4. Carson, C., S. Belongie, H. Greenspan, and J. Malik. 1997. Region-based image querying. *Proc. IEEE Workshop on Content-Based Access of Image and Video Libraries,* San Juan, Puerto Rico.
5. Del Bimbo, A., E. Vicario, and D. Zingoni. 1993. Sequence retrieval by contents through spatio temporal indexing. *IEEE Symp. Visual Lang.,* 88–92.
6. Del Bimbo, A., P. Pala, and S. Santini. 1994. Visual image retrieval by elastic deformation of object sketches. *IEEE Symp. Visual Lang.,* 216–223.
7. Fleck, M. M., D. A. Forsyth, and C. Bregler. 1996. Finding naked people. *Proc. Euro. Conf. Comput. Vision.* Springer-Verlag, New York, 593–602.
8. Forsyth, D. A., J. Malik, M. M. Fleck, H. Greenspan, T. Leung, S. Belongie, C. Carson, and C. Bregler. 1996. Finding pictures of objects in large collections of images. *Proc. 2nd Inter. Workshop on Object Representation in Comput. Vision* (April 1996).
9. Kato, T., T. Kurita, N. Otsu, and K. Hirata. 1992. A sketch retrieval method for full color image database. *11th Inter. Conf. Pattern Recog.,* 530–533.
10. Ma, W. Y., and B. S. Manjunath. 1999. Netra: a toolbox for navigating large image databases. *Multimedia Systems,* v. 7(3):184–198.

11. Minka, T. P., and R. W. Picard. 1996. Interactive learning with a society of models. *Proc. CVPR-96,* 447–452.
12. Niblack, W., and others. 1993. The QBIC project: Querying images by content using color, texture, and shape. *SPIE Proc. Storage and Retrieval for Image and Video Databases,* 173–187.
13. Rowley, H., S. Baluja, and T. Kanade. 1996. *Human Face Detection in Visual Scenes.* Carnegie-Mellon University, Pittsburgh, PA.
14. Samet, H. 1990. *The Design and Analysis of Spatial Data Structure.* Addison-Wesley, Reading, MA.

9

Motion from 2D Image Sequences

A changing scene may be observed via a sequence of images. One might learn a golf swing by observing the motions of an expert with a video camera, or better understand root growth by observing the root using many images taken hours apart. Action phenomena observed over time can be due to the motion of objects or the observer, or both. Changes in an image sequence provide features for detecting objects that are moving or for computing their trajectories. In case the viewer is moving through a relatively static world, image changes allow for computing the motion of the viewer in the world.

Similarly changing pixels in an image provide an important feature for object detection and recognition. Motion can reveal the shape of an object as well as other characteristics, such as speed or function. Analysis of object motion over time may be the ultimate goal; for instance, in controlling traffic flow or in analyzing the gait of a person with a new prosthesis. Today, a huge amount of videos are made to record events and structure in the world. It is necessary to have methods of segmenting these image sequences into meaningful events or scenes for easy access, analysis, or editing.

This chapter concentrates on detection of motion from 2D images and video sequences and the image analysis used to extract features. Methods for solution of the application problems just mentioned are discussed. Analysis of 3D structure and motion derived from 2D images is discussed in Chapter 13.

9.1 MOTION PHENOMENA AND APPLICATIONS

It is useful to consider the various cases of motion observable in an image sequence and the several important related applications. The problems to be solved range from mere detection of a moving object to analyzing the related motion and shape of multiple moving objects.

We identify the following four general cases of motion. We use the term *camera* to be interchangable with the term *observer*.

- Still camera, single moving object, constant background,
- Still camera, several moving objects, constant background,
- Moving camera, relatively constant scene,
- Moving camera, several moving objects.

The simplest case occurs when a still sensor stares at a relatively constant background. Objects moving across that background will result in changes to the image pixels associated with the object. Detection of these pixels can reveal the shape of the object as well as its speed and path. Such sensors are commonly used for safety and security. It is common for homes to use such sensors to automatically switch on a light upon detection of significant motion, which might be due to the owner coming home or to an unwelcome intruder. These simple motion sensors can also be used in manufacturing to detect the presence of a part fed into a workspace or in traffic control systems that detect moving vehicles.

A staring camera can also provide data for analysis of the movements of one or several objects. Moving objects must be tracked over time to produce a trajectory or path, which in turn can reveal the behavior of the object. For example, a camera might be used to analyze the behavior of people who enter the lobby of some business or who use some workspace. Several cameras can be used to produce different views of the same object, thus enabling the computation of paths in 3D. This is often done in the analysis of the motion of athletes or patients in rehabilitation. A system currently under development tracks the players and ball in a tennis match and provides an analysis of the elements of the game.

A moving camera creates image change due to its own motion, even if the 3D environment is unchanging. This motion has several uses. First, it may create more observations of the environment than available from a single viewpoint—this is the case when a panning camera is used to provide a wider (panoramic) view of the scene. Second, it can provide for computation of relative depth of objects since the images of close objects change faster than the images of remote objects. Third, it can provide for perception and/or measurement of the 3D shape of nearby objects—the multiple viewpoints allow for a triangulating computation similar to binocular stereo. In processing or analyzing video or film content, it is often important to detect points in time when the camera is panned or zoomed: In this case, we may not be interested in the contents of the scene but rather in the manner in which the scene was viewed.

The most difficult motion problems involve moving sensors and scenes containing so many moving objects that it is difficult to identify any constant background. Such a case arises when a robot vehicle navigates through heavy traffic. Another interesting case might be using communicating cameras to make correspondences in their observations in order to track several moving objects in the workspace.

The next sections examine various image analysis methods which analyze a sequence of two or more images in order to detect changes due to motion, or to analyze the objects themselves or their motion.

Exercise 9.1

Locate a motion detector in your neighborhood that is used to switch on a light. These devices are commonly used near a garage or house entry. Verify that quickly entering the area switches on a light. (a) Experiment by walking very slowly. Can you fool the motion detector into missing you? (b) What does this experiment tell you about how the motion detector works? (c) How does this relate to the Tyrannosaurus rex in the movie *Jurassic Park?*

9.2 IMAGE SUBTRACTION

Image subtraction was introduced in Chapter 1 as a means of detecting an object moving across a constant background. Suppose a video camera provides thirty frames per second of a conveyor belt that creates a uniform dark background. As brighter objects move across the camera view, the forward and rear edges of the object advance only a few pixels per frame. By subtracting the image I_t from the previous image I_{t-1}, these edges should be evident as the only pixels significantly different from zero.

Figure 9.1 shows the results of differencing over an interval of a few seconds for the purpose of monitoring a workspace (surveillance). A background image, which is nonuniform in this case, is derived from many video frames. A person who enters the workspace changes a region of the image, which can be detected by image subtraction as shown in Figure 9.1. The bounding box delimits a rectangular region where the change is detected. Further analysis of this bounding box might reveal object shape and even object type. The center image of Figure 9.1 actually shows three separate regions of change corresponding to (1) the person; (2) the door that the person opened; and (3) a computer monitor. A surveillance system may be provided the knowledge of the location of such objects and might even be primed to observe them or to ignore them. For example, the door might be carefully monitored while the CRTs are ignored. Related applications include monitoring and inventory of parking lots and monitoring the flow of vehicles on streets or people in rooms.

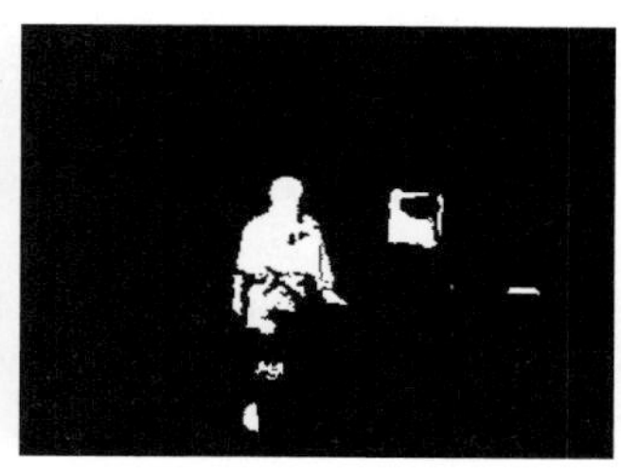

Figure 9.1 (left) A person appears in a formerly unoccupied workspace; (center) image substraction reveals changed regions where the person occludes the background and at the door and a CRT; and (right) the change due to the person is deemed significant while the other two are expected and hence ignored. (Images courtesy of S.-W. Chen.)

Detect changes between two images

Input $\mathbf{I_t}[\mathbf{r}, \mathbf{c}]$ and $I_{t-\Delta}[\mathbf{r}, \mathbf{c}]$: two monochrome input images taken Δ seconds apart.
Input τ is an intensity threshold.
$\mathbf{I_{out}}[\mathbf{r}, \mathbf{c}]$ is the binary output image; **B** is a set of bounding boxes.

1. For all pixels $[\mathbf{r},\mathbf{c}]$ in the input images,
   set $\mathbf{I_{out}}[\mathbf{r}, \mathbf{c}] = 1$ if $(|\mathbf{I_t}\ [\mathbf{r}, \mathbf{c}] - \mathbf{I_{t-\Delta}}\ [\mathbf{r}, \mathbf{c}]| > \tau)$
   set $\mathbf{I_{out}}[\mathbf{r}, \mathbf{c}] = 0$ otherwise.
2. Perform connected components extraction on $\mathbf{I_{out}}$.
3. Remove small regions assuming they are noise.
4. Perform a closing of $\mathbf{I_{out}}$ using a small disk to fuse neighboring regions.
5. Compute the bounding boxes of all remaining regions of changed pixels.
6. Return $\mathbf{I_{out}}[\mathbf{r}, \mathbf{c}]$ and the bounding boxes **B** of regions of changed pixels.

Algorithm 9.1 Detection of change via image subtraction.

A sketch of the steps in change detection via image subtraction is given as Algorithm 9.1. Steps of this algorithm are operations given in Chapter 3.

Exercise 9.2

This exercise requires a workstation equipped with an attached camera and software to access the frames from the camera. Write a program that monitors the top of your desk (next to the workstation). The program should successively acquire frames, compute a histogram for each, and sound an alarm whenever there is a significant change in the histogram. Test your program on various still scenes and by moving various objects on and off the desk.

9.3 COMPUTING MOTION VECTORS

Motion of 3D scene points results in motion of the image points to which they project. Figure 9.2 shows three typical cases. Zooming out can be performed by reducing the focal length of a still camera or by backing away from the scene while keeping the focal length fixed. The optical axis points toward a scene point whose image does not move: this is the *focus of contraction. Zooming in* is performed by increasing the focal length of a still camera or by moving toward a particular scene point whose image does not change (*focus of expansion*). Panning a camera or turning our heads causes the images of the 3D scene points to translate, as shown at the right in Figure 9.2.

70 Definition. A 2D array of 2D vectors representing the motion of 3D scene points (as shown in Figure 9.2) is called the **motion field.** The motion vectors in the image represent the displacements of the images of moving 3D points. Each motion

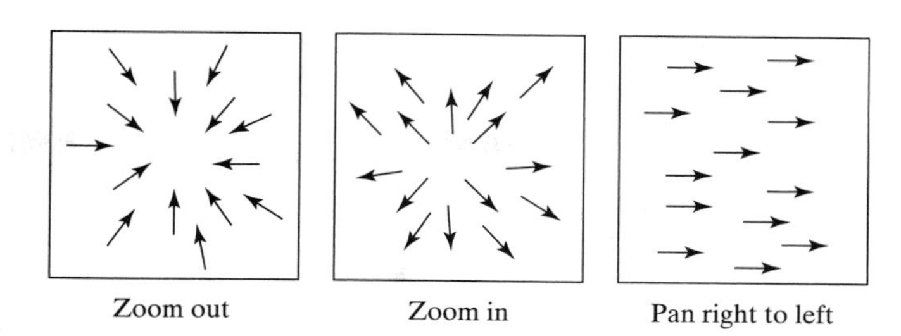

Figure 9.2 Effects of zooming and panning on imaged features. The effect of zoom in is similar to that observed when we move forward in a scene, and the effect of panning is similar to that observed when we turn.

vector might be formed with its tail at an imaged 3D point at time t and its head at the image of that same 3D point imaged at time $t + \Delta$. Alternatively, each motion vector might correspond to an instantaneous velocity estimate at time t.

71 Definition. The **focus of expansion (FOE)** is that image point from which all motion field vectors diverge. The FOE is typically the image of a 3D scene point toward which the sensor is moving. The **focus of contraction (FOC)** is that image point toward which all motion vectors converge, and is typically the image of a 3D scene point from which the sensor is receding.

Computation of the motion field can support both the recognition of objects and an analysis of their motion. One of two—not too constraining—assumptions is usually made in order to compute motion vectors. First, we can assume that the intensity of a 3D scene point P and that of its neighbors remain nearly constant during the time interval (t_1, t_2) over which the motion estimate for P is made. Alternatively, we can assume that the intensity differences observed along the images of the edges of objects are nearly constant during the time interval (t_1, t_2).

72 Definition. **Image flow** is the motion field computed under the assumption that image intensity near corresponding points is relatively constant.

Two methods for computing image flow are given next. Before developing these we describe a video game application that uses motion fields.

9.3.1 The Decathlete Game

Researchers at Mitsubishi Electric in Sagamihara, Japan, and at (Mitsubishi Electric Research Laboratory) MERL in Cambridge, Massachusetts have reported results of using motion analysis to control the Sega Saturn *Decathlete* game. They replaced a keypad interface with control through image flow computation from a low-resolution camera. The actual motion of the player of the game was used to control the motion of the player's avatar in the simulation. In the example shown here, the avatar is running the hurdles against another simulated runner. In Figure 9.3 the man at the left is the player and is making running motions with his arms and hands. The faster the motions, the faster his avatar runs. The avatar

Figure 9.3 The man at the left is making running motions with his arms and hands to control the game of running the hurdles. The game display is shown at the right. In the lower right corner, a video camera (Mitsubishi Electric's CMOS image sensor with on-chip image processing) observes the motion of the player, which is used to control the running speed and jumping of the hurdling avatar. (Reprinted from *IEEE Computer Graphics,* vol. 18, no. 3 (May–June 1998) by permission of IEEE.)

must also jump the hurdles at the proper instants of time: The player jumps by raising both fists upward. The *Decathlete* display is shown on the monitor at the right in Figure 9.3. In the lower right corner of that figure, one can see the video camera that observes the motion of the player. The two persons in the middle of the picture are watching the fun.

Figure 9.4 illustrates the motion analysis used to control the hurdling game. Figure 9.4(a) gives a snapshot of the motion analysis, while Figure 9.4(b) provides an explanatory map of the content of (a). Note that the top left of (a) shows a video frame of the *running* player seen by the camera, while the middle left of (a) shows motion vectors extracted from multiple frames. The bottom left of (a) shows a time history of the average horizontal motion over the video frame; the middle of (a) shows the average vertical motion.

The camera must be set up to view the player's two hands as the player "runs" and "jumps." A running motion causes alternations in the horizontal average motion. The frequency of those alternations indicates how fast the player is running in place, which controls the runner's speed. A jumping motion causes the average vertical velocity to exceed a threshold, which sends a jump command to the game. The spatial resolution is very low but the temporal resolution is high.

The type of coarse motion analysis used in the decathalon game might provide a general gesture interface to computers. For example, future computers might provide for input using American Sign Language (ASL) or some smaller gesture language.

9.3.2 Using Point Correspondences

A sparse motion field can be computed by identifying pairs of points that correspond in two images taken at times t_1 and $t_1 + \Delta$. The points we use must be distinguished in some way so that they can be identified and located in both images. Detecting corner points or

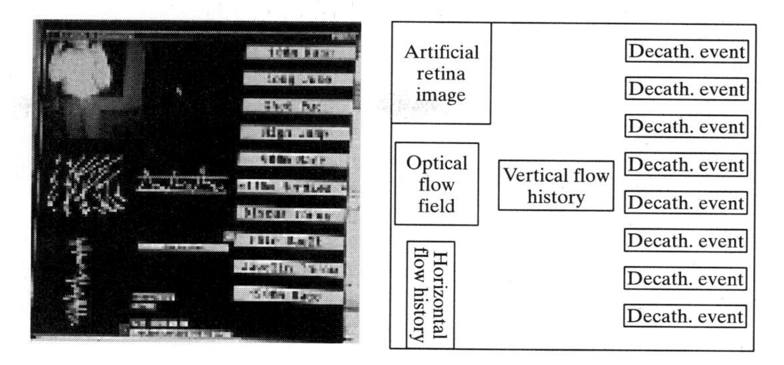

Figure 9.4 Illustration of the motion analysis used to control the hurdling game. The top left shows a video frame of the running player, while the middle left shows motion vectors extracted from multiple frames. The jumping of the hurdles is indicated by the vertical motion plot shown in the middle. (Reprinted from *IEEE Computer Graphics,* vol. 18, no. 3 (May–June 1998) with permission of IEEE.)

high interest points should work for both color and monochrome images. Alternatively, centroids of persistent moving regions from segmented color images might be used. Corner points can be detected by using masks such as the Kirsch edge operator or the ripple masks of the Frie-Chen operator set (Chapter 5). Alternatively, an *interest operator* can be used. The operator computes intensity variance in the vertical, horizontal, and diagonal directions through the neighborhood centered at a pixel **P**. Only if the minimum of these four variances exceeds a threshold is **P** passed as an interesting point. These operations are sketched in Algorithm 9.2. An alternative operator based on texture is developed in Exercise 9.3.

Exercise 9.3: Texture-based interest operator.

Experiment with the following interest operator, which is based on the texture of an entire $n \times n$ neighborhood. First, compute the gradient magnitude for the entire input image using a 3×3 or 2×2 mask. Second, threshold the magnitude image to produce a binary image. A pixel $[r, c]$ in the original image is interesting only if there is significant variation in each of the four main directions in the $n \times n$ neighborhood of $B[r, c]$ in the binary image. Variation in direction $[\Delta r, \Delta c] = [0, 1], [1, 0], [1, 1], [1, -1]$ is just the sum of $B[r, c] \otimes B[r + \Delta r, c + \Delta c]$ for all pixels in the $n \times n$ neighborhood centered at $B[r, c]$. $\otimes$ is the **exclusive or** operator that returns 1 if and only if the two inputs are different. An *interest image* is formed by assigning IN[r, c] the minimum of the four variations in $B[r, c]$ computed as above. Try your operator on a few monochrome images, including a checkerboard.

Once a set of interesting points $\{P_j\}$ is identified in the image I_1 taken at time t, corresponding points must be identified in the image I_2 taken at time $t + \Delta$. Rather than

```
Find interesting points of a given input image.

procedure detect_corner_points(I, V);
{
        "I[r, c] is an input image of MaxRow rows and MaxCol columns"
        "V is an output set of interesting points from I."
        "τ is a threshold on the interest operator output"
        "w is the halfwidth of the neighborhood for the interest operator"
        for r := 0 to MaxRow - 1
           for c := 0 to MaxCol - 1
              {
              if I[r,c] is a border pixel then break;
              else if (interest_operator (I, r, c, w ) ≥ τ1) then add
              [(r,c),(r,c)] to set V;
        "The second (r, c) is a place holder in case vector tip found later."
              }
}
real procedure interest_operator (I, r, c, w)
{
        "w is the halfwidth of operator window"
        "See alternate texture-based interest operator in the exercises."
        v1 := variance of intensity of horizontal pixels I1[r, c − w] ... I1[r, c + w];
        v2 := variance of intensity of vertical pixels I1[r − w, c] ... I1[r + w, c];
        v3 := variance of intensity of diagonal pixels
                                    I1[r − w, c − w] ... I1[r + w, c + w];
        v4 := variance of intensity of diagonal pixels
                                    I1[r − w, c + w] ... I1[r + w, c − w];
        return minimum {v1, v2, v3, v4};
}
```

Algorithm 9.2 Algorithm for detecting interesting image points.

extract points from I_2 in the same manner and then attempt to make correspondences, we can directly search I_2 to determine the new location of each point from I_1. This can be done by using the cross-correlation method described in Chapter 5. Given an interesting point P_j from I_1, we take its neighborhood in I_1 and find the best correlating neighborhood in I_2 under the assumption that the amount of movement is limited. Figure 9.5 sketches how to search frame I_2 for a good match to the neighborhood of point P_j in frame I_1. The center $P_k = [P_{kr}, P_{kc}]$ of the best correlating neighborhood in $\mathbf{I}_2$ is taken to be the corresponding point and will become the tip of a motion vector with $P_j = [P_{jr}, P_{jc}]$ being the tail. The search for P_k is limited to a rectangular $C \times R$ region in image rows $P_{jr} - R \ldots P_{jr} + R$ and image columns $P_{jc} - C \ldots P_{jc} + C$. A small search region speeds up the search for a match and also reduces ambiguity, but is only useful when there is a justifiable assumption on the limit to the velocity of objects. The resulting algorithm is given as Algorithm 9.3.

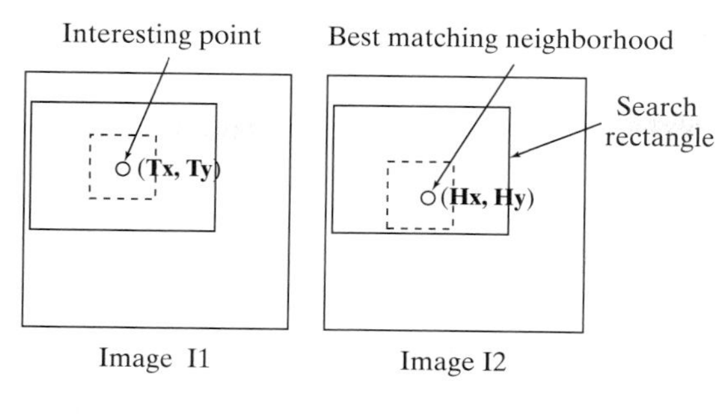

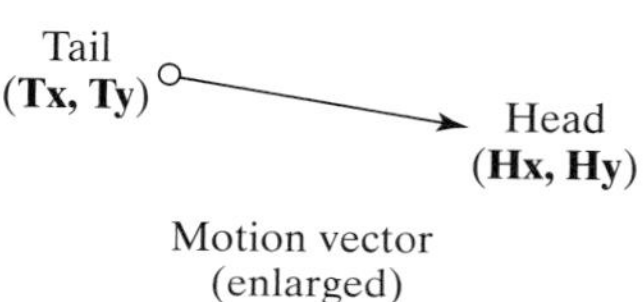

Figure 9.5 For each interesting point **(Tx, Ty)** in image $\mathbf{I_1}$ a rectangular region of image $\mathbf{I_2}$ is searched for the best match to a small neighborhood of **(Tx, Ty)**. If the best match is good, then it becomes the head **(Hx, Hy)** of a motion vector.

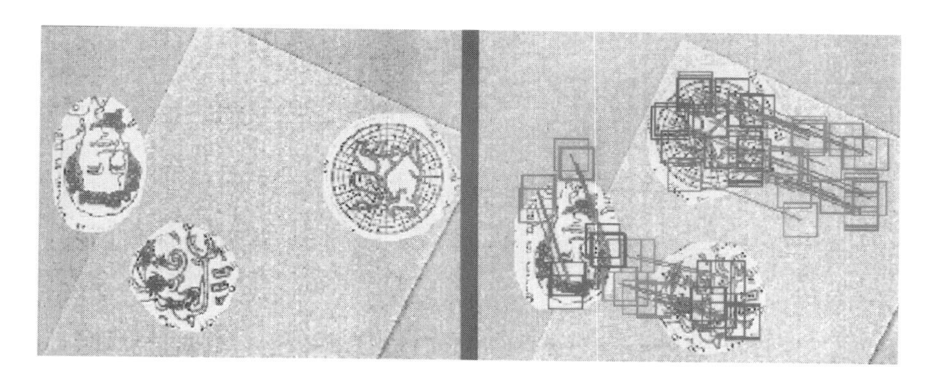

Figure 9.6 Results of applying Algorithm 9.3. At the left is the image at time t_1. At the right is the image at time t_2 with the motion analysis overlaid. Red squares indicate the location of the original neigborhoods detected by the interest operator in the image at the left. Blue squares indicate the best matches to these neighborhoods in the image at the right. There are three coherent sets of motion vectors (green lines) corresponding to the three moving objects. The leftmost object moved down and slightly right. The lowest object moves right and slightly down; the rightmost object moves left and slightly up. (Analysis courtesy of Adam T. Clark.) See colorplate.

Results of applying this algorithm are shown in Figure 9.6. The test imagery was created by moving three highly textured cutouts across a lightly textured background.

Algorithm 9.3 can be controlled to iterate through pairs of frames so that feature points can be continuously tracked over many frames. The corner points identified in frame $t + \Delta$ can replace those formerly identified in frame t and the new, possibly changed, neighborhoods used for cross-correlation. In this manner, significant points can be tracked in a dynamic scene provided that their neighborhoods change in a gradual manner. In general, we must also provide for the disappearance of corner points due to occlusion and the appearance of new unoccluded corner points. These ideas are discussed in Section 9.4.

From two input images, derive motion vectors for interesting points.

$\mathbf{I_1[r, c]}$ and $\mathbf{I_2[r, c]}$ are input images of **MaxRow** rows and **MaxCol** columns.
V is the output set of motion vectors $\{[(T_x, T_y), (H_x, H_y)]_i\}$
where (T_x, T_y) is the tail of a motion vector and (H_x, H_y) is its head.

```
procedure extract_motion_field(I1, I2, V)
{
"Detect matching corner points and returning motion vectors V"
"τ2 is a threshold on neighborhood cross-correlation"
      detect_corner_points(I1, V);
      for all vectors [(Tx, Ty), (Ux, Uy)] in V
         match := best_match( I1, I2, Tx, Ty, Hx, Hy);
         if (match < τ2) then delete [(Tx, Ty), (Ux, Uy)] from V;
         else replace [(Tx, Ty), (Ux, Uy)] with [(Tx, Ty), (Hx, Hy)] in V;
}
real procedure best_match( I1, I2, Tx, Ty, Hx, Hy);
"(Hx, Hy) is returned as the center of the neighborhood in I2 that matches best"
"to the neighborhood centered at (Tx, Ty) in I1."
"sh and sw define search rectangle: h and w define neighborhood size."
{
      "first indicate that a good match has not been found"
      Hx := -1; Hy   := -1;   best := 0.0;
      for r := Ty - sh to Ty + sh
         for c := Tx - sw to Tx + sw
            {
               "cross correlate N in I1 with N in I2 as in Chapter 5"
               match := cross_correlate(I1, I2, Tx, Ty, r, c, h, w);
               if ( match > best) then
                  {
                     Hy := r;   Hx := c;   best := match;
                  }
            }
}
```

Algorithm 9.3 Compute sparse set of motion vectors from a pair of input images.

Exercise 9.4

Consider the image of a standard checkerboard: (a) Design a corner detector that will respond only at the corners of four of the squares and not within squares or along the edges between two squares; (b) take several images by slowly moving a checkerboard in front of a stationary camera; (c) test your corner detector on these images and report on the number of correct and incorrect detections in each of the images; and (d) implement and test Algorithm 9.3 on several pairs of the images with a small amount of motion between them.

9.3.3 MPEG Compression of Video

MPEG video compression uses complex operations to compress a video stream up to 200:1. We note the similarity between MPEG motion compensation and Algorithm 9.3. The subgoal of MPEG is not to compute a motion field but to compress the size of an image sequence by predicting the content of some frames from other frames. It is not important that the motion vectors be correct representations of moving objects, but only that they provide for good approximations of one image neighborhood from another. An MPEG encoder replaces an entire 16×16 image block in one frame with a motion vector defining how to locate the best matching 16×16 block of intensities in some previous frame. Figure 9.7 illustrates the use of motion estimation in the MPEG compression scheme. Details of the scheme are given in the figure caption. Distinguished image points are not used; instead a uniform grid of blocks is used and a match of each block is sought by searching a previous image of the video sequence. The figure only shows how a few of the many blocks are computed. Ideally, each block B_k of 16×16 pixels can be replaced by a single vector $[V_x, V_y]_k$, which the encoder found to locate the best block of matching intensities in a previous frame. If there is a difference in the two intensity arrays, then this difference can be represented by a small number of bits and transmitted also.

Although the MPEG motion vectors are designed for the purpose of compression and not motion analysis, researchers have begun to experiment with using them for the purpose of providing a motion field. The advantage is that MPEG encoders can now compute these

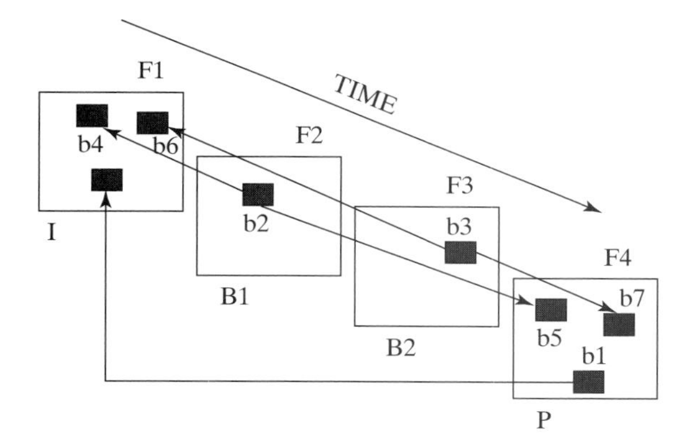

Figure 9.7 A coarse view of the MPEG use of motion vectors to compress the video sequence of four frames F1, F2, F3, F4. F1 is coded as an *independent* (I) frame using the JPEG scheme for single still images. F4 is a P frame predicted from F1 using motion vectors together with block differences: 16×16 pixel blocks (b1) are located in frame F1 using a motion vector and a block of differences to be added. Between frames B1 and B2 are determined entirely by interpolation using motion vectors: 16×16 blocks (b2) are reconstructed as an average of blocks (b4) in frame F1 and (b5) in frame F4. Between frames F2 and F3 can only be decoded after predicted frame F4 has been decoded even though these images were originally created before F4. Between frames yield the most compression since each 16×16 pixel block is represented by only two motion vectors. I frames yield the least compression.

vectors in real time and they are already present in the video stream. Options in future codecs may indeed provide useful motion fields for motion analysis applications.

Exercise 9.5

Assume a video sequence of frames that each have 320×240 8-bit monochrome pixels. (a) What is the representation output by an MPEG-type encoder for a between frame? (b) How many bytes are needed for the representation? (c) What is the compression ratio for the between frame relative to an original image?

9.3.4 Computing Image Flow*

Methods have been developed for estimating image flow at all points of an image and not just at interesting points. We study a classical method that combines spatial and temporal gradients computed from at least two consecutive frames of imagery. Figure 9.8 shows an ideal example of what a camera might observe when an object moves in its field of view. The image (a) from time t_1 shows a triangular object in the lower left corner, while (b) shows that the triangle has moved upward by time t_2. This simple example serves to illustrate some of the assumptions that we will make to develop a mathematical model of image flow.

- We assume that the object reflectivity and the illumination of the object do not change during the interval $[t_1, t_2]$.
- We assume that the distances of the object from the camera or light sources do not vary significantly over this interval.
- We shall also assume that each small intensity neighborhood $N_{x,y}$ at time t_1 is observed in some shifted position $N_{x+\Delta x, y+\Delta y}$ at time t_2.

These assumptions do not hold tight in real imagery, but in some cases they can lead to useful computation of image flow vectors. Having motivated the theory with a simple discrete case, we now proceed with the development of the image flow equation for the case of a continuous intensity function $f(x, y)$ of continuous spatial parameters.

$(a)\ t_1$	$(b)\ t_2$
3333333333	3333333333
3333333333	3333333333
3333333333	3373333333
3373333333	3397533333
3397533333	3399753333
3399753333	3399975333
3399975333	3333333333
3333333333	3333333333

Figure 9.8 An example of image flow. A brighter triangle moves one pixel upward from time t_1 to time t_2. Background intensity is 3 while object intensity is 9.

Exercise 9.6

Refer to Figure 9.8. The intensity function is $f(x, y, t)$. Consider the topmost pixel at time t_1 with intensity 7 within the context 9 7 5 (its spatial coordinates are $x = y = 4$). Estimate the spatial derivatives of the image function $\partial f/\partial x$ and $\partial f/\partial y$ at $x = y = 4; t = t_1$ using a 3×3 neighborhood. Estimate the temporal derivative $\partial f/\partial t$ at $x = y = 4; t = t_1$: what method is appropriate?

9.3.5 The Image Flow Equation*

Using the previous assumptions, we now derive what is called the *image flow equation* and show how it can be used to compute image flow vectors. Using the continuous model of the intensity function $f(x, y, t)$, we apply its Taylor series representation in a small neighborhood of an arbitrary point (x, y, t).

$$f(x + \Delta x, y + \Delta y, t + \Delta t) = f(x, y, t) + \frac{\partial f}{\partial x}\Delta x + \frac{\partial f}{\partial y}\Delta y + \frac{\partial f}{\partial t}\Delta t + h.o.t. \quad (9.1)$$

Note that Equation 9.1 is a multivariable version of the very intuitive approximation for the one variable case: $f(x + \Delta x) \approx f(x) + f'(x)\Delta x$. For small neighborhoods of (x, y, t) we ignore the higher order terms $h.o.t.$ of Equation 9.1 and work only with the linear terms. Our next crucial step is illustrated in Figure 9.9. The image flow vector $\mathbf{V} = [\Delta x, \Delta y]$ for which we want to solve carries the intensity neighborhood N_1 of (x, y) at time t_1 to an identical intensity neighborhood N_2 of $(x + \Delta x, y + \Delta y)$ at time t_2. This assumption means that

$$f(x + \Delta x, y + \Delta y, t + \Delta t) = f(x, y, t) \quad (9.2)$$

We obtain the image flow equation by combining Equations 9.1 and 9.2 and ignoring the higher order terms.

$$-\frac{\partial f}{\partial t}\Delta t = \frac{\partial f}{\partial x}\Delta x + \frac{\partial f}{\partial y}\Delta y = \left[\frac{\partial f}{\partial x}, \frac{\partial f}{\partial y}\right] \circ [\Delta x, \Delta y] = \nabla f \circ [\Delta x, \Delta y] \quad (9.3)$$

The image flow equation does not give a unique solution for the flow vector $\mathbf{V}$, but imposes a linear constraint on it. In fact, there may be many neighborhoods N_2 that have the same intensities as N_1. Figure 9.10 shows how multiple possibilities exist for the flow vector when restricted to a small neighborhood, or *aperture,* about point (x, y). Observing only the small aperture about point $\boldsymbol{P}$, it is possible that $\boldsymbol{P}$ moves to $\boldsymbol{R}$ or $\boldsymbol{Q}$ or some other

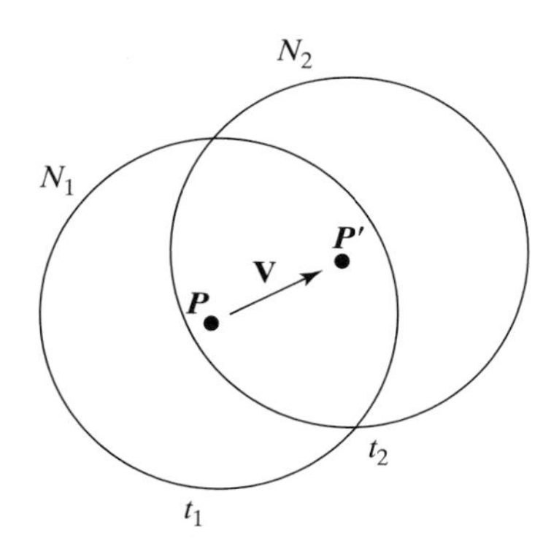

Figure 9.9 Due to motion in direction **v**, the intensities of neighborhood N_1 at time t_1 are the same as the intensities of neighborhood N_2 at time t_2.

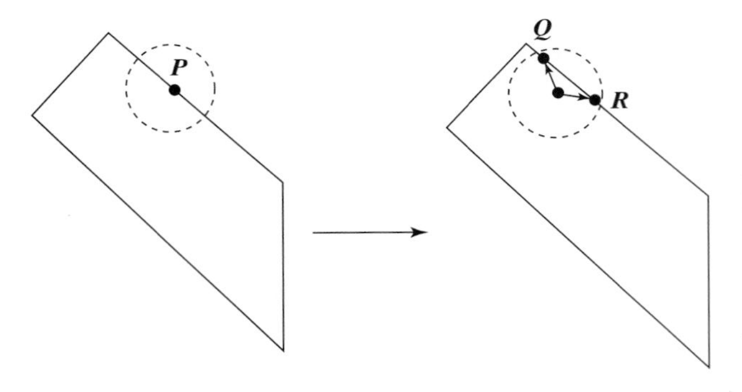

Figure 9.10 An intensity edge moves toward the right from time t_1 to time t_2. However, due to the limited size of the neighborhood, or *aperture,* used for matching, the location of the displaced point ***P*** could be ***R*** or ***Q*** or some other point along the edge segment determined by them.

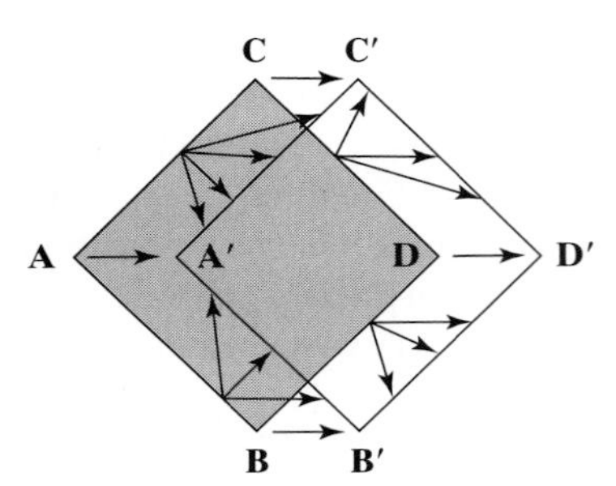

Figure 9.11 A square object is moving toward the right. Motion vectors with their tails on the edge at time t_1 are constrained by a linear relationship that puts their heads on an edge at time t_2. Common constraints at the corners A, B, C, D force the *move right* interpretation, which can then be propagated to all edge points by enforcing consistency along the entire boundary.

point along the segment ***QR***. Figure 9.11 shows four different edge cases from a square object. In general, we may not have pronounced object edges; however, Figure 9.9 still applies to curved isobrightness contours. The edges shown in the picture would be the tangents to the contour that approximate the contour locally.

We can interpret Figure 9.10 as follows. A change is observed at point ***P*** and can be measured by $-\frac{\partial f}{\partial t}\Delta t$. This change equals the dot product of the spatial gradient ∇f and the flow vector $\mathbf{V}$. $|\mathbf{V}|$ can be as small as the perpendicular distance to the new edge location, or it can be much larger in case the flow vector is in a direction much different from the spatial gradient. The latter case would result when a rope being pulled very fast vertically vibrates slightly horizontally resulting in small changes of image edge position over time.

9.3.6 Solving for Image Flow by Propagating Constraints*

The image flow equation provides a constraint that can be applied at every pixel position. By assuming coherence, neighboring pixels are constrained to have similar flow vectors. Figure 9.12 shows how neighboring constraints can be used to reduce the ambiguity in the direction of motion. Figure 9.12(*b*) shows a blowup of the neighborhood of corner A of the moving square shown in (*a*). The image flow equation constrains the direction θ_x of motion at point X to be in the interval between $5\pi/4$ and $\pi/4$. The direction θ_y of motion at point Y is constrained to the interval between $-\pi/4$ and $3\pi/4$. If points X and Y are assumed to be on the same rigid object, then the flow vectors at X and Y are now constrained to the intersection of these constraints, which is the interval $-\pi/4$ and $\pi/4$.

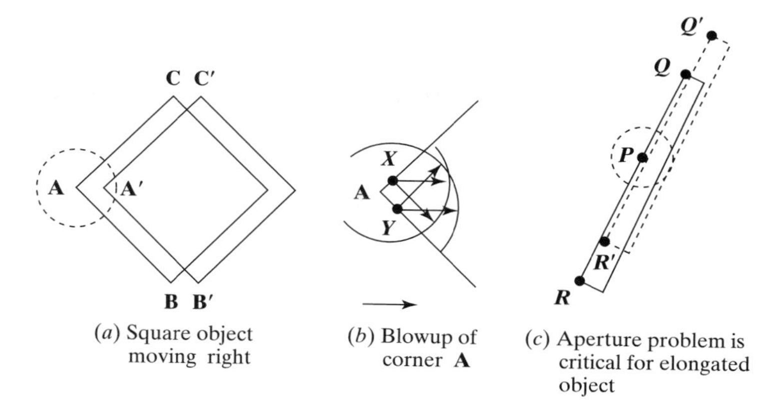

Figure 9.12 (*a*) A square object moves right; (*b*) a blowup of corner ***A*** shows how the constraints from two neighboring image flow equations can reduce the ambiguity of direction to $\pi/2$, and (*c*) an extreme aperture problem: an elongated object moves in the direction of its length; from the aperture around point ***P*** and its neighbors, the ambiguity in the direction of motion is π.

Figures 9.11 and 9.12 emphasize two points. First, only at the interesting corner points can image flow be safely computed using small apertures. Second, constraints on the flow vectors at the corners can be propagated down the edges; however, as Figure 9.12(c) shows, it might take many iterations to reach an interpretation for edge points, such as P, that are distant from any corner. Some experiments in flow computations have been performed using random pixel images. Our development shows that such images are probably easier than highly structured images because neighborhoods are more likely to be unique. Relaxation methods in 2D are studied in Chapter 11. Use of differential equations to solve for image flow can be found in the paper by Horn and Schunck (1981) cited in the references in Section 9.6.

9.4 COMPUTING THE PATHS OF MOVING POINTS

The previous sections discussed methods for identifying interesting points of a frame at time t_1 and for locating that same object point in a following frame at time t_2 close in time to t_1. If the intensity neighborhood of each point is uniquely textured, then we should be able to track the point over time using normalized cross-correlation. Also, domain knowledge might make it easy to track an object in an image sequence, as in the case of tracking an orange tennis ball in a tennis match, or a pink face looking at a workstation.

We now consider a more general situation where moving points are not tagged with unique texture or color information. In this case, the characteristics of the motion itself must be used to collect points into trajectories. Figure 9.13 shows the smooth trajectories of three objects over six instances of time. We mention three concrete problems before considering the abstract general case. For one situation, consider a box of many tennis balls dumped out on the ground: the problem is to compute the path of each ball from a video sequence.

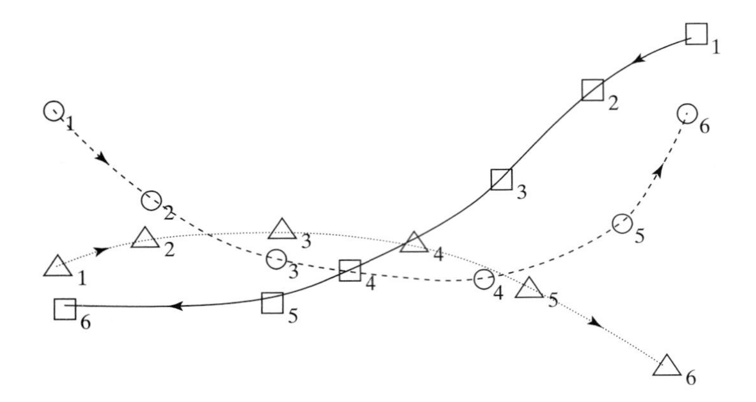

Figure 9.13 Trajectories of three objects, ○, △, □ are shown: the location of each object is shown for six instants of time. ○ and △ are generally moving from left to right while □ is moving right to left

In a second situation, we want to study fluid flow through a chamber by mixing fluorescent particles with the fluid and observing their motion over time. We assume that individual particles appear to be the same in an image. In a third case, we want to compute the paths of people walking in a mall. Clothing may make the images of some people unique, but surely there would be a high probability that some sets of individuals would have similar appearances in the images.

We can exploit the following general assumptions that hold for physical objects in 3D space.

1. The location of a physical object changes smoothly over time.
2. The velocity of a physical object changes smoothly over time: This includes both its speed and direction.
3. An object can be at only one location in space at a given time.
4. Two objects cannot occupy the same location at the same time.

The first three assumptions hold for the 2D projections of 3D space; smooth 3D motion creates smooth 2D trajectories. The fourth assumption may be violated under projection, since one object might occlude another, and this will create problems for a single observer. Experiments with humans have shown that humans can recognize objects and analyze their motion when presented with many frames of the moving object points. In a well-known experiment by Johansson (1976), lights were affixed to various points of a human body. Observers of only the lights on a still person could not recognize that it was a person; however, when the person moved, the observers could easily recognize that they were seeing a person.

We now present an algorithm, developed by Sethi and Jain (1987), that uses these four assumptions to compute a smoothest set of paths through points observed in a sequence of frames. First, we give a mathematical definition for the smoothness of a single path. Second, we define the smoothest set of m paths as that set of paths with the optimal sum of the m

smoothness values. Finally, we define the *Greedy Exchange Algorithm,* which iteratively extends the m paths from time t_1 through time t_n by making an optimal set of assignments at each time instant.

73 Definition. If an object i is observed at time instants $t = 1, 2, \ldots, n$, then the sequence of image points $T_i = (p_{i,1}, p_{i,2}, \ldots, p_{i,t}, \ldots, p_{i,n})$ is called the **trajectory of i.**

Between any two points of a trajectory, we can define their difference vector

$$V_{i,t} = p_{i,t+1} - p_{i,t} \tag{9.4}$$

We can define a smoothness value at a trajectory point $p_{i,t}$ in terms of the difference of vectors reaching and leaving that point. Smoothness of direction is measured by their dot product, while smoothness of speed is measured by comparing the geometric mean of their magnitudes to their average magnitude.

$$S_{i,t} = w\left(\frac{V_{i,t-1} \circ V_{i,t}}{|V_{i,t-1}|\,|V_{i,t}|}\right) + (1-w)\left(\frac{2\sqrt{|V_{i,t-1}|\,|V_{i,t}|}}{|V_{i,t-1}| + |V_{i,t}|}\right) \tag{9.5}$$

The weight w of the two factors is such that $0 \leq w \leq 1$, which yields $0 \leq S_{i,t} \leq 1$ (See the exercises in this section.) Note that for a straight trajectory with equally spaced points all the difference vectors are the same and Equation 9.5 yields 1.0, which is the optimal point smoothness value. Change of direction or speed decreases the value of $S_{i,t}$. We assume that m unique points are extracted from each of the n frames, although we can relax this later. Points of the first frame are labeled $i = 1, 2, \ldots, m$. The problem is to construct the m trajectories T_i with the maximum total smoothness value. Total smoothness is defined in Equation 9.6 as a sum of smoothness of all the interior points of all the m paths.

$$\textit{total smoothness } T_s = \sum_{i=1}^{m}\sum_{t=2}^{n-1} S_{i,t} \tag{9.6}$$

Exercise 9.7

Assume that $w = 0.5$ so that direction and speed are weighted equally. (a) Show that point smoothness at each vertex of a regular hexagon with unit sides is 0.75. (b) What is the point smoothness at the vertices of a square?

Exercise 9.8

(a) Refer to the Cauchy-Schwartz inequality (Chapter 5) and show that the factor $\frac{V_{i,t-1} \circ V_{i,t}}{|V_{i,t-1}|\,|V_{i,t}|}$ of $S_{i,t}$ is between 0 and 1. (b) Show that for two positive numbers x and y, their geometric mean $\sqrt{xy}$ never exceeds their arithmetic mean $(x + y)/2$. Use this to show that the factor $\frac{2\sqrt{|V_{i,t-1}|\,|V_{i,t}|}}{|V_{i,t-1}| + |V_{i,t}|}$ is between 0 and 1. (c) Now show that $S_{i,t}$ in Equation 9.5 is between 0 and 1 provided that w is between 0 and 1.

Exercise 9.9

What is the total smoothness of these two paths: a 4-point trajectory along four sides of an octagon with sides s and a 4-point trajectory along a square with sides s?

Algorithm 9.4 develops a set of m trajectories over n frames. It is not guaranteed to produce the minimum possible T_s, but experiments have shown it to work well. First, we intuitively examine some of its operations with reference to the simple example in Figure 9.14. Table 9.1 provides a record of the smoothness of some paths considered. We can assign point labels arbitrarily in the first frame; for example, object $\square_1 \equiv 1 = T[1, 1]$ and object $\bigcirc_1 \equiv 2 = T[2, 1]$. The trajectories can then be extended to the nearest point in subsequent frames: $T[1, 2] = \bigcirc_2$, the closest point, and $T[2, 2] = \square_2$ by process of elimination. We have made a mistake by switching the actual trajectories. We compute the

Input sets of 2D points over time and compute smooth paths.

P[i, t] holds $i = 1, 2, \ldots, m$ 2D points from frames $t = 1, 2, \ldots, n$;
T[i, t] is the output set of trajectories of m rows and n columns.
T[i, t] = k means that object i is observed as the k-th point of frame t.

1. **initialize:** create m complete paths by linking nearest neighbors
   - **(a)** First frame: for all i, set object labels **T[i, 1]** $= i$;
   - **(b)** Other frames: for $t = 2, 3, \ldots, n$, assign $\mathbf{T}[i, t] = k$ where point $\mathbf{P}[k,t]$ is the nearest point to point $\mathbf{P}[\mathbf{T}[i,t\text{-}1],t\text{-}1]$ which is not already assigned.
2. **Exchange Loop:** for $t := 2$ to $n - 1$
   - **(a)** for all pairs (j,k) with $j \neq k$, compute the increase of smoothness due to exchanging the assignments $\mathbf{T}[j, t]$ and $\mathbf{T}[k, t]$;
   - **(b)** make the exchange that produces the maximum increase in smoothness; make no exchange if none would increase total smoothness;
   - **(c)** set the exchange flag on only if an exchange was made;
3. **Test Termination:** If an exchange was made in the above loop, reset the exchange flag to off and repeat the exchange loop.

Algorithm 9.4 Greedy-Exchange Algorithm.

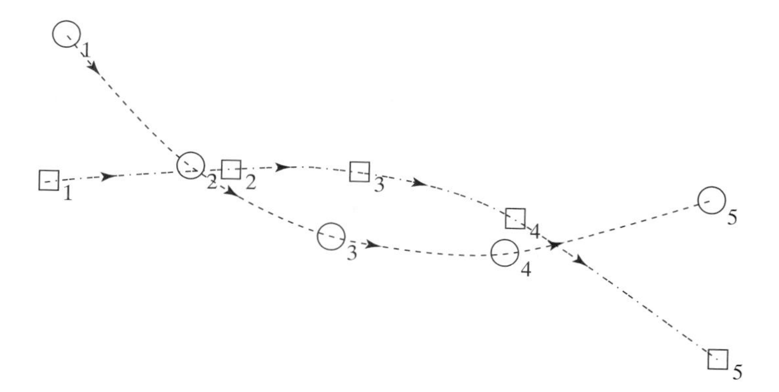

Figure 9.14 Trajectories of two objects, ○ and □ are shown along with the image flow vector at each of the first five points. A tracker would consider $\square_2$ as a likely successor to $\bigcirc_1$ and $\bigcirc_5$ to be a possible ending of the sequence $\square_1, \square_2, \square_3, \square_4$.

TABLE 9.1 SMOOTHNESS FOR PATHS OF FIGURE 9.14

t = 1	t = 2	t = 3	t = 4	t = 5	smoothness
$\bigcirc_1$(112 262)	$\square_2$(206 185)	$\bigcirc_3$(250 137)			0.97
$\square_1$(106 175)	$\bigcirc_2$(180 188)	$\square_3$(280 185)			0.98
$\bigcirc_1$(112 262)	$\bigcirc_2$(180 188)	$\bigcirc_3$(250 137)			0.99
$\square_1$(106 175)	$\square_2$(206 185)	$\square_3$(280 185)			0.99
$\bigcirc_1$(112 262)	$\bigcirc_2$(180 188)	$\bigcirc_3$(250 137)	$\bigcirc_4$(360 137)		1.89
$\square_1$(106 175)	$\square_2$(206 185)	$\square_3$(280 185)	$\square_4$(365 156)		1.96
$\bigcirc_1$(112 262)	$\bigcirc_2$(180 188)	$\bigcirc_3$(250 137)	$\bigcirc_4$(360 137)	$\square_5$(482 80)	2.84
$\square_1$(106 175)	$\square_2$(206 185)	$\square_3$(280 185)	$\square_4$(365 156)	$\bigcirc_5$(478 170)	2.91
$\bigcirc_1$(112 262)	$\bigcirc_2$(180 188)	$\bigcirc_3$(250 137)	$\bigcirc_4$(360 137)	$\bigcirc_5$(478 170)	2.89
$\square_1$(106 175)	$\square_2$(206 185)	$\square_3$(280 185)	$\square_4$(365 156)	$\square_5$(482 80)	2.94

total smoothness for these two paths after looking ahead to the nearest neighbor assignments at time $t = 3$. From the first two rows of Table 9.1 we see that the total smoothness for these two paths is $0.97 + 0.98 = 1.95$. If the assignments $T[1, 2] = \bigcirc_2$ and $T[2, 2] = \square_2$ are exchanged, a better smoothness value of $0.99 + 0.99 = 1.98$ is achieved. After this exchange, the two trajectories up to time $t = 2$ are $(\square_1, \square_2)$ and $(\bigcirc_1, \bigcirc_2)$. Nearest point initial assignments will give the best smoothness values at times $t = 3, 4$ and no exchanges are needed. However, at time $t = 5$ the nearest point assignments yield trajectories $(\square_1, \square_2, \square_3, \square_4, \bigcirc_5)$ and $(\bigcirc_1, \bigcirc_2, \bigcirc_3, \bigcirc_4, \square_5)$. Computing total smoothness with these last two assignments exchanged improves total smoothness from $2.84 + 2.91 = 5.75$ to $2.89 + 2.94 = 5.83$ over the interior 3 trajectory points so the final result will be correct according to the labels in Figure 9.14.

Algorithm 9.4 initializes m complete paths of n points each before it applies any smoothness criteria. A variable number of exchange loops attempt to improve smoothness by exchanging points between two paths. If an improvement is made (and it is always

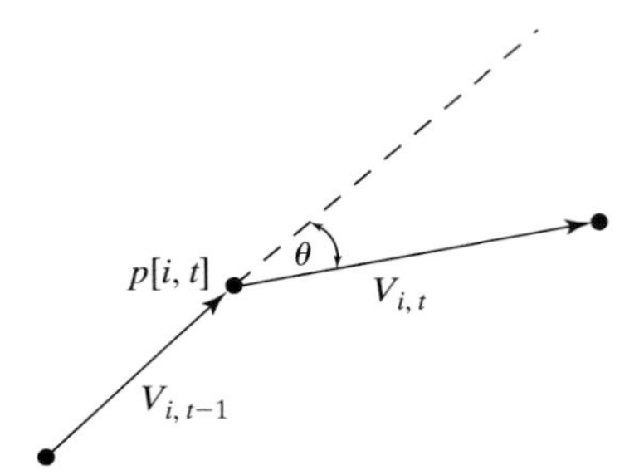

Figure 9.15 Vectors entering and leaving trajectory point $\boldsymbol{p}[i, t]$.

the biggest improvement) by an exchange at any time t, then the entire exchange loop is repeated. In general, $\binom{m}{2}$ possible exchanges must be considered at each frame t. The algorithm requires at least $(n-2)\binom{m}{2}$ operations; more effort is required for any additional exchange loops. Total smoothness cannot increase beyond the value $1.0m(t - 2)$ so the number of times that it can increase is limited and the algorithm must terminate.

Given that the assignments at frame $t = 1$ are arbitrary, there are m^{n-1} possible paths to consider overall—an exhorbitant number to evaluate. The Greedy-Exchange Algorithm does not consider exchanges of more than one pair of assignments at a time and hence might not obtain a global minimum. Algorithm 9.4 may be modified so that it is initialized using only one or three frames and continues as each new frame is sensed and processed for feature points. If all points of all frames are available, the algorithm can be improved by using both forward and backward processing in the exchange loop. Also, the algorithm has been extended to handle cases where points may appear or disappear between frames, primarily due to occlusions of one moving object by another. *Ghost points* can be used to represent points in frames containing fewer than m points.

Exercise 9.10

The following sets of three points were extracted from six frames of video and correspond to the data shown in Figure 9.13. Identify the smoothest set of three trajectories according to the Greedy-Exchange Algorithm.

t = 1	t = 2	t = 3	t = 4	t = 5	t = 6
(483 270)	(155 152)	(237 137)	(292 128)	(383 117)	(475 220)
(107 225)	(420 237)	(242 156)	(358 125)	(437 156)	(108 108)
(110 133)	(160 175)	(370 180)	(310 145)	(234 112)	(462 75)

Exercise 9.11

Would you expect the Greedy-Exchange Algorithm to succeed in constructing trajectories from points in image sequences in the following cases? Explain why or why not. (a) The video is of a rotating carrousel with wooden horses that move up and down. (b) The video is of a street taken from the sidewalk: two cars pass just in front of the camera going 35 MPH in opposite directions. (c) High speed film is taken of the collison of two billiard balls. The moving white ball hits a still red ball: After the collision, the white ball is still and the red ball has taken all of its momentum.

9.4.1 Integrated Problem-Specific Tracking

Use of Algorithm 9.4 has demonstrated the power of using only the general constraints of smoothness. In specific applications much more information might be available to increase both the robustness and speed of tracking. If features are available for each of the m points, then feature matching can be included in the smoothness computations. Going even further, fitting of the current partial trajectory up to time t can be used to predict the location of the next trajectory point in frame $t + 1$, which can greatly reduce the effort of cross-correlation used to find that point. Algorithms incorporating these methods can be found in the recent research literature. Maes and others (1996) have tracked human motion by computing the trajectories of the hands, feet, and head, which are identified as high-curvature protrusions from the silhouette of the moving human form. Bakic and Stockman (1999) track the human face, eyes, and nose using a workstation camera in order to move the mouse cursor. Figure 9.16 shows a sample screen exhibiting the features detected in the current frame and

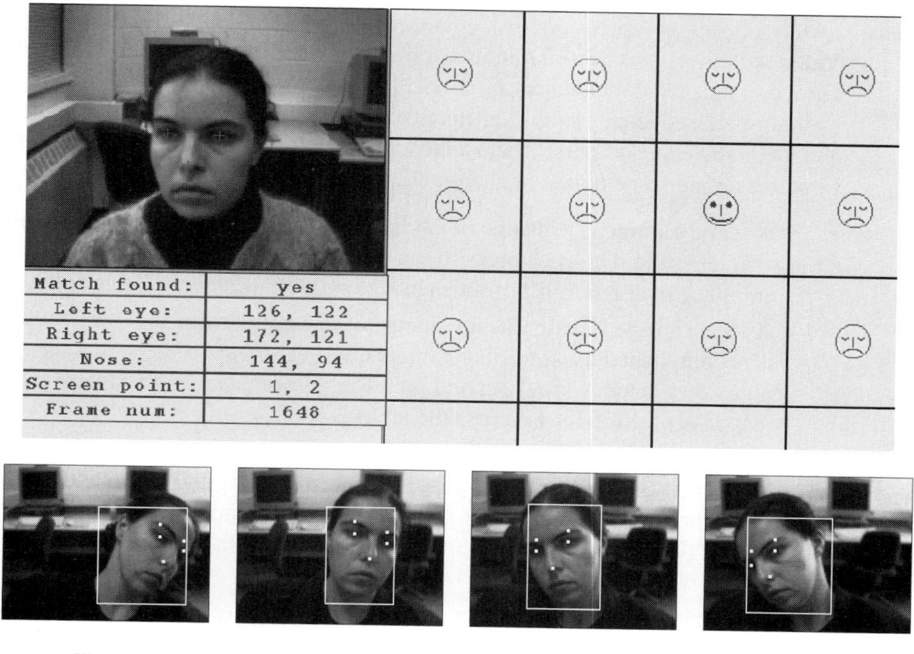

Figure 9.16 Tracking of the eyes and nose of a workstation user enables movement of the cursor without using the mouse. (top) Face pose determines selection in menu; (bottom) A sequence of face images showing tracking of the eyes and nose. (Images courtesy of Vera Bakic.)

the resulting position of the cursor in an array of 8×8 menu choices. The smiling face in the second row, third column indicates the user's selection. Processing can be done at 15 or more frames per second because of the integrated use of domain knowledge. The knowledge of face color is used to find the face in the image and knowledge of the structure of a face is used to locate the eyes and nose. Moreover, trajectories of the eyes and nose are used to predict where those features should be found in the next incoming frame: when they are found in the predicted neighborhood, global face detection is not done.

The making of the movie *Titanic* is one of many examples where integrated computer graphics and computer vision techniques were used to combine real imagery and synthetic imagery. Imagery was shot of a model ship and was later augmented by placing moving models on the deck of the ship. An actress was used to capture the motion of a live human wearing a flowing dress typical of the early twentieth century. Many small lights were attached to a real dress so that feature points could easily be detected in a motion sequence. The trajectories of these moving points were then used to orchestrate the motion of model people and model clothes, which were added to various positions in the images taken of the model ship. Many minutes of computer and operator time are used per frame of such a blockbuster movie, so not all steps need to be fully automatic.

9.5 DETECTING SIGNIFICANT CHANGES IN VIDEO

Video sequences may record minutes or hours of surveillance, different takes of a TV news crew, or a finished documentary or movie. It is becoming increasingly important to segment and store subsequences in digital libraries for random access. Some important concepts and methods related to parsing and analyzing video sequences are discussed in this section. First, we define some points of change found in videos and some other image sequences.

- A **scene change** is a change of environment; for example, from a restaurant scene to a street scene. Gross changes in the background are expected. Often, scene changes are made over ten to fifty frames using one of the camera effects below.
- A **shot change** is a significant change of camera view of the same scene. Often, this is accomplished by switching cameras. For example, in the restaurant scene camera one views actor A speaking and then frames are taken from camera two to view the response of the actor B across the table.
- A **camera pan** is used to sweep a horizontal view of the scene. If the camera pans from right to left, objects appear to enter at the left and move across the images to the right, finally exiting right. Motion vectors computed from consecutive frames of such a panning sequence of a static scene will be in the same direction from left to right.
- **Camera zoom** changes the focal length over time to expand the image of some part of the scene (zoom in) or to reduce the image of a scene part and include more adjacent background (zoom out).
- **Camera effects: fade, dissolve, and wipe** are used for transitions from one source of imagery to a different source of imagery. A *fade out* is a continuous transition from

one video source to black or to white frames, whereas a *fade in* is a transition from black or from white frames to some video source. A transition from video source A to video source B can be achieved by fading out A and then fading in B. A *dissolve* changes the pixels of A into pixels of B over several frames. One type of dissolve weights the pixels of A by $(1 - t/T)$ and the pixels of B by t/T over the frames $t = 0, \ldots, T$. A *wipe* makes the transition from source A to B by changing the size of the regions where A and B appear in the frame. One can imagine a windshield wiper crossing our window with source A displayed on one side of the wiper and source B on the other. Wipes can be done using a vertical, horizontal, or diagonal boundary between the two regions. Or, source B can appear within a small circular region that grows to cover the entire frame.

Exercise 9.12

Construct a pseudo-code algorithm that blends video source A into video source B using a wipe. Source A is the image sequence $A_t[r, c]$ and source B is the image sequence $B_t[r, c]$. (a) Suppose the wipe is accomplished from time t_1 to time t_2 by using a diagonal line of slope 1 originating through pixel [0, 0] (top left) at time t_1 and ending through pixel $[M - 1, N - 1]$ at time t_2. (b) Suppose the wipe is accomplished by a growing circular region at the frame center. At time t_1 the radius of the circle is 0 and at time t_2 the radius is large enough such that the circle circumscribes the entire frame.

9.5.1 Segmenting Video Sequences

The goal of the analysis is to parse a long sequence of video frames into subsequences representing single shots or scenes. As an example, consider a 30-minute video of a TV news program. There will be several 10 to 15 second segments where the camera shot is of a newscaster reporting from a desk: the background is a constant office background, but there may be zooming of the camera. After such a segment, it is common to transition to another source documenting another event, perhaps frames of a flood, an interesting play in sports, a meeting, or a government official jogging. Often there are several different shots of the event being reported with transitions between them. The transitions can be used to segment the video and can be detected by large changes in the features of the images over time.

One obvious method of computing the difference between two frames I_t and $I_{t+\Delta}$ of a sequence is to compute the average difference between corresponding pixels as in Equation 9.7. Depending on the camera effect being detected, the time interval Δ may be one or more frames.

$$d_{pixel}(I_t, I_{t+\Delta}) = \frac{\sum_{r=0}^{MaxRow-1} \sum_{c=0}^{MaxCol-1} | I_t[r, c] - I_{t+\Delta}[r, c] |}{MaxRow \times MaxCol} \tag{9.7}$$

Equation 9.7 is likely to mislead us by yielding a large difference when there is even a small amount of camera pan or some object motion in an otherwise stable shot. A more robust variation, proposed by Kasturi and Jain (1991), is to break the image into larger blocks and test to see if a majority of the blocks are essentially the same in both images. A likelihood ratio test defined in Equation 9.8 was proposed to evaluate whether or not there was significant change in the intensities of corresponding blocks. Let block B_1 in image I_1 have intensity mean and variance u_1 and v_1 and block B_2 in image I_2 have intensity mean and variance u_2 and v_2. The block difference is defined in terms of the likelihood ratio in Equation 9.8. If a sufficient number of the blocks have zero difference then the decision is that the two images are from essentially the same shot. Clearly, Equation 9.8 will be more stable than Equation 9.7 when the images are highly textured and are not stabilized from frame to frame to remove the effects of small motions of the camera.

$$r = \frac{\left[\frac{v_1+v_2}{2} + \left(\frac{u_1-u_2}{2}\right)^2\right]^2}{v_1 v_2} \tag{9.8}$$

$$d_{block}(B_1, B_2) = 1 \quad \mathit{if}\ r > \tau_r$$

$$= 0 \quad \mathit{if}\ r \leq \tau_r$$

$$d(I_1, I_2) = \sum_{B_{1i} \in I_1;\, B_{2i} \in I_2} d_{block}(B_{1i}, B_{2i})$$

The difference between two images can be represented in terms of the difference between their histograms as was done in $d_{hist}(I, Q)$ of Chapter 8. In our current discussion, I is image I_1 and Q is image I_2. A 64-level histogram can provide enough levels. For color video frames, a value in the interval [0, 63] can be obtained by concatenating the higher order two bits of the red, green, and blue color values. Histogram comparison can be faster than the previous methods and is potentially a better representative of the general features of the scene. Since it totally avoids any spatial coherence checking, it can be completely fooled when two images have similar histograms but totally different spatial distributions, and in fact are from two different shots.

Figure 9.17 shows four frames from the same documentary video. The top two frames occur prior to a scene break and the bottom two frames occur after the scene break. Figure 9.18 shows the histograms computed from the first three frames shown in Figure 9.17. The top two histograms are similar: this means that the two frames from which they were derived are likely to be from the same shot. The bottom histogram differs significantly from the first two, and thus the third frame from Figure 9.17 is likely to be from a different shot.

9.5.2 Ignoring Certain Camera Effects

We do not want to segment the video sequence when adjacent frames differ significantly due only to certain camera effects such as pan or zoom. Transitions detected as in Section 9.5.1 can be subject to some simple motion analysis to determine if they are effects to be ignored. Panning can be detected by computing motion vectors and determining if the motion vectors closely cluster around some modal direction and magnitude. We can do this by simple analysis of the set **V** output from Algorithm 9.3. Zooming can be detected by examining the motion vectors at the periphery of the motion field. Zooming in or out is indicated by the

Figure 9.17 Four frames from a documentary video. The top two and the bottom two are separated across a camera break. (Reprinted from Zhang and others (1993) with permission from Springer-Verlag.)

motion vectors at the periphery approximately summing to zero. Using only the periphery of the motion field accounts for cases where the FOE or FOC is not near the center of the field. Suppose that motion vectors are computed using MPEG-type block matching techniques so that there are motion vectors for blocks along the top and bottom of the motion field determined by I_1 and I_2. The difference between the vertical components of motion vectors in corresponding positions across the image should be greater than that of either the top or bottom motion vector as shown in Figure 9.19. The relationship is similar for the horizontal components of motion vectors horizontally related across the image. Both zoom in and out can be detected reasonably well using these heuristics. The quality of the motion field derived from block matching will deteriorate, however, due to change of scale as the zooming speeds up.

Exercise 9.13

Obtain two consecutive frames of a video of the same scene. (a) Compute the average pixel difference as defined in Equation 9.7. (b) Partition the image into $2 \times 2 = 4$ blocks and compute the sum of their block differences as defined in Equation 9.8.

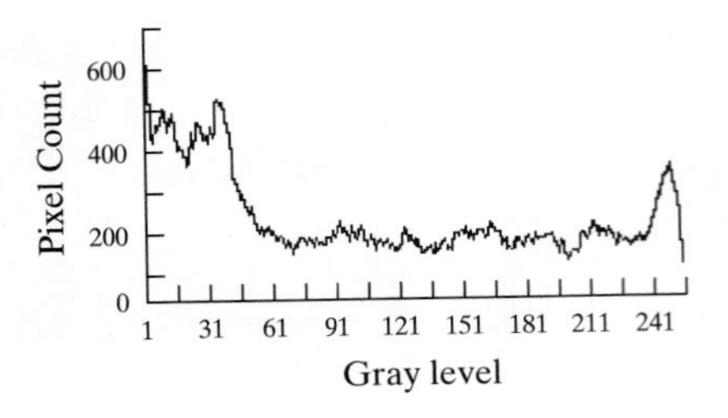

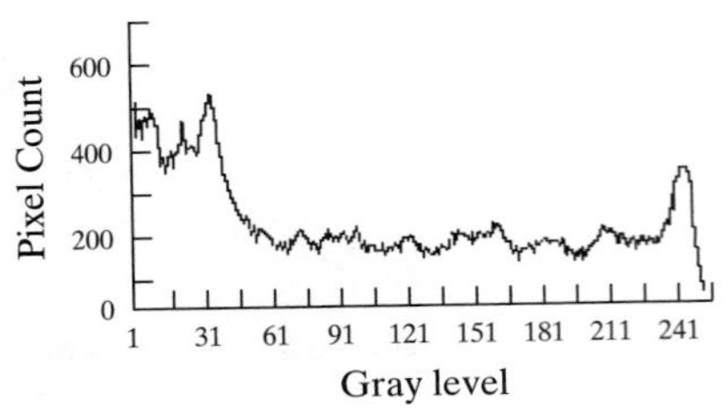

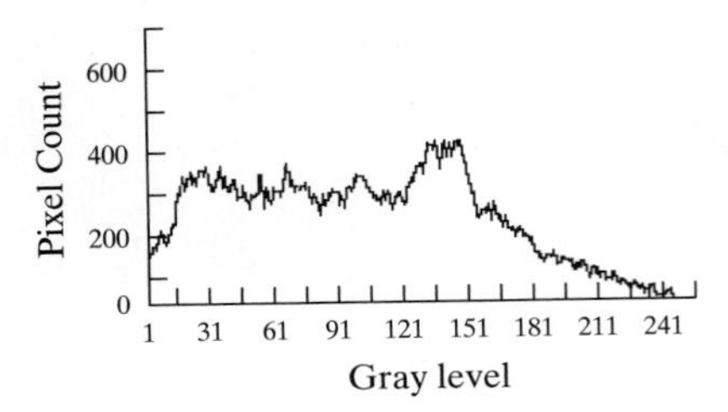

Figure 9.18 Histograms from the first three frames shown in Figure 9.17. The top two are similar as are the frames from which they were derived. The bottom histogram is significantly different from the top two, indicating that the frame from which is was derived is different from the first two. (Reprinted from Zhang and others (1993) with permission from Springer-Verlag.)

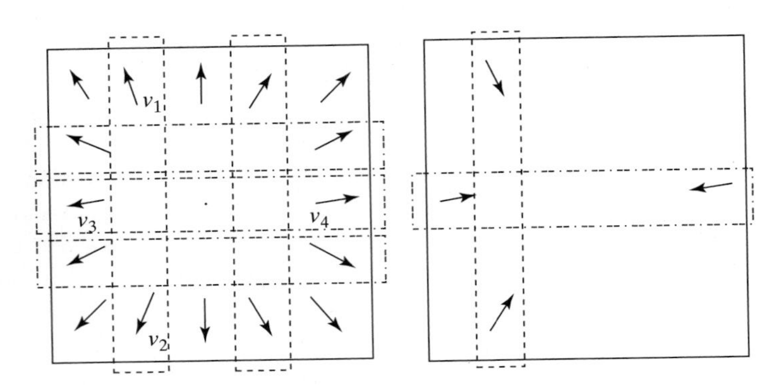

Figure 9.19 Heuristics for detection of camera zoom by comparing motion vectors across the periphery of the motion field. The difference of vertical components across the field exceeds the vertical component of either vector: $|v_{1r} - v_{2r}| > max\{|v_{1r}|, |v_{2r}|\}$. Similarly, $|v_{3c} - v_{4c}| > max\{|v_{3c}|, |v_{4c}|\}$ for horizontally opposed motion vectors. These relations hold for both zoom in (left) and zoom out (right).

9.5.3 Storing Video Subsequences

Once a long video sequence is parsed into meaningful subsequences, these subsequences can be stored in a video database for query and retrieval. They can be accessed using some of the same methods discussed in Chapter 8. Certain *key frames* can be identified and used for access in the same manner as in Chapter 8. In the future, we will probably be able to go much farther; for example, automatic motion analysis might be performed to assign symbolic action labels such as *running, fighting,* or *debating*. Face recognition might be performed in order to label famous people, or general object recognition might provide labels such as *horse* or *house*. Although the use of many frames implies more computational effort with video compared to still images, the information provided by motion analysis should increase both the cability to segment objects from background and the capability to classify them.

Exercise 9.14

Consider the application, discussed earlier in this chapter, of computing an analysis of a tennis match from a video of the match. (a) What are the actions or events that an analysis program should report? (b) What quantitative data should be reported?

9.6 REFERENCES

The discussion of tracking the players and ball in a tennis match is based upon the recent work of Pingali, Jean, and Carlbom (1998) at Bell Laboratories. The paper by Freeman and others (1998) describes the results of several experiments that have used computer vision techniques to create a gesture interface for interacting with several existing applications. Included is a description of a fast motion estimation algorithm. The details of the fast motion estimation algorithm, and much detail about the game interface are given in Kage and others (1999). Our treatment of video parsing and indexing followed the work of Zhang and others (1993) and Smolier and Zhang (1996). The treatment of computing smooth trajectories from many frames of featureless points was based on the work of Sethi and Jain (1987), which was done in an era of computer vision where it was common to study what could be computed from a small set of general assumptions. More recent work, such as that reported by Maes and others (1996), Darrell and others (1998) and Bakic and Stockman (1999) integrates problem-specific knowledge in order to speed up the process and make it more robust. The work by Ayers and Shah (1998) shows how to interpret motion and change in terms of the semantics relevant to a surveillance application.

1. Ayers, D., and M. Shah. 1998. Recognizing human actions in a static room. *Proc. 4th IEEE Workshop on Applications of Computer Vision,* Princeton, NJ (19–21 Oct. 1998), 42–47.
2. Bakic, V., and G. Stockman. 1999. Menu selection by facial aspect. *Proc. Vision Interface '99,* Quebec, Canada (18–21 May 1999),18–21.

3. Darrell, T. 1998. A radial cumulative similarity transform for robust image correspondence. *Proc. IEEE CVPR,* Santa Barbara, CA (June 1998), 656–662.
4. Darrell, T., G. Gordon, M. Harville, and J. Woodfill. 1998. Integrated person tracking using stereo, color, and pattern detection. *Proc. IEEE CVPR,* Santa Barbara, CA (June 1998), 601–608.
5. Freeman, W., D. Anderson, P. Beardsley, C. Dodge, M. Roth, C. Weissman, W. Yerazunis, H. Kage, K. Kyuma, Y. Miyake, and K. Tanaka. 1998. Computer vision for interactive computer graphics. *IEEE Comput. Graphics and Applications,* v. 18(3) (May–June 1998), 42–53.
6. Horn, B., and B. Schunck. 1981. Determining optical flow. *Artificial Intelligence,* v. 17:185–203.
7. Johansson, G. 1964. Perception of motion and changing form. *Scandanavian J. Psychology,* v. 5:181–208.
8. Kage, H., W. T. Freeman, Y. Miyake, E. Funatsu, K. Tanaka, and K. Kyuma. 1999. Artificial retina chips as on-chip image processors and gesture-oriented interfaces. *Optical Engineering,* 38(12):1979–1988.
9. Kasturi, R., and R. Jain. 1991. Dynamic vision. In *Computer Vision Principles,* R. Kasturi and R. Jain, eds. IEEE Computer Society Press, Washington, D.C., 469–480.
10. Maes, P., T. Darrell, B. Blumberg, and A. Pentland, 1996, *The ALIVE System: Wireless, Full-Body, Interaction with Autonomous Agents.* ACM Multimedia Systems: Special Issue on Multimedia and Multisensory Virtual Worlds, Sprint.
11. Pingali, G., Y. Jean, and I. Carlbom. 1998. Real time tracking for enhanced tennis broadcasts. *Proc. IEEE CVPR,* Santa Barbara, CA (June 1998), 260–265.
12. Salari, V., and I. Sethi. 1990. Correspondence of feature points in presence of occlusion. *IEEE Trans. on Pattern Analysis and Machine Intelligence,* v. 12(1):87–91.
13. Sethi, I., and R. Jain. 1987. Finding trajectories of feature points in a monocular image sequence. *IEEE Trans. on Pattern Analysis and Machine Intelligence,* v. 9(1):56–73.
14. Smolier, S., and H-J Zhang. 1996. Video indexing and retrieval. In *Multimedia Systems and Techniques,* B. Furht, ed. Kluwer Academic Publishers, Boston, 293–322.
15. Zhang, H-J., A. Kankanhalli, and S. Smoliar. 1993. Automatic partitioning of full-motion video. *Multimedia Systems,* v. 1(1):10–28.

10

Image Segmentation

The term *image segmentation* refers to the partition of an image into a set of regions that cover it. The goal in many tasks is for the regions to represent meaningful areas of the image, such as the crops, urban areas, and forests of a satellite image. In other analysis tasks, the regions might be sets of border pixels grouped into such structures as line segments and circular arc segments in images of 3D industrial objects. Regions may also be defined as groups of pixels having both a border and a particular shape such as a circle, ellipse, or polygon. When the interesting regions do not cover the whole image, we can still talk about segmentation, into foreground regions of interest and background regions to be ignored.

Segmentation has two objectives. The first objective is to decompose the image into parts for further analysis. In simple cases, the environment might be well enough controlled so that the segmentation process reliably extracts only the parts that need to be analyzed further. For example, in the chapter on color, an algorithm was presented for segmenting a human face from a color video image. The segmentation is reliable, provided that the person's clothing or room background does not have the same color components as a human face. In complex cases, such as extracting a complete road network from a gray-scale aerial image, the segmentation problem can be very difficult and might require application of a great deal of domain knowledge.

The second objective of segmentation is to perform a change of representation. The pixels of the image must be organized into higher-level units that are either more meaningful or more efficient for further analysis (or both). A critical issue is whether or not segmentation can be performed for many different domains using general bottom-up methods that do not use any special domain knowledge. This chapter presents segmentation methods that have potential use in many different domains. Both region-based and curve-based units are discussed in the following sections. The prospects of having a single segmentation system

Figure 10.1 (left) Football image and (right) segmentation into regions. Each region is a set of connected pixels that are similar in color. See colorplate.

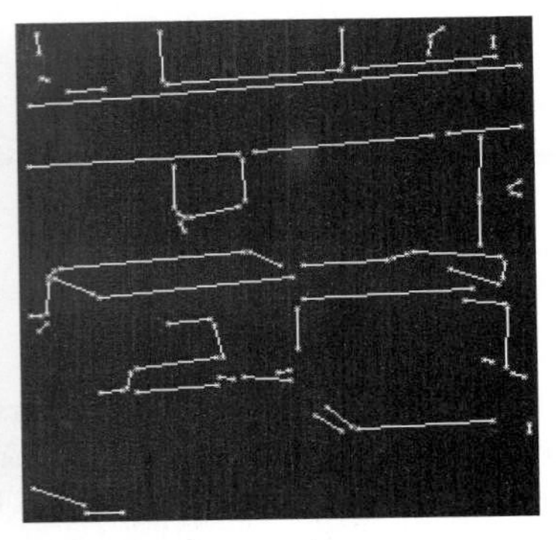

Figure 10.2 (left) Blocks image and (right) extracted set of straight line segments. The line segments were extracted by the Object Recognition Toolkit (ORT). (Courtesy of John Illingworth and Ata Etamadi.)

work well for all problems appear to be dim. Experience has shown that an implementor of machine vision applications must be able to choose from a toolset of methods and perhaps tailor a solution using knowledge of the application.

This chapter discusses several different kinds of segmentation algorithms including the classical region growers, clustering algorithms, and line and circular arc detectors. Figure 10.1 illustrates the segmentation of a colored image of a football game into regions of near-constant color. Figure 10.2 shows the line segments extracted from an image of toy blocks. In both cases, note that the results are far from perfect by human standards. However, these segmentations might provide useful input for higher-level automated processing, for example, identifying players by number or recognizing a part to be assembled.

10.1 IDENTIFYING REGIONS

- Regions of an image segmentation should be uniform and homogeneous with respect to some characteristic, such as gray level, color, or texture.
- Region interiors should be simple and without many small holes.
- Adjacent regions of a segmentation should have significantly different values with respect to the characteristic on which they are uniform.

- Boundaries of each segment should be smooth, not ragged, and should be spatially accurate.

Achieving all these desired properties is difficult because strictly uniform and homogeneous regions are typically full of small holes and have ragged boundaries. Insisting that adjacent regions have large differences in values can cause regions to merge and boundaries to be lost. In addition, the regions that humans see as homogeneous may not be homogeneous in terms of the low-level features available to the segmentation system, so higher-level knowledge may have to be used. The goal of this chapter is to develop algorithms that will apply to a variety of images and serve a variety of higher-level analyses.

10.1.1 Clustering Methods

Clustering in pattern recognition is the process of partitioning a set of pattern vectors into subsets called *clusters*. For example, if the pattern vectors are pairs of real numbers as illustrated by the point plot of Figure 10.3, clustering consists of finding subsets of points that are "close" to each other in Euclidean two-space.

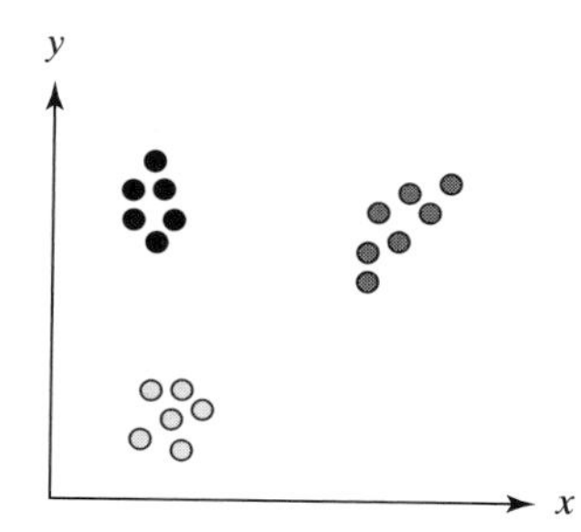

Figure 10.3 Set of points in a Euclidean measurement space that can be separated into three clusters of points. Each cluster consists of points that are in some sense close to one another. Clusters are designated by the fill patterns inside the circles.

The general term clustering refers to a number of different methods. We will look at several different types of clustering algorithms that have been found useful in image segmentation. These include classical clustering algorithms, simple histogram-based methods, Ohlander's recursive histogram-based technique, and Shi's graph-partitioning technique.

Classical Clustering Algorithms The general problem in clustering is to partition a set of vectors into groups having similar values. In image analysis, the vectors represent pixels or sometimes small neighborhoods around pixels. The components of these vectors can include:

1. intensity values,
2. RGB values and color properties derived from them,
3. calculated properties, and
4. texture measurements.

Any feature that can be associated with a pixel can be used to group pixels. Once pixels have been grouped into clusters based on these *measurement-space* values, it is easy to find connected regions using connected components labeling as in Chapter 3.

In traditional clustering, there are K clusters $C_1, C_2, \ldots, C_K$ with means $m_1, m_2, \ldots, m_K$. A *least squares error measure* can be defined as

$$D = \sum_{k=1}^{K} \sum_{x_i \in C_k} \|x_i - m_k\|^2.$$

which measures how close the data are to their assigned clusters. A least-squares clustering procedure could consider all possible partitions into K clusters and select the one that minimizes D. Since this is computationally infeasible, the popular methods are approximations. One important issue is whether or not K is known in advance. Many algorithms expect K as a parameter from the user. Others attempt to find the best K according to some criterion, such as keeping the variance of each cluster less than a specified value.

Iterative K-Means Clustering The *K-means* algorithm is a simple, iterative hill-climbing method. It can be expressed as:

Form K-means clusters from a set of n-dimensional vectors.

1. Set ic (iteration count) to 1.
2. Choose randomly a set of K means $m_1(1), m_2(1), \ldots, m_K(1)$.
3. For each vector x_i compute $D(x_i, m_k(ic))$ for each $k = 1, \ldots, K$ and assign x_i to the cluster C_j with the nearest mean.
4. Increment ic by 1 and update the means to get a new set $m_1(ic), m_2(ic), \ldots, m_K(ic)$.
5. Repeat steps 3 and 4 until $C_k(ic) = C_k(ic + 1)$ for all k.

Algorithm 10.1 K-Means Clustering.

This algorithm is guaranteed to terminate, but it may not find the global optimum in the least squares sense. Step 2 may be modified to partition the set of vectors into K random clusters and then compute their means. Step 5 may be modified to stop after the percentage of vectors that change clusters in a given iteration is small. Figure 10.4 illustrates the application of the K-means clustering algorithm in RGB space to the original football image of Figure 10.1.

Isodata Clustering *Isodata clustering* is another iterative algorithm that uses a split-and-merge technique. Again assume that there are K clusters $C_1, C_2, \ldots, C_K$ with means $m_1, m_2, \ldots, m_K$, and let Σ_k be the covariance matrix of cluster k (as defined next). If the x_i's are vectors of the form

$$x_i = [v_1, v_2, \ldots, v_n]$$

Figure 10.4 (left) Football image and (right) $K = 6$ clusters resulting from a K-means clustering procedure shown as distinct gray tones. The six clusters correspond to the six main colors in the original image: dark green, medium green, dark blue, white, silver, and black. See colorplate.

then each mean m_k is a vector

$$m_k = [m_{1k}, m_{2k}, \ldots, m_{nk}]$$

and Σ_k is defined by

$$\Sigma_k = \begin{bmatrix} \sigma_{11} & \sigma_{12} & \ldots & \sigma_{1n} \\ \sigma_{12} & \sigma_{22} & \ldots & \sigma_{2n} \\ \vdots & \vdots & \vdots & \vdots \\ \sigma_{1n} & \sigma_{2n} & \ldots & \sigma_{nn} \end{bmatrix} \tag{10.1}$$

where $\sigma_{ii} = \sigma_i^2$ is the variance of the *ith* component v_i of the vectors and $\sigma_{ij} = \rho_{ij}\sigma_i\sigma_j$ is the covariance between the *ith* and *jth* components of the vectors. (ρ_{ij} is the correlation coefficient between the *ith* and *jth* components, σ_i is the standard deviation of the *ith* component, and σ_j is the standard deviation of the *jth* component.)

Figure 10.5 illustrates the application of the isodata clustering algorithm (given in Algorithm 10.2) in RGB space to the original football image of Figure 10.1. This cluster image was the input to a connected components procedure which produced the segmentation

Figure 10.5 (left) Football image and (right) $K = 5$ clusters resulting from an isodata clustering procedure shown as distinct gray tones. The five clusters correspond to five color groups: green, dark blue, white, silver, and black. See colorplate.

shown in Figure 10.1. The threshold τ_v for the isodata clustering was set to a size of 10 percent of the side of the RGB color-space cube.

Exercise 10.1: Isodata vs. K-means clustering.

The isodata algorithm gave better results than the K-means algorithm on the football images in that it correctly grouped the dark green areas at the top of the image with those near the bottom. Why do you think the isodata algorithm was able to perform better than the K-means algorithm?

Simple Histogram-Based Methods Iterative partition rearrangement schemes have to go through the image data set many times. Because they require only one pass through the data, histogram methods probably involve the least computation time of the measurement-space clustering techniques.

Histogram mode seeking is a measurement-space clustering process in which it is assumed that homogeneous objects in the image manifest themselves as the clusters in measurement space, that is, on the histogram. Image segmentation is accomplished by mapping the clusters back to the image domain where the maximal connected components of the cluster labels constitute the image segments. For gray-tone images, the measurement-space clustering can be accomplished by determining the valleys in the histogram and declaring the clusters to be the interval of values between valleys. A pixel whose value is in the *ith* interval is labeled with index i and the segment to which it belongs is one of the connected components of all pixels whose label is i. The automatic thresholding technique discussed in Chapter 3 is an example of histogram mode seeking with a bimodal histogram.

Form isodata clusters from a set of n-dimensional vectors.

1. assign x_i to the cluster l that minimizes
$$D_\Sigma = [x_i - m_l]' \Sigma_l^{-1} [x_i - m_l].$$
2. Merge clusters i and j if
$$|m_i - m_j| < \tau_v$$
where τ_v is a variance threshold.
3. Split cluster k if the maximum eigenvalue of Σ_k is larger than τ_v.
4. Stop when
$$|m_i(t) - m_i(t+1)| < \epsilon$$
for every cluster i or when the maximum number of iterations has been reached.

Algorithm 10.2 Isodata Clustering.

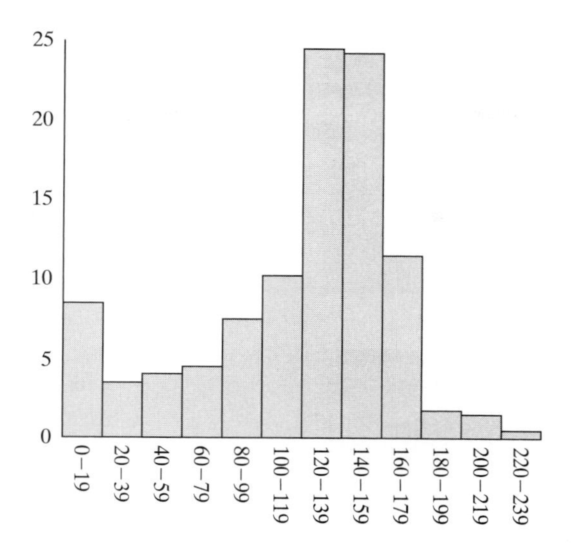

Figure 10.6 Histogram of the blocks image of Figure 10.2.

Exercise 10.2: Histogram mode seeking.

Write a program that finds the modes of a multimodal histogram by first breaking it into two parts, as does the Otsu method in Chapter 3, and then recursively trying to break each part into two more parts, if possible. Test it on gray-tone and color images.

In general, a gray-tone image will have a multimodal histogram, so that any automatic thresholding technique will have to look for significant peaks in the image and the valleys that separate them. This task is easier said than done. Figure 10.6 shows the histogram of the gray-tone blocks image. A naïve valley-seeking algorithm might judge it to be bimodal and place the single threshold somewhere between 39 and 79. Trial-and-error threshold selection, however, produced three thresholds yielding the four thresholded images of Figure 10.7, which show some meaningful regions of the image. This motivates the need for *knowledge-directed thresholding* techniques where the thresholds chosen depend on both the histogram and the quality/usefulness of the resulting regions.

Ohlander's Recursive Histogram-Based Technique Ohlander, Price, and Reddy (1978) refine the histogram-based clustering idea in a recursive way. The idea is to perform histogram mode seeking first on the whole image and then on each of the regions obtained from the resultant clusters, until regions are obtained that can be decomposed no further. They begin by defining a mask selecting all pixels in the image. Given any mask, a histogram of the masked portion of the image is computed. Measurement-space clustering is applied to this histogram, producing a set of clusters. Pixels in the image are then identified with the cluster to which they belong. If there is only one measurement-space cluster, then the mask is terminated. If there is more than one cluster, then the connected components

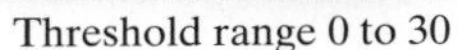

Threshold range 0 to 30

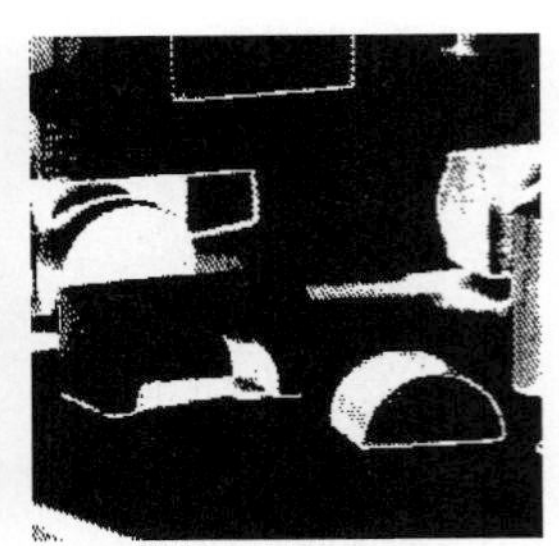

Threshold range 31 to 100

Threshold range 101 to 179

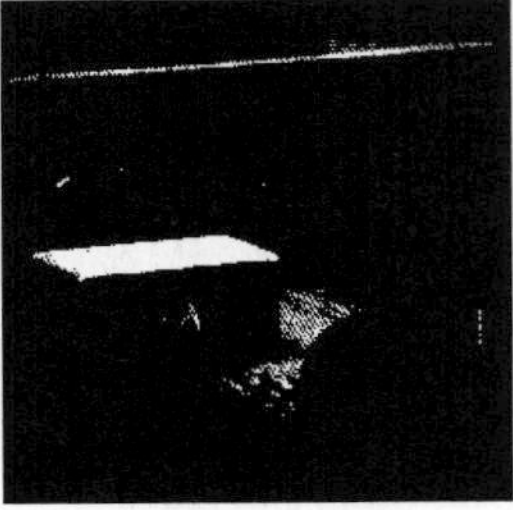

Threshold range 180 to 239

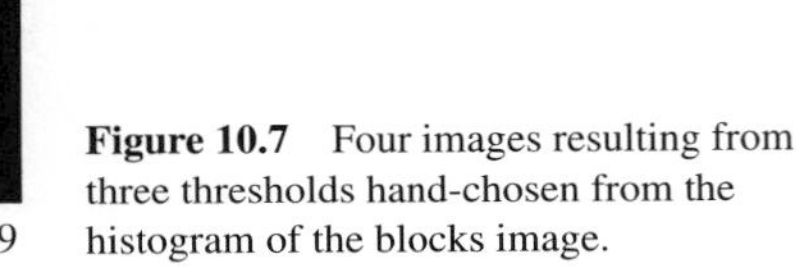

Figure 10.7 Four images resulting from three thresholds hand-chosen from the histogram of the blocks image.

operator is applied to each cluster, producing a set of connected regions for each cluster label. Each connected component is then used to generate a new mask which is placed on a mask stack. The masks on the stack represent regions that are candidates for further segmentation. During successive iterations, the next mask in the stack selects pixels in the histogram computation process. Clustering is repeated for each new mask until the stack is empty. Figure 10.8 illustrates this process, which we call recursive histogram-directed spatial clustering.

For ordinary color images, Ohta, Kanade, and Sakai (1980) suggested that histograms not be computed individually on the red, green, and blue (RGB) color variables, but on a set of variables closer to what the Karhunen-Loeve (principal components) transform would suggest: $(R + G + B)/3$, $(R - B)/2$, and $(2G - R - B)/4$.

Shi's Graph-Partitioning Technique* The Ohlander and Ohta algorithms work well on reasonably simple color scenes with man-made objects and single-color regions, but do not extend well to complex images of natural scenes, where they give lots of tiny regions in textured areas. Shi and Malik (1997) developed a method that can use color, texture, or any combination of these and other attributes. They formulated the segmentation problem as a graph-partitioning problem and developed a new graph-partitioning method that reduced to solving an eigenvector and eigenvalue problem as follows.

Let $G = (V, E)$ be a graph whose nodes are points in measurement space and whose edges each have a weight $w(i, j)$ representing the similarity between nodes i and j. The goal in segmentation is to find a partition of the vertices into disjoint sets $V_1, V_2, \ldots, V_m$ so that the similarity within the sets is high and across different sets is low.

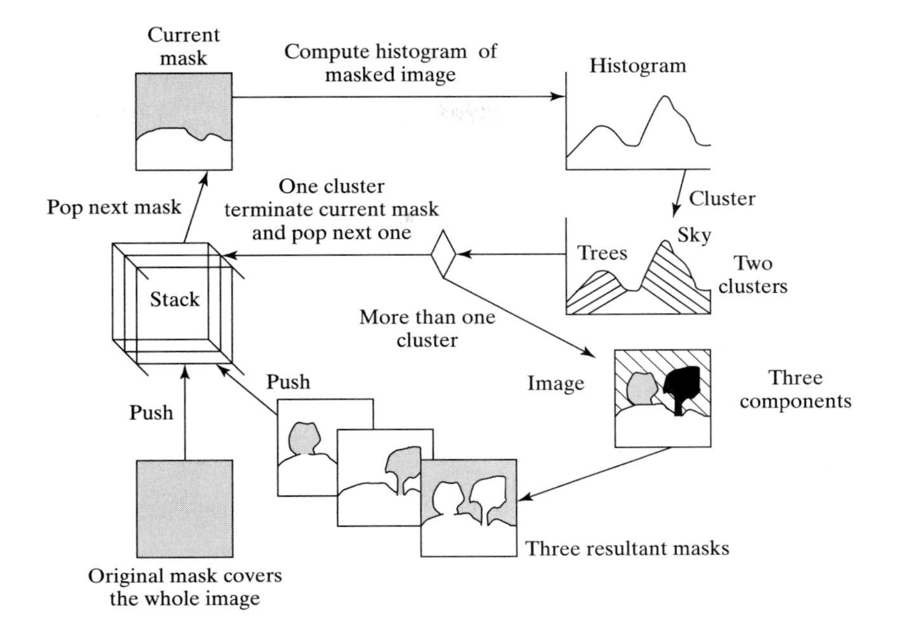

Figure 10.8 Recursive histogram-directed spatial-clustering scheme. The original image has four regions: grass, sky, and two trees. The current mask (shown at upper left) identifies the region containing the sky and the trees. Clustering its histogram leads to two clusters in color space, one for the sky and one for the trees. The sky cluster yields one connected component, while the tree cluster yields two. Each of the three connected components become masks that are pushed onto the mask stack for possible further segmentation.

A graph $G = (V, E)$ can be partitioned into two disjoint graphs with node sets A and B by removing any edges that connect nodes in A with nodes in B. The degree of dissimilarity between the two sets A and B can be computed as the sum of the weights of the edges that have been removed; this total weight is called a *cut*.

$$cut(A, B) = \sum_{u \in A, v \in B} w(u, v) \tag{10.2}$$

One way of formulating the segmentation problem is to look for the *minimum cut* in the graph, and to do so recursively until the regions are uniform enough. The minimum cut criterion, however, favors cutting small sets of isolated nodes, which is not useful in finding large uniform color or texture regions. Shi proposed the *normalized cut* (*Ncut*) defined in terms of $cut(A, B)$ and the *association* of A and the full vertex set V defined by:

$$asso(A, V) = \sum_{u \in A, t \in V} w(u, t) \tag{10.3}$$

The definition of normalized cut is then

$$Ncut(A, B) = \frac{cut(A, B)}{asso(A, V)} + \frac{cut(A, B)}{asso(B, V)} \tag{10.4}$$

With this definition, the cut that partitions out small isolated point sets will not have small *Ncut* values, and the partitions that do produce small $Ncut$ values are more likely to be useful in image segmentation. Furthermore, the related measure for total normalized *association* given by

$$Nasso(A, B) = \frac{asso(A, A)}{asso(A, V)} + \frac{asso(B, B)}{asso(B, V)} \tag{10.5}$$

measures how tightly the nodes within a given set are connected to one another. It is related to the *Ncut* by the relationship

$$Ncut(A, B) = 2 - Nasso(A, B) \tag{10.6}$$

so that either of them can be used, as convenient, by a partitioning procedure.

Given the definitions for $Ncut$ and $Nasso$, we need a procedure that segments an image by partitioning the pixel set. Shi's segmentation procedure is given in Algorithm 10.3.

Shi ran this algorithm to segment images based on brightness, color, or texture information. The edge weights $w(i, j)$ were defined by

$$w(i, j) = e^{\frac{-\|F(i)-F(j)\|_2}{\sigma_I}} * \begin{cases} e^{\frac{-\|X(i)-X(j)\|_2}{\sigma_X}} & \text{if } \|X(i) - X(j)\|_2 < r \\ 0 & \text{otherwise} \end{cases} \tag{10.11}$$

where

- $X(i)$ is the spatial location of node i.
- $F(i)$ is the feature vector based on intensity, color, or texture information and is defined by
  - $F(i) = I(i)$, the intensity value, for segmenting intensity images.
  - $F(i) = [v, v \cdot s \cdot sin(h), v \cdot s \cdot cos(h)](i)$, where h, s, and v are the HSV values, for color segmentation.
  - $F(i) = [|I * f_1|, \ldots, |I * f_n|](i)$, where the f_i are difference of difference of Gaussian (DOOG) filters at various scales and orientations, for texture segmentation.

Note that the weight $w(i, j)$ is set to 0 for any pair of nodes i and j that are more than a prespecified number r of pixels apart.

Algorithm 10.3 leads to very good segmentations of images via color and texture. Figure 10.9 illustrates the performance of the algorithm on a sample image of a natural scene. While the segmentation is very good, the complexity of the algorithm makes it unsuitable for use in a real-time system.

Perform a graph-theoretic clustering on a graph whose nodes are pixels and whose edges represent the similarities between pairs of pixels.

1. Set up a weighted graph $G = (V, E)$ whose nodeset V is the set of pixels of the image and whose edgeset E is a set of weighted edges with $w(i, j)$, the weight on the edge between node i and j, computed as the similarity between the measurement-space vector of i and the measurement-space vector of j. Let N be the size of nodeset V. Define the vector d with $d(i)$ given by

$$d(i) = \sum_j w(i, j) \tag{10.7}$$

so that $d(i)$ represents the total connection from node i to all other nodes. Let D be an $N \times N$ diagonal matrix with d on its diagonal. Let W be an $N \times N$ symmetrical matrix with $W(i, j) = w(i, j)$.

2. Let x be a vector whose components are defined by

$$x_i = \begin{cases} 1 & \text{if node } i \text{ is in } A \\ -1 & \text{otherwise} \end{cases} \tag{10.8}$$

and let y be the continuous approximation to x defined by

$$y = (1 + x) - \frac{\sum_{x_i > 0} d_i}{\sum_{x_i < 0} d_i}(1 - x). \tag{10.9}$$

Solve the system of equations

$$(D - W)y = \lambda D y \tag{10.10}$$

for the eigenvectors y and eigenvalues λ.

3. Use the eigenvector with the second smallest eigenvalue to bipartition the graph to find the splitting point such that $Ncut$ is minimized.[1]
4. Decide if the current partition should be subdivided further by checking the stability of the cut and making sure that $Ncut$ is below a pre-specified threshold value.
5. Recursively repartition the segmented parts if necessary.

Algorithm 10.3 Shi's Clustering Procedure.

10.1.2 Region Growing

Instead of partitioning the image, a *region grower* begins at one position in the image (often the top left corner) and attempts to grow each region until the pixels being compared are too dissimilar to the region to add them. Usually a statistical test is performed to decide if this

[1] Shi showed that the second smallest eigenvector of the generalized eigensystem is the real-valued solution to the normalized cut problem.

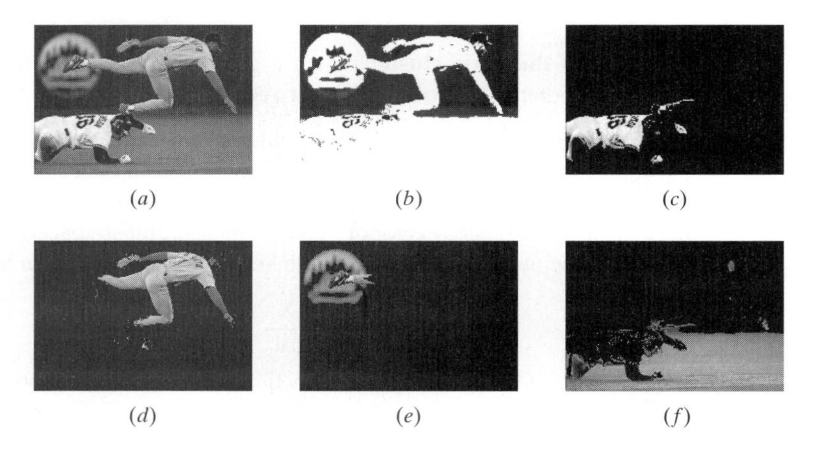

Figure 10.9 Original gray-tone image (*a*) and regions produced by the Shi segmentation method (*b*)-(*f*). In result image (*b*), the selected region is the dark background region, and it is shown in black. In all other results, the selected region is shown in its original gray tones, with the remainder of the image shown in black. (Courtesy of Jianbo Shi.)

is the case. Haralick and Shapiro (1985) proposed the following region-growing technique called the Haralick region-growing procedure, which assumes that a region is a set of connected pixels with the same population mean and variance.

Let R be a region of N pixels neighboring a pixel with gray tone intensity y. Define the region mean $\overline{X}$ and scatter S^2 by

$$\overline{X} = \frac{1}{N} \sum_{[r,c]\in R} I[r,c] \tag{10.12}$$

and

$$S^2 = \sum_{[r,c]\in R} (I[r,c] - \overline{X})^2. \tag{10.13}$$

Under the assumption that all the pixels in R and the test pixel y are independent and identically distributed normals, the statistic

$$T = \left[\frac{(N-1)N}{(N+1)} (y - \overline{X})^2 / S^2 \right]^{\frac{1}{2}} \tag{10.14}$$

has a T_{N-1} distribution. If T is small enough, y is added to region R and the mean and scatter are updated using y. The new mean and scatter are given by

$$\overline{X}_{\text{new}} \leftarrow (N\overline{X}_{\text{old}} + y)/(N+1) \tag{10.15}$$

and

$$S^2_{\text{new}} \leftarrow S^2_{\text{old}} + (y - \overline{X}_{\text{new}})^2 + N(\overline{X}_{\text{new}} - \overline{X}_{\text{old}})^2. \tag{10.16}$$

If T is too high the value y is not likely to have arisen from the population of pixels in R. If y is different from all of its neighboring regions then it begins its own region. A slightly stricter linking criterion can require that not only must y be close enough to the

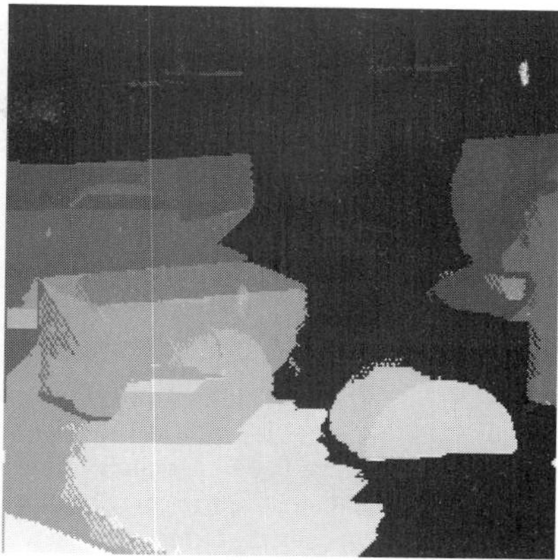

Figure 10.10 (left) The blocks image and (right) a segmentation resulting from application of the Haralick region-growing procedure. (Blocks image courtesy of John Illingworth and Ata Etamadi. Segmentation performed by the GIPSY image processing system.)

mean of the neighboring region, but that a neighboring pixel in that region must have a close enough value to y.

To give a precise meaning to the notion of too high a difference, we can use an α level statistical significance test. The fraction α represents the probability that a T statistic with $N-1$ degrees of freedom will exceed the value $t_{N-1}(\alpha)$. If the observed T is larger than $t_{N-1}(\alpha)$, then we declare the difference to be significant. If the pixel and the segment really come from the same population, the probability that the test provides an incorrect answer is α.

The significance level α is a user-provided parameter. The value of $t_{N-1}(\alpha)$ is higher for small degrees of freedom and lower for larger degrees of freedom. Thus, region scatters considered to be equal, the larger a region is, the closer a pixel's value has to be to the region's mean in order to merge into the region. This behavior tends to prevent an already large region from attracting to it many other additional pixels and tends to prevent the drift of the region mean as the region gets larger. Figure 10.10 illustrates the operation of the Haralick region-growing procedure.

Exercise 10.3: Region Growing.

Implement the Haralick region-growing operator as a program and use it to segment gray-tone images.

10.2 REPRESENTING REGIONS

Each algorithm that produces a set of image regions has to have a way to store them for future use. There are several possibilities including overlays on the original images, labeled images, boundary encodings, quadtree data structures, and property tables. Labeled images are the most commonly used representation. We describe each representation next.

(*a*) Region-border overlay

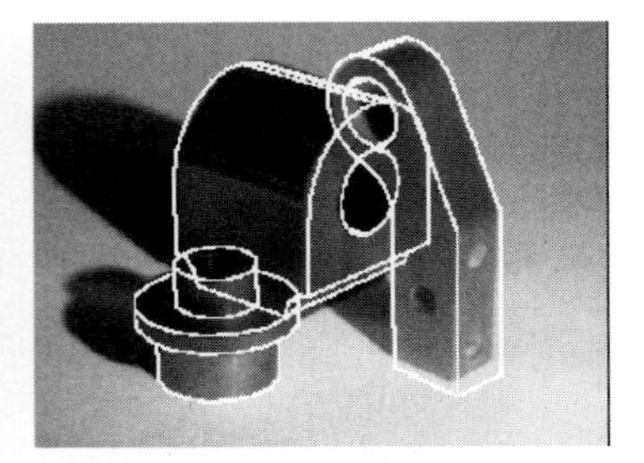

(*b*) Wire-frame-model overlay

Figure 10.11 Examples of overlays. (*a*) overlay of selected region borders onto a football image. (*b*) overlay of wire-frame 3D object models onto an industrial parts image ((*b*) courtesy of Mauro Costa).

10.2.1 Overlays

An overlay is a method of showing the regions computed from an image by overlaying some color or colors on top of the original image. Many image processing systems provide this operation as part of their image-output procedures. Usually, the original image is a gray-tone image and the overlay color is something that stands out well on the gray tones, such as red or white. To show a region segmentation, one could convert the pixels of the region borders to white and display the transformed gray-tone image. Sometimes more than one pixel in width is used to make the region borders stand out. Figure 10.11(*a*) shows the borders of selected dark regions, including dark blue referees' jackets and players' numbers, overlayed on a gray-tone football image. Another use for overlays is to highlight certain features of an image. Figure 10.11(*b*) reprints an industrial part image from Chapter 1 in which projections of the recognized object models are overlayed on the original gray-tone image.

10.2.2 Labeled Images

Labeled images are good intermediate representations for regions that can also be used in further processing. The idea is to assign each detected region a unique identifier (usually an integer) and create an image where all the pixels of a region will have its unique identifier as their pixel value. Most connected components operators (see Chapter 3) produce this kind of output. A labeled image can be used as a kind of mask to identify pixels of a region in some operation that computes region properties, such as area or length of major axis of best-fitting ellipse. It can also be displayed in gray-tone or pseudo-color. If the integer labels are small integers that would all look black in gray tone, the labeled image can be stretched or histogram-equalized to get a better distribution of gray tones. The segmentations of football images earlier in this chapter are labeled images shown in gray tone.

10.2.3 Boundary Coding

Regions can also be represented by their boundaries in a data structure instead of an image. The simplest form is just a linear list of the border pixels of each region. (See the border procedure later in this chapter, which extracts the region borders from a labeled

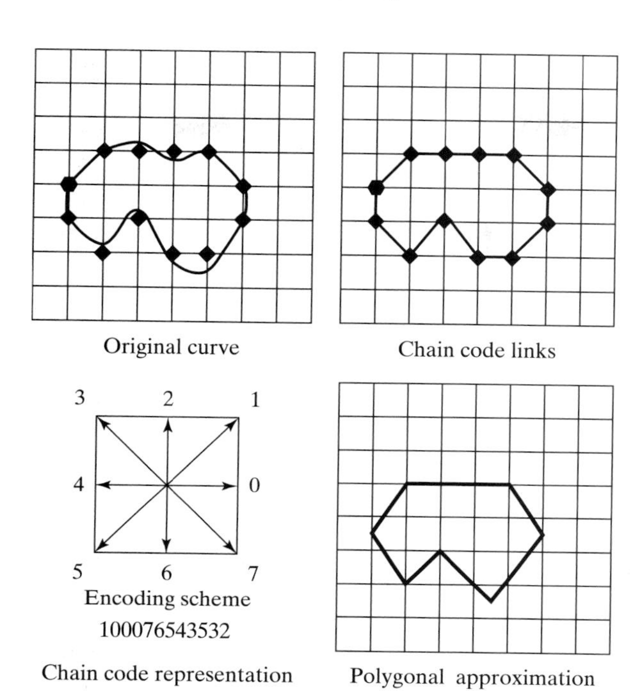

Figure 10.12 Two boundary encodings: chain-code and polygonal approximation. The chain-code boundary encoding uses an 8-symbol code to represent 8 possible angles of line segments that approximate the curve on a square grid. The polygonal approximation uses line segments fit to the original curve; the end points of the line segments have real-valued coordinates and are not constrainted to the original grid points.

image.) A variation of the list of points is the Freeman *chain code,* which encodes the information from the list of points at any desired quantization and uses less space than the original point list. Conceptually, a boundary to be encoded is overlaid on a square grid whose side length determines the resolution of the encoding. Starting at the beginning of the curve, the grid intersection points that come closest to it are used to define small line segments that join each grid point to one of its neighbors. The directions of these line segments are then encoded as small integers from zero to the number of neighbors used in the encoding. Figure 10.12 illustrates this encoding with an eight-neighbor chain code. The line segments are encoded as 0 for a 0° segment, 1 for a 45° segment, up to 7 for a 315° segment. In the figure, a hexagon symbol marks the beginning of the closed curve, and the rest of the grid intersection points are shown with diamonds. The coordinates of the beginning point plus the chain code are enough to reproduce the curve at the resolution of the selected grid. The chain code not only saves space, but can also be used in subsequent operations on the curve itself, such as shape-based object recognition. When a region has not only an external boundary, but also one or more hole boundaries, it can be represented by the chain codes for each of them.

When the boundary does not have to be exact, the boundary pixels can be approximated by straight line segments, forming a *polygonal approximation* to the boundary, as shown at the bottom right of Figure 10.12. This representation can save space and simplify algorithms that process the boundary.

10.2.4 Quadtrees

The *quadtree* is another space-saving region representation that encodes the whole region, not just its border. In general, each region of interest would be represented by a quadtree structure. Each node of a quadtree represents a square region in the image and can have one of three labels: *full, empty,* or *mixed*. If the node is labeled full, then every pixel of the square region it represents is a pixel of the region of interest. If the node is labeled empty, then there is no intersection between the square region it represents and the region of interest. If the node is labeled mixed, then some of the pixels of the square region are pixels of the region of interest and some are not. Only the mixed nodes in a quadtree have children. The full nodes and empty nodes are leaf nodes. Figure 10.13 illustrates a quadtree representation of an image region. The region looks blocky, because the resolution of the image is only 8×8, which leads to a four-level quadtree. Many more levels would be required to produce a reasonably smoothly curved boundary. Quadtrees have been used to represent map regions in geographic information systems.

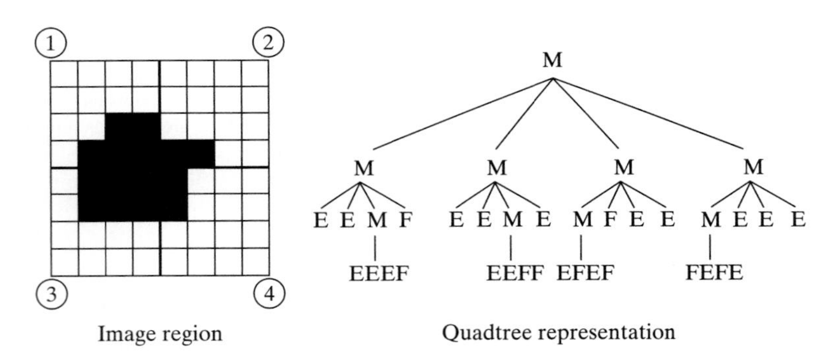

Figure 10.13 A quadtree representation of an image region. The four children of each node correspond to the upper left, upper right, lower left, and lower right quadrants, as illustrated by the numbers in circles for the first level of the tree. M = mixed; E = empty; and F = full.

10.2.5 Property Tables

Sometimes we want to represent a region by its extracted properties rather than by its pixels. In this case, the representation is called a *property table*. It is a table in the relational database sense that has a row for each region in the image and a column for each property of interest. Properties can represent the size, shape, intensity, color, or texture of the region. The features described in Chapters 3, 6, and 7 are all possibilities. For example, in a content-based image retrieval system, regions might be described by area, ratio of minor-to-major axis of the best-fitting ellipse, two main colors, and one or more texture measures. Property tables can be augmented to include or to point to the chain-code encoding or quadtree representation of the region.

Exercise 10.4: Computing area and perimeter.

Consider an image region represented by (a) a labeled image and (b) a chain code representation.

1. Give algorithms for computing the area and the perimeter of the region.
2. Give the running times of your algorithms.

10.3 IDENTIFYING CONTOURS

While some image analysis applications work directly with regions, others need the borders of these regions or various other structures, such as line and circular arc segments. This section discusses the extraction of these structures from images.

Exercise 10.5: Testing for pixels in a region.

Consider an image region represented by (a) a labeled image and (b) a polygonal approximation to the boundary.

1. In each case, give an algorithm for testing if an arbitrary pixel $[r, c]$ is in that region.
2. Give the running times of your algorithms in terms of the appropriate parameters, that is, number of pixels in the region or number of segments in the polygonal approximation.

10.3.1 Tracking Existing Region Boundaries

Once a set of regions has been determined by a procedure such as segmentation or connected components, the boundary of each region may be extracted. *Boundary extraction* can be done simply for small-sized images. Scan through the image and make a list of the first border pixel for each connected component. Then for each region, begin at its first border pixel and follow the border of the connected component around in a clockwise direction until the tracking returns to the first border pixel. For large-sized images that do not reside in memory, this simple border tracking algorithm results in excessive I/O to the mass storage device on which the image resides.

We will describe an algorithm called *border* which can extract the boundaries for all regions in one left-right, top-bottom scan through the image. *Border* inputs a labeled image and outputs, for each region, a clockwise ordered list of the coordinates of its border pixels. The algorithm is flexible in that it can be easily modified to select the borders of specified regions.

The input to *border* is a labeled image whose pixel values denote region labels. It is assumed that there is one background label that designates those pixels in part of a possibly disconnected background region whose borders do not have to be found. Rather than tracing all around the border of a single region and then moving on to the next region, the border algorithm moves in a left-right, top-bottom scan down the image collecting chains of border pixels that form connected sections of the borders of regions. At any given time during execution of the algorithm, there is a set of current regions whose borders have

Find the borders of every region of a labeled image S.
S[R, C] is the input labeled image.
NLINES is the number of rows in the image.
NPIXELS is the number of pixels per row.
NEWCHAIN is a flag that is true when a pixel starts a new chain and false when a new pixel is added to an existant chain.

```
procedure border(S);
{
  for R:= 1 to NLINES
    {
      for C:= 1 to NPIXELS
        {
        LABEL:= S[R, C];
        if new_region(LABEL) then add(CURRENT,LABEL);
        NEIGHB:= neighbors(R,C,LABEL);
        T:= pixeltype(R,C,NEIGHB);
        if T == 'border'
        then for each pixel N in NEIGHB
          {
          CHAINSET:= chainlist(LABEL);
          NEWCHAIN:= true;
          for each chain X in CHAINSET while NEWCHAIN
            if N==rear(X)
            then {add(X,[R, C]); NEWCHAIN:= false}
          if NEWCHAIN
          then make_new_chain(CHAINSET,[R,C],LABEL);
          }
        }
      for each region REG in CURRENT
        if complete(REG)
        then {connect_chains(REG); output(REG); free(REG)}
    }
}
```

Algorithm 10.4 Finding the Borders of Labeled Regions.

been partially scanned, but not yet output, a set of *past regions* that have been completely scanned and their borders output, and a set of *future regions* that have not yet been reached by the scan.

The data structures contain the chains of border pixels of the current regions. Since there may be a huge number of region labels in the image, but only at most 2 × *number_of_columns* may be active at once, a hash table can be used as the device to allow rapid access to the chains of a region given the label of the region. (2 × *number_of_columns* is a safe upper bound; the actual number of regions will be smaller.) When a region is completed and output, it is removed from the hash table. When a new region is encountered during the scan, it is added to the hash table. The hash table entry for a region points to a linked list of chains that have been formed so far for that region. Each chain is a linked list of pixel positions that can be grown from the beginning or the end.

The tracking algorithm examines three rows of the labeled image at a time: The current row being processed; the row above it; and the row below it. Two dummy rows of background pixels are appended to the image, one on top and one on the bottom, so that all rows can be treated alike. The algorithm for an NLINES by NPIXELS labeled image **S** is as follows.

In this procedure, **S** is the name of the labeled image; thus **S[R, C]** is the value (**LABEL**) of the current pixel being scanned. If this is a new label, it is added to the set **CURRENT** of current region labels. **NEIGHB** is the list of neighbors of pixel **[R, C]** which have label **LABEL.** The function *pixeltype* looks at the values of **[R, C]** and its neighbors to decide if **[R, C]** is a nonbackground border pixel. If so, the procedure searches for a chain of the region with label **LABEL** that has a neighbor of **[R, C]** at its rear, and, if it finds one, appends **[R, C]** to the end of the chain by the procedure *add* whose first argument is a chain and whose second argument is **[R, C]**. If no neighbor of **[R, C]** is at the rear of a chain of this region, then a new chain is created containing **[R, C]** as its only element by the procedure *make_new_chain* whose first argument is the set of chains in which a new chain is being added whose sole element is the location **[R, C]**, which is its second argument. Its third argument is the label **LABEL** to be associated with the chain.

After each row **R** is scanned, the chains of those current regions whose borders are now complete are merged into a single border chain which is output. The hash table entrees and list elements associated with those regions are then freed. Figure 10.14 shows a labeled image and its output from the border procedure.

Exercise 10.6: Limitations of the border tracking algorithm.

The border tracking algorithm makes certain assumptions about the regions that it is tracking. Under what conditions could it fail to properly identify the border of a region?

10.3.2 The Canny Edge Detector and Linker

The Canny edge detector and linker extracts boundary segments of an intensity image. It was briefly introduced in Chapter 5 along with other edge detectors. The Canny operator is often used and recent work comparing edge operators justifies its popularity. Examples of its use

	1	2	3	4	5	6	7
1	0	0	0	0	0	0	0
2	0	0	0	0	2	2	0
3	0	1	1	1	2	2	0
4	0	1	1	1	2	2	0
5	0	1	1	1	2	2	0
6	0	0	0	0	2	2	0
7	0	0	0	0	0	0	0

(*a*) A labeled image with two regions

Region	Length	List
1	8	(3, 2)(3, 3)(3, 4)(4, 4)(5, 4)(5, 3)(5, 2)(4, 2)
2	10	(2, 5)(2, 6)(3, 6)(4, 6)(5, 6)(6, 6)(6, 5)(5, 5)(4, 5)(3, 5)

(*b*) The output of the border procedure for the labeled image

Figure 10.14 Action of the border procedure on a labeled image.

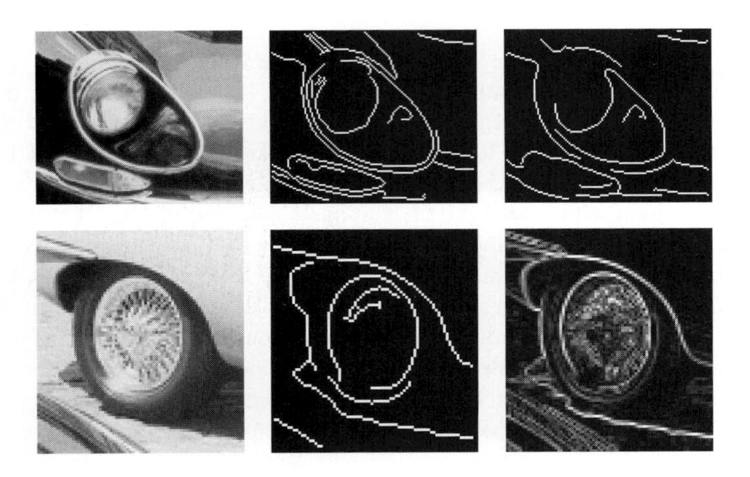

Figure 10.15 (top left) Image of headlight of a black car; (top center) results of Canny operator with $\sigma = 1$; (top right) results of Canny operator with $\sigma = 4$; (bottom left) image of car wheel; (bottom center) results of Canny operator with $\sigma = 1$; and (bottom right) results of Roberts operator. Note in the top row how specular reflection at the top left distracts the edge detector from the boundary of the chrome headlight rim. In the bottom row, note how the shadow of the car connects to the tire which connects to the fender: Neither the tire nor the spokes are detected well.

were provided in Chapter 5: Figure 10.15 shows two examples of images of car parts taken from a larger image shown in Chapter 2. Both of these show well-known problems with all edge detection and boundary following algorithms: The contour segments from actual object parts erroneously merge with contour segments from illumination or reflectance boundaries. Such contours are difficult to analyze in a bottom-up manner by a general object recognition system. However, top-down matching of such representations to models of specific objects can be done successfully, as we shall see in subsequent chapters. Thus, the quality of these edge representations of images depends upon their use in the overall machine vision system.

The Canny edge detection algorithm defined in Algorithm 10.5 produces thin fragments of image contours and is controlled by the single smoothing parameter σ. The image is first smoothed with a Gaussian filter of spread σ and then gradient magnitude and direction are computed at each pixel of the smoothed image. Gradient direction is used to thin edges by suppressing any pixel response that is not higher than the two neighboring pixels on either side of it along the direction of the gradient. This is called *nonmaximum suppression,* a good operation to use with any edge operator when thin boundaries are wanted. The two 8-neighbors of a pixel $[x, y]$ that are to be compared are found by rounding off the computed gradient direction to yield one neighbor on each side of the center pixel. Once the gradient magnitudes are thinned, high magnitude contours are tracked. In the final aggregation phase, continuous contour segments are sequentially followed. Contour following is initiated only on edge pixels where the gradient magnitude meets a high threshold; however, once started, a contour may be followed through pixels whose gradient magnitude meet a lower threshold, usually about half of the higher starting threshold.

Image regions can sometimes be detected when boundary segments close on themselves. Examples of this are shown in Figures 10.16 and 10.17. Such segments can be

Figure 10.16 Identifying regions corresponding to symbols on surfaces is often easy because they are created with good contrast. These results were obtained by applying only the Canny operator: (left set) character carefully written with ink on paper; (right set) weathered signage on a brick wall. (Left set of images courtesy of John Weng.)

Figure 10.17 Image of Mao's Tomb and results of applying the Canny operator with $\sigma = 1$ and $\sigma = 2$. Several objects are detected very well, but so are some shadows.

Compute thin connected edge segments in the input image.

```
I[x, y] : input intensity image; σ : spread used in Gaussian smoothing;
E[x, y] : output binary image;
IS[x, y] : smoothed intensity image;
Mag[x, y] : gradient magnitude; Dir[x, y] : gradient direction;
T_low is low intensity threshold; T_high is high intensity threshold;

procedure Canny( I[], σ);
{
      IS[] = image I[] smoothed by convolution with Gaussian G_σ(x, y);
      use Roberts operator to compute Mag [x, y] and Dir[x, y] from IS[];
      Suppress_Nonmaxima(Mag[], Dir[], T_low, T_high);
      Edge_Detect(Mag[], T_low, T_high, E[]);
}
procedure Suppress_Nonmaxima(Mag[], Dir[]);
{
      define +Del[4] = (1,0), (1,1), (0,1) (−1,1);
      define −Del[4] = (−1,0), (−1−,1), (0,−1) (1,−1);
        for x := 0 to MaxX-1;
        for y := 0 to MaxY-1;
        {
          direction := (Dir[x, y] + π/8) modulo π/4;
          if (Mag[x, y] ≤ Mag[(x, y)+Del[direction] ]) then Mag[x, y] := 0;
          if (Mag[x, y] ≤ Mag[(x, y)+−Del[direction] ]) then Mag[x, y] := 0;
        }
} procedure Edge_Detect(Mag[], T_low, T_high, E[]);
{
      for x := 0 to MaxX - 1;
      for y := 0 to MaxY - 1;
        {
          if (Mag[x, y] ≥ T_high) then Follow_Edge(x, y, Mag[], T_low, T_high, E[]);
        };
}
procedure Follow_Edge(x, y, Mag[], T_low, T_high, E[]);
{
      E [x, y] := 1;
      while Mag[u, v] > T_low for some 8-neighbor [u, v] of [x, y]
        {
          E[u, v] := 1;
          [x, y] := [u, v];
        };
}
```

Algorithm 10.5 Canny Edge Detector.

further analyzed by segmenting the set of boundary pixels into straight or circular sides, etc. For example, the boundary of a rectangular building might result in four straight line segments. Straight line segments can be identified by the Hough transform or by direct fitting of a parameteric line model.

10.3.3 Aggregating Consistent Neighboring Edgels into Curves

The border-tracking algorithm in Section 10.3.1 required as input a labeled image denoting a set of regions. It tracked along the border of each region as it scanned the image, row-by-row. Because of the assumption that each border bounded a closed region, there was never any point at which a border could be split into two or more segments. When the input is instead a labeled edge image with a value of 1 for edge pixels and 0 for non-edge pixels, the problem of tracking edge segments is more complex. Here it is not necessary for edge pixels to bound closed regions and the segments consist of connected edge pixels which go from end point, corner, or junction to endpoint, corner, or junction with no intermediate junctions or corners. Figure 10.18 illustrates such a labeled edge image. Pixel [3, 3] of the image is a *junction* pixel where three different edge (line) segments meet. Pixel [5, 3] is a corner pixel and may be considered a segment end point as well, if the application requires ending segments at corners. An algorithm that tracks segments like these has to be concerned with the following tasks:

1. starting a new segment,
2. adding an interior pixel to a segment,
3. ending a segment,
4. finding a junction, and
5. finding a corner.

	1	2	3	4	5
1	**1**	0	0	0	**1**
2	0	**1**	0	**1**	0
3	0	0	**1**	0	0
4	0	0	**1**	0	0
5	0	0	**1**	**1**	**1**

Figure 10.18 Labeled edge image containing a junction of three line segments at pixel [3, 3] and a potential corner at pixel [5, 3].

Exercise 10.7

Consider the contour following phase of the Canny edge-detection algorithm. When following an image contour by tracing pixels of high gradient magnitude, would it be a good idea to select the next possible pixel only from the two neighbors that are perpendicular to the gradient direction? Why or why not? Show specific cases to support your answer.

Exercise 10.8: Measuring across Canny edges.

Perform the following experiment. Obtain a program for the Canny edge detector or some image tool that contains it. Obtain some flat objects with precise parallel edges, such as razor blades, and some rounded objects, such as the shanks of drill bits. Image several of these objects in several different orientations: Use high resolution scanning, if possible. Apply the Canny edge detector and study the quality of edges obtained, including the repeatability of the distance across parallel edges. Is there any difference between measuring the razor blades, which have sharp edges and measuring the drill bits, which may have soft edges in the image.

As in border tracking, efficient data structure manipulation is needed to manage the information at each step of the procedure. The data structures used are very similar to those used in the border algorithm. Instead of past, current, and future regions, there are past, current, and future segments. Segments are lists of edge points that represent straight or curved lines on the image. Current segments are kept in internal memory and accessed by a hash table. Finished segments are written out to a disk file and their space in the hash table freed. The main difference is the detection of junction points and the segments entering them from above or the left and the segments leaving them from below or the right. We will assume an extended neighborhood operator called *pixeltype* that determines if a pixel is an isolated point, the starting point of a new segment, an interior pixel of an old segment, an ending point of an old segment, a junction or a corner. If the pixel is an interior or end point of an old segment, the segment ID of the old segment is also returned. If the pixel is a junction or a corner point, then a list (INLIST) of segment IDs of incoming segments plus a list (OUTLIST) of pixels representing outgoing segments are returned. A procedure for tracking edges in a labeled image is given in Algorithm 10.6. Figure 10.19 gives the results of its application on the labeled image of Figure 10.18.

Segment ID	Length	List
1	3	(1, 1)(2, 2)(3, 3)
2	3	(1, 5)(2, 4)(3, 3)
3	3	(3, 3)(4, 3)(5, 3)
4	3	(5, 3)(5, 4)(5, 5)

Figure 10.19 Output of the *edge_track* procedure on the image of Fig. 10.18, assuming the point (5, 3) is judged to be a corner point. If corner points are not used to terminate segments, then segement 3 would have length 5 and list ((3, 3)(4, 3)(5, 3) (5, 4)(5, 5)).

The details of keeping track of segment IDs entering and leaving segments at a junction have been supressed. This part of the procedure can be very simple and assume every pixel adjacent to a junction pixel is part of a different segment. In this case, if the segments are more than one-pixel wide, the algorithm will detect a large number of small segments that are really not new line segments at all. This can be avoided by applying a connected shrink operator to the edge image. Another alternative would be to make the pixeltype operator even smarter. It can look at a larger neighborhood and use heuristics to decide if this is just a thick part of the current segment, or a new segment is starting. Often the application will dictate what these heuristics should be.

Find the line segments of binary edge image S.
S[R, C] is the input labeled image.
NLINES is the number of rows in the image.
NPIXELS is the number of pixels per row.
IDNEW is the ID of the newest segment.
INLIST is the list of incoming segment IDs returned by pixeltype.
OUTLIST is the list of outgoing segment IDs returned by pixeltype.

```
procedure edge_track(S);
{
IDNEW := 0;
for R := 1 to NLINES
  for C := 1 to NPIXELS
    if S[R,C] ≠ background pixel
    {
    NAME := address(R, C); NEIGHB := neighbors(R, C);
    T := pixeltype(R,C,NEIGHB,ID,INLIST,OUTLIST);
    case
      T = isolated point : next;
      T = start point of new segment: {
        IDNEW := IDNEW + 1;
        make_new_segment(IDNEW,NAME);};
      T = interior point of old segment : add(ID,NAME);
      T = end point of old segment : {
        add(ID,NAME);
        output(ID); free(ID)};
      T = junction or corner point:
        for each ID in INLIST {
          add(ID,NAME);
          output(ID); free(ID);};
        for each pixel in OUTLIST {
          IDNEW := IDNEW + 1;
          make_new_segment(IDNEW,NAME);};
    }
}
```

Algorithm 10.6 Tracking Edges of a Binary Edge Image.

10.3.4 Hough Transform for Lines and Circular Arcs

The *Hough transform* is a method for detecting straight lines and curves in gray-tone (or color) images. The method is given the family of curves being sought and produces the set of curves from that family that appear on the image. In this section we describe the Hough transform technique, and show how to apply it to finding straight line segments and circular arcs in images.

Exercise 10.9: Determining the type of a pixel.

Give the code for the operator pixeltype, using a 3×3 neighborhood about a pixel to classify it as one of the types: isolated, start or end, interior, junction, and corner.

The Hough Transform Technique The Hough transform algorithm requires an accumulator array whose dimension corresponds to the number of unknown parameters in the equation of the family of curves being sought. For example, finding line segments using the equation $y = mx + b$ requires finding two parameters for each segment: m and b. The two dimensions of the accumulator array for this family would correspond to quantized values for m and quantized values for b. The accumulator array accumlates evidence for the existence of the line $y = mx + b$ in bin **A[M, B]** where **M** and **B** are quantizations of m and b, respectively.

Using an accumulator array **A**, the Hough procedure examines each pixel and its neighborhood in the image. It determines if there is enough evidence of an edge at that pixel, and if so calculates the parameters of the specified curve that passes through this pixel. In the straight line example with equation $y = mx + b$, it would estimate the m and the b of the line passing through the pixel being considered if the measure of edge strength (such as gradient) at that pixel were high enough. Once the parameters at a given pixel are estimated, they are quantized to corresponding values **M** and **B** and the accumulator **A[M, B]** is incremented. Some schemes increment by one and some by the strength of the gradient at the pixel being processed. After all pixels have been processed, the accumulator array is searched for peaks. The peaks indicate the parameters of the most likely lines in the image.

Although the accumulator array tells us the parameters of the infinite lines (or curves), it does not tell us where the actual segments begin and end. In order to have this information, we can add a parallel structure called **PTLIST. PTLIST[M, B]** contains a list of all the pixel positions that contributed to the sum in accumulator **A[M, B]**. From these lists the actual segments can be determined.

This description of the Hough method is general; it leaves out the details needed for an implementation. We will now discuss algorithms for straight line and circle finding in detail.

Finding Straight Line Segments The equation $y = mx + b$ for straight lines does not work for vertical lines. A better model is the equation $d = xcos\theta + ysin\theta$ where d is the perpendicular distance from the line to the origin and θ is the angle the perpendicular makes with the x-axis. We will use this form of the equation but convert to row (r) and column (c) coordinates. Since the column coordinate c corresponds to x and the row coordinate r corresponds to $-y$, our equation becomes

$$d = ccos\theta - rsin\theta \tag{10.17}$$

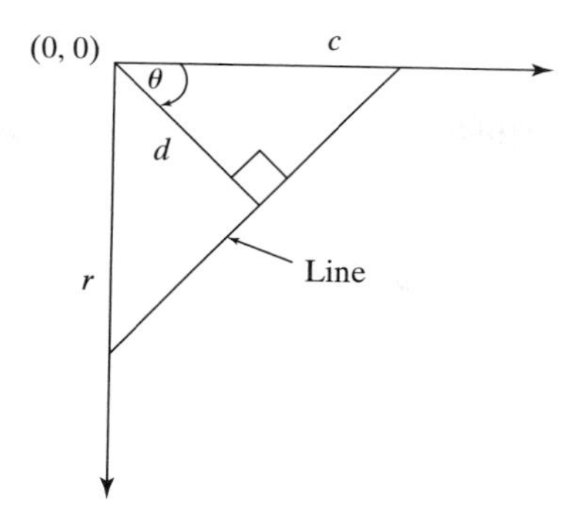

Figure 10.20 The parameters d and θ used in the equation $d = -r sin\theta + c cos\theta$ of a straight line.

where d is the perpendicular distance from the line to the origin of the image (assumed to be at upper left), and θ is the angle this perpendicular makes with the c (column) axis. Figure 10.20 illustrates the parameters of the line segment. Suppose the point where the perpendicular from the origin intersects the line is [50, 50] and that $\theta = 315°$. Then we have

$$d = 50cos(315) - 50sin(315) = 50(.707) - 50(-.707) \approx 70$$

The accumulator **A** has subscripts that represent quantized values of d and θ. O'Gorman and Clowes (1976) quantized the values of d by 3s and θ by 10° increments in their experiments on gray level images of puppet objects. An accumulator array quantized in this fashion is illustrated in Fig. 10.21. The O'Gorman and Clowes algorithm for filling the accumulator **A** and parallel list array **PTLIST** is given in procedure *accumulate_lines* below.

The algorithm is expressed in (row,column) space. The functions *row_gradient* and *column_gradient* are neighborhood functions that estimate the row and column components of the gradient, while the function *gradient* combines the two to get the magnitude. The function *atan2* is the standard scientific library function that returns the angle in the correct quadrant given the row and column components of the gradient. We assume here that *atan2* returns a value between 0° and 359°. Many implementations return the angle in radians which would have to be converted to degrees. If the distance d comes out negative (for

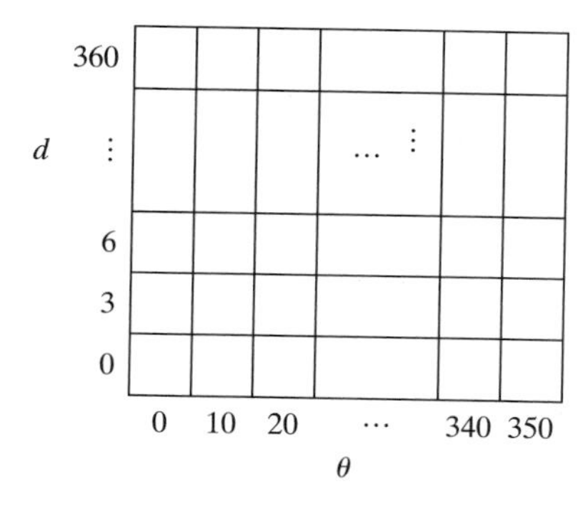

Figure 10.21 The accumulator array for finding straight line segments in images of size 256 × 256.

example, for $\theta = 135°$), its absolute value gives the distance to the line. The actions of the procedure are illustrated in Fig. 10.22. Notice that with a 3×3 gradient operator, the lines are two pixels wide. Notice also that counts appear in other accumulators than the two correct ones.

Procedure *accumulate_lines* is O'Gorman's and Clowes's version of the Hough method. Once the accumulator and list arrays are filled, though, there is no standard method for extracting the line segments. Their ad hoc procedure, *find_lines,* which illustrates some of the problems that come up in this phase of the line segment extraction process, is expressed as Algorithm 10.8:

Accumulate the straight line segments in gray-tone image S to accumulator A.
S[R, C] is the input gray-tone image.
NLINES is the number of rows in the image.
NPIXELS is the number of pixels per row.
A[DQ, THETAQ] is the accumulator array.
DQ is the quantized distance from a line to the origin.
THETAQ is the quantized angle of the normal to the line.

```
procedure accumulate_lines(S,A);
{
A := 0;
PTLIST := NIL;
for R := 1 to NLINES
   for C := 1 to NPIXELS
      {
      DR := row_gradient(S,R,C);
      DC := col_gradient(S,R,C);
      GMAG := gradient(DR,DC);
      if GMAG > gradient_threshold
         {
         THETA := atan2(DR,DC);
         THETAQ := quantize_angle(THETA);
         D := abs(C*cos(THETAQ) – R*sin(THETAQ));
         DQ := quantize_distance(D);
         A[DQ,THETAQ] := A[DQ,THETAQ] + GMAG;
         PTLIST(DQ,THETAQ) := append(PTLIST(DQ,THETAQ),[R,C])
         }
      }
}
```

Algorithm 10.7 Hough Transform for Finding Straight Lines.

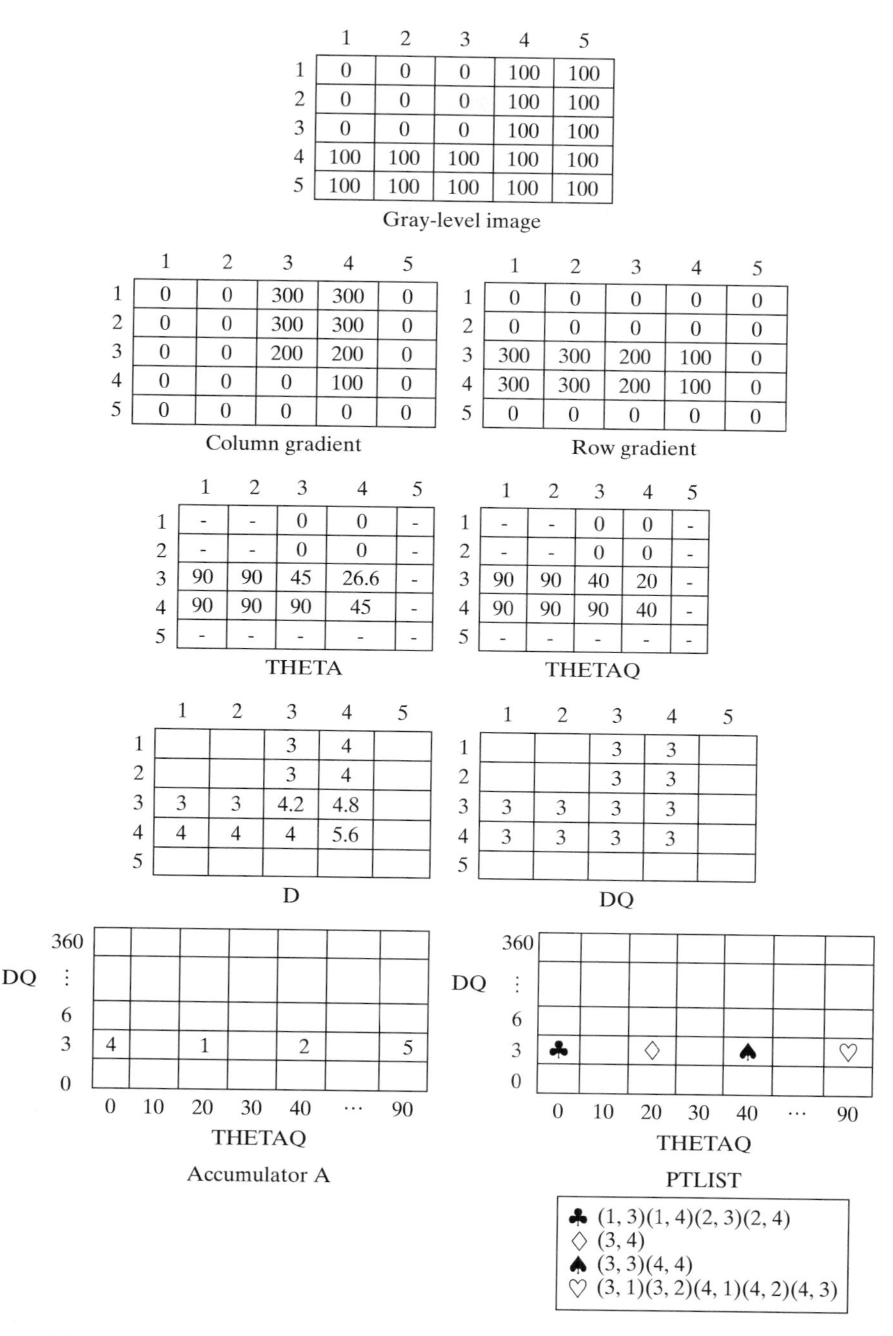

Figure 10.22 The results of the operation of procedure *accumulate* on a simple gray-level image using Prewitt masks. For this small example, the evidence for correct detections is not much larger than that for incorrect ones, but in real images with long segments, the difference will be much more pronounced.

```
Find the point lists corresponding to separate line segments.
A[DQ, THETAQ] is the accumulator array from accumulate_lines.
DQ is the quantized distance from a line to the origin.
THETAQ is the quantized angle of the normal to the line.

   procedure find_lines;
   {
   V := pick_greatest_bin(A,DQ,THETAQ);
   while V > value_threshold
     {
     list_of_points := reorder(PTLIST[DQ,THETAQ]);
     for each point [R, C] in list_of_points
       for each neighbor [R',C'] of [R, C] not in list_of_points
         {
         DPRIME := D[R',C'];
         THETAPRIME := THETA[R',C'];
         GRADPRIME := GRADIENT[R',C'];
         if GRADPRIME > gradient_threshold
           and abs(THETAPRIME–THETAQ)≤ 10
         then {
                merge(PTLIST[DQ,THETAQ],PTLIST[DPRIME,
                  THETAPRIME]);
                set_to_zero[A,DPRIME,THETAPRIME];
              }
         }
     final_list_of_points := PTLIST[DQ,THETAQ];
     create_segments(final_list_of_points);
     set_to_zero[A,DQ,THETAQ];
     V := pick_greatest_bin[A,DQ,THETAQ];
     }
   }
```

Algorithm 10.8 O'Gorman and Clowes Method for Extracting Straight Lines.

The function *pick_greatest_bin* returns the value in the largest accumulator while setting its last two parameters, **DQ** and **THETAQ,** to the quantized d and θ values for that bin. The *reorder function* orders the list of points in a bin by column coordinate for $\theta < 45$ or $\theta > 135$ and by row coordinate for $45 \leq \theta \leq 135$. The arrays **D** and **THETA** are expected to hold the quantized **D** and **THETA** values for a pixel that were computed during the accumulation. Similarly the array **GRADIENT** is expected to contain the computed gradient magnitude. These can be saved as intermediate images. The *merge* procedure merges the list of points from a neighbor of a pixel in with the list of points for that pixel, keeping the spatial ordering. The *set_to_zero* procedure zeroes out an accumulator so that it will not be reused. Finally, the procedure *create_segments* goes through the final ordered set of points searching for gaps longer than one pixel. It creates and saves a set of line segments terminating at gaps. For better accuracy, a least-squares procedure can be used to fit lists of points to line segments. It is important to mention that the Hough procedure can gather strong evidence from broken or virtual lines such as a row of stones or a road broken by overhanging trees.

Exercise 10.10

This exercise follows the work of Kasturi and others (1990): The problem is to apply the Hough transform to identify lines of text. Apply existing programs or tools and write new ones as needed to perform the following experiment. (a) Type or print a few lines of text in various directions and binarize the image. Add some other objects, such as blobs or curves. (b) Apply connected components processing and output the centroids of all objects whose bounding boxes are appropriate for characters. (c) Input the set of all selected centroids to a Hough line detection procedure and report on how well the text lines can be detected.

Finding Circles The Hough transform technique can be extended to circles and other parametrized curves. The standard equation of a circle has three parameters. If a point **[R, C]** lies on a circle then the gradient at **[R, C]** points to the center of that circle as shown in Fig. 10.23. So if a point **[R, C]** is given, a radius d is selected, and the direction of the vector from **[R, C]** to the center is computed, the coordinates of the center can be found. The radius, d, the row-coordinate of the center, r_o, and the column-coordinate of the center, c_o, are the three parameters used to vote for circles in the Hough algorithm. In row-column coordinates, circles are represented by the equations

$$r = r_o + d\sin\theta \tag{10.18}$$

$$c = c_o - d\cos\theta \tag{10.19}$$

With these equations, the accumulate algorithm for circles becomes algorithm *accumulate_circles* on the next page.

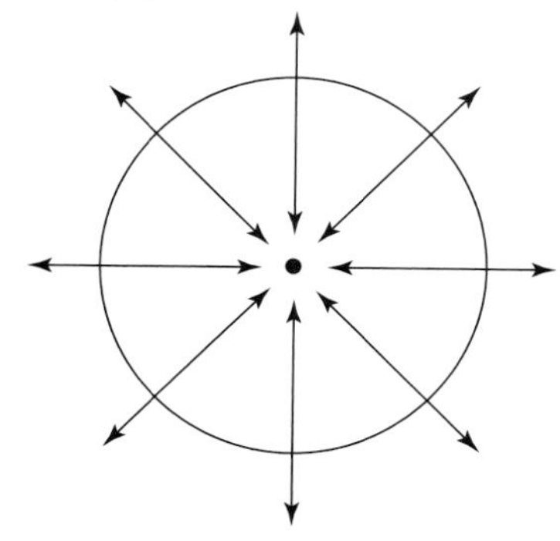

Figure 10.23 Illustrates the direction of the gradient at the boundary points of a circle. The inward pointing gradients are the ones that will accumulate evidence for the center of the circle.

Accumulate the circles in gray-tone image S to accumulator A.
S[R, C] is the input gray-tone image.
NLINES is the number of rows in the image.
NPIXELS is the number of pixels per row.
A[R, C, RAD] is the accumulator array.
R is the row index of the circle center.
C is the column index of the circle center.
RAD is the radius of the circle.

```
procedure accumulate_circles(S,A);
{
A := 0;
PTLIST := 0;
for R := 1 to NLINES
  for C := 1 to NPIXELS
    for each possible value RAD of radius
      {
      THETA := compute_theta(S,R,C,RAD);
      R0 := R – RAD*cos(THETA);
      C0 := C + RAD*sin(THETA);
      A[R0,C0,RAD] := A[R0,C0,RAD]+1;
      PTLIST[R0,C0,RAD] := append(PTLIST[R0,C0,RAD],[R,C])
      }
}
```

Algorithm 10.9 Hough Transform for Finding Circles.

This procedure can easily be modified to take into account the gradient magnitude as did the procedure for line segments. The results of applying it to a technical document image are shown in Figure 10.24.

Extensions The Hough transform method can be extended to any curve with analytic equation of the form $f(\mathbf{x}, \mathbf{a}) = 0$ where $\mathbf{x}$ denotes an image point and $\mathbf{a}$ is a vector of parameters. The procedure is as follows:

1. Initialize accumulator array **A[a]** to zero.
2. For each edge pixel $\mathbf{x}$ determine each $\mathbf{a}$ such that $f(\mathbf{x}, \mathbf{a}) = 0$ and set **A[a]** := **A[a]** + 1.
3. Local maxima in **A** correspond to curves of f in the image.

If there are m parameters in $\mathbf{a}$, each having M discrete values, then the time complexity is $O(M^{m-2})$. The Hough transform method has been further generalized to arbitrary shapes specified by a sequence of boundary points (Ballard, 1981). This is known as the *generalized Hough transform.*

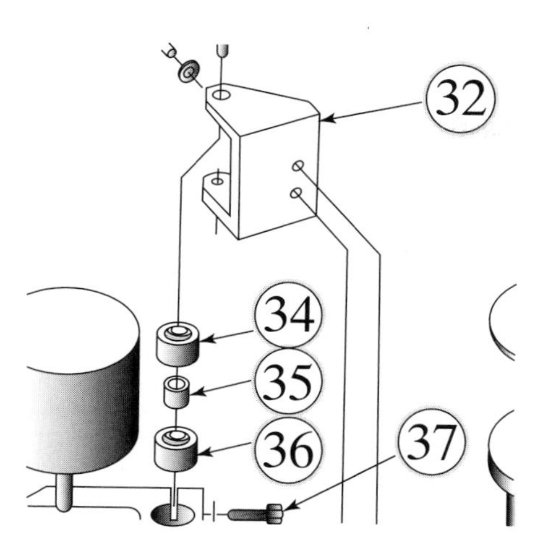

Figure 10.24 Circles detected by the Hough transform on a section of a technical drawing, shown by overlaying an extra circle of slightly larger radius on each detected circle.

The Burns Line Finder A number of hybrid techniques exist that use some of the principles of the Hough transform. The Burns line finder (Burns and others, 1986) was developed to find straight lines in complex images of outdoor scenes. The Burns method can be summarized as follows:

1. Compute the gradient magnitude and direction at each pixel.
2. For points with high enough gradient magnitude, assign two labels representing two different quantizations of the gradient direction. (For example, for eight bins, if the first quantization is 0 to 44, 45 to 90, 91 to 134, etc., then the second can be −22 to 22, 23 to 67, 68 to 112, etc.) The result is two symbolic images.
3. Find the connected components of each symbolic image and compute line length for each component.
   - Each pixel is a member of two components, one from each symbolic image.
   - Each pixel votes for its *longer* component.
   - Each component receives a count of pixels that voted for it.
   - The components (line segments) that receive the majority support are selected.

The Burns line finder takes advantage of two powerful algorithms: The Hough transform and the connected components algorithm. It attempts to get rid of the quantization problems that forced O'Gorman and Clowes to search neighboring bins by the use of two separate quantizations. In practice, it suffers from a problem that will affect any line finder that estimates angle based on a small neighborhood around a pixel. Digital lines are not straight. Diagonal lines are really a sequence of horizontal and vertical steps. If the angle detection technique uses too small a neighborhood, it will end up finding a lot of tiny horizontal and vertical segments instead of a long diagonal line. Thus in practice, the Burns line finder and any other angle-based line finder can break up lines that a human would like to detect as a connected whole.

Exercise 10.11: Burns compared to Hough.

Implement both the Hough transform and the Burns operator for line finding and compare the results on real-world images having a good supply of straight lines.

Exercise 10.12: Line Detection.

Implement the following approach to detecting lines in a gray-tone image **I**.

> **for** all image pixels **I[R,C]**
> {
> compute the gradient $\mathbf{G_{mag}}$ and $\mathbf{G_{dir}}$
> **if** $\mathbf{G_{mag}}$ > threshold
> **then** output [$\mathbf{G_{mag}}$,$\mathbf{G_{dir}}$] to set **H**
> }
> detect clusters in the set **H**;

The significant clusters will correspond to the significant line segments in **I**.

10.4 FITTING MODELS TO SEGMENTS

Mathematical models that fit data not only reveal important structure in the data, but also can provide efficient representations for further analysis. A straight line model might be used for the edge pixels of a building or a planar model might apply to surface data from the face of a building. Convenient mathematical models exist for circles, cylinders, and many other shapes.

Next, we present the *method of least squares* for determining the parameters of the best mathematical model fitting observed data. This data might be obtained using one of the region or boundary segmentation methods described previously; for example, we have already mentioned that we could fit a straight line model to all the pixels $[r, c]$ voting for a particular line hypothesis **A[THETAQ, DQ]** in the Hough accumulator array. In order to apply least squares, there must be some way of establishing which model or models should be tried out of an infinite number of possibilities. Once a model is fit and its parameters determined, it is possible to determine whether or not the model fits the data acceptably. A good fit might mean that an object of some specified shape has been detected; or, it might just provide a more compact representation of the data for further analysis.

Fitting a Straight Line We introduce the least-squares theory by way of a simple example. One straight line model is a function with two parameters: $y = f(x) = c_1 x + c_0$. Suppose we want to test whether or not a set of observed points $\{(x_j, y_j), j = 1, n\}$ form a line. To do this, we determine the best parameters c_1 and c_0 of the linear function and then examine how close the observed points are to the function. Different criteria can be used to quantify how close the observations are to the model. Figure 10.25 shows the fitting of a line to six data points. Surely, we could move the line slightly and we would have a

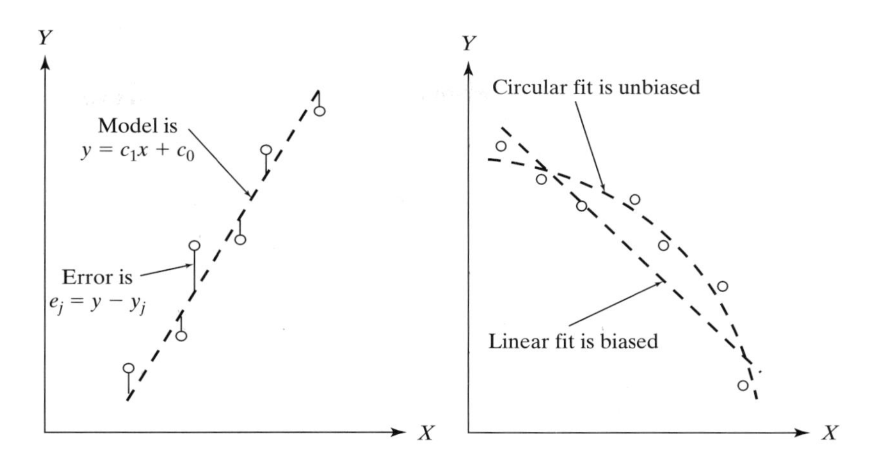

Figure 10.25 (left) Fit of model $y = f(x)$ to six data points; and (right) competing straight line and circular models: The signs of the residual errors show that the line fit is biased and the circular fit is unbiased.

different line that still fits well. The *least-squares criteria* defines a best line according to the Definition 74.

74 Definition. Least-Squares Error Criteria: The measure of how well a model $y = f(x)$ fits a set of n observations $\{(x_j, y_j), j = 1, n\}$ is

$$LSE = \sum_{j=1}^{n} (f(x_j) - y_j)^2$$

The best model $y = f(x)$ is the model with the parameters minimizing this criteria.

75 Definition. The **root-mean-square error, or RMSE**, is the average difference of observations from the model:

$$RMSE = \left[\sum_{j=1}^{n} (f(x_j) - y_j)^2)/n \right]^{1/2}$$

Note that for the straight line fit, this difference is not the Euclidean distance of the observed point from the line, but the distance measured parallel to the y-axis as shown in Figure 10.25.

76 Definition. Max-Error Criteria: The measure of how well a model $y = f(x)$ fits a set of n observations $\{(x_j, y_j), j = 1, n\}$ is

$$MAXE = max(\{|(f(x_j) - y_j)|\}_{j=1,n})$$

Note that this measure depends only on the worst fit point, whereas the RMS error depends on the fit of all of the points.

TABLE 10.1 LEAST-SQUARES FIT OF DATA GENERATED USING $y = 3x - 7$ PLUS NOISE GIVES FITTED MODEL $y = 2.971x - 6.962$

Data Pts (x_j, y_j)	(0.0, −6.8)	(1.0, −4.1)	(2.0, −1.1)	(3.0, 1.8)	(4.0, 5.1)	(5.0, 7.9)
Residuals $y - y_j$:	−0.162	0.110	0.081	0.152	−0.176	−0.005

Closed Form Solutions for Parameters The least-squares criteria is popular for two reasons; first, it is a logical choice when a Gaussian noise model holds, second, derivation of a closed form solution for the parameters of the best model is easy. We first derive the closed form solution for the parameters of the best-fitting straight line. Development of other models follows the same pattern. The least-squares error for the straight line model can be explicitly written as follows: Note that the observed data x_j, y_j are regarded as constants in this formula.

$$LSE = \varepsilon(c_1, c_0) = \sum_{j=1}^{n} (c_1 x_j + c_0 - y_j)^2 \tag{10.20}$$

The error function ε is a smooth non-negative function of the two parameters c_1 and c_0 and will have a global minimum at the point (c_1, c_0) where $\partial\varepsilon/\partial c_1 = 0$ and $\partial\varepsilon/\partial c_0 = 0$. Writing out these derivatives from the formula in Equation 10.20 and using the fact that the derivative of a sum is the sum of the derviatives yields the following derivation.

$$\partial\varepsilon/\partial c_1 = \sum_{j=1}^{n} 2(c_1 x_j + c_0 - y_j)x_j = 0 \tag{10.21}$$

$$= 2\left(\sum_{j=1}^{n} x_j^2\right) c_1 + 2\left(\sum_{j=1}^{n} x_j\right) c_0 - 2\sum_{j=1}^{n} x_j y_j \tag{10.22}$$

$$\partial\varepsilon/\partial c_0 = \sum_{j=1}^{n} 2(c_1 x_j + c_0 - y_j) = 0 \tag{10.23}$$

$$= 2\left(\sum_{j=1}^{n} x_j\right) c_1 + 2\sum_{j=1}^{n} c_0 - 2\sum_{j=1}^{n} y_j \tag{10.24}$$

These equations are nicely represented in matrix form. The parameters of the best line are found by solving the equations. The general case representing an arbitrary polynomial fit results in a highly patterned set of equations called the *normal equations*.

$$\begin{bmatrix} \sum_{j=1}^{n} x_j^2 & \sum_{j=1}^{n} x_j \\ \sum_{j=1}^{n} x_j & \sum_{j=1}^{n} 1 \end{bmatrix} \begin{bmatrix} c_1 \\ c_0 \end{bmatrix} = \begin{bmatrix} \sum_{j=1}^{n} x_j y_j \\ \sum_{j=1}^{n} y_j \end{bmatrix} \tag{10.25}$$

Exercise 10.13: Fitting a line to 3 points.

Using Equation 10.25, compute the parameters c_1 and c_0 of the best line through the points $(0, -7)$, $(2, -1)$ and $(4, 5)$.

Exercise 10.14: Normal equations.

(a) Derive the matrix form for the equations constraining the 4 parameters for fitting a cubic polynomial $c_3x^3 + c_2x^2 + c_1x + c_0$ to observed data (x_j, y_j), $j = 1, n$. (b) From the pattern of the matrix elements, predict the matrix form for fitting a polynomial of degree four.

Empirical Interpretation of the Error Empirical interpetation of the error and individual errors is often straightforward in machine vision problems. For example, we might accept the fit if all observed points are within a pixel or two of the model. In a controlled 2D imaging environment, one could image many straight-sided objects and study the variation of detected edge points off the ideal line. If individual points are far from the fitted line (these are called *outliers*), they could indicate feature detection error, an actual defect in the object, or that a different object or model exists. In these cases, it is appropriate to delete the outliers from the set of observations and repeat the fit so that the model is not pulled off by points which it should not model. All the original points can still be interpreted relative to the updated model. If the model fitting is being used for curve segmentation, it is typically the extreme points that are deleted, because they are actually part of a differently shaped object or part.

Statistical Interpretation of the Error* Error can be interpreted relative to a formal statistical hypothesis. The common assumption is that the observed value of y_j is just the model value $f(x_j)$ plus (Gaussian) noise from the normal distribution $N(0, \sigma)$, where σ is known from the analysis of measurement error, which could be done empirically as above. It is also assumed that the noise in any individual observation j is independent of the noise in any other observation k. It follows that the variable $S_{sq} = \sum_{j=1}^{n}((f(x_j) - y_j)^2)/\sigma^2)$ is χ^2 distributed, so its likelihood can be determined by formula or table lookup. The number of degrees of freedom is $n - 2$ for the straight line fit since two parameters are estimated from the n observations. If 95 percent of the χ^2 distribution is below our observed S_{sq}, then perhaps we should reject the hypothesis that this model fits the data. Other confidence levels can be used. The χ^2 test is not only useful for accepting/rejecting a given hypothesis, but it is also useful for selecting the most likely model from a set of competing alternatives. For example, a parabolic model may compete with the straight line model. Note that in this case, the parabolic model $y = c_2x^2 + c_1x + c_0$ has three parameters so the χ^2 distribution would have $n - 3$ degrees of freedom.

Intuitively, we should not be too comfortable in assuming that error in observation j is independent of the error in observations $j - 1$ or $j + 1$. For example, an errant manufacturing process might distort an entire neighborhood of points from the ideal model. The independence hypothesis can be tested using a *run-of-signs* test, which can detect systematic bias in the error, which in turn indicates that a different shape model will fit better. If the noise is truly random, then the signs of the error should be random and hence fluctuate frequently. Figure 10.25 (right) shows a biased linear fit competing with an unbiased circular fit. The signs of the errors indicate that the linear fit is biased. Consult the references at the end of the chapter for more reading on statistical hypothesis-testing for evaluating fit quality.

Exercise 10.15: Fitting a planar equation to 3D points.

(a) Solve for the parameters a, b, c of the model $z = f(x, y) = ax + by + c$ of the least-squares plane through the five surface points (20, 10, 130), (25, 20, 130), (30, 15, 145), (25, 10, 140), (30, 20, 140). (b) Repeat part (a) after adding a random variation to each of the three coordinates of each of the five points. Flip a coin to obtain the variation: if the coin shows heads, then add 1, if it shows tails then subtract 1.

Exercise 10.16: Prewitt operator is *optimal.*

Show that the Prewitt gradient operator from Chapter 5 can be obtained by fitting the least-squares plane through the 3×3 neighborhood of the intensity function. To compute the gradient at $I[x, y]$, fit the nine points $(x + \Delta x, y + \Delta y, I[x + \Delta x, y + \Delta y])$ where Δx and Δy range through $-1, 0, +1$. Having the planar model $z = ax + by + c$ that best fits the intensity surface, show that using the two Prewitt masks actually compute a and b.

Problems in Fitting It is important to consider several kinds of problems in fitting.

Outliers Since every observation affects the RMS error, a large number of outliers may render the fit worthless: The initial fit may be so far off the ideal that it is impossible to identify and delete the real outliers. Methods of robust statistics can be applied in such cases: Consult the Boyer, Mirza, and Ganguly (1994) reference cited at the end of the chapter.

Error Definition The mathematical definition of error as a difference along the y-axis is not a true geometric distance; thus the least squares fit does not necessarily yield a curve or surface that best approximates the data in geometric space. The rightmost data point at the right of Figure 10.25 illustrates this problem—that point is geometrically very close to the circle, but the functional difference along the y-axis is rather large. This effect is even more pronounced when complex surfaces are fit to 3D points. While geometric distance is usually more meaningful than functional difference, it is not always easy to compute. In the case of fitting a straight line, when the line gets near to vertical, it is better to use the best axis computation given in Chapter 3 rather than the least squares method presented here. The best axis computation is formulated based on minimizing the geometric distances between line and points.

Nonlinear Optimization Sometimes, a closed form solution to the model parameters is not available. The error criteria can still be optimized, however, by using a technique that **searches parameter space** for the best parameters. Hill-climbing, gradient-based search, or even exhaustive search can be used for optimization. See the works by Chen and Medioni (1994) and Sullivan, Sandford, and Ponce (1994), which address this and the previous issue.

High Dimensionality When the dimensionality of the data and/or the number of model parameters is high, both empirical and statistical interpretation of a fit can be difficult. Moreover, if a search technique is used to find parameters it may not even be known whether or not these parameters are optimal or just result from a local minima of the error criteria.

Fit Constraints Sometimes, the model being fit must satisfy additional constraints. For example, we may need to find the best line through observations that is also perpendicular to another line. Techniques for constrained optimization can be found in the references.

Segmenting Curves via Fitting The model fitting method and theory presented above assumes that both a model hypothesis and set of observations are given. Boundary tracking can be done in order to obtain long strings of boundary points which can then be segmented as follows. First, high curvature points or cusps can be detected in the boundary sequence in order to segment it. Then, model hypotheses can be tested against the segments between breakpoints. The result of this process is a set of curve segments and the mathematical model and parameters which characterize the shape of each segment. An alternative method is to use the model-fitting process in order to segment the original boundary curve. In the first stage, each subsequence of k consecutive points is fit with each model. The χ^2 value of the fit is stored in a set with each acceptable fit. The second stage attempts to extend acceptable fits by repetitively adding another endpoint to the subsequence. Fitted segments are grown until addition of an endpoint decreases the χ^2 value of the fit. The result of this process is a set of possibly overlapping subsequences, each with a model and the χ^2 value of the model fit. This set is then passed to higher-level processes which can construct domain objects from the available detected parts. This process is similar in spirit to the region-grower described in Section 10.1.2, which has been successfully used to grow line segments using the direction estimate at each edge pixel as the critical property instead of the gray-tone property used in growing regions.

10.5 IDENTIFYING HIGHER-LEVEL STRUCTURE

Analysis of images often requires the combination of segments. For example, quadrilateral regions and straight edge segments might be combined as evidence of a building or intersecting edge segments might accurately define the corner of a building, or a green region inside a blue region might provide evidence of an island. The methods for combining segments are limitless. Below, we look at just two general cases of combining edge segments to form more informative structures: These are the *ribbon* and the *corner*.

10.5.1 Ribbons

A very general type of image segment is the ribbon. Ribbons are commonly produced by imaging elongated objects in 2D or in 3D; for example, by imaging a conduction path on a printed circuit board, the door of a house, a pen on a table, or a road through fields. In these examples, the sides of the ribbons are approximately parallel to each other, but not necessarily straight. Although we limit ourselves to straight-sided ribbons in this text, ribbons can have more general shape, such as that of a wine bottle or ornate lampost, where the shape of the silhouette is some complex curve with reflective symmetry relative to the axis of the ribbon. An electric cord, a rope, a meandering stream, or road each produce a ribbon in an image, as will the shadow of a rope or lampost. Chapter 14 discusses 3D object parts called *generalized cylinders,* which produce ribbons when viewed. At the left

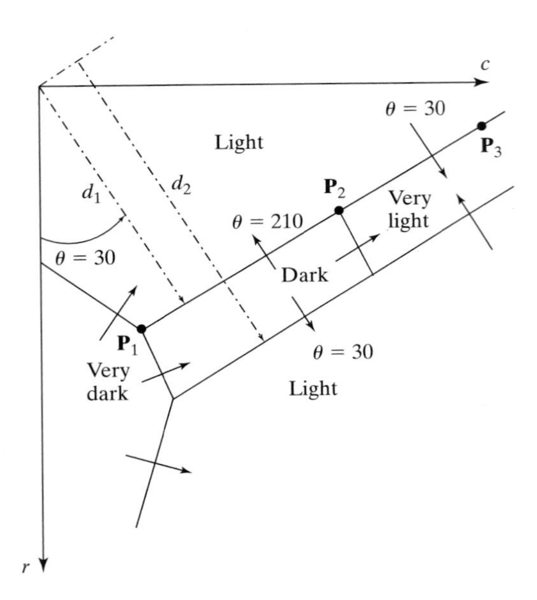

Figure 10.26 The Hough transform can encode the location and orientation of an edge and its gradient direction. The transition from a dark region to a lighter one gives the opposite gradient direction from the transition from the lighter region to the darker one, along the same image line.

in Figure 10.16 is a symbol that is well represented by four ribbons, two of which are highly curved. We leave extraction of general ribbons to future study and concentrate on those with straight sides.

77 Definition. A **ribbon** is an elongated region that is approximately symmetrical about its major axis. Often, but not always, the edges of a ribbon contrast symmetrically with its background.

Figure 10.26 shows how the Hough Transform can be extended slightly to encode the gradient direction across an edge in addition to its orientation and location. As was shown in Chapter 5, and earlier in this chapter, gradient direction θ at a pixel $[r, c]$ that has significant gradient magnitude can be computed in the interval $[0, 2\Pi)$ using operators such as the Sobel operator. The vector from the image origin to the pixel is $[r, c]$: We project this vector onto the unit vector in the direction θ to obtain a signed distance d.

$$d = [r, c] \circ [-sin\theta, cos\theta] = -r\ sin\theta + c\ cos\theta \tag{10.26}$$

A positive value for d is the same as that obtained in the usual polar coordinate representation for pixel $[r, c]$. However, a negative d results when the direction from the origin to the edge is opposite to the gradient direction: This will result in two separate clusters for every line on a checkerboard for example. Figure 10.26 illustrates this idea. Consider the edge P_2P_3 in the figure. Pixels along this edge should all have gradient direction approximately 30 degrees. The perpendicular direction from the origin to P_2P_3 is in the same direction, so pixels along P_2P_3 will transform to (approximately) $[d_1, 30°]$ in the Hough parameter space. Pixels along line segment P_1P_2 however, have a gradient direction of 210 degrees, which is opposite to the direction of the perpendicular from the origin to P_1P_2. Thus, pixels along segment P_1P_2 will transform to (approximately) $[-d_1, 210°]$.

Exercise 10.17

Figure 10.27 shows a dark ring on a light background centered at the image origin. Sketch the parameter space that would result when such an image is transformed using the Hough Transform as shown in Figure 10.26.

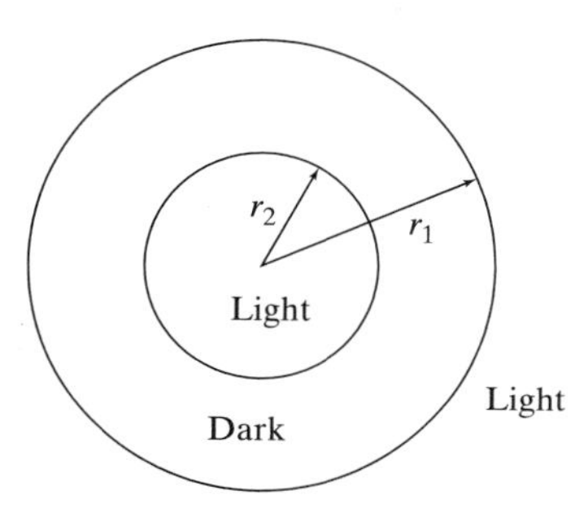

Figure 10.27 Dark ring centered at the origin on light background. Dark region is outside the smaller circle and inside the larger one.

Detecting Straight Ribbons By using the Hough parameters along with the point lists obtained by Algorithm *accumulate_lines,* more complex image structure can be detected. Two edges whose directions differ by 180° provide evidence of a possible ribbon. If in addition, the point lists are located near each other, then there is evidence of a larger linear feature that reverses gradient, such as the columns as in Figure 10.17.

Figure 10.28 shows an image of part of a white house containing a downspout. The image was taken in strong sunlight and this resulted in hard shadows. By using a gradient operator and then collecting pixels on edge segments using *accumulate_lines,* there is strong evidence of a bright ribbon on a dark background corresponding to the downspout (sides are AB and ED). The shadow of the downspout **S** also creates evidence of a dark ribbon on a bright background.

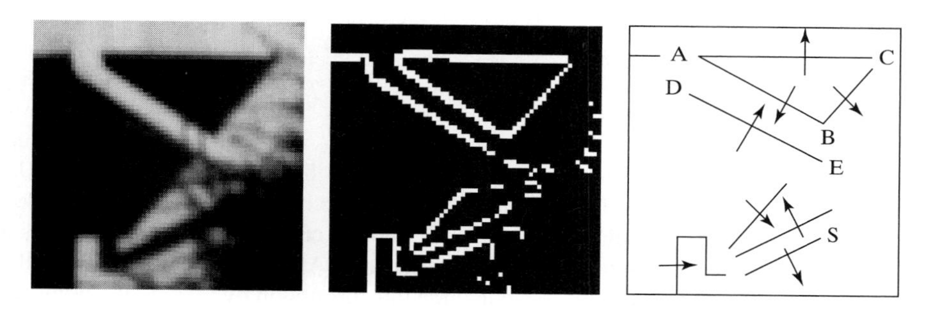

Figure 10.28 (left) Region of an image of a house showing a downspout and strong shadows; (center) the highest 10 percent gradient magnitudes computed by the Prewitt 3 × 3 operator; and (right) sketch of ribbons and corners evident.

Exercise 10.18

Write a computer program to study the use of the Hough Transform to detect ribbons. (a) Use the Sobel operator to extract the gradient magnitude and direction at all pixels. Then transform only pixels of high magnitude. (b) Detect any clusters in $[d, \theta]$-space. (c) Detect pairs of clusters, $([d_1, \theta_1], [d_2, \theta_2])$, where θ_1 and θ_2 are π apart. (d) Delete pairs that are not approximately symmetrical across an axis between them.

10.5.2 Detecting Corners

Significant region corners can be detected by finding pairs of detected edge segments E_1 and E_2 in the following relationship.

1. Lines fit to edge point sets E_1 and E_2 intersect at point $[u, v]$ in the real image coordinate space.
2. Point $[u, v]$ is close to extreme points of both sets E_1 and E_2.
3. The gradient directions of E_1 and E_2 are symmetric about their axis of symmetry.

This definition models only corners of type 'L': constraint (2) rules outs those of type 'T', 'X' and 'Y'. The computed intersection $[u, v]$ will have subpixel accuracy. Figure 10.29 sketches the geometry of a corner structure. Edge segments can be identified intitially by using the Hough transform or by boundary following and line fitting or by any other appropriate algorithm. For each pair $([d_1, \theta_1], [d_2, \theta_2])$ satisfying the above criteria, add the quad $([d_1, \theta_1], [d_2, \theta_2], [u, v], \alpha)$ to a set of candidate corners. The angle α is formed at the corner. This set of corner features can be used for building higher level descriptions, or can be used directly in image matching or warping methods as shown in Chapter 11.

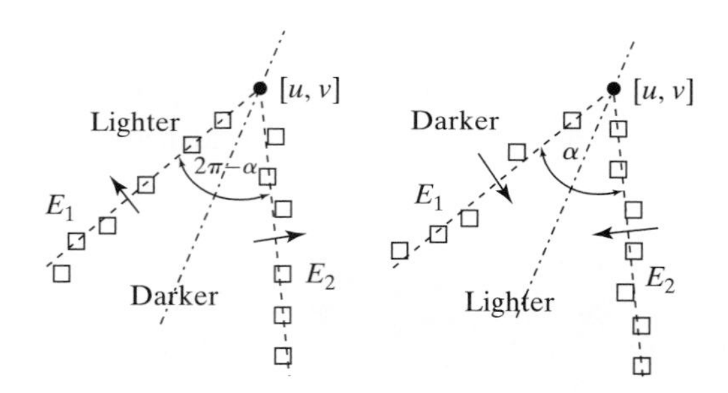

Figure 10.29 Corners are detected as pairs of detected edge segments appropriately related.

Several corners can easily be extracted from the blocks image in Figure 10.2; however, most of them are due to viewpoint dependent occlusions of one object crease by another and not by the joining of two actual 3D object creases. Four actual corners are evident for the top of the arch. The corners of triangle ABC in Figure 10.28 are all artifacts of the lighting and viewpoint. We conclude this discussion by making the point that although representations using edge segments are often used in specific problem domains, they may be highly ambiguous when used in general. Usually, problem-specific knowledge is needed in order to interpret higher-level structure.

Exercise 10.19

Describe how to change the ribbon detection algorithm so that (a) it only detects ribbons that are nearly vertical, (b) it detects ribbons that are no wider than W.

10.6 SEGMENTATION USING MOTION COHERENCE

As we have seen, motion is important for determining scene content and action. Chapter 9 presented methods for detecting change in a scene and for tracking motion over many frames.

10.6.1 Boundaries in Space-Time

The contours of moving objects can be identified by using both spatial and temporal contrast. Our previous examples have used only spatial contrast of some property such as intensity or texture in a single image. Spatial and temporal gradients can be computed and combined if we have two images **I[x, y, t]** and **I[x, y, t + Δt]** of the scene. We can define a *spatio-temporal gradient magnitude* as the product of the spatial gradient magnitude and the temporal gradient magnitude as in Equation 10.27. Once an image **STG**[] is computed, it is amenable to all the contour extraction methods already discussed. The contours that are extracted will be the boundaries of moving objects and not static ones, however.

$$STG[x, y, t] = Mag[x, y, t]\,(|I[x, y, t] - I[x, y, t + \Delta t]|) \tag{10.27}$$

10.6.2 Aggregrating Motion Trajectories

Assume that motion vectors are computed across two frames of an image sequence. This can be done using special interest points or regions as described in Chapter 9. Region segmentation can be performed on the motion vectors by clustering according to image position, speed, and direction as shown in Figure 10.30. Clustering should be very tight for a translating object, because points of the object should have the same velocity. Through more complex analysis, objects that rotate and translate can also be detected.

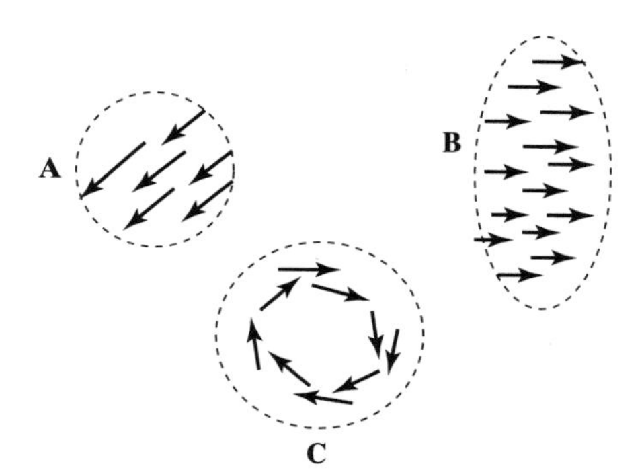

Figure 10.30 Aggregation of vectors from the motion field via compatible location, velocity and direction: translating objects (**A,B**) are easier to detect than rotating objects (**C**).

Exercise 10.20

Given two lines parameterized by $([d_1, \theta_1], [d_2, \theta_2])$, (a) derive a formula for their intersection point $[x, y]$ and (b) derive a formula for their axis of symmetry $[d_a, \theta_a]$.

Exercise 10.21

Obtain two successive images of a scene with moving objects and compute a spatio-temporal image from them using Equation 10.27. (Two frames from a Motion JPEG video would be good. Or, one could digitize some dark cutouts on a flatbed scanner, moving them slightly for the second image.)

Figure 10.31 shows processing from an application where the purpose of motion is communication: The goal is to input to a machine via American Sign Language (ASL). The figure shows only a sample of frames from a sequence representing about two seconds of gesturing by the human signer. The results shown in Figure 10.31 were produced using both color segmentation within a single frame and motion segmentation across pairs of frames. A sketch of steps of an algorithm is given in Algorithm 10.10: For details of the actual algorithm producing Figure 10.31, consult the reference by Yang and Ahuja (1999). The first several steps of the algorithm can be generally applied to many different kinds of

Input a video sequence of a person signing in ASL.
Output motion trajectories of the two palms.

1. Segment each frame $\mathbf{I}_t$ of the sequence into regions using color.
2. Match the regions of each pair of images $(\mathbf{I}_t, \mathbf{I}_{t+1})$ by color and neighborhood.
3. Compute the affine transformation matching each region of $\mathbf{I}_t$ to the corresponding region of $\mathbf{I}_{t+1}$.
4. Use the transformation matching regions to guide the computation of motion vectors for individual pixels.
5. Segment the motion field derived above using motion coherence and image location.
6. Identify two hand regions and the face region using a skin color model.
7. Merge adjacent skin colored regions that were fragmented previously.
8. Find an ellipse approximating each hand and the face.
9. Create motion trajectories by tracking each ellipse center over the entire sequence.
10. (Recognize the gesture using the trajectories of the two hands.)

Algorithm 10.10 Algorithm using color and motion to track ASL gestures (Motivated by Yang and Ahuja (1999)).

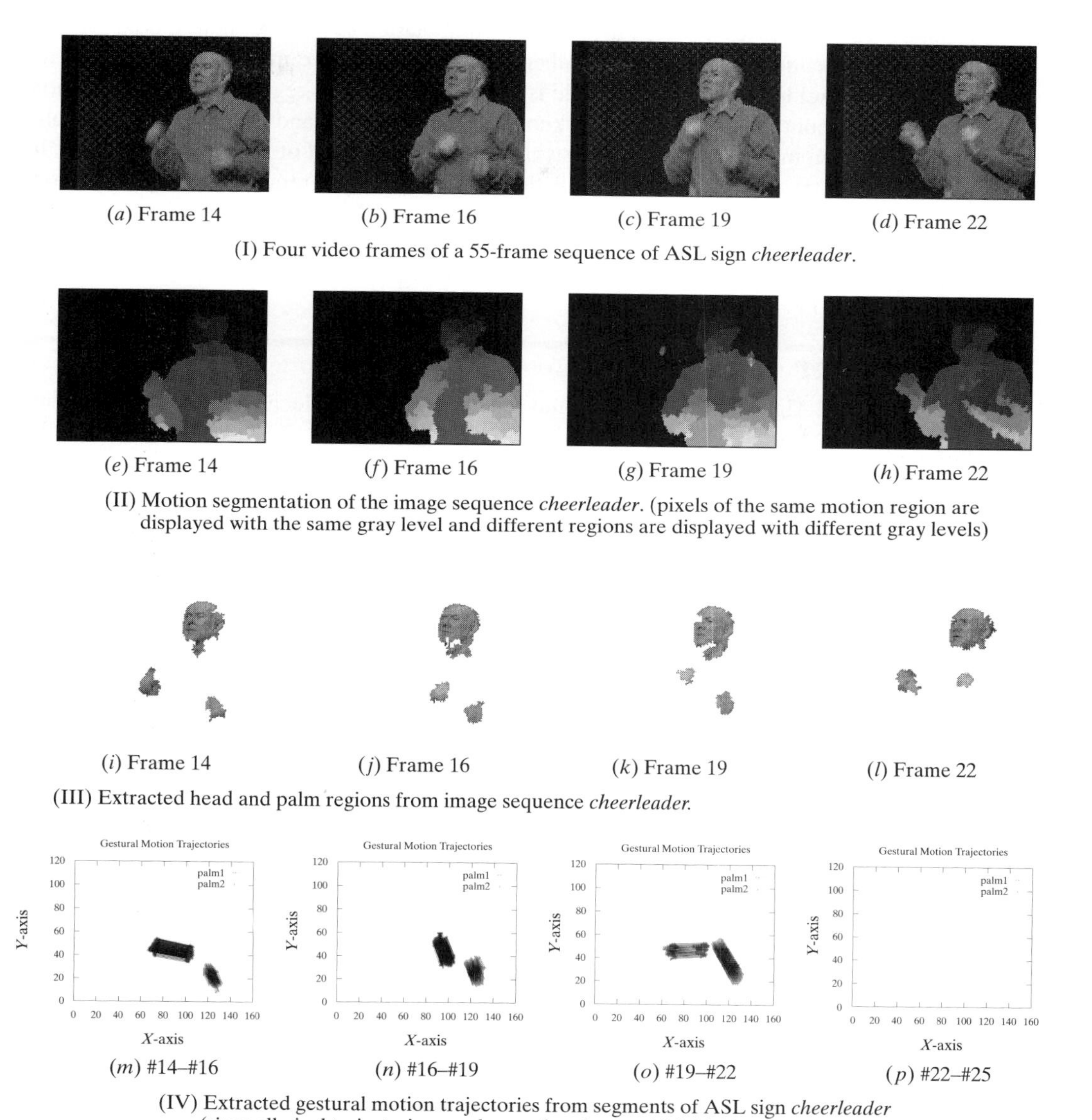

(*a*) Frame 14 (*b*) Frame 16 (*c*) Frame 19 (*d*) Frame 22

(I) Four video frames of a 55-frame sequence of ASL sign *cheerleader*.

(*e*) Frame 14 (*f*) Frame 16 (*g*) Frame 19 (*h*) Frame 22

(II) Motion segmentation of the image sequence *cheerleader*. (pixels of the same motion region are displayed with the same gray level and different regions are displayed with different gray levels)

(*i*) Frame 14 (*j*) Frame 16 (*k*) Frame 19 (*l*) Frame 22

(III) Extracted head and palm regions from image sequence *cheerleader.*

(*m*) #14–#16 (*n*) #16–#19 (*o*) #19–#22 (*p*) #22–#25

(IV) Extracted gestural motion trajectories from segments of ASL sign *cheerleader* (since all pixel trajectories are shown, they form a thick blob)

Figure 10.31 Extraction of motion trajectories from image sequence. (I) Sample frames from video sequence; (II) frames segmented using motion; (III) head and palm regions extracted by color and size; and (IV) motion trajectories for points on the palms. (Figures courtesy of Ming-Hsuan Yang and Narendra Ahuja.)

sequences. Color segmentation is done for each image and the segments are matched across frames. These matches are used to guide the computation of a dense motion field for each pair of consecutive images, which is then segmented to derive motion trajectories at the individual pixel level. This motion field is then segmented into regions of uniform motion. Only at this point do we add domain knowledge to identify hands and face: A skin color model, as seen in Chapter 6, identifies skin regions, the largest of which is taken to be the face. The center of each hand region is then tracked over all frames: The two trajectories can then be used to recognize the sign made. Yang and Ahuja (1999) reported recognition rates of over 90 percent in experiments with many instances of a set of 40 American Sign Language signs.

Exercise 10.22

Describe how Algorithm 10.10 can be modified to make it simpler and faster by specializing all of its steps for the ASL application.

Exercise 10.23

Suppose we have two motion trajectories $\mathbf{P}_j$, $j = 1, N$ and $\mathbf{Q}_k$, $k = 1, M$, where $\mathbf{P}_j$ and $\mathbf{Q}_k$ are points in 2D in proper time sequence. Devise an algorithm for matching two such trajectories, such that 1.0 is output when both trajectories are identical and 0.0 is output when they are very different. Note the M and N may not be equal.

10.7 REFERENCES

Segmentation is one of the oldest, and still unsolved, areas of computer vision. The 1985 survey by Haralick and Shapiro gives a good overview of the early work, most of which used gray-tone images. The first useful segmentation work with natural color images was done by Ohlander, Price, and Reddy in 1978. Only in recent years has the area become fruitful again. The work of Shi and Malik on normalized cuts—starting in 1997—can be credited with being the catalyst for newer work in which segmentations of arbitrary color images from large image collections is being undertaken. In line-drawing analysis, Freeman first proposed his chain code in the 1960s; his 1974 article discusses its use. While the Hough Transform was published only as a patent, it was popularized and expanded by Duda and Hart (1972), and its use is nicely illustrated for line segments in O'Gorman's and Clowes's 1976 paper and for circles in the Kimme, Ballard, and Sklansky 1975 work. The Burns line finder, published ten years later is an improvement to the technique to make it more robust and reliable. Samet's 1990 book on spatial data structures is an excellent reference on quad trees.

Boyer and others (1994) show how to use robust statistics to fit models to data for segmentation. Any anticipated model can be fit to all of the image: Robust fitting can eliminate a huge number of outliers, resulting in a segment of the image where the particular model fits well. An image can be said to be segmented when all anticipated models have been fitted: The segments are comprised of the points that have been fitted.

1. Ballard, D. H., 1981, Generalizing the Hough transform to detect arbitrary shapes. *Pattern Recog.,* v. 13(2):111–122.
2. Boyer, K., K. Mirza, and G. Ganguly. 1994. The robust sequential estimator: a general approach and its application to surface organization in range data, *IEEE Trans. Pattern Analysis and Machine Intelligence,* v. 16(10) (Oct. 1994), 987–1001.
3. Burns, J. R., A. R. Hanson, and E. M. Riseman. 1986. Extracting straight lines. *IEEE Trans. Pattern Analysis and Machine Intelligence*, v. PAMI-8:425–455.
4. Chen, Y., and G. Medioni. 1994. Surface description of complex object from multiple range images. *Proc. IEEE Conf. Comput. Vision and Pattern Recog.,* Seattle, WA (June 1994), 513–518.
5. Duda, R. O., and P. E. Hart. 1972. Use of the Hough transform to detect lines and curves in pictures. *Communications of the ACM,* v. 15:11–15.
6. Freeman, H. 1974. Computer processing of line-drawing images. *Computing Surveys,* v. 6:57–97.
7. Haralick, R. M., and L. G. Shapiro. 1985. Image segmentation techniques. *Comput. Vision, Graphics, and Image Proc.,* v. 29(1) (January 1985), 100–132.
8. Kasturi, R., S. Bow, W. El-Masri, J. Shah, J. Gattiker, and U. Mokate. 1990. A system for interpretation of line drawings. *IEEE Trans. Pattern Analysis and Machine Intelligence,* v. PAMI-12:978–992.
9. Kimme, C., D. Ballard, and J. Sklansky. 1975. Finding circles by an array of accumulators. *Communications of the ACM,* v. 18:120–122.
10. O'Gorman, F., and M. B. Clowes. 1976. Finding picture edges through collinearity of feature points. *IEEE Trans. Comput.,* v. C-25:449–454.
11. Ohlander, R., K. Price, and D. R. Reddy. 1978. Picture segmentation using a recursive region splitting method. *Comput. Graphics and Image Proc.,* v. 8:313–333.
12. Ohta, Y., T. Kanade, and T. Sakai. 1980. Color information for region segmentation. *Comput. Graphics and Image Proc.,* v. 13:222–241.
13. Rao, K. 1988. *Shape Description from Sparse and Imperfect Data.* Ph.D. thesis, Univ. of Southern California.
14. Samet, H. 1990. *Design and Analysis of Spatial Data Structures.* Addison-Wesley, Reading, MA.
15. Shi, J., and J. Malik. 1997. Normalized cuts and image segmentation. *IEEE Conf. Comput. Vision and Pattern Recog.,* 731–737.
16. Sullivan, S., L. Sandford, and J. Ponce. 1994. Using geometric distance for 3D object modeling and recognition. *IEEE Trans. Pattern Analysis and Machine Intelligence,* v. 16(12) (Dec. 1994), 1183–1196.
17. Yang, M.-H., and N. Ahuja. 1999. Recognizing hand gesture using motion trajectories. *Proc. IEEE Conf. Comput. Vision and Pattern Recog. 1999,* Ft. Collins, CO (23–25 June 1999), 466–472.

11

Matching in 2D

This chapter explores how to make and use correspondences between images and maps, images and models, and images and other images. All work is done in two dimensions; methods are extended in Chapter 14 to 3D-2D and 3D-3D matching. There are many immediate applications of 2D matching which do not require the more general 3D analysis.

Consider the problem of taking inventory of land use in a township for the purpose of planning development or collecting taxes. A plane is dispatched on a clear day to take aerial images of all the land in the township. These pictures are then compared to the most recent map of the area to create an updated map. Moreover, other databases are updated to record the presence of buildings, roads, oil wells, etc., or perhaps the kind of crops growing in the various fields. This work can all be done by hand, but is now commonly done using a computer. A second example is from the medical area. The bloodflow in a patient's heart and lungs is to be examined. An X-ray image is taken under normal conditions, followed by one taken after the injection into the bloodstream of a special dye. The second image should reveal the bloodflow, except that there is a lot of noise due to other body structures such as bone. Subtracting the first image from the second will reduce noise and artifact and emphasize only the changes due to the dye. Before this can be done, however, the first image must be *geometrically transformed* or *warped* to compensate for small motions of the body due to body positioning, heart motion, or breathing, etc.

11.1 REGISTRATION OF 2D DATA

A simple general mathematical model applies to all the cases in this chapter and many others not covered. Equation 11.1 and Figure 11.1 show an invertible mapping between points of a model **M** and points of an image **I**. Actually, **M** and **I** can each be any 2D coordinate space

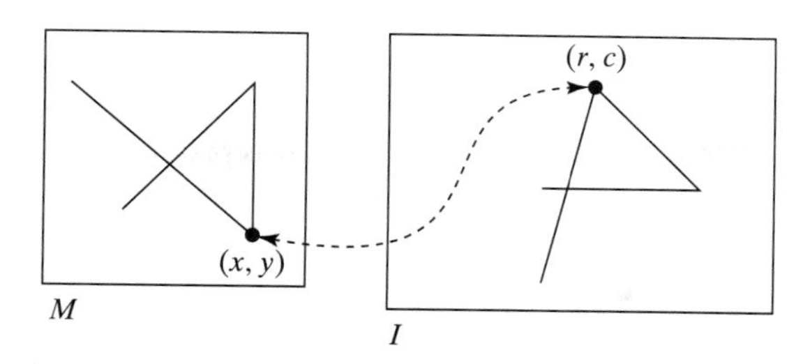

Figure 11.1 A mapping between 2D spaces M and I. M may be a model and I an image, but in general any 2D spaces are possible.

and can each represent a map, model, or image.

$$\begin{aligned} M[x, y] &\cong I[g(x, y), h(x, y)] \\ I[r, c] &\cong M[g^{-1}(r, c), h^{-1}(r, c)] \end{aligned} \tag{11.1}$$

78 Definition. The mapping from one 2D coordinate space to another is called a **2D transformation.**

The type of transformation defined in Equation 11.1 is sometimes called a spatial transformation, geometric transformation, or warp (although to some the term warp is reserved for only nonlinear transformations). The functions g and h create a correspondence between model points $[x, y]$ and image points $[r, c]$ so that a point feature in the model can be located in the image: We assume that the mapping is invertible so that we can go in the other direction using their inverses. Having such mapping functions in the tax record problem allows one to transform property boundaries from a map into an aerial image. The region of the image representing a particular property can then be analyzed for new buildings or crop type, etc. (Currently, the analysis is likely to be done by a human using an interactive graphics workstation.) Having such functions in the medical problem allows the radiologist to analyze the difference image $I_2[r_2, c_2] - I[g(r_2, c_2), h(r_2, c_2)]$: Here the mapping functions register like points in the two images.

79 Definition. **Image registration** is the process by which points of two images from similar viewpoints of essentially the same scene are geometrically transformed so that corresponding feature points of the two images have the same coordinates after transformation.

Another common and important application, although not actually a matching operation, is creation of a new image by collecting sample pixels from another image. For example we might want to cut out a subimage I_2 from an image I_1 as shown in Figure 11.2. Although the content of the new image I_2 is a subset of the original image I_1, it is possible that I_2 can have the same number of pixels as I_1 or even more.

There are several issues of this theory which have practical importance. What is the form of the functions g and h, are they linear, continuous, etc? Are straight lines in one space mapped into straight or curved lines in the other space? Are the distances between point pairs the same in both spaces? More important, how do we use the properties of different functions to achieve the mappings needed? Is the 2D space of the model or image continuous or discrete? If at least one of the spaces is a digital image then

Figure 11.2 (left) Image of a scene containing signage used in Chapter 8 and (right) a new image cut out of the original using a sampling transformation.

quantization effects will impact both accuracy and visual quality. (The quantization effects have deliberately been kept in the right image of Figure 11.2 to demonstrate this point.)

Exercise 11.1

Describe how to enhance the right image of Figure 11.2 to lessen the quantization or aliasing effect.

11.2 REPRESENTATION OF POINTS

In this chapter, we work specifically with points from 2D spaces. Extension of the definitions and results to 3D is done later in Chapter 13; most, but not all, extensions are straightforward. It is good for the student to master the basic concepts and notation before handling the increased complexity sometimes present in 3D. A 2D point has two coordinates and is conveniently represented as either a row vector $\mathbf{P} = [x, y]$ or column vector $\mathbf{P} = [x, y]^t$. The column vector notation will be used in our equations here to be consistent with most engineering books which apply transformations $\mathbf{T}$ on the left to points $\mathbf{P}$ on the right. For convenience, in our text we will often use the row vector form and omit the formal transpose notation t. Also, we will separate coordinates by commas, something that is not needed when a column vector is displayed vertically.

$$\mathbf{P} = [x, y]^t = \begin{bmatrix} x \\ y \end{bmatrix}$$

Sometimes we will have need to label a point according to the type of feature from which it was determined. For example, a point could be the center of a hole, a vertex of a polygon, or the computed location where two extended line segments intersect. Point type will be used to advantage in the automatic matching algorithms discussed later in the chapter.

Reference Frames The coordinates of a point are always relative to some coordinate frame. Often, there are several coordinate frames needed to analyze an environment, as discussed at the end of Chapter 2. When multiple coordinate frames are in use, we may use a special superscript to denote which frame is in use for the coordinates being given for the point.

80 Definition. If $\mathbf{P}_j$ is some feature point and $\mathbf{C}$ is some reference frame, then we denote the **coordinates** of the point relative to the coordinate system as ${}^c\mathbf{P}_j$.

Homogeneous Coordinates As will soon become clear, it is often convenient notationally and for computer processing to use *homogeneous coordinates* for points, especially when affine transformations are used.

81 Definition. The **homogeneous coordinates** of a 2D point $\mathbf{P} = [x, y]^t$ are $[sx, sy, s]^t$, where s is a scale factor, commonly 1.0.

Finally, we need to note the conventions of coordinate systems and programs that display pictures. The coordinate systems in the drawn figures of this chapter are typically plotted as they are in mathematics books with the first coordinate (x or u or even r) increasing to the right from the origin and the second coordinate (y or v or even c) increasing upward from the origin. However, our image display programs display an image of n rows and m columns with the first row (row $r = 0$) at the top and the last row (row $r = n - 1$) at the bottom. Thus r increases from the top downward and c increases from left to right. This presents no problem to our algebra, but may give our intuition trouble at times since the displayed image needs to be mentally rotated counterclockwise 90 degrees to agree with the conventional orientation in a math book.

11.3 AFFINE MAPPING FUNCTIONS

A large class of useful spatial transformations can be represented by multiplication of a matrix and a homogeneous point. Our treatment here is brief but fairly complete: for more details, consult one of the computer graphics texts or robotics texts listed in the references. Properties of vector spaces can be reviewed in Chapter 5.

Scaling A common operation is scaling. Uniform scaling changes all coordinates in the same way, or equivalently changes the size of all objects in the same way. Figure 11.3 shows a 2D point $\mathbf{P} = [1, 2]$ scaled by a factor of 2 to obtain the new point $\mathbf{P}' = [2, 4]$. The same scale factor is applied to the three vertices of the triangle yielding a triangle twice as large. Scaling is a *linear transformation,* meaning that it can be easily represented in terms of the scale factor applied to the two basis vectors for 2D Euclidean space. For example, $[1, 2] = 1[1, 0] + 2[0, 1]$ and $2[1, 2] = 2(1[1, 0] + 2[0, 1]) = 2[1, 0] + 4[0, 1] = [2, 4]$. Equation 11.2 shows how scaling of a 2D point is conveniently represented using multiplication by a simple matrix containing the scale factors on the diagonal. The second case is the general case where the x and y unit vectors are scaled differently and is given in Equation 11.3. Recall the five coordinate frames introduced in Chapter 2: The change of coordinates of real image points expressed in *mm* units to pixel image points expressed in row and column units is one such scaling. In the case of a square pixel camera, $c_x = c_y = c$,

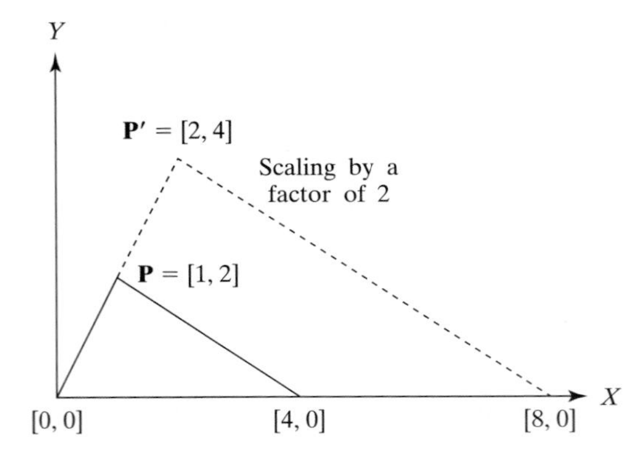

Figure 11.3 Scaling both coordinates of a 2D vector by scale factor 2.

but these constants will be in the ratio of 4/3 for cameras built using TV standards.

$$\begin{bmatrix} x' \\ y' \end{bmatrix} = \begin{bmatrix} c & 0 \\ 0 & c \end{bmatrix} \begin{bmatrix} x \\ y \end{bmatrix} = \begin{bmatrix} cx \\ cy \end{bmatrix} = c \begin{bmatrix} x \\ y \end{bmatrix} \tag{11.2}$$

$$\begin{bmatrix} x' \\ y' \end{bmatrix} = \begin{bmatrix} c_x & 0 \\ 0 & c_y \end{bmatrix} \begin{bmatrix} x \\ y \end{bmatrix} = \begin{bmatrix} c_x x \\ c_y y \end{bmatrix} \tag{11.3}$$

Exercise 11.2: Scaling for a non-square pixel camera.

Suppose a square CCD chip has side 0.5 inches and contains 480 rows of 640 pixels each on this active area. Give the scaling matrix needed to convert pixel coordinates $[r, c]$ to coordinates $[x, y]$ in inches. The center of pixel [0, 0] corresponds to [0, 0] in inches. Using your conversion matrix, what are the integer coordinates of the center of the pixel in row 100, column 200?

Rotation A second common operation is rotation about a point in 2D space. Figure 11.4 (left) shows a 2D point $\mathbf{P} = [x, y]$ rotated by angle θ counterclockwise about the origin to obtain the new point $\mathbf{P'} = [x', y']$. Equation 11.4 shows how rotation of a 2D point about the origin is conveniently represented using multiplication by a simple matrix. As for any linear transformation, we take the columns of the matrix to be the result of the transformation applied to the basis vectors (Figure 11.4 (right)); transformation of any other vector can be expressed as a linear combination of the basis vectors.

$$\begin{aligned} R_\theta([x, y]) &= R_\theta(x[1, 0] + y[0, 1]) \\ &= xR_\theta([1, 0]) + yR_\theta([0, 1]) = x[cos\theta, sin\theta] + y[-sin\theta, cos\theta] \\ &= [xcos\theta - ysin\theta,\ xsin\theta + ycos\theta] \end{aligned}$$

$$\begin{bmatrix} x' \\ y' \end{bmatrix} = \begin{bmatrix} cos\theta & -sin\theta \\ sin\theta & cos\theta \end{bmatrix} \begin{bmatrix} x \\ y \end{bmatrix} = \begin{bmatrix} xcos\theta - ysin\theta \\ xsin\theta + ycos\theta \end{bmatrix} \tag{11.4}$$

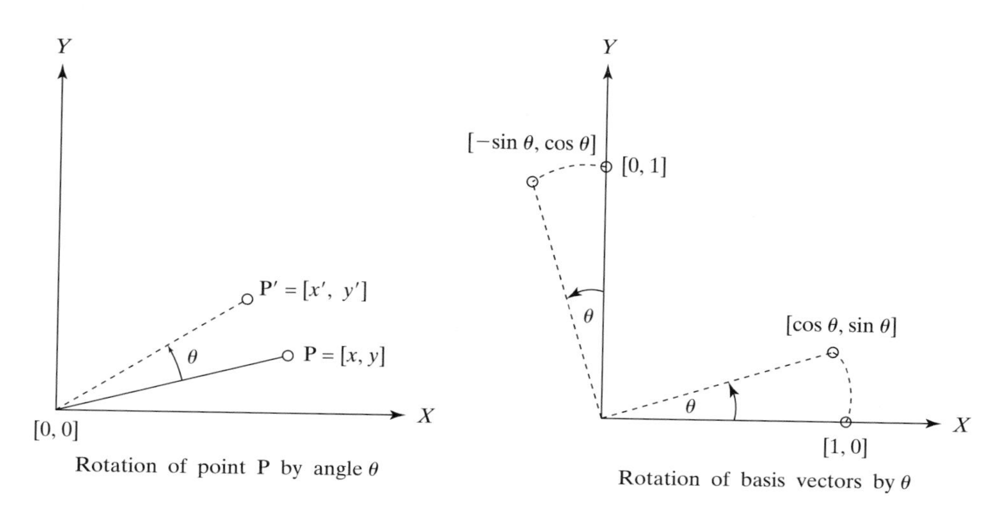

Figure 11.4 Rotation of any 2D point in terms of rotation of the basis vectors.

2D rotations can be made about some arbitrary point in the 2D plane, which need not be the origin of the reference frame. Details are left for a guided exercise later in this section.

Exercise 11.3

(a) Sketch the three points [0, 0], [2, 2], and [0, 2] using an XY coordinate system. (b) Scale these points by 0.5 using Equation 11.2 and plot the results. (c) Using a new plot, plot the result of rotating the three points by 90 degrees about the origin using Equation 11.4. (d) Let the scaling matrix be S and the rotation matrix be R. Let SR be the matrix resulting from multiplying matrix S on the left of matrix R. Is there any difference if we transform the set of three points using SR and RS?

Orthogonal and Orthonormal Transforms*

82 Definition. A set of vectors is said to be **orthogonal** if all pairs of vectors in the set are perpendicular; or equivalently, have scalar product of zero.

83 Definition. A set of vectors is said to be **orthonormal** if it is an orthogonal set and if all the vectors have unit length.

A rotation preserves both the length of the basis vectors and their orthogonality. This can be seen both intuitively and algebraically. As a direct result, the distance between any two transformed points is the same as the distance between the points before transformation. A *rigid transformation* has this same property: A *rigid transformation* is a composition of rotation and translation. Rigid transformations are commonly used for rigid objects or for change of coordinate system. A uniform scaling that is not 1.0 does not preserve length;

however, it does preserve the angle between vectors. These issues are important when we seek properties of objects that are invariant to how they are placed in the scene or how a camera views them.

Translation Often, point coordinates need to be shifted by some constant amount, which is equivalent to changing the origin of the coordinate system. For example, row-column coordinates of a pixel image might need to be shifted to transform to latitude-longitude coordinates of a map. Since translation does not map the origin [0, 0] to itself, we cannot model it using a simple 2×2 matrix as has been done for scaling and rotation: In other words, it is not a linear operation. We can extend the dimension of our matrix to 3×3 to handle translation as well as some other operations: Accordingly, another coordinate is added to our point vector to obtain homogeneous coordinates. Typically, the appended coordinate is 1.0, but other values may sometimes be convenient.

$$\mathbf{P} = [x, y] \simeq [wx, wy, w] = [x, y, 1] \quad \textit{for } w = 1$$

The matrix multiplication shown in Equation 11.5 can now be used to model the translation $\mathbf{D}$ of point $[x, y]$ so that $[x', y'] = \mathbf{D}([x, y]) = [x + x_0, y + y_0]$.

$$\begin{bmatrix} x' \\ y' \\ 1 \end{bmatrix} = \begin{bmatrix} 1 & 0 & x_0 \\ 0 & 1 & y_0 \\ 0 & 0 & 1 \end{bmatrix} \begin{bmatrix} x \\ y \\ 1 \end{bmatrix} = \begin{bmatrix} x + x_0 \\ y + y_0 \\ 1 \end{bmatrix} \tag{11.5}$$

Exercise 11.4: Rotation about a point.

Give the 3×3 matrix that represents a $\pi/2$ rotation of the plane about the point [5, 8]. *Hint:* First derive the matrix $\mathbf{D}_{-5,-8}$ that translates the point [5, 8] to the origin of a new coordinate frame. The matrix which we want will be the combination $\mathbf{D}_{5,8}\ \mathbf{R}_{\pi/2}\ \mathbf{D}_{-5,-8}$. Check that your matrix correctly transforms points [5, 8], [6, 8], and [5, 9].

Exercise 11.5: Reflection about a coordinate axis.

A *reflection* about the y-axis maps the basis vector [1, 0] onto [−1, 0] and the basis vector [0, 1] onto [0, 1]. (a) Construct the matrix representing this reflection. (b) Verify that the matrix is correct by transforming the three points [1, 1], [1, 0], and [2, 1].

Rotation, Scaling and Translation Figure 11.5 shows a common situation: An image **I[r, c]** is taken by a square-pixel camera looking perpendicularly down on a planar workspace **W[x, y]**. We need a formula that converts any pixel coordinates **[r, c]** in units of rows and columns to, say, mm units [x, y]. This can be done by composing a rotation **R**, a scaling **S**, and a translation **D** as given in Equation 11.6 and denoted ${}^w\mathbf{P}_j = \mathbf{D}_{x_0,y_0}\ \mathbf{S}_s\ \mathbf{R}_\theta\ {}^i\mathbf{P}_j$. There are four parameters that determine the mapping between row-column coordinates and x-y coordinates on the workbench; angle of rotation θ, scale factor s that converts pixels to *mm*, and the two displacements x_0, y_0. These four parameters can be obtained from the coordinates of two *control points* $\mathbf{P}_1$ and $\mathbf{P}_2$. These points are determined by clearly marked

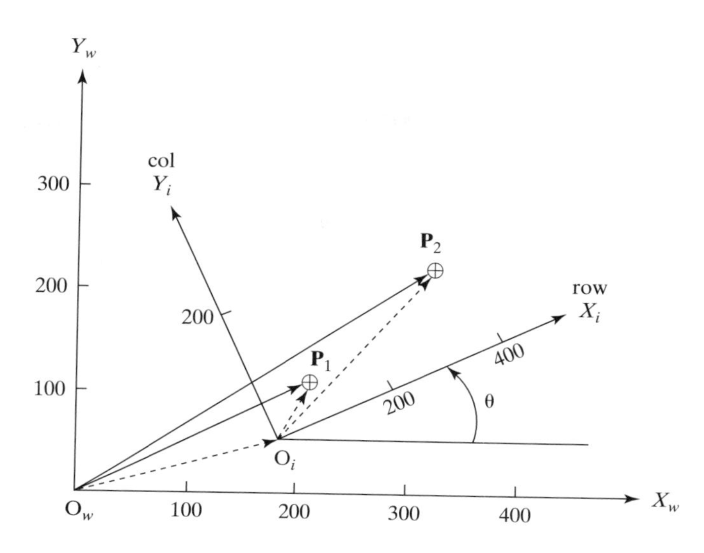

Figure 11.5 Image from a square-pixel camera looking vertically down on a workbench: Feature points in image coordinates need to be rotated, scaled, and translated to obtain workbench coordinates.

and easily measured features in the workspace that are also readily observed in the image—'+' marks, for example. In the land use application, road intersections, building corners, sharp river bends, etc. are often used as control points. It is important to emphasize that the same point feature, say $\mathbf{P}_1$, can be represented by two (or more) distinct vectors, one with row-column coordinates relative to **I** and one with mm x-y coordinates relative to **W**. We denote these representations as ${}^i\mathbf{P}_1$ and ${}^w\mathbf{P}_1$ respectively. For example, in Figure 11.5 we have ${}^i\mathbf{P}_1 = [100, 60]$ and ${}^w\mathbf{P}_1 = [200, 100]$.

84 Definition. Control points are clearly distinguishable and easily measured points used to establish known correspondences between different coordinate spaces.

Given coordinates for point $\mathbf{P}_1$ in both coordinate systems, matrix Equation 11.6 yields two separate equations in the four unknowns.

$$\begin{bmatrix} x_w \\ y_w \\ 1 \end{bmatrix} = \begin{bmatrix} 1 & 0 & x_0 \\ 0 & 1 & y_0 \\ 0 & 0 & 1 \end{bmatrix} \begin{bmatrix} s & 0 & 0 \\ 0 & s & 0 \\ 0 & 0 & 1 \end{bmatrix} \begin{bmatrix} cos\theta & -sin\theta & 0 \\ sin\theta & cos\theta & 0 \\ 0 & 0 & 1 \end{bmatrix} \begin{bmatrix} x_i \\ y_i \\ 1 \end{bmatrix} \tag{11.6}$$

$$x_w = x_i\, s\, cos\,\theta - y_i\, s\, sin\,\theta + x_0 \tag{11.7}$$

$$y_w = x_i\, s\, sin\,\theta + y_i\, s\, cos\,\theta + y_0 \tag{11.8}$$

Using point $\mathbf{P}_2$ yields two more equations and we should be able to solve the system to determine the four parameters of the conversion formula. θ is easily determined independent of the other parameters as follows: First, the direction of the vector $\mathbf{P}_1\mathbf{P}_2$ in **I** is determined as $\theta_i = arctan((^iy_2 - {}^iy_1)/(^ix_2 - {}^ix_1))$. Then, the direction of the vector in $\mathbf{W}$ is determined as $\theta_w = arctan((^wy_2 - {}^wy_1)\,/(^wx_2 - {}^wx_1))$. The rotation angle is just the difference of these two angles: $\theta = \theta_w - \theta_i$. Once θ is determined, all the *sin* and *cos* elements are known: There are 3 equations and 3 unknowns which can easily be solved for s and x_0, y_0. The reader should complete this solution via Exercise 11.6.

Exercise 11.6: Converting image coordinates to workbench coordinates.

Assume an environment as in Figure 11.5. (Perhaps the vision system must inform a pick-and-place robot of the locations of objects.) Give, in matrix form, the transformation that relates image coordinates $[x_i, y_i, 1]$ to workbench coordinates $[x_w, y_w, 1]$. Compute the four parameters using these control points: ${}^i\mathbf{P}_1 = [100, 60]$, ${}^w\mathbf{P}_1 = [200, 100]$; ${}^i\mathbf{P}_2 = [380, 120]$, ${}^w\mathbf{P}_2 = [300, 200]$.

An Example Affine Warp It is easy to extract a parallelogram of data from a digital image by selecting three points. The first point determines the origin of the output image to be created, while the second and third points determine the extreme point of the parallelogram sides. The output image will be a rectangular pixel array of any size constructed from samples from the input image. Figure 11.6 shows results of using a program based on this idea. To create the center image of the figure, the three selected points defined nonorthogonal axes, thus creating shear in the output; this shear was removed by extracting a third image from the center image and aligning the new sampling axes with the skewed axes. Figure 11.3 shows another example: A distorted version of Andrew Jackson's head has been extracted from an image of a U.S. $20 bill (see Figure 11.7). In both of these examples, although only a portion of the input image was extracted, the output image contains the same number of pixels.

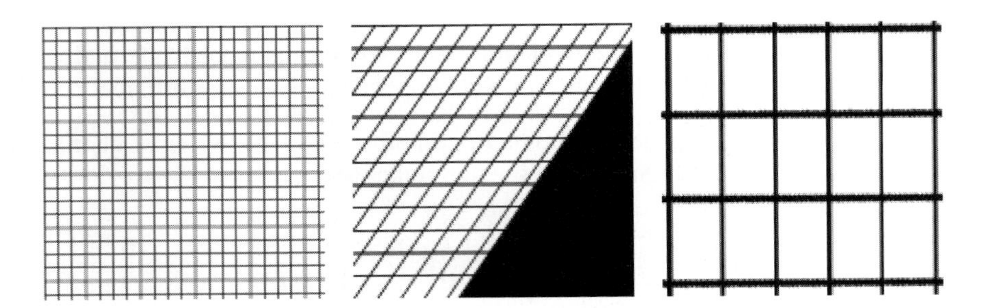

Figure 11.6 (left) 128 × 128 digital image of grid; (center) 128 × 128 image extracted by an affine warp defined by three points in the left image; and (right) 128 × 128 rectified version of part of the center image.

The program that created Figure 11.7 used the three user selected points to transform the input parallelogram defined at the left of the figure. The output image was $n \times m$ or 512×512 pixels with coordinates $[r, c]$; for each pixel $[r, c]$ of the output, the input image value was sampled at pixel $[x, y]$ computed using the transformation in Equation 11.9. The

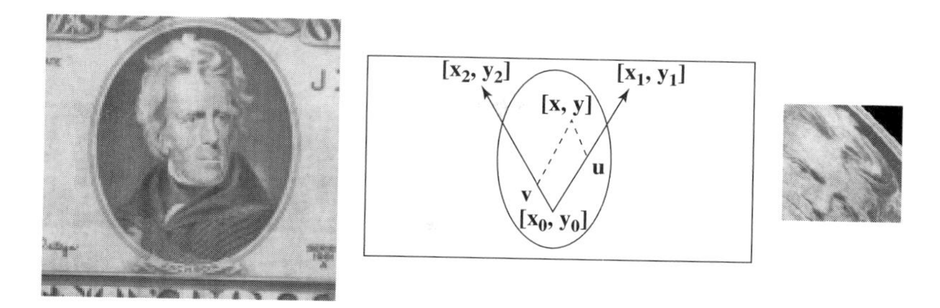

Figure 11.7 Distorted face of Andrew Jackson extracted from a U.S. \$20 bill by defining an affine mapping with shear.

first form of the equation is the intuitive construction in terms of basis vectors, while the second is its equivalent in standard form.

$$\begin{bmatrix} x \\ y \end{bmatrix} = \begin{bmatrix} x_0 \\ y_0 \end{bmatrix} + \frac{r}{n}\left(\begin{bmatrix} x_1 \\ y_1 \end{bmatrix} - \begin{bmatrix} x_0 \\ y_0 \end{bmatrix}\right) + \frac{c}{m}\left(\begin{bmatrix} x_2 \\ y_2 \end{bmatrix} - \begin{bmatrix} x_0 \\ y_0 \end{bmatrix}\right)$$

$$\begin{bmatrix} x \\ y \\ 1 \end{bmatrix} = \begin{bmatrix} (x_1 - x_0)/n & (x_2 - x_0)/m & x_0 \\ (y_1 - y_0)/n & (y_2 - y_0)/m & y_0 \\ 0 & 0 & 1 \end{bmatrix} \begin{bmatrix} r \\ c \\ 1 \end{bmatrix} \tag{11.9}$$

Conceptually, the point $[x, y]$ is defined in terms of the new unit vectors along the new axes defined by the user selected points. The computed coordinates $[x, y]$ must be rounded to get integer pixel coordinates to access digital image ${}^1\mathbf{I}$. If either x or y are out of bounds, then the output point is set to black, in this case ${}^2\mathbf{I}[r, c] = 0$; otherwise ${}^2\mathbf{I}[r, c] = {}^1\mathbf{I}[x, y]$. One can see a black triangle at the upper right of Jackson's head because the sampling parallelogram protrudes above the input image of the \$20 bill image.

Object Recognition and Location Example Consider the example of computing the transformation matching the model of an object shown at the left in Figure 11.8 to the object in the image shown at the right of the figure. Assume that automatic feature extraction produced only three of the holes in the object. The spatial transformation will map points $[x, y]$ of the model to points $[u, v]$ of the image. Assume that we have a controlled imaging environment and the known scale factor has already been applied to the image coordinates to produce the u-v coordinates shown. Only two image points are needed in order to derive the rotation and translation that will align all model points with corresponding image points. Point locations in the model and image and interpoint distances are shown in Tables 11.1 and 11.2. We will use the hypothesized correspondences (A, H_2) and (B, H_3) to deduce the transformation. Note that these correspondences are consistent with the known interpoint distances. We will discuss algorithms for making such hypotheses in Section 11.5.

The direction of the vector from A to B in the model is $\theta_1 = arctan(9.0/8.0) = 0.844$ and the heading of the corresponding vector from H_2 to H_3 in the image is $\theta_2 = arctan(12.0/0.0) = \pi/2 = 1.571$. The rotation is thus $\theta = 0.727$ radians. Using Equation 11.6 and substituting the known matching coordinates for points A in the model and H_2 in the image, we obtain the following system, where u_0, v_0 are the unknown translation

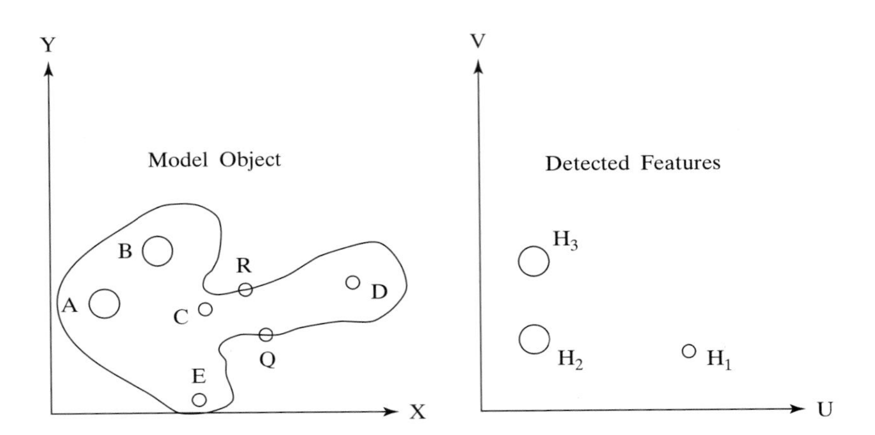

Figure 11.8 (left) Model object and (right) three holes detected in an image.

TABLE 11.1 MODEL POINT LOCATIONS AND INTERPOINT DISTANCES († COORDINATES ARE FOR THE CENTER OF HOLES)

Point	Coordinates †	to A	to B	to C	to D	to E
A	(8,17)	0	12	15	37	21
B	(16,26)	12	0	12	30	26
C	(23,16)	15	12	0	22	15
D	(45,20)	37	30	22	0	30
E	(22,1)	21	26	15	30	0

TABLE 11.2 IMAGE POINT LOCATIONS AND INTERPOINT DISTANCES († COORDINATES ARE FOR THE CENTER OF HOLES)

Point	Coordinates †	to H_1	to H_2	to H_3
H_1	(31,9)	0	21	26
H_2	(10,12)	21	0	12
H_3	(10,24)	26	12	0

components in the image plane. Note that the values of $sin\,\theta$ and $cos\,\theta$ are actually known since θ has been computed.

$$\begin{bmatrix} u \\ v \\ 1 \end{bmatrix} = \begin{bmatrix} 10 \\ 12 \\ 1 \end{bmatrix} = \begin{bmatrix} cos\theta & -sin\theta & u_0 \\ sin\theta & cos\theta & v_0 \\ 0 & 0 & 1 \end{bmatrix} \begin{bmatrix} 8 \\ 17 \\ 1 \end{bmatrix} \tag{11.10}$$

The two resulting linear equations readily produce $u_0 = 15.3$ and $v_0 = -5.95$. As a check, we can use the matching points B and H_3, obtaining a similar result. Each distinct pair of points will result in a slightly different transformation. Methods of using many more points to obtain a transformation that is more accurate across the entire 2D space are discussed later in this chapter. Having a complete spatial transformation, we can now compute the location of any model points in the image space, including the grip points $R = [29, 19]$ and $Q = [32, 12]$. As shown next, model point R transforms to image point ${}^iR = [24.4, 27.4]$: Using $Q = [32, 12]$ as the input to the transformation outputs the image location ${}^iQ = [31.2, 24.2]$ for the other grip point.

$$\begin{bmatrix} u_R \\ v_R \\ 1 \end{bmatrix} = \begin{bmatrix} 24.4 \\ 27.4 \\ 1 \end{bmatrix} = \begin{bmatrix} cos\theta & -sin\theta & 15.3 \\ sin\theta & cos\theta & -5.95 \\ 0 & 0 & 1 \end{bmatrix} \begin{bmatrix} 29 \\ 19 \\ 1 \end{bmatrix} \quad (11.11)$$

Given this knowledge, a robot could grasp the real object being imaged, provided that it had knowledge of the transformation from image coordinates to coordinates of the workbench supporting the object. Of course, a robot gripper would be opened a little wider than the transformed length obtained from ${}^iR{}^iQ$ indicates, to allow for small effects such as distortions in imaging, inaccuracies of feature detection, and computational errors. The required gripping action takes place on a real continuous object and real number coordinates make sense despite the fact that the image holds only discrete spatial samples. The image data itself is only defined at integer grid points. If our action were to verify the presence of holes ${}^iC$ and ${}^iD$ by checking for a bright image pixel, then the transformed model points should be rounded to access an image pixel. Or, perhaps an entire digital neighborhood containing the real transformed coordinates should be examined. With this example, we have seen the potential for recognizing 2D objects by aligning a model of the object with important feature points in its image.

85 Definition. Recognizing an object by matching transformed model features to image features via a rotation, scaling, and translation (RST) is called **recognition-by-alignment.**

Exercise 11.7: Are transformations commutative?

Suppose we have matrices for three primitive transformations: $\mathbf{R}_\theta$ for a rotation about the origin, $\mathbf{S}_{s_x,s_y}$ for a scaling, and $\mathbf{D}_{x_0,y_0}$ for a translation. (a) Do scaling and translation commute; that is, does $\mathbf{S}_{s_x,s_y}\ \mathbf{D}_{x_0,y_0} = \mathbf{D}_{x_0,y_0}\ \mathbf{S}_{s_x,s_y}$? (b) Do rotation and scaling commute; that is, does $\mathbf{R}_\theta\ \mathbf{S}_{s_x,s_y} = \mathbf{S}_{s_x,s_y}\ \mathbf{R}_\theta$? (c) Same question for rotation and translation. (d) Same question for scaling and translation. Do both the algebra and intuitive thinking to derive your answers and explanations.

Exercise 11.8

Construct the matrix for a reflection about the line $y = 3$ by composing a translation with $y_0 = -3$ followed by a reflection about the x-axis. Verify that the matrix is correct by transforming the three points [1, 1], [1, 0], and [2, 1] and plotting the input and output points.

Exercise 11.9

Verify that the product of the matrices $\mathbf{D}_{x_0,y_0}$ and $\mathbf{D}_{-x_0,-y_0}$ is the 3×3 identity matrix. Explain why this should be the case.

General Affine Transformations* We have already covered affine transformation components of rotation, scaling, and translation. A fourth component is shear. Figure 11.9 shows the effect of shear. Using u-v coordinates, point vectors move along the v-axis in proportion to their distance from the v-axis. The point $[u, v]$ is transformed to $[u, e_u u + v]$ with v-axis shear and to $[u + e_v v, v]$ with u-axis shear. The matrix equations are given in Equation 11.12 and Equation 11.13. Recall that the column vectors of the shear matrix are just the images of the basis vectors under the transformation.

$$\begin{bmatrix} x \\ y \\ 1 \end{bmatrix} = \begin{bmatrix} 1 & 0 & 0 \\ e_u & 1 & 0 \\ 0 & 0 & 1 \end{bmatrix} \begin{bmatrix} u \\ v \\ 1 \end{bmatrix} \tag{11.12}$$

$$\begin{bmatrix} x \\ y \\ 1 \end{bmatrix} = \begin{bmatrix} 1 & e_v & 0 \\ 0 & 1 & 0 \\ 0 & 0 & 1 \end{bmatrix} \begin{bmatrix} u \\ v \\ 1 \end{bmatrix} \tag{11.13}$$

Reflections are a fifth type of component. A reflection about the u-axis maps the basis vectors $[1, 0], [0, 1]$ onto $[1, 0], [0, -1]$ respectively, while a reflection about the v-axis maps $[1, 0], [0, 1]$ onto $[-1, 0], [0, 1]$. The 2×2 or 3×3 matrix representation is straightforward. Any affine transformation can be constructed as a composition of any of the component types—rotations, scaling, translation, shearing, or reflection. Inverses of these components exist and are of the same type. Thus, it is clear that the matrix of a general affine transformation composed of any components has six parameters as shown in Equation 11.14. These six parameters can be determined using 3 sets of noncolinear points known to be in correspondence by solving 3 matrix equations of this type. We have already

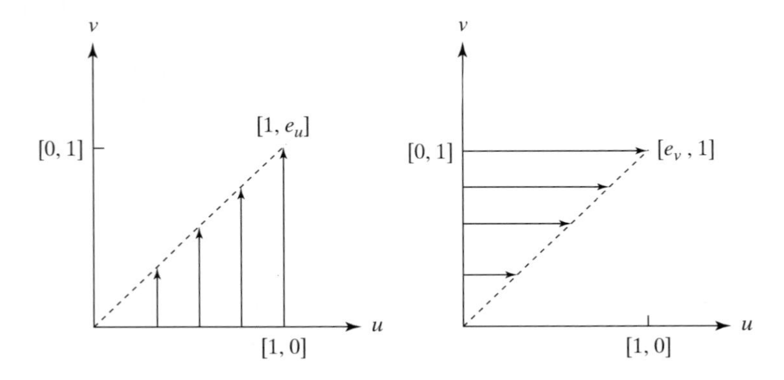

Figure 11.9 (Left) v-axis shear and (right) u-axis shear.

seen an example using shear in the case of the slanted grid in Figure 11.6.

$$\begin{bmatrix} x \\ y \\ 1 \end{bmatrix} = \begin{bmatrix} a_{11} & a_{12} & a_{13} \\ a_{21} & a_{22} & a_{23} \\ 0 & 0 & 1 \end{bmatrix} \begin{bmatrix} u \\ v \\ 1 \end{bmatrix} \tag{11.14}$$

11.4 A BEST 2D AFFINE TRANSFORMATION*

A general affine transformation from 2D to 2D as in Equation 11.15 requires six parameters and can be computed from only 3 matching pairs of points $([x_j, y_j], [u_j, v_j])_{j=1,3}$.

$$\begin{bmatrix} u \\ v \\ 1 \end{bmatrix} = \begin{bmatrix} a_{11} & a_{12} & a_{13} \\ a_{21} & a_{22} & a_{23} \\ 0 & 0 & 1 \end{bmatrix} \begin{bmatrix} x \\ y \\ 1 \end{bmatrix} \tag{11.15}$$

Error in any of the coordinates of any of the points will surely cause error in the transformation parameters. A much better approach is to use many more pairs of matching control points to determine a least-squares estimate of the six parameters. We can define an error criteria function similar to that used for fitting a straight line in Chapter 10.

$$\begin{aligned} \varepsilon(a_{11}, a_{12}, a_{13}, a_{21}, a_{22}, a_{23}) = \sum_{j=1}^{n} ((a_{11}x_j + a_{12}y_j + a_{13} - u_j)^2 \\ + (a_{21}x_j + a_{22}y_j + a_{23} - v_j)^2) \end{aligned} \tag{11.16}$$

Taking the six partial derivatives $\partial\varepsilon/a_{ij}$ of the error function with respect to each of the six variables a_{ij} and setting this expression to zero gives us the six equations represented in matrix form in Equation 11.17.

$$\begin{bmatrix} \Sigma x_j^2 & \Sigma x_j y_j & \Sigma x_j & 0 & 0 & 0 \\ \Sigma x_j y_j & \Sigma y_j^2 & \Sigma y_j & 0 & 0 & 0 \\ \Sigma x_j & \Sigma y_j & \Sigma 1 & 0 & 0 & 0 \\ 0 & 0 & 0 & \Sigma x_j^2 & \Sigma x_j y_j & \Sigma x_j \\ 0 & 0 & 0 & \Sigma x_j y_j & \Sigma y_j^2 & \Sigma y_j \\ 0 & 0 & 0 & \Sigma x_j & \Sigma y_j & \Sigma 1 \end{bmatrix} \begin{bmatrix} a_{11} \\ a_{12} \\ a_{13} \\ a_{21} \\ a_{22} \\ a_{23} \end{bmatrix} = \begin{bmatrix} \Sigma u_j x_j \\ \Sigma u_j y_j \\ \Sigma u_j \\ \Sigma v_j x_j \\ \Sigma v_j y_j \\ \Sigma v_j \end{bmatrix} \tag{11.17}$$

Exercise 11.10

Solve Equation 11.17 using the following three pairs of matching control points: ([0, 0], [0, 0]), ([1, 0],[0, 2]), ([0, 1], [−2, 0]). Do your computations give the same answer as reasoning about the transformation using basis vectors?

Exercise 11.11

Solve Equation 11.17 using the following three pairs of matching control points: ([0, 0], [1, 2]), ([1, 0],[3, 2]), ([0, 1], [1, 4]). Do your computations give the same answer as reasoning about the transformation using basis vectors?

It is common to use many control points to put an image and map or two images into correspondence. Figure 11.10 shows two images of approximately the same scene. Eleven pairs of matching control points are given at the bottom of the figure. Control points are

```
======Best 2D Affine Fit Program======
Matching control point pairs are:
288 210  31 160    232 288  95 205    195 372 161 229    269 314 112 159
203 424 199 209    230 336 130 196    284 401 180 124    327 428 198  69
284 299 100 146    337 231  45 101    369 223  38  64

The Transformation Matrix is:

[ -0.0414 ,  0.773 , -119
  -1.120  , -0.213 ,  526 ]

Residuals (in pixels) for 22 equations are as follows:

0.18  -0.68  -1.22  0.47  -0.77  0.06   0.34  -0.51   1.09   0.04  0.96
1.51  -1.04  -0.81  0.05   0.27  0.13  -1.12   0.39  -1.04  -0.12  1.81

======Fitting Program Complete======
```

Figure 11.10 Images of same scene and best affine mapping from the left image into the right image using 11 control points. $[x, y]$ coordinates for the left image with x increasing downward and y increasing to the right; $[u, v]$ coordinates for the right image with u increasing downward and v toward the right. The 11 clusters of coordinates directly below the images are the matching control points x, y, u, v. Can you match features across the two images? (Images courtesy of Oliver Faugeras.)

corners of objects that are uniquely identifiable in both images (or map). In this case, the control points were selected using a display program and mouse. The list of residuals shows that, using the derived transformation matrix, no u or v coordinate in the right image will be off by two pixels from the transformed value. Most residuals are less than one pixel. Better results can be obtained by using automatic feature detection which locates feature points with subpixel accuracy: Control point coordinates are often off by one pixel when chosen using a computer mouse and the human eye. Using the derived affine transformation, the right image can be searched for objects known to be in the left image. Thus we have reached the point of understanding how the tax-collector's map can be put into correspondence with an aerial image for the purpose of updating its inventory of objects.

Exercise 11.12

Take three pairs of matching control points from Figure 11.10 (for example, ([288, 210, 1], [31, 160, 1])) and verify that the affine transformation matrix maps the first into the second.

11.5 2D OBJECT RECOGNITION VIA AFFINE MAPPING

In this section we study a few methods of recognizing 2D objects through mappings of model points onto image points. We have already introduced one method of recognition-by-alignment in the section on affine mappings. The general methods work with general point features; however, each application domain will present distinguishing features which allow labels to be attached to feature points. Thus we might have corners or centers of holes in a part sorting application, or intersections and high curvature land and water boundary points in a land use application.

Figure 11.11 illustrates the overall model-matching paradigm. Figure 11.11(a) is a boundary model of an airplane part. Feature points that may be used in matching are indicated with small black circles. Figure 11.11(b) is an image of the real airplane part in approximately the same orientation as the model. Figure 11.11(c) is a second image of the real part rotated about 45 degrees. Figure 11.11(d) is a third image of the real part in which the camera angle causes a large amount of skewing in the resultant image. The methods described in this section are meant to determine if a given image, such as those of Figures 11.11(b), 11.11(c), and 11.11(d) contains an instance of an object model such as that of Figure 11.11(a) and to determine the *pose* (position and orientation) of the object with respect to the camera.

Local-Feature-Focus Method The local-feature-focus method uses local features of an object and their 2D spatial relationships to recognize the object. In advance, a set of object models is constructed, one for each object to be recognized. Each object model contains a set of *focus features,* which are major features of the object that should be easily detected if they are not occluded by some other object. For each focus feature, a set of its nearby features is included in the model. The nearby features can be used to verify that the correct focus feature has been found and to help determine the pose of the object.

In the matching phase, feature extraction is performed on an image of one or more objects. The matching algorithm looks first for focus features. When it finds a focus feature

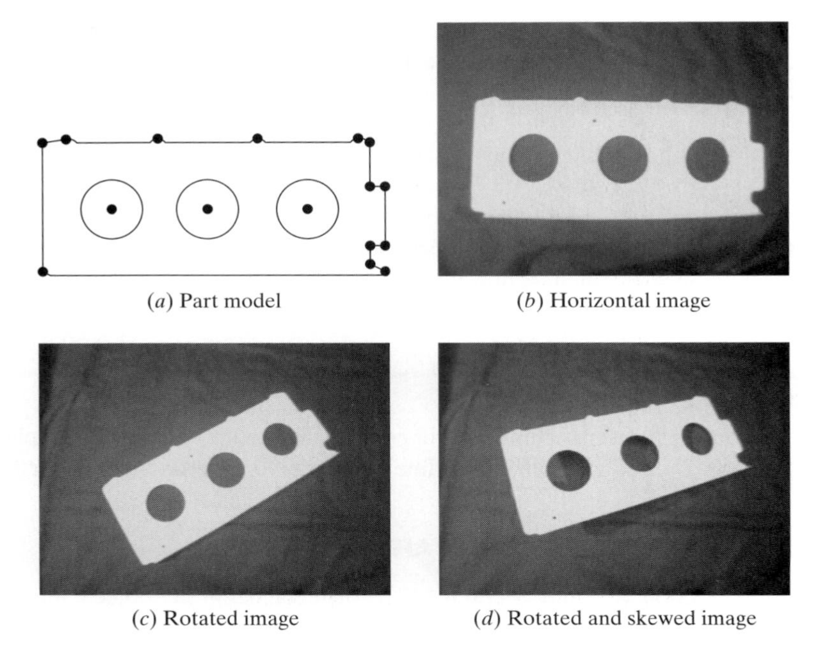
(*a*) Part model
(*b*) Horizontal image
(*c*) Rotated image
(*d*) Rotated and skewed image

Figure 11.11 A 2D model and 3 matching images of an airplane part.

belonging to a given model, it looks for a cluster of image features near the focus feature that match as many as possible of the required nearby features for that focus feature in that model. Once such a cluster has been found and the correspondences between this small set of image features and object model features have been determined, the algorithm hypothesizes that this object is in the image and uses a verification technique to decide if the hypothesis is correct.

The verification procedure must determine whether there is enough evidence in the image that the hypothesized object is in the scene. For polyhedral objects, the boundary of the object is often used as suitable evidence. The set of feature correspondences is used to determine a possible affine transformation from the model points to the image points. This transformation is then used to transform each line segment of the boundary into the image space. The transformed line segments should approximately line up with image line segments wherever the object is unoccluded. Due to noise in the image and errors in feature extraction and matching, it is unlikely that the transformed line segments will exactly align with image line segments, but a rectangular area about each transformed line segment can be searched for evidence of possibly matching image segments. If sufficient evidence is found, that model segment is marked as verified. If enough model segments are verified, the object is declared to be in the image at the location specified by the computed transformation.

The local-feature-focus algorithm for matching a given model $\mathbf{F}$ to an image is given next. The model has a set $\{\mathbf{F}_1, \mathbf{F}_2, \ldots, \mathbf{F}_M\}$ of focus features. For each focus feature $\mathbf{F}_m$, there is a set $\mathbf{S}(\mathbf{F}_m)$ of nearby features that can help to verify this focus feature. The image

Find the transformation from model features to image features using the local-feature-focus approach.

$\mathbf{G_i}$, $\mathbf{i = 1, I}$ is the set of the detected image features.
$\mathbf{F_m}$, $\mathbf{m = 1, M}$ is the set of focus features of the model.
$\mathbf{S(f)}$ is the set of nearby features for any feature $\mathbf{f}$.

```
procedure local_feature_focus(G, F);
{
for each focus feature F_m
  for each image feature G_i of the same type as F_m
    {
      Find the maximal subgraph S_m of S(F_m) that
        matches a subgraph S_i of S(G_i);
      Compute the transformation T that maps the points of
        each feature of S_m to the corresponding feature of S_i;
      Apply T to the boundary segments of the model;
      if enough of the transformed boundary segments find
        evidence in the image then return(T);
    };
}
```

Algorithm 11.1 Local-Feature-Focus Method.

has a set $\{\mathbf{G}_1, \mathbf{G}_2, \ldots, \mathbf{G}_I\}$ of detected image features. For each image feature $\mathbf{G}_i$, there is a set of nearby image features $\mathbf{S}(\mathbf{G}_i)$.

Figure 11.12 illustrates the local-feature-focus method with two models, E and F, and an image. The detected features are circular holes and sharp corners. Local feature F1 of model F has been hypothesized to correspond to feature G1 in the image. Nearby features

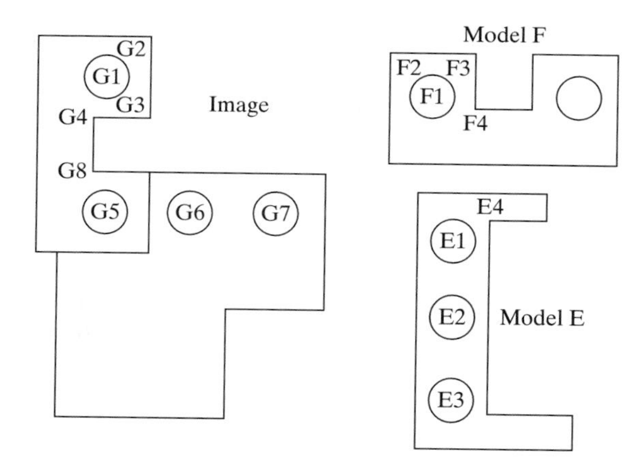

Figure 11.12 The Local-Feature-Focus Method. The image shows an instance of Model F on top of another object.

F2, F3, and F4 of the model have been found to correspond well to nearby features G2, G3, and G4, respectively, of the image. The verification step will show that model F is indeed in the image. Considering the other model E, feature E1 and the set of nearby features E2, E3, and E4 have been hypothesized to correspond to features G5, G6, G7, and G8 in the image. However, when verification is performed, the boundary of model E will not line up well with the image segments, and this hypothesis will be discarded.

Pose Clustering We have seen that an alignment between model and image features using an RST transformation can be obtained from two matching control points. The solution can be obtained using Equation 11.6 once two control points have been matched between image and model. Obtaining the matching control points automatically may not be easy due to ambiguous matching possibilities. The pose-clustering approach computes an RST alignment for all possible control point pairs and then checks for a cluster of similar parameter sets. If indeed there are many matching feature points between model and image, then a cluster should exist in the parameter space. A pose-clustering algorithm is sketched below.

86 Definition. Let T be a spatial transformation aligning model M to an object O in image I. The **pose** of object O is its location and orientation as defined by the parameters α of T.

Find the transformation from model features to image features using pose clustering.

$\mathbf{P_i}$, $\mathbf{i = 1, D}$ is the set of detected image features.
$\mathbf{L_j}$, $\mathbf{j = 1, M}$ is the set of stored model features.

```
procedure pose_clustering(P, L);
{
for each pair of image feature points (P_i, P_j)
   for each pair of model feature points (L_m, L_n) of same type
      {
         compute parameters α of RST mapping
            pair (L_m, L_n) onto (P_i, P_j);
         contribute α to the cluster space;
      };
examine space of all candidates α for clusters;
verify every large cluster by mapping all
   model feature points and checking the image;
return(verified {α_k});
}
```

Algorithm 11.2 Pose-Clustering Algorithm.

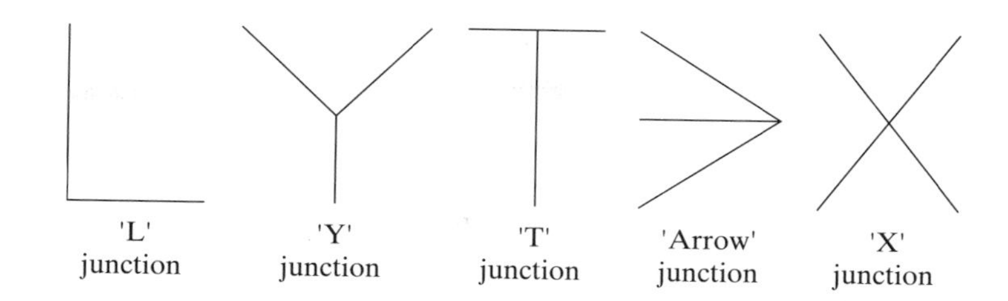

Figure 11.13 Common line-segment junctions used in matching.

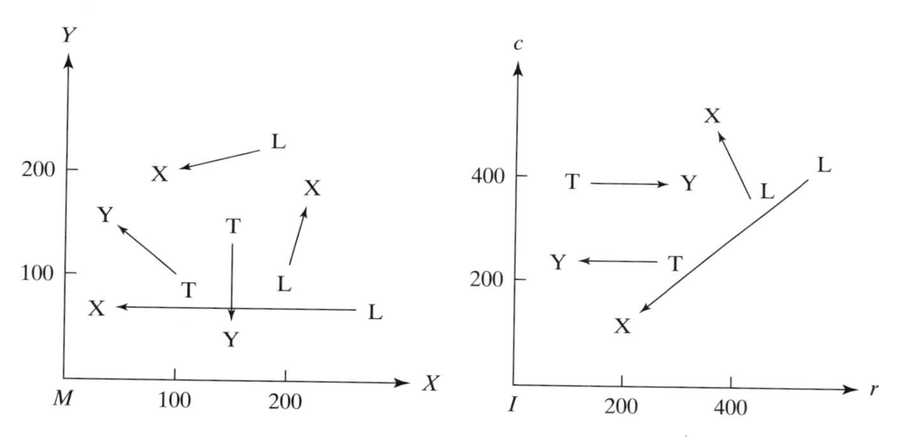

Figure 11.14 Example pose detection problem with 5 model feature point pairs and 4 image feature point pairs.

Using all possible pairs of feature points would provide too much redundancy. An application in matching aerial images to maps can use intersection points detected on road networks or at the corners of regions such as fields. The degree of the intersection gives it a type to be used in matching; for example, common intersections have type 'L', 'Y', 'T', 'Arrow', and 'X' as shown in Figure 11.13. Assume that we use only the pairs of combined type LX or TY. Figure 11.14 shows an example with 5 model pairs and 4 image pairs. Although there are $4 \times 5 = 20$ possible pairings, only 10 of them have matching types for both points. The transformations computed from each of those are given in Table 11.3. The 10 transformations computed have scattered and inconsistent parameter sets, except for 3 of them indicated by asterisks (*) in the last column of the table. These 3 parameter sets form a cluster whose average parameters are $\theta = 0.68, s = 2.01, u_0 = 233, v_0 = -41$. While one would like less variance than this for correct matches, this variance is typical due to slight errors in point feature location and small nonlinear distortions in the imaging process. If the parameters of the RST mapping are inaccurate, they can be used to verify matching points which can then be used as control points to find a nonlinear mapping or affine mapping (with more parameters) that is more accurate in matching control points.

TABLE 11.3 CLUSTER SPACE FORMED FROM 10 POSE COMPUTATIONS FROM FIGURE 11.14

Model Pair	Image Pair	θ	s	u_0	v_0	
L(170,220),X(100,200)	L(545,400),X(200,120)	0.403	6.10	118	−1240	
L(170,220),X(100,200)	L(420,370),X(360,500)	5.14	2.05	−97	514	
T(100,100),Y(40,150)	T(260,240),Y(100,245)	0.663	2.05	225	−48	*
T(100,100),Y(40,150)	T(140,380),Y(300,380)	3.87	2.05	166	669	
L(200,100),X(220,170)	L(545,400),X(200,120)	2.53	6.10	1895	200	
L(200,100),X(220,170)	L(420,370),X(360,500)	0.711	1.97	250	−36	*
L(260, 70),X(40, 70)	L(545,400),X(200,120)	0.682	2.02	226	−41	*
L(260, 70),X(40, 70)	L(420,370),X(360,500)	5.14	0.651	308	505	
T(150,125),Y(150, 50)	T(260,240),Y(100,245)	4.68	2.13	3	568	
T(150,125),Y(150, 50)	T(140,380),Y(300,380)	1.57	2.13	407	60	

Pose-clustering can work using low-level features; however, both accuracy and efficiency are improved when features can be filtered by type. Clustering can be performed by a simple $O(n^2)$ algorithm: For each parameter set α, count the number of other parameter sets α_i that are close to it using some permissible distance. This requires $n-1$ distance computations for each of the n parameter sets in cluster space. A faster, but less flexible, alternative is to use binning. Binning has been the traditional approach reported in the literature and was discussed in Chapter 10 with respect to the Hough transform. Each parameter set produced is contributed to a bin in parameter space, after which all bins are examined for significant counts. A cluster can be lost when a set of similar α_i spreads over neighboring bins.

The clustering approach has been used to detect the presence of a particular model of airplane from an aerial image, as shown in Figure 11.15. Edge and curvature features are extracted from the image using the methods of Chapters 5 and 10. Various overlapping windows of these features are matched against the model shown in part (*b*) of the figure. Part (*c*) of the figure shows the edges detected in one of these windows where many of the features aligned with the model features using the same transformation parameters.

Geometric Hashing Both the local-feature-focus method and the pose-clustering algorithm were designed to match a single model to an image. If several different object models were possible, then these methods would try each model, one at a time. This makes them less suited for problems in which a large number of different objects can appear. Geometric hashing was designed to work with a large database of models. It trades a large amount of offline preprocessing and a large amount of space for a potentially fast online object recognition and pose determination.

Suppose we are given

1. a large database of models, and
2. an unknown object whose features are extracted from an image and which is known to be an affine transformation of one of the models,

and we wish to determine which model it is and what pose transformation was applied.

(a) Original airfield image

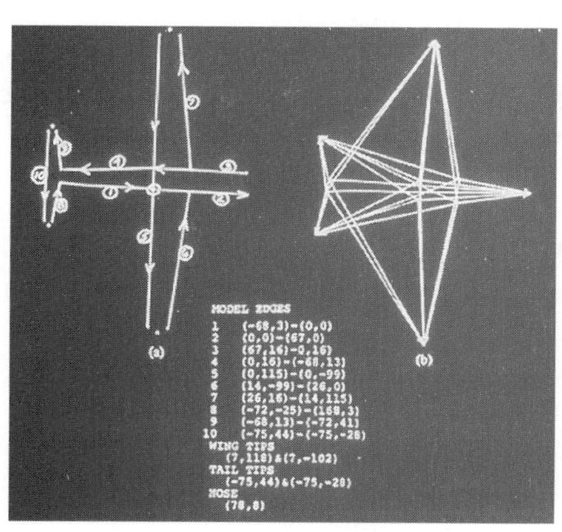

(b) Model of object

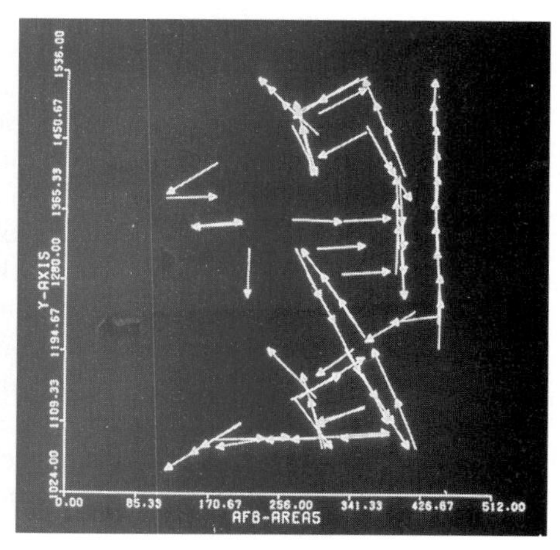

(c) Detections matching model

Figure 11.15 Pose-clustering applied to detection of a particular airplane. (a) Aerial image of an airfield; (b) object model in terms of real edges and abstract edges subtending one corner and one curve tip point; and (c) image window containing detections that match many model parts via the same transformation. Reprinted by permission of IEEE.

Consider a model $\mathbf{M}$ to be an ordered set of feature points. Any subset of three non-collinear points $E = \{e_{00}, e_{01}, e_{10}\}$ of $\mathbf{M}$ can be used to form an affine basis set, which defines a coordinate system on $\mathbf{M}$, as shown in Figure 11.16(a). Once the coordinate system is chosen, any point $x \in \mathbf{M}$ can be represented in affine coordinates (ξ, η) where

$$x = \xi(e_{10} - e_{00}) + \eta(e_{01} - e_{00}) + e_{00}$$

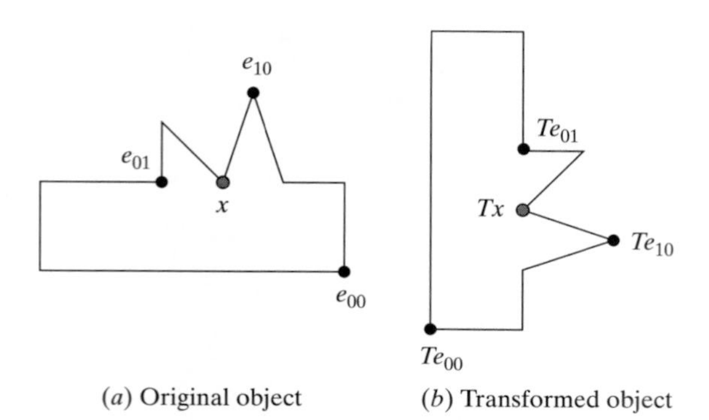

Figure 11.16 The affine transformation of a point with respect to an affine basis set.

Furthermore, if we apply an affine transform T to point x, we get

$$Tx = \xi(Te_{10} - Te_{00}) + \eta(Te_{01} - Te_{00}) + Te_{00}$$

Thus Tx has the same affine coordinates (ξ, η) with respect to $(Te_{00}, Te_{01}, Te_{10})$ as x has with respect to (e_{00}, e_{01}, e_{10}). This is illustrated in Figure 11.16(b).

Offline Preprocessing The offline preprocessing step creates a hash table containing all of the models in the database. The hash table is set up so that the pair of affine coordinates (ξ, η) indexes a bin of the hash table that stores a list of model-basis pairs **(M, E)** where some point x of model **M** has affine coordinates (ξ, η) with respect to basis **E**. The offline preprocessing algorithm is given in Algorithm 11.3.

Online Recognition The hash table created in the preprocessing step is used in the online recognition step. The recognition step also uses an accumulator array **A** indexed by model-basis pairs. The bin for each **(M, E)** is initialized to zero and used to vote for the hypothesis that there is a transformation **T** that places **(M, E)** in the image. Computation of the actual transformations is done only for those model-basis pairs that achieve a high number of votes and is part of the verification step that follows the voting. The online recognition and pose estimation algorithm is given below.

Suppose that there are s models of approximately n points each. Then the preprocessing step has complexity $O(sn^4)$ which comes from processing s models, $O(n^3)$ triples per model, and $O(n)$ other points per model. In the matching, the amount of work done depends somewhat on how well the feature points can be found in the image, how many of them are occluded, and how many false or extra feature points are detected. In the best case, the first triple selected consists of three real feature points all from the same model, this model gets a large number of votes, the verification procedure succeeds, and the task is done. In this best case, assuming the average list in the hash table is of a small constant size and that hashing time is approximately constant, the complexity of the matching step is approximately $O(n)$. In the worst case, for instance when the model is not in the database at all, every triple is tried, and the complexity is $O(n^4)$. In practice, although it would be rare to try all bases, it is also rare for only one basis to succeed. A number of different things can go wrong:

1. feature point coordinates have some error,
2. missing and extra feature points,

Set up the hash table for matching to a database of models using geometric hashing.

D is the database of models.
H is an initially empty hash table.

```
procedure GH_Preprocessing(D, H);
{
for each model M
  {
  Extract the feature point set F_M of M;
  for each noncollinear triple E of points from F_M
    for each other point x of F_M
      {
      Calculate (ξ, η) for x with respect to E;
      Store (M, E) in hash table H at index (ξ, η);
      };
  };
}
```

Algorithm 11.3 Geometric Hashing Offline Preprocessing.

3. occlusion, multiple objects,
4. unstable bases, and
5. weird affine transforms on a subset of the points.

In particular, the algorithm can hallucinate a transformation based on a subset of the points that passes the point verification tests, but gives the wrong answer. Figure 11.17 illustrates this point. Pose-clustering and focus feature methods are also susceptible to this same phenomenon.

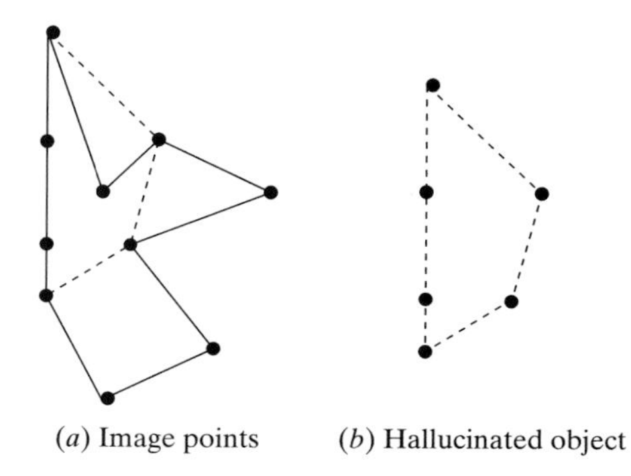

Figure 11.17 The geometric hashing algorithm can hallucinate that a given model is present in an image. In this example, 60 percent of the feature points (left) led to the verified hypothesis of an object (right) that was not actually present in the image.

Use the hash table to find the correct model and transformation that maps image features to model features.

H is the hash table created by the preprocessing step.
A is an accumulator array indexed by **(M, E)** pairs.
I is the image being analyzed.

```
procedure GH_Recognition(H, A, I);
{
Initialize accumulator array A to all zeroes;
Extract feature points from image I;
for each basis triple F
  {
  for each other point v
    {
    Calculate (ξ, η) for v with respect to F;
    Retrieve the list L of model-basis pairs from the
      hash table H at index (ξ, η);
    for each pair (M, E) of L
      A[M, E] = A[M, E] + 1;
    };
  Find the peaks in accumulator array A;
  for each peak (M, E)
    {
    Calculate T such that F = TE;
    if enough of the transformed model points of M find
      evidence on the image then return(T);
    };
  };
}
```

Algorithm 11.4 Geometric Hashing Online Recognition.

11.6 2D OBJECT RECOGNITION VIA RELATIONAL MATCHING

We have previously described three useful methods for matching observed image points to model points: These were local-feature-focus, pose clustering, and geometric hashing. In this section, we examine three simple general paradigms for object recognition within the context given in this chapter. All three paradigms view recognition as a mapping from model structures to image structures: A consistent labeling of image features is sought in terms of model features, recognition is equivalent to mapping a sufficient number of features from a single model to the observed image features. The three paradigms differ in how the mapping is developed.

Four concepts important to the matching paradigms are *parts, labels, assignments, and relations*.

- A **part** is an object or structure in the scene such as region segment, edge segment, hole, corner or blob.
- A **label** is a symbol assigned to a part to identify and, or recognize it at some level.
- An **assignment** is a mapping from parts to labels. If P_1 is a region segment and L_1 is the lake symbol and L_2 the field symbol, an assignment may include the pair (P_1, L_2) or perhaps $(P_1, \{L_1, L_2\})$ to indicate remaining ambiguity. A pairing (P_1, NIL) indicates that P_1 has no interpretation in the current label set. An interpretation of the scene is just the set of all pairs making up an assignment.
- A **relation** is the formal mathematical notion. Relations will be discovered and computed among scene objects and will be stored for model objects. For example, $R4(P_1, P_2)$ might indicate that region P_1 is *adjacent to* region P_2.

Given these four concepts, we can define a consistent labeling.

87 Definition. Given a set of parts P, a set of labels for those parts L, a relation R_P over P, and a second relation R_L over L, a **consistent labeling** f is an assignment of labels to parts that satisfies:

$$\text{If } (p_i, p_{i'}) \in R_P, \quad \text{then } (f(p_i), f(p_{i'})) \in R_L.$$

For example, suppose we are trying to find a match between two images: For each image we have a set of extracted line segments and a connection relation that indicates which pairs of line segments are connected. Let P be the set of line segments and R_P be the set of pairs of connecting line segments from the first image, $R_P \subseteq P \times P$. Similarly, let L be the set of line segments and R_L be the set of pairs of connecting line segments from the second image, $R_L \subseteq L \times L$. Figure 11.18 illustrates two sample images and the sets P, R_P, L, and R_L. Note that both R_P and R_L are symmetric relations; if (Si, Sj) belongs to such a relation, then so does (Sj, Si). In our examples, we list only tuples (Si, Sj) where $i < j$, but the mirror image tuple (Sj, Si) is implicitly present.

A consistent labeling for this problem is the mapping f given below:

$$\begin{array}{ll} f(S1) = Sj & f(S7) = Sg \\ f(S2) = Sa & f(S8) = Sl \\ f(S3) = Sb & f(S9) = Sd \\ f(S4) = Sn & f(S10) = Sf \\ f(S5) = Si & f(S11) = Sh \\ f(S6) = Sk & \end{array}$$

For another example, we return to the recognition of the kleep object shown in Figure 11.8 and defined in the associated tables. Our matching paradigms will use the distance relation defined for any two points: Each pair of holes is related by the distance between them. Distance is invariant to rotations and translations but not scale change. Abusing

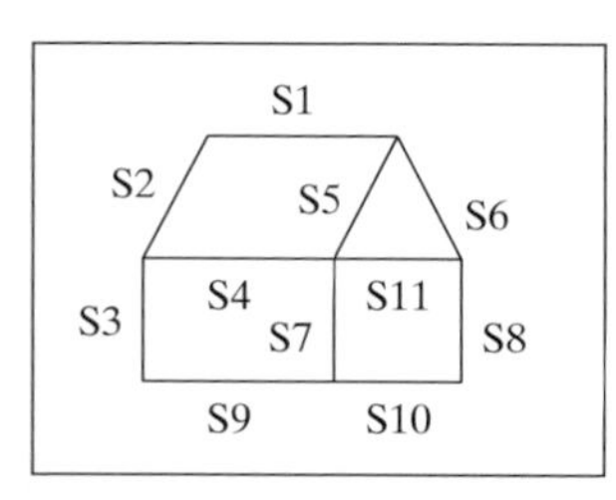

Image 1

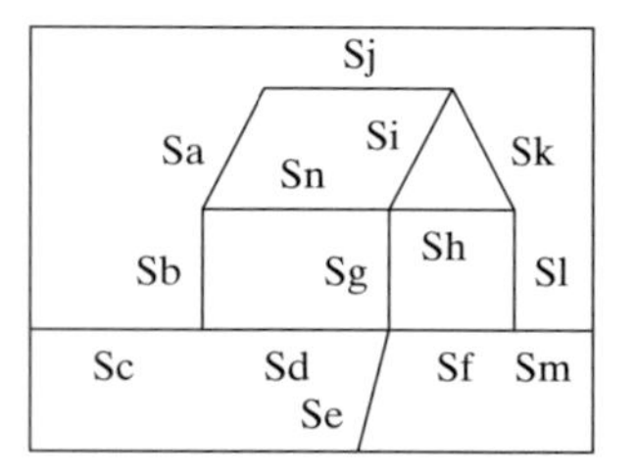

Image 2

$P = \{$S1, S2, S3, S4, S5, S6, S7, S8, S9, S10, S11$\}$.
$L = \{$Sa, Sb, Sc, Sd, Se, Sf, Sg, Sh, Si, Sj, Sk, Sl, Sm$\}$.

$R_P = \{$ (S1, S2), (S1, S5), (S1, S6), (S2, S3), (S2, S4), (S3, S4), (S3, S9), (S4, S5), (S4, S7), (S4, S11), (S5, S6), (S5, S7), (S5, S11), (S6, S8), (S6, S11), (S7, S9), (S7, S10), (S7, S11), (S8, S10), (S8, S11), (S9, S10) $\}$.

$R_L = \{$ (Sa, Sb), (Sa, Sj), (Sa, Sn), (Sb, Sc), (Sb, Sd), (Sb, Sn), (Sc, Sd), (Sd, Se), (Sd, Sf), (Sd, Sg), (Se, Sf), (Se, Sg), (Sf, Sg), (Sf, Sl), (Sf, Sm), (Sg, Sh), (Sg, Si), (Sg, Sn), (Sh, Si), (Sh, Sk), (Sh, Sl), (Sh, Sn), (Si, Sj), (Si, Sk), (Si, Sn), (Sj, Sk), (Sk, Sl), (Sl, Sm) $\}$.

Figure 11.18 Example of a Consistent Labeling Problem.

notation, we write $12(A, B)$ and $12(B, C)$ to indicate that points A and B are distance 12 apart in the model, similarly for points B and C. $12(C, D)$ does NOT hold, however, as we see from the distance tables. To allow for some distortion or detection error, we might allow that $12(C, D)$ is true even when the distance between C and D is actually $12 \pm \Delta$ for some small amount Δ.

Exercise 11.13: Consistent Labeling Problem.

Show that the labeling f given above is a consistent labeling. Because the relations are symmetric, the following modified constraint must be satisfied:

$$\text{If } (p_i, p_{i'}) \in R_P, \quad \text{then } (f(p_i), f(p_{i'})) \in R_L \quad \text{or} \quad (f(p_{i'}), f(p_i)) \in R_L.$$

The Interpretation Tree

88 Definition. An interpretation tree (IT) is a tree that represents all possible assignments of labels to parts. Every path in the tree is terminated either because it represents a complete consistent assignment, or because the partial assignment it represents fails some relation.

A partial interpretation tree for the image data of Figure 11.8 is shown in Figure 11.19. The tree has three levels, each to assign a label to one of the three holes H_1, H_2, H_3 observed in the image. No inconsistencies occur at the first level since there are no distance constraints to check. However, most label possibilities at level 2 can be immediately terminated using one distance check. For example, the partial assignment $\{(H_1, A), (H_2, A)\}$ is inconsistent

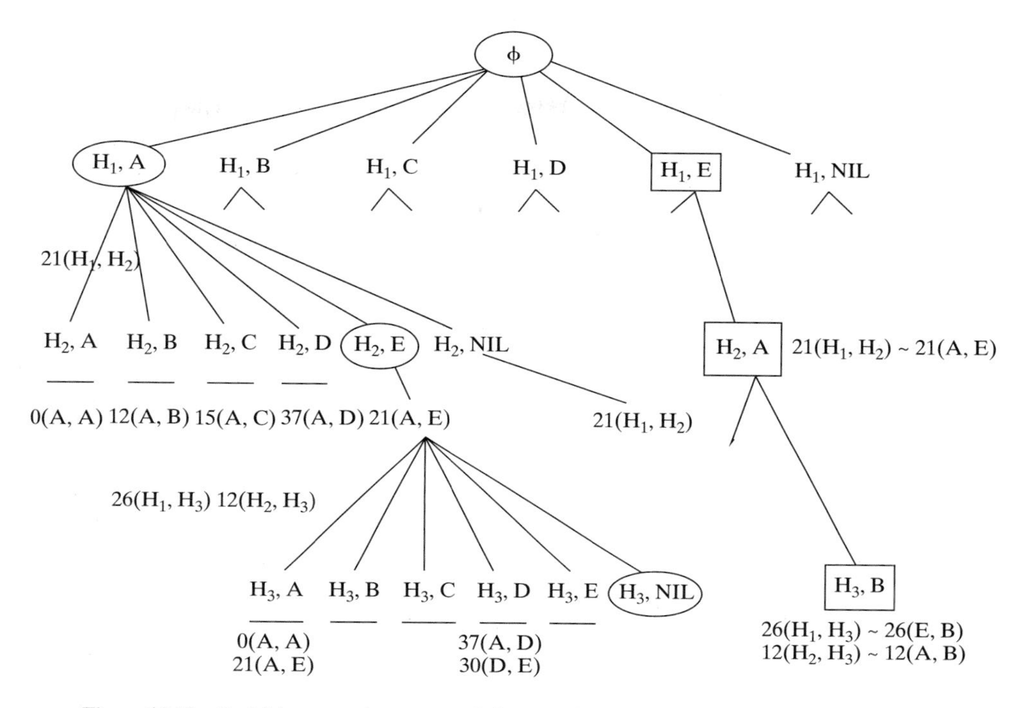

Figure 11.19 Partial interpretation tree search for a consistent labeling of the kleep parts in Figure 11.8 (right).

because the relation $21(H_1, H_2)$ is violated by $0(A, A)$. Many paths are not shown due to lack of space. The path of labels denoted by the boxes yields a complete and consistent assignment. The path of labels denoted by ellipses is also consistent; however, it contains one NIL label and thus has fewer constraints to check. This assignment has the first two pairs of the complete (boxed) assignment reversed in labels and the single distance check is consistent. Multiple paths of an IT can succeed due to symmetry. Although the IT potentially contains an exponential number of paths, it has been shown that most paths will terminate by level 3 due to the relational constraints. Use of the label NIL allows for detection of artifacts or the presence of features from another object in the scene.

The IT can easily be developed using a recursive backtracking process that develops paths in a depth-first fashion. At any instantiation of the procedure, the parameter f, which is initially NIL, contains the consistent partial assignment. Whenever a new labeling of a part is consistent with the partial assignment, the algorithm goes deeper in the tree by hypothesizing another label for an unlabeled part; if an inconsistency is detected, then the algorithm backs up to make an alternate choice. As coded, the algorithm returns the first completed path, which may include *NIL* labels if that label is explicitly included in L. An improvement would be to return the completed path with the most non*NIL* pairs, or perhaps all completed paths.

Find the mapping from model features to image features that satisfies the model relationships by a tree search.

P is the set of detected image features.
L is the set of stored model features.
$\mathbf{R_P}$ is the relationship over the image features.
$\mathbf{R_L}$ is the relationship over the model features.
f is the consistent labeling to be returned, initially NIL.

```
procedure Interpretation_Tree_Search(P, L, R_P, R_L, f);
{
p := first(P);
for each l in L
  {
  f' = f ∪ {(p, l)}; /* add part-label to interpretation */
  OK = true;
  for each N-tuple (p_1, ... ,p_N) in R_P containing component p
    and whose other components are all in domain(f)
    /* check on relations */
    if (f'(p_1), ... ,f'(p_N)) is not in R_L then
      {
      OK: = false;
      break;
      }
  if OK then
    {
    P' = rest(P);
    if isempty(P') then output(f');
    else Interpretation_Tree_Search(P', L, R_P, R_L, f');
    }
  }
}
```

Algorithm 11.5 Interpretation Tree Search.

The recursive interpretation tree search algorithm is defined to be general and to handle arbitrary N-ary relations, R_P and R_L, rather than just binary relations. R_P and R_L can be single relations, such as the connection relation in our first example, or they can be unions of a number of different relations, such as connection, parallel, and distance.

Discrete Relaxation Relaxation uses only local constraints rather than all the constraints available; for example, all constraints from matches on a single path of the

IT. After N iterations, local constraints from one part neighborhood can propagate across the object to another part on a path N edges distant. Although the constraints used in one iteration are weaker than those available using the IT search, they can be applied in parallel, thus making faster and simpler processing possible.

Initially, a part can be labeled by any label permitted by its type; suppose we assign it a set of all these possible labels. Discrete relaxation examines the relations between a particular part and all others, and by doing so, reduces the possible labels for that particular part. In the related problem of character recognition, if it is known that the following letter cannot be 'U', then we can conclude that the current letter cannot be 'Q'. In yet a different domain, if it is known that an image region is not water, then an object in it is not a ship. Discrete relaxation was popularized by David Waltz, who used it to constrain the labels assigned to edges of line drawings. (Waltz filtering is discussed in the text by Winston (1977).) Waltz used an algorithm with some sequential character; here we present a parallel approach.

Each part $\mathbf{P_i}$ is initially assigned the entire **set** of possible labels $\mathbf{L_j}$ according to its type. Then, all relations are checked to see if some labels are impossible: Inconsistent labels are removed from the set. The label sets for each part can be processed in parallel through passes. If, after any pass some labels have been filtered out of some sets, then another pass is executed; if no labels have changed, then the filtering is completed. There may be no interpretations left possible, or there may be several. The following example should be instructive. To keep it simple, we assume that there are no extra features detected that are not actual parts of the model; as before, we assume that some features may have been missed.

We now match the data in Tables 11.1 and 11.2. The filtering process begins with all 5 labels possible for each of the 3 holes H_1, H_2, H_3. To add interest and to be more practical, *we will allow a tolerance of* ± 1 *on distance matches*. Table 11.4 shows the 3 label sets at some point midway through the first pass. Each cell of the table gives the reason why a label must be deleted or why it survives. A is deleted from the label set for H_1 because the relation $26(H_1, H_3)$ cannot be explained by any label for H_3. The label A

TABLE 11.4 MIDWAY THROUGH FIRST PASS OF RELAXATION LABELING

	A	B	C	D	E
H_1	no $N \ni$ $d(A, N) = 26$	no $N \ni$ $d(B, N) = 21$	no $N \ni$ $d(C, N) = 26$	no $N \ni$ $d(D, N) = 26$	$21(H_1, H_2)$ $A \in L(H_2)$ $26(H_1, H_3)$ $B \in L(H_3)$
H_2	$21(H_2, H_1)$ $E \in L(H_1)$ $12(H_2, H_3)$ $B \in L(H_3)$	no $N \ni$ $d(B, N) = 21$	$21(H_2, H_1)$ $D \in L(H_1)$ $12(H_2, H_3)$ $B \in L(H_3)$		
H_3	no $N \ni$ $d(A, N) = 26$	$12(H_3, H_2)$ $A \in L(H_2)$ $26(H_3, H_1)$ $E \in L(H_1)$			

survives for H_2 because there is label $E \in L(H_1)$ to explain the relation $21(H_2, H_1)$ and label $B \in L(H_3)$ to explain relation $12(H_2, H_3)$. The label C survives for H_2, because $d(H_2, H_1) = 21 \approx 22 = d(C, D)$.

At the end of the first pass, as Table 11.5 shows, there are only two labels possible for H_2, only label E remains for H_1, and only label B remains for H_3. At the end of pass 1 the reduced label sets are made available for the parallel processing of pass 2, where each label set is further filtered in asynchronous parallel order.

TABLE 11.5 AFTER COMPLETION OF THE FIRST PASS OF RELAXATION LABELING

	A	B	C	D	E
H_1	no	no	no	no	possible
H_2	possible	no	possible	no	no
H_3	no	possible	no	no	no

Exercise 11.14

Give detailed justification for each of the labels being in or out of each of the label sets after pass 1 as shown in Table 11.5.

Pass 2 deletes label C from $L(H_2)$ because the relation $21(H_1, H_2)$ can no longer be explained by using D as a label for H_1. After pass 3, additional passes cannot change any label set so the process has converged. In this case, the label sets are all singletons representing a single assignment and interpretation. A high-level sketch of the algorithm is given in Algorithm 11.6. Although a simple and potentially fast procedure, relaxation labeling sometimes leaves more ambiguity in the interpretation than does IT search because constraints are only applied pairwise. Relaxation labeling can be applied as preprocessing for IT search: It can substantially reduce the branching factor of the tree search.

Continuous Relaxation* In exact consistent labeling procedures, such as tree search and discrete relaxation, a label l for a part p is either possible or impossible at any stage of the process. As soon as a part-label pair (p, l) is found to be incompatible with some already instantiated pair, the label l is marked as illegal for part p. This property of calling a label either possible or impossible in the preceding algorithms makes them *discrete* algorithms. In contrast, we can associate with each part-label pair (p, l) a real

TABLE 11.6 AFTER COMPLETION OF THE SECOND PASS OF RELAXATION LABELING

	A	B	C	D	E
H_1	no	no	no	no	possible
H_2	possible	no	no	no	no
H_3	no	possible	no	no	no

TABLE 11.7 AFTER COMPLETION OF THE THIRD PASS OF RELAXATION LABELING

	A	B	C	D	E
H_1	no	no	no	no	possible
H_2	possible	no	no	no	no
H_3	no	possible	no	no	no

Remove incompatible labels from the possible labels for a set of detected image features.

$\mathbf{P_i}$, $\mathbf{i} = 1, D$ is the set of detected image features.
$\mathbf{S(P_i)}$, $\mathbf{i} = 1, D$ is the set of initially compatible labels.
R is a relationship over which compatibility is determined.

```
procedure Relaxation_Labeling(P, S, R);
{
repeat
  for each (P_i, S(P_i))
  {
    for each label L_k ∈ S(P_i)
      for each relation R(P_i, P_j) over the image parts
        if ∃ L_m ∈ S(P_j) with R(L_k, L_m) in model
          then keep L_k in S(P_i)
          else delete L_k from S(P_i)
  }
until no change in any set S(P_i)
return(S);
}
```

Algorithm 11.6 Discrete Relaxation Labeling.

number representing the probability or certainty that part p can be assigned label l. In this case the corresponding algorithms are called *continuous*. In this section we will look at a labeling algorithm called *continuous relaxation* for symmetric binary relations.

A continuous relaxation labeling problem is a 6-tuple $CLRP = (P, L, R_P, R_L, PR, C)$. As before, P is a set of parts, L is a set of labels for those parts, R_P is a relationship over the parts, and R_L is a relationship over the labels. L is usually given as the union over all parts i of L_i, the set of allowable labels for part i. Suppose that $|P| = n$. Then PR is a set of n functions $PR = \{pr_1, \ldots, pr_n\}$ where $pr_i(l)$ is the *a priori* probability that label l is valid for part i. C is a set of n^2 compatibility coefficients $C = \{c_{ij}\}, i = 1, \ldots, n; j = 1, \ldots, n$. c_{ij} can be thought of as the influence that part j has on the labels of part i. Thus, if we view the constraint relation R_P as a graph, we can view c_{ij} as a weight on the edge between part i and part j.

Instead of using R_P and R_L directly, we combine them to form a set R of n^2 functions $R = \{r_{ij}\}, i = 1, \ldots, n; j = 1, \ldots, n$, where $r_{ij}(l, l')$ is the compatibility of label l for part i with label l' for part j. In the discrete case, $r_{ij}(l, l')$ would be 1, meaning $((i, l), (j, l'))$ is allowed or 0, meaning that combination is incompatible. In the continuous case, $r_{ij}(l, l')$ can be any value between 0 and 1, indicating how compatible the relationship between parts i and j is with the relationship between labels l and l'. This information can come from R_P and R_L—which may be themselves simple, binary relations or may be attributed binary relations—where the attribute associated with a pair of parts (or pair of labels) represents the likelihood that the required relationship holds between them. The solution of a continuous relaxation labeling problem, like that of a consistent labeling problem, is a mapping $f : P \rightarrow L$ that assigns a label to each unit. Unlike the discrete case, there is no external definition stating what conditions such a mapping f must satisfy. Instead, the definition of f is implicit in the procedure that produces it. This procedure is known as *continuous relaxation*.

As discrete relaxation algorithms iterate to remove possible labels from the label set L_i of a part i, continuous relaxation iterates to update the probabilities associated with each part-label pair. The initial probabilities are defined by the set *PR* of functions defining *a priori* probabilities. The algorithm starts with these initial probabilities at step 0. Thus we define the probabilities at step 0 by

$$pr_i^0(l) = pr_i(l) \tag{11.18}$$

for each part i and label l. At each iteration k of the relaxation, a new set of probabilities $\{pr_i^k(l)\}$ is computed from the previous set and the compatibility information. In order to define $pr_i^k(l)$ we first introduce a piece of it, $q_i^k(l)$ defined by

$$q_i^k(l) = \sum_{\{j|(i,j)\in R_P\}} c_{ij} \left[\sum_{l'\in L_j} r_{ij}(l, l')pr_j^k(l') \right] \tag{11.19}$$

The function $q_i^k(l)$ represents the influence that the current probabilities associated with labels of other parts constrained by part i have on the label of part i. With this formulation, the formula for updating the pr_i^k's can be written as

$$pr_i^{k+1}(l) = \frac{pr_i^k(l)\left(1 + q_i^k(l)\right)}{\sum_{l'\in L_i} pr_i^k(l')\left(1 + q_i^k(l')\right)} \tag{11.20}$$

The numerator of the expression allows us to add to the current probability $pr_i^k(l)$ a term that is the product $pr_i^k(l)q_i^k(l)$ of the current probability and the opinions of other related parts, based on the current probabilities of their own possible labels. The denominator normalizes the expression by summing over all possible labels for part i.

Exercise 11.15: Continous Relaxation.

Figure 11.20 shows a model and an image, each composed of line segments. Two line segments are said to be in the relationship *closadj* if their endpoints either coincide or are close to each other. (a) Construct the attributed relation R_P over the parts of the model defined by $R_P = \{(p_i, p_j, d) \mid p_i \textit{ closadj } p_j\}$ and the attributed relation R_L over the labels of the image defined by $R_L = \{(l_i, l_j) \mid l_i \textit{ closadj } l_j\}$. (b) Define the compatibility coefficients by $c_{ij} = 1 \textit{ if } (p_i, p_j) \in R_P \textit{ else } 0$. Use R_P and R_L together to define R in a manner of your choosing. Let $pr_i(l_j)$ be given as 1 if p_i has the same orientation as l_j, 0 if they are perpendicular, and 0.5 if one is diagonal and the other is horizontal or vertical. Define pr for the parts of the model and labels of the image. (c) Apply several iterations of continuous relaxation to find a probable labeling from the model parts to the image labels.

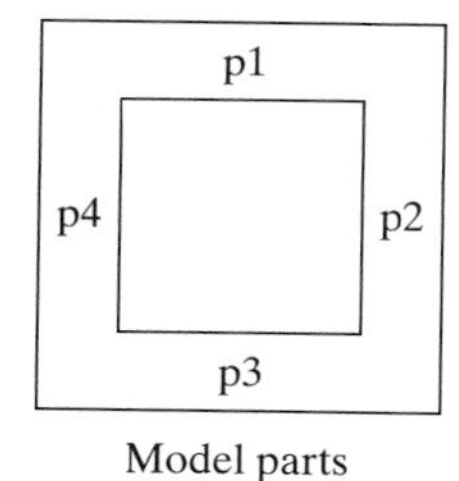

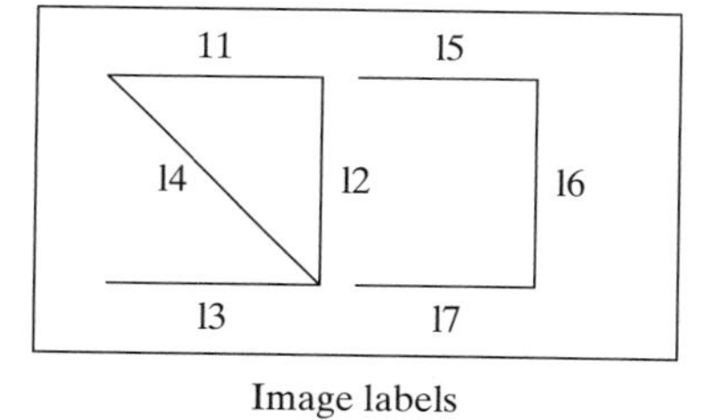

Figure 11.20 A model and image for continuous relaxation.

Relational Distance Matching A fully consistent labeling is unrealistic in many real applications. Due to feature extraction errors, noise, and occlusion of one object by another, an image may have missing and extra parts, and required relationships may not hold. Continous relaxation may be used in these cases, but it is not guaranteed to find the best solution. In problems where finding an optimal solution is important, we can perform a search to find the best mapping f from P to L, in the sense that it preserves the most relationships and, or minimizes the number of NIL labels. The concept of *relational distance* as originally defined by Haralick and Shapiro (1981) allows us to define the best mapping in the general case where there may be any number of relations with possibly different dimensions. To do this we first need the concept of a *relational description* of an image or object.

> **89 Definition.** A **relational description** D_P is a sequence of relations $D_X = \{R_1, \ldots, R_I\}$ where for each $i = 1, \ldots, I$, there exists a positive integer n_i with $R_i \subseteq P^{n_i}$ for some set P. P is a set of the parts of the entity being described and the relations R_i indicate various relationships among the parts.

A relational description is a data structure that may be used to describe two-dimensional shape models, three-dimensional object models, regions on an image, and so on.

Let $D_A = \{R_1, \ldots, R_I\}$ be a relational description with part set A and $D_B = \{S_1, \ldots, S_I\}$ be a relational description with part set B. We will assume that $|A| = |B|$; if this is not the case, we will add enough dummy parts to the smaller set to make it the case. The assumption is made in order to guarantee that the relational distance is a metric.

Let f be any one-one, onto mapping from A to B. For any $R \subseteq A^N$, N a positive integer, the *composition* $R \circ f$ of relation R with function f is given by

$$R \circ f = \{(b_1, \ldots, b_N) \in B^N \mid \textit{there exists}(a_1, \ldots, a_N) \in R \\ \textit{with } f(a_n) = b_n, n = 1, \ldots, N\} \tag{11.21}$$

This composition operator takes N-tuples of R and maps them, component by component, into N-tuples of B^N.

The function f maps parts from set A to parts from set B. The *structural error* of f for the ith pair of corresponding relations (R_i and S_i) in D_A and D_B is given by

$$E_S^i(f) = |R_i \circ f - S_i| + |S_i \circ f^{-1} - R_i|. \tag{11.22}$$

The structural error indicates how many tuples in R_i are not mapped by f to tuples in S_i and how many tuples in S_i are not mapped by f^{-1} to tuples in R_i. The structural error is expressed with respect to only one pair of corresponding relations.

The *total error* of f with respect to D_A and D_B is the sum of the structural errors for each pair of corresponding relations. That is,

$$E(f) = \sum_{i=1}^{I} E_S^i(f). \tag{11.23}$$

The total error gives a quantitative idea of the difference between the two relational descriptions D_A and D_B with respect to the mapping f.

The *relational distance* $GD(D_A, D_B)$ between D_A and D_B is then given by

$$GD(D_A, D_B) = \min_{\substack{1-1 \\ f:A \to B \\ \text{onto}}} E(f). \tag{11.24}$$

That is, the relational distance is the minimal total error obtained for any one-one, onto mapping f from A to B. We call a mapping f that minimizes total error a best mapping from D_A to D_B. If there is more than one best mapping, additional information that is outside the pure relational paradigm can be used to select the preferred mapping. More than one best mapping will occur when the relational descriptions involve certain kinds of symmetries.

We illustrate the relational distance with several examples. Figure 11.21 shows two digraphs, each having four nodes. A best mapping from $A = \{1, 2, 3, 4\}$ to $B = \{a, b, c, d\}$

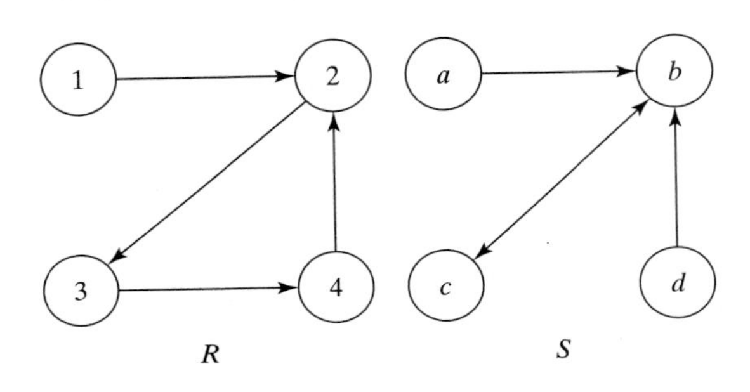

Figure 11.21 Two digraphs whose relational distance is 3.

is $\{f(1) = a,\ f(2) = b,\ f(3) = c,\ f(4) = d\}$. For this mapping we have

$$
\begin{aligned}
|R \circ f - S| &= |\{(1,2)(2,3)(3,4)(4,2)\} \circ f - \{(a,b)(b,c)(c,b)(d,b)\}| \\
&= |\{(a,b)(b,c)(c,d)(d,b)\} - \{(a,b)(b,c)(c,b)(d,b)\}| \\
&= |\{(c,d)\}| \\
&= 1 \\
|S \circ f^{-1} - R| &= |\{(a,b)(b,c)(c,b)(d,b)\} \circ f^{-1} - \{(1,2)(2,3)(3,4)(4,2)\}| \\
&= |\{(1,2)(2,3)(3,2)(4,2)\} - \{(1,2)(2,3)(3,4)(4,2)\}| \\
&= |\{(3,2)\}| \\
&= 1 \\
E(f) &= |R \circ f - S| + |S \circ f^{-1} - R| \\
&= 1 + 1 \\
&= 2
\end{aligned}
$$

Since f is a best mapping, the relational distance is also 2.

Figure 11.22 gives a set of object models M_1, M_2, M_3, and M_4 whose primitives are image regions. Two relations are shown in the figure: The connection relation and the parallel relation. Both are binary relations over the set of primitives. Consider the first two models,

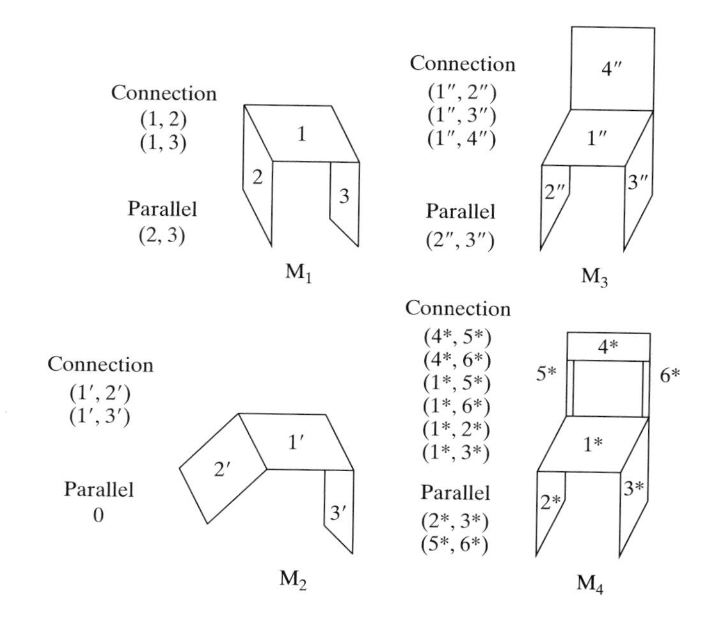

Figure 11.22 Four object models. The relational distance of model M_1 to M_2 and M_1 to M_3 is 1. The relational distance of model M_3 to M_4 is 6.

M_1 and M_2. The best mapping f maps primitive 1 to 1′, 2 to 2′, and 3 to 3′. Under this mapping the connection relations are isomorphic. The parallel relationship (2, 3) in model M_1 does not hold between 2′ and 3′ in model M_2. Thus the relational distance between M_1 and M_2 is exactly 1. Now consider models M_1 and M_3. The best mapping maps 1 to 1″, 2 to 2″, 3 to 3″, and a dummy primitive to 4″. Under this mapping, the parallel relations are now isomorphic but there is one more connection in M_3 than in M_2. Again the relational distance is exactly 1.

Finally consider models M_3 and M_4. The best mapping maps 1″ to 1*, 2″ to 2*, 3″ to 3*, 4″ to 4*, 5_d to 5*, and 6_d to 6*. (5_d and 6_d are dummy primitives.) For this mapping we have

$$\begin{aligned}
|R_1 \circ f - S_1| &= |\{(1'', 2'')(1'', 3'')(1'', 4'')\} \circ f \\
&\quad - \{(4^*, 5^*)(4^*, 6^*)(1^*, 5^*)(1^*, 6^*)(1^*, 2^*)(1^*, 3^*)\}| \\
&= |\{(1^*, 2^*)(1^*, 3^*)(1^*, 4^*)\} \\
&\quad - \{(4^*, 5^*)(4^*, 6^*)(1^*, 5^*)(1^*, 6^*)(1^*, 2^*)(1^*, 3^*)\}| \\
&= |\{(1^*, 4^*)\}| \\
&= 1 \\
|S_1 \circ f^{-1} - R_1| &= |\{(4^*, 5^*)(4^*, 6^*)(1^*, 5^*)(1^*, 6^*)(1^*, 2^*)(1^*, 3^*)\} \circ f^{-1} \\
&\quad - \{(1'', 2'')(1'', 3'')(1'', 4'')\}| \\
&= |\{(4'', 5_d)(4'', 6_d)(1'', 5_d)(1'', 6_d)(1'', 2'')(1'', 3'')\} \\
&\quad - \{1'', 2'')(1'', 3'')(1'', 4'')\}| \\
&= |\{(4'', 5_d)(4'', 6_d)(1'', 5_d)(1'', 6_d)\}| \\
&= 4 \\
|R_2 \circ f - S_2| &= |\{(2'', 3'')\} \circ f - \{2^*, 3^*)(5^*, 6^*)\}| \\
&= |\{(2^*, 3^*)\} - \{(2^*, 3^*)(5^*, 6^*)\}| \\
&= |\emptyset| \\
&= 0 \\
|S_2 \circ f^{-1} - R_2| &= |\{(2^*, 3^*)(5^*, 6^*)\} \circ f^{-1} - \{(2'', 3'')\}| \\
&= |\{(2'', 3'')(5_d, 6_d)\} - \{(2'', 3'')\}| \\
&= |\{(5_d, 6_d)\}| \\
&= 1 \\
E_S^1(f) &= 1 + 4 = 5 \\
E_S^2(f) &= 0 + 1 = 1 \\
E(f) &= 6
\end{aligned}$$

Exercise 11.16: Relational Distance Tree Search.

Modify the algorithm for Interpretation Tree Search to find the relational distance between two structural descriptions and to determine the best mapping in the process of doing so.

Exercise 11.17: One-Way Relational Distance.

The definition of relational distance in Equation 11.24 uses a two-way mapping error, which is useful when comparing two objects that stand alone. When matching a model to an image, we often want to use only a one-way mapping error, checking how many relationships of the model are in the image, but not vice versa. Define a modified one-way relational distance that can be used for model-image matching.

Exercise 11.18: NIL Mappings in Relational Distance.

The definition of relational distance in Equation 11.24 does not handle NIL labels explicitly. Instead, if part j has a NIL label, then any relationship (i, j) will cause an error, since $(f(i), NIL)$ will not be present. Define a modified relational distance that counts NIL labels as errors only once and does not penalize again for missing relationships caused by NIL labels.

Exercise 11.19: Attributed Relational Distance.

The definition in Equation 11.24 also does not handle attributed relations in which each tuple, in addition to a sequence of parts, contains one or more attributes of the relation. For example, a connection relation for line segments might have as an attribute the angle between the connecting segments. Formally, an attributed n-relation R over part set P and attribute set A is a set $R \subseteq P_n \times A_m$ for some nonnegative integer m that specifies the number of attributes in the relation. Define a modified relational distance in terms of attributed relations.

Relational Indexing Sometimes a tree search even with relaxation filtering is too slow, especially when an image is to be compared to a large database of models. For structural descriptions in terms of labeled relations, it is possible to approximate the relational distance with a simpler voting scheme. Intuitively, suppose we observe two concentric circles and two 90 degree corners connected by an edge. We would like to quickly find all models that have these structures and match them in more detail. To achieve this, we can build an index that allows us to look up the models given the partial graph structure. Given two concentric circles, we look up all models containing these related features and give each of those models one vote. Then, we look up all models having connected 90 degree corners: Any models repeating from the first test will now have two votes. These lookups can be done rapidly provided that an index is built offline before recognition by extracting significant binary relations from each model and recording each in a lookup table.

Let $DB = \{M_1, M_2, \ldots, M_T\}$ be a database of T object models. Each object model M_t consists of a set of attributed parts P_t plus a labeled relation R_t. For simplicity of explanation, we will assume that each part has a single label, rather than a vector of attributes and that the relation is a binary relation, also with a single label attached to each tuple. In this case, a model is represented by a set of *2-graphs* each of which is a graph with two nodes and two directed edges. Each node represents a part, and each edge represents a directed binary relationship. The value in the node is the *label* of the part, rather than a unique identifier. Similarly, the value in an edge is the *label* of the relationship. For example, one node could represent an ellipse and another could represent a pair of parallel lines. The edge from the parallel lines node to the ellipse node could represent the relationship encloses, while the edge in the opposite direction represents the relationship is enclosed by.

Relational indexing requires a preprocessing step in which a large hash table is set up. The hash table is indexed by a string representation of a 2-graph. When it is completed, one can look up any 2-graph in the table and quickly retrieve a list of all models containing that particular 2-graph. In our example, all models containing an ellipse between two parallel line segments can be retrieved. During recognition of an object from an image, the features are extracted and all the 2-graphs representing the image are computed. A set of accumulators, one for each model in the database are all set to zero. Then each 2-graph in the image is used to index the hash table, retrieve the list of associated models, and vote for each one. The discrete version of the algorithm adds one to the vote; a probabilistic algorithm would add a probability value instead. After all 2-graphs have voted, the models with the largest numbers of votes are candidates for verification.

11.7 NONLINEAR WARPING

Nonlinear warping functions are also important. We may need to rectify nonlinear distortion in an image; for example, radial distortion from a fisheye lens. Or, we may want to distort an image in creative ways. Figure 11.23 shows a nonlinear warp which maps a regular grid onto a cylinder. The effect is the same as if we wrapped a flat image around a cylinder and then viewed the cylinder from afar. This same warp applied to a twenty-dollar-bill is shown in Figure 11.24. Intuitively, we need to choose some image axis corresponding to the center of a cylinder and then use a formula which models how the pixels of the input image will *compress* off the cylinder axis in the output image. Figure 11.24 shows two warps; the rightmost one uses a cylinder of smaller radius than the center one.

Figure 11.25 shows how to derive the cylindrical warp. We choose an axis for the warp (determined by x_0) and a width W. W corresponds to one quarter of the circumference

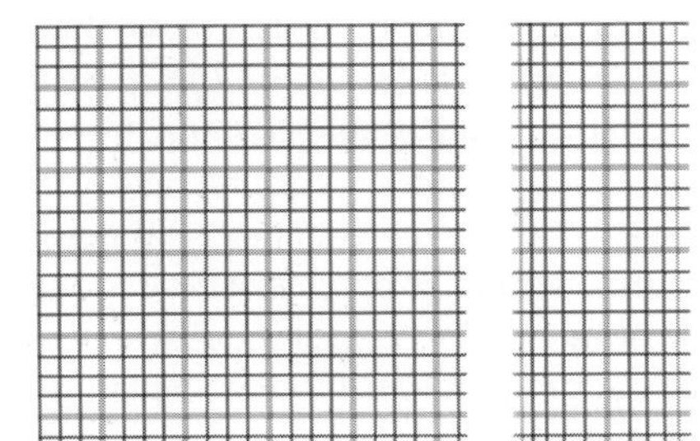

Figure 11.23 (left) Regular grid of lines; and (right) grid warped by wrapping it around a cylinder.

Figure 11.24 (left) Image of center of U.S.$20 bill; (center) image of Andrew Jackson wrapped around a cylinder of circumference 640 pixels; and (right) same as center except circumference is 400 pixels.

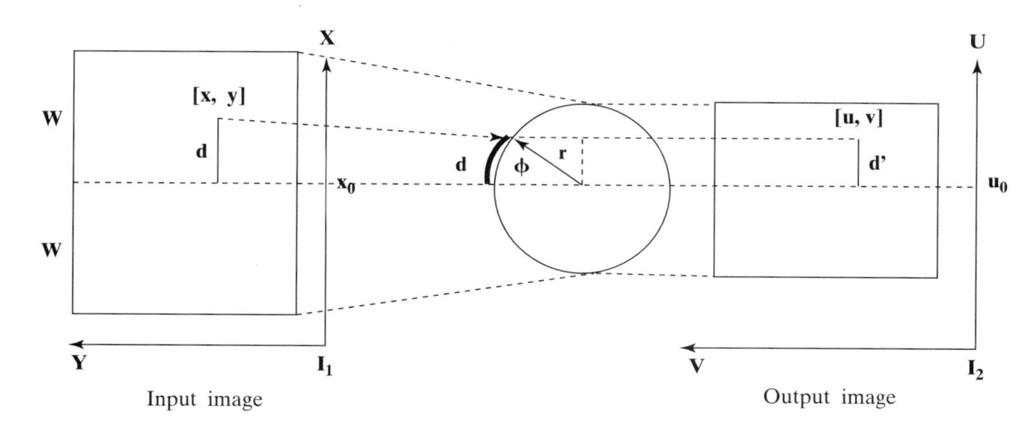

Figure 11.25 The output image at the right is created by *wrapping* the input image at the left around a cylinder (center): distance **d** in the input image becomes distance **d**′ on output.

of the cylinder. Any length d in the input image is *wrapped around the cylinder* and then projected to the output image. Actually, d corresponds to the length $x - x_0$ where x_0 is the x-coordinate of the axis of the cylinder. The y coordinate of the input image point is preserved by the warp, so we have $v = y$. From the figure, the following equations are seen to hold. First, $W = (\pi/2)r$ since W accounts for one quarter of the circumference. d accounts for a fraction of that: $d/W = \phi/(\pi/2)$ and $sin\phi = d'/r$. Combining these yields $d = x - x_0 = (2W/\pi)arcsin((\pi/2W)(u - u_0))$. Of course, $d' = u - u_0 = u - x_0$.

We now have a formula for computing input image coordinates $[x, y]$ from output coordinates $[u, v]$ and the warp parameters x_0 and W. This seems to be backwards; why not take each pixel from the input image and transform it to the output image? Were we to do this, there would be no guarantee that the output image would have each pixel set. For a digital image, we would like to generate each and every pixel of output exactly once and obtain the pixel value from the input image as Algorithm 11.7 shows. Moreover, using this approach it is easily possible for the output image to have more or fewer pixels than the

Perform a cylindrical warp operation.

$^1\mathbf{I}[\mathbf{x}, \mathbf{y}]$ is the input image.
$\mathbf{x_0}$ is the axis specification.
$\mathbf{W}$ is the width.
$^2\mathbf{I}[\mathbf{u}, \mathbf{v}]$ is the output image.

```
procedure Cylindrical_Warp(¹I[x, y])
{
r = 2W/π;
for u:=0, Nrows-1
   for v:=0, Ncols-1
   {
      ²I[u, v] = 0; // set as background
      if (|u − u₀| ≤ r)
         {
           x = x₀ + r arcsin((u − x₀)/r);
           y = v;
           ²I[u, v] = ¹I[round(x), round(y)];
         }
   }
return(²I[u, v]);
}
```

Algorithm 11.7 Cylindrical Warp of Image Region.

input image. The concept is that in generating the output image, we *map back* into the input image and sample it.

Exercise 11.20

(a) Determine the transformation that maps a circular region of an image onto a hemisphere and then projects the hemisphere onto an image. The circular region of the original image is defined by a center (x_c, y_c) and radius r_0. (b) Develop a computer program to carry out the mapping.

Rectifying Radial Distortion Radial distortion is present in most lenses: It might go unnoticed by human interpreters, but sometimes can produce large errors in photometric measurements if not corrected. Physical arguments deduce that radial distortion in the location of an imaged point is proportional to the distance from that point to the optical axis. Figure 11.26 shows two common cases of radial distortion along with the desirable rectified image. If we assume that the optical axis passes near the image center, then the

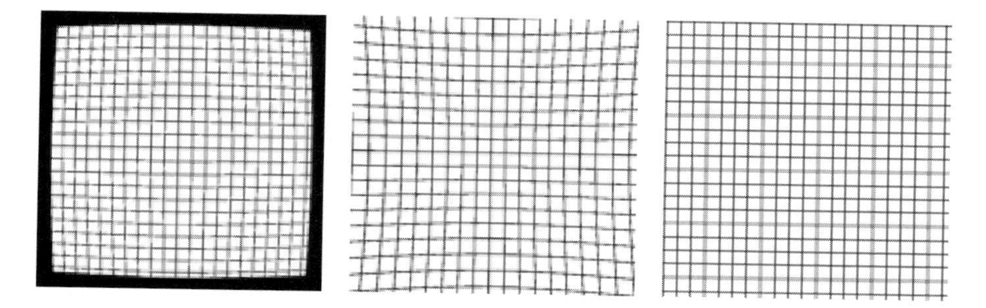

Figure 11.26 Two types of radial distortion, (left) barrel and (center) pincushion which can be removed by warping to produce a (right) rectified image.

distortion can be corrected by displacing all image points either toward or away from the center by a displacement proportional to the square of the distance from the center. This is not a linear transformation since the displacement varies across the image. Sometimes, more even powers of the radial distance are used in the correction, as the mathematical model in Equation 11.25 shows. Let $[x_c, y_c]$ be the image center which we are assuming is also where the optical axis passes through the image. The corrections for the image points are as follows, assuming we need the first two even powers of the radial distance to compute the radial distortion. The best values for the constants c_2 and c_4 can be found by empirical study of the radial displacements of known control points or by formally using least-squares fitting in a calibration process.

$$\begin{aligned} R &= \sqrt{((x - x_c)^2 + (y - y_c)^2)} \\ D_r &= (c_2 R^2 + c_4 R^4) \\ x &= x_c + (x - x_c) D_r \\ y &= y_c + (y - y_c) D_r \end{aligned} \tag{11.25}$$

Polynomial Mappings Many small global distortion factors can be rectified using polynomial mappings of maximum degree two in two variables as defined in Equation 11.26. Twelve different coefficients must be estimated in order to adapt to the different geometric factors. To estimate these coefficients, we need the coordinates of at least six control points before and after mapping; however, many more points are used in practice. (Each such control point yields two equations.) Note that if only the first three terms are used in Equation 11.26 the mapping is an affine mapping.

$$\begin{aligned} u &= a_{00} + a_{10}x + a_{01}y + a_{11}xy + a_{20}x^2 + a_{02}y^2 \\ v &= b_{00} + b_{10}x + b_{01}y + b_{11}xy + b_{20}x^2 + b_{02}y^2 \end{aligned} \tag{11.26}$$

Exercise 11.21

Show that radial distortion in Equation 11.25 with $c_4 = 0$ can be modeled exactly by a polynomial mapping of the form shown in Equation 11.26.

11.8 SUMMARY

Multiple concepts have been discussed in this chapter under the theme of 2D matching. One major theme was 2D mapping using transformations. These could be used as simple image processing operations which could extract or sample a region from an image, register two images together in the same coordinate system, or remove or creatively add distortion to 2D images. Algebraic machinery was developed for these transformations and various methods and applications were discussed. This development is continued in Chapter 13 as it relates to mapping points in 3D scenes and 3D models. The second major theme of this chapter was the interpretation of 2D images through correspondences with 2D models. A general paradigm is *recognition-by-alignment:* The image is interpreted by discovering a model and an RST transformation such that the transformation maps known model structures onto image structures. Several different algorithmic approaches were presented, including pose-clustering, interpretation tree search, and the local-feature-focus method. Discrete relaxation and relational matching were also presented: These two methods can be applied in very general contexts even though they were introduced here within the context of constrained geometric relationships. Relational matching is potentially more robust than rigid alignment when the relations themselves are more robust than those depending on metric properties. Image distortions caused by lens distortions, slanted viewing axes, quantification effects, etc., can cause metric relations to fail; however, topological relationships such as cotermination, connectivity, adjacency, and insideness are usually invariant to such distortions. A successful match using topological relationships on image and, or model parts might then be used to find a large number of matching points, which can then be used to find a mapping function with many parameters that can adapt to the metric distortions. The methods of this chapter are directly applicable to many real world applications. Chapter 14 extends the methods to 3D.

11.9 REFERENCES

The paper by Van Wie and Stein (1977) discusses a system to automatically bring satellite images into registration with a map. An approximate mapping is known using the time at which the image is taken. This mapping is used to do a refined search for control points using templates: The control points are then used to obtain a refined mapping. The book by Wolberg (1990) gives a complete account of image warping including careful treatment of sampling the input image and lessening aliasing by smoothing. The treatment of 2D matching via pose clustering was drawn from Stockman and others (1982), which contained the airplane detection example provided. A more general treatment handling the 3D case is given in Stockman (1987). The paper by Grimson and Lozano-Perez (1984) demonstrates how distance constraints can be used in matching model points to observed data points. Least-squares fitting is treated very well in other references and is a topic of much depth. Least-squares techniques are in common use for the purpose of estimating transformation parameters in the presence of noise by using many more than the minimal number of control points. The book by Wolberg (1990) treats several least-squares methods within the context of warping while the book by Daniel and Wood (1971) treats the general fitting problem. In some problems, it is not possible to find a good geometric transformation that can be applied globally to the entire image. In such cases, the image can be partitioned into a number of regions, each with its own control points, and separate warps can be applied to each region. The warps from neighboring regions must smoothly agree along the boundary between them. The paper by Goshtasby (1988) presents a flexible way of doing this.

Theory and algorithms for the consistent labeling problems can be found in the papers by Haralick and Shapiro (1979, 1980). Both discrete and continuous relaxation are defined in the paper by Rosenfeld, Hummel, and Zucker (1976), and continuous relaxation is further analyzed in the paper by Hummel and Zucker (1983). Methods for matching using structural descriptions have been derived from Shapiro and Haralick (1981); relational indexing using 2-graphs can be found in Costa and Shapiro (1995). Use of invariant attributes of structural parts for indexing into models can be found in Chen and Stockman (1996).

1. Chen, J. L., and G. Stockman. 1996. Indexing to 3D model aspects using 2D contour features. *Proc. Int. Conf. Comput. Vision and Pattern Recog. (CVPR),* San Francisco, CA (June 18–20), expanded paper to appear in the journal *CVIU.*
2. Clowes, M. 1971. On seeing things. *Artifical Intelligence,* v. 2:79–116.
3. Costa, M. S., and L. G. Shapiro. 1995. Scene analysis using appearance-based models and relational indexing. *IEEE Symposium on Comput. Vision* (Nov. 1995), 103–108.
4. Daniel, C., and F. Wood. 1971. *Fitting Equations to Data.* John Wiley & Sons, Inc., New York.
5. Goshtasby, A. 1988. Image registration by local approximation methods. *Image and Vision Computing,* v. 6(4):255–261.
6. Grimson, W., and T. Lozano-Perez. 1984. Model-based recognition and localization from sparse range or tactile data. *Int. J. Robotics Research,* v. 3(3):3–35.
7. Haralick, R, and L. Shapiro. 1979. The consistent labeling problem I. *IEEE Trans.,* v. PAMI-1:173–184.
8. Haralick, R., and L. Shapiro. 1980. The consistent labeling problem II. *IEEE Trans.,* v. PAMI-2:193–203.
9. Hummel, R., and S. Zucker. 1983. On the foundations of relaxation labeling processes. *IEEE Trans.,* v. PAMI-5:267–287.
10. Lamden, Y., and H. Wolfson. 1988. Geometric hashing: a general and efficient model-based recognition scheme. *Proc. 2nd Int. Conf. Comput. Vision,* Tarpon Springs, FL (Nov. 1988), 238–249.
11. Rogers, D., and J. Adams. 1990. *Mathematical Elements for Computer Graphics,* 2nd ed. McGraw-Hill, New York.
12. Rosenfeld, A., R. Hummel, and S. Zucker. 1976. Scene labeling by relaxation operators. *IEEE Trans. Systems, Man, and Cybern.,* v. SMC-6:420–453.
13. Shapiro, L. G., and R. M. Haralick. 1981. Structural descriptions and inexact matching. *IEEE Trans. Pattern Recog. and Machine Intelligence,* v. PAMI-3(5):504–519.
14. Stockman, G., S. Kopstein, and S. Benett. 1982. Matching images to models for registration and object detection via clustering. *IEEE Trans. PAMI,* v. PAMI-4(3): 229–241.

15. Stockman, G. 1987. Object recognition and localization via pose clustering. *Comput. Vision, Graphics and Image Proc.,* v. 40:361–387.
16. Van Wie, P., and M. Stein. 1977. A LANDSAT digital image rectification system. *IEEE Trans. Geosci. Electron.,* v. GE-15 (July 1977).
17. Winston, P. 1977. *Artificial Intelligence.* Addison-Wesley.
18. Wolberg, G. 1990. *Digital Image Warping.* IEEE Computer Society Press, Los Alamitos, CA.

12

Perceiving 3D from 2D Images

This chapter investigates phenomena that allow 2D image structure to be interpreted in terms of 3D scene structure. Humans have an uncanny ability to perceive and analyze the structure of the 3D world from visual input. Humans operate effortlessly and often have little idea what the mechanisms of visual perception are. Before proceeding, three points must be emphasized. First, while the discussion appeals to analytical reasoning, humans readily perceive structure without conscious reasoning. Also, many aspects of human vision are still not clearly understood. Second, although we can nicely model several vision cues separately, interpretation of complex scenes surely involves competitive and cooperative processes using multiple cues simultaneously. Finally, our interest need not be in explaining human visual behavior at all, but instead be in solving a particular application problem in a limited domain, which allows us to work with simpler sets of cues.

The initial approach used in this chapter is primarily descriptive. The next section discusses the *intrinsic image,* which is an intermediate 2D representation that stores important local properties of the 3D scene. Then we explore properties of texture, motion, and shape that allow us to infer properties of the 3D scene from the 2D image. Although the emphasis of this chapter is more on identifying sources of information rather than mathematically modeling them, the final sections do treat mathematical models. Models are given for perspective imaging, computing depth from stereo, and for relating field of view to resolution and blur via the thin lens equation. Other mathematical modeling is left for Chapter 13.

12.1 INTRINSIC IMAGES

It is convenient to think of a 3D scene as composed of object surface elements that are illuminated by light sources and that project as regions in a 2D image. Boundaries between 3D surface elements or changes in the illumination of these surface elements result in

Figure 12.1 (left) Intensity image of three blocks and (right) result of 5 × 5 Prewitt edge operator. (Images courtesy of Deborah Trytten.)

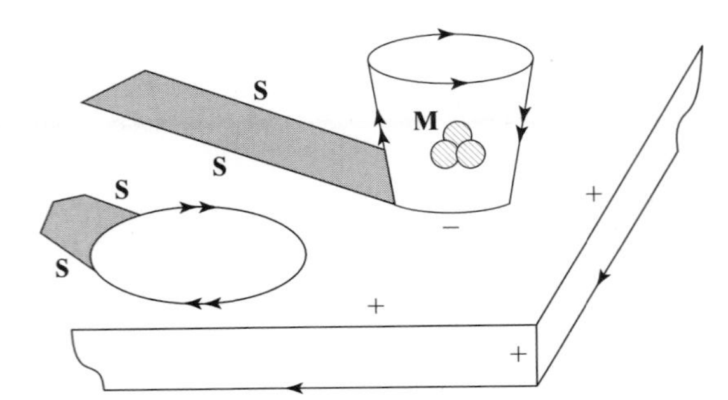

Figure 12.2 A 2D image with contour labels relating 2D contrasts to 3D phenomena such as surface orientation and lighting. Surface creases are indicated by + or −, an arrowhead (>) indicates a blade formed by the surface to its right while a double arrowhead indicates a smooth limb to its right, shadow boundaries are indicated by **S**, and reflectance boundaries are indicated by **M**.

contrast edges or *contours* in the 2D image. For simple scenes, such as those shown in Figures 12.1 and 12.2, all surface elements and their lighting can be represented in a description of the scene. Some scientists believe that the major purpose of the lower levels of the human visual system is to construct some such representation of the scene as the base for further processing. This is an interesting research question, but we do not need an answer in order to proceed with our work. Instead, we will use such a representation for scene and image description and machine analysis without regard to whether or not such a representation is actually computed by the human visual system.

Figure 12.2 shows an egg and an empty thin cup near the corner of a table. For this viewpoint, both the egg and cup occlude the planar table. Arrows along the region edges show which surface element occludes the other. The direction of the arrow is used to indicate which is the occluding surface; by convention, it is to the right as the edge is followed in the direction of the arrow. A single arrowhead (>) indicates a *blade,* such as the blade of a knife, where the orientation of the occluding surface element does not change much as the edge is approached; as the edge is crossed, the orientation of the occluded surface has no relation to that of the occluding surface. All of the object outlines in the right image of Figure 12.1 are due to blades. In Figure 12.2 a blade is formed at the lower table edge because the table edge, a narrow planar patch, occludes an unknown background. The top edge of the (thin) paper cup is represented as a blade because that surface occludes the background and has a consistent surface orientation as the boundary is approached. A more interesting case is the blade representing the top front cup surface occluding the cup interior.

A *limb* (≫) is formed by viewing a smooth 3D object, such as the limb of the human body; when the edge of a limb boundary is approached in the 2D image, the orientation of the corresponding 3D surface element changes and approaches the perpendicular to the line of sight. The surface itself is *self-occluding,* meaning that its orientation continues to change smoothly as the 3D surface element is followed behind the object and out of the 2D view. A blade indicates a real edge in 3D whereas a limb does not. All of the boundary of the image of the egg is a limb boundary, while the cup has two separate limb boundaries. As artists know, the shading of an object darkens when approaching a limb away from the direction of lighting. Blades and limbs are often called *jump edges:* There is an indefinite jump in depth (range) from the occluding to occluded surface behind. Looking ahead to Figure 12.10 one can see a much more complex scene with many edge elements of the same type as in Figure 12.2. For example, the lightpost and light have limb edges and the rightmost edge of the building at the left is a blade.

Exercise 12.1

Put a cup on your desk in front of you and look at it with one eye closed. Use a pencil touching the cup to represent the normal to the surface and verify that the pencil is perpendicular to your line of sight.

Creases are formed by abrupt changes to a surface or the joining of two different surfaces. In Figure 12.2, creases are formed at the edge of the table and where the cup and table are joined. The surface at the edge of the table is convex, indicated by a '+' label, whereas the surface at the join between cup and table is concave, indicated by a '−' label. Note that a machine vision system analyzing bottom-up from sensor data would not know that the scene contained a cup and table; nor would we humans know whether or not the cup were glued to the table, or perhaps even have been cut from the same solid piece of wood, although our experience biases such top-down interpretations! Creases usually, but not always, cause a significant change of intensity or contrast in a 2D intensity image because one surface usually faces more directly toward the light than does the other.

Exercise 12.2

The triangular block viewed in Figure 12.1 results in six contour segments in the edge image. What are the labels for these six segments?

Exercise 12.3

Consider the image of the three machine parts from Figure 1.7, Chapter 1. (Most, but not all, of the image contours are highlighted in white.) Sketch all of the contours and label each of them. Do we have enough labels to interpret all of the contour segments? Are all of our available labels used?

Two other types of image contours are not caused by 3D surface shape. The *mark* ('M') is caused by a change in the surface albedo; for example, the logo on the cup in Figure 12.2 is a dark symbol on lighter cup material. Illumination boundaries ('I'), or shadows ('S'), are caused by a change in illumination reaching the surface, which may be due to some shadowing by other objects.

We summarize the surface structure that we are trying to represent with the following definitions. It is very important to understand that we are representing 3D scene structure as seen in a particular 2D view of it. These 3D structures usually, but not always produce detectable contours in case of a sensed intensity image.

90 Definition. A **crease** is an abrupt change to a surface or a join between two different surfaces. While the surface points are continuous across the crease, the surface normal is discontinuous. The surface geometry of a crease may be observed from an entire neighborhood of viewpoints where it is visible.

91 Definition. A **blade** corresponds to the case where one continuous surface occludes another surface in its background: the normal to the surface changes smoothly and continues to face the view direction as the boundary of the surface is approached. The contour in the image is a smooth curve.

92 Definition. A **limb** corresponds to the case where one continuous surface occludes another surface in its background: the normal to the surface is smooth and becomes perpendicular to the view direction as the contour of the surface is approached, thus causing the surface to occlude itself as well. The image of the boundary is a smooth curve.

93 Definition. A **mark** is due to a change in reflectance of the surface material; for example, due to paint or the joining of different materials.

94 Definition. An **illumination boundary** is due to an abrupt change in the illumination of a surface, due to a change in lighting or shadowing by another object.

95 Definition. A **jump edge** is a limb or blade and is characterized by a depth discontinuity across the edge(contour) between an occluding object surface and the background surface that it occludes.

Exercise 12.4: Line labeling of the image of a cube.

Draw a cube in general position so that the picture shows 3 faces, 9 line segments, and 7 corners. (a) Assuming that the cube is floating in the air, assign one of the labels from $\{+, -, >, \gg\}$ to each of the 9 line segments, which gives the correct 3D interpretation for the phenomena creating it. (b) Repeat (a) with the assumption that the cube lies directly on a planar table. (c) Repeat (a) assuming that the cube is actually a thermostat attached to a wall.

Exercise 12.5: Labeling images of common objects.

Label the line segments shown in Figure 12.3: An unopened can of brand X soda and an open and empty box are lying on a table.

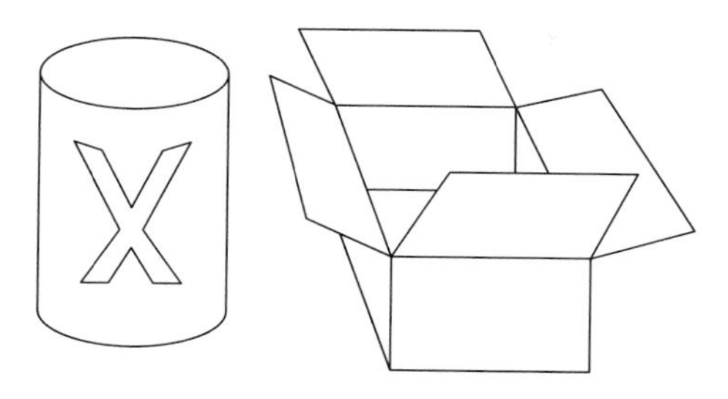

Figure 12.3 (left) An unopened can of Brand X Soda, which is a solid blue can with a single large orange block character 'X'; (right) an empty box with all four of its top flaps open, so one can see part of the box bottom that is not occluded by the box sides.

In Chapter 5 we studied methods of detecting contrast points in intensity images. Methods of tracking and representing contours were given in Chapter 10. Unfortunately, several 3D phenomena can cause the same kind of effect in the 2D image. For example, given a 2D contour tracked in an intensity image, how do we decide if it is caused by viewing an actual object or another object's shadow? Consider, for example, an image of a grove of trees taken on a sunny day (or, refer to the image of the camel on the beach toward the end of Chapter 5, where the legs of the camel provide the phenomena). The shadows of the trees on the lawn ('S') may actually be better defined by our edge detector than are the limb boundaries (≫) formed by the tree trunks. In interpreting the image, how do we tell the difference between the image of the shadow and the image of the tree; or, between the image of the shadow and the image of a sidewalk?

Exercise 12.6

Relating our work in Chapter 5 to our current topic, explain why the shadow of a tree trunk might be easier to detect in an image compared to the tree trunk itself.

Some researchers have proposed developing a sensing system that would produce an *intrinsic image*. An intrinsic image would contain four intrinsic scene values in each pixel.

- **range** or **depth** to the scene surface element imaged at this pixel
- **orientation** or **surface normal** of the scene element imaged at this pixel
- **illumination** received by the surface element imaged at this pixel
- **albedo** or surface reflectance of the surface element imaged at this pixel

Humans are good at making such interpretations for each pixel of an image given their surrounding context. Automatic construction of an intrinsic image is still a topic of research, but it is not being pursued as intensively as in the past. Many image analysis tasks do not need an intrinsic image. Chapter 13 will treat some methods useful for constructing

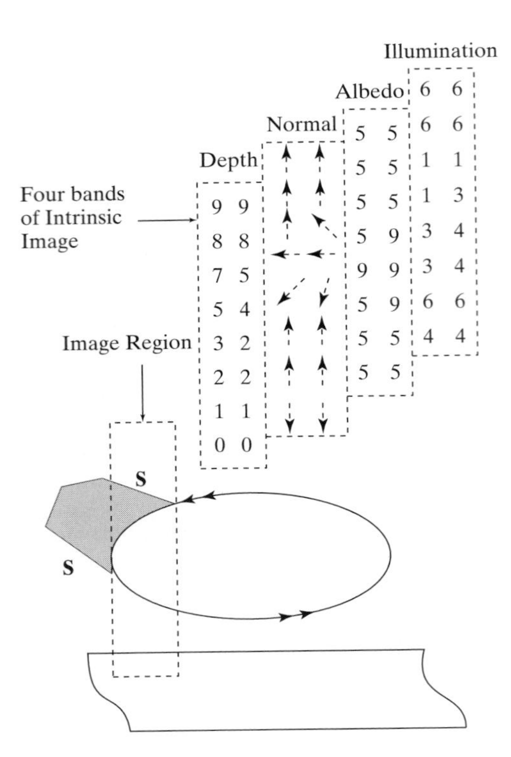

Figure 12.4 Intrinsic image corresponding to a small band across the egg of Figure 12.2. Each pixel contains four values representing surface depth, orientation, illumination, and albedo. See text for details.

intrinsic images or partial ones. An example of the intrinsic image corresponding to the image in Figure 12.2 is shown in Figure 12.4. The figure shows only the information from a small band of the intrinsic image across the end of the egg. The depth values show a gradual change across the table, except that at the edge of the table the change is more rapid, and there is a jump where the surface of the egg occludes the table. The orientation, or normal, of the table surface is the same at all the points of the table top; and, there is an abrupt change at the edge of the table. The orientation of the surface of the egg changes smoothly from one point to the next. The albedo values show that the table is a darker (5) material than the egg (9). The illumination values record the difference between table pixels that are in shadow (1) versus those that are not. Similarly, pixels from the egg surface, that is curving away from the illumination direction—assumed to be from the upper right—appear darker (3) than those directly facing the light because they receive less light energy per unit area.

Exercise 12.7: Line labeling an image of an outdoor scene.

Refer to the picture taken in Quebec City shown in Chapter 2. Sketch some of the major image contours visible in this image and label them using the label set $\{I/S, M, +, -, >, \gg\}$

12.2 LABELING OF LINE DRAWINGS FROM BLOCKS WORLD

The structure of contours in an image is strongly related to the structure of 3D objects. In this section, we demonstrate this in a microworld containing restricted objects and viewing conditions. **We assume that the universe of 3D objects are those with trihedral corners: all surface elements are planar faces and all corners are formed by the intersection of exactly three faces.** The block in Figure 12.5 is one such object. We use the terms *faces, creases* and *corners* for the 3D structures and we use the terms *regions, edges,* and *junctions* for the images of those structures in 2D. A 2D image of the 3D blocks world is assumed to be a line drawing consisting of regions, edges, and junctions. Moreover, we make the assumption that small changes in the viewpoint creating the 2D image cause no changes in the topology of this line drawing; that is, no new faces, edges, or junctions can appear or disappear. Often it is said that the object is *in general position.*

Although our blocks microworld is so limited that it is unrealistic, the methods developed in this context have proven to be useful in many real domains. Thus we will add to the set of algorithms developed in Chapter 11 for matching and interpretation. Also, the blocks domain has historical significance and supports an intuitive development of the new methods.

From the previous section, we already know how to label the image edges using the labels $\{+, -, >\}$ to indicate which are creases or blades according to our interpretation of the 3D structure. No limbs are used since they do not exist in the blocks world. About thirty years ago, it was discovered that the possible combinations of line labels forming junctions is strongly constrained. In total, there are only 16 such combinations possible: these are shown in Figure 12.6. Figure 12.5 shows how these junction configurations appear in two distinct 3D interpretations of the same 2D line drawing.

There are four types of junctions according to the number of edges joining and their angles: for obvious reasons, they're called *Ls, arrows, forks, and Ts* from top to bottom by rows in Figure 12.6. Figure 12.5 shows an example with all four junction types. The

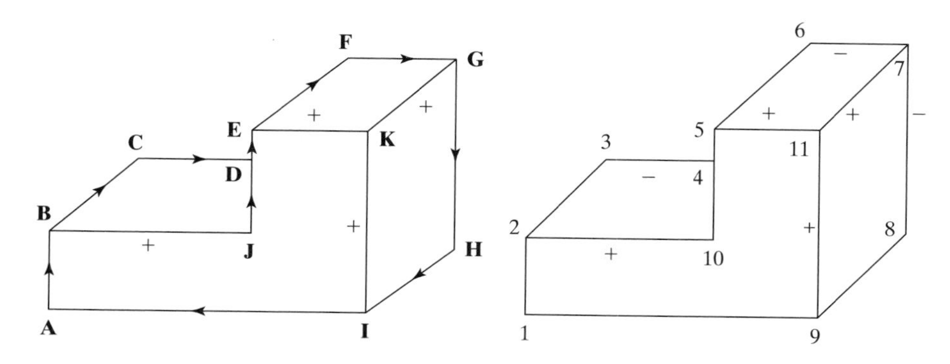

Figure 12.5 Two different interpretations for the same line drawing: (left) block floating in space and (right) block glued to back wall. The blade labels, omitted from the right figure, are the same as on the left.

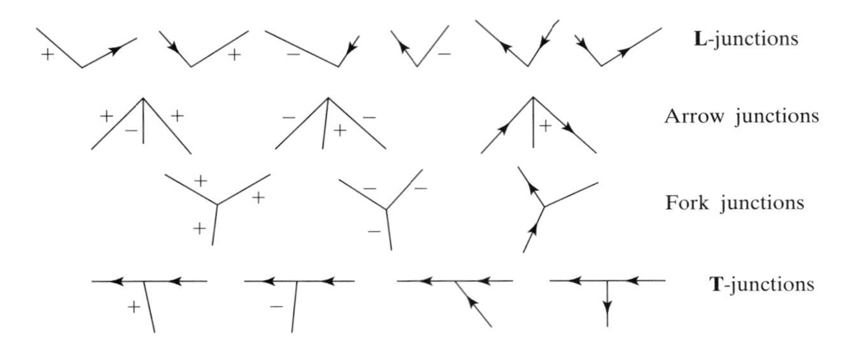

Figure 12.6 The only 16 topologically possible line junctions for images of trihedral blocks world (all 3D corners are formed by intersecting 3 planes and the object is viewed in general position). Junction types are, from top to bottom, **L**-junctions, arrows, forks, and **T**-junctions.

junction marked **J** is an instance of the leftmost **L**-junction shown at the top of the catalog in Figure 12.6, whereas the junction marked **C** is an instance of the second **L**-junction from the top right. **G** is the rightmost arrow-junction in the second row of Figure 12.6. There is only one **T**-junction, marked **D** in the figure. Note that, as Figure 12.6 shows, the occluding edge (cross) of the **T**-junction places no constraint on the occluded edge; all four possibilities remain, as they should. The four arrows in the block at the left (**B, E, G, I**) all have the same (convex) structure; however, the block at the right in Figure 12.5 has one other (concave) type of arrow-junction (**7**), indicating the convexity formed by the join of the block and wall.

Before proceeding, the reader should be convinced that all of the 16 junctions are, in fact, derivable from projections of 3D blocks. It is more difficult to reason that there are no other junctions possible: this has been proven, but for now, the reader should just verify that no others can be found while doing Exercises 12.8 and 12.9.

Exercise 12.8

Label the lines in the right part of Figure 12.1 according to how your visual system interprets the scene at the left.

Exercise 12.9

Try to label all edges of all the blocks in Figure 12.7 as creases or blades. Every junction must be from the catalog in Figure 12.6. (a) Which line drawings have consistent labelings? (b) Which drawings seem to correspond to a real object but cannot be labeled: why does the labeling fail? (c) Which drawings seem to correspond to impossible objects? Can any of these drawings be labeled consistently?

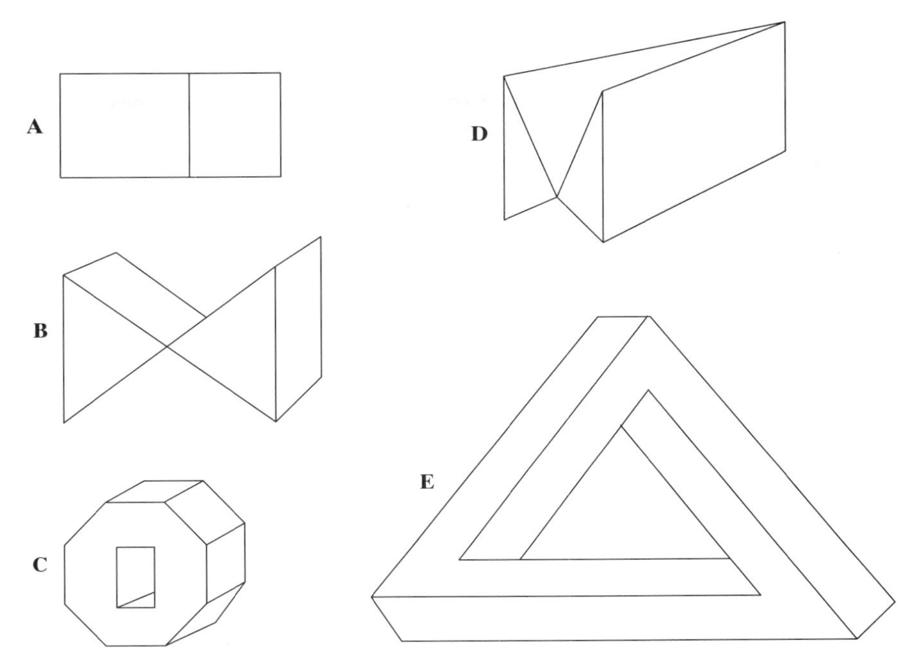

Figure 12.7 Line drawings which may or may not have 3D interpretations in our limited blocks world: which do and which do not and why?

Exercise 12.10

Sketch and label the line drawing of a real scene that yields at least two different instances of each of the four junction types. Create your own scene: you may use any structures from the several figures in this section.

Two algorithmic approaches introduced in Chapter 11 can be used to automatically label such line drawings; one is sequential backtracking and the other is parallel relaxation labeling. We first formalize the problem to be solved: **Given a 2D line drawing with a set of edges P_i (the observed objects), assign each edge a label L_j (the model objects) which interprets its 3D cause, and such that the combinations of labels formed at the junctions belong to the junction catalog.** The symbols P and L have been used to be consistent with Chapter 11, which should be consulted for algorithm details. Coarse algorithm designs are given below to emphasize certain points. Both algorithms often produce many interpretations unless some external information is provided. A popular choice is to label all edges on the convex hull of the drawing as $>$ such that the hull is to the right.

If possible, the edges with the most constrained labels should be assigned first: it may even be that outside information (say stereo) already indicates that the edge corresponds to a 3D crease, for example. Some preprocessing should be done to determine the type of each junction according to the angles and number of incident edges. Other versions of this approach assign catalog interpretations to the junction labels and proceed to eliminate those that are inconsistent across an edge with the neighboring junction interpretations. Figure 12.8 shows an interpretation tree for interpreting the line drawing of a pyramid with four faces. The search space is rather small, demonstrating the strong constraints in trihedral blocks world.

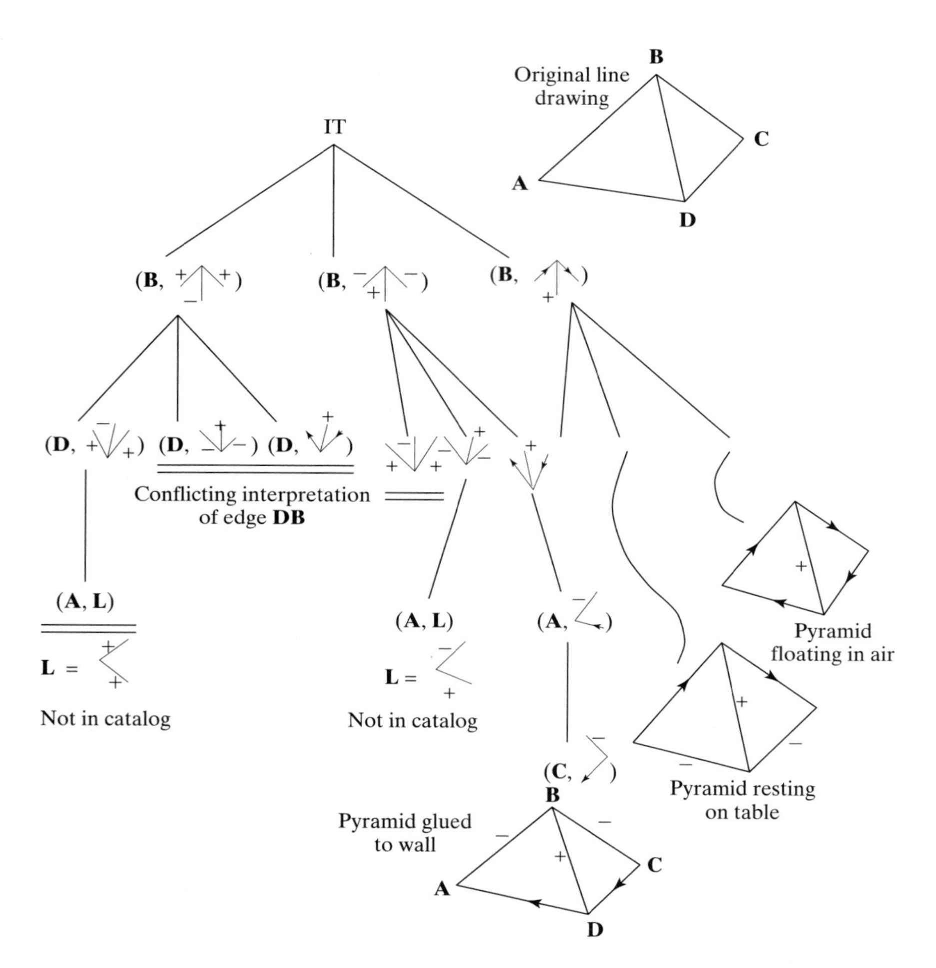

Figure 12.8 Interpretation Tree for the line drawing of a pyramid shown at the top right. At each level of the tree, one of the four junctions is labeled with one of the 16 junction labels from the catalog in Figure 12.6. At the first level of the tree, an interpretation is assigned to junction **B**; subsequent levels assign interpretations to junctions **D**, **A**, and **C**. Three paths complete to yield the three interpretations shown at the bottom right.

Assign consistent interpretations to all edges of a scene graph.

Input: a graph representing edges $\mathbf{E}$ and junctions $\mathbf{V}$.
Output: a mapping of edge set $\mathbf{E}$ onto label set $\mathbf{L} = \{+, -, >, <\}$.

- Assume some order on the set of edges, which may be arbitrary: $\mathbf{E} = \{P_1, P_2, \ldots, P_n\}$.
- At forward stage i, try labeling edge P_i using the next untried label from label set $\mathbf{L} = \{+, -, >, <\}$.
- Check the consistency of the new label with all other edges that are adjacent to it via a junction in $\mathbf{V}$.
- If a newly assigned label produces a junction that is not in the catalog, then backtrack; otherwise, try the next forward stage.

Algorithm 12.1 Labeling the edges of Blocks via Backtracking.

Exercise 12.11

Complete the IT shown in Figure 12.8 by providing all the edges and nodes omitted from the right side of the tree.

Exercise 12.12

Construct the 5-level IT to assign consistent labels to all the edges of the pyramid shown in Figure 12.8. First, express the problem using the consistent labeling formalism from Chapter 11; define P, L, R_P, R_L, using the 5 observed edges and the 4 possible edge labels. Secondly, sketch the IT. Are there three completed paths corresponding to the three completed paths in Figure 12.8?

Line Labeling via Relaxation As we have studied in Chapter 11, a discrete relaxation algorithm can also be used to constrain the interpretations of the parts of the line drawing. Here we assign labels to the edges of the line drawing: a similar procedure can be used to assign labels to the junctions.

Algorithm 12.2 is a simple representative of a large area of work with many variations. Because its simple operations can be done in any order—or even in parallel within each stage—the paradigm makes an interesting model for what might be happening in the human neural network downstream from the retina. Researchers have studied how to incorporate constraints from intensity and how to work at multiple resolutions. The blocks world work is very interesting and has been fruitful in its extensions. However, in the toy form presented here, it is useless for almost all real scenes, because (a) most 3D objects do not satisfy the assumptions and (b) actual 2D image representations are typically quite far from the

Assign consistent interpretations to all edges of a scene graph.

Input: a graph representing edges **E** and junctions **V**.
Output: a mapping of edge set **E** onto *subsets of* label set $L = \{+, -, >, <\}$.

- Initially assign all labels $\{+, -, >, <\}$ to the label set of every edge P_i.
- At every stage, filter out possible edge labels by working on all edges as follows:
  - If a label L_j cannot form a legal junction using the possible labels for the edges connected to edge P_i, then eliminate label L_j from the label set of P_i.
- Stop iterating when no more label sets decrease in size.

Algorithm 12.2 Labeling the edges of Blocks via Discrete Relaxation.

line drawings required. Extensions, such as extending the label and junction catalogues to handle objects with curved surfaces and fixing errors in the line drawing, have been made and are mentioned in the references.

Exercise 12.13: Demonstrating the Necker Phenomena.

This exercise is a variation on a previous one concerning the labeling of the edges of the image of a pyramid in general position (see previous exercise). Figure 12.9 shows a wireframe drawing of a cube: There is no occlusion and all 12 creases are visible as edges in the image. (a) Stare at the leftmost of the three drawings. Does your visual system produce a 3D interpretation of the line drawing? Does this interpretation change over a few minutes of staring? (b) Label the center drawing so that junction G is the image of a front corner. Cross out junction H and the three edges incident there so that a solid cube is represented. (c) Repeat (b) except this time have junction H be the image of a front corner and delete the edges incident on G. *Note that the 3D wireframe object is NOT a legal object in the*

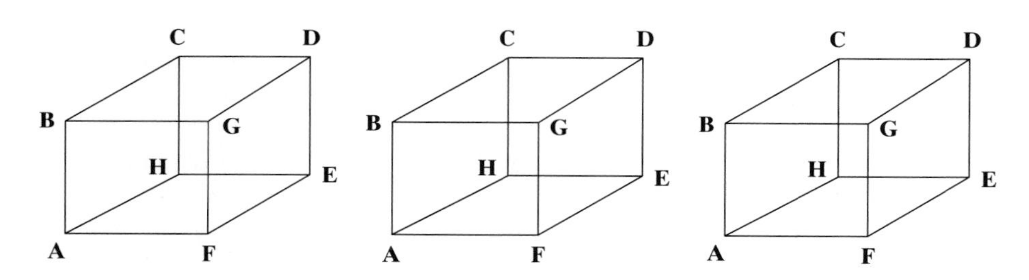

Figure 12.9 See the related Exercise 12.13. The *Necker Cube* has multiple interpretations: A human staring at one of these figures typically experiences changing interpretations. The interpretations of the two fork junctions in the center flip flop between the image of a front corner and the image of a back corner.

blocks world we defined. However, we can use the same reasoning in interpreting small neighborhoods about any of the cube corners, which do belong to the junction catalog.

Exercise 12.14

Apply the interpretation tree procedure of Chapter 11 to the line drawings in Figure 12.7. Show any correct labelings resulting from completed paths.

12.3 3D CUES AVAILABLE IN 2D IMAGES

An image is a 2D projection of the world. However, anyone who appreciates art or the movies knows that a 2D image can evoke rich 3D perceptions. There are many cues used to make 3D interpretations of 2D imagery.

Several depth cues can be seen in the image of Figure 12.10. Two sleeping persons occlude the bench, which occludes a post, which occludes an intricate railing, which occludes the trees, which occlude the buildings with the spires, which occlude the sky. The sun is off to the far right as is indicated by the shadow of the rightmost lamp post and its brighter right side. Also, the unseen railing at the right casts its intricate shadow on the

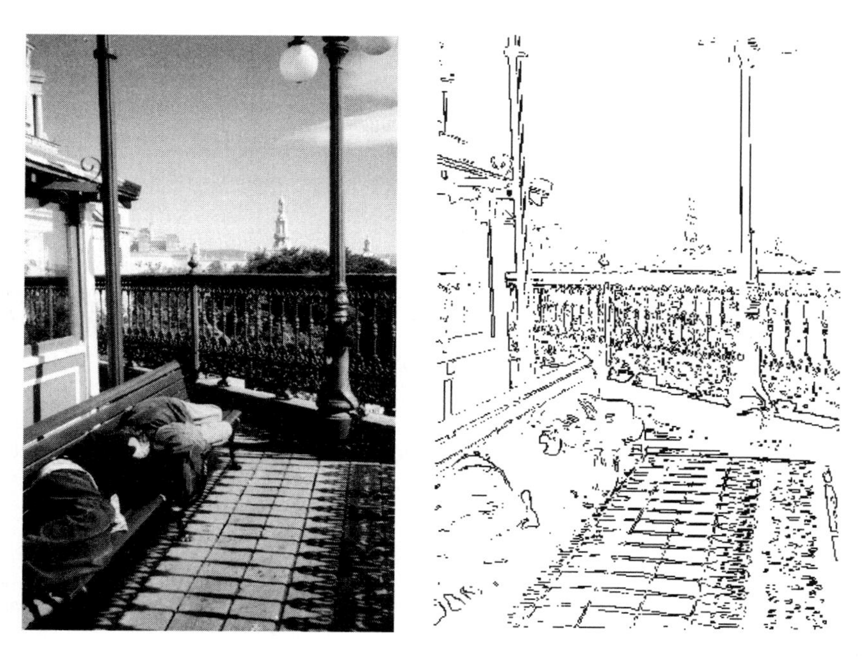

Figure 12.10 In Quebec City above the cliff above the great St. Lawrence River: (left) image showing many depth cues and (right) result of Roberts edge operator followed by thresholding to pass 10 percent of the pixels.

planks, giving the false appearance of a tiled patio. The texture on the ground indicates a planar surface and the shrinking of the texture indicates the gradual regress of the ground away from the viewer. The orientation of the building wall at the left is obvious to a human interpreter from the orientation of the edges on it. The image of the railing tapers from right to left giving a strong cue that its depth is receding in 3D. Similarly, the boards of the bench taper from left to right. The images of the lamp post and the people are much larger than the images of the spires, indicating that the spires are far away.

96 Definition. **Interposition** occurs when one object occludes another object, thus indicating that the occluding object is closer to the viewer than the occluded object.

Exercise 12.15

Apply the relaxation labeling procedure of Chapter 11 to the line drawings in Figure 12.7. If the label set for any edge becomes NULL, then there is no consistent interpretation. If any edges have more than one label in the final label set, then there are ambiguous interpretations as far as this algorithm is concerned. In such cases, use the multiple labels on the line drawing and verify which are actually realizable.

Exercise 12.16

Find all T-junctions in the line segments of the box in Figure 12.3. Does each truly indicate occlusion of one surface by another?

Object interposition gives very strong cues in interpreting the image of Figure 12.10 as discussed above. Clearly, the bench is closer than the post that it occludes, and the lamp post is closer than the railing. Recognition of individual objects may help in using this cue; however, it is not necessary. T-junctions formed in the image contours give a strong local cue. See Figure 12.11. Note that in the edge image at the right of Figure 12.10 the building edge is the bar of a T-junction formed with the top of the bench and the railing forms the bar of a T-junction with the right edge of the lamp post. A matching pair of T-junctions is an even stronger cue because it indicates a continuing object passing behind another. This edge image is difficult as is often the case with outdoor scenes: for simpler situations, consider the next exercise. **Interposition of recognized objects or surfaces can be used to compute the relative depth of these objects.**

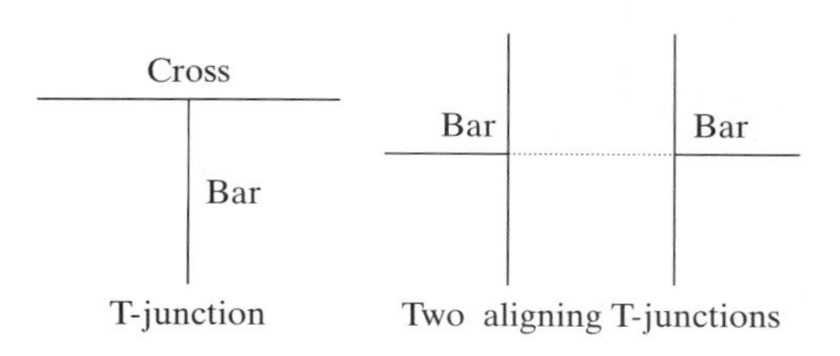

Figure 12.11 T-junctions indicate occlusion of one object by another: The inference is that the edge of the cross corresponds to the occluding object, while the edge of the bar corresponds to the occluded object. Two aligning T-junctions provide much stronger evidence of occlusion than just one.

97 Definition. **Perspective scaling** indicates that the distance to an object is inversely proportional to its image size. The term *scaling* is reserved for comparing object dimensions that are parallel to the image plane.

Once we recognize the spires in Figure 12.10, we know they are quite distant because their image size is small. The vertical elements of the railing decrease in size as we see them recede in distance from right to left. Similarly, when we look down from a tall building to the street below, our height is indicated by how small the people and cars appear. **The size of an object recognized in the image can be used to compute the depth of that object in 3D.**

98 Definition. **Foreshortening** of an object's image is due to viewing the object at an acute angle to its axis and gives another strong cue of how the 2D view relates to the 3D object.

As we look at the bench in Figure 12.10 and the people on it, their image length is shorter than it would be were the bench at a consistent closest distance to the viewer. Similarly, the vertical elements of the railing get closer together in the image as they get further away in the scene. This foreshortening would not occur were we to look perpendicularly at the railing. Texture gradient is a related phenomena. Elements of a texture are subject to scaling and foreshortening, so these changes give a viewer information about the distance and orientation of a surface containing texture. This effect is obvious when we look up at a brick building, along a tiled floor or railroad track, or out over a field of corn or a stadium crowd. Figure 12.12 demonstrates this. Texture gradients also tell us something about the shape of our friends' bodies when they wear clothes with regular textured patterns. A sketch of a simple situation creating a texture gradient is given in Figure 12.13: The texels, or images of the dots, move closer and closer together toward the center of the image corresponding to the increasing distance in 3D. Figure 12.14 shows how a texture is formed when objects in a scene are illuminated by a regular grid of light stripes. This artificial (structured) lighting not only gives humans more information about the shape of the surfaces, but also allows

Figure 12.12 Image of a cornfield shows multiple textures (corn plants and rows of corn plants) and texture gradients. Texture becomes more dense from bottom to top in the image because each square centimeter of image contains more corn leaves. (Image courtesy of John Gerrish.)

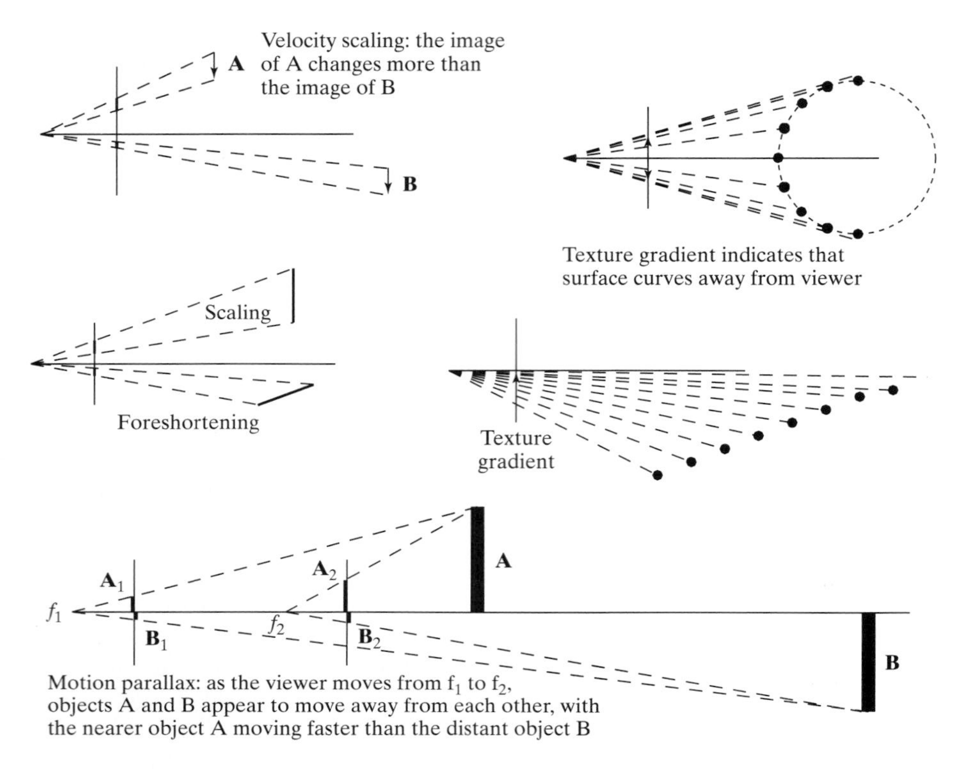
Velocity scaling: the image of A changes more than the image of B
A
B
Texture gradient indicates that surface curves away from viewer
Scaling
Foreshortening
Texture gradient
A
A_1
A_2
f_1
f_2
B_1
B_2
B
Motion parallax: as the viewer moves from f_1 to f_2, objects A and B appear to move away from each other, with the nearer object A moving faster than the distant object B

Figure 12.13 Sketches of the effects of scaling, foreshortening, texture gradient, and motion parallax. In each case, the front image plane is represented by a vertical line segment; objects are toward the right.

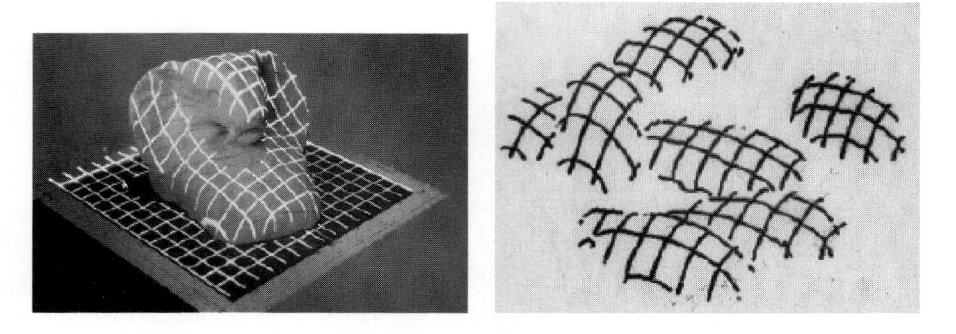

Figure 12.14 The texture formed by light stripes projected onto objects reveals their 3D surface shape: (left) a grid of light projected on a smooth sculpture on a planar background; (right) stripes on objects with planar background removed—what are the objects? (Images courtesy of Gongzhu Hu.)

automatic computation of surface normal or even depth as the next chapter will show. Change of texture in an image can be used to compute the orientation of the 3D surface yielding that texture.

99 Definition. **Texture gradient** is the change of image texture (measured or perceived) along some direction in the image, often corresponding to either a change in distance or surface orientation in the 3D world containing the objects creating the texture.

Regular textured surfaces in 3D viewed nonfrontally create texture gradients in the image, but the reverse may not be true. Certainly, artists create the illusion of 3D surfaces by creating texture gradients on a single 2D sheet of paper.

100 Definition. **Motion parallax** gives a moving observer information about the depth to objects, as even stationary objects appear to move relative to each other: the images of closer objects will move faster than the images of distant objects.

Although motion parallax is the result of viewer motion, a similar effect results if the viewer is stationary and the objects are moving. Figure 12.13 relates the several effects under discussion to the perspective viewing projection. When we walk down a street (assume one eye closed), the image of objects we pass, such as trash cans or doorways, move much faster across our retina than do the images of the same kind of objects one block ahead of us. When driving in a car, oncoming vehicles in the distance present stable images which ultimately speed up to pass rapidly by our side window. Similarly, the images of cars passing by a standing person change much faster than the images of cars one block away. Motion parallax is related to scaling and foreshortening by the same mathematics of perspective projection.

There are even more 3D cues available from single 2D images than what we have discussed. For example, distant objects may appear slightly more blueish than closer objects. Or, their images may be less crisp due to scattering effects of the air in between these objects and the viewer. Also, depth cues can be obtained from focusing, as will be discussed in Chapter 13. Moreover, we have not discussed some other assumptions about the real world; for example, we have not assumed a ground plane or a world with gravity defining a special vertical direction, two conditions under which the human visual system evolved.

Exercise 12.17

Examine a pencil with one eye closed. Keep it parallel to the line between your eyeballs and move it from your face to arms length. Is the change in image size due to scaling, foreshortening, or both? Now hold the pencil at its center and maintain a fixed distance between your eye and its center: Rotate the pencil about its center and observe the change in its image. Is the change due to scaling, foreshortening, or both? Write an approximate trigonometric formula for image size as a function of rotation angle.

Exercise 12.18

Hold a finger vertically and close in front of your nose and alternately open one eye only for two seconds. Observe the apparent motion of your finger (which should not be moving). Move your finger further away and repeat. Move your finger to arms length and repeat again. (It might help to line up your finger tip with a door knob or other object much farther away.) Describe the amount of apparent motion relative to the distance of finger to nose.

12.4 OTHER PHENOMENA

In Chapter 10, we discussed principles, such as the Gestalt principles, for grouping image features to make larger structures. These principles are often fruitful in inverting the image to obtain a 3D interpretation—and sometimes are deceiving, in the sense that they stimulate incorrect interpretations for some sets of conditions. Some additional important phenomena for interpreting 3D structure from 2D image features are briefly discussed next.

12.4.1 Shape from X

The 1980s saw a flurry of work on computational models for computing surface shape from different image properties. The research usually concentrated on using a single image property rather than combinations of them. Some of the mathematical models are examined in Chapter 13, while the phenomena used are discussed in more detail below. Here we just introduce the property X used as a 3D shape cue.

Shape from Shading Use of shading is taught in art class as an important cue to convey 3D shape in a 2D image. Smooth objects, such as an apple, often present a highlight at points where a reception from the light source makes equal angles with reflection toward the viewer. At the same time, smooth objects get increasingly darker as the surface normal becomes perpendicular to rays of illumination. Planar surfaces tend to have a homogeneous appearance in the image with intensity proportional to the angle made between the normal to the plane and the rays of illumination. Computational formulas have been developed for computing surface normal from image intensity; however, most methods require calibration data in order to model the relationship between intensity and normal and some methods require multiple cameras. Recall that computational procedures must in general use a model formula relating several parameters—the reflected energy, the incident energy, the direction of illumination, the direction of reflection, and the orientation and reflectance of the surface element. With so many parameters, SFS can only be expected to work well by itself in highly controlled environments. Figure 12.15 shows an image of a uniform cylinder with a grid of lines illuminated from a single direction, while Figure 12.16 shows two images of buff objects yielding good shading information for perceiving shape.

Shape from Texture Whenever texture is assumed to lie on a single 3D surface and to be uniform, the notion of texture gradient in 2D can be used to compute the 3D orientation of the surface. The concept of texture gradient has already been described. Figure 12.18 shows a uniform texture on a 3D surface viewed at an angle so that a texture

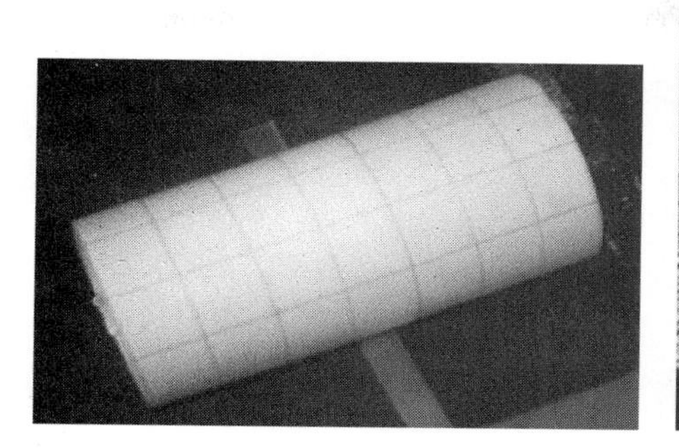

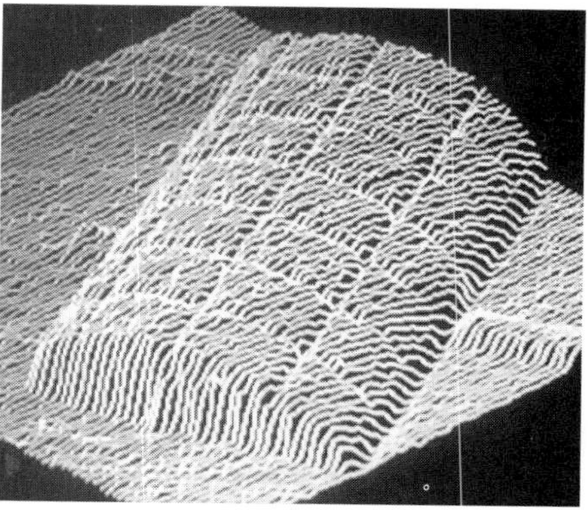

Figure 12.15 (left) Image of a carefully illuminated cylinder formed by wrapping a gridded paper around a can and (right) a 3D plot of the intensity function from a slightly different viewpoint. Note how the shape of the cylinder is well represented by the intensities.

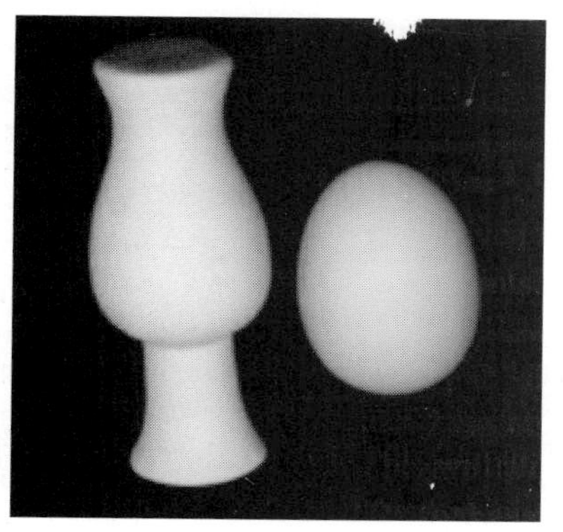

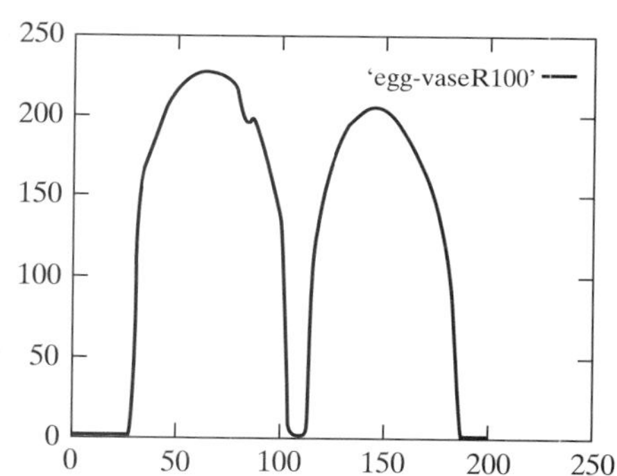

Figure 12.16 Intensity image of smooth buff objects—a vase and an egg—and a plot of intensities across the highlighted row. Note how the intensities are closely related to the object shape. (Courtesy of D. Trytten.)

gradient is formed in the 2D image. There are two angles specially defined to relate the orientation of the surface to the viewer.

101 Definition. The **tilt** of a planar surface is defined as the direction of the surface normal projected in the image. The **slant** of a surface is defined to be the angle made between the surface normal and the line of sight. See Figure 12.18.

Consider a person standing erect and looking ahead in a flat field of wheat. Assuming that the head is vertical, the tilt of the field is 90 degrees. If the person is looking far away, then the slant is close to 90 degrees; if the person is looking just beyond her feet, then the slant is close to zero. Now, if the person rotates her head 45 degrees to the left, then the tilt of the field becomes 45 degrees, while a 45 degree tilt of the head toward the right would

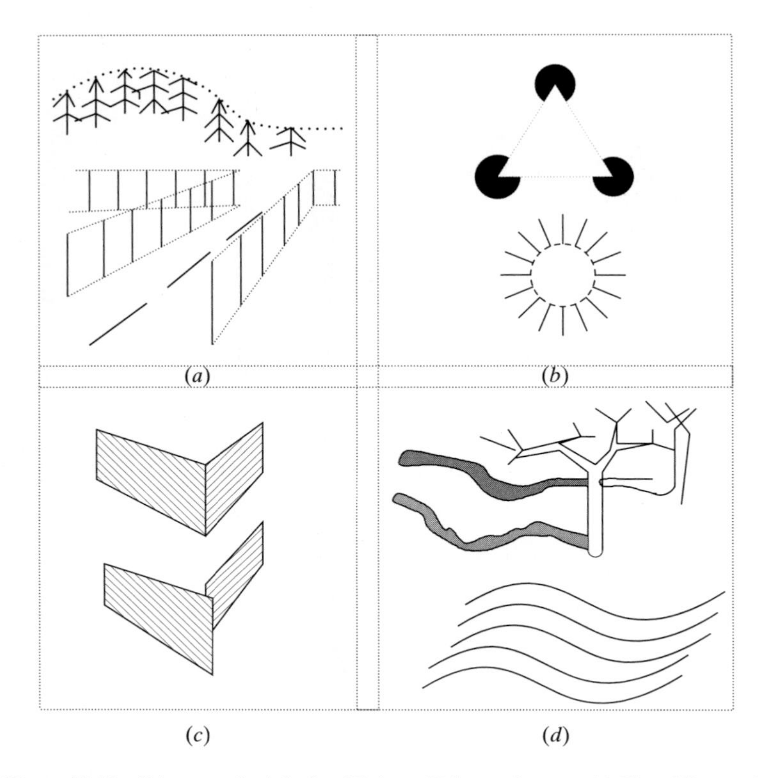

Figure 12.17 Other cues for inferring 3D from 2D image features: (*a*) Virtual lines and curves can be formed by grouping of similar features; (*b*) virtual boundaries can deceive humans into perceiving interposing objects with intensity different from the background; (*c*) alignments in 2D usually, but not always, indicate alignment in 3D; and (*d*) 2D image curves induce perception of 3D surface shape.

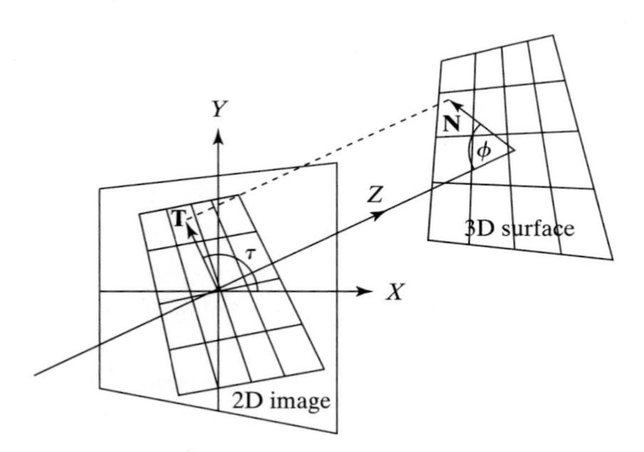

Figure 12.18 Tilt and slant of a surface are determined from the orientation of the surface normal **N** relative to the viewing coordinate system. *Tilt* τ is the direction of **N** projected into the image (**T**). *Slant* ϕ is the angle between **N** and the line of sight.

Figure 12.19 (left) Image containing many textures and (right) result of 5 × 5 Prewitt edge operator. The walk has tilt of 90 degrees and slant of about 75 degrees: the brick wall has tilt of about 170 degrees and slant of about 70 degrees.

produce a tilt of 135 degrees for the field surface. Figure 12.19 is a picture containing two major planar surfaces, a walkway on the ground plane and a terrace wall. The walkway has tilt of 90 degrees and slant of about 75 degrees. (The path inclines 15 degrees uphill.) The terrace wall has a tilt of about 170 degrees and slant of about 70 degrees. The concepts of tilt and slant apply to any surfaces, not just those near the ground plane; for example, outside or inside walls of buildings, the faces of boxes or trucks. In fact, these concepts apply to the elements of curved surfaces as well; however, the changing surface normal makes it more difficult to compute the texture gradient in the image.

Exercise 12.19

(a) Give the tilt and slant for each of the four faces of the object in Figure 12.5. (b) Do the same for the faces of the objects in Figure 12.1.

Shape from Boundary Humans infer 3D object shape from the shape of the 2D boundaries in the image. Given an ellipse in the image, immediate 3D interpretations are disk or sphere. If the circle has uniform shading or texture the disk will be favored, if the shading or texture changes appropriately toward the boundary, the sphere will be favored. Cartoons and other line drawings often have no shading or texture, yet humans can derive 3D shape descriptions from them.

Computational methods have been used to compute surface normals for points within a region enclosed by a smooth curve. Consider the simple case of a circle: The smoothness assumption means that in 3D, the surface normals on the object limb are perpendicular to both the line of sight and the circular cross section seen in the image. This allows us to assign a unique normal to the boundary points in the image. These normals are in opposite directions for boundary points that are the endpoints of diameters of the circle. We can then interpolate smooth changes to the surface normals across an entire diameter, making sure that the middle pixel has a normal pointing right at the viewer. Surely, we need to make an additional assumption to do this, because if we were looking at the end of an ellipse, the surface would be different from a sphere. Moreover, we could be looking at the inside of a half spherical shell! Assumptions can be used to produce a unique surface, which may be the wrong one. Shading information can be used to constrain the propagation of surface normals. Shading can differentiate between an egg and a ball, but maybe not between the outside of a ball and the inside of a ball.

Exercise 12.20

Find a cartoon showing a smooth human or animal figure. (a) Is there any shading, shadows, or texture to help with perception of 3D object shape? If not, add some yourself by assuming a light bulb at the front top right. (b) Trace the object boundary onto a clean sheet of paper. Assign surface normals to 20 or so points within the figure boundary to represent the object shape as it would be represented in an intrinsic image.

12.4.2 Vanishing Points

The perspective projection distorts parallel lines in interesting ways. Artists and drafts persons have used such knowledge in visual communication for centuries. Figure 12.20 shows two well-known phenomena. First, a 3D line skew to the optical axis will appear to vanish at a point in the 2D image called the *vanishing point*. Secondly, a group of parallel lines will have the same vanishing point as shown in the figure. These effects are easy to show from the algebraic model of projective projection. *Vanishing lines* are formed by the vanishing points from different groups of lines parallel to the same plane. In particular, a *horizon line* is formed from the vanishing points of different groups of parallel lines on the ground plane. In Figure 12.20, points $\mathbf{V_1}$ and $\mathbf{V_3}$ form a horizon for the ground plane formed by the rectangular tiled surface. Note that the independent bundle of three parallel lines (highway) vanishes at point $\mathbf{V_2}$ which is on the same horizon line formed by the rectangular texture. Using these properties of perspective, camera models can be deduced from imagery taken from ordinary uncalibrated cameras. Recently, systems have been constructed using these principles that can build 3D models of scenes from an ordinary video taken from several viewpoints in the scene.

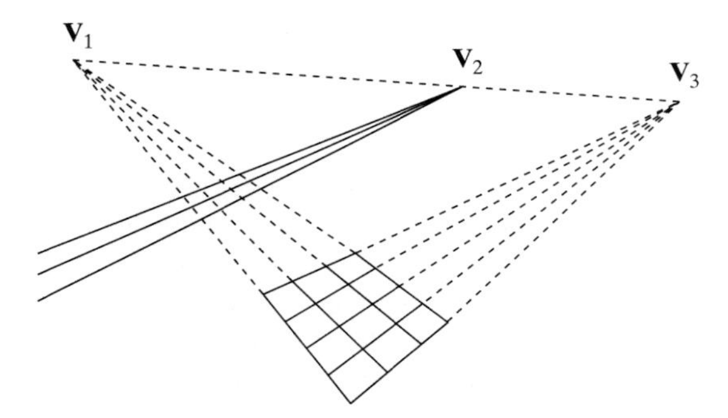

Figure 12.20 Under perspective projection, lines in 3D slanted with respect to the optical axis appear to vanish at some point in the 2D image, and parallel lines appear to intersect at the same *vanishing point*.

12.4.3 Depth from Focus

A single camera can be used to compute the depth to the surface imaged at a pixel as can the human eye. Muscles in the human eye change its shape and hence the effective focal length causing the eye to focus on an object of attention. By bringing an object, or object edges, into focus, the sensor obtains information on the range to that object. Devices operating on this principle have been built, including the control for automatic focus cameras. Being brief here, we can imagine that the focal length of the camera is smoothly changed over a range of values; for each value of f an edge detector is applied to the resulting image. For each pixel, the value of f which produces the sharpest edge is stored and then used to determine the range to the 3D surface point imaged at that pixel. Many image points will not result from a contrasting neighborhood in 3D and hence will not produce usable edge sharpness values. Lenses with short focal length, for example, $f < 8mm$, are known to have a good *depth of field,* which means that an object remains in sharp focus over a wide range from the camera. Short focal lengths would not be good for determining an accurate range from focus; longer focal lengths do a better job. In Section 12.7, we look at how this follows from the lens equation from physics.

Use of Intensity and Shadows It has already been mentioned that structured illumination is useful for producing features on otherwise uniform surfaces. Shadows can be helpful in a similar manner. Humans and machines can infer the existence and shape of a surface from any pattern on it. Consider the bottom right example in Figure 12.17. The pattern of curves at the bottom stimulate a 3D interpretation of an undulating surface. The shadows from the trees on snow-covered terrain are helpful to skiers who can easily lose their balance from even a six-inch mistake in judging ground elevation. A similar situation is shown in Figure 12.14, where the pattern of projected light stripes reveals the ellipsoidal shape of some potatoes.

12.4.4 Motion Phenomena

Motion parallax has already been considered. When a moving visual sensor pursues an object in 3D, points on that object appear to expand in the 2D image as the sensor closes in on the object. (Points on that object would appear to contract, not expand, if that object were escaping faster than the pursuer.) The point which is the center of pursuit is called the *focus of expansion*. The same phenomena results if the object moves toward the sensor: In this case, the rapid expansion of the object image is called *looming*. Chapter 9 treated these ideas in terms of optical flow. Quantitative methods exist to relate the image flow to the distance and speed of the objects or pursuer.

12.4.5 Boundaries and Virtual Lines

Boundaries or curves can be *virtual,* as shown in Figure 12.17. At the top left, image curves are formed by the ends of the fence posts, the tips of the trees, and the painted dashes on the highway. The top right shows two famous examples from psychology: Humans see a brighter triangular surface occluding three dark circles, and a brighter circular surface occluding a set of rays radiating from a central point. It has been hypothesized that once the human visual system perceives (incorrectly) that there is an interposing object, it then must reject the interpretation that that object accidentally has the same reflectance as the background it occludes. So strong is this perception that the dotted virtual curves provided in Figure 12.17 need not be provided to a human at all. A machine vision system will not be fooled into perceiving that the central region is brighter than the background because it has objective pixel intensities.

102 Definition. **Virtual lines** or curves are formed by a compelling grouping of similar points or objects along an image line or curve.

Exercise 12.21

Carefully create two separate white cards containing the illusory figures at the top right of Figure 12.17. Show each to five human subjects to determine if they do see a brighter central region. You must not ask them this question directly: Instead, just ask them general questions to get them to describe what they perceive. Ask *What objects do you perceive?* Ask *Please describe them in terms of their shape and color.* Summarize your results.

12.4.6 Alignments are Non-accidental

Humans tend to reject interpretations that imply *accidental alignments* of objects in space or of objects and the viewer. Instead, alignments in the 2D image are usually assumed to have some cause in 3D. For example, a human viewing the two quadrilateral regions at the top of the panel in the lower left of Figure 12.17 will infer that two [rectangular] surfaces meet in 3D at the image edge, which is perceived as a crease due to the foreshortening cues. The bottom of the panel shows how the image might appear after a small change in viewpoint: the fork and arrow junctions from the top have become T-junctions, and the perception is now of one surface occluding another. Perception of virtual curves is another form of this same principle. Actually, all of the four panels of Figure 12.17 could be included under the same principle. As proposed in Irving Rock's 1983 treatise, the human visual system seems to accept the simplest hypothesis that explains the image data. (This viewpoint explains many experiments and makes vision similar to reasoning. However, it also seems to contradict some experimental data and may make vision programming very difficult.)

Some of the heuristics that have been proposed for image interpretation follow. None of these give correct interpretations in all cases: counterexamples are easy to find. We use the term *edge* here also to mean crease, mark, or shadow in 3D in addition to its 2D meaning.

- From a straight edge in the image, infer a straight edge in 3D.
- From edges forming a junction in the 2D image, infer edges forming a corner in 3D. (More generally, from coincidence in 2D infer coincidence in 3D.)
- From similar objects on a curve in 2D, infer similar objects on a curve in 3D.
- From a 2D polygonal region, infer a polygonal face in 3D.
- From a smooth curved boundary in 2D, infer a smooth object in 3D.
- Symmetric regions in 2D correspond to symmetric objects in 3D.

12.5 THE PERSPECTIVE IMAGING MODEL

We now derive the algebraic model for perspective imaging. The derivation is rather simple when done by relating points in the camera coordinate frame **C** to the real image coordinate frame **R**. First, consider the simple 1D case shown in Figure 12.21, which is an adequate model for problems such as imaging a flat surface of earth from an airplane looking directly downward. The sensor observes a point **B** which projects to image point **E**. The center of the sensor coordinate system is point **O** and the distance from **O** to **B** measured along the optical axis from **O** to **A** is z_c. Point **B** images a distance x_i from the image center. f is the focal length. Using similar triangles, we obtain Equation 12.1. This says that the real 2D image coordinate (or size) is just the 3D coordinate (or size) scaled by the ratio of the focal length to the distance. Provided that all points being imaged lie on the same plane at the same distance from the sensor, the image is just a scaled down version of the 2D world. This model is applicable to real applications such as analyzing microscope, areal images, or scanned documents.

$$x_i/f = x_c/z_c \quad \text{or} \quad x_i = (f/z_c)\, x_c \tag{12.1}$$

It is convenient to think of a *front image plane* rather than the actual image plane so that the image is not inverted in direction relative to the world. The front image plane is an abstract image plane f units on the world side of the optical center: Objects on the front image plane have the same proportions as those on the actual image plane and the same sense of direction as those in the real world. Points **C** and **D** are the front image plane points corresponding to points **F** and **E** on the real image plane. The perspective imaging equation holds for points on the front image plane. From now on, the front image plane will be used for our illustrations.

The case of perspective projection from 3D to 2D is shown in Figure 12.22 and modeled by the Equations 12.2. The equations for both x and y dimensions are obtained using the same derivation from similar triangles as in the 1D case. Note that, as a projection

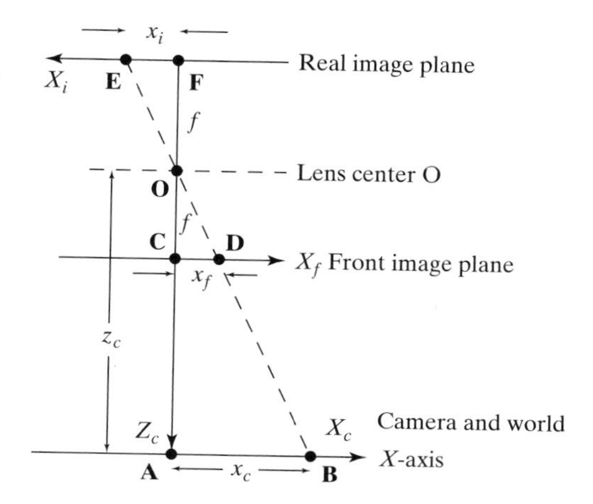

Figure 12.21 Simple perspective model with actual and front image planes. Object size in the world x_c is related to its image size x_i via similar triangles: $x_i/f = x_c/z_c$.

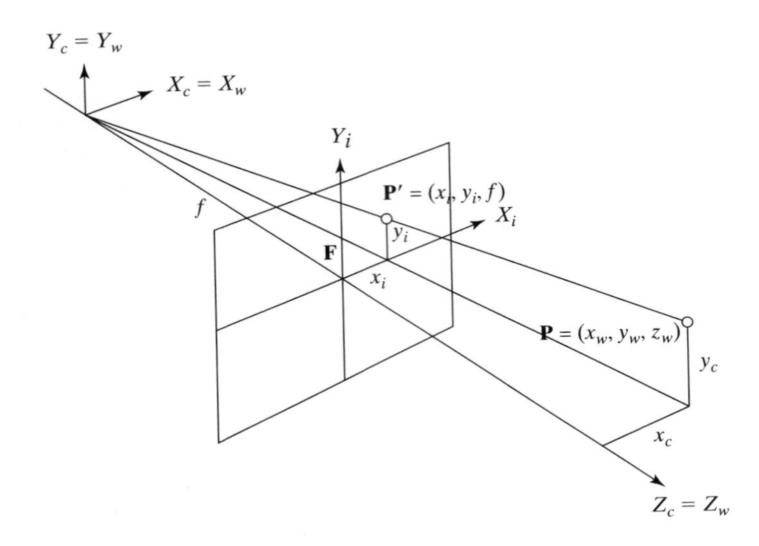

Figure 12.22 General model of perspective projection to a 2D image.

of 3D to 2D, this is a many-to-one mapping. All 3D points along the same ray from an image point out into the 3D space image to the same 2D image point and thus a great deal of 3D information is lost in the process. Equation 12.2 provides the algebraic model which computer algorithms must use to model the set of all 3D points on the ray from image point (x_i, y_i) out into the world. This algebra is of fundamental importance in the 3D work of the text. Before leaving this topic, we emphasize that the simple Equation 12.2 only relates points in the 3D camera frame to points in the 2D real image frame. Relating points from object coordinate frames or a world coordinate frame requires the algebra of transformations, which we will take up in Chapter 13. If the camera views planar material a constant distance $z_c = c_1$ away, then the image is a simple scaling of that plane. Setting $c_2 = f/c_1$ yields the simple relations $x_i = c_2 x_c$ and $y_i = c_2 y_c$. Thus, we can simply work in image coordinates, knowing that the image is a scaled version of the world.

$$x_i/f = x_c/z_c \quad or \quad x_i = (f/z_c)\, x_c$$
$$y_i/f = y_c/z_c \quad or \quad y_i = (f/z_c)\, y_c \qquad (12.2)$$

Exercise 12.22: Uniform scaling property.

A camera looks vertically down at a table so that the image plane is parallel to the table top (similar to a photo enlarging station) as in Figure 12.21. Prove that the image of a 1-inch-long nail (line segment) has the same length regardless of where the nail is placed on the table within the FOV.

Exercise 12.23: A vision-guided tractor.

See Figure 12.23. Suppose a forward-looking camera is used to guide the steering of a farm tractor and its application of weed control and fertilizer. The camera has a focal length of 100*mm* and is positioned 3,000*mm* above the ground plane as shown in the figure. It has an angular field of view of 50 degrees and its optical axis makes an angle of 35 degrees with the ground plane. (a) What is the length of the FOV along the ground ahead of the tractor?

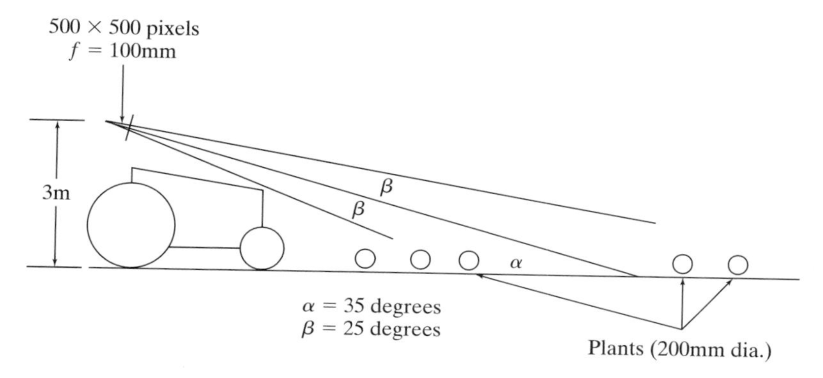

Figure 12.23 Sketch of camera on a vision-guided farm tractor: See Exercise 12.23.

(b) Assuming that plants are spaced 500*mm* apart, what is their spacing in the image when they are at the extremes of the FOV? (Image spacing is different for near and far plants.) (c) If plants are approximated as spheres 200*mm* in diameter and the image is 500×500 pixels, what is their image diameter in pixels? (Again, different answer, for near and far plants.) (d) Will successive plants overlap each other in the image or will there be space in between them?

12.6 DEPTH PERCEPTION FROM STEREO

Only simple geometry and algebra is needed to understand how 3D points can be located in space using a stereo sensor as is shown in Figure 12.24. Assume that two cameras, or eyes, are carefully aligned so that their X-axes are collinear and their Y-axis and Z-axis are parallel. The Y-axis is perpendicular to the page and is not actually used in our derivations. The origin—or center of projection—of the right camera is offset by b, which is the *baseline* of the stereo system. The system observes some object point $\mathbf{P}$ in the left image point $\mathbf{P_l}$ and the right image point $\mathbf{P_r}$. Geometrically, we know that the point $\mathbf{P}$ must be located at the intersection of the ray $\mathbf{LP_l}$ and the ray $\mathbf{RP_r}$.

From similar triangles, Equations 12.3 are obtained.

$$z/f = x/x_l \tag{12.3}$$

$$z/f = (x - b)/x_r$$

$$z/f = y/y_l = y/y_r$$

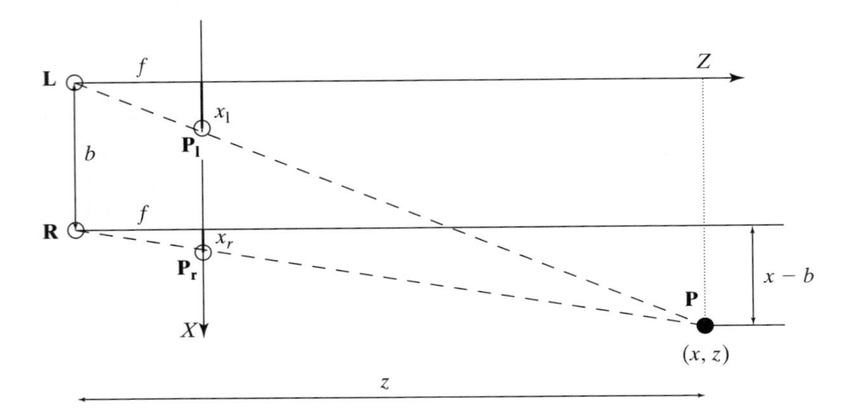

Figure 12.24 Geometric model for a simple stereo system: The sensor coordinate system is established in the left eye **L** (or camera) and the *baseline* is b; all quantities are measured relative to **L**, except for x_r which is measured relative to **R**.

Note that by construction the image coordinates y_l and y_r may be assumed to be identical. A few substitutions and manipulations yields a solution for the two unknown coordinates x and z of the point **P**.

$$z = fb/(x_l - x_r) = fb/d \tag{12.4}$$

$$x = x_l z/f = b + x_r z/f$$

$$y = y_l z/f = y_r z/f$$

In solving for the depth of point **P**, we have introduced the notion of *disparity* d in Equations 12.4, which is the difference between the image coordinates x_l and x_r in the left and right images. Solution of these equations yields all three coordinates completely locating point **P** in 3D space. Equation 12.4 clearly shows that the distance to point **P** increases as the disparity decreases and decreases as the disparity increases. Distance goes to infinity as disparity goes to zero. By construction of this simple stereo imaging system, there is no disparity in the two y image coordinates.

103 Definition. **Disparity** refers to the difference in the image location of the same 3D point when projected under perspective to two different cameras.

Figure 12.24 shows a single point **P** being located in 3D space so there is no problem identifying the matching image points $\mathbf{P_l}$ and $\mathbf{P_r}$. Determining these corresponding points can be very difficult for a real 3D scene containing many surface points because it is often unclear which point in the left image corresponds to which point in the right image. Consider a stereo pair of images from a cornfield as in Figure 12.12. There would be many similar edge points across the image rows. Often, stereo cameras are very precisely aligned so that the search for corresponding points can be constrained to the same rows of the two images. Although many constraints are known and used, problems still remain. One obvious example is the case where a point **P** is not visible in both images. While the busy texture of the cornfield presents problems in handling too many feature points, the opposite problem of having too few points is also common. Too few points are available for smooth objects with no texture, such as a marble statue or a snow-covered hill. In an industrial setting, it is easy to add artificial points using illumination as is shown in Figure 12.14. More details are given below.

Despite the difficulties mentioned above, research and development has produced several commercially available stereo systems. Some designs use more than two cameras. Some systems can produce depth images at nearly the frame rate of video cameras. Chapter 16 discusses use of stereo sensing at an ATM to identify people.

Exercise 12.24

Perform the following informal stereo experiment. (a) View a book frontally about $30cm$ in front of your nose. Use one eye for two seconds, then the other eye, in succession. Do you observe the disparity of the point features, say the characters in the title, in the left and right images? (b) Repeat the experiment holding the book at arm's length. Are the disparities larger or smaller? (c) Rotate the book significantly. Can you find an orientation such that the right eye sees the cover of the book but the left eye cannot?

Exercise 12.25: On the error in stereo computations.

Assume that stereo cameras with baseline $b = 10cm$ and focal lengths $f = 2cm$ view a point at $\mathbf{P} = (10cm, 1{,}000cm)$. Refer to Figure 12.24: Note that point **P** lies along the optical axis of the right camera. Suppose that due to various errors, the image coordinate x_l is 1 percent smaller than its true value, while the image coordinate x_r is perfect. What is the error in depth z, in centimeters, computed by Equations 12.4?

Stereo Displays Stereo displays are generated by computer graphics systems in order to convey 3D shape to an interactive user. The graphics problem is the inverse of the computer vision problem: All the 3D surface points (x, y, z) are known and the system must create the left and right images. By rearranging the Equations 12.4 we arrive at Equations 12.5, which give formulas for computing image coordinates (x_l, y_l) and (x_r, y_r) for a given object point (x, y, z) and fixed baseline b and focal length f. Thus, given a computer model of an object, the graphics system generates two separate images. These two images are presented to the user in one of two ways: (a) one image is piped to the left eye and one is piped to the right eye using a special helmet or (b) the two images are presented alternately on a CRT using complementary colors which the user views with different filters on each eye. There is an inexpensive third method if motion is not needed: Humans can actually fuse a stereo pair of images printed side-by-side on plain paper. (For example, stare at the stereogram that is Figure 12.25 of the paper by Tanimoto (1998) cited in the references.)

$$x_l = xf/z \tag{12.5}$$

$$x_r = f(x - b)/z$$

$$y_l = y_r = yf/z$$

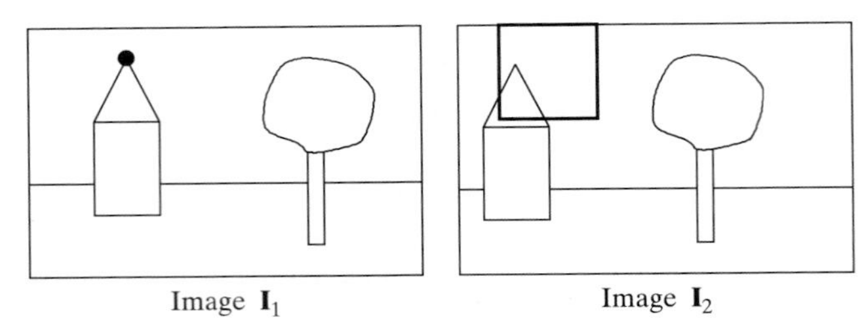

Figure 12.25 The cross-correlation technique for finding correspondences in a stereo pair of images.

Chapter 15 discusses in more detail how stereo displays are used in virtual reality systems, which engage users to an increased degree as a result of the 3D realism. Also, they can be useful in conveying to a radiologist the structure of 3D volumetric data from an MRI device.

12.6.1 Establishing Correspondences

The most difficult part of a stereo vision system is not the depth calculation, but the determination of the correspondences used in the depth calculation. If any correspondences are incorrect, they will produce incorrect depths, which can be just a little off or totally wrong. In this section we discuss the major methods used for finding correspondences and some helpful constraints.

Cross Correlation The oldest technique for finding the correspondence between pixels of two images uses the cross-correlation operator described in Chapter 5. The assumption is that for a given point $\mathbf{P}_1$ in image $\mathbf{I}_1$ (the first image of a stereo pair), there will be a fixed region of image $\mathbf{I}_2$ (the second image of the pair) in which the point $\mathbf{P}_2$ that corresponds to $\mathbf{P}_1$ must be found. The size of the region is determined by information about the camera setup that is used to take the images. In industrial vision tasks, this information is readily available from the camera parameters obtained by the calibration procedure (see Chapter 13). In remote sensing and other tasks, the information may have to be estimated from training images and ground truth. In any case, for pixel $\mathbf{P}_1$ of image $\mathbf{I}_1$, the selected region of $\mathbf{I}_2$ is searched, applying the cross-correlation operator to the neighborhoods about $\mathbf{P}_1$ and $\mathbf{P}_2$. The pixel that maximizes the response of the cross correlation operator is deemed the best match to $\mathbf{P}_1$ and used to find the depth at the corresponding 3D point. The cross-correlation technique has been used very successfully to find correspondences in satellite and aerial imagery. Figure 12.25 illustrates the cross-correlation technique. The black dot in image $\mathbf{I}_1$ indicates a point whose correspondence is sought. The square region in image $\mathbf{I}_2$ is the region to be searched for a match.

Symbolic Matching and Relational Constraints A second common approach to finding correspondences is to look for a feature in one image that matches a feature in the other. Typical features used are junctions, line segments, or regions. This type of matching can use the consistent labeling formalism defined in Chapter 11. The part set $\mathbf{P}$ is the set of features of the first image $\mathbf{I}_1$. The label set L is the set of features of the second image $\mathbf{I}_2$. If features can have more than one type, then the label for a part must be of the same type as that part. (Note that T-junctions are often avoided since they usually are due to

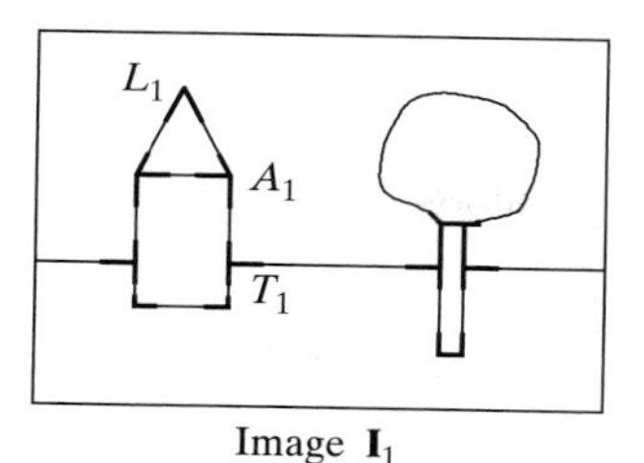

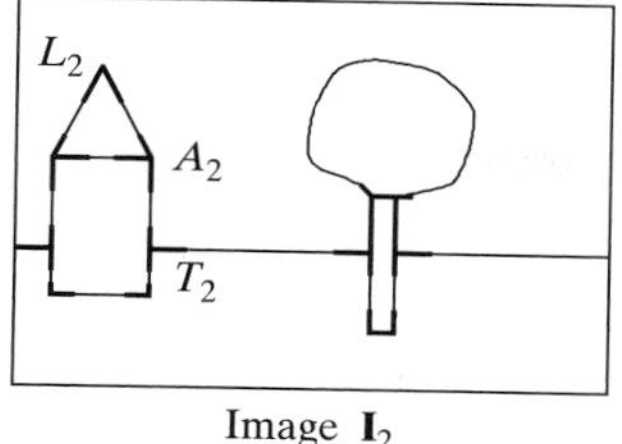

Figure 12.26 Symbolic matching technique for finding correspondences in a stereo pair of images. The L- and A- (arrow) junctions shown are potential matches. The T-junctions probably should be avoided since they usually result from occlusion rather than real features in 3D.

the occlusion of one edge by another and not to the structure of one 3D object.) Furthermore, the spatial relationships R_P that hold over $\mathbf{P}$ should be the same as the spatial relationships R_L that hold over L. For instance, if the features to be matched are junctions, as shown in Figure 12.26, then corresponding junctions should have the same types (an L-junction maps to another L-junction) and if two junctions are connected by a line segment in the first image (L and A for example), the corresponding junctions should be connected by a line segment in the second image. If the features to be matched are line segments, such relationships as parallel and collinear can be used in matching. For region matching, the region adjacency relationship can be used.

This brings up one problem that can occur in any kind of stereo matching. Not every feature of the first image will be detected in the second. Some features are just not there, due to the viewpoint. Some features appear in one image, but are occluded in the other. Some features may be misdetected or just missed, and extraneous features may be found. So the symbolic matching procedure cannot look for a perfect consistent labeling, but instead must use some inexact version, either looking for a least-error mapping or applying continuous relaxation to achieve an approximate answer.

Once a mapping has been found from features of the first image to features of the second, we are not yet done. The correspondence of junctions produces a sparse depth map with the depth known only at a small set of points. The correspondence of line segments can lead to correspondences between their endpoints or, in some work, between their midpoints. The correspondences between regions still requires some extra work to decide which pixels inside the regions correspond. The sparse depth maps obtained can then be augmented through linear interpolation between the known values. As you can see, there is a lot of room for error in this process and that is probably why cross correlation is still widely used in practice, especially where the images are natural—rather than industrial—scenes.

The Epipolar Constraint Stereo matching can be greatly simplified if the relative orientation of the cameras is known, as the two-dimensional search space for the point in one image that corresponds to a given point in a second image is reduced to a one-dimensional search by the so called *epipolar geometry* of the image pair. Figure 12.27 shows the epipolar geometry in the simple case where the two image planes are identical and are parallel to the baseline. In this case, given a point $\mathbf{P}_1 = (x_1, y_1)$ in image $\mathbf{I}_1$, the corresponding point $\mathbf{P}_2 = (x_2, y_2)$ in image $\mathbf{I}_2$ is known to lie on the same scan line; that is, $y_1 = y_2$. We call this a normal image pair.

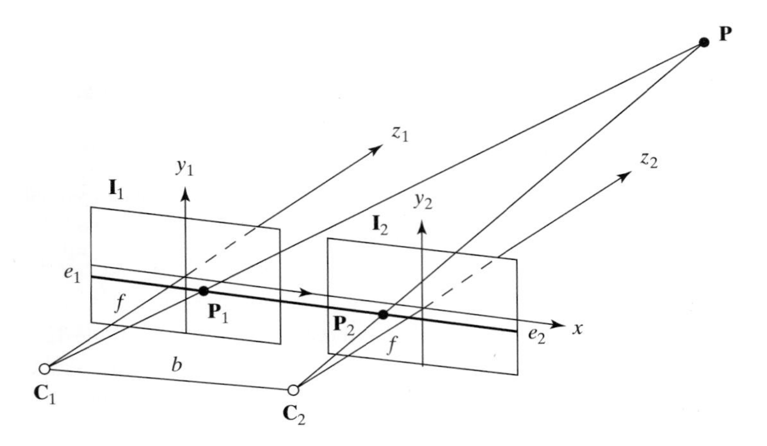

Figure 12.27 Epipolar geometry of a normal image pair: Point **P** in 3D projects to point $\mathbf{P}_1$ on image $\mathbf{I}_1$ and point $\mathbf{P}_2$ on image $\mathbf{I}_2$, which share the same image plane that is parallel to the baseline between the cameras. The optical axes are perpendicular to the baseline and parallel to each other.

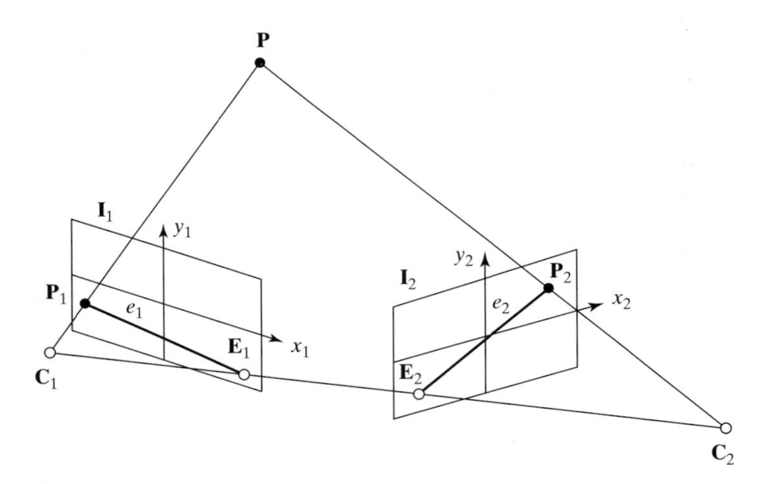

Figure 12.28 Epipolar geometry of a general image pair: Point **P** in 3D projects to point $\mathbf{P}_1$ on image $\mathbf{I}_1$ and point $\mathbf{P}_2$ on image $\mathbf{I}_2$; these image planes are not the same. The epipolar line that $\mathbf{P}_1$ lies on in image $\mathbf{I}_1$ is line e_1, and the corresponding epipolar line that $\mathbf{P}_2$ lies on in image $\mathbf{I}_2$ is line e_2. $\mathbf{E}_1$ is the epipole of image $\mathbf{I}_1$, and $\mathbf{E}_2$ is the epipole of image $\mathbf{I}_2$.

While the normal setup makes the geometry simple, it is not always possible to place cameras in this position, and it may not lead to big enough disparities to calculate accurate depth information. The general stereo setup has arbitrary positions and orientations of the cameras, each of which must view a significant subvolume of the object. Figure 12.28 shows the epipolar geometry in the general case.

104 Definition. The plane that contains the 3D point **P**, the two optical centers (or cameras) $\mathbf{C}_1$ and $\mathbf{C}_2$, and the two image points $\mathbf{P}_1$ and $\mathbf{P}_2$ to which **P** projects is called the **epipolar plane**.

105 Definition. The two lines e_1 and e_2 resulting from the intersection of the epipolar plane with the two image planes $\mathbf{I}_1$ and $\mathbf{I}_2$ are called **epipolar lines**.

Given the point $\mathbf{P}_1$ on epipolar line e_1 in image $\mathbf{I}_1$ and knowing the relative orientations of the cameras (see Chapter 13), it is possible to find the corresponding epipolar line e_2 in image $\mathbf{I}_2$ on which the corresponding point $\mathbf{P}_2$ must lie. If another point $\mathbf{P}_1{}'$ in image $\mathbf{I}_1$ lies in a different epipolar plane from point $\mathbf{P}_1$, it will lie on a different epipolar line.

106 Definition. The **epipole** of an image of a stereo pair is the point at which all its epipolar lines intersect.

Points $\mathbf{E}_1$ and $\mathbf{E}_2$ are the epipoles of images $\mathbf{I}_1$ and $\mathbf{I}_2$, respectively.

The Ordering Constraint Given a pair of points in the scene and their corresponding projections in each of the two images, the ordering constraint states that if these points lie on a continuous surface in the scene, they will be ordered in the same way along the epipolar lines in each of the images. This constraint is more of a heuristic than the epipolar constraint, because we do not know at the time of matching whether or not two image points lie on the same 3D surface. Thus it can be helpful in finding potential matches, but it may cause correspondence errors if it is strictly applied.

Error versus Coverage In designing a stereo system, there is a trade off between scene coverage and error in computing depth. When the baseline is short, small errors in location of image points $\mathbf{P}_1$ and $\mathbf{P}_2$ will propagate to larger errors in the computed depth of the 3D point **P** as can be inferred from the figures. Increasing the baseline improves accuracy. However, as the cameras move further apart, it becomes more likely that image point correspondences are lost due to increased effects of occlusion. It has been proposed that an angle of $\pi/4$ between optical axes is a good compromise.

12.7 THE THIN LENS EQUATION*

The principle of the thin lens equation is shown in Figure 12.29. A ray from object point **P** parallel to the optical axis passes through the lens and focal point $\mathbf{F_i}$ in route to image point $\mathbf{p}'$. Other rays from **P** also reach $\mathbf{p}'$ because the lens is a light collector. A ray through the optical center **O** reaches $\mathbf{p}'$ along a straight path. A ray from image point $\mathbf{p}'$ parallel to the optical axis passes through the lens and second focal point $\mathbf{F_j}$.

The thin lens equation can be derived from the geometry in Figure 12.29. Since distance X is the same as the distance from **R** to **O**, similar triangles $\mathbf{ROF_i}$ and $\mathbf{Sp'F_i}$ give the following equation.

$$\frac{X}{f} = \frac{x'}{z'} \tag{12.6}$$

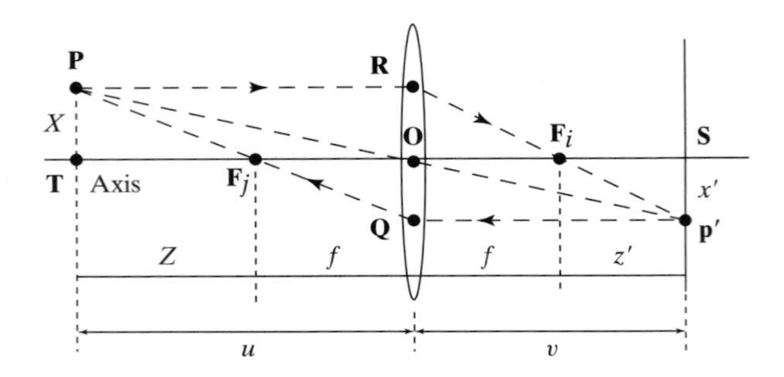

Figure 12.29 Principle of the thin lens. Ray from object point **P** parallel to the optical axis passes through the lens and focal point $\mathbf{F_i}$ in route to image point $\mathbf{p'}$. Ray from image point $\mathbf{p'}$ parallel to the optical axis passes through the lens and focal point $\mathbf{F}_j$.

Using similar triangles **POT** and **p′OS** we obtain a second equation.

$$\frac{X}{f+Z} = \frac{x'}{f+z'} \tag{12.7}$$

Substituting the value of X from Equation 12.6 into Equation 12.7 yields

$$f^2 = Zz' \tag{12.8}$$

Substituting $u - f$ for Z and $v - f$ for z' yields

$$uv = f(u+v) \tag{12.9}$$

and finally dividing both sides by (uvf) yields the most common form of the lens equation, which relates the focal length to the object distance u from the lens center and the image distance v from the lens center.

$$\frac{1}{f} = \frac{1}{u} + \frac{1}{v} \tag{12.10}$$

Focus and Depth of Field Assuming that a point **P** is in perfect focus as shown in Figure 12.29, the point will be out of focus if the image plane is moved, as shown in Figure 12.30. The lens equation, which held for v, is now violated for the new distance v'. Similarly, if the image plane remains constant but point **P** is moved, distance u changes and the lens equation is violated. In either case, instead of a crisp point on the image plane, the image of a world point spreads out into a circle of diameter b. We now relate the size of this circle to the resolution of the camera and its depth of field.

Assume that a blur b the size of a pixel can be tolerated. From this assumption, we will compute the nearest and farthest that a point **P** can be away from the camera and have a blur within this limit. We fix distance u as the nominal depth at which we want to sense; v as the ideal image distance according to the lens equation; a as the aperture of the lens; and f as the focal length used. The situation is sketched in Figure 12.30. Assuming that

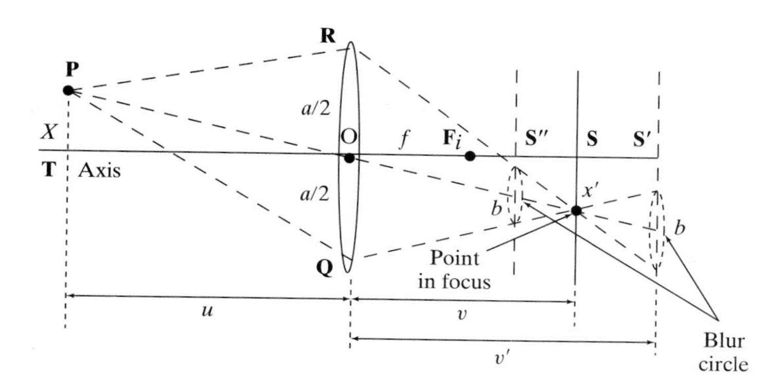

Figure 12.30 Point **P** will be out of focus if either the depth of the point or the position of the image plane violates the lens equation. If **S** is the image location where **P** is in focus, then **P** will be out of focus as the image plane is moved to **S′** or **S″**. The image of **P** will be a disk of diameter b.

perfect focus is achieved for the above conditions, we now investigate how much u can be varied keeping the blur within the limit of b.

Using the extreme cases of v' in Figure 12.30, similar triangles yield

$$v' = \frac{a+b}{a} v \quad : in\ case\ v' > v$$

$$v' = \frac{a-b}{a} v \quad : in\ case\ v' < v \tag{12.11}$$

Note that for $v' > v$, the lens equation shows that u' will be closer to the camera than u and for $v' < v$, u' will be farther. We compute the nearest point u_n that will result in the blur of b shown in the figure, using the lens equation several times to relate the parameters u, v, f and using the relation for $v' > v$ from Equation 12.11.

$$\begin{aligned} u_n &= \frac{fv'}{v'-f} = \frac{f\frac{(a+b)}{a}v}{\frac{(a+b)}{a}v - f} \\ &= \frac{f\frac{(a+b)}{a}\frac{uf}{(u-f)}}{\frac{(a+b)}{a}\frac{uf}{(u-f)} - f} \\ &= \frac{uf(a+b)}{af+bu} = \frac{u(a+b)}{a+\frac{bu}{f}} \end{aligned} \tag{12.12}$$

Similarly, by using $v' < v$ and the same steps given above, we obtain the location u_r of the far plane.

$$u_r = \frac{uf(a-b)}{af-bu} = \frac{u(a-b)}{a-\frac{bu}{f}} \tag{12.13}$$

107 Definition. The **depth of field** is the difference between the far and near planes for the given imaging parameters and limiting blur b.

Since $u > f$ in typical situations, one can see from the final expression in Equation 12.12 that $u_n < u$. Moreover, holding everything else constant, using a shorter focal length f will bring the near point closer to the camera. The same kind of argument shows that $u_r > u$ and that shortening f will push the far point farther from the camera. Thus, shorter focal length lenses have a larger depth of field than lenses with longer focal lengths. (Unfortunately, shorter focal length lenses often have more radial distortion.)

Relating Resolution to Blur A CCD camera with ideal optics can at best resolve $n/2$ distinct lines across n rows of pixels, assuming that a spacing of one pixel between successive pairs of lines is necessary. A 512×512 CCD array can detect a grid of 256 dark lines separated by one-pixel wide rows of bright pixels. (If necessary, we can move the camera perpendicular to the optical axis slightly until the line pattern aligns optimally with the rows of pixels.) Blur larger than the pixel size will fuse all the lines into one gray image. The formulas given above enable one to engineer the imaging equipment for a given detection problem. Once it is known what features must be detected and what standoff is practical, one can then decide on a detector array and lens.

108 Definition. The **resolving power** of a camera is defined as $R_p = 1/(2\Delta)$ in units of lines per inch (*mm*) where Δ is the pixel spacing in inches (*mm*).

For example, if a CCD array $10mm$ square has 500×500 pixels, the resolving power is $1/(2 \times 2 \times 10^{-2} mm/line)$ or 25 lines per *mm*. If we assume that black and white film comprises silver halide molecules $5 \times 10^{-3} mm$ apart, the resolving power is 100 lines per *mm* or 2,500 lines per inch. The cones that sense color in the human eye are packed densely in an area called the *fovea,* perhaps with a spacing of $\Delta = 10^{-4}$ inches. This translates to a resolving power of 5×10^3 lines per inch on the retina. Using $20mm = 0.8in$ as the diameter of an eye, we can compute the subtended angle shown in Figure 12.31 as $\theta \approx sin(\theta) = 2\Delta/0.8in = 2.5 \times 10^{-4}$ radians, which is roughly one minute of arc. This means that a human should be able to detect a line made by a $0.5mm$ lead pencil on a wall 2 meters away.

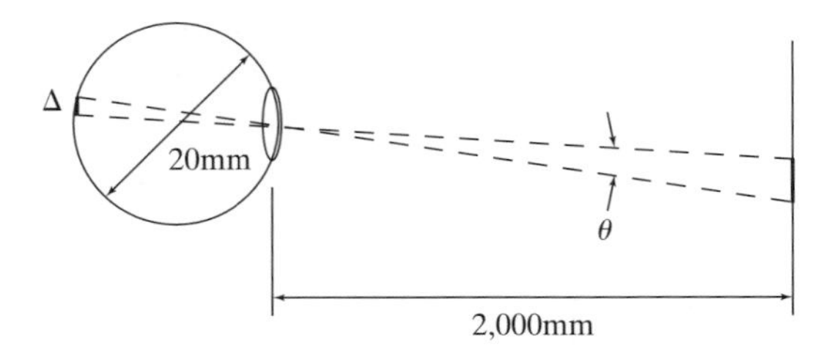

Figure 12.31 A small object images on the human retina.

12.8 CONCLUDING DISCUSSION

This chapter examined a number of relations between 2D image structure and the surfaces and objects in the 3D world. Humans use many of these relations in combination to perceive the world and to navigate within it. Although we only took a careful look at the relationship of depth to stereo and focus, quantitative models have been formulated for many phenomena, including shape from shading and shape from texture. These models are important tools for artists, especially when computer graphics is used, to convey 3D structure using a 2D canvas or 2D graphics display. Chapter 13 shows how some of these methods can be employed for automatic recognition of 3D structure from 2D images. We caution the reader that some of these models are too fragile and inaccurate to be used by themselves except in controlled environments. Combination of computational methods to provide real-time vision for an outdoor navigating robot is still a difficult problem and the subject of much current research activity.

Exercise 12.26

Locate a painting of a plaza or acropolis in an art book and make a copy of it. Mark vanishing points and vanishing lines that the artist may have used in creating the picture.

12.9 REFERENCES

The early book by visual psychologist J. J. Gibson (1950) is a classic in the study of visual information cues. Much of the work of computer vision researchers of the 1980s can be traced to roots in Gibson's book. Many experiments were influenced by the approach of David Marr (1982), whose *information processing paradigm* held that one should first isolate the information being used to make a decision or perception, second explore mathematical models and finally pursue possible implementations. Marr also believed that the human visual system actually constructed fairly complete descriptions of the surfaces of the scene, a belief that is less common today. The 1983 book by visual psychologist Irvin Rock contains a retrospective view of many years of experiments and reaches the conclusion that visual perception requires intelligent operations and shares many properties of reasoning: this book is both a resource for properties of human vision and also for the methodology used to study it. The notion of the intrinsic image was introduced by Barrow and Tenenbaum in (1978). Their proposal is roughly equivalent to the 2 1/2 D sketch proposed by David Marr and appearing in his 1982 book. Our treatment of the intrinsic image has been strongly influenced by the development in Chapter 3 of the text by Charniak and McDermott (1985).

Huffman (1971) and Clowes (1971) are jointly credited with discovery of the junction constraints in blocks world. The extension of that work by Waltz in 1975 to handle shadows and nontrihedral corners enlarged the junction catalogue to thousands of cases—probably too much for a consciously reasoning human to manipulate, but no trouble for a computer. Waltz developed an efficient algorithm to discard possible line labels, which is offen dubbed *Waltz filtering*. The 1977 *A I* text by Winston is a good source of details on the topic of shape interpretation in blocks world: it also contains a treatment of how the junction catalogues are derived.

The parallel relaxation approach which we gave here is derived from a large amount of similar work by Rosenfeld and others (1976), and others, some of which grew from

Waltz's results. Consult Sugihara (1986) for additional geometric constraints which prevent interpretation of line drawings which cannot be formed by actually imaging a blocks world object. The paper by Malik (1987) shows how to extend the catalogue of line and junction labels to handle a large class of curved objects. Work by Stockman and others (1990) shows how sparse depth samples can be used to reconstruct and interpret incomplete line drawings of scenes that are much more general than trihedral blocks world. The two-volume set by Haralick and Shapiro (1992/93) contains a large collection of information about the perspective transformation.

In a popular paper, Marr and Poggio (1979) characterized the human stereo system in terms of information processing and proposed an implementation similar to relaxation. The paper by Tanimoto (1998) includes a section on how stereograms can be used to motivate mathematics; in particular, it includes some color stereograms which humans can view to perceive 3D shape. Creation of stereo material for human viewing is an active area of current research; for example, Peleg and Ben-Ezra (1999) have created stereo scenes of historical places using a single moving camera. Automatic focusing mechanisms abound in the commercial camera market: It is clear that depth to scene surfaces can be computed by fast cheap mechanisms. The work of Krotkov (1987), Nayar and others (1992), and Subbarao and Tyan (1998) provides background in this area.

1. Barrow, H., and J. Tenenbaum. 1978. Recovering intrinsic scene characteristics from images. In *Computer Vision Systems,* A. Hansom and E. Riseman, eds. Academic Press, New York.
2. Charniak, E., and D. McDermott. 1985. *Artifical Intelligence.* Addison-Wesley, Reading, MA.
3. Clowes, M. 1971. On seeing things. *Artificial Intelligence,* v. 2:79–116.
4. Gibson, J. J. 1950. *The Perception of the Visual World.* Houghton-Mifflin, Boston.
5. Haralick, R., and L. Shapiro. 1992/3. *Computer and Robot Vision, Volumes I and II.* Addison-Wesley, Reading, MA.
6. Huffman, D. 1971. Impossible objects as nonsense sentences. In *Machine Intelligence,* v. 6, B. Meltzer and D. Michie, eds. Elsevier, New York, 295–323.
7. Kender, J. 1980. *Shape from Texture,* Ph.D. dissertation. Dept. of Computer Science, Carnegie Mellon Univ., Pittsburgh, PA.
8. Koenderink, J. 1984. What does the occluding contour tell us about solid shape? *Perception,* v. 13.
9. Krotkov, E. 1987. Focusing, *Int. J. Comput. Vision,* v. 1:223–237.
10. Malik, J. 1987. Interpreting line drawings of curved objects. *Int. J. Comput. Vision,* v. 1(1).
11. Marr, D., and T. Poggio. 1979. A computational theory of human stereo vision. *Proc. Royal Society,* v. B 207:207–301.
12. Marr, D. 1982. *Vision: A Computational Investigation into the Human Representation and Processing of Visual Information.* W. H. Freeman and Co., New York.

13. Nayar, S. 1992. Shape from focus system. *Proc. Comput. Vision and Pattern Recog.*, Champaign, Illinois (June 1992), 302–308.
14. Peleg, S., and M. Ben-Ezra. 1999. Stereo panorama with a single camera. *Proc. Comput. Vision and Pattern Recog.*, Fort Collins, CO (23–25 June 1999), v. I:395–401.
15. Rock, I. 1983. *The Logic of Perception.* A Bradford Book, MIT Press, Cambridge, MA.
16. Rosenfeld, A., R. Hummel, and S. Zucker. 1976. Scene labeling by relaxation processes. *IEEE Trans.* SMC, v. 6.
17. Stockman, G., G. Lee, and S. W. Chen. 1990. Reconstructing line drawings from wings: the polygonal case. *Proc. of Int. Conf. Comput. Vision 3,* Osaka, Japan.
18. Subbarao, M., and J-K. Tyan. 1998. Selecting the optimal focus measure for autofocusing and depth-from-focus. *IEEE-T-PAMI,* v. 20(8):864–870.
19. Sugihara, K. 1986. *Machine Interpretation of Line Drawings.* MIT Press, Cambridge, MA.
20. Tanimoto, S. 1998. Connecting middle school mathematics to computer vision and pattern recognition. *Int. J. Pattern Recog. and Artificial Intelligence,* v. 12(8):1053–1070.
21. Waltz, D. 1975. Understanding line drawings of scenes with shadows. In *The Psychology of Computer Vision,* P. Winston, ed. McGraw-Hill, New York, 19–91.
22. Winston, P. 1977. *Artificial Intelligence.* Addison-Wesley, Reading, MA.

13

3D Sensing and Object Pose Computation

The main concern of this chapter is the quantitative relationship between 2D image structures and their corresponding 3D real-world structures. The previous chapter investigated relationships between image and world by primarily studying the qualitative phenomena. Here, we show how to make measurements needed for recognition and inspection of 3D objects and for robotic manipulation or navigation.

Consider Figure 13.1, for example. The problem is to measure the body posture of a driver operating a car: the ultimate purpose is to design a better driving environment. In another application, shown in Figure 13.2, the camera system must recognize 3D objects

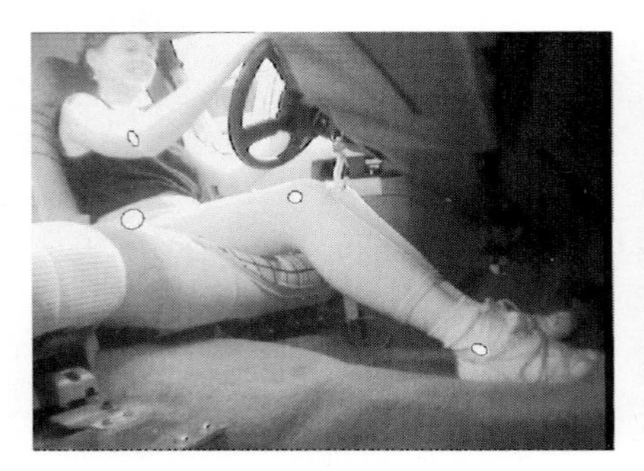

Figure 13.1 Two images of the driver of a car taken from two of four onboard cameras. A multiple camera measurement system is used to compute the 3D location of certain body points (identified by the bright ellipses) which are then used to compute posture. (Images courtesy of Herbert Reynolds, Michigan State University Ergonomics Lab.)

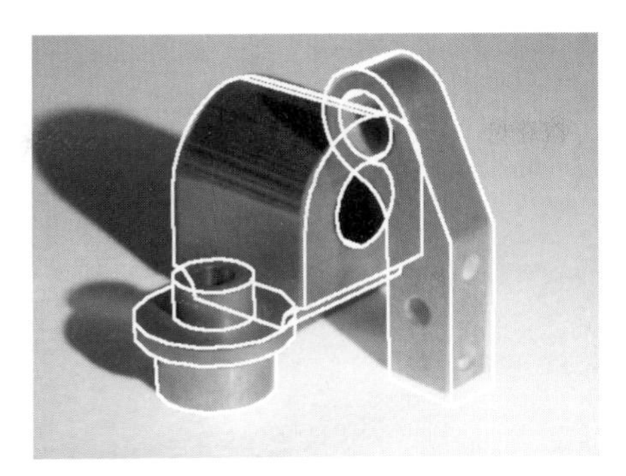

Figure 13.2 The overlay of graphics on the image of three 3D objects; computer vision has been used to recognize and localize the objects. In order to do this, the recognition system has matched 2D image parts to 3D model parts and has computed the geometrical 3D transformation needed to create the observed image from the model objects. The robot controller can then be told of the identity and pose of each part. (Image courtesy of Mauro Costa.)

and their pose so that a parts handling robot can grasp them. For this application, the camera system and the robot arm communicate in terms of 3D world coordinates.

This chapter treats some of the engineering and mathematics for sensing in 3D. Problems are formulated in terms of the intuitive geometry and then the mathematical models are developed. The algebra of transformations in 3D is central to the mathematics. The role of 3D object models is described as well as different sensor configurations and their calibration procedures are discussed.

13.1 GENERAL STEREO CONFIGURATION

Figure 13.3 shows a general configuration of two cameras viewing the same 3D workspace. Often in computer graphics, a right-handed coordinate system is used with the negative z-axis extending out from the camera so that points farther from the camera have more negative depth cordinates. We keep depth positive in most of the models in this chapter but sometimes use another system to be consistent with a published derivation. Figure 13.3 shows a general stereo configuration that does not have the special alignment of the cameras as assumed in Chapter 12. The cameras both view the same workpiece on a worktable: The worktable is the entire 3D world in this situation and it has its own global coordinate system **W** attached to it. Intuitively, we see that the location of 3D point ${}^{\mathbf{w}}\mathbf{P} = [{}^{w}P_x, {}^{w}P_y, {}^{w}P_z]^t$ in the workspace can be obtained by simply determining the intersection of the two imaging rays ${}^{\mathbf{w}}\mathbf{P}^{1}\mathbf{O}$ and ${}^{\mathbf{w}}\mathbf{P}^{2}\mathbf{O}$. We shall derive the algebra for this computation in Section 13.3.3: It is straighforward but there are some complications due to measurement error.

In order to perform the general stereo computation illustrated in Figure 13.3, the following items must be known.

- We must know the pose of camera $\mathbf{C_1}$ in the workspace **W** and some camera internals, such as focal length. All this information will be represented by a *camera matrix,* which algebraically defines a ray in 3D space for every image point ${}^{1}\mathbf{P}$. Sections 13.3 and 13.7 describe camera calibration procedures by which this information can be obtained.
- Similarly, we must know the pose of camera $\mathbf{C_2}$ in the workspace **W** and its internal parameters; equivalently, we need its camera matrix.

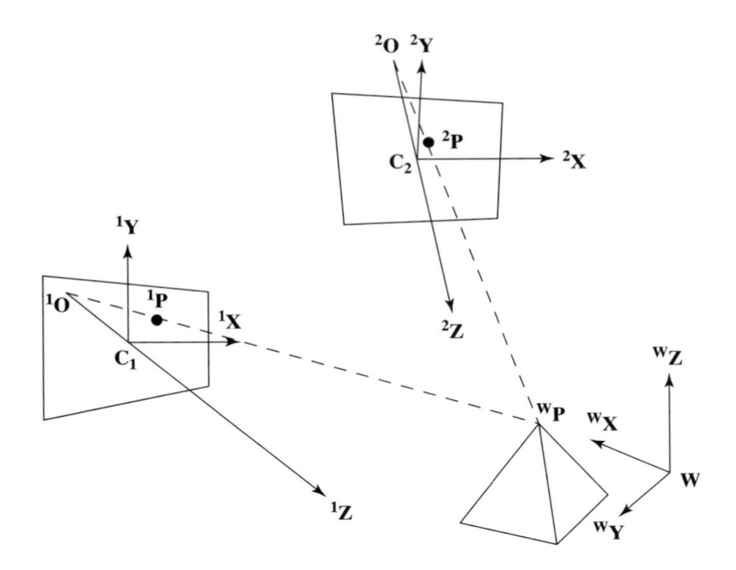

Figure 13.3 Two cameras $\mathbf{C_1}$ and $\mathbf{C_2}$ view the same 3D workspace. Point **P** on a workpiece is imaged at point $^1\mathbf{P}$ on the first image plane and at point $^2\mathbf{P}$ on the second image plane.

- We need to identify the correspondence of the 3D point to the two 2D image points ($^w\mathbf{P}$, $^1\mathbf{P}$, $^2\mathbf{P}$).
- We need a formula for computing $^w\mathbf{P}$ from the two imaging rays $^w\mathbf{P}^1\mathbf{O}$ and $^w\mathbf{P}^2\mathbf{O}$.

Before addressing these items, we take the opportunity to describe three important variations on the configuration shown in Figure 13.3.

- **The configuration shown in Figure 13.3 consists of two cameras calibrated to the world coordinate space.** Coordinates of 3D point features are computed by intersecting the two imaging rays from the corresponding image points.
- **One of the cameras can be replaced by a projector** which illuminates one or more surface points using a beam of light or a special pattern such as a crosshair. This is shown in Figure 13.4. As we shall see below, the projector can be calibrated in much the same manner as a camera: The projected ray of light has the same algebraic representation as the ray imaging to a camera. Using a projector has advantages when surface point measurements are needed on a surface that has no distinguishing features.
- **Knowledge of the model object can replace one of the cameras.** Assume that the height of the pyramid in Figure 13.3 is known; thus, we already know coordinate ^wP_z, which means that point **P** is constrained to lie on the plane $z = {}^wP_z$. The other two coordinates are easily found by intersecting the imaging ray from the single camera $\mathbf{C_1}$ with that plane. In many cases model information adds enough constraint so that a single camera is sufficient.

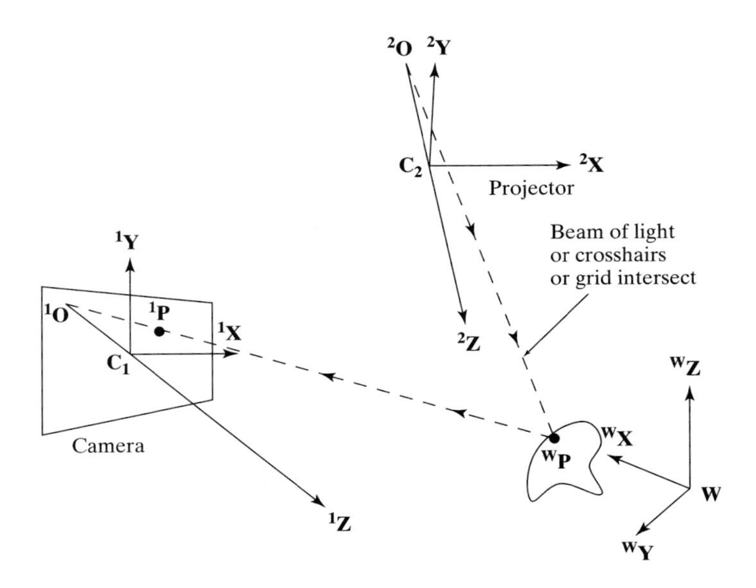

Figure 13.4 A projector can replace one camera in the general stereo configuration. The same geometric and algebraic constraints hold as in Figure 13.3; however, the projector can add surface features to an otherwise featureless surface.

13.2 3D AFFINE TRANSFORMATIONS

Affine transformations of 2D spaces were treated in Chapter 11. In this chapter, we make the extension to 3D. These transformations are very important not only for 3D machine vision, but also for robotics and virtual reality. Basic transformations are translation, rotation, scaling, and shear. These primitive transformations extend in a straightforward manner; however, some are more difficult to visualize. Once again we use the convenience of homogeneous coordinates, which extends a 3D point $[P_x, P_y, P_z]$ to have the four coordinates $[sP_x, sP_y, sP_z, s]$, where s is a nonzero scale factor. (As before, points are column vectors but we will sometimes omit the transpose symbol as there is no ambiguity caused by this.) In this chapter, we often use superscripts on point names because we need to name more coordinate systems than we did in Chapter 11. We will be adding perspective, orthographic, and weak perspective projections from 3D space to 2D space to our set of transformations.

13.2.1 Coordinate Frames

Coordinate frames or *coordinate systems* are needed in order to quantitatively locate points in space. Figure 13.5 shows a scene with four different relevant coordinate systems. Point **P**, the apex of a pyramid, has four different coordinate representations. First, the point is represented in a CAD model as $^{\mathbf{M}}\mathbf{P} = [^{M}P_x, {^{M}P_y}, {^{M}P_z}] = [b/2, b/2, \frac{\sqrt{2}}{2}b]$, where b is the size of its base. Second, an instance of this CAD model is posed on the workbench as

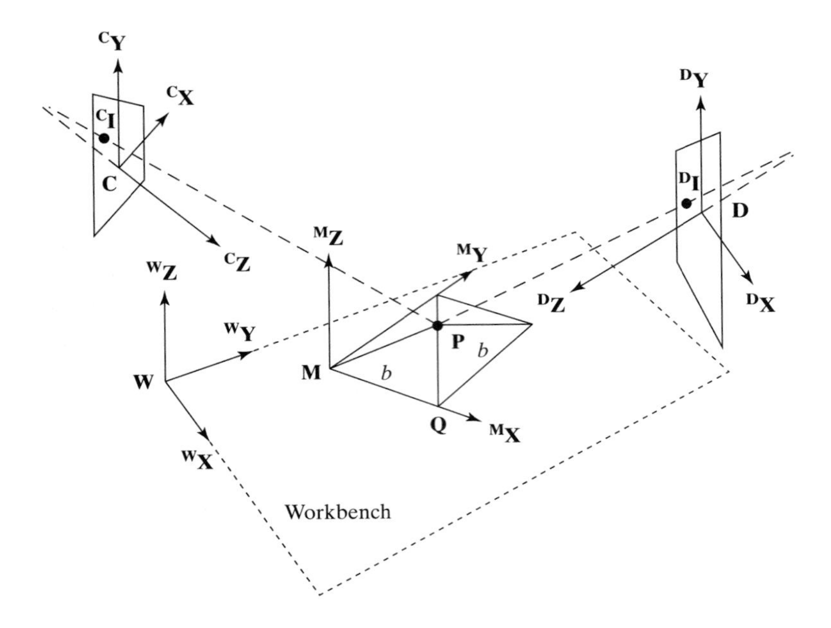

Figure 13.5 Point **P** can be represented in terms of coordinates relative to four distinct coordinate frames—(1) the model coordinate system **M**; (2) the world or workbench coordinate system **W**; (3) sensor **C**; and sensor **D**. Coordinates change with the coordinate frame; for example, point **P** appears to be left of **Q** to sensor **C**, but to the right of **Q** to **D**.

shown. The representation of the pyramid apex relative to the workbench frame is

$$^{\mathbf{W}}\mathbf{P} = \left[{}^{W}P_x, {}^{W}P_y, {}^{W}P_z\right] = \mathbf{TR}\left[\frac{b}{2}, \frac{b}{2}, \frac{\sqrt{2}}{2}b\right], \tag{13.1}$$

where **TR** is the combined rotation and translation of coordinate frame **M** relative to coordinate frame **W**. Finally, if two sensors **C** and **D** (or persons) view the pyramid from opposite sides of the workbench, the left-right relationship between points **P** and **Q** is reversed: the coordinate representations are different. $^{\mathbf{C}}\mathbf{P} = [{}^{C}P_x, {}^{C}P_y, {}^{C}P_z] \neq {}^{\mathbf{D}}\mathbf{P} = [{}^{D}P_x, {}^{D}P_y, {}^{D}P_z]$.

In order to relate sensors to each other, to 3D objects in the scene, and to manipulators that operate on them, we will develop mathematical methods to relate their coordinate frames. These same methods also enable us to model the motion of an object in space. Sometimes we will use a convenient notation to emphasize the coordinate frame in which a point is represented and the direction of the coordinate transformation. We denote the transformation $^{\mathbf{W}}_{\mathbf{M}}\mathbf{T}$ of a model point $^{\mathbf{M}}\mathbf{P}$ from model coordinates to workbench coordinates $^{\mathbf{W}}\mathbf{P}$ as follows:

$$^{\mathbf{W}}\mathbf{P} = {}^{\mathbf{W}}_{\mathbf{M}}\mathbf{T}\ {}^{\mathbf{M}}\mathbf{P} \tag{13.2}$$

This notation is developed in the robotics text by Craig (1986): It can be very helpful when reasoning about motions of objects or matching of objects. In simple situations where the coordinate frame is obvious, we use simpler notation. We now proceed with the study of transformations.

13.2.2 Translation

Translation adds a translation vector of three coordinates x_0, y_0, z_0 to point ${}^1\mathbf{P}$ in coordinate frame **1** to get point ${}^2\mathbf{P}$ in coordinate frame **2**. In the example in Figure 13.5 some translation (and rotation) is necessary in order to relate a point in model coordinates to its pose on the workbench.

$$
{}^2\mathbf{P} = \mathbf{T}(x_0, y_0, z_0)\, {}^1\mathbf{P}
$$

$$
{}^2\mathbf{P} = \begin{bmatrix} {}^2P_x \\ {}^2P_y \\ {}^2P_z \\ 1 \end{bmatrix} = \begin{bmatrix} 1 & 0 & 0 & x_0 \\ 0 & 1 & 0 & y_0 \\ 0 & 0 & 1 & z_0 \\ 0 & 0 & 0 & 1 \end{bmatrix} \begin{bmatrix} {}^1P_x \\ {}^1P_y \\ {}^1P_z \\ 1 \end{bmatrix} \tag{13.3}
$$

13.2.3 Scaling

A 3D scaling matrix can apply individual scale factors to each of the coordinates. Sometimes, all scale factors will be the same, as in the case of a change of measurement units or a uniform scaling in instantiating a model to a certain size.

$$
{}^2\mathbf{P} = \mathbf{S}\, {}^1\mathbf{P} = \mathbf{S}(s_x, s_y, s_z)\, {}^1\mathbf{P}
$$

$$
\begin{bmatrix} {}^2P_x \\ {}^2P_y \\ {}^2P_z \\ 1 \end{bmatrix} = \begin{bmatrix} s_x\, {}^2P_x \\ s_y\, {}^2P_y \\ s_z\, {}^2P_z \\ 1 \end{bmatrix} = \begin{bmatrix} s_x & 0 & 0 & 0 \\ 0 & s_y & 0 & 0 \\ 0 & 0 & s_z & 0 \\ 0 & 0 & 0 & 1 \end{bmatrix} \begin{bmatrix} {}^1P_x \\ {}^1P_y \\ {}^1P_z \\ 1 \end{bmatrix} \tag{13.4}
$$

13.2.4 Rotation

Creation of a matrix representing a primitive rotation about a coordinate axis is especially easy; all we need do is write down the column vectors of the matrix to be the transformed values of the unit vectors under the rotation. (Recall that any 3D linear transformation is completely characterized by how that transformation transforms the three basis vectors.) The transformation about the z-axis is actually the same as the 2D transformation done in Chapter 11, except that it now carries along a copy of the z-coordinate of the 3D point. Figure 13.6 shows how the basis vectors are transformed by the primitive rotations.

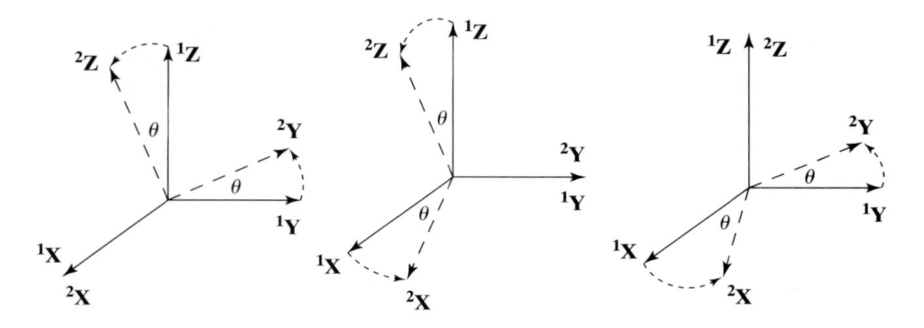

Figure 13.6 Rotations by angle θ about the (left) x-axis, (center) y-axis, and (right) z-axis.

Rotation of θ about the *X*-axis:

$$
{}^2\mathbf{P} = \mathbf{R}({}^1X, \theta)\ {}^1\mathbf{P}
$$

$$
\begin{bmatrix} {}^2P_x \\ {}^2P_y \\ {}^2P_z \\ 1 \end{bmatrix} = \begin{bmatrix} 1 & 0 & 0 & 0 \\ 0 & cos\theta & -sin\theta & 0 \\ 0 & sin\theta & cos\theta & 0 \\ 0 & 0 & 0 & 1 \end{bmatrix} \begin{bmatrix} {}^1P_x \\ {}^1P_y \\ {}^1P_z \\ 1 \end{bmatrix} \tag{13.5}
$$

Rotation of θ about the *Y*-axis:

$$
{}^2\mathbf{P} = \mathbf{R}({}^1Y, \theta)\ {}^1\mathbf{P}
$$

$$
\begin{bmatrix} {}^2P_x \\ {}^2P_y \\ {}^2P_z \\ 1 \end{bmatrix} = \begin{bmatrix} cos\theta & 0 & sin\theta & 0 \\ 0 & 1 & 0 & 0 \\ -sin\theta & 0 & cos\theta & 0 \\ 0 & 0 & 0 & 1 \end{bmatrix} \begin{bmatrix} {}^1P_x \\ {}^1P_y \\ {}^1P_z \\ 1 \end{bmatrix} \tag{13.6}
$$

Rotation of θ about the *Z*-axis:

$$
{}^2\mathbf{P} = \mathbf{R}({}^1Z, \theta)\ {}^1\mathbf{P}
$$

$$
\begin{bmatrix} {}^2P_x \\ {}^2P_y \\ {}^2P_z \\ 1 \end{bmatrix} = \begin{bmatrix} cos\theta & -sin\theta & 0 & 0 \\ sin\theta & cos\theta & 0 & 0 \\ 0 & 0 & 1 & 0 \\ 0 & 0 & 0 & 1 \end{bmatrix} \begin{bmatrix} {}^1P_x \\ {}^1P_y \\ {}^1P_z \\ 1 \end{bmatrix} \tag{13.7}
$$

Exercise 13.1

Verify that the columns of the matrices representing the three primitive rotations are orthonormal. Same question for the rows.

Exercise 13.2

Construct the rotation matrix for a counterclockwise rotation of $\Pi/4$ about the axis defined by the origin and the point $[1, 1, 0]^t$.

Exercise 13.3

Show how to make the general construction of the rotation matrix given a rotation angle of θ radians and the direction cosines $[c_x, c_y, c_z]^t$ of the axis.

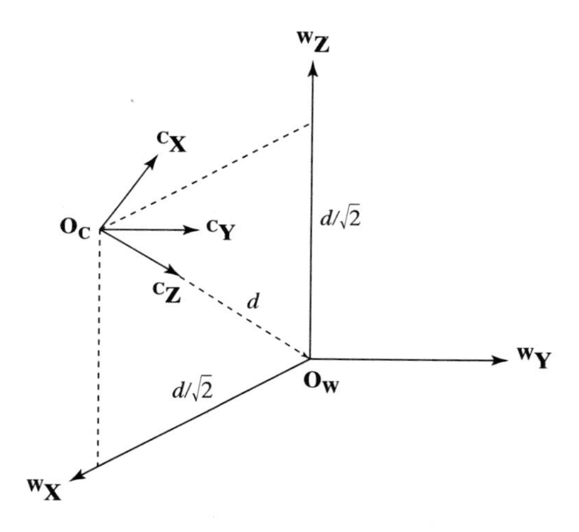

Example: Derive the combined rotation and translation needed to transform world coordinates **W** into camera coordinates **C**.

To construct the rotation matrix **R**, we write the coordinates of the basis vectors of **W** in terms of those of **C** so that any point in **W** coordinates can be transformed into **C** coordinates.

$$
\begin{aligned}
{}^{W}\mathbf{X} &= \frac{-\sqrt{2}}{2}\,{}^{C}\mathbf{X} + 0\,{}^{C}\mathbf{Y} + \frac{-\sqrt{2}}{2}\,{}^{C}\mathbf{Z} \\
{}^{W}\mathbf{Y} &= 0\,{}^{C}\mathbf{X} + 1\,{}^{C}\mathbf{Y} + 0\,{}^{C}\mathbf{Z} \\
{}^{W}\mathbf{Z} &= \frac{\sqrt{2}}{2}\,{}^{C}\mathbf{X} + 0\,{}^{C}\mathbf{Y} + \frac{-\sqrt{2}}{2}\,{}^{C}\mathbf{Z}
\end{aligned}
\qquad (13.8)
$$

These three vectors will be the three columns of the rotation matrix encoding the orientation of camera frame **C** relative to the world frame **W**. Once the camera is rotated, world points must be translated by d along the z-axis so that the world origin will be located at coordinate

$[0, 0, d]^t$ relative to **C**. The final change of coordinate transformation is

$$ {}^{\mathbf{C}}_{\mathbf{W}}\mathbf{TR} = \begin{bmatrix} \frac{-\sqrt{2}}{2} & 0 & \frac{\sqrt{2}}{2} & 0 \\ 0 & 1 & 0 & 0 \\ \frac{-\sqrt{2}}{2} & 0 & \frac{-\sqrt{2}}{2} & d \\ 0 & 0 & 0 & 1 \end{bmatrix} \tag{13.9}$$

Check that ${}^{\mathbf{C}}_{\mathbf{W}}\mathbf{TR}\,{}^{\mathbf{W}}\mathbf{O}_{\mathbf{w}} = {}^{\mathbf{C}}_{\mathbf{W}}\mathbf{TR}[0, 0, 0, 1]^t = [0, 0, d, 1]^t = {}^{\mathbf{C}}\mathbf{O}_{\mathbf{w}}$, and ${}^{\mathbf{C}}_{\mathbf{W}}\mathbf{TR}\,{}^{\mathbf{W}}\mathbf{O}_{\mathbf{c}} = {}^{\mathbf{C}}_{\mathbf{W}}\mathbf{TR}[d\frac{\sqrt{2}}{2}, 0, d\frac{\sqrt{2}}{2}, 1]^t = [0, 0, 0, 1]^t = {}^{\mathbf{C}}\mathbf{O}_{\mathbf{c}}$.

Exercise 13.4

Consider the environment of the previous example. Place a unit cube at the world origin $\mathbf{O_W}$. Transform its corners K_j into camera coordinates. Verify that four of the edges have unit length by computing $\| K_i - K_j \|$ using camera coordinates.

13.2.5 Arbitrary Rotation

Any rotation can be expressed in the form shown in Equation 13.10. The matrix of coefficients r_{ij} is an orthonormal matrix: all the columns (rows) are unit vectors that are mutually orthogonal. All of the primitive rotation matrices given above have this property. Any rigid rotation of 3D space can be represented as a rotation of some angle θ about a single axis **A**. **A** need not be one of the coordinate axes; it can be any axis in 3D space. To see this, suppose that the basis vector ${}^1\mathbf{X}$ is transformed into a different vector ${}^2\mathbf{X}$. The axis of rotation **A** can be found by taking the cross product of ${}^1\mathbf{X}$ and ${}^2\mathbf{X}$. In case ${}^1\mathbf{X}$ is invariant under the rotation, then it is itself the axis of rotation.

$$ {}^2\mathbf{P} = \mathbf{R}(A, \theta)\ {}^1\mathbf{P} $$

$$ \begin{bmatrix} {}^2P_x \\ {}^2P_y \\ {}^2P_z \\ 1 \end{bmatrix} = \begin{bmatrix} r_{11} & r_{12} & r_{13} & 0 \\ r_{21} & r_{22} & r_{23} & 0 \\ r_{31} & r_{32} & r_{33} & 0 \\ 0 & 0 & 0 & 1 \end{bmatrix} \begin{bmatrix} {}^1P_x \\ {}^1P_y \\ {}^1P_z \\ 1 \end{bmatrix} \tag{13.10}$$

As a consequence, *the result of the motion of a rigid object moving from time t_1 to time t_2 can be represented by a single translation vector and a single rotation matrix, regardless of its actual trajectory during this time period.* Only one homogeneous matrix is needed to store both the translation and rotation: there are six parameters, three for rotation and three for translation.

$$ \begin{bmatrix} {}^2P_x \\ {}^2P_y \\ {}^2P_z \\ 1 \end{bmatrix} = \begin{bmatrix} r_{11} & r_{12} & r_{13} & t_x \\ r_{21} & r_{22} & r_{23} & t_y \\ r_{31} & r_{32} & r_{33} & t_z \\ 0 & 0 & 0 & 1 \end{bmatrix} \begin{bmatrix} {}^1P_x \\ {}^1P_y \\ {}^1P_z \\ 1 \end{bmatrix} \tag{13.11}$$

Exercise 13.5

Refer to Figure 13.7. Give the homogeneous transformation matrix which maps all corners of the block at the origin onto corresponding corners of the other block. Assume that corner $^1\mathbf{O}$ maps onto $^2\mathbf{O}$ and corner $^1\mathbf{P}$ maps onto $^2\mathbf{P}$ and that the transformation is a rigid transformation.

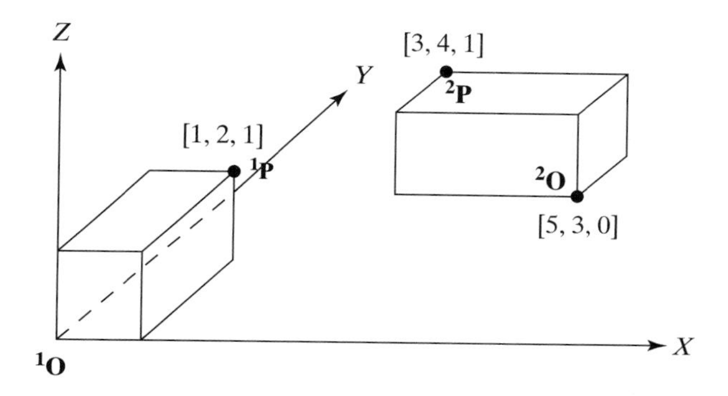

Figure 13.7 Two instances of the same block model.

Exercise 13.6: Inverse of a rotation matrix.

Argue that a rotation matrix always has an inverse. What is the inverse of the rotation $\mathbf{R}(A, \theta)$ of Equation 13.10?

13.2.6 Alignment via Transformation Calculus

Here we show how to align a model triangle with a sensed triangle. This should be a convincing example of the value of doing careful calculus with transformations. More important, it provides a basis for aligning any rigid model via correspondences between three model points with three sensed points. The development is done only with algebra using the basic transformation units already studied. Figures 13.8 and 13.9 illustrate the steps with a sketch of the geometry.

The problem is to derive the transformation $^W_M\mathbf{T}$ that maps the vertices A, B, C of a model triangle onto the vertices D, E, F of a congruent triangle in the workspace. In order to achieve this, we transform both triangles so that side AB lies along the ^WX-axis and so that the entire triangle lies in the XY-plane of $\mathbf{W}$. In such a configuration, it must be that coordinates of A and D are the same; also B and E, and C and F. The transformation equation derived to do this can be rearranged to produce the desired transformation $^W_M\mathbf{T}$ mapping each point $^M\mathbf{P_i}$ onto the corresponding $^W\mathbf{P_i}$. The operations are given in Algorithm 13.1. It should be clear that each of these operations can be done and that the inverse of each

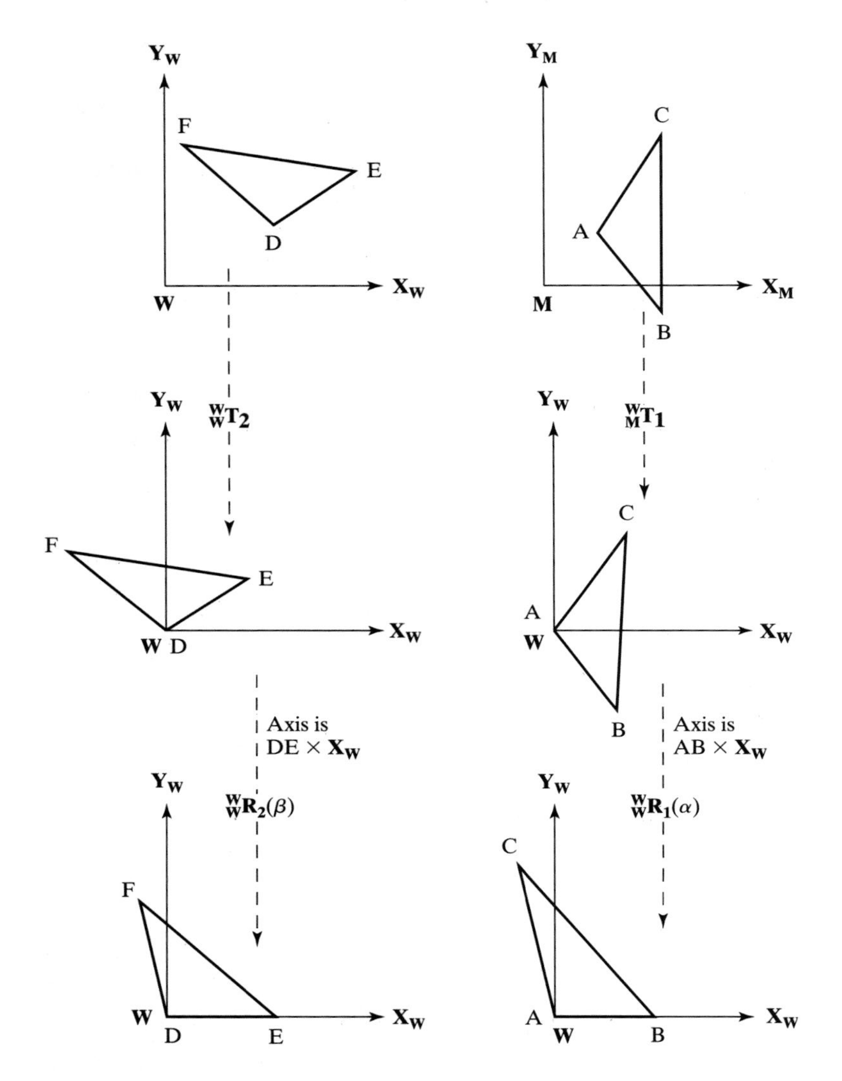

Figure 13.8 (part I) Alignment of two congruent triangles. $\triangle$ABC is the *model* triangle while $\triangle$DEF is the sensed triangle. First, object points are translated such that A and D move to the origin. Second, each triangle is rotated so that rays AB and DE lie along the X-axis.

exists. The family of operations requires careful programming and data structure design, however. This is a very important operation; in theory, any two congruent rigid objects can be aligned by aligning a subset of three corresponding points. In practice, measurement and computational errors are likely to produce significant error in aligning points far away from the $\triangle$ABC used. An optimization procedure using many more points is usually needed.

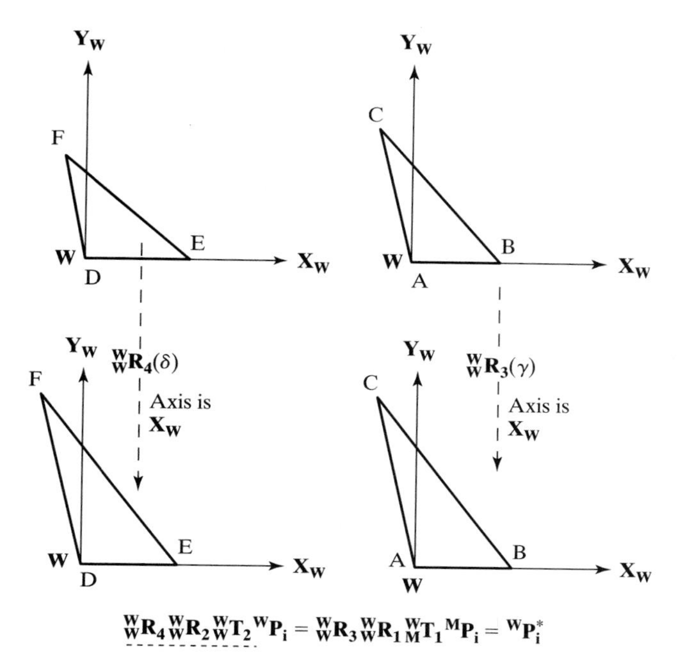

Figure 13.9 (part II) Alignment of two congruent triangles continued. Each triangle is rotated into the *XY*-plane. The axis of rotation is the *X*-axis; the angles of rotation are determined by rays AC and DF relative to the *XY*-plane.

Compute rigid transformation ${}^{W}_{M}\mathbf{T}$ that aligns model points A, B, C with world points D, E, F

1. Input the three 3D model points A, B, C and the corresponding three 3D world points D, E, F.
2. Construct translation ${}^{W}_{M}\mathbf{T}_1$ to shift points so that model point A maps to the world origin. Construct translation ${}^{W}_{W}\mathbf{T}_2$ to shift world points so that D maps to the world origin. This will align only points A and D in **W**.
3. Construct rotation ${}^{W}_{W}\mathbf{R}_1$ so that side AB is along the X-axis. Construct rotation ${}^{W}_{W}\mathbf{R}_2$ so that side DE is along the X-axis. This will align sides AB and DE in **W**.
4. Construct rotation ${}^{W}_{W}\mathbf{R}_3$ about the X-axis so that point C maps into the XY-plane. Construct rotation ${}^{W}_{W}\mathbf{R}_4$ about the X-axis so that point F maps into the XY-plane. Now all three points are aligned in **W**.
5. The model triangle and world triangle are now aligned, as represented by the following equation.

$$ {}^{W}_{W}\mathbf{R}_3\, {}^{W}_{W}\mathbf{R}_1\, {}^{W}_{M}\mathbf{T}_1\, {}^{M}\mathbf{P}_i = {}^{W}_{W}\mathbf{R}_4\, {}^{W}_{W}\mathbf{R}_2\, {}^{W}_{W}\mathbf{T}_2\, {}^{W}\mathbf{P}_i \qquad (13.12) $$

$$ {}^{W}\mathbf{P}_i = \left(\mathbf{T}_2^{-1}\, \mathbf{R}_2^{-1}\, \mathbf{R}_4^{-1}\, \mathbf{R}_3\, \mathbf{R}_1\, \mathbf{T}_1\right) {}^{M}\mathbf{P}_i \qquad (13.13) $$

6. Return ${}^{W}_{M}\mathbf{T} = (\mathbf{T}_2^{-1}\, \mathbf{R}_2^{-1}\, \mathbf{R}_4^{-1}\, \mathbf{R}_3\, \mathbf{R}_1\, \mathbf{T}_1)$

Algorithm 13.1 Derive the rigid transformation needed to align a model triangle with a congruent world triangle.

13.3 CAMERA MODEL

Our goal in this section is to show that the camera model $\mathbf{C}$ in Equation 13.14 is the appropriate algebraic model for perspective imaging and then to show how to determine the matrix elements from a fixed camera setup. The matrix elements are then used in computer programs that perform 3D measurements.

$$^{\mathbf{I}}\mathbf{P} = {}^{\mathbf{I}}_{\mathbf{W}}\mathbf{C}\,{}^{\mathbf{W}}\mathbf{P} \tag{13.14}$$

$$\begin{bmatrix} s\,{}^IP_r \\ s\,{}^IP_c \\ s \end{bmatrix} = {}^{\mathbf{I}}_{\mathbf{W}}\mathbf{C} \begin{bmatrix} {}^WP_x \\ {}^WP_y \\ {}^WP_z \\ 1 \end{bmatrix} = \begin{bmatrix} c_{11} & c_{12} & c_{13} & c_{14} \\ c_{21} & c_{22} & c_{23} & c_{24} \\ c_{31} & c_{32} & c_{33} & 1 \end{bmatrix} \begin{bmatrix} {}^WP_x \\ {}^WP_y \\ {}^WP_z \\ 1 \end{bmatrix}$$

$${}^IP_r = \frac{[c_{11}\ c_{12}\ c_{13}\ c_{14}] \circ \left[{}^WP_x\ {}^WP_y\ {}^WP_z\ 1\right]}{[c_{31}\ c_{32}\ c_{33}\ 1] \circ \left[{}^WP_x\ {}^WP_y\ {}^WP_z\ 1\right]}$$

$${}^IP_c = \frac{[c_{21}\ c_{22}\ c_{23}\ c_{24}] \circ \left[{}^WP_x\ {}^WP_y\ {}^WP_z\ 1\right]}{[c_{31}\ c_{32}\ c_{33}\ 1] \circ \left[{}^WP_x\ {}^WP_y\ {}^WP_z\ 1\right]}$$

Exercise 13.7: Inverse of a translation matrix.

Argue that a translation matrix always has an inverse. What is the inverse of the translation $\mathbf{T}(t_x, t_y, t_z)$ of Equation 13.3?

Exercise 13.8

Clearly, Algorithm 13.1 must fail if $\|A - B\| \neq \|D - E\|$. Insert the appropriate tests and error returns to handle the case when the triangles are not congruent.

Exercise 13.9: Program the triangle alignment.*

Using the method sketched in Algorithm 13.1, write and test a program to align a model triangle with a congruent triangle in world coordinates. Write separate functions to perform each basic operation.

Exercise 13.10

(a) Argue that the transformation returned by Algorithm 13.1 does map model point A to world point D, model point B to world E, and model C to world F. (b) Argue that any other model point will maintain the same distances to A, B, C when transformed. (c) Argue that if a rigid n-vertex polyhedral model can be aligned with an object in the world using a rigid transformation, then the transformation can be obtained by aligning just two triangles using Algorithm 13.1.

We will justify next that the perspective imaging transformation is represented by this 3×4 camera matrix ${}^{\mathbf{I}}_{\mathbf{W}}\mathbf{C}_{3\times4}$, which projects 3D world point ${}^{\mathbf{W}}\mathbf{P} = [{}^{W}P_x, {}^{W}P_y, {}^{W}P_z]^t$ to image point ${}^{\mathbf{I}}\mathbf{P} = [{}^{I}P_r, {}^{I}P_c]^t$. Using this matrix, there are enough parameters to model the change of coordinates between world **W** and camera **C** and to model all the scaling needed for both perspective transformation and the scaling of the real image coordinates into row and column of the image array. The matrix equation uses homogeneous coordinates: Removal of the scale factor s using the dot product is also shown in Equation 13.14. We will now derive the parameters of the camera matrix ${}^{\mathbf{I}}_{\mathbf{W}}\mathbf{C}$.

13.3.1 Perspective Transformation Matrix

The algebra for the perspective tranformation was given in Chapter 12: the resulting equations are rephrased in Equation 13.15. Recall that these equations are derived from the simple case where the world and camera coordinates are identical. Moreover, the image coordinates $[{}^{F}P_x, {}^{F}P_y]$ are in the same (real) units as the coordinates in the 3D space and not in pixel coordinates. (Think of the superscript F as denoting a *floating point number* and not focal length, which is the parameter f of the perspective transformation.)

$$\begin{aligned} {}^{F}P_x/f &= {}^{C}P_x/{}^{W}P_z \quad or \quad {}^{F}P_x = \left(f/{}^{C}P_z\right)\,{}^{C}P_x \\ {}^{F}P_y/f &= {}^{C}P_y/{}^{W}P_z \quad or \quad {}^{F}P_y = \left(f/{}^{C}P_z\right)\,{}^{C}P_y \end{aligned} \tag{13.15}$$

A pure perspective transformation shown in Figure 13.10 is defined in terms of the single parameter f, the focal length. The matrix ${}^{\mathbf{F}}_{\mathbf{C}}\mathbf{\Pi}(f)$ is shown in Equation 13.16 in its 4×4 form so that it can be combined with other transformation matrices; however, the third row producing ${}^{F}P_z = f$ is not needed and is ultimately ignored and often not given.

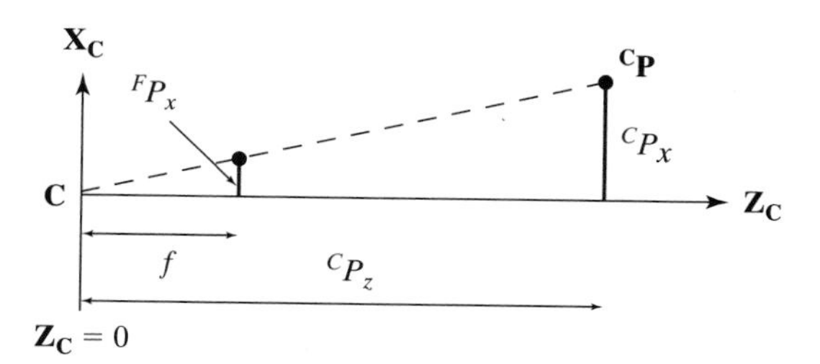

Figure 13.10 With the camera frame origin at the center of projection, ${}^{F}P_z = f$ always.

Note that the matrix is of rank 3, not 4, and hence an inverse does not exist.

$$\mathbf{^FP} = {}^{\mathbf{F}}_{\mathbf{C}}\mathbf{\Pi}(f)\ {}^{\mathbf{C}}\mathbf{P}$$

$$\begin{bmatrix} s\ {}^FP_x \\ s\ {}^FP_y \\ s\ {}^FP_z \\ s \end{bmatrix} = \begin{bmatrix} 1 & 0 & 0 & 0 \\ 0 & 1 & 0 & 0 \\ 0 & 0 & 1 & 0 \\ 0 & 0 & 1/f & 0 \end{bmatrix} \begin{bmatrix} {}^CP_x \\ {}^CP_y \\ {}^CP_z \\ 1 \end{bmatrix} \tag{13.16}$$

An alternative perspective transformation can be defined by placing the camera origin at the center of the image so that ${}^FP_z = 0$ as shown in Figure 13.11. The projection matrix would be as in Equation 13.17. (An advantage of this formulation is that it is obvious that we get orthographic projection as $f \to \infty$.)

$$\mathbf{^FP} = {}^{\mathbf{F}}_{\mathbf{C}}\mathbf{\Pi}(f)\ {}^{\mathbf{C}}\mathbf{P}$$

$$\begin{bmatrix} s\ {}^FP_x \\ s\ {}^FP_y \\ s\ {}^FP_z \\ s \end{bmatrix} = \begin{bmatrix} 1 & 0 & 0 & 0 \\ 0 & 1 & 0 & 0 \\ 0 & 0 & 0 & 0 \\ 0 & 0 & 1/f & 1 \end{bmatrix} \begin{bmatrix} {}^CP_x \\ {}^CP_y \\ {}^CP_z \\ 1 \end{bmatrix} \tag{13.17}$$

In the general case, as in Figure 13.3, the world coordinate system **W** is different from the camera coordinate system **C**. A rotation and translation are needed to convert world point $\mathbf{^WP}$ into camera coordinates $\mathbf{^CP}$. Three rotation parameters and three translation parameters are needed to do this, but they are combined in complex ways to yield transformation matrix elements as given in the previous sections.

$$\mathbf{^CP} = \mathbf{T}(t_x, t_y, t_z)\mathbf{R}(\alpha, \beta, \gamma)\ \mathbf{^WP}$$

$$\mathbf{^CP} = {}^{\mathbf{C}}_{\mathbf{W}}\mathbf{TR}(\alpha, \beta, \gamma, t_x, t_y, t_z)\ \mathbf{^WP}$$

$$\begin{bmatrix} {}^cP_x \\ {}^cP_y \\ {}^cP_z \\ 1 \end{bmatrix} = \begin{bmatrix} r_{11} & r_{12} & r_{13} & t_x \\ r_{21} & r_{22} & r_{23} & t_y \\ r_{31} & r_{32} & r_{33} & t_z \\ 0 & 0 & 0 & 1 \end{bmatrix} \begin{bmatrix} {}^wP_x \\ {}^wP_y \\ {}^wP_z \\ 1 \end{bmatrix} \tag{13.18}$$

We can compose transformations in order to model the change of coordinates from **W** to **C** followed by the perspective transformation of $\mathbf{^CP}$ onto the real image plane yielding $\mathbf{^FP}$. The third row of the matrix is dropped because it just yields the constant value for ${}^FP_z$. $\mathbf{^FP}$ is on the *real image plane* and a scaling transformation is needed to convert to

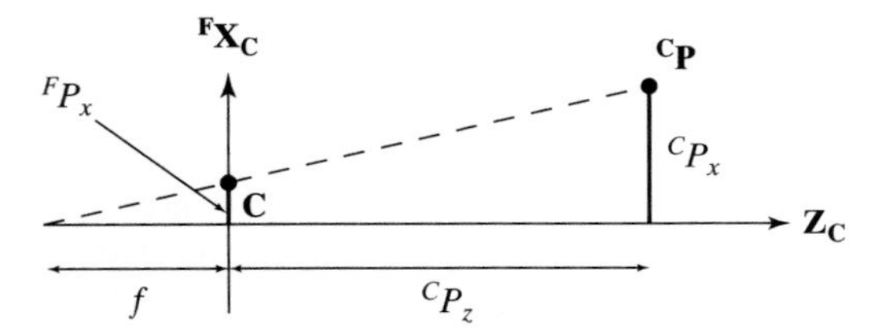

Figure 13.11 With the camera frame origin at the center of the image, ${}^FP_z = 0$ always.

the pixel row and column coordinates of $^{\mathbf{I}}\mathbf{P}$. Recall that matrix multiplication representing composition of linear transformations is associative.

$$\begin{aligned} {}^{\mathbf{F}}\mathbf{P} &= {}^{\mathbf{F}}_{\mathbf{C}}\mathbf{\Pi}(f)\ {}^{\mathbf{C}}\mathbf{P} \\ &= {}^{\mathbf{F}}_{\mathbf{C}}\mathbf{\Pi}(f)\left({}^{\mathbf{C}}_{\mathbf{W}}\mathbf{TR}(\alpha, \beta, \gamma, t_x, t_y, t_z)\ {}^{\mathbf{W}}\mathbf{P}\right) \\ &= \left({}^{\mathbf{F}}_{\mathbf{C}}\mathbf{\Pi}(f){}^{\mathbf{C}}_{\mathbf{W}}\mathbf{TR}(\alpha, \beta, \gamma, t_x, t_y, t_z)\right)\ {}^{\mathbf{W}}\mathbf{P} \end{aligned}$$

$$\begin{bmatrix} s\ {}^F P_x \\ s\ {}^F P_y \\ s \end{bmatrix} = \begin{bmatrix} d_{11} & d_{12} & d_{13} & d_{14} \\ d_{21} & d_{22} & d_{23} & d_{24} \\ d_{31} & d_{32} & d_{33} & 1 \end{bmatrix} \begin{bmatrix} {}^W P_x \\ {}^W P_y \\ {}^W P_z \\ 1 \end{bmatrix} \tag{13.19}$$

The matrix Equation 13.19 uses elements d_{ij} and not c_{ij} because it is not quite the camera matrix, whose derivation was our goal. This is because our derivation thus far has used the same units of real space, say *mm* or *inches,* and has not included the scaling of image points into pixel rows and columns. Scale factors converting *mm* to pixels for the image rows and columns are easily combined with Equation 13.19 to obtain the full camera matrix **C**. Suppose that d_x is the horizontal size and d_y is the vertical size of a pixel in real-valued units. Instead of the real-valued coordinates $[{}^F P_x\ {}^F P_y]$ whose reference coordinate system has [0.0, 0.0] at the lower left of the image, we want to go one step further to the integer-valued coordinates $[r, c]$ that refer to the row and column coordinates of a pixel of the image array whose reference coordinate system has [0, 0] at the top left pixel of the image. The transformation from the real numbers to pixels, including the reversal of direction of the vertical axis is given by

$${}^{\mathbf{I}}\mathbf{P} = \begin{bmatrix} s\ r \\ s\ c \\ s \end{bmatrix} = {}^{\mathbf{I}}_{\mathbf{F}}\mathbf{S} \begin{bmatrix} s\ {}^F P_x \\ s\ {}^F P_y \\ s \end{bmatrix} \tag{13.20}$$

where ${}^{\mathbf{I}}_{\mathbf{F}}\mathbf{S}$ is defined by

$${}^{\mathbf{I}}_{\mathbf{F}}\mathbf{S} = \begin{bmatrix} 0 & -\frac{1}{d_y} & 0 \\ \frac{1}{d_x} & 0 & 0 \\ 0 & 0 & 1 \end{bmatrix} \tag{13.21}$$

The final result for the full camera matrix that transforms 3D points in the real world to pixels in an image is given by

$${}^{\mathbf{I}}\mathbf{p} = \left({}^{\mathbf{I}}_{\mathbf{F}}\mathbf{S}\ {}^{\mathbf{F}}_{\mathbf{C}}\Pi(f)\ {}^{\mathbf{C}}_{\mathbf{W}}\mathbf{TR}(\alpha, \beta, \gamma, t_x, t_y, t_z)\right)\ {}^{\mathbf{W}}\mathbf{P}$$

$$\begin{bmatrix} s\ {}^I P_r \\ s\ {}^I P_c \\ s \end{bmatrix} = \begin{bmatrix} c_{11} & c_{12} & c_{13} & c_{14} \\ c_{21} & c_{22} & c_{23} & c_{24} \\ c_{31} & c_{32} & c_{33} & 1 \end{bmatrix} \begin{bmatrix} {}^W P_x \\ {}^W P_y \\ {}^W P_z \\ 1 \end{bmatrix} \tag{13.22}$$

which is the full camera matrix of Equation 13.14.

Let us review intuitively what we have just done to model the viewing of 3D world points by a camera sensor. First, we have placed the camera so that its coordinate system is completely aligned with the world coordinate system. Next, we rotated the camera ($^{\mathbf{W'}}_{\mathbf{W}}\mathbf{R}$) into its final orientation relative to $\mathbf{W}$. Then, we translated the camera ($^{\mathbf{C}}_{\mathbf{W'}}\mathbf{T}$) to view the workspace from the appropriate position. Now all 3D points can be projected to the image plane of the camera using the model of the perspective projection ($^{\mathbf{F}}_{\mathbf{C}}\Pi(\mathbf{f})$). Finally, we need to scale the real image coordinates $[^FP_x, {}^FP_y]$ and reverse the direction of the vertical axis to get the pixel coordinates $[^iP_r, {}^iP_c]$. We have used our transformation notation fully to account for all the steps. It is usually difficult to execute this procedure with enough precision in practice using distance and angle measurements to obtain a sufficiently accurate camera matrix $\mathbf{C}$ to use in computations. Instead, a camera calibration procedure is used to obtain the camera matrix. While the form of the camera matrix is justified by the above reasoning, the actual values of its parameters are obtained by fitting to control points as described in Section 13.4. Before treating calibration, we show some important uses of the camera matrix obtained by it.

Exercise 13.11

From the arguments of this section, it is easy to see that the form of the camera matrix is as follows.

$$\begin{bmatrix} s & {}^IP_r \\ s & {}^IP_c \\ & s \end{bmatrix} = \begin{bmatrix} c_{11} & c_{12} & c_{13} & c_{14} \\ c_{21} & c_{22} & c_{23} & c_{24} \\ c_{31} & c_{32} & c_{33} & c_{34} \end{bmatrix} \begin{bmatrix} {}^WP_x \\ {}^WP_y \\ {}^WP_z \\ 1 \end{bmatrix} \tag{13.23}$$

Show that we can derive the 11 parameter form of Equation 13.22 from this 12 parameter form just by scaling the camera matrix by $1/c_{34}$. Verify that the 11 parameter form performs the same mapping of 3D scene points to 2D image points.

13.3.2 Orthographic and Weak Perspective Projections

The orthographic projection of $^{\mathbf{C}}\mathbf{P}$ just drops the z-coordinate of the world point: This is equivalent to projecting each world point parallel to the optical axis and onto the image plane. Figure 13.12 compares perspective and orthographic projections. Orthographic projection can be viewed as a perspective projection where the focal length f has gone to infinity as shown in Equation 13.24. Orthographic projection is often used in computer graphics to convey true scale for the cross section of an object. It is also used for development of computer vision theory, because it is simpler than perspective, yet in many cases a good enough approximation to it to test the theory.

$$^{\mathbf{F}}\mathbf{P} = {}^{\mathbf{F}}_{\mathbf{C}}\mathbf{\Pi}(\infty)\ {}^{\mathbf{C}}\mathbf{P}$$

$$\begin{bmatrix} {}^FP_x \\ {}^FP_y \end{bmatrix} = \begin{bmatrix} 1 & 0 & 0 & 0 \\ 0 & 1 & 0 & 0 \end{bmatrix} \begin{bmatrix} {}^CP_x \\ {}^CP_y \\ {}^CP_z \\ 1 \end{bmatrix} \tag{13.24}$$

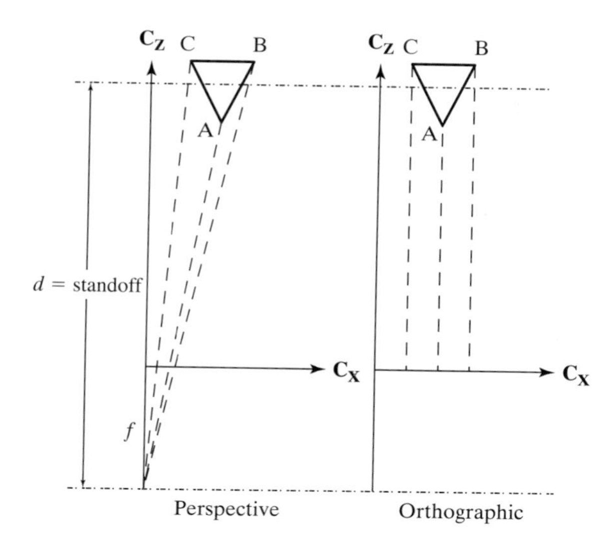

Figure 13.12 (left) Perspective projection versus (right) orthographic.

Often, perspective transformation can be nicely approximated by an orthographic projection followed by a uniform scaling in the real image plane. Projecting away the z-coordinate of a point and applying uniform scaling has been called *weak perspective*. A good scale factor is the ratio of the standoff of the object to the focal length of the camera ($s = f/d$ in Figure 13.12).

$$^{\mathbf{F}}\mathbf{P} = {}^{\mathbf{F}}_{\mathbf{C}}\mathbf{\Pi}(s)\ {}^{\mathbf{C}}\mathbf{P}$$

$$\begin{bmatrix} {}^{F}P_x \\ {}^{F}P_y \end{bmatrix} = \begin{bmatrix} s & 0 & 0 & 0 \\ 0 & s & 0 & 0 \end{bmatrix} \begin{bmatrix} {}^{C}P_x \\ {}^{C}P_y \\ {}^{C}P_z \\ 1 \end{bmatrix} \tag{13.25}$$

A rule of thumb is that the approximation will be acceptable if the standoff of the object is twenty times the size of the object. The approximation also depends upon how far the object is off the optical axis; the closer the better. As the triangular object of Figure 13.12 moves farther to the right of the optical axis, the perspective images of points A and B will crowd closer together until B will be occluded by A; however, the orthographic images of A and B will maintain their distance. Most robotic or industrial vision systems will attempt to center the object of attention in the field of view. This might also be true of aerial imaging platforms. In these cases, weak perspective might be an appropriate model.

Exercise 13.12

Derive Equation 13.24 from Equation 13.16 using $f \to \infty$.

Mathematical derivations and algorithms are usually more easily developed using weak perspective rather than true perspective. The approximation will usually be good

TABLE 13.1 WEAK PERSPECTIVE VERSUS TRUE PERSPECTIVE

${}^wP_x$	$f = 5mm$	$s = 5/1000$	$f = 20mm$	$s = 20/1000$	$f = 50$	$s = 50/1000$
0	0.000	0.000	0.000	0.000	0.000	0.000
10	0.051	0.050	0.204	0.200	0.510	0.500
20	0.102	0.100	0.408	0.400	1.020	1.000
50	0.255	0.250	1.020	1.000	2.551	2.500
100	0.510	0.500	2.041	2.000	5.102	5.000
200	1.020	1.000	4.082	4.000	10.204	10.000
500	2.551	2.500	10.204	10.000	25.510	25.000
1000	5.102	5.000	20.408	20.000	51.020	50.000

Comparison of values ${}^iP_x$ computed by the perspective transformation ${}^{\mathbf{F}}_{\mathbf{C}}\Pi_{\mathbf{pers}}(\mathbf{f})$ versus transformation using weak perspective ${}^{\mathbf{F}}_{\mathbf{C}}\Pi_{\mathbf{weak}}(\mathbf{s})$ with scale $s = f/1{,}000$ for 3D points $[{}^cP_x, 0, 980]^t$. Focal lengths for perspective transformation are 5, 20, and 50mm. Weak perspective scale is set at $f/1{,}000$ for a nominal standoff of 1,000mm.

enough to do the matching required by recognition algorithms. Moreover, a closed form weak perspective solution might be a good starting point for a more complex iterative algorithm using true perspective. Huttenlocher and Ullman (1988) have published some fundamental work on this issue. Table 13.1 gives some numerical comparisons between true perspective and weak perspective: the data shows a good approximation within the range of the table.

By substituting into the definition given in Equation 13.25, a weak perspective transformation is defined by eight parameters as follows:

$$ {}^{\mathbf{F}}\mathbf{P} = {}^{\mathbf{F}}_{\mathbf{C}}\mathbf{\Pi}_{\mathbf{weak}} \quad {}^{\mathbf{C}}_{\mathbf{W}}\mathbf{TR} \quad {}^{\mathbf{W}}\mathbf{P} $$

$$ \begin{bmatrix} {}^FP_x \\ {}^FP_y \end{bmatrix} = \begin{bmatrix} s & 0 & 0 & 0 \\ 0 & s & 0 & 0 \end{bmatrix} \begin{bmatrix} r_{11} & r_{12} & r_{13} & t_x \\ r_{21} & r_{22} & r_{23} & t_y \\ r_{31} & r_{32} & r_{33} & t_x \\ 0 & 0 & 0 & 1 \end{bmatrix} \begin{bmatrix} {}^WP_x \\ {}^WP_y \\ {}^WP_z \\ 1 \end{bmatrix} \tag{13.26} $$

$$ = \begin{bmatrix} c_{11} & c_{12} & c_{13} & c_{14} \\ c_{21} & c_{22} & c_{23} & c_{24} \end{bmatrix} \begin{bmatrix} {}^WP_x \\ {}^WP_y \\ {}^WP_z \\ 1 \end{bmatrix} \tag{13.27} $$

Exercise 13.13

While Equation 13.27 shows eight parameters in the weak perspective transformation, there are only seven independent parameters. What are these seven?

13.3.3 Computing 3D Points Using Multiple Cameras

We show how to use two camera models to compute the unknown 3D point $[x, y, z]$ from its two images $[r_1, c_1]$ and $[r_2, c_2]$ from two calibrated cameras. Since the coordinate system for our points is now clear, we drop the superscripts from our point notation. Figure 13.3 sketches the environment and Equation 13.14 gives the model for each of the cameras. The following imaging equations yield four linear equations in the three unknowns x, y, z.

$$\begin{bmatrix} sr_1 \\ sc_1 \\ s \end{bmatrix} = \begin{bmatrix} b_{11} & b_{12} & b_{13} & b_{14} \\ b_{21} & b_{22} & b_{23} & b_{24} \\ b_{31} & b_{32} & b_{33} & 1 \end{bmatrix} \begin{bmatrix} x \\ y \\ z \\ 1 \end{bmatrix}$$

$$\begin{bmatrix} tr_2 \\ tc_2 \\ t \end{bmatrix} = \begin{bmatrix} c_{11} & c_{12} & c_{13} & c_{14} \\ c_{21} & c_{22} & c_{23} & c_{24} \\ c_{31} & c_{32} & c_{33} & 1 \end{bmatrix} \begin{bmatrix} x \\ y \\ z \\ 1 \end{bmatrix} \tag{13.28}$$

Eliminating the homogeneous coordinates s and t from Equations 13.28, we obtain the following four linear equations in the three unknowns.

$$\begin{aligned} r_1 &= (b_{11} - b_{31}r_1)x + (b_{12} - b_{32}r_1)y + (b_{13} - b_{33}r_1)z + b_{14} \\ c_1 &= (b_{21} - b_{31}c_1)x + (b_{22} - b_{32}c_1)y + (b_{23} - b_{33}c_1)z + b_{24} \\ r_2 &= (c_{11} - c_{31}r_2)x + (c_{12} - c_{32}r_2)y + (c_{13} - c_{33}r_2)z + c_{14} \\ c_2 &= (c_{21} - c_{31}c_2)x + (c_{22} - c_{32}c_2)y + (c_{23} - c_{33}c_2)z + c_{24} \end{aligned} \tag{13.29}$$

Any three of these four equations could be solved to obtain the point $[x, y, z]$; however, each subset of three equations would yield slightly different coordinates. Together, all four simultaneous equations are typically inconsistent; due to approximation errors in the camera model and image points, the two camera rays will not intersect in mathematical 3D space. The better solution is to compute the closest approach of the two skew rays—or equivalently—the shortest line segment connecting them. If the length of this segment is suitably short, then we assign its midpoint as the intersection of the rays, which is the point $[x, y, z]$ as shown in Figure 13.13. If the shortest connecting segment is too long, then we assume that there was some mistake in corresponding the image points $[r_1, c_1]$ and $[r_2, c_2]$.

Applying the orthogonality constraints to the shortest connecting segment yields the

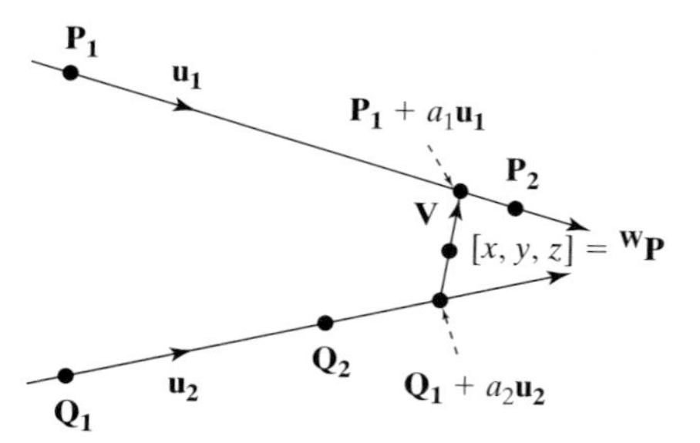

$\mathbf{P_1}$ and $\mathbf{P_2}$ are points on one line, while $\mathbf{Q_1}$ and $\mathbf{Q_2}$ are points on the other line. $\mathbf{u_1}$ and $\mathbf{u_2}$ are unit vectors along these lines. Vector $\mathbf{V} = \mathbf{P_1} + a_1\mathbf{u_1} - (\mathbf{Q_1} + a_2\mathbf{u_2})$ is the shortest (free) vector connecting the two lines, where a_1, a_2 are two scalars to be determined. a_1, a_2 can be determined using calculus to minimize the length of $\mathbf{V}$; however, it is easy to compute them using the constraints that $\mathbf{V}$ must be orthogonal to both $\mathbf{u_1}$ and $\mathbf{u_2}$.

Figure 13.13 The shortest distance between two skew lines is measured along a connecting line segment that is orthogonal to both lines.

following two linear equations in the two unknowns a_1, a_2.

$$\begin{aligned}((\mathbf{P_1} + a_1\mathbf{u_1}) &- (\mathbf{Q_1} + a_2\mathbf{u_2})) \circ \mathbf{u_1} = 0 \\ ((\mathbf{P_1} + a_1\mathbf{u_1}) &- (\mathbf{Q_1} + a_2\mathbf{u_2})) \circ \mathbf{u_2} = 0\end{aligned} \tag{13.30}$$

$$\begin{aligned}1a_1 - (\mathbf{u_1} \circ \mathbf{u_2})a_2 &= (\mathbf{Q_1} - \mathbf{P_1}) \circ \mathbf{u_1} \\ (\mathbf{u_1} \circ \mathbf{u_2})a_1 - 1a_2 &= (\mathbf{Q_1} - \mathbf{P_1}) \circ \mathbf{u_2}\end{aligned} \tag{13.31}$$

These equations are easily solved for a_1, a_2 by elimination of variables or by the method of determinants to obtain these solutions.

$$\begin{aligned}a_1 &= \frac{(\mathbf{Q_1} - \mathbf{P_1}) \circ \mathbf{u_1} - ((\mathbf{Q_1} - \mathbf{P_1}) \circ \mathbf{u_2}) \circ (\mathbf{u_1} \circ \mathbf{u_2})}{1 - (\mathbf{u_1} \circ \mathbf{u_2})^2} \\ a_2 &= \frac{((\mathbf{Q_1} - \mathbf{P_1}) \circ \mathbf{u_1})(\mathbf{u_1} \circ \mathbf{u_2}) - (\mathbf{Q_1} - \mathbf{P_1}) \circ \mathbf{u_2}}{1 - (\mathbf{u_1} \circ \mathbf{u_2})^2}\end{aligned} \tag{13.32}$$

Provided that $\|s\mathbf{V}\|$ is less than some threshold, we report the intersection of the two lines as $[x, y, z]^t = (1/2)[(\mathbf{P_1} + a_1\mathbf{u_1}) + (\mathbf{Q_1} + a_2\mathbf{u_2})]$. It is important to go back to the beginning and realize that all the computations depend upon being able to define each of the lines by a pair of points (that is, $\mathbf{P_1}, \mathbf{P_2}$) on the line. Often, each ray is determined by the optical center of the camera and the image point. If the optical center is not known, a point on the first camera ray can be found by choosing some value $z = z_1$ and then solving the two Equations 13.29 for coordinates x and y. If the ray is nearly parallel with the z-axis, then $x = x_0$ should be chosen. In this manner, the four needed points can be selected.

Exercise 13.14

Implement and test a function in your favorite programming language to determine both the distance between two skew lines and the midpoint of the shortest connecting line segment. The function should take four 3D points as input and execute the mathematical formulas in this section.

Exercise 13.15

Use of Cramer's rule to solve Equations 13.31 for a_1, a_2 requires that the determinant of the matrix of coefficients be nonzero. Argue why this must be the case when two cameras view a single point.

We can use one camera and one projector to sense 3D surfaces. The geometry and mathematics is the same as in the case of two cameras. The biggest advantage is that a projector can artificially create surface texture on smooth surfaces so that feature points may be defined and put into correspondence. Use of structured light is treated below, after we show how to obtain camera models or projector models via calibration.

13.4 BEST AFFINE CALIBRATION MATRIX

The problem of camera calibration is to relate the locations of the pixels in the image array of a given camera to the real-valued points in the 3D scene being imaged. This process generally precedes any image analysis procedures that involve the computation of 3D location and orientation of an object or measurements of its dimensions. It is also needed for the stereo triangulation procedure described in Section 13.3.3.

It was shown in Section 13.3 that the eleven parameter camera matrix of Equation 13.14 was an appropriate mathematical model. We now show how to derive the values of the eleven parameters using least-squares fitting. The camera view and focus are fixed and a calibration object, or *jig*, with known measurements, is placed in the scene as in Figure 13.14. A set of n data tuples $\langle {}^{\mathbf{I}}\mathbf{P}_j,\ {}^{\mathbf{W}}\mathbf{P}_j \rangle$ are then taken: ${}^{\mathbf{I}}\mathbf{P}_j = [{}^{I}P_r, {}^{I}P_c]$ is the pixel in the image where 3D point ${}^{\mathbf{W}}\mathbf{P}_j = [{}^{W}P_x, {}^{W}P_y, {}^{W}P_z]$ is observed. The number of points n must be at least 6, but results are better with $n = 25$ or more.

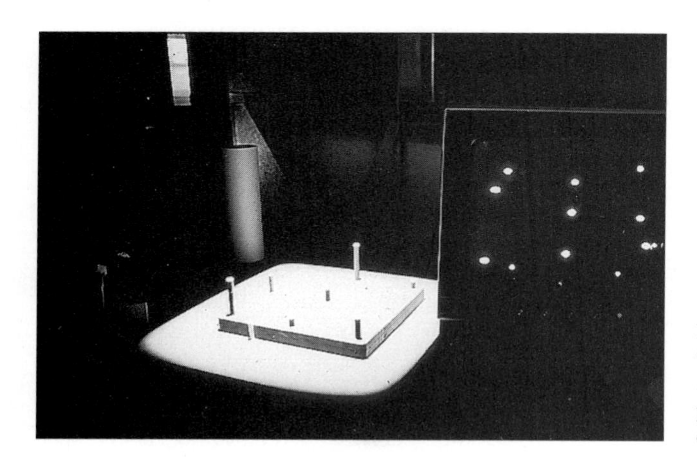

Figure 13.14 (left) The calibration jig has 9 dowels of random height (it can be rotated 3 times to obtain 25 unique calibration points) and (right) the image of the jig on the display.

13.4.1 Calibration Jig

The purpose of a *calibration jig* is to conveniently establish some well-defined 3D points. Figures 13.14, 13.18 and 13.22 show three different jigs. The jig is carefully located in the world coordinate system **W**, or perhaps defines the world coordinate system itself, such that the 3D coordinates $[{}^{W}P_x, {}^{W}P_y, {}^{W}P_z]$ of its features are readily known. The camera then views these features and obtains their 2D coordinates $[{}^{I}P_r, {}^{I}P_c]$. Other common types of jigs are rigid frames with wires and beads or rigid boards with special markings.

13.4.2 Defining the Least-Squares Problem

Equation 13.33 was obtained by eliminating the homogeneous scale factor s from the imaging model. We thus have two linear equations modeling the geometry of each imaging ray, and we have one ray for each calibration point. To simplify notation and also to avoid confusion of symbols we use $[x_j, y_j, z_j]$ for the world point ${}^{\mathbf{W}}\mathbf{P}_j = [{}^{W}P_x, {}^{W}P_y, {}^{W}P_z]$, and $[u_j, v_j]$ for the image point ${}^{\mathbf{I}}\mathbf{P}_j = [{}^{I}P_r, {}^{I}P_c]$. For each calibration point, the following two

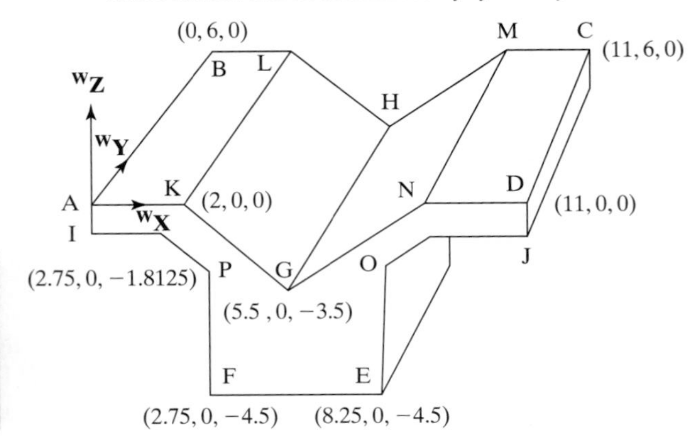

Figure 13.15 Precisely machined jig with many corners; the jig is 11 inches long, 6 inches wide, and 4.5 inches high. All 3D corner coordinates are given in Figure 13.18.

equations are obtained.

$$u_j = (c_{11} - c_{31}u_j)x_j + (c_{12} - c_{32}u_j)y_j + (c_{13} - c_{33}u_j)z_j + c_{14}$$
$$v_j = (c_{21} - c_{31}v_j)x_j + (c_{22} - c_{32}v_j)y_j + (c_{23} - c_{33}v_j)z_j + c_{24} \tag{13.33}$$

We rearrange the equations separating the knowns from the unknowns into vectors. All the entities on the left are known from the calibration tuples, while all the c_{km} on the right are unknowns to be determined.

$$\begin{bmatrix} x_j, y_j, z_j, 1, 0, 0, 0, 0, -x_ju_j, -y_ju_j, -z_ju_j \\ 0, 0, 0, 0, x_j, y_j, z_j, 1, -x_jv_j, -y_jv_j, -z_jv_j \end{bmatrix} \begin{bmatrix} c_{11} \\ c_{12} \\ c_{13} \\ c_{14} \\ c_{21} \\ c_{22} \\ c_{23} \\ c_{24} \\ c_{31} \\ c_{32} \\ c_{33} \end{bmatrix} = \begin{bmatrix} u_j \\ v_j \end{bmatrix} \tag{13.34}$$

Since each imaging ray gives two such equations, we obtain $2n$ linear equations from n calibration points, which can be compactly represented in the classic matrix form where $\mathbf{x}$ is the column vector of unknowns and $\mathbf{b}$ is the column vector of image coordinates.

$$\mathbf{A}_{2n\times 11}\mathbf{x}_{11\times 1} \approx \mathbf{b}_{2n\times 1} \tag{13.35}$$

Since there are 11 unknowns and 12 or more equations, this is an overdetermined system. There is no vector of parameters $\mathbf{x}$ for which all the equations hold: A least-squares solution is appropriate. We want the set of parameters such that the sum (over all equations) of the squared differences between the observed coordinate and the coordinate predicted by the camera matrix is minimized. Figure 13.16 shows four of these differences for two calibration points. These differences are called *residuals* as in Chapter 11. Figure 13.17 gives an abstract representation of the least-squares solution: We want to compute the c_{km} that give a linear combination of the columns of **A** closest to **b**. The key to solving this problem is the observation that the residual vector $\mathbf{r} = \mathbf{b} - \mathbf{Ax}$ is orthogonal to the column space of **A** yielding $\mathbf{A^t r} = \mathbf{0}$. Substituting $\mathbf{b} - \mathbf{Ax}$ for **r**, we obtain $\mathbf{A^t A x} = \mathbf{A^t b}$. $\mathbf{A^t A}$ is symmetric and positive definite, so it has an inverse, which can be used to solve for $\mathbf{x} = (\mathbf{A^t A})^{-1}\mathbf{A^t b}$. Several common libraries of numerical methods are available to solve this. (Using MATLAB, the solution is invoked using the simple statement `X = A \ B`. Once the least-squares solution `X` is found, the vector `R` of residuals is computed as `R` = `B` − `AX`.) Consult the Heath (1997) reference or the users manual for your local linear algebra library.)

Figure 13.18 shows example results of computing the camera matrix for a camera viewing the calibration jig shown in Figure 13.15. The corners of the jig are labeled 'A'

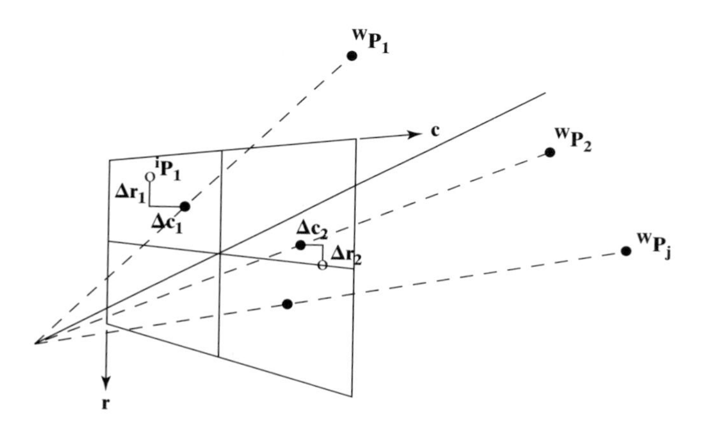

Figure 13.16 Image plane residuals are the differences between the actual observed image points (open dots) and the points computed using the camera matrix of Equation 13.14 (filled dots).

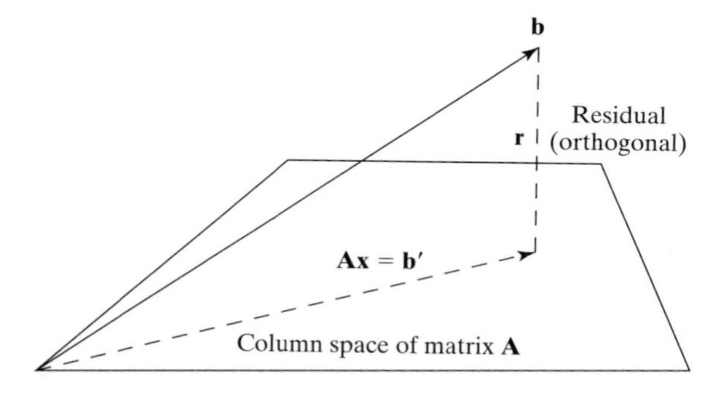

Figure 13.17 Least-squares solution of the system $\mathbf{Ax} \approx \mathbf{b}$. The plane represents the 11-dimensional column space of matrix $\mathbf{A}_{2n\times 11}$. All linear combinations **Ax** must be in this space, but $\mathbf{B}_{2n\times 1}$ will not be in it. The least-squares solution computes $\mathbf{b}'$ as the projection of **b** onto the 11D space, making $\mathbf{b}'$ the closest point in the space to **b**.

```
#
#  IMAGE: g1view1.ras
#
#                 INPUT DATA                               |             OUTPUT  DATA
#                                                          |
Point  Image 2-D (U,V)    3-D Coordinates (X,Y,Z)          |   2-D Fit Data        Residuals X Y
                                                           |
 A      95.00   336.00     0.00      0.00      0.00  |   94.53   337.89 |    0.47   -1.89
 B                         0.00      6.00      0.00  |                  |
 C                        11.00      6.00      0.00  |                  |
 D     592.00   368.00    11.00      0.00      0.00  |  592.21   368.36 |   -0.21   -0.36
 E     472.00   168.00     8.25      0.00     -4.50  |  470.14   168.30 |    1.86   -0.30
 F     232.00   155.00     2.75      0.00     -4.50  |  232.30   154.43 |   -0.30    0.57
 G     350.00   205.00     5.50      0.00     -3.50  |  349.17   202.47 |    0.83    2.53
 H     362.00   323.00     5.00      6.00     -3.50  |  363.44   324.32 |   -1.44   -1.32
 I      97.00   305.00     0.00      0.00     -0.75  |   97.90   304.96 |   -0.90    0.04
 J     592.00   336.00    11.00      0.00     -0.75  |  591.78   334.94 |    0.22    1.06
 K     184.00   344.00     2.00      0.00      0.00  |  184.46   343.40 |   -0.46    0.60
 L     263.00   431.00     2.00      6.00      0.00  |  261.52   429.65 |    1.48    1.35
 M                         9.00      6.00      0.00  |                  |
 N     501.00   363.00     9.00      0.00      0.00  |  501.16   362.78 |   -0.16    0.22
 O     467.00   279.00     8.25      0.00     -1.81  |  468.35   281.09 |   -1.35   -2.09
 P     224.00   266.00     2.75      0.00     -1.81  |  224.06   266.43 |   -0.06   -0.43


#              CALIBRATION  MATRIX

     44.84            29.80         -5.504        94.53

      2.518           42.24          40.79       337.9

     -0.0006832        0.06489      -0.01027       1.000
```

Figure 13.18 Camera calibration output using the jig in Figure 13.15.

to 'P' and their world coordinates [X, Y, Z] are given in Figure 13.18 alongside the image coordinates [U,V] where the camera images them. Corner points 'B', 'C', and 'M' are occluded in this view, so no image coordinates are available. The camera matrix obtained by fitting the 13 correspondences is given at the bottom of the figure and the residuals are shown at the right. 16 of the 26 coordinates computed by the camera matrix applied to the 3D points are within one pixel of the observed image coordinate, 10 are more than one pixel different, but only 2 are off by two pixels. This example supports the validity of the affine camera model, but also exhibits the errors due to corner location and the distortion of a short focal length lens.

Exercise 13.16: Replication of camera model derivation.

(a) Locate software to perform least-squares fitting. Enter the point correspondences from Figure 13.18 and compute the camera matrix: compare it to the matrix shown in Figure 13.18. (b) Delete three points with the largest residuals and derive a new camera matrix. Are any of the new residuals worse than 2 pixels? (c) Define the 3D coordinates of a $1 \times 1 \times 1$

cube that would rest on one of the top horizontal surfaces of the jig shown in Figure 13.15. Transform each of the eight corners of the cube into image coordinates by using the derived camera matrix. Also transform the four points defining the surface on which the cube rests. Plot the images of the transformed points and draw the connecting edges. Does the image of the cube make sense?

Exercise 13.17: Subpixel accuracy.

Refer to Figure 13.14. The center of the dowels can be computed to subpixel accuracy using the methods of Chapter 3. How? Can we use these noninteger coordinates for $[{}^I P_r, {}^I P_c]$ for the calibration data? How?

Exercise 13.18: Best weak perspective camera model.

Find the best weak perspective camera matrix to fit the data at the left in Figure 13.18. Carefully go back over the derivation of the simpler imaging equations to develop a new system of equations $\mathbf{Ax} = \mathbf{b}$. After obtaining the best camera matrix parameters, compare its residuals with those at the right in Figure 13.18.

Exercise 13.19: Camera calibration.

Table 13.2 shows the image points recorded for 16 3D corner points of the jig from Figure 13.15. In fact, coordinates in two separate images are recorded. Using the affine calibration procedure, compute a camera matrix using the 5-tuples from table columns 2–6.

TABLE 13.2 3D FEATURE POINTS OF JIG IMAGED WITH TWO CAMERAS

Point	${}^w\mathbf{x}$	${}^w\mathbf{y}$	${}^w\mathbf{z}$	${}^1\mathbf{u}$	${}^1\mathbf{v}$	${}^2\mathbf{u}$	${}^2\mathbf{v}$
A	0.0	0.0	0.0	167	65	274	168
B	0.0	6.0	0.0	96	127	196	42
C	11.0	6.0	0.0	97	545	96	431
D	11.0	0.0	0.0	171	517	154	577
E	8.25	0.0	−4.5	352	406	366	488
F	2.75	0.0	−4.5	347	186	430	291
G	5.5	0.0	−3.5	311	294	358	387
H	5.5	6.0	−3.5	226	337	NA	NA
I	0.0	0.0	−0.75	198	65	303	169
J	11.0	0.0	−0.75	203	518	186	577
K	2.0	0.0	0.0	170	143	248	248
L	2.0	6.0	0.0	96	198	176	116
M	9.0	6.0	0.0	97	465	114	363
N	9.0	0.0	0.0	173	432	176	507
O	8.25	0.0	−1.81	245	403	259	482
P	2.75	0.0	−1.81	242	181	318	283

3D world coordinates ${}^w\mathbf{x}$, ${}^w\mathbf{y}$, ${}^w\mathbf{z}$ are in inches. Coordinates of image 1 are ${}^1\mathbf{u}$ and ${}^1\mathbf{v}$ and are in row and column units. Coordinates of image 2 are ${}^2\mathbf{u}$ and ${}^2\mathbf{v}$.

Exercise 13.20: Stereo computation.

(a) Using the data in Table 13.2, compute two calibration matrices, one from columns 2–6 and one from columns 2–4 and 7–8. (b) Using the method of Section 13.3.3, compute the 3D coordinates of point **A** using only the two camera matrices and the image coordinates in columns 5–8 of the Table. Compare your result to the coordinates in columns 2–4 of the table. (c) Consider the obtuse corner point between corner points **I** and **P** of the jig; call it point **Q**. Suppose point **Q** images at [196, 135] and [281, 237] respectively. Use the stereo method to compute the 3D coordinates of world point **Q** and verify that your results are reasonable.

13.4.3 Discussion of the Affine Method

The major issue relates to whether or not there really are 11 camera model parameters to estimate. As we have seen, there are only 3 independent parameters of a rotation matrix and 3 translational parameters defining the camera pose in world coordinates. There are 2 scaling factors relating real image coordinates to pixel rows and columns and focal length **f**. Thus, not all 11 parameters are independent. Treating them as independent means that the rotation parameters will not define an orthonormal rotation matrix. In case of a precisely built camera, we waste constraints; however, if the image plane might not be perpendicular to the optical axis, the increased parameters can produce a better model. We need more calibration points to estimate more free parameters, and the parameters that are derived do not explicitly yield the intrinsic properties of the camera. The affine method has several advantages, however—it works well even with skew between image rows and columns or between image plane and optical axis, it works with either pixel coordinates or real image coordinates, and the solution can be quickly computed without iteration. In Section 13.7, we will introduce a different calibration method that uses more constraints and overcomes some of the problems mentioned.

Exercise 13.21: Calibrate your home camera.

If you do not have a simple film camera, borrow one or buy a cheap disposable one. Obtain a rigid box to use as a jig: draw a few Xs on each face. Measure the $[x, y, z]$ coordinates of each corner and each X relative to a RH coordinate system with origin at one corner of the box and axes along its edges. Take and develop a picture of the box. Identify and mark 15 corners and Xs in the picture. Measure the coordinates of these points in inches using a rule. Following the example of this section, derive the camera matrix and residuals and report on these results. Check how well the camera matrix works for some points that were not used in deriving it. Repeat the experiment using *mm* units for the picture coordinates.

Exercise 13.22: *Texture map* your box.

Texture map your picture onto the box of the previous exercise. (a) First, create a .pgm image file of your box as above. (b) Using the methods of Chapter 11, create a mapping from one face of the box (in 2D coordinates) to an image array containing your own picture.

(c) Update the `.pgm` image file of your box by writing pixels from your picture into it. Hint: Mapping two triangles rather than one parallelogram will produce a better result. Why?

13.5 USING STRUCTURED LIGHT

Sensing via *structured light* was motivated in the first section and illustrated in Figure 13.4. We now have all the mathematical tools to implement it. Figure 13.19 gives a more detailed view. Object surfaces are illuminated by a pattern of light: In this case a slide projector projects a regular grid of bright lines on surfaces. The camera then senses the result as a grid distorted by the surface structure and its pose. Since the grid of light is highly structured, the imaging system has strong information about which projector rays created the intersections that it sensed. Assume for the moment that the imaging system knows that grid intersect ${}^{G}P_{lm}$ is imaged at ${}^{I}P_{uv}$. There are then four linear equations to use to solve for the 3D surface point ${}^{W}P_{lm}$ being illuminated by the special light pattern. We must have both the camera calibration matrix $\mathbf{C}$ and the projector calibration matrix $\mathbf{D}$. The solution of the system $\mathbf{D}^{W}P_{lm} = {}^{G}P_{lm}$ and $\mathbf{C}^{W}P_{lm} = {}^{I}P_{uv}$ is the same as previously given for the general case of two camera stereo in Section 13.3.3.

A projector can be calibrated in much the same way as a camera. The projector is turned on so that it illuminates the planer surface of the worktable. Precise blocks are located on the table so that one corner exactly intercepts one of the projected grid intersects. A calibration tuple $\langle [{}^{G}P_l, {}^{G}P_m], [{}^{W}P_x, {}^{W}P_y, {}^{W}P_z] \rangle$ is obtained, where ${}^{G}P_l, {}^{G}P_m$ are the ordinals (integers) of the grid lines defining the intersection and ${}^{W}P_x, {}^{W}P_y, {}^{W}P_z$ are the measured world coordinates of the corner of the block. We note in passing that by using the affine calibration procedure, we can simply order the grid lines of the slide as $m = 1, 2, \ldots$ or $l = 1, 2, \ldots$ because the procedure can adapt to any scale factor.

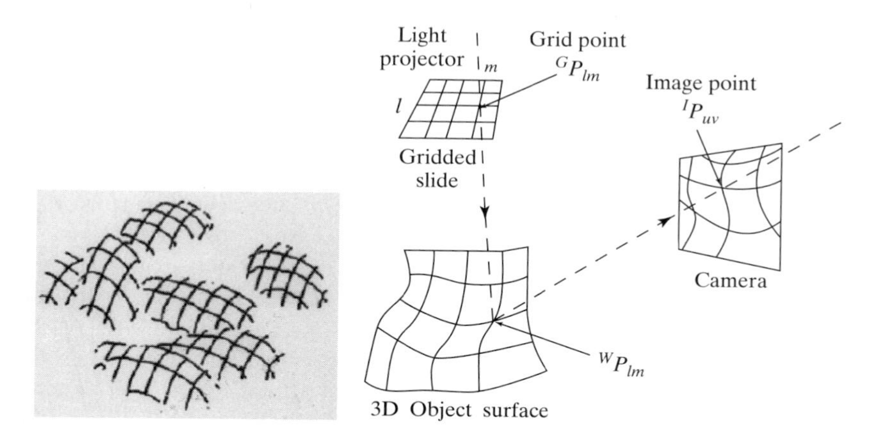

Figure 13.19 (left) Potatoes illuminated by a slide containing a grid of lines and (right) structured light concept: If the imaging system can deduce which ray ${}^{G}P_{lm}$ created a certain bright feature imaged at ${}^{I}P_{uv}$, then four equations are available to solve for 3D surface point ${}^{W}P_{lm}$. Camera and projector matrices must both be available.

Compute mesh of 3D points from image of scene with light stripes.
Offline Procedure:

1. Calibrate the camera to obtain camera matrix **C**.
2. Calibrate the projector to obtain projector matrix **D**.

Online Procedure:

1. Input camera and projector matrices **C**, **D**.
2. Input image of scene surface with light stripes.
3. Extract grid of bright stripes and intersections.
4. Determine the labels l, m of all grid intersections.
5. For each projected point P_{lm} imaged at P_{uv} compute 3D surface point P using **C**, **D** and stereo equations.
6. Output mesh as a graph with vertices as 3D points and edges as their connections by bright stripes.

Algorithm 13.2 Computing 3D surface coordinates using calibrated camera and projector.

Exercise 13.23

Create and calibrate a structured light projector in the following manner: Using your favorite image editing tool, create a digital image of a bright grid on a dark background; or, draw a pattern on paper and digitize it on a scanner. Connect a laptop to a projector and display your digital image; this should project your grid into space. Pose the projector so that it illuminates a tabletop workspace. Place precise blocks on the table to obtain calibration points and perform the affine calibration procedure. Report on the results.

The correspondence problem is still present as in two-camera stereo, although not quite as severe. Referring to the image of the potatoes (see Figure 13.19), it is clear that there is a problem for the imaging system to determine the exact grid intersections seen. If only surface shape is needed and not location, it usually only matters that grid intersections are identified consistently relative to each other. See Figure 13.20 and the references by Hu and Stockman (1989) and by Shrikhande and Stockman (1989). Various engineering solutions have been implemented; for example, coding grid lines using variations of shape or color. Another solution is to change the grid pattern rapidly over time with the imaging system taking multiple images from which grid patterns can be uniquely determined. White light projectors have a limted depth-of-field where the grid pattern is sharp. Laser light projectors do not suffer from this limitation but the reflection off certain objects will be poor due to the low power and limited spectrum of a typical laser. In many controlled sensing environments structured light sensors are highly effective. Off-the-shelf sensor

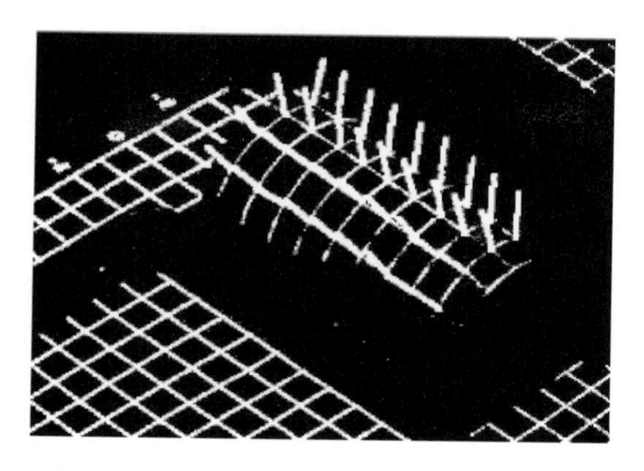

Figure 13.20 Surface normals computed using a projected grid: Normals can be computed by assuming the grid lines are in sequence but exact grid lines are not needed. (Weak perspective models were used for both projector and camera.) (Courtesy of Shrikhande and Stockman, 1989.)

units can be purchased from several companies; some may have only one ray, one stripe, or perhaps two orthogonal stripes.

13.6 A SIMPLE POSE ESTIMATION PROCEDURE

We want to use cameras to compute object geometry and pose. In this section we study a simple method of computing the pose of an object from three points of the image of that object. It is assumed that a geometric model of the object is known and that the focal length f of the camera is known. We study this simple method because it not only gives a practical means of computing pose, but because it also introduces some important concepts in a simple context. One of these is the idea of *inverse perspective*—computing 3D properties from properties in the 2D image plane of a perspective transformation. Another is the idea of using optimization to find a best set of parameters that will match 3D object points to 2D image points. Our most important simplifying assumption is that the world coordinate system is the same as the camera coordinate system, so we omit superscripts of points because they are not needed to indicate coordinate system (${}^{\mathbf{W}}\mathbf{P}_j = {}^{\mathbf{C}}\mathbf{P}_j \equiv P_j$). Another simplification is that we will work only in real geometrical space—we use no image quantization or pixel coordinates.

The environment for the *Perspective 3 Point Problem* (**P3P**) is shown in Figure 13.21. Three 3D scene points P_i are observed in the u-v image plane as Q_i. The coordinates of the points P_i are the unknowns for which we want to solve. Since we assume that we know what points of an object model we are observing (a big assumption), we do know the distances between pairs of points: For a rigid object, these three distances do not vary as the object moves in space. In an application of human-computer interaction (HCI), the points P_i could be facial features of a specific person, such as the eyes and tip of the nose. The face can be measured for the required distances. Computing the pose of the face will reveal where the person is looking. In a navigation application, the three points P_i could be geographic landmarks whose locations are known in a map. A navigating robot or drone can compute its own position relative to the landmarks using the following method.

The image locations of the observed points Q_i are known. Let $\mathbf{q}_i$ be the unit vector from the origin in the direction of Q_i. The 3D point P_i is also in this same direction.

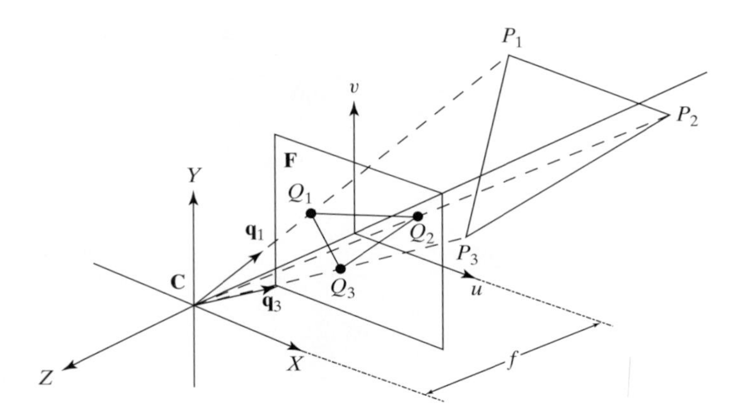

Figure 13.21 A simple case for pose estimation: A triangle $P_1 P_2 P_3$ with sides of known length in 3D is observed in the u-v image plane as triangle $Q_1 Q_2 Q_3$. The 3D positions of points P_j can be computed from the observed image points Q_j. The pose of an object containing the P_j can then be determined relative to the camera coordinate system. The focal length f is the distance from the center of projection to the image plane. Coordinates of the image points are $Q_j = [u_j, v_j, -f]$ relative to the camera coordinate system **C**.

Therefore, we can solve for the locations of P_i from Q_i if we can compute the three scalars a_i such that

$$P_i = a_i \mathbf{q}_i \tag{13.36}$$

From the three equations represented in Equation 13.36 we derive three others relating the distances between points, which are known from the model.

$$d_{mn} = \| P_m - P_n \| \quad (m \neq n) \tag{13.37}$$

We rewrite the P_i in terms of the observed Q_i and compute the 3D distances, using the dot product and the fact that $\mathbf{q_i} \circ \mathbf{q_i} = \mathbf{1}$.

$$\begin{aligned} {d_{mn}}^2 &= \|a_m \mathbf{q_m} - a_n \mathbf{q_n}\|^2 \\ &= (a_m \mathbf{q_m} - a_n \mathbf{q_n}) \circ (a_m \mathbf{q_m} - a_n \mathbf{q_n}) \\ &= {a_m}^2 - 2a_m a_n (\mathbf{q_m} \circ \mathbf{q_n}) + {a_n}^2 \end{aligned} \tag{13.38}$$

We now have three quadratic equations in the three unknowns a_i. The three left sides ${d_{mn}}^2$ are known from the model and the three $\mathbf{q_m} \circ \mathbf{q_n}$ are known from the image points Q_i. Our P3P problem of computing the positions of the three points P_i is now reduced to solving three quadratic equations in three unknowns. In theory, there can be 8 different triples $[a_1, a_2, a_3]$ that will satisfy Equations 13.38. Extending Figure 13.21, it is easy to see that for each pose of the three points on one side of the coordinate system, there is a mirror set on the other side with parameters $[-a_1, -a_2, -a_3]$, and clearly if one triple satisfies the equations, then so must the other. Thus, we can have at most four actual poses possible in a real situation because the object must be on one side of the camera. In Fischler and Bolles (1981) it is shown that four poses are possible in special cases; however, the common case is for two solutions.

We now show how to solve for the unknowns a_i (and hence P_i) using nonlinear optimization. This will give insight to other optimizations used in later sections. Mathematically, we want to find three roots of the following functions of the a_i.

$$
\begin{aligned}
f(a_1, a_2, a_3) &= a_1^2 - 2a_1a_2(\mathbf{q_1} \circ \mathbf{q_2}) + a_2^2 - d_{12}^2 \\
g(a_1, a_2, a_3) &= a_2^2 - 2a_2a_3(\mathbf{q_2} \circ \mathbf{q_3}) + a_3^2 - d_{23}^2 \\
h(a_1, a_2, a_3) &= a_1^2 - 2a_1a_3(\mathbf{q_1} \circ \mathbf{q_3}) + a_3^2 - d_{13}^2
\end{aligned} \tag{13.39}
$$

Suppose that we are near a root $[a_1, a_2, a_3]$ but that $f(a_1, a_2, a_3) \neq 0$. We want to compute some changes $[\Delta_1, \Delta_2, \Delta_3]$ so that, ideally, $f(a_1 + \Delta_1, a_2 + \Delta_2, a_3 + \Delta_3) = 0$, and in reality, it moves toward 0. We can linearize f in the neighborhood of $[a_1, a_2, a_3]$ and then compute the changes $[\Delta_1, \Delta_2, \Delta_3]$ needed to produce 0.

$$
f(a_1 + \Delta_1, a_2 + \Delta_2, a_3 + \Delta_3) = f(a_1, a_2, a_3) + \begin{bmatrix} \frac{\partial f}{\partial a_1} & \frac{\partial f}{\partial a_2} & \frac{\partial f}{\partial a_3} \end{bmatrix} \begin{bmatrix} \Delta_1 \\ \Delta_2 \\ \Delta_3 \end{bmatrix} + \mathbf{h.o.t.} \tag{13.40}
$$

By ignoring the higher order terms in Equation 13.40 and setting the left side to zero, we obtain one linear equation in the unknowns $[\Delta_1, \Delta_2, \Delta_3]$. Using the same concept for the functions g and h, we arrive at the following matrix equation:

$$
\begin{bmatrix} 0 \\ 0 \\ 0 \end{bmatrix} = \begin{bmatrix} f(a_1, a_2, a_3) \\ g(a_1, a_2, a_3) \\ h(a_1, a_2, a_3) \end{bmatrix} + \begin{bmatrix} \frac{\partial f}{\partial a_1} & \frac{\partial f}{\partial a_2} & \frac{\partial f}{\partial a_3} \\ \frac{\partial g}{\partial a_1} & \frac{\partial g}{\partial a_2} & \frac{\partial g}{\partial a_3} \\ \frac{\partial h}{\partial a_1} & \frac{\partial h}{\partial a_2} & \frac{\partial h}{\partial a_3} \end{bmatrix} \begin{bmatrix} \Delta_1 \\ \Delta_2 \\ \Delta_3 \end{bmatrix} \tag{13.41}
$$

The matrix of partial derivatives is the Jacobian matrix $\mathbf{J}$: If it is invertible at the point $[a_1, a_2, a_3]$ of our search then we obtain the following solution for the changes to these parameters:

$$
\begin{bmatrix} \Delta_1 \\ \Delta_2 \\ \Delta_3 \end{bmatrix} = -\mathbf{J}^{-1}(a_1, a_2, a_3) \begin{bmatrix} f(a_1, a_2, a_3) \\ g(a_1, a_2, a_3) \\ h(a_1, a_2, a_3) \end{bmatrix} \tag{13.42}
$$

We can compute a new vector of parameters by adding these changes to the previous parameters. We use A^k to denote the parameters $[a_1, a_2, a_3]$ at the *kth* iteration and arrive at a familiar form of Newton's method. $\mathbf{f}$ represents the vector of values computed using functions f, g, and h.

$$
A^{k+1} = A^k - \mathbf{J}^{-1}(A^k)\mathbf{f}(A^k) \tag{13.43}
$$

Exercise 13.24: Defining the Jacobian.

Show that the Jacobian for the function $\mathbf{f}(a_1, a_2, a_3)$ is as follows, where t_{mn} denotes the dot product $\mathbf{q}_m \circ \mathbf{q}_n$.

$$\mathbf{J}(a_1, a_2, a_3) \equiv \begin{bmatrix} J_{11} & J_{12} & J_{13} \\ J_{21} & J_{22} & J_{23} \\ J_{31} & J_{32} & J_{33} \end{bmatrix}$$

$$= \begin{bmatrix} (2a_1 - 2t_{12}a_2) & (2a_2 - 2t_{12}a_1) & 0 \\ 0 & (2a_2 - 2t_{23}a_3) & (2a_3 - 2t_{23}a_2) \\ (2a_1 - 2t_{31}a_3) & 0 & (2a_3 - 2t_{31}a_1) \end{bmatrix}$$

Exercise 13.25: Computing the inverse of the Jacobian.

Using the symbols J_{ij} from the previous exercise, derive a representation for the inverse Jacobian $\mathbf{J}^{-1}$.

Algorithm 13.3 summarizes the method for computing the positions of the three 3D model points in terms of the camera coordinate system. Experience has shown that the algorithm converges within 5 to 10 iterations under typical situations. However, it is not clear how to control the algorithm so that multiple solutions can be obtained. Nonlinear optimization is sometimes as much art as algorithm and the reader should consult the references for its many nuances. Table 13.3 shows critical data during the iterations of a

TABLE 13.3 ITERATIONS OF P3P SOLUTION

It. k	$\lvert f(A^k)\rvert$	$\lvert g(A^k)\rvert$	$\lvert h(A^k)\rvert$	a_1	a_2	a_3
1	6.43e+03	3.60e+03	1.09e+04	1.63e+02	1.65e+02	1.63e+02
2	1.46e+03	8.22e+02	2.48e+03	1.06e+02	1.08e+02	1.04e+02
3	2.53e+02	1.51e+02	4.44e+02	8.19e+01	9.64e+01	1.03e+02
...		...			...	
8	2.68e+00	6.45e−01	5.78e+00	8.414e+01	9.127e+01	8.926e+01
9	5.00e−02	3.87e−02	1.71e−01	8.414e+01	9.126e+01	8.925e+01

It. k	P_{1x}	P_{1y}	P_{1z}	P_{2x}	P_{2y}	P_{2z}	P_{3x}	P_{3y}	P_{3z}
1	−36.9	−58.4	147.6	−34.4	−14.4	160.7	0.0	−14.5	162.4
2	−24.0	−38.0	96.0	−22.5	−9.3	105.2	0.0	−9.3	103.6
...	...			...			...		
8	−19.1	−30.2	76.2	−19.1	−7.9	88.9	0.0	−7.9	88.9
9	−19.1	−30.2	76.2	−19.1	−7.9	88.9	0.0	−7.9	88.9

Using focal length $f = 30$, simulated image coordinates Q_j were computed from projecting $P_1 = [-19.05, -30.16, 76.20]$, $P_2 = [-19.05, -7.94, 88.90]$, $P_3 = [0.00, -7.94, 88.90]$. Beginning parameters were set to $A^0 = [300, 300, 300]$ and $\Delta = 0.2$. In nine iterations the P3P program converged to the initial P_j accurate to two decimal places.

Compute the position of 3 3D points in space from 3 image points.
Input three pairs $({}^{\mathbf{M}}\mathbf{P}_i, {}^{\mathbf{F}}\mathbf{Q}_i)$ of corresponding points from 3D and 2D.
Each ${}^{\mathbf{M}}\mathbf{P}_i$ is in model coordinates; ${}^{\mathbf{F}}\mathbf{Q}_i$ is in real image coordinates.
Input the focal length f of camera and tolerance Δ on distance.
Output the positions ${}^{\mathbf{C}}\mathbf{P}_i$ of the three model points relative to the camera.

1. **initialize:**
   - **(a)** From model points ${}^{\mathbf{M}}\mathbf{P}_i$ compute squared distances ${d_{mn}}^2$
   - **(b)** From image points ${}^{\mathbf{F}}\mathbf{Q}_i$ compute unit vectors $\mathbf{q_i}$ and dot products $2\mathbf{q_m} \circ \mathbf{q_n}$
   - **(c)** Choose a starting parameter vector $\mathbf{A^1} = [a_1, a_2, a_3]$ (How?)
2. **iterate:** until $\mathbf{f}(\mathbf{A^k}) \approx 0$
   - **(a)** $\mathbf{A^{k+1}} = \mathbf{A^k} - \mathbf{J}^{-1}(\mathbf{A^k})\mathbf{f}(\mathbf{A^k})$
     - **i.** $\mathbf{A^k} = \mathbf{A^{k+1}}$
     - **ii.** compute $\mathbf{J}^{-1}(\mathbf{A^k})$ if $\mathbf{J}^{-1}$ exists
     - **iii.** evaluate $\mathbf{f}(\mathbf{A^k}) = [f(a_1^k, a_2^k, a_3^k), g(a_1^k, a_2^k, a_3^k), h(a_1^k, a_2^k, a_3^k)]^t$
   - **(b)** stop when $\mathbf{f}(\mathbf{A^{k+1}})$ is within error tolerance Δ of 0
     or stop if number of iterations exceeds limit
3. **compute pose:** From $\mathbf{A^{k+1}}$ compute each ${}^{\mathbf{C}}\mathbf{P}_i = {a_i}^{k+1}\mathbf{q_i}$

Algorithm 13.3 Iterative P3P Solution.

P3P solution. Simulated P_i were projected using focal length $f = 30$ to obtain simulated image coordinates Q_i. Beginning parameters were set well away from the true values. After taking some crude steps, the algorithm homes into the neighborhood of the true solution and outputs the same P_i as input, accurate to within two decimal places. As shown in columns 3–5 of Table 13.3, after the 9th iteration, the difference between the model side length and the computed side length is less than 0.2 units. Starting with each $a_i \approx 100$ halves the number of iterations. If the standoff of an object from the camera is approximately known, then this is a good starting value for each parameter.

Ohmura and others (1988) built a system that could compute the position of a person's head 10 times-per-second. For the model feature points ${}^{\mathbf{M}}\mathbf{P}_j$, blue dots were placed on the face just left of the left eye, just right of the right eye, and just under the nose. (These points will not deform much with various facial expressions.) With the blue makeup the image points ${}^{\mathbf{F}}\mathbf{Q}_j$ could be detected rapidly and robustly. The pose of the face could then be defined by the computed points ${}^{\mathbf{C}}\mathbf{P}_j$ and the transformation mapping the ${}^{\mathbf{M}}\mathbf{P}_j$ to the ${}^{\mathbf{C}}\mathbf{P}_j$ (using Algorithm 13.1). Ballard and Stockman (1995) developed a system that located the eyes and nose on a face without any makeup, but performance was much slower due to the processing needed to identify the eyes and nose. Both groups reported that the error in the normal vector of the plane formed by the three points is of the order of a few degrees. If the plane of the three points ${}^{\mathbf{C}}\mathbf{P}_j$ is nearly normal to the image plane, then small errors in the observed image coordinates ${}^{\mathbf{F}}\mathbf{Q}_j$ can produce large errors in the computed orientation of the plane in 3D. In order to counter this effect, Ohmura and others (1988) oriented the camera axis about 20 degrees off to the side of the general direction of the face.

Equations 13.38 should always have a solution; thus we can compute the pose of an airplane from three points in the image of a frog! Choosing a good candidate model is important: This might be done by knowing that airplanes are not green or cannot be present, etc. Verification of a model is also important: This can be done by projecting more object model points and verifying them in the image. For example, to distinguish between two possible poses of a face, we might look for an ear, chin, and eyebrow after computing pose from the eyes and nose. We examine verification more in the following sections. Also, we will take into consideration that digital image points are identified in pixel coordinates and that radial lens distortion must be modeled along with the perspective projection.

Exercise 13.26

Discuss how the **P5P** problem can be solved using the **P3P** solution procedure.

13.7 AN IMPROVED CAMERA CALIBRATION METHOD*

We now describe a calibration method, developed by Roger Tsai (1987), that has been widely used in industrial vision applications. It has been reported that with careful implementation this procedure can support 3D measurements with a precision of 1 part in 4,000, which is quite good. Since the general idea of calibration has been carefully developed in Section 13.3, we proceed with a simpler notation.

- $\mathbf{P} = [x, y, z]$ is a point in the 3D world coordinate system.
- $\mathbf{p} = [u, v]$ is a point in the real image plane. (One can think of the u axis as horizontal and pointing to the right and the v axis as vertical and pointing upward.)
- $\mathbf{a} = [r, c]$ is a pixel in the image array expressed by two integers; r is the row coordinate and c is the column coordinate of the pixel. (Relative to convention for u and v above, the r axis is vertical and points downward. The c axis is horizontal and points to the right.)

Camera calibration is conceived to be parameter recovery; we are solving for the *camera parameters* that characterize camera geometry and pose. There are two different types of parameters to be recovered:

1. intrinsic parameters, and
2. extrinsic parameters.

Exercise 13.27: WP3P problem.*

Acquire a copy of the 1988 paper by Huttenlocher and Ullman, which describes a solution for computing the pose of a rigid configuration of three points from a single weak perspective projection. The solution method is closed form and explicitly produces two solutions, unlike the P3P solution in this chapter. Program the solution method and test it on simulated data obtained by mathematically projecting 3-point configurations to obtain your three correspondences $\langle P_j, Q_j \rangle$.

13.7.1 Intrinsic Camera Parameters

The *intrinsic* parameters are truly camera parameters, as they depend on the particular device being used. They include the following parameters:

- principal point $[u_0, v_0]$: the intersection of the optical axis with the image plane.
- scale factors $\{d_x, d_y\}$ for the x and y pixel dimensions.
- aspect distortion factor τ_1: a scale factor used to model distortion in the aspect ratio of the camera.
- focal length f: the distance from the optical center to the image plane.
- lens distortion factor (κ_1): a scale factor used to model radial lens distortion.

These definitions refer to the optical center of the camera lens. The camera origin is at this point. The optical axis is the perpendicular to the image plane going through the optical center. The principal point is often, but not always, in the center pixel of the image. The scale factors d_x and d_y represent the horizontal size and vertical size of a single pixel in real world units, such as millimeters. We will assume that u_0, v_0, d_x, d_y, and the aspect distortion factor τ_1 are known for a particular camera. Thus only the focal length f and the lens distortion factor κ_1 will be computed during the calibration process.

13.7.2 Extrinsic Camera Parameters

The extrinsic parameters describe the position and orientation (pose) of the camera system in the 3D world. They include:

- translation:

$$\mathbf{t} = [t_x \quad t_y \quad t_z]^T \tag{13.44}$$

- rotation:

$$\mathbf{R} = \begin{bmatrix} r_{11} & r_{12} & r_{13} & 0 \\ r_{21} & r_{22} & r_{23} & 0 \\ r_{31} & r_{32} & r_{33} & 0 \\ 0 & 0 & 0 & 1 \end{bmatrix} \tag{13.45}$$

The translation parameters describe the position of the camera in the world coordinate system, and the rotation parameters describe its orientation. We emphasize at the outset that **there are only three independent rotation parameters and not nine.**

The calibration method developed below is autonomous, reasonably accurate, efficient, and flexible [see Tsai (1987)]. Furthermore, it can be used with any off-the-shelf camera and lens, although it may not perfectly model the specific lens chosen. Figure 13.22 shows a calibration object, which was also used in a 3D object reconstruction system to be

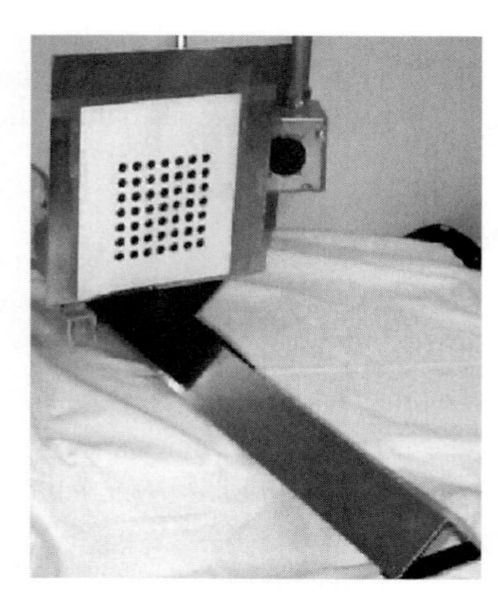

Figure 13.22 A calibration object that uses a moving 2D pattern to create many feature points in 3D space.

discussed below. The device is a metal plate onto which has been painted a 7×7 array of black circles. The centers of the circles are used as the distinguished points. The object is mounted on a horizontal rail. It is perpendicular to the rail and can be moved along it in $10mm$ steps. The position of the rail defines the 3D world coordinate system.

In the system shown in Figure 13.22, several images are taken at different positions along the rail corresponding to different distances from the camera. The camera must not be moved nor focused during the time it is being calibrated or used. In each image, the circles are detected and their centers computed. The result of the image processing is a set of correspondences between known points in the 3D world and corresponding points in the 2D images. $n > 5$ correspondences are required; we refer to them as

$$\{([x_i, y_i, z_i], [u_i, v_i]) \mid i = 1, \dots, n\}.$$

The real-valued image coordinates $[u, v]$ are computed from their pixel position $[r, c]$ by the formulas

$$u = \tau_1 d_x (c - u_0) \tag{13.46}$$

$$v = -d_y (r - v_0) \tag{13.47}$$

where d_x and d_y are the center-to-center distances between pixels in the horizontal and vertical directions, and τ_1 is the scale factor for distortions in the aspect ratio of the camera.

Figure 13.23 illustrates the geometry assumed by the procedure. The point $\mathbf{P_i} = [x_i, y_i, z_i]$ is an arbitrary point in the 3D world. The corresponding point on the image plane is labeled $\mathbf{p_i}$. Vector $\mathbf{r_i}$ is from the point $[0, 0, z_i]$ on the optical axis to the 3D point $\mathbf{P_i}$. Vector $\mathbf{s_i}$ is from the principal point $\mathbf{p_0}$ to the image point $\mathbf{p_i}$. Vector $\mathbf{s_i}$ is parallel to vector $\mathbf{r_i}$. Any radial distortion due to the lens is along $\mathbf{s_i}$.

Tsai made the following observation, which allows most of the extrinsic parameters to be computed in a first step. Since the radial distortion is along vector $\mathbf{s_i}$, the rotation matrix can be determined without considering it. Also, t_x and t_y can be computed without

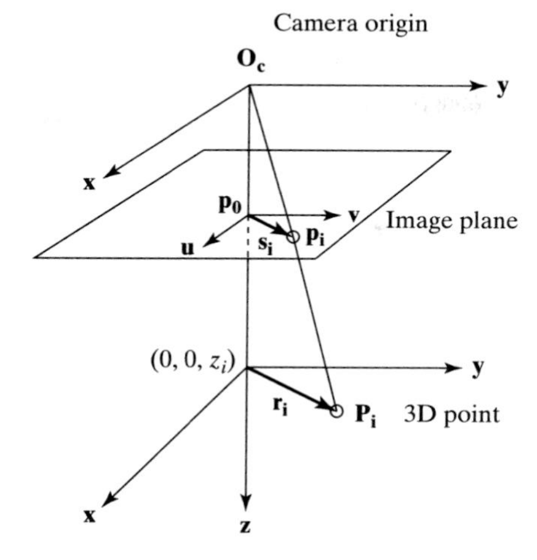

Figure 13.23 The geometric model for the Tsai calibration procedure. Point $\mathbf{p_i} = [u_i, v_i]$ on the image corresponds to point $\mathbf{P_i} = [x_i, y_i, z_i]$ on the calibration object. Point $\mathbf{p_0} = [u_0, v_0]$ is the principal point. Any radial distortion must displace image point $\mathbf{p_i}$ along direction $\mathbf{p_0} - \mathbf{p_i}$ in the image.

knowing κ_1. Computation of t_z must be done in a second step since change in t_z has an image effect similar to κ_1.

Instead of directly solving for all unknowns, we first solve for a set $\boldsymbol{\mu}$ of computed parameters from which the extrinsic parameters can be derived. Given the n point correspondences between $[x_i, y_i, z_i]$ and $[u_i, v_i]$ for $i = 1$ to n, $n > 5$, a matrix $\mathbf{A}$ is formed with rows a_i;

$$a_i = [v_i x_i,\ v_i y_i,\ -u_i x_i,\ -u_i y_i,\ v_i]. \tag{13.48}$$

Let $\boldsymbol{\mu} = [\mu_1, \mu_2, \mu_3, \mu_4, \mu_5]$ be a vector of unknown (computed) parameters defined with respect to the rotation parameters r_{11}, r_{12}, r_{21}, and r_{22} and the translation parameters t_x and t_y as

$$\mu_1 = \frac{r_{11}}{t_y} \tag{13.49}$$

$$\mu_2 = \frac{r_{12}}{t_y} \tag{13.50}$$

$$\mu_3 = \frac{r_{21}}{t_y} \tag{13.51}$$

$$\mu_4 = \frac{r_{22}}{t_y} \tag{13.52}$$

$$\mu_5 = \frac{t_x}{t_y} \tag{13.53}$$

Let the vector $\mathbf{b} = [u_1, u_2, \ldots, u_n]$ contain the u_i image coordinates of the n correspondences. Since $\mathbf{A}$ and $\mathbf{b}$ are known, the system of linear equations

$$\mathbf{A}\boldsymbol{\mu} = \mathbf{b} \tag{13.54}$$

can be solved for the unknown parameter vector $\boldsymbol{\mu}$. (See Johnson, Riess, and Arnold (1989) for techniques on solving linear systems of equations.) Now $\boldsymbol{\mu}$ can be used to compute the rotation and translation parameters as follows.

1. Let $U = \mu_1^2 + \mu_2^2 + \mu_3^2 + \mu_4^2$. Calculate the square of the y component of translation t_y as

$$t_y^2 = \begin{cases} \dfrac{U - [U^2 - 4(\mu_1\mu_4 - \mu_2\mu_3)^2]^{1/2}}{2(\mu_1\mu_4 - \mu_2\mu_3)^2} & \text{if } (\mu_1\mu_4 - \mu_2\mu_3) \neq 0 \\ \dfrac{1}{\mu_1^2 + \mu_2^2} & \text{if } \left(\mu_1^2 + \mu_2^2\right) \neq 0 \\ \dfrac{1}{\mu_3^2 + \mu_4^2} & \text{if } \left(\mu_3^2 + \mu_4^2\right) \neq 0 \end{cases} \tag{13.55}$$

2. Let $t_y = (t_y^2)^{1/2}$ (the positive square root) and compute four of the rotation parameters and the x-translation t_x from the known computed value of $\boldsymbol{\mu}$:

$$r_{11} = \mu_1 t_y \tag{13.56}$$

$$r_{12} = \mu_2 t_y \tag{13.57}$$

$$r_{21} = \mu_3 t_y \tag{13.58}$$

$$r_{22} = \mu_4 t_y \tag{13.59}$$

$$t_x = \mu_5 t_y \tag{13.60}$$

3. To determine the true sign of t_y, select an object point $\mathbf{P}$ whose image coordinates $[u, v]$ are far from the image center (to avoid numerical problems). Let $\mathbf{P} = [x, y, z]$, and compute

$$\xi_x = r_{11}x + r_{12}y + t_x \tag{13.61}$$

$$\xi_y = r_{21}x + r_{22}y + t_y \tag{13.62}$$

This is like applying the computed rotation parameters to the x and y coordinates of $\mathbf{P}$. If ξ_x has the same sign as u and ξ_y has the same sign as v, then t_y already has the correct sign, else negate it.

4. Now the remaining rotation parameters can be computed as

$$r_{13} = \left(1 - r_{11}^2 - r_{12}^2\right)^{1/2} \tag{13.63}$$

$$r_{23} = \left(1 - r_{21}^2 - r_{22}^2\right)^{1/2} \tag{13.64}$$

$$r_{31} = \frac{1 - r_{11}^2 - r_{12}r_{21}}{r_{13}} \tag{13.65}$$

$$r_{32} = \frac{1 - r_{21}r_{12} - r_{22}^2}{r_{23}} \tag{13.66}$$

$$r_{33} = (1 - r_{31}r_{13} - r_{32}r_{23})^{1/2} \tag{13.67}$$

The orthonormality constraints of the rotation matrix **R** have been used in deriving these equations. The signs of r_{23}, r_{31}, and r_{32} may not be correct due to the ambiguity of the square root operation. At this step, r_{23} should be negated if the sign of

$$r_{11}r_{21} + r_{12}r_{22}$$

is positive, so that the orthogonality of the rotation matrix is preserved. The other two may need to be adjusted after computing the focal length.

5. The focal length f and the z component of translation t_z are now computed from a second system of linear equations. First a matrix $\mathbf{A}'$ is formed whose rows are given by

$$a_i' = (r_{21}x_i + r_{22}y_i + t_y, v_i) \tag{13.68}$$

Next a vector $\mathbf{b}'$ is constructed with rows defined by

$$b'_i = (r_{31}x_i + r_{32}y_i)vi. \tag{13.69}$$

We solve the linear system

$$\mathbf{A}'\mathbf{v} = \mathbf{b}' \tag{13.70}$$

for $\mathbf{v} = (f, t_z)^T$. We obtain only estimates for f and t_z at this point.

6. If $f < 0$ then change the signs of r_{13}, r_{23}, r_{31}, r_{32}, f, and t_z. This is to force the use of a right-handed coordinate system.
7. The estimates for f and t_z can be used to compute the lens distortion factor κ_1 and to derive improved values for f and t_z. The simple distortion model used here is that the true image coordinates $[\hat{u}, \hat{v}]$ are obtained from the measured ones by the following equations:

$$\hat{u} = u(1 + \kappa_1 r^2) \tag{13.71}$$

$$\hat{v} = v(1 + \kappa_1 r^2) \tag{13.72}$$

where the radius of distortion r is given by

$$r = (u^2 + v^2)^{1/2} \tag{13.73}$$

Using the perspective projection equations, modified to include the distortion, we derive a set of nonlinear equations of the form

$$\left\{ v_i(1 + \kappa_1 r^2) = f \frac{r_{21}x_i + r_{22}y_i + r_{23}z_i + t_y}{r_{31}x_i + r_{32}y_i + r_{33}z_i + t_z} \right\} \quad i = 1, \ldots, n \tag{13.74}$$

Solving this system by nonlinear regression gives the values for f, t_z, and κ_1.

13.7.3 Calibration Example

To see how this camera calibration procedure works, we will go through an example. The following table gives five point correspondences input to the calibration system.

The units for both the world coordinate system and the **u-v** image coordinate system are centimeters.

i	x_i	y_i	z_i	u_i	v_i
1	0.00	5.00	0.00	−0.58	0.00
2	10.00	7.50	0.00	1.73	1.00
3	10.00	5.00	0.00	1.73	0.00
4	5.00	10.00	0.00	0.00	1.00
5	5.00	0.00	0.00	0.00	−1.00

Figure 13.24 shows the five calibration points in the 3D world and approximately how they will look on the image plane as viewed by the camera, whose position, orientation, and focal length are unknown and to be computed. Figure 13.25 shows the image points in the continuous **u-v** coordinate system.

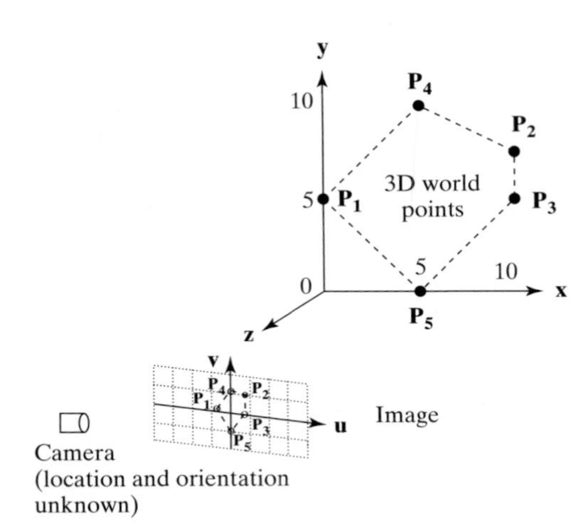

Figure 13.24 The 3D world points and corresponding 2D image points that are input to the calibration procedure, which will compute the camera parameters including position, orientation, and focal length.

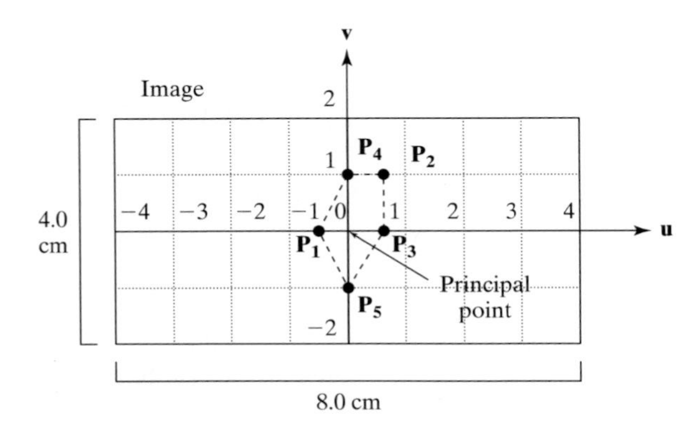

Figure 13.25 The image points in the **u-v** image coordinate system.

Using the five correspondences, the matrix **A** and vector **b** of Equation 13.54 are given by

$$\mathbf{A} = \begin{array}{c} \begin{array}{ccccc} v_i x_i & v_i y_i & -u_i x_i & -u_i y_i & v_i \end{array} \\ \left[\begin{array}{ccccc} 0.00 & 0.00 & 0.00 & 2.89 & 0.00 \\ 10.00 & 7.50 & -17.32 & -12.99 & 1.00 \\ 0.00 & 0.00 & -17.32 & -8.66 & 0.00 \\ 5.00 & 10.00 & 0.00 & 0.00 & 1.00 \\ -5.00 & 0.00 & 0.00 & 0.00 & -1.00 \end{array}\right] \end{array}$$

and

$$\mathbf{b} = \begin{array}{c} u_i \\ \left[\begin{array}{c} -.58 \\ 1.73 \\ 1.73 \\ 0.00 \\ 0.00 \end{array}\right] \end{array}$$

Solving $\mathbf{A}\boldsymbol{\mu} = \mathbf{b}$ yields the vector $\boldsymbol{\mu}$ given by

$$\boldsymbol{\mu} = \begin{array}{c} \mu_i \\ \left[\begin{array}{c} -0.17 \\ 0.00 \\ 0.00 \\ -0.20 \\ 0.87 \end{array}\right] \end{array}$$

The next step is to calculate U and use it to solve for t_y^2 in Equation 13.55. We have

$$U = \mu_1^2 + \mu_2^2 + \mu_3^2 + \mu_4^2 = .07$$

Using the first formula in Equation 13.55, we have

$$t_y^2 = \frac{U - [U^2 - 4(\mu_1\mu_4 - \mu_2\mu_3)^2]^{1/2}}{2(\mu_1\mu_4 - \mu_2\mu_3)^2} = 25$$

When t_y is set to the positive square root (5), we have

$$\begin{aligned} r_{11} &= \mu_1 t_y = -.87 \\ r_{12} &= \mu_2 t_y = 0 \\ r_{21} &= \mu_3 t_y = 0 \\ r_{22} &= \mu_4 t_y = -1.0 \\ t_x &= \mu_5 t_y = 4.33 \end{aligned}$$

Next we compute ξ_x and ξ_y for the point $\mathbf{P_2} = (10.0,7.5,0.0)$ and corresponding image point $\mathbf{p_2} = (1.73,1.0)$, which is far from the image center, to check the sign of t_y.

$$\xi_x = r_{11}x + r_{12}y + t_x = (-.87)(10) + 0 + 4.33 = -4.37$$

$$\xi_y = r_{21}x + r_{22}y + t_y = 0 + (-1.0)(7.5) + 5 = -2.5$$

Since the signs of ξ_x and ξ_y do not agree with those of $\mathbf{p_2}$, the sign of t_y is wrong, and it is negated. This gives us

$$\begin{aligned} t_y &= -5 \\ r_{11} &= .87 \\ r_{12} &= 0 \\ r_{21} &= 0 \\ r_{22} &= 1.0 \\ t_x &= -4.33 \end{aligned}$$

Continuing, we compute the remaining rotation parameters:

$$\begin{aligned} r_{13} &= \left(1 - r_{11}^2 - r_{12}^2\right)^{1/2} = 0.5 \\ r_{23} &= \left(1 - r_{21}^2 - r_{22}^2\right)^{1/2} = 0. \\ r_{31} &= \frac{1 - r_{11}^2 - r_{12}r_{21}}{r_{13}} = 0.5 \\ r_{32} &= \frac{1 - r_{21}r_{12} - r_{22}^2}{r_{23}} = 0. \\ r_{33} &= (1 - r_{31}r_{13} - r_{32}r_{23})^{1/2} = .87 \end{aligned}$$

Checking the sign of $r_{11}r_{21} + r_{12}r_{22} = 0$ shows that it is not positive, and therefore, r_{23} does not need to change.

We now form the second system of linear equations as follows:

$$\begin{array}{cc} r_{21}x_i + & \\ r_{22}y_i + t_y & v_i \end{array}$$

$$\mathbf{A}' = \begin{bmatrix} 0.00 & 0.00 \\ 2.500 & -1.00 \\ 0.00 & 0.00 \\ 5.00 & -1.00 \\ -5.00 & 1.00 \end{bmatrix}$$

and

$$(r_{31}x_i + r_{32}y_i)v_i$$

$$\mathbf{b}' = \begin{bmatrix} 0.0 \\ 5.0 \\ 0.0 \\ 2.5 \\ -2.5 \end{bmatrix}$$

Solving $\mathbf{A}'\mathbf{v} = \mathbf{b}'$ yields the vector $\mathbf{v} = [f, t_z]$ given by

$$f = -1.0$$
$$t_z = -7.5$$

Since f is negative, our coordinate system is not right-handed. To flip the z-axis, we negate r_{13}, r_{23}, r_{31}, r_{32}, f, and t_z. The results are:

$$\mathbf{R} = \begin{bmatrix} 0.87 & 0.00 & -0.50 \\ 0.00 & -1.00 & 0.00 \\ -0.50 & 0.00 & 0.87 \end{bmatrix}$$

and

$$\mathbf{T} = \begin{bmatrix} -4.33 \\ -5.00 \\ 7.50 \end{bmatrix}$$

and $f = 1$.

Since our example includes no distortion, these are the final results of the calibration procedure. Figure 13.26 illustrates the results of the calibration procedure shown from two different views.

Exercise 13.28

Verify that the rotation matrix **R** derived in the above example is orthonormal.

Exercise 13.29: Using camera parameter f.

Using 3D Euclidean geometry and Figure 13.26(a), find the projection of P_1 and P_3 on the image plane. (All the lines in Figure 13.26(a) are coplanar.) Verify that the results agree with the image coordinates p_1 and p_3 given in the example.

Exercise 13.30: Using camera parameters **R** and **T**.

P_1 and P_3 in the example were given in the world coordinate system. Find their coordinates in the camera coordinate system

1. using Euclidean geometry and Figure 13.26(a),
2. using the camera parameters determined by the calibration procedure.

13.8 POSE ESTIMATION*

In industrial vision, especially for robot guidance tasks, it is important to obtain the pose of a 3D object in workspace coordinates. Since the pose of the camera in the workspace can be computed by calibration, the problem is reduced to that of determining the pose of the object with respect to the camera. The method for determining object pose given in this

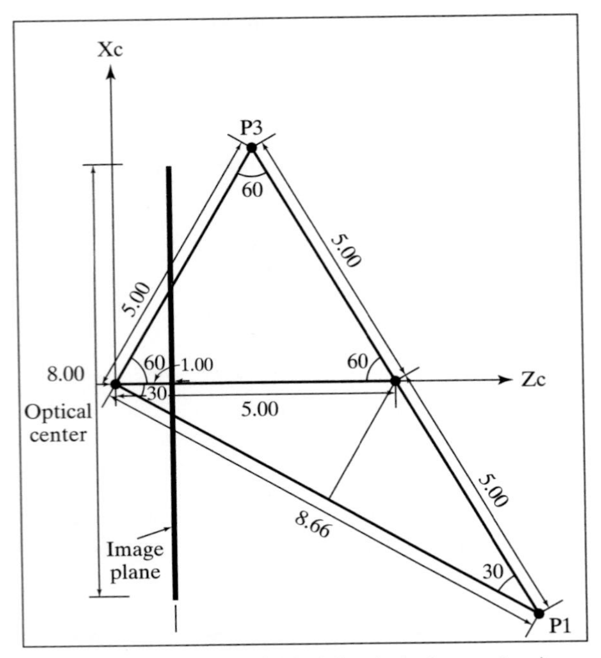

(*a*) Top view (the thick black line is the image plane)

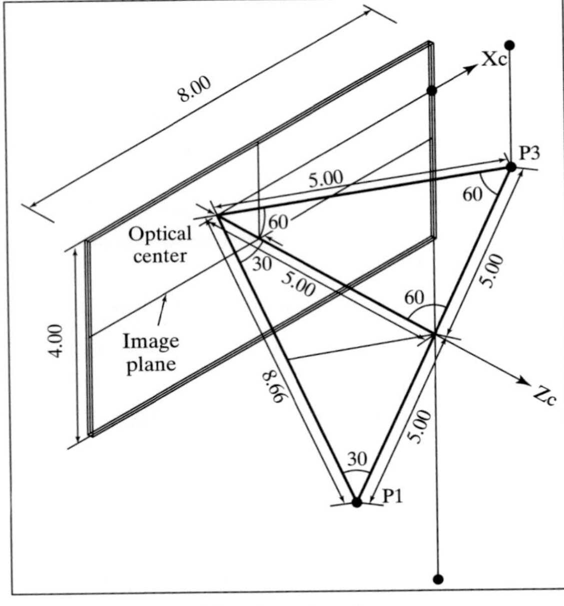

(*b*) Perspective view

Figure 13.26 Two views of the camera and image plane in the world coordinate system as computed in the example via the Tsai calibration procedure. (Courtesy of Habib Abi-Rached.)

section should have accuracy superior to the simple method given in Section 13.6. Using point-correspondences is the most basic and most-often-used method for pose computation. For use of correspondences between 2D and 3D line segments, between 2D ellipses and 3D circles, and between any combination of point pairs, line-segment pairs, and ellipse-circle pairs, see the work of Ji and others (1998).

13.8.1 Pose from 2D-3D Point Correspondences

The camera model of the previous section is used, and we assume that the camera has been calibrated to obtain the intrinsic and extrinsic parameters. The problem of determining object pose from n point correspondences between 3D model object points and 2D image points is inherently a non-linear one. Non-linear methods for estimating the pose parameters are necessary; however, under some conditions, an approximate, linear solution can be found.

Let $[x, y, z]$ be the coordinates of object point ${}^{\mathbf{M}}\mathbf{P}$ in its model coordinate system. Let the object coordinate system and the camera coordinate system be related by a transformation ${}^{\mathbf{C}}_{\mathbf{M}}\mathbf{Tr} = \{\mathbf{R}, \mathbf{T}\}$, described in the form of a rotation matrix $\mathbf{R}$ and a translation vector $\mathbf{T} = [t_x, t_y, t_z]$. Then, the perspective projection of ${}^{\mathbf{M}}\mathbf{P}$ onto the image plane yields image plane coordinates $[u, v]$, where

$$u = f\frac{r_{11}x + r_{12}y + r_{13}z + t_x}{r_{31}x + r_{32}y + r_{33}z + t_z} \qquad (13.75)$$

and

$$v = f\frac{r_{21}x + r_{22}y + r_{23}z + t_y}{r_{31}x + r_{32}y + r_{33}z + t_z} \qquad (13.76)$$

and f is the focal length of the camera, which is now known.

The transformation between object model frame and camera frame corresponds to the pose of the object with respect to the camera frame. Using our perspective imaging model as before, we have nine rotation parameters and three translation parameters involved in twelve equations of the following form.

$$\mathbf{B}\,\mathbf{w} = \mathbf{0} \qquad (13.77)$$

where

$$\mathbf{B} = \begin{pmatrix} fx_1 & fy_1 & fz_1 & 0 & 0 & 0 & -u_1x_1 & -u_1y_1 & -u_1z_1 & f & 0 & -u_1 \\ 0 & 0 & 0 & fx_1 & fy_1 & fz_1 & -v_1x_1 & -v_1y_1 & -v_1z_1 & 0 & f & -v_1 \\ fx_2 & fy_2 & fz_2 & 0 & 0 & 0 & -u_2x_2 & -u_2y_2 & -u_2z_2 & f & 0 & -u_2 \\ 0 & 0 & 0 & fx_2 & fy_2 & fz_2 & -v_2x_2 & -v_2y_2 & -v_2z_2 & 0 & f & -v_2 \\ \vdots & \vdots & \vdots & \vdots & \vdots & \vdots & \vdots & \vdots & \vdots & \vdots & \vdots & \vdots \\ fx_6 & fy_6 & fz_6 & 0 & 0 & 0 & -u_6x_6 & -u_6y_6 & -u_6z_6 & f & 0 & -u_6 \\ 0 & 0 & 0 & fx_6 & fy_6 & fz_6 & -v_6x_6 & -v_6y_6 & -v_6z_6 & 0 & f & -v_6 \end{pmatrix} \qquad (13.78)$$

and

$$\mathbf{w} = (r_{11} \quad r_{12} \quad r_{13} \quad r_{21} \quad r_{22} \quad r_{23} \quad r_{31} \quad r_{32} \quad r_{33} \quad t_x \quad t_y \quad t_z)^T . \qquad (13.79)$$

However, if one is interested in finding independent pose parameters, and not simply an affine transformation that aligns the projected model points with the image points, conditions need to be imposed on the elements of **R** such that it satisfies all the criteria a true 3D rotation matrix must satisfy. In particular, a rotation matrix needs to be orthonormal: its row vectors must have magnitude equal to one, and they must be orthogonal to each other. This can be written as:

$$\begin{aligned} \|R_1\| &= r_{11}^2 + r_{12}^2 + r_{13}^2 = 1 \\ \|R_2\| &= r_{21}^2 + r_{22}^2 + r_{23}^2 = 1 \\ \|R_3\| &= r_{31}^2 + r_{32}^2 + r_{33}^2 = 1 \end{aligned} \tag{13.80}$$

and

$$\begin{aligned} R_1 \circ R_2 &= 0 \\ R_1 \circ R_3 &= 0 \\ R_2 \circ R_3 &= 0 \end{aligned} \tag{13.81}$$

The conditions imposed on **R** turn the problem into a non-linear one. If the conditions on the magnitudes of the row vectors of **R** are imposed one at a time, and computed independently, a linear constrained optimization technique can be used to compute the constrained row vector of **R**. (See the Faugeras and others (1993) reference for a similar procedure.)

13.8.2 Constrained Linear Optimization

Given the system of Equations 13.77, the problem at hand is to find the solution vector **w** that minimizes $\|\mathbf{Bw}\|$ subject to the constraint $\|\mathbf{w}'\|^2 = 1$, where $\mathbf{w}'$ is a subset of the elements of **w**. If the constraint is to be imposed on the first row vector of **R**, then

$$\mathbf{w}' = \begin{pmatrix} r_{11} \\ r_{12} \\ r_{13} \end{pmatrix}.$$

To solve the above problem, it is necessary to rewrite the original system of equations $\mathbf{Bw} = \mathbf{0}$ in the following form

$$\mathbf{Cw}' + \mathbf{Dw}'' = \mathbf{0},$$

where $\mathbf{w}''$ is a vector with the remaining elements of **w**. Using the example above, that is, if the constraint is imposed on the first row of **R**,

$$\mathbf{w}'' = (r_{21} \quad r_{22} \quad r_{23} \quad r_{31} \quad r_{32} \quad r_{33} \quad t_x \quad t_y \quad t_z)^T.$$

The original problem can be stated as: minimize the objective function $\mathbf{O} = \mathbf{Cw}' + \mathbf{Dw}''$, that is

$$\min_{w', w''} \|\mathbf{Cw}' + \mathbf{Dw}''\|^2 \tag{13.82}$$

subject to the constraint $\|\mathbf{w}'\|^2 = 1$. Using a Lagrange multiplier technique, the above is equivalent to

$$\min_{w',w''}\left[\|\mathbf{C}\mathbf{w}' + \mathbf{D}\mathbf{w}''\|^2 + \lambda(1 - \|\mathbf{w}'\|^2)\right]. \tag{13.83}$$

The minimization problem above can be solved by taking partial derivatives of the objective function with respect to $\mathbf{w}'$ and $\mathbf{w}''$ and equating them to zero:

$$\frac{\partial \mathbf{O}}{\partial \mathbf{w}'} = 2\mathbf{C}^{\mathbf{T}}(\mathbf{C}\mathbf{w}' + \mathbf{D}\mathbf{w}'') - 2\lambda\mathbf{w}' = 0 \tag{13.84}$$

$$\frac{\partial \mathbf{O}}{\partial \mathbf{w}''} = 2\mathbf{D}^{\mathbf{T}}(\mathbf{C}\mathbf{w}' + \mathbf{D}\mathbf{w}'') = 0 \tag{13.85}$$

Equation 13.85 is equivalent to

$$\mathbf{w}'' = -(\mathbf{D}^{\mathbf{T}}\mathbf{D})^{-1}\mathbf{D}^{\mathbf{T}}\mathbf{C}\mathbf{w}'. \tag{13.86}$$

Substituting Equation 13.86 into Equation 13.84 yields

$$\lambda\mathbf{w}' = [\mathbf{C}^{\mathbf{T}}\mathbf{C} - \mathbf{C}^{\mathbf{T}}\mathbf{D}(\mathbf{D}^{\mathbf{T}}\mathbf{D})^{-1}\mathbf{D}^{\mathbf{T}}\mathbf{C}]\mathbf{w}'. \tag{13.87}$$

It can be seen that λ is an eigenvector of the matrix

$$\mathbf{M} = \mathbf{C}^{\mathbf{T}}\mathbf{C} - \mathbf{C}^{\mathbf{T}}\mathbf{D}(\mathbf{D}^{\mathbf{T}}\mathbf{D})^{-1}\mathbf{D}^{\mathbf{T}}\mathbf{C}. \tag{13.88}$$

Therefore, the solution sought for $\mathbf{w}'$ corresponds to the smallest eigenvector associated with matrix $\mathbf{M}$. The corresponding $\mathbf{w}''$ can be directly computed from Equation 13.86. It is important to notice that since the magnitude constraint was imposed only on one of the rows of $\mathbf{R}$, the results obtained for $\mathbf{w}''$ are not reliable and therefore should not be used. However, solution vector $\mathbf{w}''$ provides an important piece of information regarding the sign of the row vector on which the constraint was imposed. The constraint imposed was $\|\mathbf{w}'\|^2 = 1$, but the sign of $\mathbf{w}'$ is not restricted by this constraint. Therefore, it is necessary to check whether or not the resulting $\mathbf{w}'$ yields a solution that is physically possible. In particular, the translation t_z must be positive in order for the object to be located in front of the camera as opposed to behind it. If the element of vector $\mathbf{w}''$ that corresponds to t_z is negative, it means that the magnitude of the computed $\mathbf{w}'$ is correct, but its sign is not and must be changed. Thus, the final expression for the computed $\mathbf{w}'$ is

$$\mathbf{w}' = sign(w''_9)\mathbf{w}'. \tag{13.89}$$

13.8.3 Computing the Transformation $\mathbf{Tr} = \{\mathbf{R}, \mathbf{T}\}$

Row vector $\mathbf{R_1}$ is computed first by computing $\mathbf{w}'$ as described above, since in this case $\mathbf{R_1} = \mathbf{w}'$. Matrices $\mathbf{C}$ and $\mathbf{D}$ are

$$\mathbf{C} = \begin{pmatrix} x_1 & y_1 & z_1 \\ 0 & 0 & 0 \\ x_2 & y_2 & z_2 \\ 0 & 0 & 0 \\ \vdots & \vdots & \vdots \\ x_6 & y_6 & z_6 \\ 0 & 0 & 0 \end{pmatrix} \tag{13.90}$$

and

$$\mathbf{D} = \begin{pmatrix} 0 & 0 & 0 & -u_1x_1 & -u_1y_1 & -u_1z_1 & f & 0 & -u_1 \\ fx_1 & fy_1 & fz_1 & -v_1x_1 & -v_1y_1 & -v_1z_1 & 0 & f & -v_1 \\ 0 & 0 & 0 & -u_2x_2 & -u_2y_2 & -u_2z_2 & 0 & f & -u_2 \\ fx_2 & fy_2 & fz_2 & -v_2x_2 & -v_2y_2 & -v_2z_2 & 0 & f & -v_2 \\ \vdots & \vdots & \vdots & \vdots & \vdots & \vdots & \vdots & \vdots & \vdots \\ 0 & 0 & 0 & -u_6x_6 & -u_6y_6 & -u_6z_6 & f & 0 & -u_6 \\ fx_6 & fy_6 & fz_6 & -v_6x_6 & -v_6y_6 & -v_6z_6 & 0 & f & -v_6 \end{pmatrix}. \tag{13.91}$$

Then row vector $\mathbf{R_2}$ is computed using the same technique, except that now the constraint is imposed on its magnitude, thus $\mathbf{R_2} = \mathbf{w}'$. In this case, matrices $\mathbf{C}$ and $\mathbf{D}$ are

$$\mathbf{C} = \begin{pmatrix} 0 & 0 & 0 \\ fx_1 & fy_1 & fz_1 \\ 0 & 0 & 0 \\ fx_2 & fy_2 & fz_2 \\ \vdots & \vdots & \vdots \\ 0 & 0 & 0 \\ fx_6 & fy_6 & fz_6 \end{pmatrix} \tag{13.92}$$

and

$$\mathbf{D} = \begin{pmatrix} fx_1 & fy_1 & fz_1 & -u_1x_1 & -u_1y_1 & -u_1z_1 & f & 0 & -u_1 \\ 0 & 0 & 0 & -v_1x_1 & -v_1y_1 & -v_1z_1 & 0 & f & -v_1 \\ fx_2 & fy_2 & fz_2 & -u_2x_2 & -u_2y_2 & -u_2z_2 & f & 0 & -u_2 \\ 0 & 0 & 0 & -v_2x_2 & -v_2y_2 & -v_2z_2 & 0 & f & -v_2 \\ \vdots & \vdots & \vdots & \vdots & \vdots & \vdots & \vdots & \vdots & \vdots \\ fx_6 & fy_6 & fz_6 & -u_6x_6 & -u_6y_6 & -u_6z_6 & f & 0 & -u_6 \\ 0 & 0 & 0 & -v_6x_6 & -v_6y_6 & -v_6z_6 & 0 & f & -v_6 \end{pmatrix}. \tag{13.93}$$

$\mathbf{R_3}$ could also be computed the same way as $\mathbf{R_1}$ and $\mathbf{R_2}$ above, but that would not guarantee it to be normal to $\mathbf{R_1}$ and $\mathbf{R_2}$. Instead, $\mathbf{R_3}$ is computed as follows:

$$\mathbf{R_3} = \frac{\mathbf{R_1} \times \mathbf{R_2}}{\|\mathbf{R_1} \times \mathbf{R_2}\|}. \tag{13.94}$$

Figure 13.27 Examples of pose computed from six point correspondences using constrained linear optimization. (Courtesy of Mauro Costa.)

All the constraints on the row vectors of $\mathbf{R}$ have been satisfied, except one: There is no guarantee that $\mathbf{R_1}$ is orthogonal to $\mathbf{R_2}$. In order to solve this undesired situation, $\mathbf{R_1}$, $\mathbf{R_2}$, and $\mathbf{R_3}$ need to go through an orthogonalization process, such that the rotation matrix $\mathbf{R}$ is assured to be orthonormal. This can be accomplished by fixing $\mathbf{R_1}$ and $\mathbf{R_3}$ as computed above and recomputing $\mathbf{R_2}$ as:

$$\mathbf{R_2} = \mathbf{R_3} \times \mathbf{R_1}. \tag{13.95}$$

This way, all the rotation parameters have been calculated and they all satisfy the necessary constraints. The translation vector $\mathbf{T}$ is computed using a least-squares technique on a new, non-homogeneous, over-constrained system of twelve equations:

$$\mathbf{A}\,\mathbf{t} = \mathbf{b}, \tag{13.96}$$

where

$$\mathbf{A} = \begin{pmatrix} f & 0 & -u_1 \\ 0 & f & -v_1 \\ f & 0 & -u_2 \\ 0 & f & -v_2 \\ \vdots & \vdots & \vdots \\ f & 0 & -u_6 \\ 0 & f & -v_6 \end{pmatrix}, \tag{13.97}$$

and

$$\mathbf{b} = \begin{pmatrix} -f(r_{11}x_1 + r_{12}y_1 + r_{13}z_1) + u_1(r_{31}x_1 + r_{32}y_1 + r_{33}z_1) \\ -f(r_{21}x_1 + r_{22}y_1 + r_{23}z_1) + v_1(r_{31}x_1 + r_{32}y_1 + r_{33}z_1) \\ -f(r_{11}x_2 + r_{12}y_2 + r_{13}z_2) + u_1(r_{31}x_2 + r_{32}y_2 + r_{33}z_2) \\ -f(r_{21}x_2 + r_{22}y_2 + r_{23}z_2) + v_1(r_{31}x_2 + r_{32}y_2 + r_{33}z_2) \\ \vdots \\ -f(r_{11}x_6 + r_{12}y_6 + r_{13}z_6) + u_1(r_{31}x_6 + r_{32}y_6 + r_{33}z_6) \\ -f(r_{21}x_6 + r_{22}y_6 + r_{23}z_6) + v_1(r_{31}x_6 + r_{32}y_6 + r_{33}z_6) \end{pmatrix}. \tag{13.98}$$

13.8.4 Verification and Optimization of Pose

A quantitative measure is useful for evaluating the quality of the pose parameters. One was already used in our development of the affine calibration procedure: It was the sum of the squared distances between projected posed model points and their corresponding image points. Some correspondences must be dropped as outliers, however, due to occlusions of some object points by the same or by other objects. Other distance measures can be used; for example, the Hausdorf or modified Hausdorf distance. (See the references by Huttenlocher and others (1993) and Dubuisson and Jain (1984)). Verification can also be done using other features, such as edges, corners, or holes.

A measure of pose quality can be used to improve the estimated pose parameters. Conceptually, we can evaluate small variations of the parameters and keep only the best ones. Brute force search of 10 variations of each of six rotation and translation parameters would mean that one million sets of pose parameters would have to be evaluated—computational effort that is not often done. A nonlinear optimization method, such as Newton's method or Powell's method (see Press and others (1992)) will be much faster. Figure 13.28 shows an initial pose estimate for a single-object image as well as the final result after nonlinear optimization has been applied to that initial solution. The improved pose is clearly better for visual inspection tasks and possibly better for grasping the object, but perhaps not necessary for recognition.

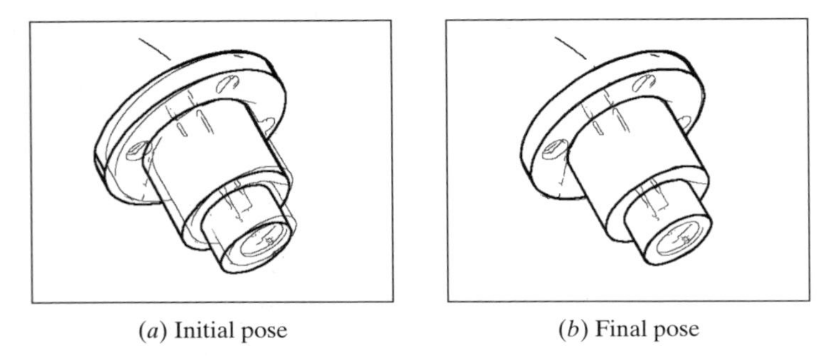

(*a*) Initial pose (*b*) Final pose

Figure 13.28 Example pose hypothesis and final pose after nonlinear optimization. (Courtesy of Mauro Costa.)

13.9 3D OBJECT RECONSTRUCTION

3D sensing is important for 3D model building. We can acquire range images of an existant object in order to create a computer model of it. This process of *3D object reconstruction* has applications in medicine and industrial vision, and is also used for producing the object models needed for virtual reality environments. By necessity, this section includes some discussion of object modeling, which is the major topic of the next chapter. The object reconstruction process has four major steps:

1. 3D data acquisition,
2. registration,

3. surface construction, and
4. optimization.

In the data acquisition phase, range data must be acquired from a set of views that cover the surface of the object. Often, 8–10 views are enough, but for complex objects or strict accuracy requirements, a larger number may be necessary. Of course, more views also mean more computation, so more is not necessarily better.

Each view obtained consists of a single range image of a portion of the object and often a registered gray-tone or color image. The range data from all of the views will be combined to produce a surface model of the object. The intensity data can be used in the registration procedure, but is really meant for use in texture mapping the objects for realistic viewing in graphics applications. The process of combining the range data by transforming them all to a single 3D coordinate system is the *registration* process.

Once the data have been registered, it is possible to view a *cloud of 3D points,* but it takes more work to create an object model. Two possible 3D representations for the object are (1) a connected mesh of 3D points and edges connecting them representing the object surface and (2) a set of 3D cubes or *voxels* representing the entire volume of the object. (See Chapter 14 for full explanations of these representations.) It is possible to convert from one representation to the other.

Exercise 13.31: Objects with hidden surfaces.

Some objects have hidden surfaces that cannot be imaged regardless of the number of views taken. Sketch one. For simplicity, you may work with a 2D model and world.

13.9.1 Data Acquisition

Range images registered to color images can be obtained from most modern commercial scanners. Here, we describe a laboratory system made with off-the-shelf components, and emphasize its fundamental operations. Figure 13.29 illustrates a specially built active stereo

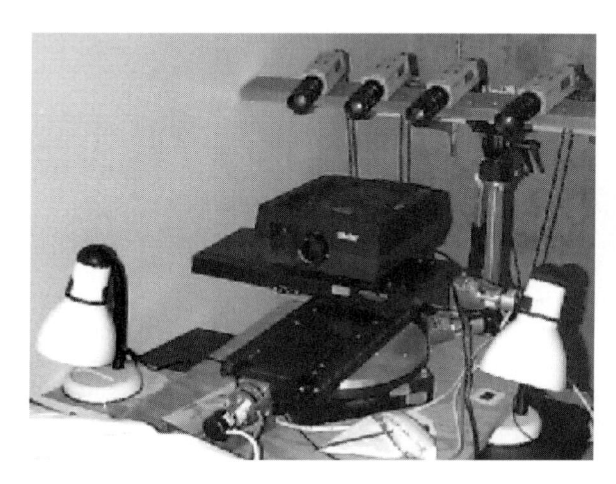

Figure 13.29 A 4-camera stereo acquisition system. (Courtesy of Kari Pulli.)

vision system that is used for acquiring range and color data. The system employs four color video cameras mounted on an aluminum bar. The cameras are connected to a digitizing board, which can switch between the four inputs under computer control, and produces images at 640×480 resolution. Below the cameras, a slide projector sits on a computer-controlled turntable. The slide projector emits a vertical stripe of white light, which is manually focused for the working volume. The light stripes are projected in a darkened room; the two side lights are turned on to take a color image after the range data has been obtained.

The cameras are calibrated simultaneously using Tsai's algorithm, described in Section 13.7. They can be used for either standard two-camera stereo or a more robust four-camera stereo algorithm. In either case, the projector is used to project a single vertical light stripe onto the object being scanned. It is controlled by computer to start at the left of the object and move, via the turntable, from left to right at fixed intervals chosen by the user to allow for either coarse or fine resolution. At each position, the cameras take a picture of the light stripe on the object in the darkened room. On each image, the intersection of the light stripe with an epipolar line provides a point for the stereo matching. Figure 13.30 illustrates the triangulation process using two cameras and a single light stripe. The two matched pixels are used to produce a point in 3D space. For a single light stripe, 3D points are computed at each pixel along that light stripe. Then the projector turns, a new light stripe is projected, new images are taken, and the process is repeated. The result is a dense range map, with 3D data associated with each pixel in the left image if that point was also visible to both the light projector and the right camera.

We can increase the reliability of the image acquisition system by using more than two cameras. One camera is our base camera in whose coordinate frame the range image will be computed. A surface point on the object must be visible from the base camera, the light projector, and at least one of the other three cameras. If it is visible in only one

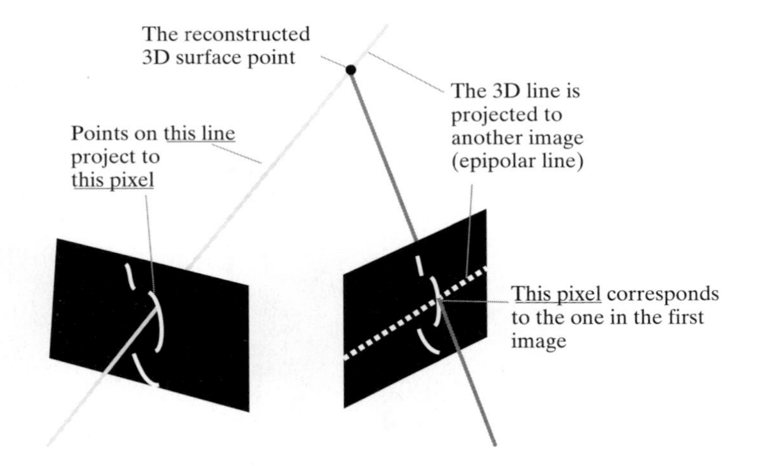

Figure 13.30 The intersections of light stripes with epipolar lines in two images gives a pair of corresponding points. (Courtesy of Kari Pulli.)

of the three, then the process reduces to two-camera stereo. If it is visible in two or three of the three, then the extra images can be used to make the process more robust. In the case of the base camera plus two more images, we have three points, so there are three different correspondences that can be triangulated to compute the 3D coordinates. It is likely that the three separate results will all be different. If they differ only by a small amount (say, they are all within a seven cubic millimeter volume), then they are all considered valid, and the average of the three 3D points can be used as the final result for that pixel. Or the pair of cameras with the widest baseline can be used as a more reliable estimate than the others. If the point is visible in all four cameras, then there are six possible combinations. Again we can check whether they all lie in a small volume, throw out outliers, and use an average value or the nonoutlier value that comes from the camera pair with the widest baseline. This procedure gives better accuracy than just using a fixed pair of cameras. (In measuring human body position in cars, as shown in Figure 13.1, the expected error was about $2mm$ in x, y, and z using this procedure.) A range image of a toy truck computed by this method is shown in Figure 13.31. The 3D truck dataset clearly shows the shape of the truck.

13.9.2 Registration of Views

In order to thoroughly cover the surface of the object, range data must be captured from multiple views. A transformation ${}^2_1\mathbf{T}$ from view 1 to view 2 is obtained either from precise mechanical movement or from image correspondence. When a highly accurate device, such as a calibrated robot or a coordinate measurement machine, is available that can move the camera system or the object in a controlled way, then the approximate transformations between the views can be obtained automatically from the system. If the movement of the cameras or object is not machine controlled, then there must be a method for detecting correspondences between views that will allow computation of the rigid transformation

Figure 13.31 A range data set for a toy truck obtained via the 4-camera active stereo system. The range points were colored with intensity data for display purposes. (Courtesy of Habib Abi-Rached.)

that maps the data from one view into that for another view. This can be done automatically using 3D features such as corner points and line segments that lead to a number of 3D-3D point correspondences from which the transformation can be computed. It can also be done interactively, allowing the user to select point correspondences in a pair of images of the object. In either case, the initial transformation is not likely to be perfect. Robots and machines will have some associated error that becomes larger as more moves take place. The automatic correspondence finder will suffer from the problems of any matching algorithm and may find false correspondences, or the features may be a little off. Even the human point clicker will make some errors, or quantization can lead to the right pixel, but the wrong transformation.

To solve these problems, most registration procedures employ an iterative scheme that begins with an initial estimate of the transformation ${}^2_1\mathbf{T}$, no matter how obtained, and refines it via a minimization procedure. For example, the iterative closest point (ICP) algorithm minimizes the sum of the distances between 3D points ${}^2_1\mathbf{T}\ {}^1\mathbf{P}$ and ${}^2\mathbf{P}$, where ${}^1\mathbf{P}$ is from one view and ${}^2\mathbf{P}$ is the closest point to it in the other view. A variation of this approach looks for a point in the second view along a normal extended from the point ${}^2_1\mathbf{T}\ {}^1\mathbf{P}$ to the surface interpolating a neighborhood in the second view. (See the references by Chen and Medioni (1992) and Dorai and others (1994), for example.) When color data is available, it is possible to project the color data from one view onto the color image from another view via the estimated transformation and define a distance measure based on how well they line up. This distance can also be iteratively minimized, leading to the best transformation from one set of 3D points to the other. Figure 13.32 illustrates the registration process for two views of a sofa using an ICP algorithm.

13.9.3 Surface Reconstruction

Once the data have been registered into a single coordinate system, reconstruction can begin. We would like the reconstructed object to come as close as possible to the shape of the actual object and to preserve its topology. Figure 13.33 illustrates problems that can occur in the reconstruction process. The registered range data is dense, but quite noisy. There are extra points outside the actual chair volume and, in particular, between the spokes of the back. The reconstruction shown in the middle is naïve in that it considered the range data only as a cloud of 3D points and did not take into account object geometry or neighbor relationships between range data points. It fails to preserve the topology of the object. The reconstruction shown on the right is better in that it has removed most of the noise and the holes between the spokes of the back have been preserved. This reconstruction was produced by a *space-carving* algorithm described next.

13.9.4 Space-Carving

The space-carving approach was developed by Curless and Levoy, and the method described here was implemented by Pulli and others (1998). Figure 13.34 illustrates the basic concept. At the left is an object to be reconstructed from a set of views. In the center, there is one camera viewing the object. The space can be partitioned into areas according to where the points lie with respect to the object and the camera. The left and bottom sides of the object

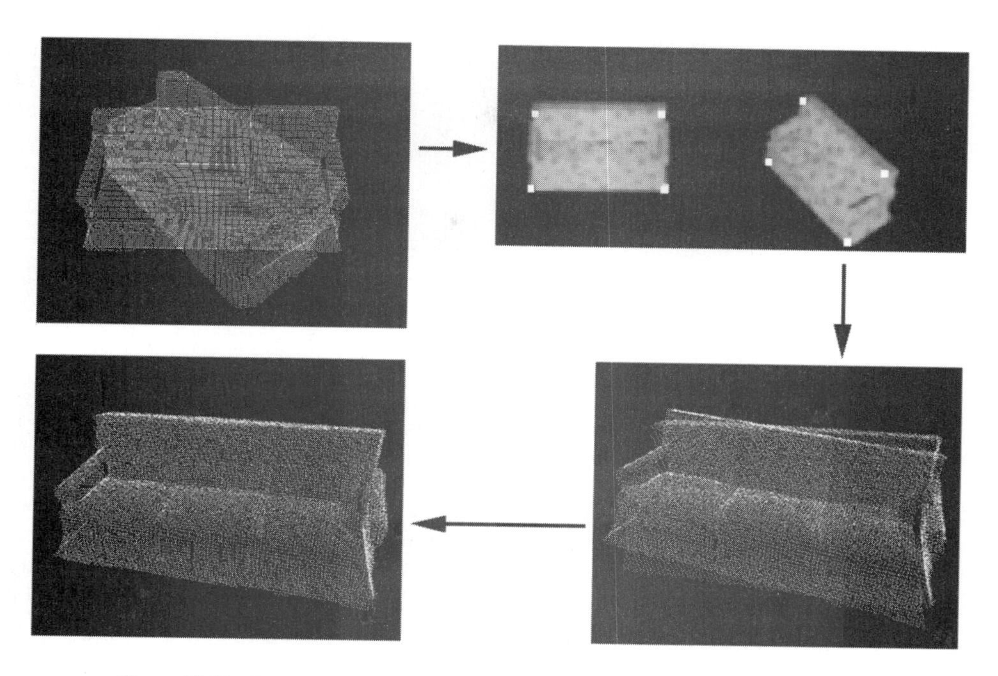

Figure 13.32 Registration of two range data sets shown at top left. The user has selected four points on intensity images corresponding to the two range views (upper right). The initial transformation is slightly off (lower right). After several iterations, the two range datasets line up well (lower left). (Courtesy of Kari Pulli.)

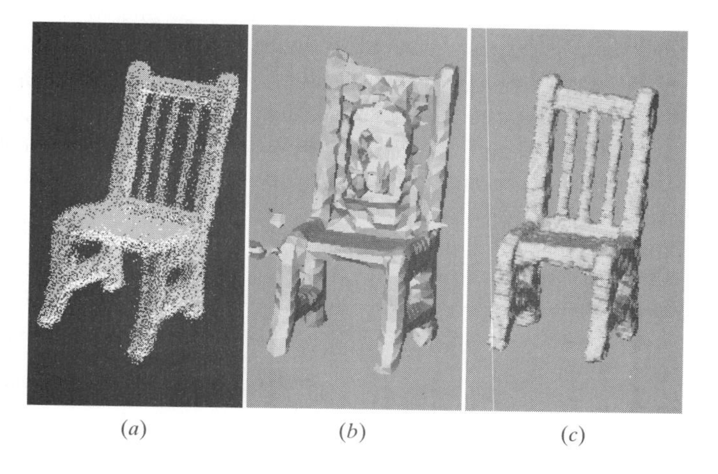

(*a*) (*b*) (*c*)

Figure 13.33 (*a*) The registered range data for a chair object; (*b*) problems that can occur in reconstruction; and (*c*) a topologically correct rough mesh model. (Courtesy of Kari Pulli.)

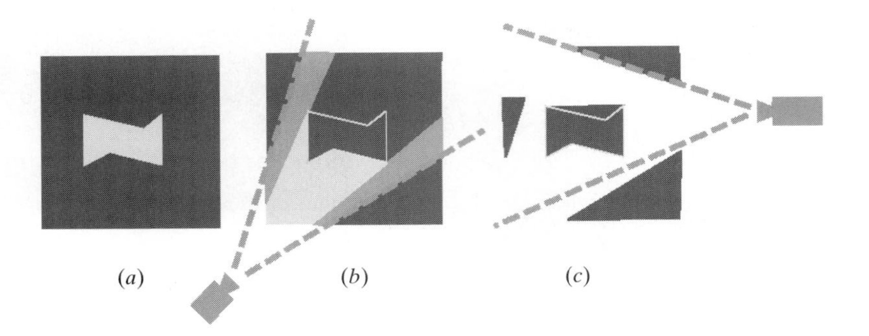

Figure 13.34 The concept of space carving: (*a*) object cross section; (*b*) camera view 1 can remove all material of light shading; (*c*) camera view 2 can remove other material. (Extra material still remains, however.) (Courtesy of Kari Pulli.)

are visible to the camera. The volume between the scanned surface and the camera (light gray) is in front of the object and can be removed. If there is data from the background, as well as the object, even more volume (darker gray) can be removed. On the other hand, points behind the object cannot be removed, because the single camera cannot tell if they are part of the object or behind it. However, another camera viewing the object (at right) can carve away more volume. A sufficient number of views will carve away most of the unwanted free space, leaving a voxel model of the object.

The space-carving algorithm discretizes space into a collection of cubes or voxels that can be processed one at a time. Figure 13.35 illustrates how the status of a single cube with respect to a single camera view is determined:

- In case (*a*) the cube lies between the range data and the sensor. Therefore the cube must lie outside of the object and can be discarded.
- In case (*b*) the whole cube is behind the range data. As far as the sensor is concerned, the cube is assumed to lie inside of the object.
- In case (*c*) the cube is neither fully in front of the data or behind the data and is assumed to intersect the object surface.

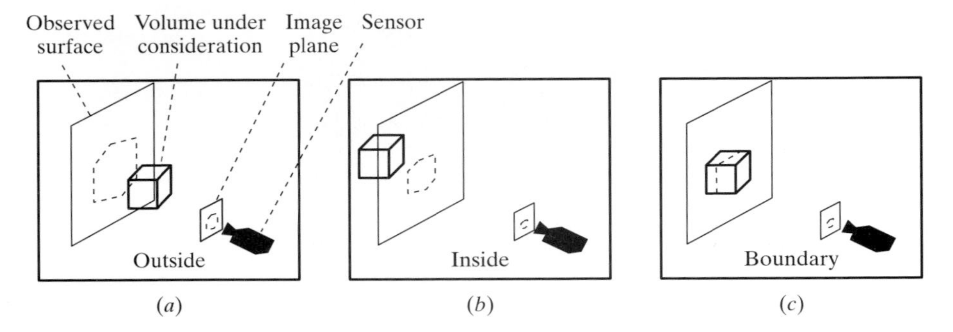

Figure 13.35 The three possible positions of a cube in space in relation to the object being reconstructed. (Courtesy of Kari Pulli.)

The cube labeling step can be implemented as follows: The eight corners of the cube are projected to the sensor's image plane, where their convex hull generally forms a hexagon. The rays from the sensor to the hexagon form a cone, which is truncated so that it just encloses the cube. If all the data points projecting onto the hexagon are behind the truncated cone (are farther than the farthest corner of the cube from the sensor), the cube is outside of the object. If all those points are closer than the closest cube corner, the cube is inside the object. Otherwise, it is a boundary cube.

So far, we have looked at one cube and one sensor. It takes a number of sensors (or views) to carve out the free space. The cube labeling step for a given cube is performed for all of the sensors. If even one sensor says that the cube is outside of the object, then it is outside. If all of the sensors say that the cube is inside of the object, then it is inside as far as we can tell. Some other view may determine that it is really outside, but we do not have that view. In the third case, if the cube is neither inside nor outside, it is a boundary cube.

Instead of using a set of fixed-size cubes, it is more efficient to perform the cube labeling in a hierarchic fashion, using an octree structure. The *octree* is described in detail in Chapter 14, but we can use it intuitively here without confusion. Initially a large cube surrounds the data. Since by definition this large cube intersects the data, it is broken into eight smaller cubes. Those cubes that are outside the object can be discarded, while those that are fully inside can be marked as part of the object. The boundary cubes are further subdivided and the process continues up to the desired resolution. The resultant octree represents the 3D object. Figure 13.36 illustrates the hierarchical space carving procedure for the chair object.

The octree representation can be converted into a 3D mesh for viewing purposes as shown in Figure 13.36. After the initial mesh is created, it can be optimized by a method that tries to simplify the mesh and better fit the data. Figure 13.37 shows the registered range data for the dog object (*a*), the initial mesh (*b*), and several steps in the optimization procedure defined by Hoppe and others (1992) (*c*)–(*f*). The final mesh (*f*) is much more concise and smoother than the initial mesh. It can now be used in a graphics system for rendering realistic views of the object as in Figure 13.38 or in model-based object recognition as in Chapter 14.

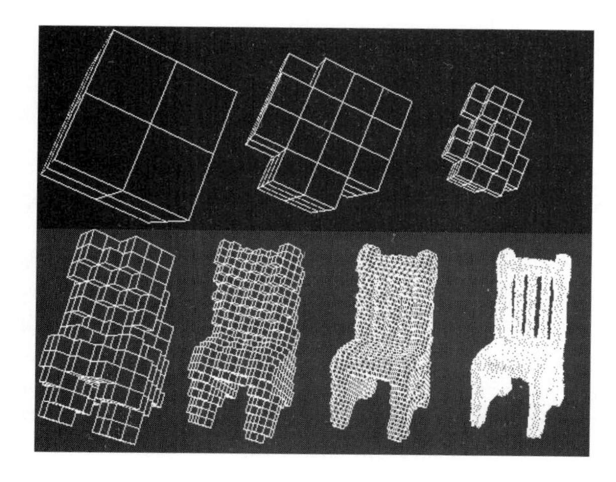

Figure 13.36 Hierarchical space carving: seven iterations to produce the chair mesh. (Courtesy of Kari Pulli.)

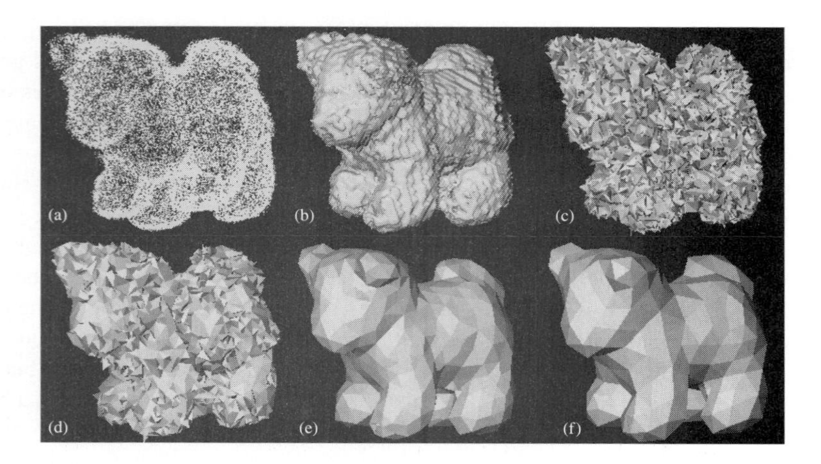

Figure 13.37 The registered range data and five steps in the creation of a dog mesh. (Courtesy of Kari Pulli.)

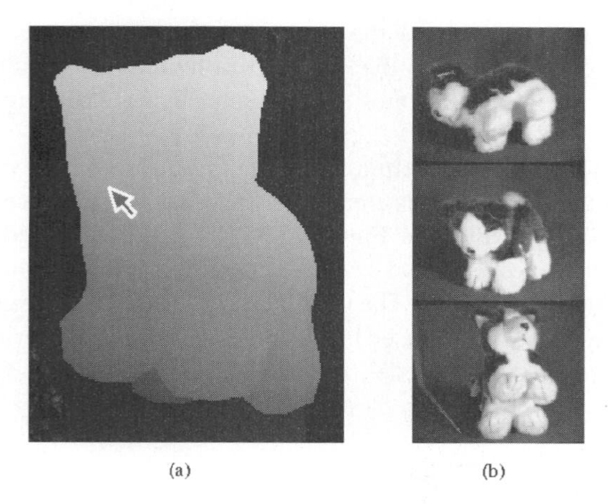

Figure 13.38 (*a*) A false color rendition of the dog model that a user can manipulate to select a desired view. (*b*) The 3D point on the model's nose marked by the arrow is projected onto 3 color images of the dog to select pixels that can be used to produce realistic color renderings of the dog. (Courtesy of Kari Pulli.)

13.10 COMPUTING SHAPE FROM SHADING

Chapters 6 and 12 both discussed how light reflecting off smooth curved objects creates a shaded image. Next, we briefly show how—under certain assumptions—the shape of the object can be computed from the shading in the image.

Humans tend to see a smoothly darkening surface as one that is turning away from the view direction. Using makeup on our face, we can change how others perceive us. Makeup that is darker than our skin color applied to our outer cheeks makes the face look narrower, because the darker shading induces the perception that the surface is turning away from the viewer faster than it actually does. Similarly, makeup lighter than our skin color will have the opposite effect, inducing the perception of a fuller face. Using the formula for Lambertian reflectance, it is possible to map an intensity value (shading) into a surface normal for the surface element being imaged. Early work by Horn and Bachman (1978) studied methods of determining the topography of the moon illuminated by the distant sun and viewed from the distant earth. The family of methods that have evolved to map back from shading in an image to a surface normal in the scene has been called *shape from shading(SFS)*.

> **109 Definition.** A **shape-from-shading** method computes surface shape $\mathbf{n} = f(x, y)$ from a shaded image, where $\mathbf{n}$ is the normal to the surface at point $[x, y]$ of the image and ${}^{F}I[x, y]$ is the intensity.

Figure 13.39 motivates shape from shading methods. At the left is an image of objects whose surfaces are approximated nicely by the Lambertian reflectance model: Image intensity is proportional to the angle between the surface normal and the direction of illumination. At the right of Figure 13.39 is a sketch of the surface normals at several points on the surfaces of the objects. Clearly, the brightest image points indicate where the surface normal points directly toward the light source: The surface normals pointing back at us appear as Xs in the figure. At limb points the surface normal is perpendicular to both the view direction and the surface boundary: This completely constrains the normal in 3D space. Using these constraints, surface normals can be propagated to all the image points. This creates a partial intrinsic image. To obtain the depth z at each image point, we can assign an arbitrary value of z_0 to one of the brightest points and then propagate depth across the image using the changes in the normals.

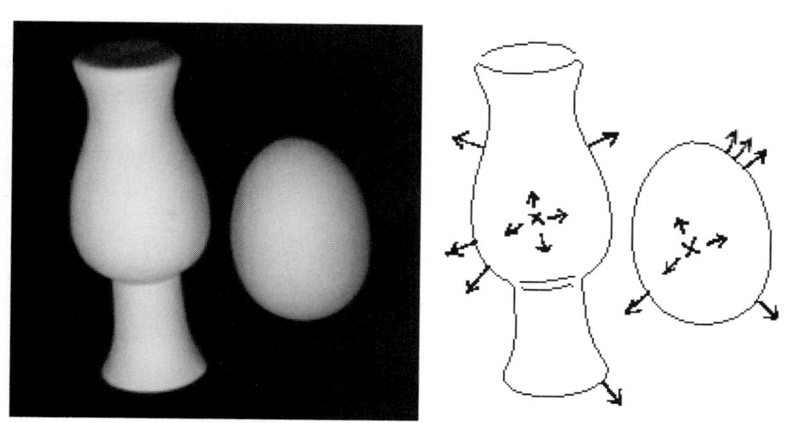

Figure 13.39 (left) Shaded image of Lambertian objects taken with light source near the camera and (right) surface normals sketched on boundaries detected by Canny operator.

Exercise 13.32

Assume a cube with a Lambertian surface posed such that all face normals make an angle of at least $\pi/6$ with the direction of illumination. Clearly, the brightest image points do not correspond to normals pointing at the light source. Why is it true for the egg and vase, but not for the cube?

Exercise 13.33

Using shading, how might it be known whether an object boundary is a limb or blade?

The Lambertian reflectance model is $i = c\,cos\theta$, where constant c results from a combination of source energy, surface albedo, and the distances between the surface element and source and sensor: All these factors are assumed to be constant. The assumption on the distances will hold when the object is many diameters away from both source and sensor. Often, it is also assumed that the illuminant direction is known, but as observed above, illuminant direction can sometimes be computed from weaker assumptions.

The orthographic projection is most convenient for the current development. Also, we use the camera frame as our only frame for 3D space $[x, y, z]$. The observed surface is $z = f(x, y)$: The problem is to compute function f at each image point from the observed intensity values ${}^{F}I[x, y]$. By rewriting and taking derivatives, we arrive at $\frac{\partial f}{\partial x}\Delta x + \frac{\partial f}{\partial y}\Delta y - \Delta z = 0$, or in vector terms $[p, q, -1] \circ [\Delta x, \Delta y, \Delta z] = 0$, where p and q denote the partial derivatives of f with respect to x and y, respectively. The last equation defines the tangent plane at the surface point $[x, y, f(x, y)]$ that has (nonunit) normal direction $[p, q, -1]$. If we know that $[x_0, y_0, z_0]$ is on the surface, and if we know p and q, then the above planar approximation allows us to compute surface point $[x_0 + \Delta x, y_0 + \Delta y, z_0 + \Delta z]$ by just moving within the approximating tangent plane. We can do this if we can estimate p and q from the intensity image and our assumptions.

We can relate surface normals to intensity by observing a known object. Whenever we know the surface orientation $[p, q, -1]$ for a point $[x, y, f(x, y)]$ observed at $I[x, y]$, we contribute tuple $\langle p, q, I[x, y]\rangle$ to a mapping so the surface orientation is associated with the shading it produces. Figure 13.40 shows how this is a many-to-one mapping: all surface orientations making a cone of angle θ with the illuminant direction will yield the same observed intensity. The best object to use for such calibration is a sphere, because (a) it exhibits all surface orientations and (b) the surface normal is readily known from the

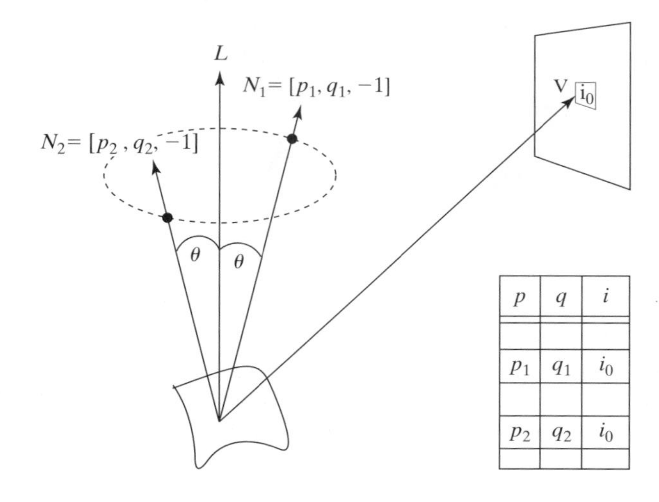

p	q	i
p_1	q_1	i_0
p_2	q_2	i_0

Figure 13.40 (left) An entire cone of possible surface normals will produce the same observed intensity and (right) a reflectance map relates surface normals to intensity values: It is a many-to-one mapping.

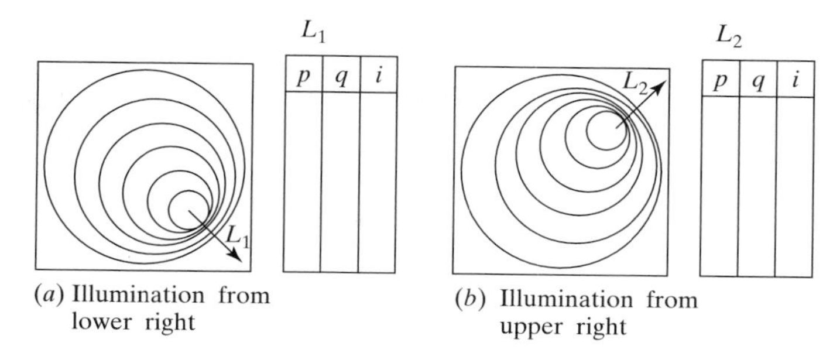

(a) Illumination from lower right

(b) Illumination from upper right

Figure 13.41 Reflectance maps can be created by using a Lambertian calibration sphere of the same material as objects to be sensed. In the observed image of the sphere, p and q are known by analytical geometry for each point $I[x, y]$: We can insert a tuple $\langle p, q, I[x, y] \rangle$ in the mapping for each image point. A different mapping is obtained for each separately used light source.

location of the image point relative to the center and radius of the sphere. Figure 13.41 shows the results of viewing a calibration sphere with two different light sources. For each light source, a reflectance map can be created that stores with each observed intensity the set of all surface orientations that produce that intensity.

Exercise 13.34

Given a calibration sphere of radius r located at $[0, 0, 100]$ in camera frame $\mathbf{C} : [x, y, z]$, derive the formulas for p and q in terms of location $[x, y]$ in the image. Recall that orthographic projection effectively drops the z-coordinate.

As Figures 13.39 and 13.41 show, image intensity gives a strong constraint on surface orientation, but not a unique surface normal. Additional constraints are needed. There are two general approaches. The first approach is to use information from the spatial neighborhood; for example, a pixel and its 4-neighborhood yield five instances of the shading equation that can be integrated to solve for a smooth surface passing through those five points. The second approach uses more than one intensity image so that multiple equations can be applied to single pixels without regard to neighbors: this has been called *photometric stereo* for obvious reasons.

13.10.1 Photometric Stereo

The photometric stereo method takes multiple images of an object illuminated in sequence by different light sources. A set of intensities is obtained for each image pixel; these can be looked up in a table to obtain the corresponding surface normal. The table is constructed by an offline photometric calibration procedure as shown in Figure 13.41. Algorithm 13.4 sketches this procedure. Photometric stereo is a fast method that has been shown to work well in a controlled environment. Ray and others (1983) reported excellent results even with specular objects using three balanced light sources. However, if the environment can be tightly controlled to support shape from shading, we can do better using structured light as demonstrated by recent trends in industry.

13.10.2 Integrating Spatial Constraints

Several different methods have been proposed for determining a smooth surface function $z = f(x, y)$ by applying the shading constraint across spatial neighborhoods. One such method is to propagate the surface from the brightest image points as mentioned previously. Minimization approaches find the best function fitting the available constraints. Figure 13.42 shows results from one such algorithm: A mesh describing the computed surface is shown for two synthetic objects and one real object. These results may or may not be good, depending on the task the data are to support. The method is not considered reliable enough to use in practice.

Shape from shading work has proven that shading information gives strong constraint on surface shape. It is a wonderful example of a pure computer vision problem—the input, output, and assumptions are very cleary defined. Many of the mathematical algorithms work well in some cases; however, none work well across a variety of scenes. The interested reader should consult the references to obtain more depth in this subject, especially for the mathematical algorithms, which were only sketched here.

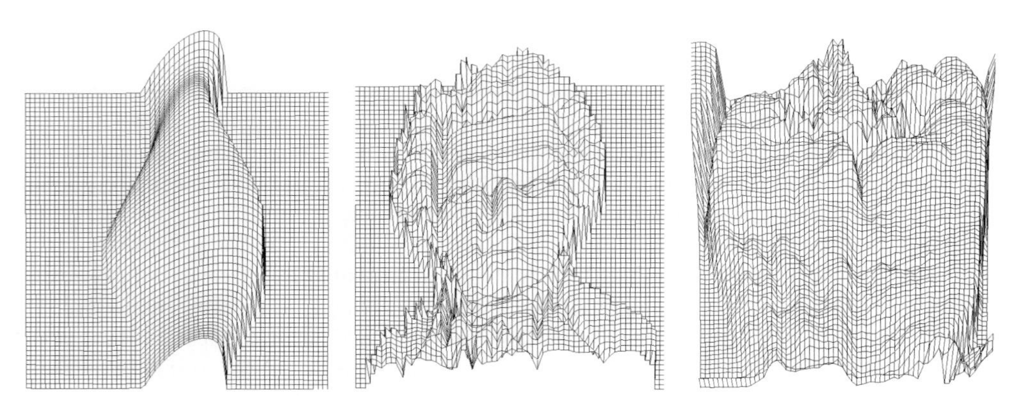

Figure 13.42 Results of the Tsai-Shah algorithm on synthetic and real images. (left) Surface obtained by algorithm from image generated applying a diffuse lighting model to a CAD model of a vase; (center) surface obtained by algorithm from a synthetic image of a bust of Mozart; and (right) surface obtained by algorithm from a real image of a green bell pepper. (Images courtesy of Mubarak Shah.)

13.11 STRUCTURE FROM MOTION

Humans perceive a great deal of information about the 3D structure of the environment by moving through it. When we move or when objects move, or both, we obtain information from images sensed over time. From flow vectors or from corresponding points, the 3D scene surfaces and corners can be reconstructed, as well as the trajectory of the sensor through the

Compute surface normals $[p, q]$ for points of a scene viewed with multiple images ${}^1\mathbf{I}, {}^2\mathbf{I}, {}^3\mathbf{I}$ using different light sources ${}^1\mathbf{L}, {}^2\mathbf{L}, {}^3\mathbf{L}$.

Offline Calibration:

1. Place calibration sphere in the center of the scene.
2. For each of the three light sources ${}^j\mathbf{L}$.
   (a) Switch on light source ${}^j\mathbf{L}$.
   (b) Record image of calibration sphere.
   (c) Create reflectance map ${}^j\mathbf{R} = \{\langle p, q, {}^j\mathbf{I}[\mathbf{x}, \mathbf{y}]\rangle\}$ where (p, q) is associated with some intensity ${}^j\mathbf{I}[\mathbf{x}, \mathbf{y}]$ of image ${}^j\mathbf{I}$.

Online Surface Sensing:

1. The object to be sensed appears in center of scene.
2. Take three separate images ${}^j\mathbf{I}$ in rapid succession using each light source ${}^j\mathbf{L}$ individually.
3. For each image point $[\mathbf{x}, \mathbf{y}]$
   (a) Use intensity $\mathbf{i_j} = {}^j\mathbf{I}[\mathbf{x}, \mathbf{y}]$ to index reflectance map ${}^j\mathbf{R}$ and access the set of tuples $\mathbf{R_j} = \{(p, q)\}$ associated with intensity $\mathbf{i_j}$.
   (b) "Intersect" the three sets: $\mathbf{S} = \mathbf{R_1} \cap \mathbf{R_2} \cap \mathbf{R_3}$.
   (c) If $\mathbf{S}$ is empty, then set $\mathbf{N}[\mathbf{x}, \mathbf{y}] = NULL$
   else set $\mathbf{N}[\mathbf{x}, \mathbf{y}]$ to the average direction vector in $\mathbf{S}$
4. Return $\mathbf{N}[\mathbf{x}, \mathbf{y}]$ as that part of the intrinsic image storing surface normals.

Algorithm 13.4 Photometric Stereo with Three Light Sources

scene. This intuitive process can be refined into an entire class of more specifically defined mathematical problems. Construction of useful computer vision algorithms to compute scene structure and object and observer motion has been difficult: progress has been steady but slow.

Figure 13.43 illustrates a general situation where both the observer and scene objects may be moving. The relative motion of objects and observer produces flow vectors in the image; these might be computable via point matching or optical flow. Figure 13.44 shows the case of two significantly different views of five 3D points. The many cases reported in the literature vary in both the problem definition and the algorithm achieved.

Exercise 13.35: Improve Algorithm 13.4.

Improve the efficiency of Algorithm 13.4 by moving all the set intersection operations into the offline procedure. Justify why this can be done. What data structure is appropriate for storing the results for use in the online procedure?

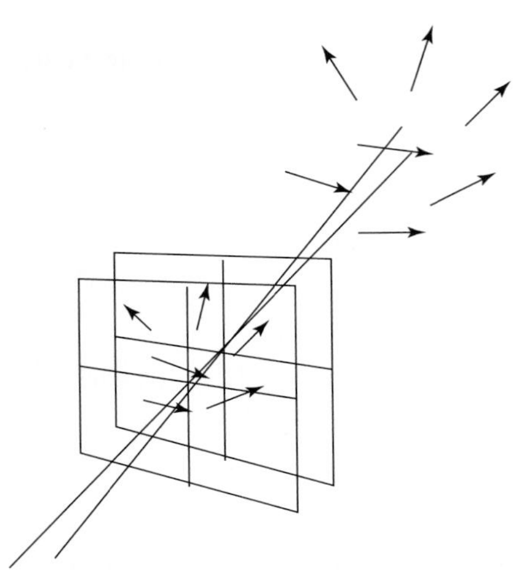

Figure 13.43 An observer moves through a scene of moving objects. The motion of a 3D point projects to a 2D flow vector spanning two images close in time.

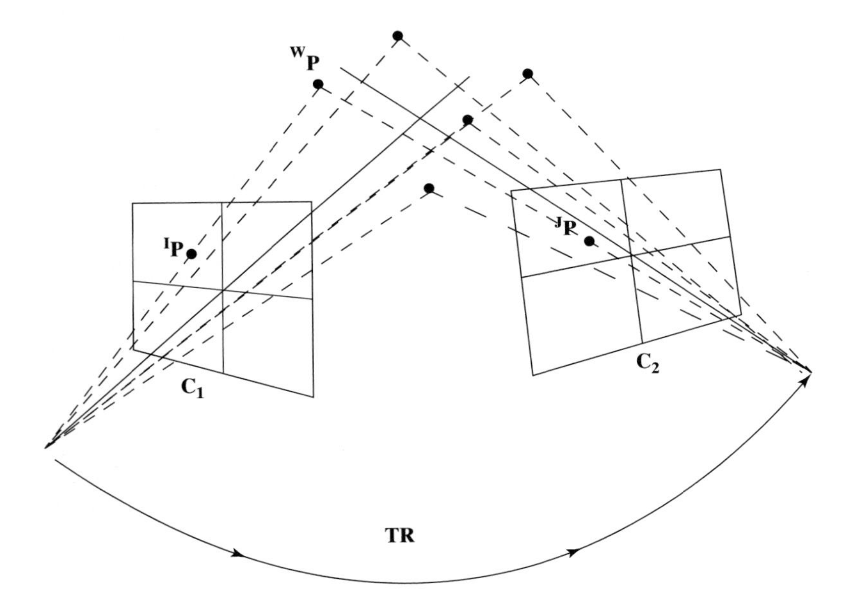

Figure 13.44 An observer moves through a scene of stationary objects. 3D object points $^{\mathbf{W}}\mathbf{P}$ project to 2D points $^{\mathbf{I}}\mathbf{P}$ and $^{\mathbf{J}}\mathbf{P}$ in two images significantly different in time and space so that point correspondence may be difficult. Given image point correspondences, the problem is to compute both the relative motion **TR** and the 3D coordinates of the points $^{\mathbf{W}}\mathbf{P}$.

The 3D objects used in the problem definition may be

- points
- lines
- planar surface patches
- curved surface patches

Given the assumptions, an algorithm should yield not only the 3D object structure, but also their motion relative to the camera coordinate system. Much algorithm development has been done assuming that the 3D object points have been reliably sensed and matched. Sensing and matching is difficult and error prone and few convincing demonstrations have been made including them. Algorithms based on image flow use small time steps between images and attempt to compute dense 3D structure, whereas algorithms based on feature correspondences can tolerate larger time steps but can only compute sparse 3D structure.

Ullman (1979) reported early results showing that the structure and motion of a rigid configuration of four points could be computed, in theory, from three orthographic projections of these four points using a stationary camera. Ten years later, Huang and Lee (1989) showed that the problem could not be solved using only two orthographic projections. While *minimalist* mathematical models for *shape-from-motion* are interesting and may be difficult to solve, they appear to be impractical because of the errors due to noise or mismatched points. Haralick and Shapiro (1993) treat several mathematical approaches and show methods for making the computations robust. Brodsky and others (1999) have recently shown good practical results for computing dense shape from a rigidly moving video camera viewing a static scene. In the first chapter of this book, we asked whether or not we might make a 3D model of Notre Dame Cathedral from a video made of it. This is one version of the structure-from-motion problem, and there is now a commercially available solution using computer vision. For a summary of the methods used, refer to the paper by Faugeras and others (1998). Having introduced the general problem of computing structure from motion and several of its aspects, we urge the reader who wants to delve deeper to consult the published literature.

13.12 REFERENCES

The method for affine camera calibration was derived from the earlier work of Ballard and Brown (1982) and Hall and others (1982). The latter article also describes a structured light system using a calibrated camera and projector. Several different viable camera calibration methods are available. For object recognition, the affine method for perspective and even weak perspective are often accurate enough. However, for inspection or accurate pose computation, methods that model radial distortion are needed: The widely used Tsai procedure was reported in Tsai (1987). Calibration is appropriate for many machine vision applications. However, much can be done with an uncalibrated camera: This is, in fact, what we have when we scan the world with our video camera. We don't know what the focal length is at every point in time and we do not know any pose parameters relative to any global coordinate system. However, humans do perceive the 3D structure of the world from such imagery. 3D structure can be computed within an unknown scale factor, assuming only that perspective projection applies. The work of Faugeras and others (1998) shows how to construct texture-mapped 3D models of buildings from image sequences. Brodsky and others (1999) show results for computing the structure of more general surfaces.

Our P3P solution followed closely the work of Ohmura and others (1988). A similar work from Linnainmaa and others (1988) appeared about the same time. However, note that Fischler and Bolles (1981) had studied this same problem and had published a closed form solution. Iterative solutions appear to have advantages when the object is being tracked in a sequence of frames, because a starting point is available, which also helps to discard a false solution. A good alternative using a weak perspective projection model is given by Huttenlocher and Ullman (1988). Their method is a fast approximation and the derivation is constructive. Fischler and Bolles (1981) were the first to formally define and study the *perspective N point* problem and gave a closed form solution for P3P. They also showed how to use it by hypothesizing N correspondences, computing object pose, and then verifying that other model points projected to correponding points in the image. They called their algorithm RANSAC since they suggested randomly choosing correspondences—something that should be avoided if properties of feature points are available.

In recent years there has been much activity in making 3D object models from multiple views in laboratory controlled environments. Many systems and procedures have been developed. The system for object reconstruction that we reported was constructed at the University of Washington by Pulli and others (1998).

1. Ballard, D., and C. Brown. 1982. *Computer Vision.* Prentice-Hall.
2. Ballard, P., and G. Stockman. 1995. Controlling a computer via facial aspect. *IEEE-Trans-SMC,* April 1995.
3. Brodsky, T., C. Fermuller, and Y. Aloimonos. 1999. Shape from video. *Proceedings of IEEE CVPR 1999,* Ft Collins, Co. (23–25 June 1999), 146–151.
4. Chen, Y., and G. Medioni. 1992. Object modeling by registration of multiple range images. *Int. J. Image and Vision Computing,* v. 10(3):145–155.
5. Craig, J. 1986. *Introduction to Robotics Mechanics and Control.* Addison-Wesley, Reading, MA.
6. Curless, B., and M. Levoy. 1996. A volumetric method for building complex models from range images. *ACM Siggraph '96,* 301–312.
7. Dorai, C., J. Weng, and A. Jain. 1994. Optimal registration of multiple range views. *Proc. 12th Int. Conf. Pattern Recognition,* Jerusalem, Israel (Oct. 1994), v. 1:569–571.
8. Dubuisson, M.-P., and A. K. Jain. 1984. A modified Hausdorff distance for object matching, *Proc. 12th Int. Conf. Pattern Recognition,* Jerusalem, Israel.
9. Faugeras, O. 1993. *Three-Dimensional Computer Vision, a Geometric Viewpoint.* The MIT Press, Cambridge, MA.
10. Faugeras, O., L. Robert, S. Laveau, G. Csurka, C. Zeller, C. Gauclin, and I. Zoghlami. 1998. 3-D reconstruction of urban scenes from image sequences. *Comput. Vision and Image Understanding,* v. 69(3):292–309.

11. Fischler, M., and R. Bolles. 1981. Random concensus: a paradigm for model fitting with applications in image analysis and automated cartography. *Communications of the ACM,* v. 24:381–395.

12. Forsyth, D., and others. 1991. Invariant descriptors for 3-d object recognition and pose. *IEEE Trans. Pattern Analysis and Machine Intelligence,* v. 13(10):971–991.

13. Hall, E., J. Tio, C. McPherson, and F. Sadjadi. 1982. Measuring curved surfaces for robot vision. *Computer,* v. 15(12):385–394.

14. Haralick, R. M., and L. G. Shapiro. 1993. *Computer and Robot Vision, Volume II.* Addison-Wesley, Reading, MA.

15. Heath, M. 1997. *Scientific Computing: An Introductory Survey.* McGraw-Hill, Inc., New York.

16. Hoppe, H., T. DeRose, T. Duchamp, J. McDonald, and W. Stuetzle. 1992. Surface reconstruction from unorganized points. *Proc. SIGGRAPH '92,* 71–78.

17. Horn, B. K. P., and B. L. Bachman. 1978. Using synthetic images to register real images with surface models. *CACM 21* v. 11:914–924.

18. Hu, G., and G. Stockman. 1989. 3-D surface solution using structured light and constraint propagation. *IEEE-TPAMI,* v. 11(4):390–402.

19. Huang, T. S., and C. H. Lee. 1989. Motion and structure from orthographic views. *IEEE Trans. Pattern Analysis and Machine Intelligence,* v. 11:536–540.

20. Huttenlocher, D., and S. Ullman. 1988. Recognizing solid objects by alignment. *Proc. DARPA Spring Meeting*, 1114–1122.

21. Huttenlocher, D. P., G. A. Klanderman, and W. J. Rucklidge. 1993. Comparing images using the Hausdorf distance. *IEEE Trans. Pattern Analysis and Machine Intelligence,* v. 15(9):850–863.

22. Ikeuchi, K., and B. K. P. Horn. 1981. Numerical shape from shading and occluding boundaries, *Artificial Intelligence,* v. 17(1–3):141–184.

23. Ji, Q., M. S. Costa, R. M. Haralick, and L. G. Shapiro. 1998. An integrated technique for pose estimation from different geometric features. *Proc. Vision Interface '98,* Vancouver (June 18–20), 77–84.

24. Johnson, L. W., R. D. Riess, and J. T. Arnold. 1989. *Introduction to Linear Algebra.* Addison-Wesley, Reading, MA.

25. Linnainmaa, S., D. Harwood, and L. Davis. 1988. Pose determination of a three-dimensional object using triangle pairs. IEEE Trans. Pattern Analysis and Machine Intelligence, v. 10(5):634–647.

26. Ohmura, K., A. Tomono, and A. Kobayashi. 1988. Method of detecting face direction using image processing for human interface. *SPIE Visual Communication and Image Processing,* v. 1001:625–632.

27. Pulli, K., H. Abi-Rached, T. Duchamp, L. G. Shapiro, and W. Stuetzle. 1998. Acquisition and visualization of colored 3D objects. *Proceedings of ICPR '98,* 11–15.

28. Ray, R., J. Birk and R. Kelley. 1983. Error analysis of surface normals determined by radiometry. *IEEE-TPAMI,* v. 5(6):631–644.

29. Shrikhande, N., and G. Stockman. 1989. Surface orientation from a projected grid. *IEEE-TPAMI,* v. 11(4):650–655.
30. Tsai, P.-S., and M. Shah. 1992. A fast linear shape from shading. *Proceedings IEEE Conf. Comput. Vision and Pattern Recognition* (June 1992), 734–736.
31. Tsai, R. 1987. A versatile camera calibration technique for high-accuracy 3D machine vision metrology using off-the-shelf cameras and lenses. *IEEE Trans. Robotics and Automation,* v. 3(4).
32. Ullman S. 1979. *The Interpretations of Visual Motion.* MIT Press, Cambridge, MA.
33. Vetterling, W. T. 1992. *Numerical Recipes in C.* Cambridge University Press, New York.
34. Zhang, R., P.-S. Tsai, J. Cryer, and M. Shah. 1999. Shape from shading: a survey. *IEEE-TPAMI,* v. 21(8):690–706.

14

3D Models and Matching

Models of 3D objects are used heavily both in computer vision and in computer graphics. In graphics, the object must be represented in a structure suitable for rendering and display. The most common such structure is the 3D mesh, a collection of polygons consisting of 3D points and the edges that join them. Graphics hardware usually supports this representation. For smoother and, or simpler surfaces, other graphics representations include quadric surfaces, B-spline surfaces, and subdivision surfaces. In addition to 3D shape information, graphics representations can contain color and texture information which is then *texture-mapped* onto the rendered object by the graphics hardware. Figure 14.1 shows a rough 3D mesh of a toy dog and a texture-mapped rendered image from the same viewpoint.

In computer vision, the object representation must be suitable for use in object recognition, which means that there must be some potential correspondence between the representation and the features that can be extracted from an image. However, there are several different types of images commonly used in 3D object recognition, in particular: Gray-scale images, color images, and range images. Furthermore, it is now common to have either gray-scale or color images registered to range data, providing recognition algorithms with a richer set of features. Most 3D object algorithms are not general enough to handle such a variety of features, but instead were designed for a particular representation. Thus it is important to look at the common representations before discussing 3D object recognition. In general categories, there are geometric representations in terms of points, lines, and surfaces; symbolic representations in terms of primitive components and their spatial relationships; and functional representations in terms of functional parts and their functional relationships. We will begin with a survey of the most common methods for representing 3D objects and then proceed to the representations required by the most common types of object recognition algorithms.

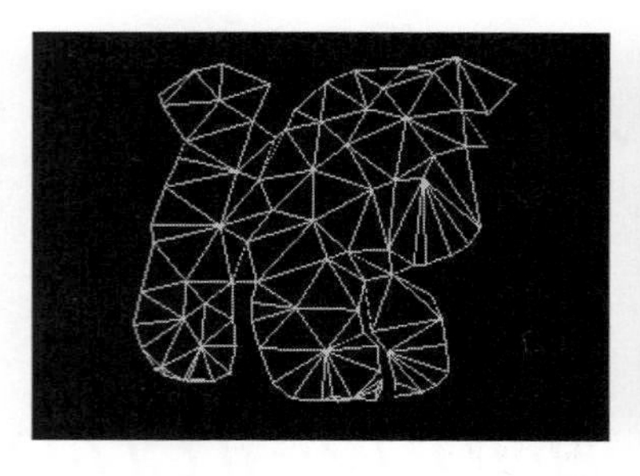
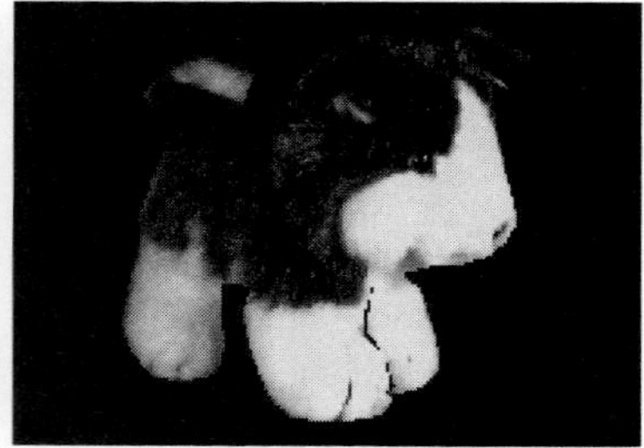

Figure 14.1 3D mesh of a toy dog and texture-mapped rendered image. (Courtesy of Kari Pulli.)

14.1 SURVEY OF COMMON REPRESENTATION METHODS

Computer vision began with the work of Roberts in 1965 on recognition of polyhedral objects, using simple wire-frame models and matching to straight line segments extracted from images. Line-segment-based models have remained popular today, but there are also a number of alternatives that attempt to more closely represent the data from objects that can have curved and even free-form surfaces. In this section, we will look at mesh models, surface-edge-vertex models, voxel and octree models, generalized-cylinder models, superquadric models, and deformable models. We will also look at the distinction between true 3D models and characteristic-view models that represent a 3D object by a set of 2D views.

14.1.1 3D Mesh Models

A 3D mesh is a very simple geometric representation that describes an object by a set of vertices and edges that together form polygons in 3D-space. An arbitrary mesh may have arbitrary polygons. A *regular mesh* is composed of polygons all of one type. One commonly used mesh is a *triangular mesh,* which is composed entirely of triangles; the mesh shown in Figure 14.1 is a triangular mesh. Meshes can represent an object at various different levels of resolution, from a coarse estimate of the object to a very fine level of detail. Figure 14.2 shows three different meshes representing different levels of resolution of the dog. They can be used both for graphics rendering or for object recognition via range data. When used for recognition, feature extraction operators must be defined to extract features from the range data that can be used in matching. Such features will be discussed in Section 14.4.1.

14.1.2 Surface-Edge-Vertex Models

Since many of the early 3D vision systems worked with polygonal objects, edges have been the main local feature used for recognition or pose estimation. A three-dimensional object model that consists of only the edges and vertices of the object is called a *wire-frame* model. The wire-frame representation assumes that the surfaces of the object are planar and that the object has only straight edges.

A useful generalization of the wire-frame model that has been heavily used in computer vision is the *surface-edge-vertex* representation. The representation is a data structure containing the vertices of the object, the surfaces of the object, the edge segments of the

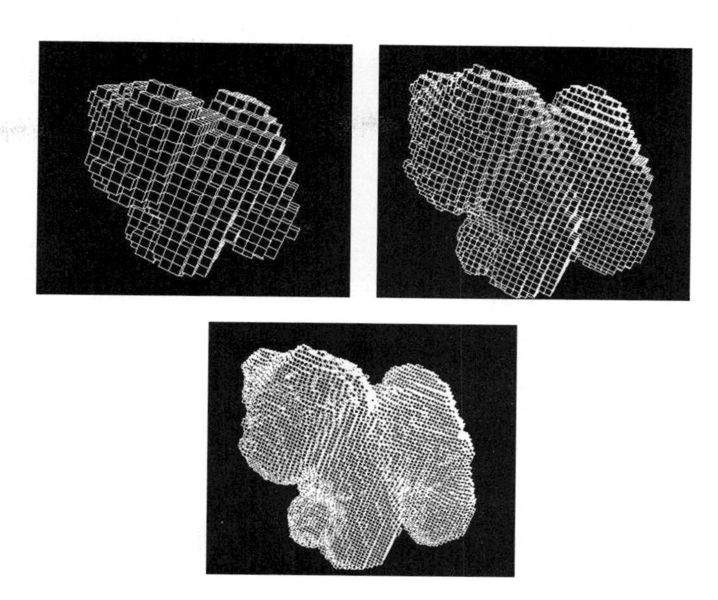

Figure 14.2 Three meshes of the dog at different resolutions. (Courtesy of Kari Pulli.)

object, and, usually, the topological relationships that specify the surfaces on either side of an edge and the vertices on either end of an edge segment. When the object is polygonal, the surfaces are planar and the edge segments are straight line segments. However, the model generalizes to include curved edge segments and/or curved surfaces.

Figure 14.3 illustrates a sample surface-edge-vertex data structure used for representing a database of object models in a 3D object recognition system. The data structure is hierarchical, beginning with the world at the top level and continuing down to the surfaces and arcs at the lowest level. In Figure 14.3 the boxes with fields labeled [name, type, ⟨entity⟩, transf] indicate the elements of a set of class ⟨entity⟩. Each element of the set has a name, a type, a pointer to an ⟨entity⟩, and a 3D transformation that is applied to the ⟨entity⟩ to obtain a potentially rotated and translated instance. For example, the world has a set called objects. In that set are named instances of various 3D object models. Any given object model is defined in its own coordinate system. The transformation allows each instance to be independently positioned in the world.

The object models each have three sets: their edges, their vertices, and their faces. A vertex has an associated 3D point and a set of edges that meet at that point. An edge has a start point, an end point, a face to its left, a face to its right, and an arc that defines its form, if it is not a straight line. A face has a surface that defines its shape and a set of boundaries including its outer boundaries and hole boundaries. A boundary has an associated face and a set of edges. The lowest-level entities—arcs, surfaces, and points—are not defined here. Representations for surfaces and arcs will depend on the application and on the accuracy and smoothness required. They might be represented by equations or further broken down into surface patches and arc segments. Points are merely vectors of (x, y, z) coordinates.

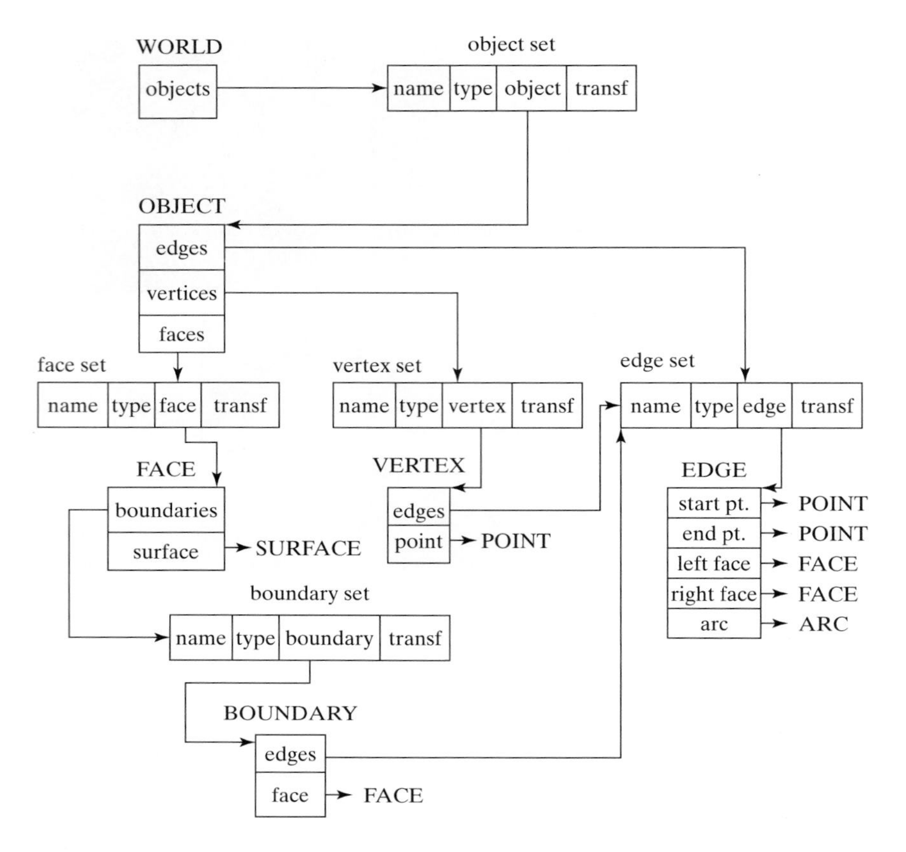

Figure 14.3 Surface-edge-vertex data structure.

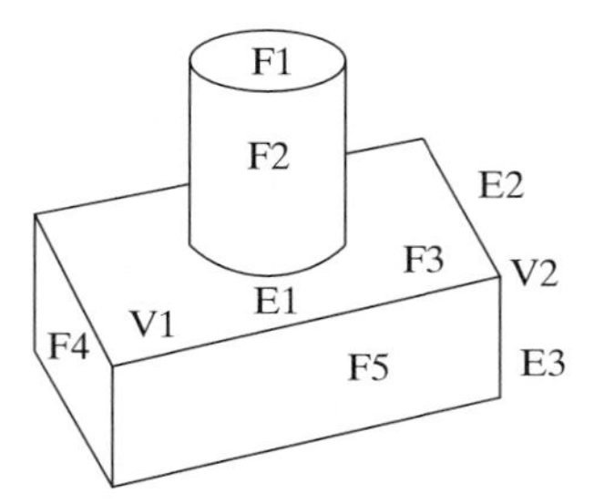

Figure 14.4 Sample 3D object with planar and cylindrical surfaces.

Figure 14.4 shows a simple 3D object that can be represented in this manner. To simplify the illustration, only a few visible surfaces and edges are discussed. The visible surfaces are F1, F2, F3, F4, and F5. F1, F3, F4, and F5 are planar surfaces, while F2 is a cylindrical surface. F1 is bounded by a single boundary composed of a single edge that can be represented by a circular arc. F2 is bounded by two such boundaries. F3 is bounded by an outer boundary composed of four straight edges and a hole boundary composed of a single circular arc. F4 and F5 are each bounded by a single boundary composed of four straight edges. Edge E1 separates faces F3 and F5. If we take vertex V1 to be its start point and V2 to be its end point, then F3 is its left face and F5 is its right face. Vertex V2 has three associated edges E1, E2, and E3.

Exercise 14.1: Surface-edge-vertex structure.

Using the representation of Figure 14.3, create a model of the entire object shown in Figure 14.4, naming each face, edge, and vertex in the full 3D object, and using these names in the structure.

14.1.3 Generalized-Cylinder Models

A *generalized cylinder* is a volumetric primitive defined by a space curve axis and a cross-section function at each point of the axis. The cross section is swept along the axis creating a solid of revolution. For example, a common circular cylinder is a generalized cylinder whose axis is a straight line segment and whose cross section is a circle of constant radius. A cone is a generalized cylinder whose axis is a straight line segment and whose cross section is a circle whose radius starts out zero at one end point of the axis and grows to its maximum at the other end point. A rectangular solid is a generalized cylinder whose axis is a straight line segment and whose cross section is a constant rectangle. A torus is a generalized cylinder whose axis is a circle and whose cross section is a constant circle.

A generalized cylinder model of an object includes descriptions of the generalized cylinders and the spatial relationships among them, plus global properties of the object. The cylinders can be described by length of axis, average cross-section width, ratio of the two, and cone angle. Connectivity is the most common spatial relationship. In addition to end-point connectivity, cylinders may be connected so that the end points of one connect to an interior point of another. In this case, the parameters of the connection—such as the position at which the cylinders touch—the inclination angle, and the girdle angle describing the rotation of one about the other may be used to describe the connection. Global properties of an object may include number of pieces (cylinders), number of elongated pieces, and symmetry of the connections. Hierarchical generalized-cylinder models, in which different levels of detail are given at different levels of the hierarchy, are also possible. For example, a person might be modeled very roughly as a stick figure (as shown in Figure 14.5) consisting of cylinders for the head, torso, arms, and legs. At the next level of the hierarchy, the torso

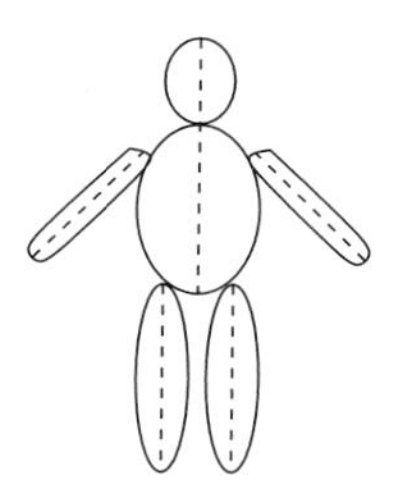

Figure 14.5 Rough generalized cylinder model of a person. The dotted lines represent the axes of the cylinders.

might be divided into a neck and lower torso, the arms into three cylinders for upper arm, lower arm, and hand, and the legs similarly. At the next level, the hands might be broken into a main piece and five fingers, and one level deeper, the fingers might be broken into three pieces and the thumb into two.

A three-dimensional generalized cylinder can project to two different kinds of two-dimensional regions on an image: ribbons and ellipses. A *ribbon* is the projection of the long portion of the cylinder, while an *ellipse* is the projection of the cross section. Of course, the cross section is not always circular, so its projection is not always elliptical, and some generalized cylinders are completely symmetric, so they have no longer or shorter parts. For those that do, algorithms have been developed to find the ribbons in images of the modelled objects. These algorithms generally look for long regions that can support the notion of an axis. Figure 14.6 shows the process of determining potential axes of generalized cylinders from a 2D shape.

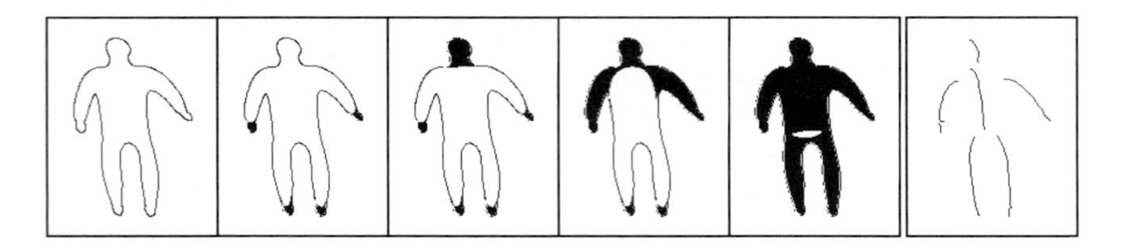

Figure 14.6 The process of constructing a generalized cylinder approximation from a 2D shape. (Example courtesy of Gerard Medioni.)

Figure 14.7 shows steps in creating a detailed model of a particular human body for the purpose of making well-fitting clothing. A special sensing environment combines input from twelve cameras. Six cameras view the human at equal intervals of a *2m* cylindrical room: there is a low set and high set so that a *2m* tall person can be viewed. As shown in Figure 14.7, silhouettes from six cameras are used to fit elliptical cross sections to obtain a cylindrical model. A light grid is also used so that triangulation can be used to compute 3D surface points in addition to points on the silhouettes. Concavities are developed using the structured light data, and ultimately a detailed mesh of triangles is computed.

Exercise 14.2: Generalized cylinder models.

Construct a generalized cylinder model of an airplane. The airplane should have a fuselage, wings, and a tail. The wings should each have an attached motor. Try to describe the connectivity relationships between pairs of generalized cylinders.

14.1.4 Octrees

An *octree* is a hierarchical 8-ary tree structure. Each node in the tree corresponds to a cubic region of the universe. The *label* of a node is either *full,* if the cube is completely enclosed

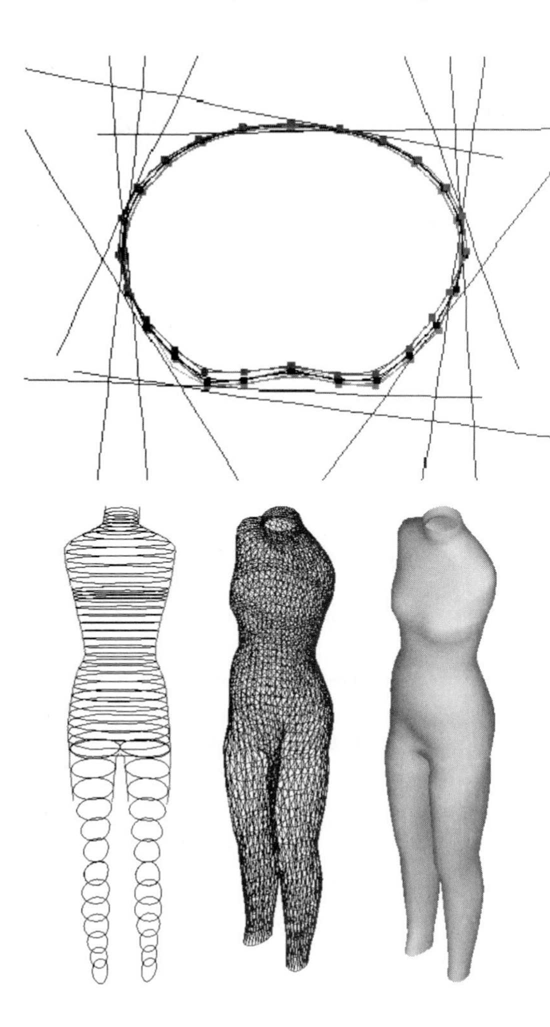

Figure 14.7 Steps in making a model of a human body for fitting clothing. (top) Three cross section curves along with cross section silhouettes from six cameras viewing the body (the straight lines project the silhouette toward the cameras). Structured light features allow 3D points to be computed in concavities; and (bottom) generalized cylinder model created by fitting elliptical cross sections to the six silhouettes, resulting triangular mesh, and shaded image. (Courtesy of Helen Shen and colleagues at the Department of Computer Science, Hong Kong University of Science and Technology: Project supported by grant AF/183/97 from the Industry and Technology Development Council of Hong Kong, SAR of China in 1997.)

by the three-dimensional object, *empty* if the cube contains no part of the object, or *partial,* if the cube partly intersects the object. A node with label *full* or *empty* has no children. A node with label *partial* has eight children representing the partition of the cube into octants.

A three-dimensional object can be represented by a $2^n \times 2^n \times 2^n$ three-dimensional array for some integer n. The elements of the array are called *voxels* and have a value of 1 (full) or 0 (empty), indicating the presence or absence of the object. The octree encoding of the object is equivalent to the three-dimensional array representation, but will generally require much less space. Figure 14.8 gives a simple example of an object and its octree encoding, using the octant numbering scheme of Jackins and Tanimoto (1980).

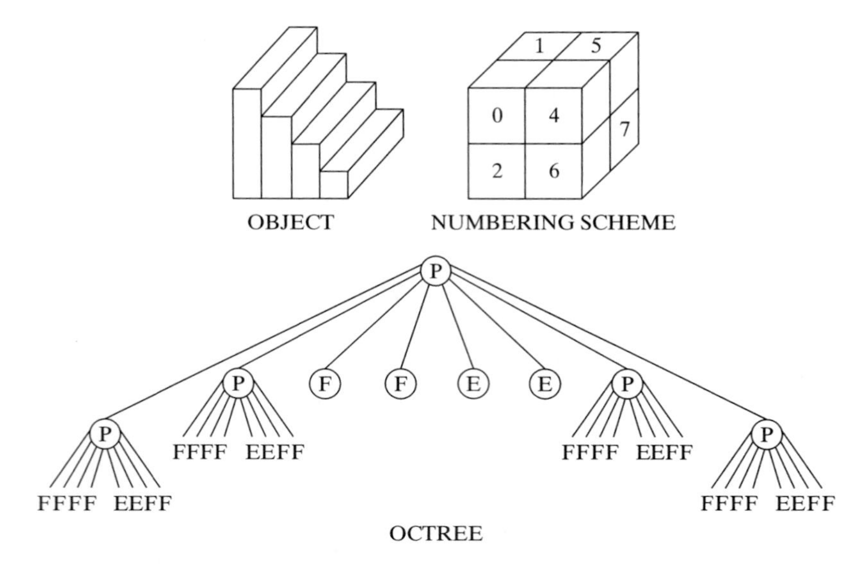

Figure 14.8 A simple three-dimensional object and its octree encoding.

Exercise 14.3: Octrees.

Figure 14.11 shows two views of a simple chair. Construct an octree model of the chair. Assume that the seat and back are both 4 voxels $\times$ 4 voxels $\times$ 1 voxel and that each of the legs is 3 voxels $\times$ 1 voxel $\times$ 1 voxel.

14.1.5 Superquadrics

Superquadrics are models originally developed for computer graphics and proposed for use in computer vision by Pentland. Superquadrics can intuitively be thought of as lumps of clay that can be deformed and glued together into object models. Mathematically, superquadrics form a parameterized family of shapes. A superquadric surface is defined by a vector **S** whose x, y, and z components are specified as functions of the angles η and ω via the equation

$$\mathbf{S}(\eta, \omega) = \begin{bmatrix} x \\ y \\ z \end{bmatrix} = \begin{bmatrix} a_1 cos^{\epsilon_1}(\eta) cos^{\epsilon_2}(\omega) \\ a_2 cos^{\epsilon_1}(\eta) sin^{\epsilon_2}(\omega) \\ a_3 sin^{\epsilon_1}(\eta) \end{bmatrix} \tag{14.1}$$

for $-\frac{\pi}{2} \leq \eta \leq \frac{\pi}{2}$ and $-\pi \leq \omega < \pi$. The parameters a_1, a_2, and a_3 specify the size of the superquadric in the x, y and z directions, respectively. The parameters ϵ_1 and ϵ_2 represent the squareness in the latitude and longitude planes.

Superquadrics can model a set of useful building blocks such as spheres, ellipsoids, cylinders, parallelepipeds, and in-between shapes. When ϵ_1 and ϵ_2 are both 1, the generated surface is an ellipsoid, and if $a_1 = a_2 = a_3$, a sphere. When $\epsilon_1 \ll 1$ and $\epsilon_2 = 1$, the surface looks like a cylinder.

The power of the superquadric representation lies not in its ability to model perfect geometric shapes, but in its ability to model deformed geometric shapes through deformations such as *tapering* and *bending*. Linear tapering along the z-axis is given by the transformation

$$x' = \left(\frac{k_x}{a_3}z + 1\right)x$$
$$y' = \left(\frac{k_y}{a_3}z + 1\right)y$$
$$z' = z$$

where k_x and k_y $(-1 \le k_x, k_y \le 1)$ are the tapering parameters with respect to the x and y planes, respectively, relative to the z direction. The bending deformation is defined by the transformation

$$x' = x + cos(\alpha)(\mathbf{R} - r),$$
$$y' = y + sin(\alpha)(\mathbf{R} - r),$$
$$z' = sin(\gamma)\left(\tfrac{1}{k} - r\right)$$

where k is the curvature, r is the projection of the x and y components onto the bending plane $z - r$ given by

$$r = cos\left(\alpha - tan^{-1}\left(\frac{y}{x}\right)\right)\sqrt{x^2 + y^2},$$

$\mathbf{R}$ is the transformation of r given by

$$\mathbf{R} = k^{-1} - cos(\gamma)(k^{-1} - r),$$

and γ is the bending angle

$$\gamma = zk^{-1}.$$

Superquadric models are mainly for use with range data and several procedures for recovering the parameters of a superquadric fit to a surface have been proposed. Figure 14.9 illustrates superquadric fits to 3D data from the left ventricle of a heart at five time points. These are extended superquadrics with parameter functions, where the parameters are not constants but functions.

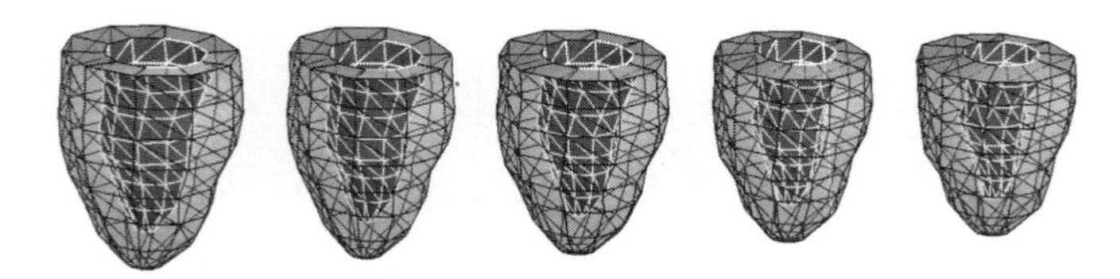

Figure 14.9 Fitted left-ventricle models at five time points during systole, using extended superquadrics with parameter functions. (Courtesy of Jinah Park and Dimitris Metaxas.)

14.2 TRUE 3D MODELS VERSUS VIEW-CLASS MODELS

All of the above object representations emphasize the three-dimensional nature of the objects, but ignore the problem of recognizing an object from a two-dimensional image taken from an arbitrary viewpoint. Most objects look different when viewed from different viewpoints. A cylinder that projects to a ribbon (see Section 14.1.3) in one set of viewpoints also projects to an ellipse in another set of viewpoints. In general, we can partition the space of viewpoints into a finite set of *view classes*[1], each view class representing a set of viewpoints that share some property. The property may be that the same surfaces of the object are visible in an image taken from that set of viewpoints, the same line segments are visible, or the relational distance (see Chapter 11) between relational structures extracted from line drawings at each of the viewpoints is small enough. Figure 14.10 shows the view classes of a cube defined by grouping together those viewpoints which produce line drawings that are topologically isomorphic. Figure 14.11 shows two views of a chair in which most, but not all, of the same surfaces are visible. These views could be grouped together via a clustering algorithm using their relational distance defined over region primitives and the region adjacency relation as a basis for closeness. They are among many different similar views that together form a view class; the number of views is potentially infinite. The main point is that once the correct view class has been determined for an object, the matching to determine the correspondences necessary for pose determination is a highly constrained, two-dimensional kind of matching.

The use of view classes was proposed by Koenderink and van Doorn (1979). The structure they proposed is called an *aspect graph*. An *aspect* is defined as a qualitatively

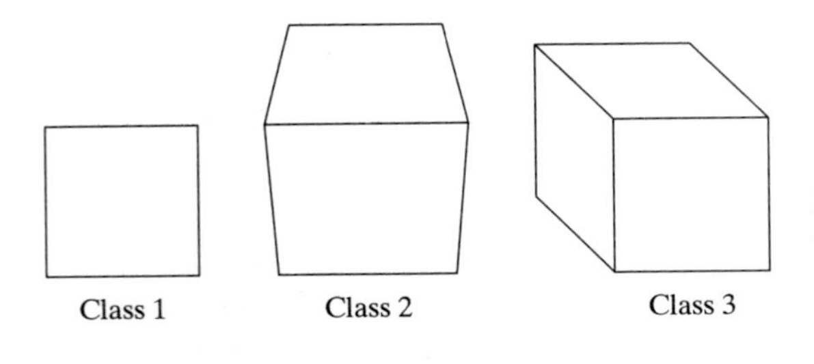

Figure 14.10 The three view classes of a cube defined by grouping together viewpoints that produce topologically isomorphic line drawings.

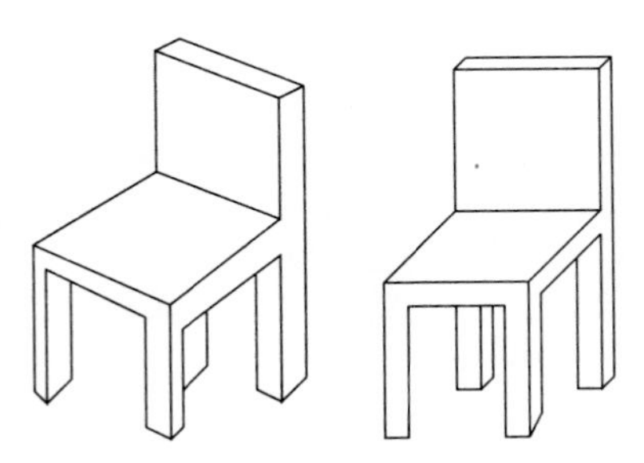

Figure 14.11 Two chair views that belong to the same view class based on their low relational distance.

[1] Also called *characteristic views*.

distinct view of an object as seen from a set of connected viewpoints. The nodes of an aspect graph represent aspects and the arcs connect adjacent aspects. The change in appearance at the boundary between two aspects is called a *visual event*. Algorithms for automatic construction of aspect graphs were developed in the late 1980s, but because the structures are very large for realistic objects, they have not been used much in object recognition. The view class or characteristic view concept has been heavily used instead.

Exercise 14.4: View-class models.

The two chairs of Figure 14.11 both belong to a single view class of a three-dimensional object. Draw pictures of three more common view-classes of the chair.

14.3 PHYSICS-BASED AND DEFORMABLE MODELS

Physics-based models can be used to model the appearance and behavior of an actual physical object being imaged. Examples are given in this section of a human heart model (Figure 14.16) and a telephone headset (Figure 14.15) model. The principles of physics can be used to model an actual physical system or instead to simulate some image analysis task. In the modeling of the heart, the objective is to model the changing shape and behavior of an object over time so that its functioning can be understood. In the modeling of the telephone handset, the objective is to obtain a good mesh model of the static measurements.

A term that is strongly related to the term *physics-based model* is *deformable model*. The latter term emphasizes that the change in object shape is to be modeled.

There has been good progress recently in physics-based and deformable modeling. Theory and applications are rich and more complex than the other areas covered in this text. The brief coverage given here is only for the purpose of introducing the topic and motivating the student to do outside reading in the rapidly developing literature.

14.3.1 Snakes: Active Contour Models

Most of us have placed a rubber band around our outstretched fingers. Our fingers are analogous to five points in 2D and the rubber band is a closed contour through the five points. The action of the rubber band can be simulated by an *active contour* which moves in an image toward a minimum energy state. A rubber band tends to contract in order to reduce its stored energy, at least until it has met with supporting forces (the fingers). Figure 14.12(right) illustrates this concept. The small dark regions are analogous to our fingers: the contraction of the rubber band is stopped by these regions. Another aspect of the analogy is that the rubber band will not have infinite curvature; even if stopped by a single point (or wire) the rubber band will smoothly wrap around that point. Preventing or punishing high curvature can be done during simulation. Figure 14.12(left) shows what would happen if a balloon were blown up inside our hand with fingers posed as if grasping a ball. Analogously, we could simulate a virtual balloon being blown up inside some image regions or points.

Figure 14.12 illustrates a critical advantage of active contours: the contour is a complete structure even though the data being fit may be badly fragmented. Moreover, other

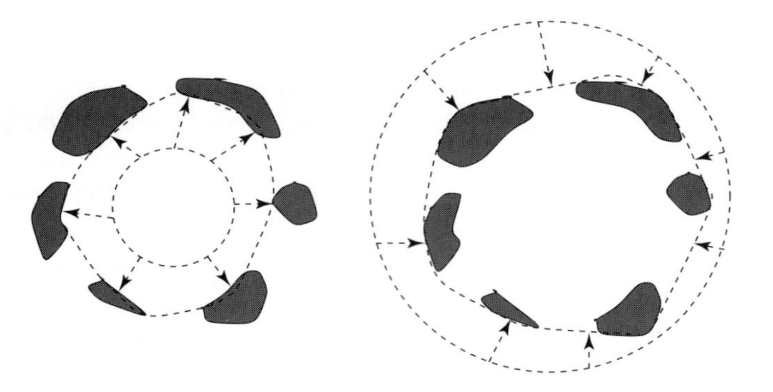

Figure 14.12 (left) 2D balloon or *active contour* being blown up to fit the 2D points; and (right) 2D rubber band stretched over the outside of the 2D points.

characteristics can be enforced top-down, such as smoothness, perimeter limits, and the property of being a simple curve. We now give a brief sketch of how the behavior of an active contour can be simulated by a computer algorithm.

To simulate the behavior of an active contour object, we first need a memory state to define the structure and location of the contour. Consider the simple case where we have a fixed set of N points at time t, each located at $P_{j,t}$ and each circularly related to two neighbors $P_{j-1,t}$ and $P_{j+1,t}$. In the case of a virtual rubber band, each point will be subject to a pulling force from the two neighbors, the result of which will accelerate the point $P_{j,t}$ to a new position $P_{j,t+1}$. Figure 14.13(left) illustrates this. We usually consider each point to have unit mass so we can easily compute acceleration in terms of force. Using acceleration we compute velocity, and using velocity we compute position. Thus, our memory state at time t should also contain the acceleration and velocity of each point; moreover, these need not be zero at the start of the simulation. One more data member is needed: we use a Boolean variable to indicate whether or not the point has stopped moving due to running into some data point (called a *hard constraint*). In addition to our active contour object, of course, we need to store the data to be modeled: this could be a gray-scale image, a set of 2D edge points, a set of 3D surface points, etc., represented in some manner as described in this chapter or in Chapters 2 or 10.

A simple algorithm for moving active contour points is sketched in Algorithm 14.1. Contour points move until they meet hard constraints or until the resultant force on them

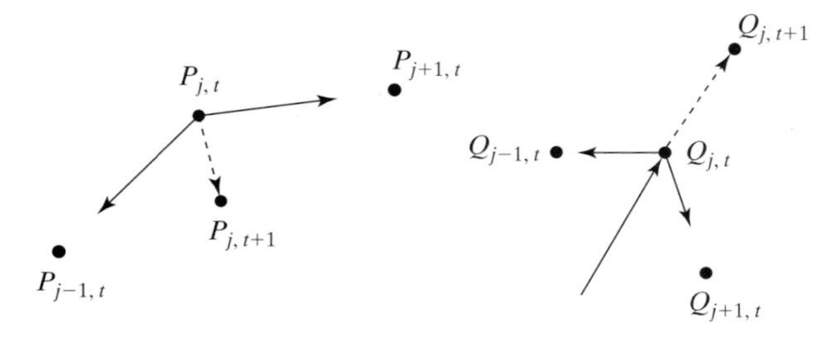

Figure 14.13 (left) Forces on a point of a stretched rubber band tend to move that point inward; and (right) an inflationary force on a point of a balloon tends to move it outward, provided the elastic forces from neighboring points are exceeded.

Move contour (snake) points $P_{j,t}$ to next position $P_{j,t+1}$.

Input: N data points at time t; each $P_{j,t}$ has velocity $V_{j,t}$ and acceleration $A_{j,t}$.

Output: N data points at time $t+1$; each $P_{j,t+1}$ has velocity $V_{j,t+1}$ and acceleration $A_{j,t+1}$.

Using time step Δt, do at each point $P_{j,t}$ that has not stopped at a hard constraint:

1. Compute the resultant force on $P_{j,t}$ using its neighbors.
2. Using the force, compute the acceleration vector $A_{j,t+1}$.
3. Compute velocity $V_{j,t+1} = V_{j,t} + A_{j,t}\Delta t$
4. Compute new position $P_{j,t+1} = P_{j,t} + V_{j,t}\Delta t$
5. If $P_{j,t+1}$ is within tolerance of some data point, freeze this position.

Algorithm 14.1 Single update stage for an active contour.

is zero. Or, perhaps the algorithm never stops, as in the case where the active contour is tracking a pair of speaking lips! Note that an initial position for the contour is required.

Algorithm 14.1 sketches a simple stage of an Euler algorithm that computes, for a small time step, acceleration from force, velocity from acceleration, and position from velocity. A point's position is frozen when it collides with a data point, edge, or surface patch. In general, this can be a costly computation that requires search through the data structure or image to find such a point.

Hooke's Law models a spring, which is a common element of physics-based models. Suppose a spring of natural length L connects points P_j and P_k. The force F acting on P_j is proportional to how the spring is stretched (or compressed) relative to its natural length.

$$F = -k_L(\|P_j - P_k\| - L)\frac{P_j - P_k}{\|P_j - P_k\|} \tag{14.2}$$

This should suffice to simulate our rubber band. A damping force should be added if it is possible that the spring system could oscillate indefinitely. A remaining problem is to determine a good length L. If we are modeling a known object, such as talking lips, we should be able to determine practical values for N, L and k_L. k_L is a stiffness parameter that relates force to deformation.

An Energy Minimizing Formulation* Although the notion of an active contour had been used previously by others, a 1987 paper by Kass, Witkin, and Terzopoulos seemed to ignite the interest of the computer vision community. Much of the discussion above was motivated by their view of *snakes,* as they were called. Fitting a snake to data was defined as an optimization problem that sought a minimum energy boundary subject to some hard constraints. A useful formulation is to consider that total energy is a sum of three components; (1) *internal contour energy* characterized by the stretching and bending of the contour itself; (2) *image energy* that characterizes how the contour fits to the image intensity or gradient; and (3) *external energy* due to constraint forces. Constraints are used to apply information from an interactive user or a higher level CV process.

A contour parameterized by $s \in [0, 1]$ is $\mathbf{v}(s) = [x(s), y(s)]$, which is just a function of real variable s. The problem is to find the function that minimizes the energy defined as follows.

$$E_{contour} = \int_0^1 (E_{internal} + E_{image} + E_{constraints})\, ds. \tag{14.3}$$

$$E_{internal} = \alpha(s)|\mathbf{v}'(s)|^2 + \beta(s)|\mathbf{v}''(s)|^2 \tag{14.4}$$

The snake can be controlled to pass near some designated points by adding in the squared distance between each point and the snake using $E_{constraints}$. E_{image} might just be the sum of the squared distances between a snake point and the closest edge point. The internal energy definition is perhaps more interesting. The first part of $E_{internal}$ punishes higher variance in the lengths of small contour segments—lower energy means small variance in their lengths. The second part punishes curvature. The weighting functions $\alpha(s)$ and $\beta(s)$ are used for blending and can also allow the process to form a sharp corner where a corner detector has found one or to take a long leap over bland texture.

The fitting of active contours to images can be related to the making of airfoils or canoes. Figure 14.14 shows what happens when a strip of wood is nailed at regular intervals to cross sections held in place by a *strongback*. The wood bends to smoothly fit the (air) space in between the cross sections making a smooth but possibly complex curve. Contact with the cross sections enforces a hard constraint. High curvature is reduced as the wood distributes the bending energy over many points. Computer algorithms can easily produce such *spline* curves—in fact, Figure 14.14 was produced by such an algorithm in the `xfig` tool!

The approaches to minimizing the energy of a contour are beyond our scope here. Careful numerical programming is needed to obtain good control of an active contour. Finite elements packages can be used. After 1987, several new works appeared that used dynamic programming, instead of the scale space approach proposed by Kass and others (1987). The interested reader can find many interesting works in the literature.

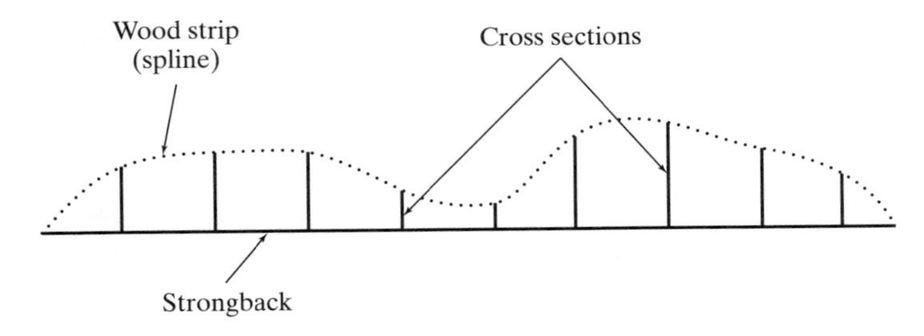

Figure 14.14 A wood strip attached to cross sections makes a *low energy* contour. (Smooth spline courtesy of `xfig`.)

14.3.2 Balloon Models for 3D

A balloon model can be a mesh made by approximating a sphere. Most soccer balls are made from 12 pentagonal and 20 hexagonal pieces, each of which can be divided into triangles. Edges of the triangles can be modeled by springs so that the shape of the entire system can deform, either by expanding or contracting. Figure 14.15 shows such a spherical model expanding inside a cloud of 3D data points taken by sensing a telephone handset. The algorithm freezes the position of a vertex when it makes contact with sensed data. An

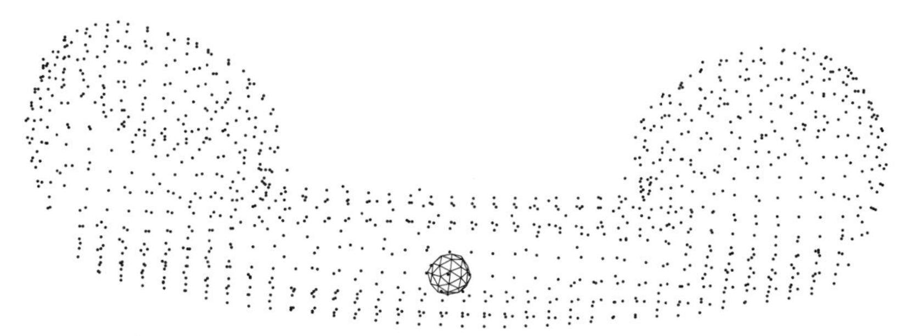

(*a*) Initialization with balloon entirely within the 3D point cloud.

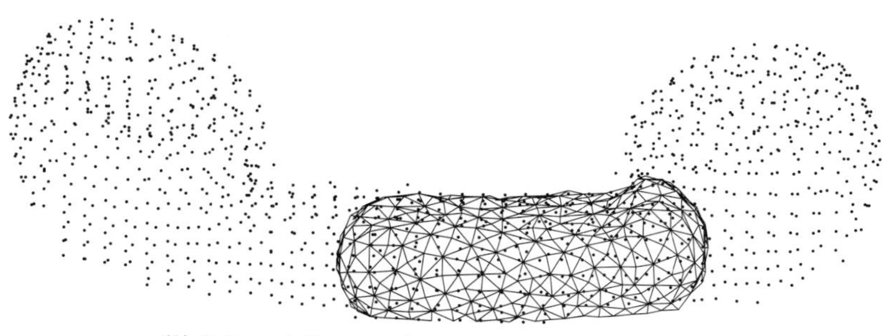

(*b*) Balloon inflated so that some triangles contact data.

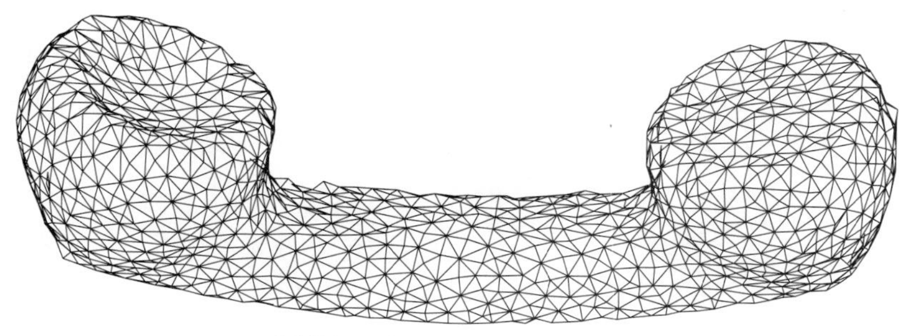

(*c*) Triangular mesh at termination.

Figure 14.15 Three snapshots of the physics-based process of inflating a mesh of triangles to fit a cloud of 3D data. (Courtesy of Yang Chen and Gerard Medioni.)

inflating force is applied to each vertex in the direction of the surface normal interpolated at that vertex. To detect contact with sensed data, the data are searched only in a direction along this normal. When an inflating triangle becomes large, the algorithm subdivides it into four triangles. In this manner, the sphere can inflate to the elongated shape of the data as shown in Figure 14.15(*b*)–(*c*). The 3D data points were obtained from several different views of the object surface using a range scanner, each rigidly transformed to a global coordinate system. Imagine the difficulty of sewing together the several different surface meshes that could have been obtained independently from the different views. The balloon model retains the correct topology and approximate uniformity of triangles as it deforms to fit the data. It is difficult to start with just a set of the 3D points and build up a good surface model.

14.3.3 Modeling Motion of the Human Heart

While triangular mesh elements are commonly used for modeling surfaces, 3D volume models can be constructed using tetrahedral elements. Each tetrahedral element has four vertices, four faces, and six edges. *Stiffness* values can be assigned to the edges based on the characteristics of the material being modeled. When forces are applied to various points of the model, the structure will deform. Figure 14.16 shows two states of a beating heart

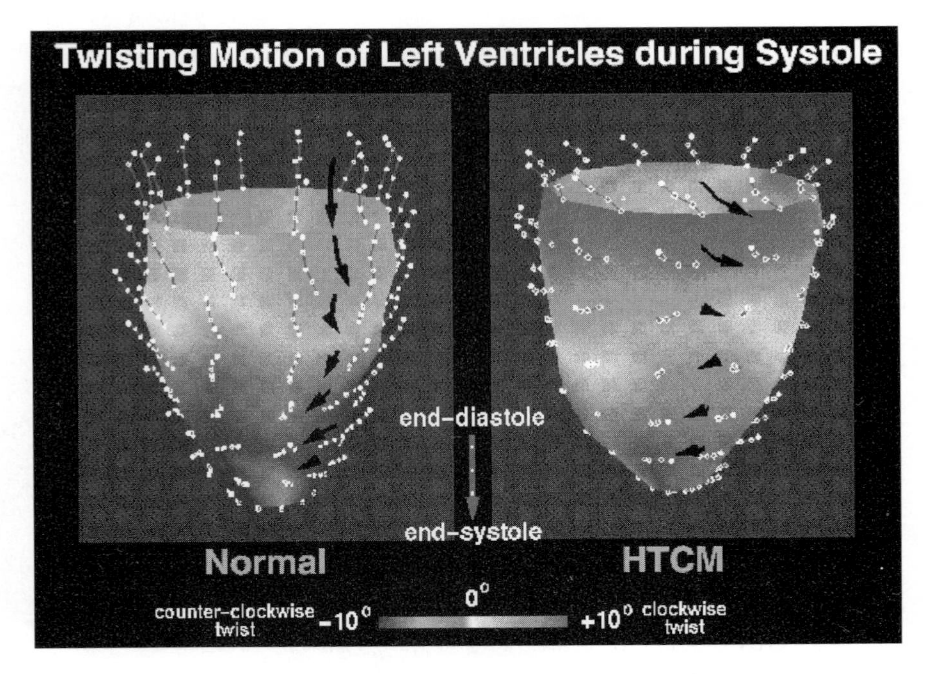

Figure 14.16 Motion of two beating hearts as computed from tagged MRI data. The sensor can tag living tissue and sense its motion in 3D. The heart model fit to the data is meant to model real physics and heart function. The motion vectors shown are different for the two hearts. (Courtesy of Jinah Park and Dimitris Metaxas.)

computed from tagged magnetic resonance imagery. The sensor can tag certain parts of living tissue so that they may be sensed in motion in 3D. The heart model fit to the data is meant to model real physics and heart function. The motion of the points of the fitted model can be used to understand how the heart is functioning. The deformation of a tetrahedral element of the model is related to the forces on it and the stiffness of the tissue that it models.

14.4 3D OBJECT RECOGNITION PARADIGMS

Having surveyed different models for 3D objects, we now discuss the most commonly used general paradigms for 3D object recognition. This is actually difficult, because the method used depends heavily on the application, the type of data, and the requirements of the recognition task. There are many dimensions along which one can classify or constrain an object recognition problem. These include:

- **Is the interest engineering or cognitive science?** In case we want to engineer a solution to an immediate practical problem, our problem may be specific enough to be simple. For example, we may have to grab a steel cylinder from a jumble of many of them. On the other hand, our interest may be in understanding human object recognition. This implies development of a very general theory which is consistent with multifarious psychological data—a much more difficult problem.
- **Does the task involve natural or manufactured objects?** Manufactured objects are usually more regular than natural ones. Moreover, there are rigid iconic prototypes for many man-made objects, making several known matching paradigms applicable. Natural objects are created by processes (geological, biological, etc.) which produce a great deal of variety that may be difficult to model. Furthermore, the context of natural objects may be less constrained and less predictable than the context in which manufactured objects are found. For example, the problem of object recognition for autonomous navigation in outdoor environments seems to be much more difficult than the problem of recognizing and determining the pose of objects for factory automation.
- **Are the object surfaces polyhedral, quadric, or free-form?** Many recognition projects have dealt only with polyhedra, which makes modeling particularly simple. Recently, researchers have turned toward use of quadric surfaces, which it is claimed can model about 85 percent of manufactured objects. The major convenience is that the modeling and the sensed data are readily described by the same primitives, possibly with some fitting of parameters. It is not clear how best to model sculpted, free-form objects even when they are rigid objects. A sculpted object, such as a sports car, turbine blade, or iceberg, may have many different smoothly blending surface features which are not easily segmented into simple primitives.
- **Is there one object in the scene or are there many?** Some object recognition schemes assume that objects to be recognized are presented in isolation. This may or may not be possible to engineer in the task domain. Multiple object environments are typically harder because object features will be both masked and intermixed. Global feature

methods work well only for single objects. The segmentation problem can be acute in multiple object environments.

- **What is the goal of the recognition?** We might need to recognize an object for inspection, grasping, or object avoidance. For inspection, we would look at the small details of at least part of the object—modeling and measurement precision must be good. Grasping an object has different requirements. Not only does the task require some rough geometrical knowledge, but it must also consider balance and strength and accessibility of the object in the workspace. A robot recognizing that an object in its path must be avoided must only have a rough idea of the size, shape, and location of that object.
- **Is the sensed data 2D or 3D data?** Humans can operate quite well with the image from only one eye. Many researchers have designed systems that use only 2D intensity images as input. 2D features from the object image are related to a 3D model via the view transformation; usually the matching process has to discover this transformation as well as the object identity. Matching is often easier if 3D data is directly available and this is the reason why many current researchers are working with range data. The belief is that the surface shape of objects and their positions can be directly sensed; this in turn provides a direct index into possible object models and also reduces the amount of ambiguity in computing the registration transformation.
- **Are object models geometric or symbolic?** Geometric models describe the exact 3D shape of an object, while symbolic models describe an entire class of objects. Geometric models are used heavily in industrial machine vision, where the objects to be recognized come from a small prespecified set. CAD data is becoming more available and will usually have all the necessary geometric detail. Symbolic models are required in tasks where there are many different varieties of the object class to be recognized. For example, in medical imaging, each organ provides a new object class and every person provides a unique variation. Many objects in our environment, for example, chairs, have many variations and require more than a geometric approach.
- **Are object models to be learned or preprogrammed?** Object models may contain a large amount of precise data which is very difficult for humans to provide. CAD data alone may not be enough; some additional organization of that data, such as emphasizing features, is often necessary. Having a system learn object geometry by presenting the object to its sensors is an attractive possibility.

14.4.1 Matching Geometric Models via Alignment

Recognition by alignment employs the same principles in 3D object recognition as in 2D matching. (See Chapter 11 for basic definitions.) The main idea is expressed in Algorithm 14.2.

We will look at both the 3D-3D and 2D-3D cases.

3D-3D Alignment Let us assume that the 3D models are, or can be converted to, collections of 3D model-point features. If the data is range data, then corresponding 3D data-point features are required for matching. The alignment procedure finds the

Determine if a set of image data points matches a 3D object model.

1. Hypothesize a correspondence between a set of model points and a set of image data points,
2. use that correspondence to determine a transformation from the model to the data,
3. apply the transformation to the model points to produce a set of transformed model points, and
4. compare the transformed model points to the data points to verify or disprove the hypothesis.

Algorithm 14.2 The Basic Alignment Algorithm.

correspondences from three chosen model-point features to three corresponding data-point features. This correspondence determines a 3D transformation consisting of a 3D rotation and a 3D translation whose application to the three model points produces the three data points. An Algorithm 13.3 to do this appeared in Chapter 13. If the point correspondence is correct and there is no noise, the correct 3D transformation can be found from the three matches. Since this is rarely the case, more robust procedures that typically use on the order of ten correspondences have been developed. In any case, once the potential transformation has been computed, it is applied to all of the model points to produce a set of transformed model points that can be directly compared to the full set of data points. As in the 2D case, a verification procedure decides how well the transformed model points line up with the data points and either declares a match or tries another possible correspondence. As in the 2D case, there are intelligent variations that select the corresponding points by the local-feature-focus method or by some other perceptual grouping technique. Figure 14.17 illustrates the 3D-3D correspondences that come about from matching a three-segment 3D junction of a chair model to a 3D mesh dataset.

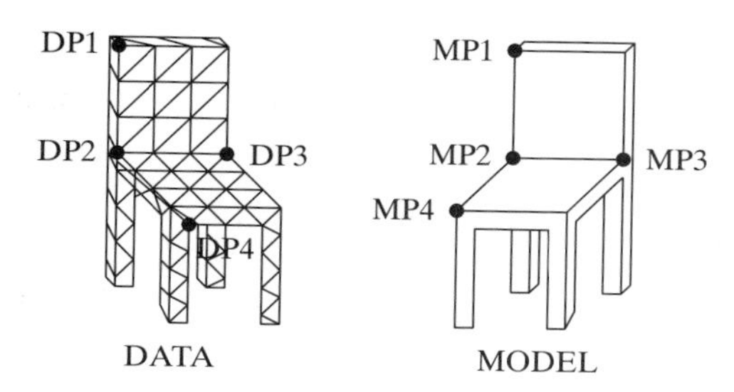

Figure 14.17 Correspondences between 3D model points and 3D mesh data points that can be used to compute the transformation from the model to the data in the 3D-3D alignment procedure.

Exercise 14.5: 3D-3D feature alignment.

Junctions of line segments are common in polyhedral objects. Consider a 3D cup object having a cylindrical part with a cylindrical cavity for the liquid and a semicircular handle.

What features of the cup might be detected in the 3D data and used for finding correspondences in matching?

Feature extraction is an important issue here. If the class of objects is such that distinguished points, such as corners points, peaks, and pits, can be easily found, then the above procedure should work well. If surfaces are smooth and distinguished points are rare or nonexistant, then a better method for finding correspondences is needed. Johnson and Hebert (1998) at CMU developed a very robust method for exactly this problem. Their 3D object representation consists of (1) a 3D mesh model of the object and (2) a set of *spin images* constructed from the mesh that characterize local shape features of the object.

Given a mesh model of a 3D object, the surface normal can be estimated at each vertex of the mesh. Then the relationship between any oriented point in 3D-space and a surface normal at a particular vertex can be represented by two distance parameters, α and β, where α is its perpendicular distance to the surface normal, and β is its signed perpendicular distance to the tangent plane at that vertex. Rotational angles are omitted from this description, because they are ambiguous.

A *spin image* is a two-dimensional histogram that can be computed at a selected vertex of the mesh. Each spin image will have a set of *contributing points* used in its construction. The size of the volume of contributing points depends on two spin-image parameters: D, the maximum distance from a contributing point to the selected vertex, and A, the angle allowed between the normal of the contributing point and the normal of the vertex. A spin image is constructed about a specified oriented point o with respect to a set of contributing points C that have been selected based on specified spin-image parameters A and D. An array of accumulators $S[\alpha, \beta]$ represents the spin image and is initially set to zero. Then for each point $c \in C$, its distance parameters α and β are computed with respect to the selected mesh vertex o, and the accumulator bin corresponding to this α and β is incremented. Note that the size of a bin in the accumulator array is on the order of the median distance between vertices in the 3D mesh. Figure 14.18 gives some examples of spin images.

Spin images are constructed at each vertex of the mesh model. This gives information on the local shape at every point of the mesh. To match two objects, the two sets of spin images are used. The spin image at each point of the first object is compared to the spin image at each point of the second object by computing the correlation coefficient of the pair. Those point pairs with high correlations form the 3D point correspondences needed for object matching. The point correspondences are grouped and outliers eliminated using geometric consistency. Then—as in alignment in general—a rigid transformation is computed and used to either verify the match or rule out the match. Figure 14.19 shows the operation of the spin-image recognition method on a difficult, cluttered image containing six different objects that correspond to models in a database of twenty object models.

2D-3D Alignment Alignment can also be applied to 2D-3D matching in which the object model is three-dimensional, but the data comes from a 2D image. In this case, the transformation from model points to data points is more complex. In addition to the 3D rotation and 3D translation, there is a perspective projection component. The full transformation can be estimated from a set of point correspondences, a set of line segment correspondences,

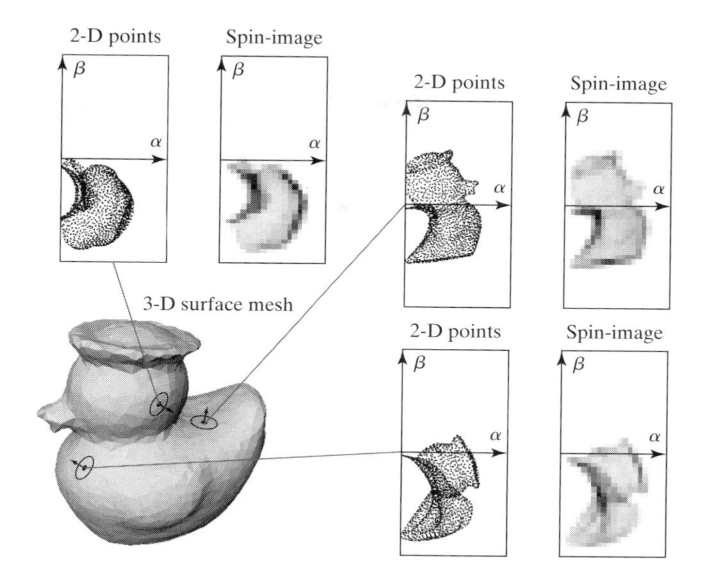

Figure 14.18 Examples of spin images. (Figure courtesy of Andrew Johnson with permission of IEEE. Reprinted from "Efficient Multiple Model Recognition in Cluttered 3-D Scenes," by A. E. Johnson and M. Hebert, *Proceedings of the IEEE Conference on Computer Vision and Pattern Recognition,* June 1998. © 1998 IEEE.)

a 2D ellipse to 3D circle plus single point correspondence, or combinations of all three types of features. This gives us a powerful tool for matching. Correspondences are hypothesized either blindly or through relational matching (see Section 14.4.2) and used to determine a potential transformation. The transformation is applied to the 3D model features to produce 2D data features. Here, a new problem emerges that did not appear in 2D-2D alignment. In any 2D perspective view of a 3D object, some of the transformed features appear on surfaces that do not face the camera and are occluded by other surfaces that are closer to the viewer. Thus, in order to accurately produce a set of transformed features to compare to image features, a hidden feature algorithm must be applied. Hidden feature algorithms are related to graphics rendering algorithms and, if applied in software, can be prohibitively slow. If appropriate mesh models and graphics hardware are available, then the full rendering is possible. Otherwise, it is common to either ignore the hidden feature problem or use an approximate algorithm that is not guaranteed to be accurate, but may be good enough for verification.

The TRIBORS object recognition system (Pulli and Shapiro, 1996) uses view-class models of polyhedral objects and finds correspondences between triplets of model line segments and triplets of 2D image line segments. Model triplets are ranked in a training phase so that triplets with high probabilities of detection are selected first in the matching phase and those with low probabilities are not at all considered. Triplets are described by a vector of nine parameters that describe the appearance of that triplet in the view class being matched. Figure 14.20 shows the parametrization of a line segment triplet. A model triplet

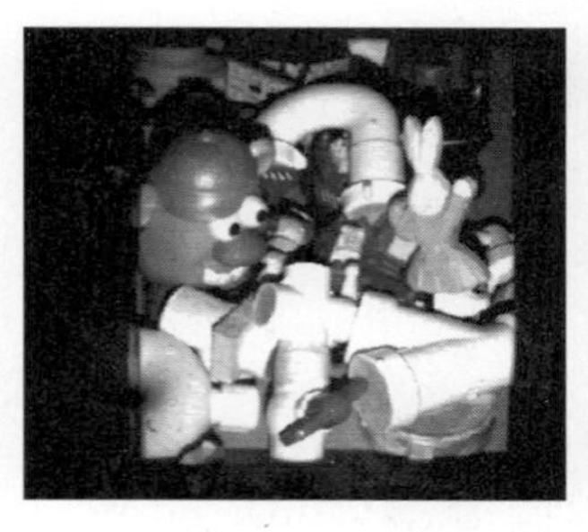

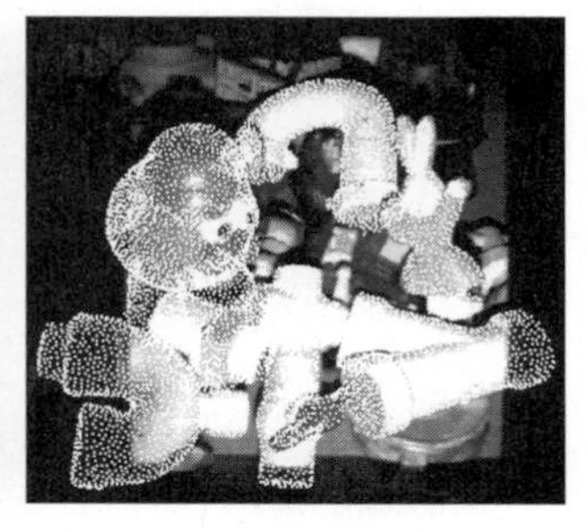

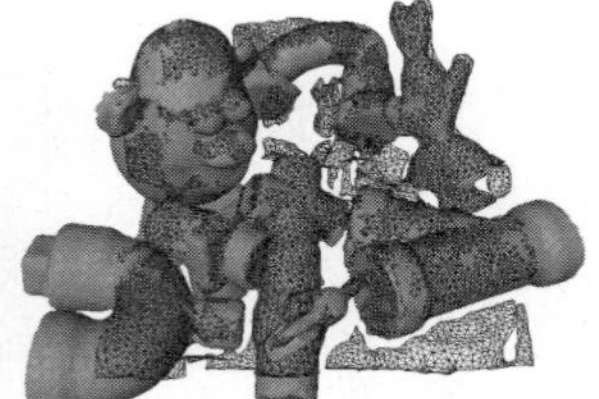

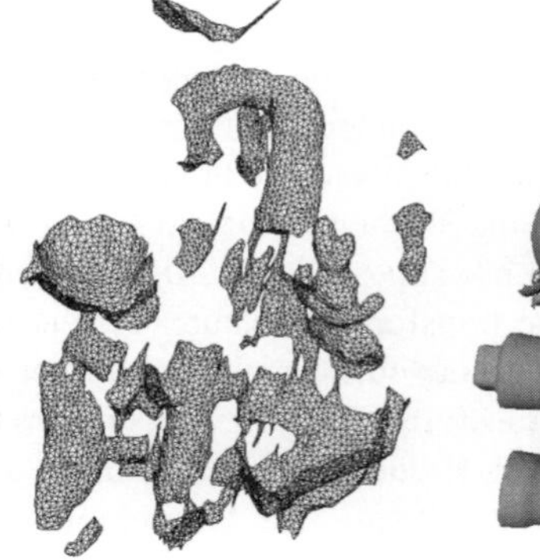

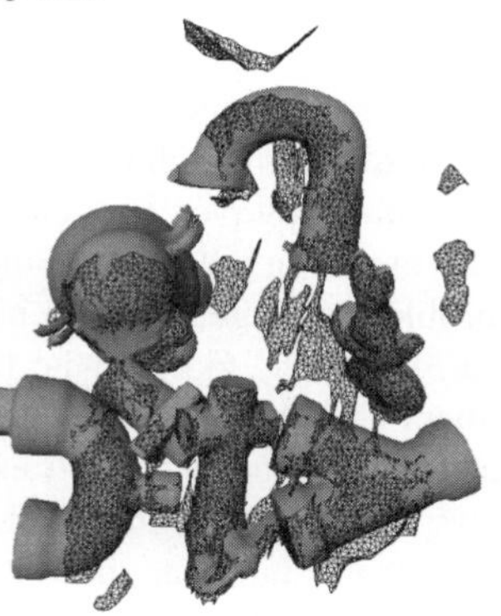

Figure 14.19 Operation of the spin-image recognition system. (Figure courtesy of Andrew Johnson with permission of IEEE. Reprinted from "Efficient Multiple Model Recognition in Cluttered 3-D Scenes," by A. E. Johnson and M. Hebert, *Proceedings of the IEEE Conference on Computer Vision and Pattern Recognition,* June 1998. © 1998 IEEE.)

is matched to an image triplet that has similar parameters. Once a match is hypothesized, the line junction points from the data triplet are paired with the hypothesized corresponding 3D vertices from the model, and an iterative point-to-point correspondence-based exterior orientation algorithm (as given in Chapter 13) is used to determine the transformation. The transformation is then applied to a wireframe model of the 3D object and the visible edges are determined through a hidden line detection algorithm. For each predicted edge, the closest image line segment is determined and verification is performed based on how similar each predicted edge is to its closest image segment. Figure 14.21 shows the operation of the TRIBORS system.

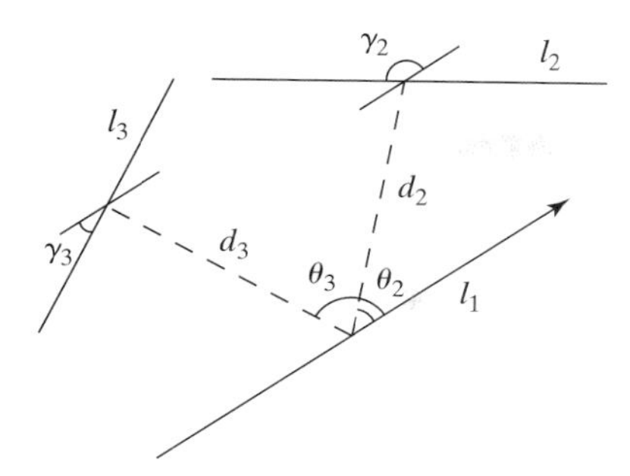

Figure 14.20 The parametrization of a line triplet in the TRIBORS system. The quantities d_2 and d_3 are the distances from the midpoint of line l_1 to the midpoints of lines l_2 and l_3, respectively. Angles γ_2 and γ_3 are the angles that line segments l_2 and l_3 make with line segment l_1. Angles θ_2 and θ_3 are the angles between line segment l_1 and the connecting lines shown to l_2 and l_3, respectively.

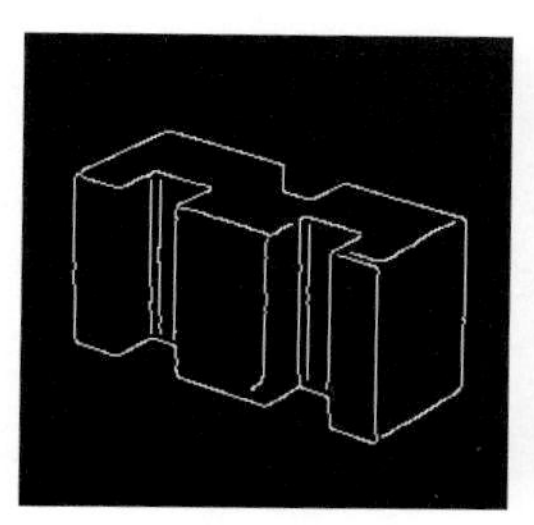
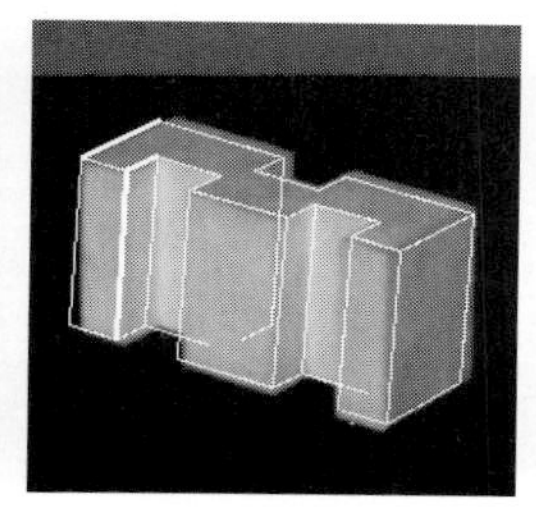
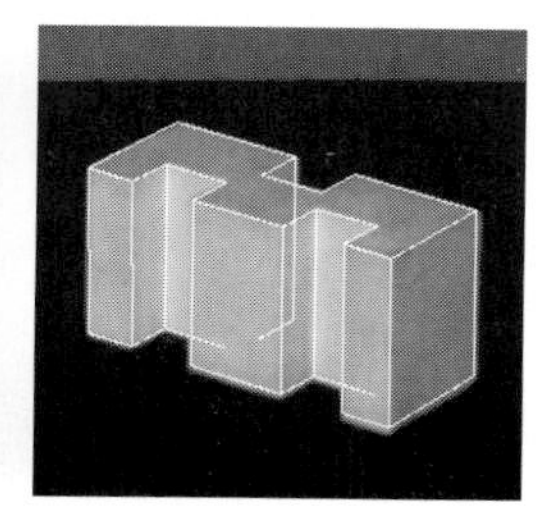

Figure 14.21 (*a*) The edges extracted from a real image; (*b*) the matched triplet of line segments (thick lines) and the initial pose estimate; and (*c*) the final match and pose. (Courtesy of Kari Pulli.)

Exercise 14.6: Matching in TRIBORS.

TRIBORS uses the 9 parameters associated with a triple of line segments to identify potential matches between a model triple and an image triple. Generate several different views of a single view class of a 3D polyhedral object, such as the chair object of Figure 14.11. Identify three major line segments that appear in all your views and compute the 9 parameters shown in Figure 14.20. How similar are the different parameter vectors that you computed? Compare the 9 parameters of these three line segments to those of a completely different set of three line segments. How well do the 9 parameters discriminate among triples of line segments?

Smooth Object Alignment We have talked about the alignment of a 3D mesh model to a 3D range image and a 3D polyhedral model to a 2D intensity image. We consider here the problem of computing the identity and pose of a free-form 3D object from a single 2D intensity image. The solution we will look at employs a view-class type of model, but the representation of a view class is quite different from the sets of triplets of line segments that TRIBORS used and matching is performed at the lowest level on edge-images.

The method discussed here is based on the work of Chen and Stockman (1996), who created a system to determine the pose of 3D objects with smooth surfaces. In this system, a 3D object is modeled by a collection of $2\frac{1}{2}$D views (called model aspects) each constructed from five images taken by rotating the viewpoint up, down, left, and right of a central

Figure 14.22 Five input images used for constructing one model aspect of a car. (Courtesy of Jin-Long Chen.)

viewpoint. Input for construction of one model aspect of a car is shown in Figure 14.22. The silhouette of the central edgemap is extracted and segmented into curve segments whose invariant features are derived for indexing to this model aspect during recognition.

Stereo-like computations are performed to compute points $[x, y, z]$ on the 3D rim for each 2D silhouette point $[u, v]$ in the central image. The up and down images are used to compute the curvature of the object rim in the y direction and the left and right images are used to compute curvature in the x direction. Similarly, stereo-like computations are used to compute the 3D locations of crease and mark points in the central edgemap. Thus, the $2\frac{1}{2}$D model aspects consist of the 3D rim, crease and mark points corresponding to the central edge image plus the x and y curvature at each of these points. Using this information, a mathematical formula can then produce an edge map for any other viewpoint in that view class, given the view parameters. A view class is described by (1) the set of 3D points and curvatures described above and (2) a set of invariant features to be used for indexing. The 3D points are derived from stereo-like correspondences between the central and adjacent edgemaps as described in Chapter 13. The invariant features are derived from the 2D edgemap of the central image as described in Chapter 10.

An image to be analyzed is processed to produce its edge map and a set of curve segments. The curve segments are used to index the database of model views to produce object-view hypotheses. The matching scheme tests hypotheses generated from the indexing scheme; each hypothesis includes both object identity and approximate pose. Verification is carried out by fitting the $2\frac{1}{2}$D aspect of each candidate model to the observed edgemap. Initially, the pose of the object is set to the pose that would have produced the central

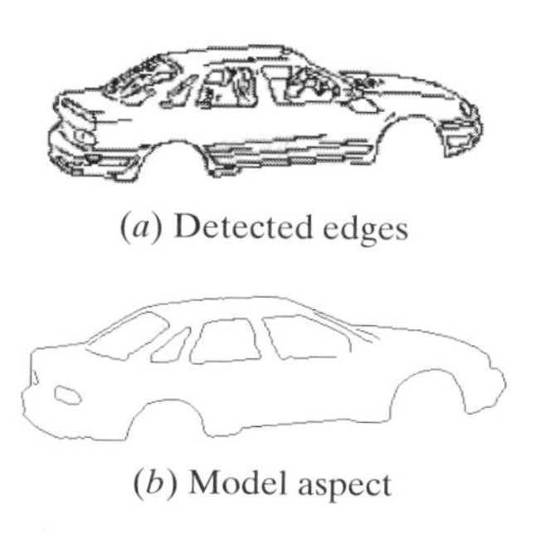

(a) Detected edges

(b) Model aspect

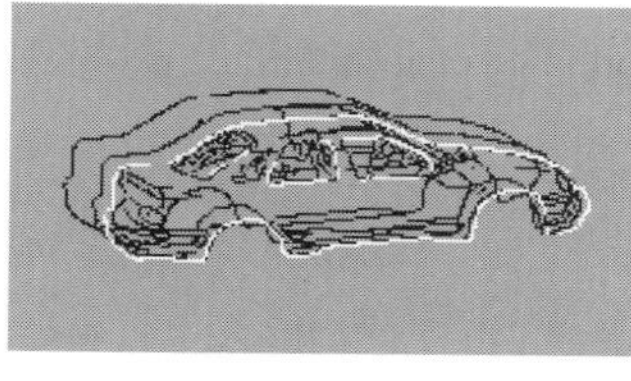

(c) Fitting steps

(d) Final fit

Figure 14.23 Matching edge maps using 1:2 scale s; (a) observed edge map; (b) the model edge map; (c) evolution of convergence in the alignment algorithm; and (d) fitted edge map shown superimposed on the original image. (Example courtesy of Jin-Long Chen.)

aspect that has been hypothesized to match. The projected edge image of that model aspect is compared to the observed edge map. In most cases, they will not align well enough. Thus the matching will proceed by refining pose parameters $\vec{\omega}$ in order to diminish the 2D distance between the projected model edgemap and the observed edgemap. Figure 14.23 illustrates the matching step. The edgemap derived from the input image is shown in (a), and a model pose hypothesized from the indexing step is shown in (b). Several iterations of model boundary generation are shown in (c), and the first acceptable match is shown in (d).

Exercise 14.7

In each case below explain how many object models and how many model aspects would be needed for the application. Also, explain what accuracy of pose might be needed. (a) In a carwash, the automatic machinery needs to reset itself according to the model of car that has entered. (b) In a parking garage, a monitoring and inventory system needs to recognize each car model that enters and exits and record the time of the event. (c) A computer vision system needs to scan an automobile graveyard to record how many wrecks are present and what kind they are.

Exercise 14.8

Make a curvature model aspect of some object and show that it can generate silhouettes of that object under small rotations. For example, consider a torus of major outside diameter

10 and minor diameter of 1. The model aspect is centered at the view perpendicular to the circle of the major diameter. Define a set of 3D points along the model silhouette together with their x and y curvatures. Then show how this silhouette changes under small rotations by creating synthetic images.

14.4.2 Matching Relational Models

As in two-dimensional matching, relational models can be used in 3D object recognition to move away from the geometric and toward the symbolic. Algorithm 14.3 summarizes the basic relational distance matching technique that was described in Chapter 11, simplified to single relations. The exact models and methods used depend on whether the image data is 3D or 2D.

3D Relational Models Three-dimensional relational models are composed of 3D primitives and 3D spatial relationships. Primitives may be volumes, surface patches, or line and/or curve features in 3D space. Generalized cylinders are commonly used as volumetric primitives along with some kind of 3D connection relationship. *Geons,* or *geometric ions,* are volumetric primitives hypothesized to be used in human vision, have also been used in 3D object recognition. Industrial objects may be represented by their planar and cylindrical surfaces and the surface adjacency relationship. Three-dimensional line and curve segments can be used with different kinds of spatial relationships, such as connections, parallel pairs, and collinear pairs.

The *sticks, plates, and blobs* models were designed to describe very rough models of 3D objects and are intended for the description and recognition of complex man-made objects, which are made up of many parts. The parts can have flat or curved surfaces, and they exist in a large variety. Instead of trying to describe each part precisely, as in the surface-edge-vertex models, for rough matching each part can be classified as a *stick,* a *plate,* or a *blob.* Sticks are long, thin parts that have only one significant dimension. Plates are flatish, wide parts with two nearly flat surfaces connected by a thin edge between them. Plates have two significant dimensions. Blobs are parts that have all three significant dimensions. All three kinds of parts are near-convex, so a stick cannot bend very much, the surfaces of a

Determine if two relational descriptions are similar enough to match.

$\mathbf{P}$ is a set of model parts.
$\mathbf{L}$ is a set of possible labels for the parts.
$\mathbf{R_P}$ is a relation over the parts.
$\mathbf{R_L}$ is a relation over the labels.

Find a mapping f from $\mathbf{P}$ to $\mathbf{L}$ that minimizes the error given by $E_S(f) = |\mathbf{R_P} \circ f - \mathbf{R_L}| + |\mathbf{R_L} \circ f^{-1} - \mathbf{R_P}|$ using an interpretation tree, discrete or probabilistic relaxation, or other methods described in Chapter 11.

Algorithm 14.3 The Basic Relational Distance Matching Technique.

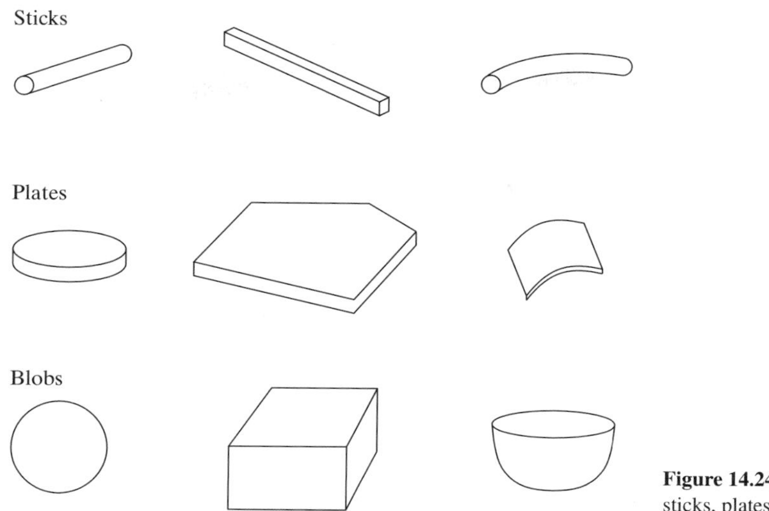

Figure 14.24 Several examples each of sticks, plates, and blobs.

plate cannot fold very much, and a blob can be bumpy, but cannot have large concavities. Figure 14.24 shows several examples of sticks, plates, and blobs.

A sticks-plates-and-blobs model describes how the sticks, plates, and blobs are put together to form an object. These descriptions are also rough; they cannot specify the physical points where two parts join. A stick has two logical end points, a logical set of interior points, and a logical center of mass that can be specified as connection points. A plate has a set of edge points, a set of surface points, and a center of mass. A blob has a set of surface points and a center of mass. Only this minimal information can be used in the object models.

The relational model for a sticks-plates-and-blobs model is a good example of a fairly detailed symbolic object model that has been used successfully in symbolic object recognition. It consists of five relations. The unary SIMPLE PARTS relation is a list of the parts of the object. Each part has several descriptive attributes including its type (stick, plate, or blob) and may also include numeric information pertaining to the size and, or shape of the part. The CONNECTS/SUPPORTS relation contains some of the most important information on the structure of the object. It consists of 6-tuples of the form $(s_1, s_2, SUPPORTS, HOW)$. The components s_1 and s_2 are simple parts, SUPPORTS is true if s_1 supports s_2 and false otherwise, and HOW describes the connection type of s_1 and s_2.

The other four relations express constraints. The TRIPLE CONSTRAINT relation has 4-tuples of the form $(s_1, s_2, s_3, SAME)$ where simple part s_2 touches both s_1 and s_3, and SAME is true if s_1 and s_3 touch s_2 on the same end (or surface) of s_2 and false otherwise. The PARALLEL relation and the PERPENDICULAR relation have pairs of the form (s_1, s_2) where simple parts s_1 and s_2 are parallel (or perpendicular) in the model. Figure 14.25 illustrates the sticks-plates-and-blobs model of a prototype chair object. All chairs with similar relations should match this model, regardless of the exact shapes of the parts.

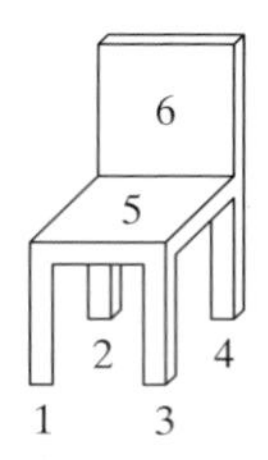

SIMPLE-PARTS

PART#	TYPE
1	Stick
2	Stick
3	Stick
4	Stick
5	Plate
6	Plate

CONNECTS-SUPPORTS

SP1	SP2	SUPPORTS	HOW
1	5	True	end-edge
2	5	True	end-edge
3	5	True	end-edge
4	5	True	end-edge
5	6	True	edge-edge

TRIPLES

SP1	SP2	SP3	SAME
1	5	2	True
1	5	3	True
1	5	4	True
1	5	6	False
2	5	3	True
2	5	4	True
2	5	6	False
3	5	4	True
3	5	6	False
4	5	6	False

PARALLEL

SP1	SP2
1	2
1	3
1	4
2	3
2	4
3	4

PERPENDICULAR

SP1	SP2
1	5
2	5
3	5
4	5
5	6

Figure 14.25 The full relational structure of the sticks-plates-and-blobs model of a chair object.

Exercise 14.9: Sticks-plates-and-blobs models.

Draw a picture of a simple polyhedral desk object and construct a full relational sticks-plates-and-blobs model for the object.

View-Class Relational Models When the data consists of 2D images, view class models can be used instead of full 3D object models. Training data derived either synthetically or from a set of real images of the object can be used in the construction of these models. Depending on the class of objects, a set of useful 2D features are extracted from images of the object. The features extracted from each training image are used to produce a relational description of that view of the object. The relational descriptions are then clustered to form the view-classes of the object. Each view class is represented by a combined relational description that includes all the features that have been detected in all views of that view class. The combined relational description is the relational model of the view class. Typically, an object will have on the order of five view classes, each with its own relational description. The view-class models can be used for full relational matching, which is expensive when there are many different models in the database or for relational indexing, as introduced in Chapter 11. For our example here, we take the relational indexing approach.

The RIO object recognition system, developed by Mauro Costa at the University of Washington, recognizes 3D objects in multi-object scenes from 2D images. Images are taken in pairs, with the camera fixed for the pair. One image has the light source at the left and the other has the light source at the right. The two images are used to determine which regions are shadows and highlights, so that a high-quality edge image of just the objects can be obtained. The edge image is used to obtain straight-line and circular-arc segments from which the recognition features are constructed. Figure 14.26 shows a sample left and right image pair and the extracted edge image obtained. Figure 14.27 shows the straight lines and circular arcs extracted from the edge image.

RIO objects can have planar, cylindrical, and threaded surfaces. This leads to a number of useful high-level features. The ten features employed by RIO are: ellipses, coaxial arcs (two, three, and multiple); parallel pairs of line segments (both close and far); triples of line segments (U-shaped and Z-shaped); and L-junctions, Y-junctions, and V-junctions. Figure 14.28 shows some of the features constructed from the line segments and arcs of Figure 14.27. The line features include two L-junctions and a pair of parallel lines. The arc cluster shows three coaxial arcs. Note that not every line segment or arc segment becomes

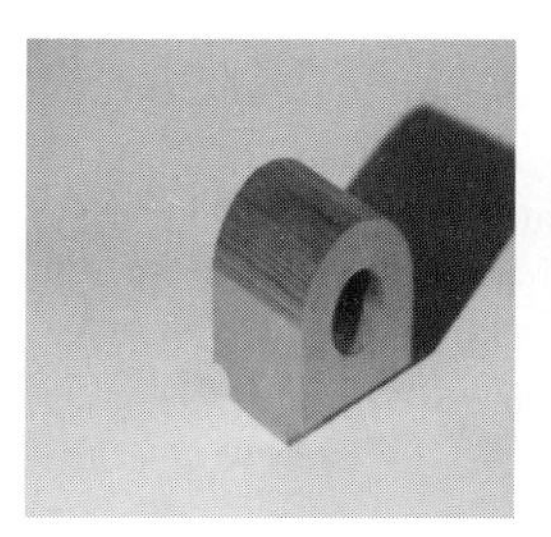
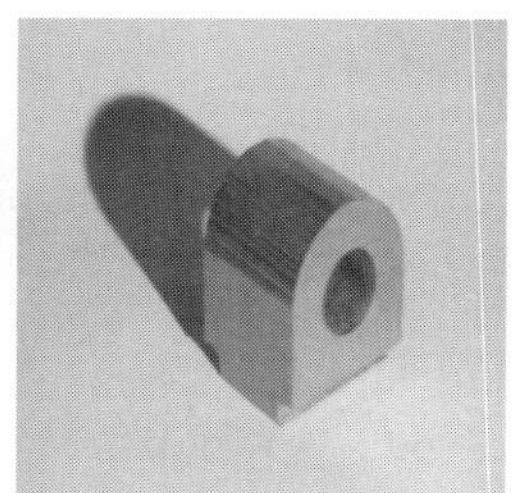
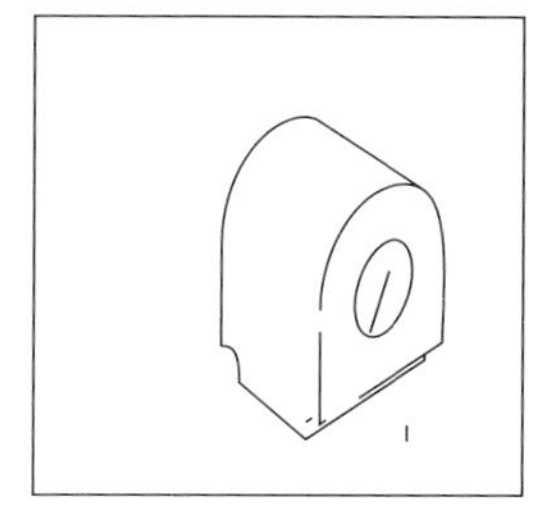

Figure 14.26 The left and right images of an industrial object and the edges extracted through an image processing step that removes shadows and most highlights. (Courtesy of Mauro Costa.)

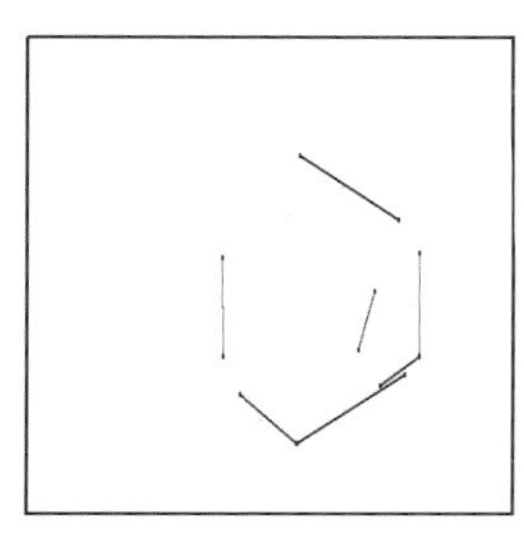
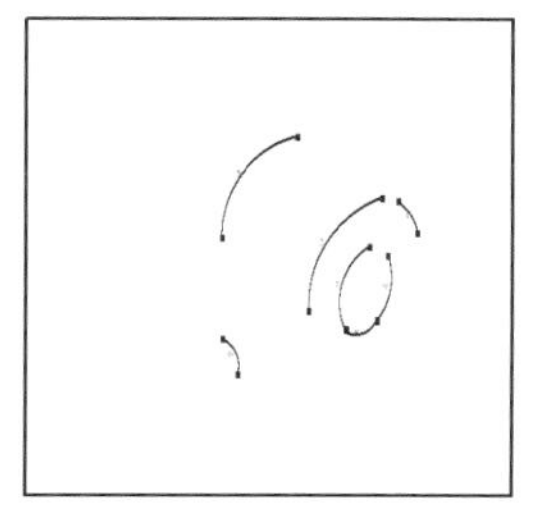

Figure 14.27 The straight line segments and circular arcs extracted from the edge image of Figure 14.26. (Courtesy of Mauro Costa.)

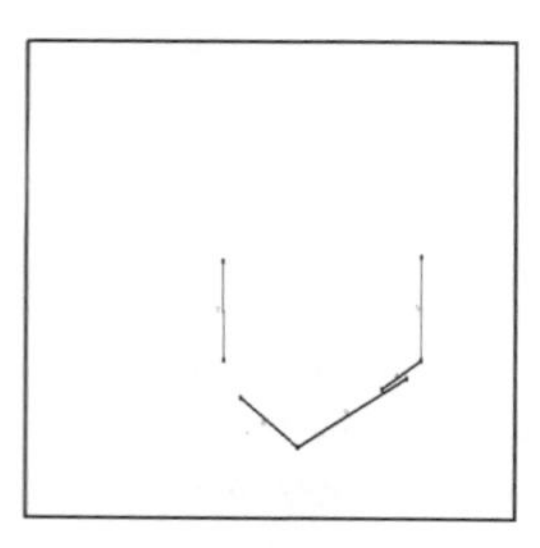
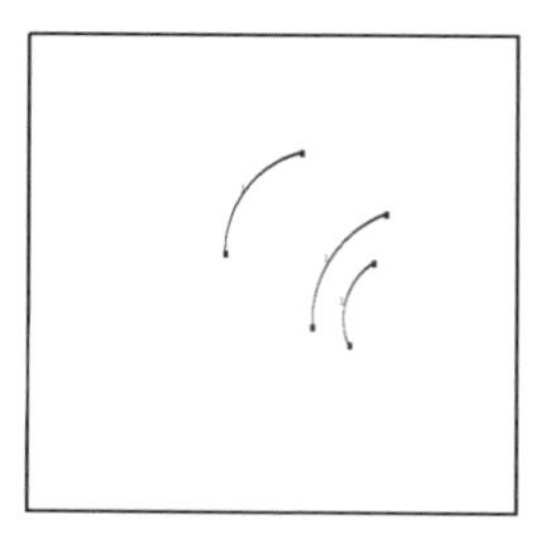
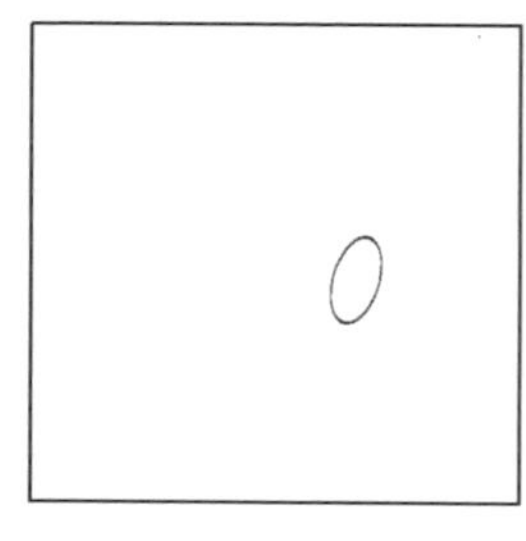

Figure 14.28 The line features, arc features and ellipse features constructed from the lines and arcs of Figure 14.27. (Courtesy of Mauro Costa.)

a part of the final features used in matching. Figure 14.29 illustrates the entire set of RIO features.

In addition to the labeled features, RIO uses labeled binary relations over the features to recognize objects. The relationships employed by RIO are: sharing one arc, sharing one line, sharing two lines, coaxiality, proximity at extremal points, and encloses/enclosed-by, as shown in Figure 14.30.

The structural description of each model-view is a graph structure whose nodes are the feature types and whose edges are the relationship types. For use in the relational indexing procedure, the graph is decomposed into a set of 2-graphs, each having two nodes and a relationship between them. Figure 14.31 shows one model-view of a hexnut object, a partial full-graph structure representing three of its features and their relationships, and the 2-graph decomposition.

Relational indexing is a procedure that matches an unknown image to a potentially large database of object-view models, producing a small set of hypotheses as to which objects are present in the image. There is an off-line preprocessing phase to set up the data structures and an on-line matching phase. The off-line phase constructs a hash table that is used by the on-line phase. The indices to the hash table are 4-tuples representing 2-graphs of a model-view of an object. The components of the 4-tuple are the types of the two nodes and the types of the two relationships. For example, the 4-tuple (ellipse, far parallel pair, enclosed-by, encloses) means that the 2-graph represents an ellipse feature and a far parallel pair feature; the ellipse is enclosed by the parallel pair of line segments and the parallel pair thus encloses the ellipse. Since most of the RIO relations are symmetric, the two relationships are often the same. For instance, the 4-tuple (ellipse, coaxial arc cluster, share an arc,share an arc) describes a relationship where an ellipse and a coaxial arc cluster share an arc segment. The symbolic components of the 4-tuples are converted to numbers for hashing. The preprocessing stage goes through each model-view in the database, encodes each of its 2-graphs to produce a 4-tuple index, and stores the name of the model-view and associated information in a list in the selected bin of the hash table.

Once the hash table is constructed, it is used in on-line recognition. Also used is a set of accumulators for voting, one for each possible model-view in the database. When a scene is analyzed, its features are extracted and a relational description in the form of a set of 2-graphs is constructed. Then, each 2-graph in the description is encoded to produce

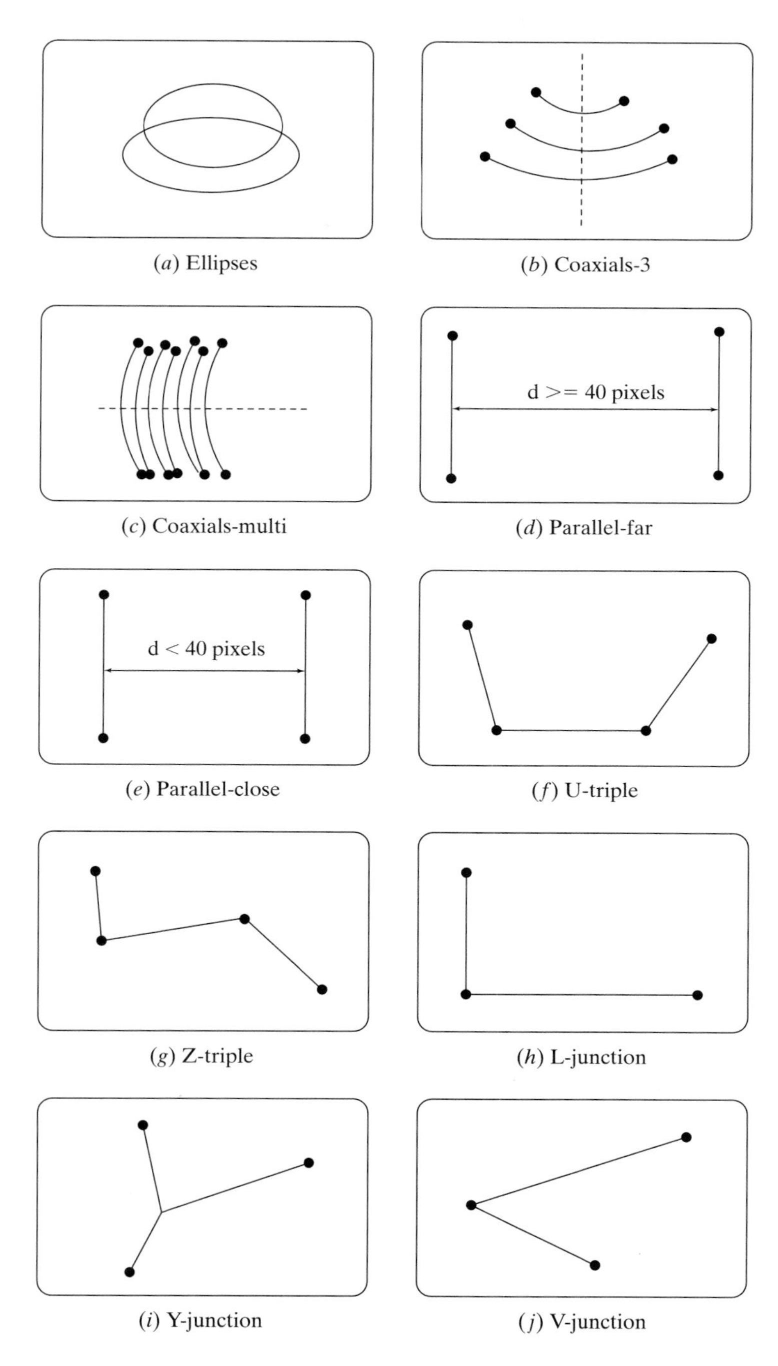

Figure 14.29 Features used in the RIO system. (Courtesy of Mauro Costa.)

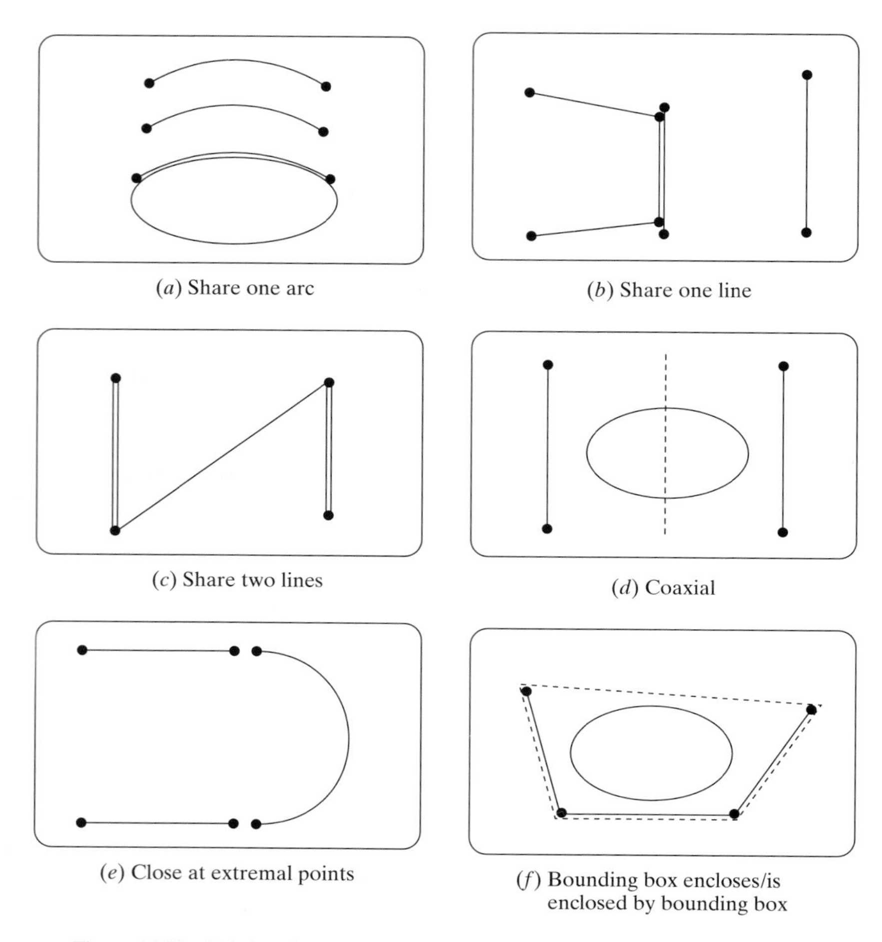

(*a*) Share one arc

(*b*) Share one line

(*c*) Share two lines

(*d*) Coaxial

(*e*) Close at extremal points

(*f*) Bounding box encloses/is enclosed by bounding box

Figure 14.30 Relations between sample pairs of features. (Courtesy of Mauro Costa.)

an index with which to access the hash table. The list associated with the selected bin is retrieved; it consists of all model-views that have this particular 2-graph. A vote is then cast for each model-view in the list. This is performed for all the 2-graphs of the image. At the end of the procedure, the model-views with the highest votes are selected as hypotheses. Figure 14.32 illustrates the on-line recognition process. The 2-graph shown in the figure is converted to the numeric 4-tuple (1, 2, 9, 9) which selects a bin in the hash table. That bin is accessed to retrieve a list of four models: M_1, M_5, M_{23}, and M_{81}. The accumulators of each of these model-views are incremented.

After hypotheses are generated, verification must be performed. The relational indexing step provides correspondences from 2D image features to 2D model features in a model-view. These 2D model features are linked with the full 3D model features of the hypothesized object. The RIO system performs verification by using corresponding 2D-3D

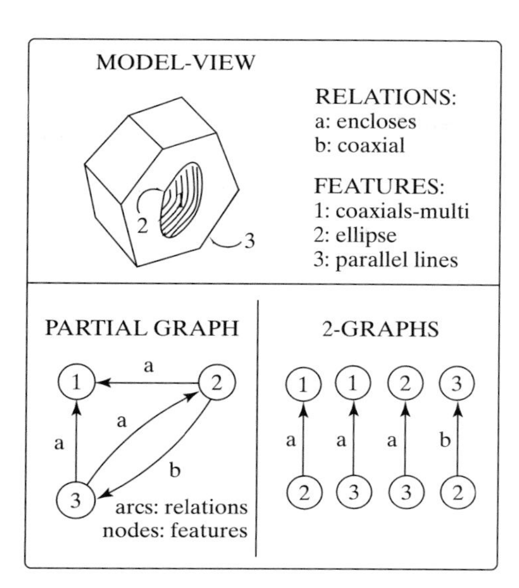

Figure 14.31 Sample graph and corresponding 2-graphs for the *hexnut* object. (Courtesy of Mauro Costa.)

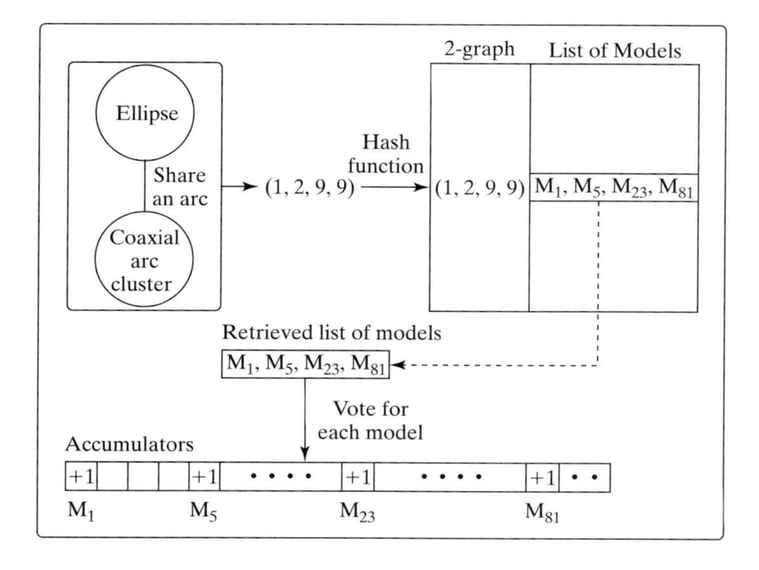

Figure 14.32 Voting scheme for relational indexing. (Courtesy of Mauro Costa.)

point pairs, 2D-3D line segment pairs, and 2D ellipse-3D circle pairs to compute an estimate of the transformation from the 3D model of the hypothesized object to the image. Line and arc segments are projected to the image plane and a distance is computed that determines if the verification is successful or if the hypothesis is incorrect. Figures 14.33 and 14.34 show a sample run of the RIO system. Figure 14.33 shows the edge image from a multi-object scene and the line features, circular arc features, and ellipses detected. Figure 14.34 shows

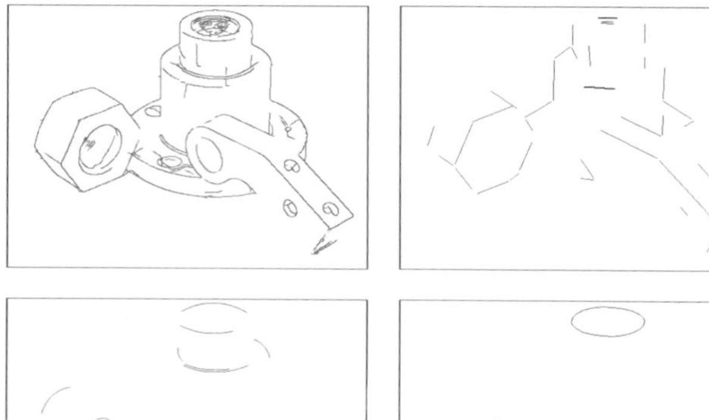

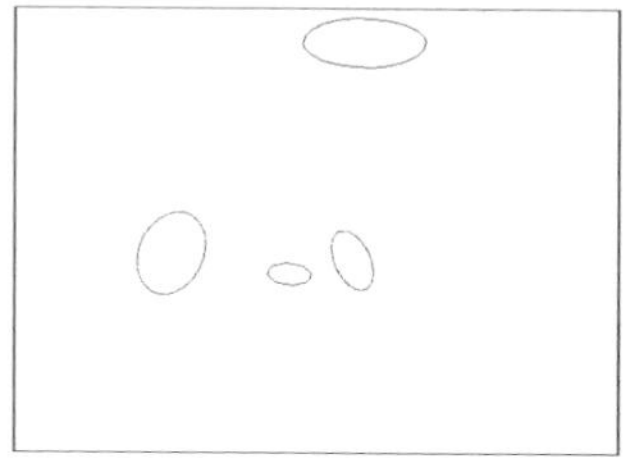

Figure 14.33 A test image and its line features, circular arc features, and ellipse features. (Courtesy of Mauro Costa.)

Figure 14.34 An incorrect hypothesis (upper left) and three correct hypotheses. (Courtesy of Mauro Costa.)

an incorrect hypothesis produced by the system, which was ruled out by the verification procedure and three correct hypotheses, which were correctly verified. The RIO pose estimation procedure for point correspondences was given in Chapter 13. Figure 14.35 shows a block diagram of the whole RIO system.

Exercise 14.10: Relational indexing.

Write a program that implements relational indexing for object matching. The program should use a stored library of object models, each represented by a set of 2-graphs. The

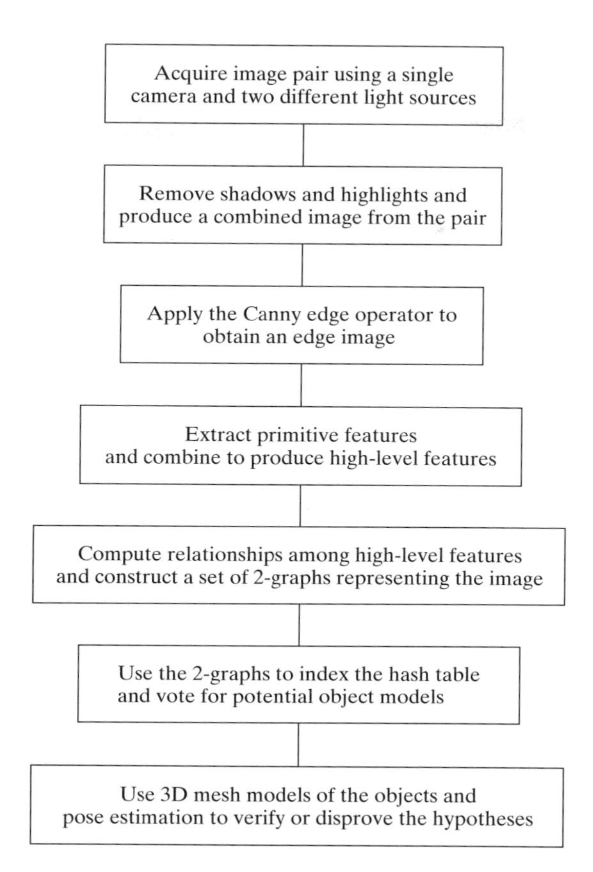

Figure 14.35 Flow diagram of the RIO object recognition system.

input to the recognition phase is a representation of a multi-object image, also in terms of a set of 2-graphs. The program should return a list of each model in the database that has at least 50 percent of its 2-graphs in the image.

14.4.3 Matching Functional Models

Geometric models give precise definitions of specific objects; a CAD model describes a single object with all critical points and dimensions spelled out. Relational models are more general in that they describe a class of objects, but each element of that class must have the same relational structure. For example, a chair might be described as having a back, a seat, and four legs attached underneath and at the corners of the seat. Another chair that has a pedestal and base instead of four legs would not match this description. The function-based approach to object recognition takes this a step further. It attempts to define classes of objects through their function. Thus, a chair is something that a human would sit on and may have many different relational structures, which all satisfy a set of functional constraints.

Function-based object recognition was pioneered by Stark and Bowyer (1996) in their Generic Object Recognition Using Form and Function (GRUFF) system. GRUFF contains three levels of knowledge:

1. the category hierarchy of all objects in the knowledge base,
2. the definition of each category in terms of functional properties, and
3. the knowledge primitives upon which the functional definitions are based

Knowledge Primitives Each knowledge primitive is a parametrized procedure that implements a basic concept about geometry, physics, or causation. A knowledge primitive takes a portion of a 3D shape description as input and returns a value that indicates how well it satisfies specific requirements. The six GRUFF knowledge primitives define the concepts of:

- relative orientation
- dimensions
- proximity
- stability
- clearance
- enclosure

The *relative orientation* primitive is used to determine how well the relative orientation of two surfaces satisfies some desired relationship. For example, the top surface of the seat of a chair should be approximately perpendicular to the adjacent surface of the back. The *dimensions* primitive performs dimensions tests for six possible dimension types: width, depth, height, area, contiguous surface, and volume. In most objects the dimensions of one part of the object constrain the dimensions of the other parts. The *proximity* primitive checks for qualitative spatial relationships between elements of an object's shape. For example, the handle of a pitcher must be situated above the average center of mass of an object to make it easy to lift.

The stability primitive checks that a given shape is stable when placed on a supporting plane in a given orientation and with a possible force applied. The *clearance* primitive checks whether a specified volume of space between parts of the object is clear of obstructions. For example, a rectangular volume above the seat must be clear for a person to sit on it. Finally, the *enclosure* primitive tests for required concavities of the object. A wine goblet, for instance, must have a concavity to hold the wine.

Functional Properties The definition of a functional object class specifies the functional properties it must have in terms of the Knowledge Primitives. The GRUFF functional categories that have been used for objects in the classes *furniture, dishes,* and *handtools* are defined by four possible templates:

- provides stable X
- provides X surface
- provides X containment
- provides X handle

where X is a parameter of the template. For example, a chair must provide stable support and a sittable surface for the person who will sit on it. A soup bowl must provide stable containment for the soup it will contain. A cup must contain a suitable handle that allows it to be picked up and fits the dimensions of its body.

The Category Hierarchy GRUFF groups all categories of objects into a category tree that lists all the categories the system can currently recognize. At the top level of the tree are very generic categories such as furniture and dishes. Each succeeding level goes into more detail; for example, furniture has specific object classes: chair, table, bench, bookshelf, and bed. Even these object classes can be divided further; chairs can be conventional chairs, lounge chairs, balans chairs, and highchairs, etc. Figure 14.36 shows portions of the GRUFF category definition tree.

Rather than recognizing an object, GRUFF uses the function-based definition of an object category to reason about whether an observed object instance (in range data) can function as a member of that category. There are two main stages of this function-based analysis process: the preprocessing stage and the recognition stage. The preprocessing stage is category independent; all objects are processed in the same manner. In this stage the 3D data is analyzed and all potential functional elements are enumerated. The recognition stage uses these elements to construct indexes that are used to rank order the object categories. An index consists of a functional element plus its area and volume. Those categories that would be impossible to match based on the index information are pruned from the search. The others are rank ordered for further evaluation. For each class hypothesis, first each

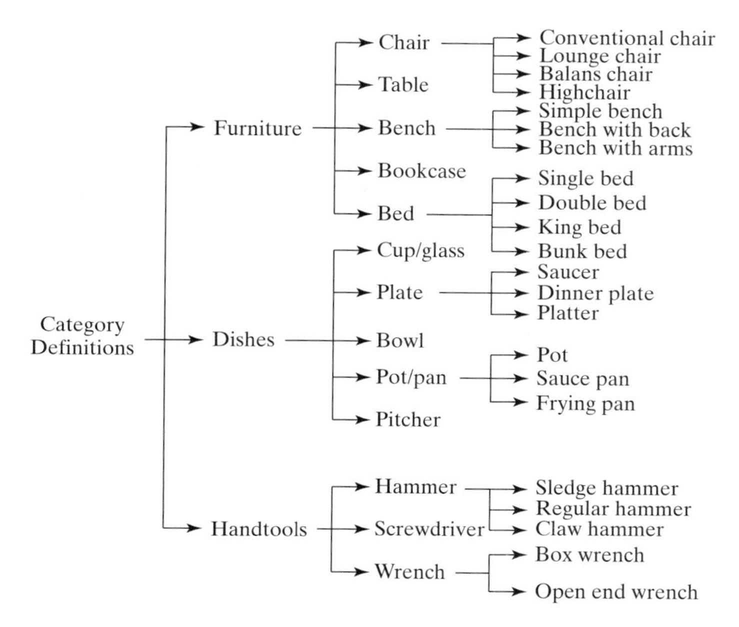

Figure 14.36 Portions of the GRUFF Category Definition Tree. (Courtesy of Louise Stark and Kevin Bowyer.)

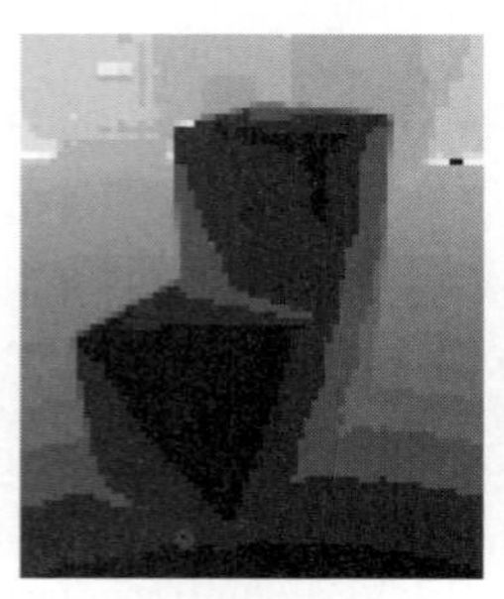
Range image

Segmented range image

Figure 14.37 Input data for the GRUFF system. (Courtesy of Louise Stark and Kevin Bowyer.)

of its knowledge primitives is invoked to measure how well a functional element from the data fits its requirements. Each knowledge primitive returns an evaluation measure. These are then combined to form a final association measure that describes how well the whole set of function elements from the data matches the hypothesized object category. Figure 14.37 shows a typical input to the GRUFF system, and Figure 14.38 shows a portion of the functional reasoning in the analysis of that data.

Exercise 14.11: Functional object recognition.

Consider two tables: one with 4 legs at the 4 corners and the other having a pedestal. What similarities between these two tables would a functional object recognition system use to classify them both as the same object?

14.4.4 Recognition by Appearance

In most 3D object recognition schemes, the model is a separate entity from the 2D images of the object. Here we examine the idea that an object can be learned by memorizing a number of 2D images of it: Recognition is performed by matching the sensed image of an unknown object to images in memory. Object representation is kept at the *signal level;* matching is done by directly comparing intensity images. Higher level features—possibly from parts extraction—are not used, and thus possibly time consuming and complex programming that is difficult to test is not needed. Several problems with this signal level approach are addressed next. The simplicity of appearance-based recognition methods have allowed them to be trained and tested on large sets of images and some impressive results have been obtained. Perhaps, the most important results have been obtained with human face recognition, which we will use here as a clarifying example.

A coarse description of recognition-by-appearance is as follows:

- During a training, or learning, phase a database of labeled images is created. $\mathbf{DB} = \{\langle I_j[\,], L_j\rangle_{j=1,k}\}$ where I_j is the *jth* training image and L_j is its label.
- An unknown object is recognized by comparing its image I_u to those in the database and assigning the object the label L_j of the closest training image I_j. The closest

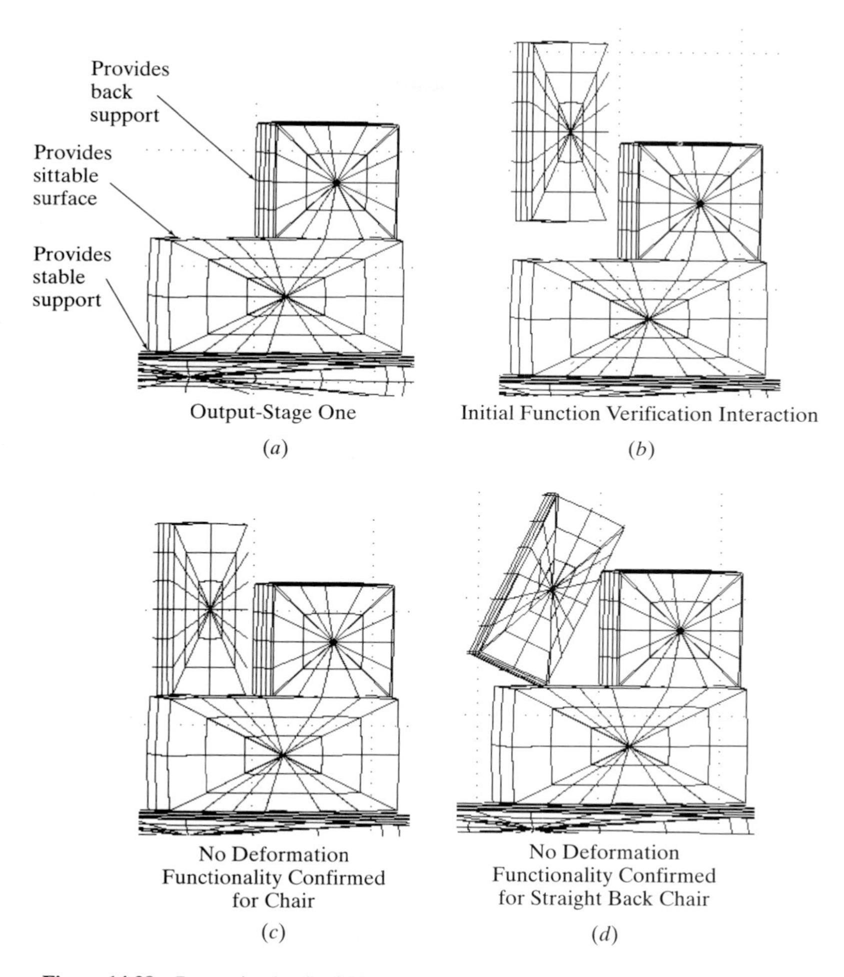

Figure 14.38 Processing by the GRUFF system. (Courtesy of Louise Stark and Kevin Bowyer.)

training image I_j can be defined by the minimum Euclidean distance $\| I_u[\,] - I_j[\,] \|$ or by the maximum dot product $I_u \circ I_j$, both of which were defined in Chapter 5.

There are, of course, complications in each step that must be addressed.

- Training images must be representative of the instances of objects that are to be recognized. In the case of human faces (and most other objects), training must include changes of expression, variation in lighting, and small rotations of the head in 2D and 3D.
- The object must be well-framed; the position and size of all faces must be roughly the same. Otherwise, a search of size and position parameters is needed.

- Since the method does not separate object from background, background will be included in the decisions and this must be carefully considered in training.
- Even if our images are as small as 100×100, which is sufficient for face recognition, the dimension of the space of all images is 10,000. It is likely that the number of training samples is much smaller than this; thus, some method of *dimensionality reduction* should be used.

While continuing, the reader should consider the case of discriminating between two classes of faces—those with glasses and those without them; or, between cars with radio antennas and those without them. Can these differences be detected among all the other variations that are irrelevant?

We now focus on the important problem of reducing the number of signal features used to represent our objects. For face recognition, it has been shown that dimensionality can be reduced from 100×100 to as little as 15 yet still supporting 97 percent recognition rates. Chapter 5 discussed using different bases for the space of $R \times C$ images, and showed how an image could be represented as a sum of meaningful images, such as step edges, ripple, etc. It was also shown that image energy was just the sum of the squares of the coefficients when the image was represented as a linear combination of orthonormal basis images.

Basis Images for the Set of Training Images Suppose for the present that a set of orthonormal basis images **B** can be found with the following properties.

1. $\mathbf{B} = \{F_1, F_2, \ldots, F_m\}$ with m much smaller than $N = R \times C$.
2. The average quality of representing the image set using this basis is satisfactory in the following sense. Over all the M images I_j in the training set, we have

$$I_j^m = a_{j1}F_1 + a_{j2}F2 + \cdots + a_{jm}F_m$$

and

$$\sum_{j=1}^{m} \left(\|I_j^m - I_j\|^2 / \|I_j\|\right)^2 < P\%.$$

I_j^m is the approximation of original image I_j using a linear combination of just the m basis images.

The top row of Figure 14.39 shows six training images from one of many individuals in the database made available by the Weizmann Institute. The middle row of the figure shows four basis images that have been derived for representing this set of faces; the leftmost is the mean of all training samples. The bottom row of the figure shows how the original six face images would appear when represented as a linear combination of just the four basis vectors. Several different research projects have shown that perhaps $m = 15$ or $m = 20$ basis images are sufficient to represent a database of face images (such as 3,000 face images in one of Pentland's studies (1986)), so that the average approximation of I_j^m is within 5 percent of I_j. Therefore, matching using the approximation will yield almost the same results as matching using the original image. It is important to emphasize that for the database illustrated by Figure 14.39, **each training image can be represented in memory by only four numbers,** which enables efficent comparison with unknown images. Provided

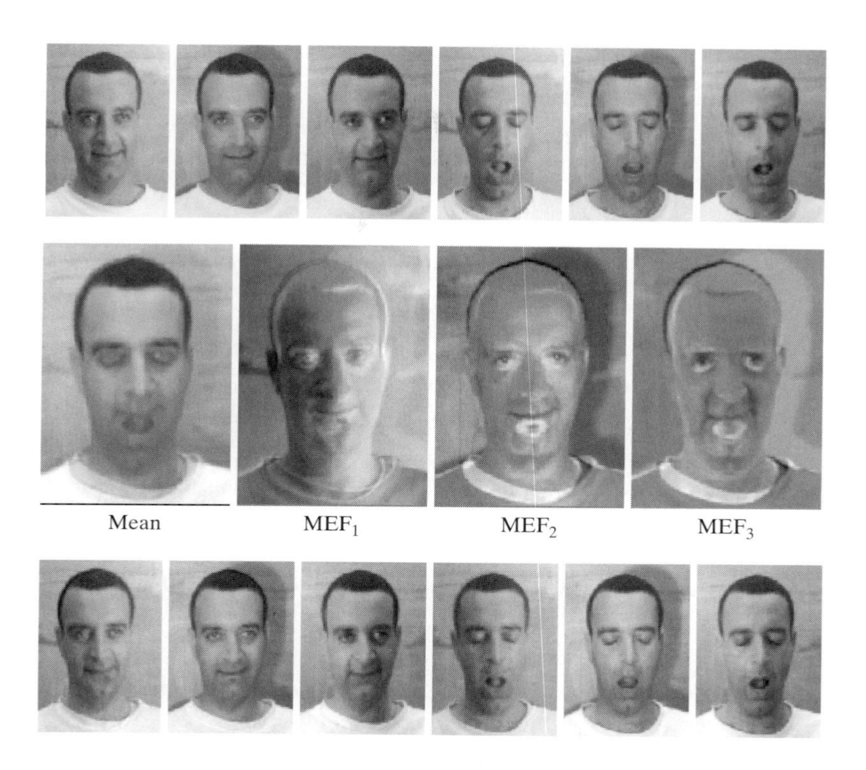

Figure 14.39 (top row) Six training images from one of many individuals in a face image database; (middle row) average training image and three most significant eigenvectors derived from the scatter matrix; and (bottom row) images of the top row represented as a linear combination of only the four images in the middle row. (Database of images courtesy of Yael Moses of The Weizmann Institute; processed images courtesy of John Weng.)

that the four basis vectors are saved in memory, a close approximation to the original face image can be regenerated when needed. (Note that the first basis vector is the mean of the original face set and not actually one of the orthonormal set.)

Computing the Basis Images Existence of the basis set **B** allows great compression in memory and speedup in computations because m is much smaller than N, the number of pixels in the original image. The basis images F_i are called *principal components* of the set of training samples. Algorithm 14.4 sketches recognition-by-appearance using these principal components. It has two parts: an offline training phase and an online recognition phase. The first step in the training phase is to compute the mean of the training images and use them to produce a set Φ of difference images, each being the difference of a training image from the mean image. If we think of each difference image Φ_i as a vector of N elements, Φ becomes an array of R rows and C columns. The next step is to compute the covariance matrix Σ_Φ of the training images. By definition, $\Sigma_\Phi[i, i]$ is the variance of

Offline Training Phase:
Input a set I of M labeled training images and produce a basis set B and a vector of coefficients for each image.

$\mathbf{I} = \{I_1, I_2, \ldots, I_M\}$ is the set of training images. (input)
$\mathbf{B} = \{F_1, F_2, \ldots, F_m\}$ is the set of basis vectors. (output)
$\mathbf{A_j} = [a_{j1}, a_{j2}, \ldots, a_{jm}]$ is the vector of coefficients for image I_j. (output)

1. $\mathbf{I_{mean}} = mean(\mathbf{I})$.
2. $\Phi = \{\Phi_i | \Phi_i = I_i - \mathbf{I_{mean}}\}$, the set of difference images
3. Σ_Φ = the covariance matrix obtained from Φ.
4. Use the principal components method to compute eigenvectors and eigenvalues of Σ_Φ. (see text)
5. Construct the vector **B** as the basis set by selecting the most significant m eigenvectors; start from the largest eigenvalue and continue in decreasing order of the eigenvalues to select the corresponding eigenvectors.
6. Represent each training image I_j by a linear combination of the basis vectors: $I_j^m = a_{j1}F_1 + a_{j2}F_2 + \cdots + a_{jm}F_m$

Online Recogniton Phase:
Input the set of basis vectors **B**, the database of coefficient sets $\{\mathbf{A_j}\}$, and a test image $\mathbf{I_u}$. Output the class label of $\mathbf{I_u}$.

1. Compute vector of coefficients $\mathbf{A_u} = [a_{u1}, a_{u2}, \ldots, a_{um}]$ for $\mathbf{I_u}$;
2. Find the h nearest neighbors of vector $\mathbf{A_u}$ in the set $\{\mathbf{A_j}\}$;
3. Decide the class of $\mathbf{I_u}$ from the labels of the h nearest neighbors (possibly reject in case neighbors are far or inconsistent in labels);

Algorithm 14.4 Recognition-by-Appearance using a Basis of Principal Components.

the *ith* pixel, while $\Sigma_\Phi[i, j]$ is the covariance of the *ith* and *jth* pixels, over all the training images. Since we have already computed the mean and difference images, the covariance matrix is defined by

$$\Sigma_\Phi = \Phi^T \Phi \tag{14.5}$$

The size of this covariance matrix is very large, $N \times N$, where N is the number of pixels in an image, typically 256×256 or even 512×512. So the computation of eigenvectors and eigenvalues in the next step of the algorithm would be extremely time-consuming if Σ_Φ were used directly. [See *Numerical Recipes in C* for the principal components algorithm (Vetterling, 1992).] Instead, we can compute a related matrix Σ'_Φ given by

$$\Sigma'_\Phi = \Phi \Phi^T \tag{14.6}$$

which is much smaller ($m \times m$). The eigenvectors and eigenvalues of Σ'_{Φ} are related to those of Σ_{Φ} as follows:

$$\Sigma_{\Phi} F = \lambda F \tag{14.7}$$

$$\Sigma'_{\Phi} F' = \lambda F' \tag{14.8}$$

$$F = \Phi^T F' \tag{14.9}$$

where λ is the vector of eigenvalues of Σ_{Φ}, F is the vector of eigenvectors of Σ_{Φ}, and F' is the vector of eigenvectors of Σ'_{Φ}.

The methods of principal components analysis discussed here have produced some impressive results in face recognition (consult the references by Kirby and Sirovich (1990), Turk and Pentland (1991), and Swets and Weng (1996)). Skeptics might argue that this method is unlikely to work for images with significant high frequency variations because autocorrelation will drop fast with small shifts in the image, thus stressing the object framing requirement. Picture functions for faces do not face this problem. Swets and Weng (1996) have shown good results with many (untextured) objects other than faces as have Murase and Nayar (1995), who were actually able to interpolate 3D object pose to an accuracy of two degrees using a training base of images taken in steps of ten degrees.

Turk and Pentland (1991) gave solutions to two of the problems noted above. First, they used motion techniques, as in Chapter 9, to segment the head from a video sequence—this enabled them to frame the face and also to normalize image size. Second, they reweighted the image pixels by filtering with a broad Gaussian that dropped the peripheral background pixels to near zero while preserving the important center face intensities.

Exercise 14.12

Obtain a set of 10 face images and 10 landscapes such that all images have the same dimensions $R \times C$. Compute the Euclidean distance between all pairs of images and display the distances in a 20×20 upper triangular matrix. Do the faces cluster close together? The landscapes? What is the ratio of the closest to farthest distance? Is the Euclidean distance promising for retrieval from an image database? Explain.

Exercise 14.13

Let I_u be an image of an unknown object and let $\mathbf{B} = \{\langle I_j, L_j \rangle\}$ be a set of labeled training images. Assume that all images are normalized so that $\| I_j[\,] \| = 1$. (a) Show that $\| I_u - I_j \|$ is minimized when $I_u \circ I_j$ is maximized. (b) Explain why the result is not true without the assumption that all images have unit size.

Better Discrimination and Faster Search of Memory The methods of prinipal components analysis allow a subspace of training patterns to be represented in a compact form. The basis that best represents the training data, computed in Algorithm 14.4, has been called the set of *most expressive features (MEFs)*. The work of John Weng has demonstrated that, while the most expressive features *represent* the subspace of training images optimally, they need not represent well the *differences between images* in different classes. Weng introduced the use of the *most discriminating features (MDFs)*, which can be derived from discriminant analysis. MDFs focus on image variance that can differentiate objects in different classes. Figure 14.40 contrasts MEFs with MDFs. The original data coordinates are (x_1, x_2). y_1 is the direction of maximum variance and y_2 is orthogonal to y_1; thus, coordinates y_1, y_2 are MEFs. The original classes of vectors are represented by the ellipses with major and minor axes aligned with y_1, y_2. (Recall that an algorithm for finding these axes in the 2D case was first presented in Chapter 3.) Thresholds on either y_1 or y_2 do not discriminate well between the two classes. MDF axes z_1, z_2, computed from discriminant analysis, allow perfect separation of the training samples based on a threshold on z_1.

A second improvement made by Weng and his colleagues to the *eigenspace recognition-by-appearance* approach is the development of a search tree construction procedure that provides $O(log_2 S)$ search time for finding the nearest neighbors in a database of S training samples. Recall that decision trees for object classification were introduced in Chapter 4. At each decision point in the tree, an unknown image is projected onto the most discriminating subspace needed to make a decision on which branch or branches to take next. The MDFs used at different nodes in the decision tree vary with the training samples from which they were derived and are tuned to the particular splitting decisions that are needed. It is best for the interested reader to consult the references to obtain more details on this recently developed theory.

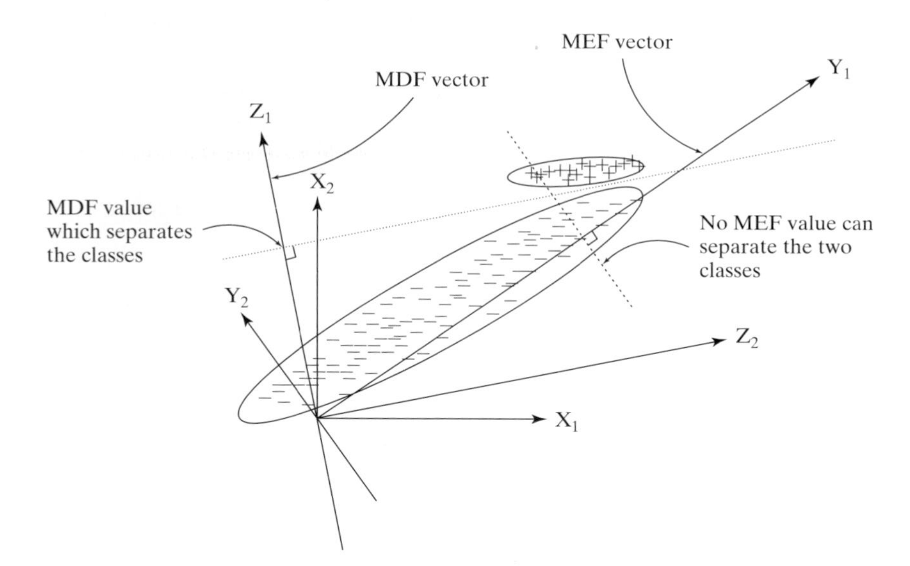

Figure 14.40 The most expressive features determined by the eignevectors of the scatter matrix represent the data well, but may not represent the differences between classes well. Discriminant analysis can be used to find subspaces that emphasize differences between classes. (Figure contributed by J. Swets and J. Weng.)

Exercise 14.14

(a) Obtain a set of 300 human face images and treat each one as a single vector of $R \times C$ coordinates. (b) Compute the scatter matrix and mean image from these 300 samples. (c) From the results in (b) compute the m largest eigenvalues of the scatter matrix and the corresponding m eigenvectors so that 95 percent of the energy in the scatter matix is represented. (d) Select 5 of the original faces at random; represent each as a linear combination of the m best eigenvectors from (c). (e) Display each of the 5 approximations derived in (d) and compare them to the original image.

14.5 REFERENCES

Mesh-models come from computer graphics, where they are usually called polygon meshes. The Foley and others (1996) graphics text is a good reference for this subject. The surface-edge-vertex representation was introduced in the VISIONS system at University of Massachusetts in the 1970s. The structure shown in this text comes from the more recent work of Camps (1992). Generalized cylinder models were first proposed by Binford and utilized by Nevatia and Binford (1977) who worked with range data. The more recent article by Rom and Medioni (1993) discusses the computation of cylinders from 2D data. Octrees were originally proposed by Hunter (1978) and developed further by Jackins and Tanimoto (1980). They are discussed in detail in Samet's book (1990). Our discussion of superquadrics comes mostly from the work of Gupta, Bogoni, and Bajcsy (1989), and the left ventricle illustrations are from the more recent work of Park, Metaxas, and Axel (1996), which is also discussed in the section on deformable models.

The introduction of the view-class concept is generally credited to Koenderink and van Doorn (1979). Camps and others (1992), Pulli (1996), and Costa (1995) used view-class models to recognize three-dimensional objects. Matching by alignment was introduced by Lowe (1987) and thoroughly analyzed by Huntenlocher and Ullmann (1990). The 3D-3D alignment discussed here comes from the work of Johnson and Hebert (1998), while the 2D-3D discussion comes from the work of Pulli and Shapiro (1996). The treatment of recognition-by-alignment of smooth objects was taken from the work of Jin-Long Chen and Stockman (1996), and is related to the original work of Basri and Ullman (1988). Matching sticks-plates-and-blobs models was described in the work of Shapiro and others (1984). Relational matching in general was discussed in Shapiro and Haralick (1981, 1985). Relational indexing can be found in the work of Costa and Shapiro (1995). Our discussion on functional object recognition comes from the work of Stark and Bowyer (1996).

Kirby and Sirovich (1990) approached the problem of face image compression, which Turk and Pentland (1991) then adopted for more efficient recognition of faces. Swets and Weng (1996) developed a general learning system, called SHOSLIF, which improved upon the principal components approach by using MDFs and by constructing a tree-structured database in order to search for nearest neighbors in $log_2 N$ time. Murase and Nayar (1994) also produced an efficient search method and showed that 3D object pose might be estimated to within 2 degrees by interpolating training views taken at 10° intervals; moreover, while working with several objects other than faces, they also found that an eigenspace of dimension 20 or less was sufficient for good performance. The coverage of recognition-by-appearance in this chapter drew heavily from the work of Swets and Weng (1996) and the frequently referenced work of Turk and Pentland (1991).

Energy minimization was used in the 1970s for smoothing contours. However, the 1987 paper by Kass, Witkin, and Terzopoulos in which the term *snake* was introduced, seemed to freshly ignite the research interest of many other workers. Applications to fitting and tracking surfaces and volumes quickly followed. Amini and others (1988) proposed dynamic programming to fit active contours to images. One of many examples of its use in medical images is Yue and others (1995). The works by Chen and Medioni (1995) and Park, Metaxas, and Axel (1996) are two good examples of rapidly developing research and applications in physics-based and deformable modeling.

1. Amini, A., S. Tehrani, and T. Weymouth. 1988. Using dynamic programming for minimizing the energy of active contours in the presence of hard constraints. *Proc. IEEE Int. Conf. Comput. Vision,* 95–99.
2. Basri, R., and S. Ullman. 1988. The alignment of objects with smooth surfaces. *Proc. 2nd Intern. Conf. Comput. Vision,* 482–488.
3. Biederman, I. 1985. Human image understanding: recent research and theory. *Comput. Vision, Graphics, and Image Proc.,* v. 32(1):29–73.
4. Camps, O. I., L. G. Shapiro, and R. M. Haralick. 1992. Image prediction for computer vision. In *Three-dimensional Object Recognition Systems,* A. Jain and P. Flynn, eds. Elsevier Science Publishers BV, Amsterdam.
5. Chen, J. L., and G. Stockman. 1996. Determining pose of 3D objects with curved surfaces. *IEEE Trans. Pattern Analysis and Machine Intelligence,* v. 18(1):57–62.
6. Chen, Y., and G. Medioni. 1995. Description of complex objects from multiple range images using an inflating balloon model. *Comput. Vision and Image Understanding,* v. 61(3):325–334.
7. Costa, M. S., and L. G. Shapiro. 1995. Scene analysis using appearance-based models and relational indexing. *IEEE Symp. Comput. Vision* (Nov. 1995), 103–108.
8. Foley, J., A. van Dam, S. Feiner, and J. Hughes. 1996. *Computer Graphics: Principles and Practice.* Addison-Wesley, Reading, MA.
9. Gupta, A., L. Bogoni, and R. Bajcsy. 1989. Quantitative and qualitative measures for the evaluation of the superquadric model. *Proc. IEEE Workshop on Interpretation of 3D Scenes,* 162–169.
10. Hunter, G. M. 1978. *Efficient Computation and Data Structures for Graphics.* Ph.D. Dissertation, Princeton University, Princeton, NJ.
11. Huttenlocher, D. P., and S. Ullman. 1990. Recognizing solid objects by alignment with an image. *Int. J. Comput. Vision,* v. 5(2):195–212.
12. Jackins, C. L., and S. L. Tanimoto. 1980. Oct-trees and their use in representing three-dimensional objects. *Comput. Graphics and Image Proc.,* v. 14:249–270.

13. Johnson, A. E., and M. Hebert. 1998. Efficient multiple model recognition in cluttered 3-D scenes. *Proc. IEEE Conf. Comput. Vision and Pattern Recognition,* 671–677.
14. Kass, M., A. Witkin, and D. Terzopoulos. 1987. Snakes: active contour models. *Proc. First Int. Conf. Comput. Vision,* London, UK, 259–269.
15. Kirby, M., and L. Sirovich. 1990. Application of the Karhunen-Loeve procedure for the characterization of human faces. *IEEE Trans. Pattern Anal. and Machine Intelligence,* v. 12(1):103–108.
16. Koenderink, J. J., and A. J. van Doorn. 1979. The internal representation of solid shape with respect to vision. *Biological Cybernetics,* v. 32:211–216.
17. Lowe, D. G. 1987. The viewpoint consistency constraint. *Int. J. Comput. Vision,* v. 1:57–72.
18. Murase, H., and S. Nayar. 1995. Parametric appearance representation. In *3D Object Representations in Computer Vision,* J. Ponce and M. Herbert, eds. Springer-Verlag.
19. Nevatia, R., and T. O. Binford. 1977. Description and recognition of curved objects. *Artificial Intelligence,* v. 8:77–98.
20. Park, J., D. Metaxas, and L. Axel. 1996. Analysis of left ventricular wall motion based on volumetric deformable models and MRI-SPAMM. *Medical Image Anal. J.,* v. 1(1):53–71.
21. Pentland, N. P. 1986. Perceptual organization and the representation of natural form. *Artificial Intelligence,* v. 28:29–73.
22. Pulli, K., and L. G. Shapiro. 1996. Triplet-based object recognition using synthetic and real probability models. *Proc. ICPR96,* v. IV:75–79.
23. Roberts, L. G. 1977. Machine perception of three-dimensional solids. In *Computer Methods in Image Analysis,* J. K. Aggarwal, R. O. Duda, and A. Rosenfeld, eds. IEEE Computer Society Press, Los Alamitos, CA.
24. Rom, H., and G. Medioni. 1993. Hierarchical decomposition and axial shape description. *IEEE Trans. Pattern Anal. and Machine Intelligence,* v. 15(10):973–981.
25. Samet, H. 1990. *Design and Analysis of Spatial Data Structures.* Addison-Wesley, Reading, MA.
26. Shapiro, L. G., J. D. Moriarty, R. M. Haralick, and P. G. Mulgaonkar. 1984. Matching three-dimensional objects using a relational paradigm. *Pattern Recog.,* v. 17(4):385–405.
27. Shapiro, L. G., and R. M. Haralick. 1981. Structural descriptions and inexact matching. *IEEE Trans. Pattern Anal. and Machine Intelligence,* v. PAMI-3(5):504–519.
28. Shapiro, L. G., and R. M. Haralick. 1985. A metric for comparing relational descriptions. *IEEE Trans. Pattern Anal. and Machine Intelligence,* v. PAMI-7(1):90–94.
29. Stark, L., and K. Bowyer. 1996. *Generic Object Recognition Using Form and Function.* World Scientific Publishing Co. Pte. Ltd., Singapore.
30. Swets, D., and J. Weng. 1996. Using discriminant eigenfeatures for image retrieval. *IEEE Trans. Pattern Anal. and Machine Intelligence,* v. 18:831–836.

31. Turk, M., and A. Pentland. 1991. Eigenfaces for recognition. *J. Cognitive Neuroscience,* v. 3(1):71–86.

32. Yue, Z., A. Goshtasby, and L. Ackerman. 1995. Automatic detection of rib borders in chest radiographs. *IEEE Trans. Medical Imaging,* v. 14(3):525, 536.

15

Virtual Reality

Suppose a surgeon needs to plan a brain operation to remove a tumor. 3D images of the patient's skull and brain are available from the diagnosis. Using *virtual reality,* the surgeon can practice with the 3D data model rather than the real object. Different entry routes can be examined and different physical operations tried in order to choose a best intervention for the real patient. Moreover, it is possible to correspond a generic brain atlas with the 3D image data so the surgeon can see it in overlay and evaluate the consequences of different surgical options, avoiding damage to critical brain structures. Virtual reality (VR) systems have been developed to facilitate such virtual surgery. Figure 15.1 shows a rendered image of a brain model that could be used in a VR system.

Virtual reality is a new field and often considered to be a subfield of computer graphics, because computer generated displays are an important component of a VR system. VR applications are important for study in their own right and VR systems also relate to the

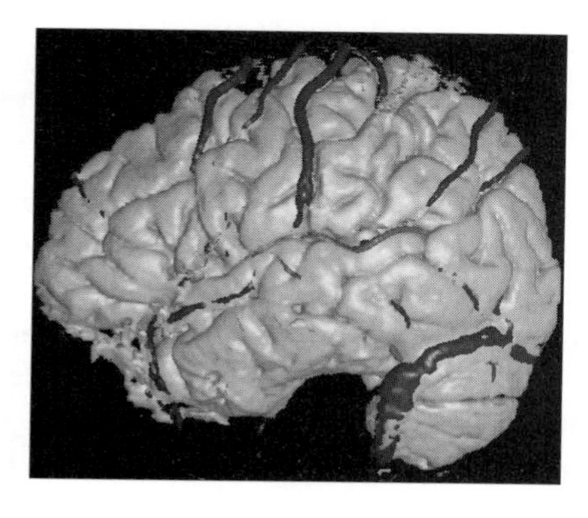

Figure 15.1 Artificial rendered image of a 3D mesh constructed from real patient MRI data. (Courtesy of the University of Washington Human Brain Project.)

purpose of this text in multiple ways: (a) real images and image processing are often needed; (b) quality stereo imagery must be produced to *immerse* the human user in the virtual environment; (c) common mathematical models are used to correspond 3D points in various real and model spaces; and (d) machine vision is sometimes used to sense the position of the user or other real objects. VR has grown from roots in engineering of simulators—particularly flight simulators—teleoperation, and computer games.

15.1 FEATURES OF VIRTUAL REALITY SYSTEMS

We identify the important features of VR systems or VEs (virtual environments), after which we cite several exciting applications.

- A human user operates with a **model** of reality and simulates many of the operations that are possible with real objects.
- The display methods are of high quality in resolution and speed, so that the user is *immersed* in the data and problem to approximate perception of the real thing.
- The user must be able to interact with and change the model environment in smoothly perceived real time.
- 3D visual feedback is of paramount importance. The VR system typically provides a way for the user to change her viewpoint of the environment or to rotate or move objects for better viewing. Although visual feedback is most important, some tactile, motion, force, or auditory feedback should also be present, so the user can feel objects or hear them collide, etc.

Figure 15.2 illustrates the usual scenario of a human operating in a real environment, while Figure 15.3 illustrates the newer scenario of a human operating in a virtual environment.

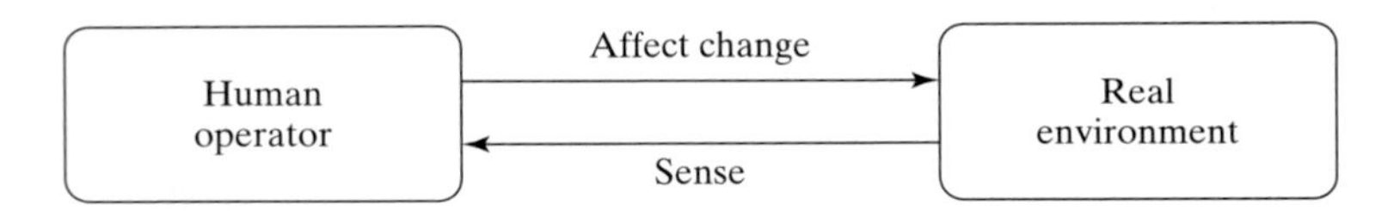

Figure 15.2 Human operator working in a real (natural) environment.

Exercise 15.1: Books and Movies.

- People can certainly become immersed in a plot when reading a book. Which of the above four features of VR systems are present and which are not? (Note that there are some books that allow the reader to select from different continuations of the plot.)
- People can also become immersed in a movie, especially when a wide screen or 3D display techniques are used. Which of the above four features of VR systems are present and which are not?
- Do you know of or play any video games that have all four features given above? Explain.

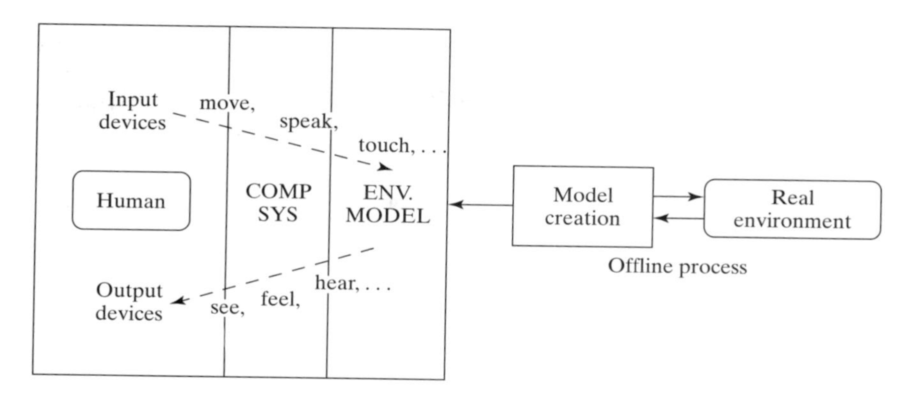

Figure 15.3 Human operator immersed in a virtual environment.

15.2 APPLICATIONS OF VR

VR is usually associated with exotic new applications enabled by expensive new hardware. However, VR existed to some degree in several commonly known older systems.

Architectural Walkthrough A user might interact with an architectural model house by taking a virtual walk through the house, perhaps looking out the virtual windows at a virtual landscape. If the house were Jefferson's Monticello, the user could view Jefferson's collection of artifacts, his unique bed, and cannonball clock. Many such historical places and *virtual museums* are currently being digitized. In simple cases, users can explore these virtual environments using the World Wide Web and an ordinary flat display. Users might not be allowed to modify a model of Monticello, but they might be able to pick up and examine artifacts inside of it. If the user is planning to build a real house, then he should be able to make changes to the architectural model, perhaps experimenting with different wall coverings or furniture placement.

Related to the *walkthrough* is the *flyby:* one can view the Grand Canyon as if flying over it. In the simple case, the view cannot be determined by the user who is a mere passenger in the plane; in the more complex case, the user is the pilot who can dynamically change course and view the scenery in many different ways.

Flight Simulation With a flight simulator, a user can practice control of an aircraft navigating various terrain and in takeoffs and landings at various airports. Thirty-year-old stories tell of the increased pulse and perspiration of humans exiting flight trainers, attesting to the depth of their immersion in the virtual environment.

Interactive Segmentation of Anatomical Structure A VR system can help medical personnel identify and extract models of anatomy from 3D sensed data; in other words, support interactive segmentation. Suppose, for example, that 3D MRI data is displayed to a user via stereo images. By interacting with the data, the user could mark a sequence of points at the center of a blood vessel, or on the boundary of a chamber of the heart. Stereo display devices and 3D input devices needed for this application are discussed in Section 15.5.

Figure 15.4 The Dynamic Virtual Playground is an experimental environment in which users can collaborate. (Courtesy of the University of Washington HIT Lab.)

Today, there are many more applications of VR systems including pain management, phobia treatment, low-vision aids, driving simulators, scientific visualizations, and virtual classrooms. Figure 15.4 illustrates a current VR project at the University of Washington: the Dynamic Virtual Playground. The virtual playground is a prototype system designed to investigate multiple simultaneous collaborations in a virtual setting. It could be used to simulate a school lab where each group of students is working on a different project.

15.3 AUGMENTED REALITY (AR)

A contractor remodeling an existing site needs to know the locations of existing water and gas pipes and electrical conduit. Such data could be available from existing maps, blueprints, or even CAD files. The following scenario is an example of *augmented reality* (AR), also called *mixed reality*. The contractor wears a head-mounted display (HMD) that overlays computer graphics on the real scene that he is viewing. When he looks at the ground, he sees blue lines where water pipes are buried and when he looks at a wall he sees red lines for electrical conduit and blue lines for water pipe. In some sense, the AR gives the contractor Superman's (TM) capability of seeing through walls!

Creation of such an AR system requires the following.

- 3D models of the objects needed to augment the real view.
- Correspondence of the user's real workspace to the 3D model data via calibration.
- Tracking of the user's pose to determine the user's viewpoint within the real workspace.
- Real time display combining the real images and the computer graphics generated from the models.
- The response time to head movements and the registration accuracy between image and graphics, which are critical to the effectiveness of the system.

An augmented reality environment is sketched in Figure 15.5 and should be compared to the other diagrams of this chapter. There are many applications of augmented reality: here are a few.

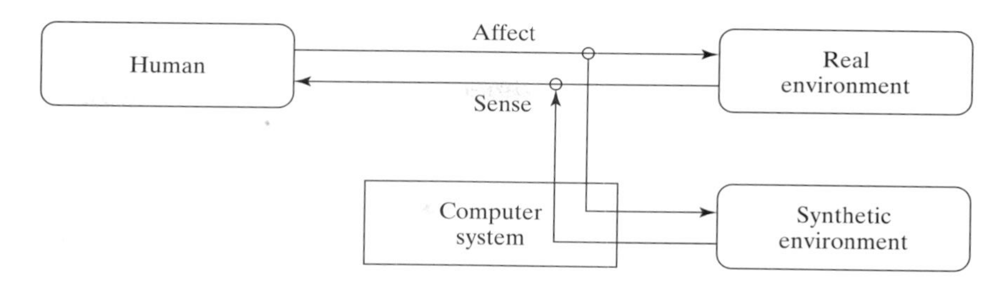

Figure 15.5 Human operator working in an augmented reality environment.

- Consider AR-assisted surgery. The surgeon operating on a real patient views CAT scan data including a path plan superimposed on a live image of the patient. (The path plan might have been obtained using a VR system as above.)
- In PC board inspection, a human inspector compares a new PC board with a CAD model and verifies the presence of all required components and leads. The board is placed accurately in a jig so that its image from a camera can be accurately registered with the CAD model and presented to the inspector via a large computer display.
- The driver of an auto views a display that shows geographic features ahead. Projectors in the dashboard project the names of buildings and streets on the inside of the windshield.
- Several people at a meeting want to talk about a computer model they are jointly creating. They want to be able to look at the model, point to and discuss its features, and still see one another and their surroundings. This scenario can be extended to teleconferencing where some of the people are located at a remote location. They may want to see not only the common computer model, but the other participants.

Figure 15.6 illustrates a potential teleconferencing application. The user of this system has placed two cards on his desk. Each card has a white background, a black square region,

Figure 15.6 An application of augmented reality techniques to teleconferencing. (Courtesy of the University of Washington HIT Lab.)

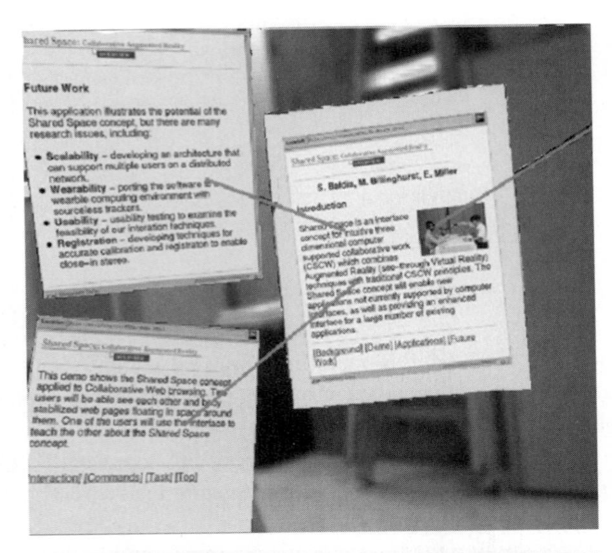

Figure 15.7 Two people wearing see-through, augmented-reality goggles (below). They can see the real world, and they can also see computer-generated displays. They are both looking at the Web pages shown above, which are perceived to be floating in space. (Courtesy of the University of Washington HIT Lab.)

and a pattern inside the black region. Computer vision techniques are used to find the cards and statistical pattern recognition techniques are used to recognize the patterns within them, thus identifying the meaning of each card. Using augmented reality techniques, an image of the remote collaborator is placed on one of the cards, while an image of a graphics model to be discussed is placed on the second.

Figure 15.7 illustrates the case of two people in the same room using see-through goggles to look at a set of Web pages they are both discussing.

15.4 TELEOPERATION

Teleoperation is an established engineering discipline which has provided much to virtual reality; in particular, the sensors and effectors needed to couple a human agent to the environment. The sketch in Figure 15.8 should be compared to the others of this chapter for similarities and differences. Using teleoperation, a human performs real work in a real environment that may be remotely located. A robot or robot-like machine carries out operations in the real environment according to the human control. Example successful application are as follows.

1. From a computer at Pathfinder mission control, a human operator gives a command to a navigating robot on Planet Mars to advance $10cm$ and take a soil sample. The human gets feedback from images taken from cameras on the nearby lander that delivered the robot to Mars. Transfer of images and commands takes about eleven minutes due to the long length of the link.
2. Communicating via radio frequency, a human operator controls a remote robot that is vacuuming up radioactive waste in a dangerous area of a nuclear power plant after a minor accident. The human wears a head-mounted display which shows the contaminated area as the robot cameras see it and uses a simulated shaft to simulate the real vacuum shaft. Sensors in the simulated shaft sense the position and motion of the shaft yielding control signals to control the motion of the remote real vacuum.
3. A surgeon performs remote surgery by sewing a synthetic object similar to a football. Sensors record the delicate sewing motion parameters which are then transmitted to a remote robot that sews up an incision in a real live dog. This experiment has already been performed and may be a step in delivering the skills of a surgeon to otherwise unreachable places.
4. (Future scenario.) A patient goes to a hospital to have arteries cleaned of plaque (arteriosclerosis). The person is placed in an MRI machine which enables real-time 3D viewing of the body. Microrobots are injected into the blood stream and allowed to disperse throughout the body. The attending doctor, aided by maps from previous diagnoses, indicates via a 3D input device what regions of the vasculature need cleaning. The MRI device then operates in alternating modes, one imaging as before,

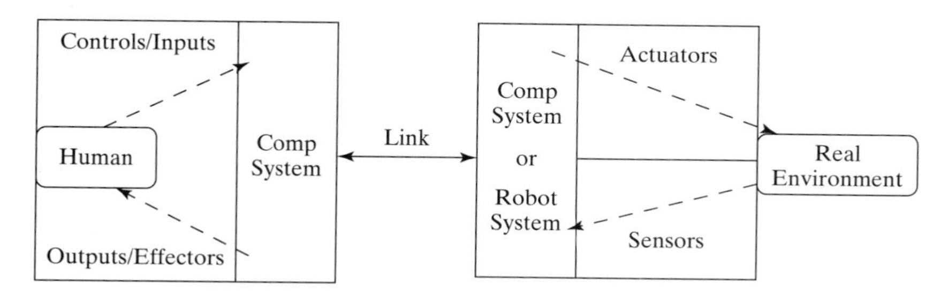

Figure 15.8 Human operator using teleoperation.

Figure 15.9 The *Hot-Line Telerobot System* from Kyushu Electric Power Company. (left) A teleoperated robot repairs a high-voltage power line; (right) the operator's interface to the system. (Courtesy of Blake Hannaford with permission of MIT Press. Reprinted from K. Goldberg, *The Robot in the Garden,* Cambridge, MA: The MIT Press, 2000.)

and the other stimulating the microrobots to perform cleaning actions in the designated areas.

Figure 15.9 shows an example of a real, working telerobot system from Kyushu Electric Power Company. The robots are used to safely repair high-voltage electrical power lines.

Before moving on to discuss the devices and mathematical models needed to implement virtual environment technology, we give a humorous and instructive anecdote. A teleoperated power shovel was devised so that a remote operator could move soil, coal, etc. The operator wore a *data glove* which could sense the position of his hand and fingers, which the system then translated into control signals for the power shovel. Two cameras on the power shovel provided left and right images for the operator's HMD. The operator would use his hand to grasp in a pile of sawdust on a table, and the remote power shovel would mimic these motions by grasping in a pile of actual coal. Once the human operator had an itchy nose and reached to scratch it with the data glove hand! Moving the controlling hand toward the nose caused the power shovel effector to move toward the cameras in the real environment. When the resulting images were transmitted back to the operator via the HMD, he perceived that he would be struck in the face by a massive shovel! This was only a virtual blow and the operator was not physically harmed; however, he was psychologically upset enough to need a break from the work. (Had the system not been carefully designed, the cameras on the real power shovel might have actually been damaged.)

110 Definition. Using **teleoperation,** a human operates a real device in a remote real environment: The feedback from the remote environment and the controls of the device provide some illusion that the operator is present in the real environment.

111 Definition. **Virtual reality** is a synthetic reality provided by a computer system to a human user via rich models of reality and immersive input/output devices: The human operator has the illusion of working with real objects when none are actually present. The illusory environment which the operator perceives and changes is called a virtual environment (VE).

112 Definition. An **augmented reality** or **mixed reality** is created by combination of a real environment and a virtual environment: Synthetic outputs from a computer system are combined with sensed data from a real environment to augment the human's perception of reality.

113 Definition. A **synthetic environment (SE)** is an environment provided to a human operator by a computer system and immersive I/O devices via teleoperation, augmented reality, or virtual reality. (Sometimes, the term virtual environment (VE), is also used to cover all of these cases.)

15.5 VIRTUAL REALITY DEVICES

Several devices are commonly used to couple a human operator to a synthetic environment. Figure 15.10 guides our discussion: It shows a previously discussed AR application, but the devices are used in all three types of SE applications. A contractor walks about in a building to be remodeled and marks the walls where water pipes and electrical conduit are present. The contractor sees real walls via a see-through HMD containing optics to overlay computer generated images showing the pipes and conduits.

The contractor (a) interacts with a real environment (b) augmented by a computer system (c) which adds otherwise invisible features to the images seen (d, e, f) and uses speech communication (g, h) with the contractor, who is free to walk around and mark the walls (i). As the human moves about, the image of the real world is seen through a beam-splitter (d). A pose sensor (e) transmits the orientation and location of the head to the computer, which then uses these parameters to generate a computer image from a 3D CAD model of the pipes and conduits. The generated image is projected onto the inside of the one-side mirror through which the operator is looking (f) to augment the view of the real world.

The Head-mounted Display A schematic of the *see-through* HMD is given at the left of Figure 15.11. A beam-splitter allows light from the real-world to pass through from the outside, but acts as a mirror on the inside in reflecting the light from a computer-generated display, resulting in the operator seeing the augmented image. Note that the graphics display is a miniature display with small optics residing in the head-mounted display or helmet. An alternative design, the opaque HMD, is shown at the right of Figure 15.11. Note that all elements reside in the HMD and move with it—the mirror, the camera, and the graphics display. Here the synthetic image is a mixture of a digital image from the camera and a

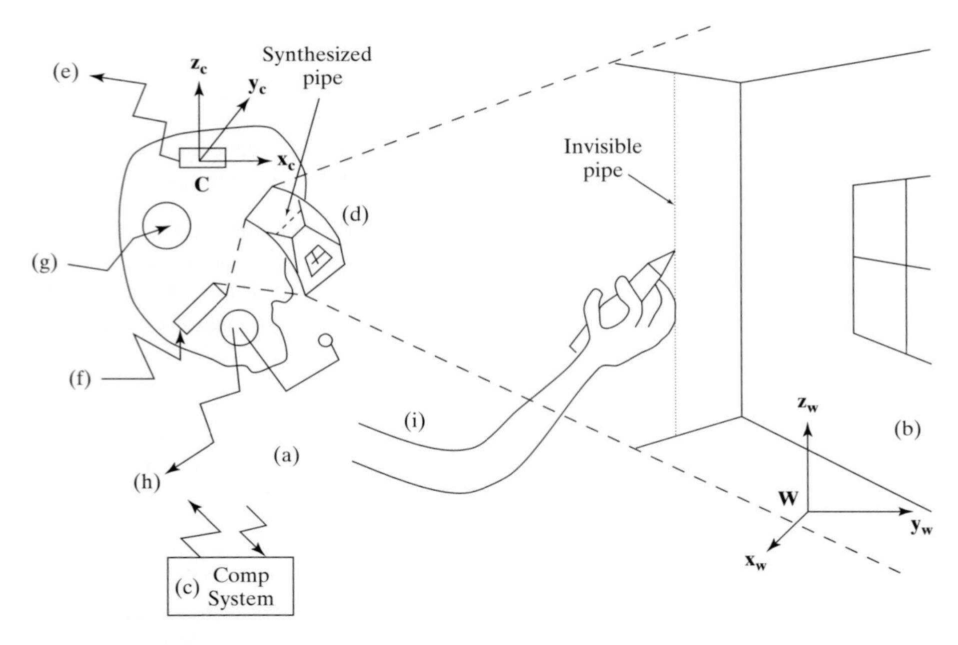

Figure 15.10 Devices used by a human operator in an AR application. The operator marks the location of pipes in a wall: Location of the invisible pipes is known from a CAD model of the house.

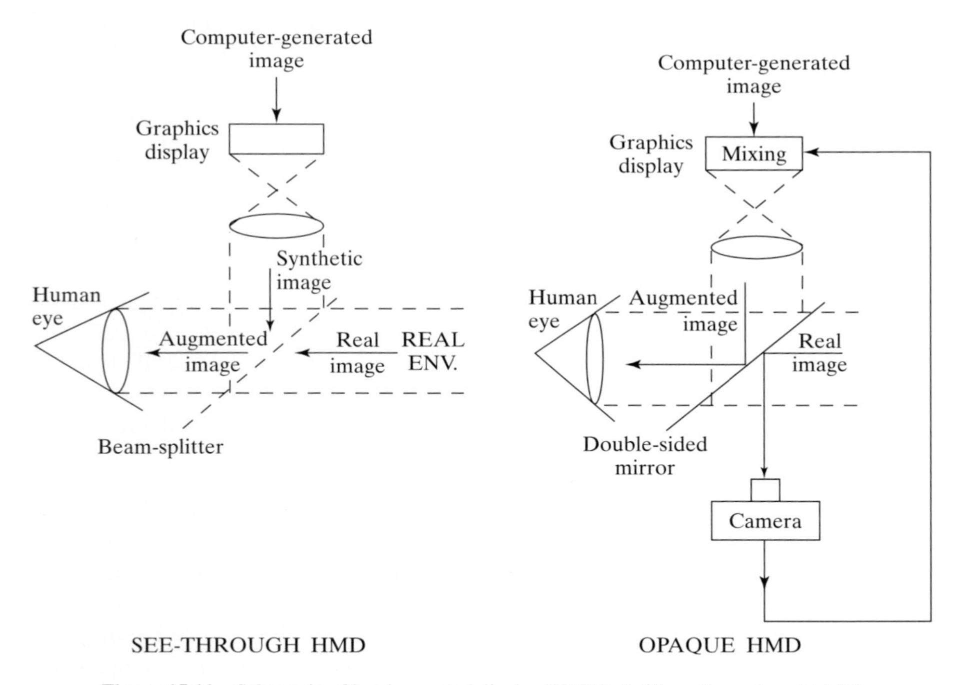

Figure 15.11 Schematic of head-mounted display (HMD): (left) see-through and (right) opaque.

computer generated image. The see-through design can provide higher resolution images, while the opaque design provides more options for controlling what the user sees.

Exercise 15.2: On registration accuracy for AR systems.

Refer to Figure 15.10. (a) Suppose that the pose sensor makes an error of two degrees in pan angle when reporting the pose of the operator's head. What would be the error (in cm) in the real world between where the operator would mark a vertical pipe versus its real location? (b) Suppose that the visual display has an FOV of 120 degrees mapped across 500 pixels. (If needed, you may assume a focal length of $2.0cm$ for the imaging system. Also, you may assume the standoff from the wall being observed is $3m$.) What would be the horizontal distance, in pixels, between the projection of the pipe in the augmented image and the true location of the pipe in the image?

Exercise 15.3: On registration accuracy for AR systems.

This problem relates to the previous one and uses the same devices. The problem is inspection of automobile instrument panels via human comparison of a real instrument panel with a CAD model using an AR system. Unlike the previous problem, all CAD features should be visible to the human operator in the real image. The operator is to check the existence and correct functioning of the instruments, such as odometer, oil pressure guage, radio, etc. The AR computer system also drives the test equipment and gives input to the operator prompting the operator on what to look for. Assume an FOV of 60 degrees and a standoff of about $60cm$. As in the problem above, suppose the pose sensor has an error of two degrees in the pan angle. (a) If the augmented image has a red circle where a radio knob should be, what is the horizontal error in registration caused by the error in pan angle? (b) Assuming misregistration error is bad, describe a method to automatically reduce it using the computer system. Would either type of HMD work as well for your method? (c) Assuming that the operator could do better inspections by being able to control small misregistrations, describe a method to automatically produce small misregistrations using the computer system. Would either type of HMD work as well for your method? (d) Is an AR system needed for this problem or could it be solved with a completely automatic system? Explain.

Exercise 15.4: Multiple operator AR.

Suppose a team of several surgeons will do surgery on a patient. Is it possible for each to wear an HMD to view the surgery plans and anatomical structures overlaid on their view of the real patient? Explain how it might be done or why it can't be done.

Dextrous Virtual Work A VR system supporting dextrous handwork on models is sketched in Figure 15.12. The human operator views the synthetic workspace via a stereo display projected onto a mirror from above. This allows normal freedom of hand movements in a real 3D workspace below the mirror. The operator manipulates a tool below the mirror;

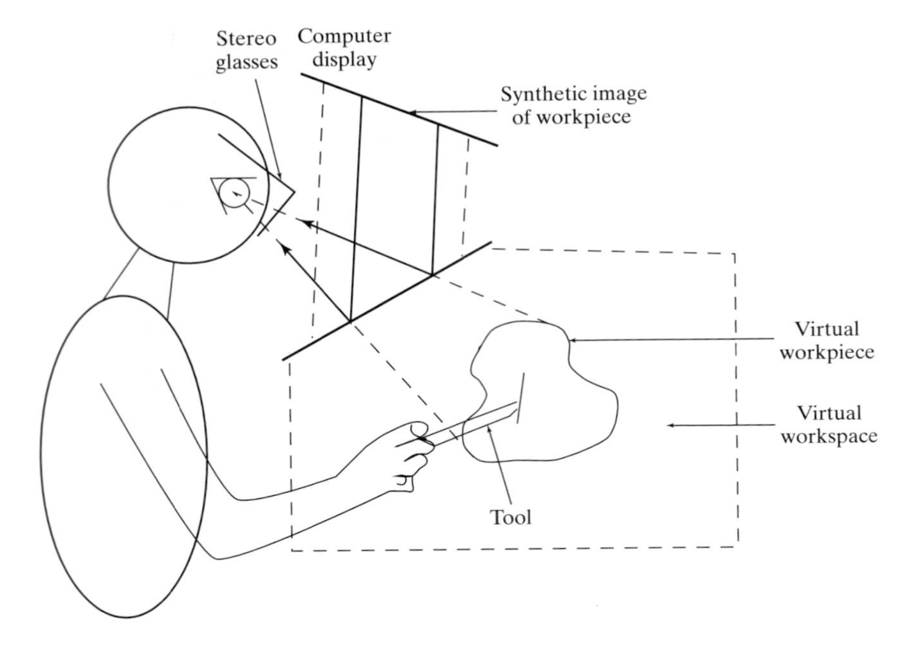

Figure 15.12 Schematic of workbench for dextrous virtual handwork.

the pose of the tool is carefully tracked and its image is projected back onto the mirror so that the operator receives visual feedback. The figure shows an operator practicing making an incision on a virtual organ. Different technologies available for the 3D tool are given next. Clearly, accuracy of 3D pose and speed of sensing and update of the display are critical to the system. Besides obvious applications for the practice or plannning of surgery, such a system might be useful for an artist in digitally sculpting a 3D model.

Stereoscopic Display Devices Stereo vision is perhaps the most important visual cue for sensing 3D objects that are within 10*m* and is the predominant means of feedback from VR systems. There are two common designs for stereo displays. With opaque HMDs, separate images can be presented to the two eyes; the left and right images are synthesized with the proper disparity from the object model using the mathematical model of Chapter 12. Opaque HMDs can produce a wide FOV on an infinite virtual world and hence high degree of presence and immersion; however, there are many design constraints in creating devices that will comfortably fit different human operators. Note that multiple users can be immersed in the same virtual world: each user has an HMD with individual pose sensor, providing an individual immersive display of the VE from the user's viewpoint. An alterative is to use an ordinary graphics display as shown in Figure 15.12 which displays the left and right images interleaved over time. The operator uses shuttered glasses that are synchronized in time with the display, so that the left eye sees only the odd frames and the right eye sees only the even frames, or vice versa. The perceptual system is able to fuse this time-varying input to perceive 3D. This design permits an inexpensive system that is easy to use. However, the degree of immersion is limited because all imagery is limited to the display screen (called *fishtank virtual reality* for obvious reasons): the view of the virtual world is limited by being clipped to the display screen (*fishtank*). Moreover, since there is only one physical display synched to one user, other users viewing the display do not have their own viewpoint and do not benefit from their own motion. The fishtank effect can be reduced by surrounding the user with displays, thus creating a *CAVE*, but the full head-coupled stereoscopic effect is still unavailable to multiple users.

15.6 SUMMARY OF SENSING DEVICES FOR VR

Visual Visual output from a VR system, as described in the previous section, typically consists of stereoscopic displays to the user, either via the HMD or standard computer display viewed through shuttered glasses. This output is coupled to the user via sensors that sense user pose in the VE so that the display portrays the appropriate viewpoint.

Visual input to a VR system, if present, usually consists of tracking the users eye[s], head, or body parts and providing pose as inputs to the activity being modeled. Eye trackers can provide gaze direction. Devices can be incorporated into HMDs or can be totally separate from the operator. Tracking head pose using HMDs is possible using a camera that tracks special feature points on the HMD. Recent work in VEs has demonstrated the capability of tracking the hands, head, and feet, or limbs using the best views from multiple cameras observing the human. One advantage of visual input devices is that they need not be another constraining wearable device; however, although the user's movement may be unconstrained by the device, the devices still have limited work volumes. Some eye trackers depend on special IR illumination and some body trackers depend on controlled background colors.

Auditory Speech input to a computer has been available for over fifteen years and is now appearing in many interfaces, including telephone systems and home PC applications. Speech input has the advantage of being natural, that is, not requiring special learning, and may be necessary when an operator's eyes and hands are involved in other tasks and unavailable. Similarly, speech output is a convenient communication channel independent of the display.

Auditory output might enhance any interface—for example, a metallic banging sound confirms that a file has been deleted when a folder icon has been dragged to the trash can icon. It also enhances the degree of immersion in a VE: The operator of a virtual vehicle can hear the sound of the engine and the wheels squealing to a stop. Or, a teleoperator can sense the amount of radioactivity present via the degree of frenzy encoded in music being played.

114 Definition. **Sonification** is the encoding into sound of data or control information that is not naturally sound.

Pose 3D position sensors sense the position and orientation of human body parts or tools that are held. Six degree-of-freedom sensors include the Polyhemus sensor commonly used on HMDs, joysticks, and newer devices such as the *bat* described by Green and Halliday (1996). x-y-z position sensors include the sparking styllus and various mechanical devices. It is also possible, although not common, for computers to output pose to a human by positioning mechanical joints that are attached to the human. This is related to force feedback, which is described below.

Haptic Humans connect to the real world via their touch, force, and motion. Sense of touch is due to nerves in the skin that can sense temperature, firmness, and smoothness of surfaces. Nerves in the limbs and muscles can sense limb pose and muscle tension and, or length and their changes. Nerves in the vestibular system can sense body motion.

115 Definition. The human **haptic sense** involves the sense of touch (tactile sense) and the senses of body position, forces, or motion (kinesthetic sense).

A variety of electro-mechanical devices are available to provide input and output of forces.

Motion Human perception of motion is affected by the integration of many sensing systems. Visual displays, as in the movies or simulators, can provide enough stimulation to actually induce motion sickness. Kinesthetic stimulation via treadmills, robotic linkages, or platforms or centrifuges that actually move the operator is employed in various VE systems to increase the degree of immersion above that provided by the visual display. Alternatively, the vestibular system might be artificially stimulated using a stream of cool air or water. Computer vision can play a strong role in motion sensing. Ideally, a human operator would move freely in a real environment and analysis of images from several tracking cameras would provide all the needed interpretation of motion and perhaps map it onto a computer model of the human body. Commercial systems are available for tracking the moving human body; typically, they depend upon placement of specially reflective targets at special locations on the body in order to simplify image segmentation and feature extraction. Such systems are used in various sports' studies and in prescribing orthopedic devices. Research systems have demonstrated tracking of the human head, hands, and feet without use of special targets on the body.

15.7 RENDERING SIMPLE 3D MODELS

In order to create virtual scenes, we need the tools for constructing object models and for producing images of them. The object models used can be complex mesh models as discussed in Chapters 13 and 14, or they can be simpler wire-frame models, which can be created using Computer-Aided Design (CAD) software packages, such as AUTOCAD. Figure 15.13 shows a wireframe model of a car created by a user interacting with a CAD package.

Once we have the 3D model, we would like to be able to display it from arbitrary viewpoints and with various different lightings.

116 Definition. **Rendering** is the process of creating images from models.

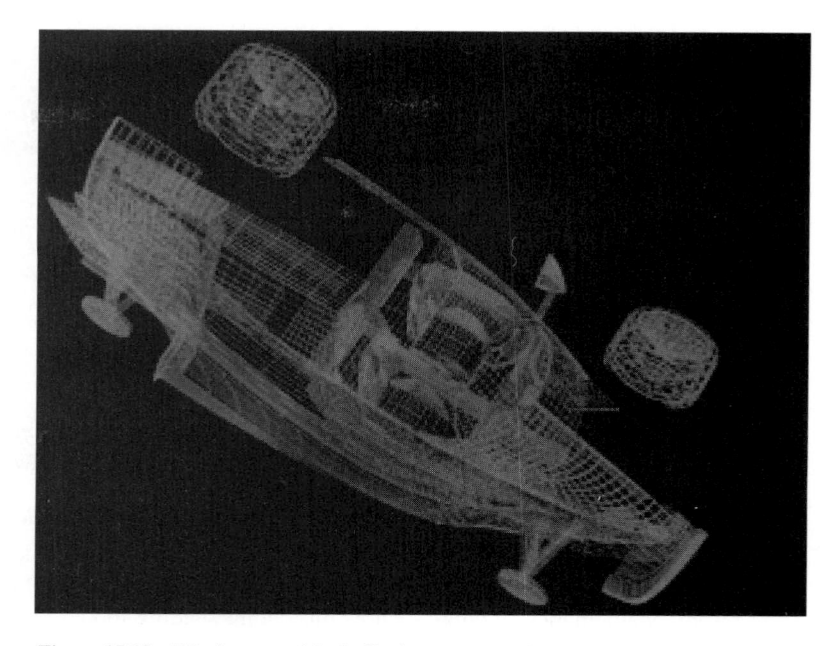

Figure 15.13 Wireframe model of a Ford concept car. (Courtesy of Ford Motor Company.)

Figure 15.14 Rendered image of the car alone. (Courtesy of Ford Motor Company.)

Rendering can be thought of in two steps:

1. Determine which surfaces of the model are visible to the selected viewpoint, and
2. determine the corresponding pixel values of the created image.

Conceptually, Step 1 is achieved by constructing a ray from the viewing point in the desired direction to the object. The first intersection of the ray with the object is the surface

Figure 15.15 Scenes using a car and other models. (Courtesy of Habib Abi-Rached.)

point that will be visible along that ray. This concept is called *ray tracing,* and there are many software algorithms that perform it. Today's computers use a hardware mechanism called a *z-buffer* to carry out this step rapidly.

Step 2 can be simple or very complex. In the simple case, the object has a particular color and is made of a particular material whose reflectance properties are known. The light comes from a point light source in a particular direction. Some amount of ambient light is also present. Mathematical models such as the Phong shading model given in Chapter 6 are then used to determine the pixel color corresponding to a small area of the object. Figure 15.14 shows a simple rendering of the wireframe car model. More realistic images can be created (at the cost of speed) by adding such factors as multiple light sources, areal light sources, shadows, transparent surfaces, and interreflections.

Figure 15.15 shows two rendered images using several different vehicle models. While some of the objects in these images were rendered as described previously, some of them have textured surfaces involving complex patterns that would be very slow to render. (See the building on the far left of the left image and the sidewalk next to it.) These surfaces have been *texture mapped* instead of being rendered. Texture mapping will be discussed in the next section.

15.8 COMPOSING REAL AND SYNTHETIC IMAGERY

Synthetic rendering can only go so far. Not only is it not sufficiently realistic, but attempts at realism make it very time-consuming. Instead, existing images of complex textures can be used to both improve realism and speed up the rendering. For this purpose, a *texture* can be an artificially generated pattern or a (portion of a) real image. In the process of rendering a surface, instead of coloring it with a single color value, we would like to paste or paint a given texture onto it. This leads to the process of *texture mapping,* where the final values of the visible pixels are selected from the pixels of a given texture.

117 Definition. **Texture mapping** is the process of painting a texture onto a smooth surface, thus creating a textured image of that surface.

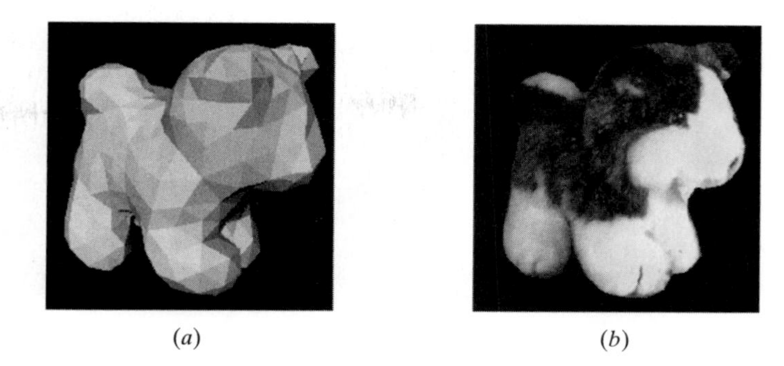

(*a*) (*b*)

Figure 15.16 Rough mesh model of a dog and texture-mapped image. (Courtesy of Kari Pulli.)

The surfaces of Figure 15.15 that were texture mapped were planar polygons, the easiest surfaces to texture-map. Texture mapping can also be done on more complex curved surfaces, such as painting a rough *peel* texture on an orange. When the object is free form and represented by a mesh, we can do this in a piecewise fashion. But for difficult objects, we need more sophisticated methods. A recently developed technique in computer graphics uses real images of the object to provide the needed textures. Figure 15.16(*a*) shows a rough mesh model of the reconstructed dog from Chapter 13, and Figure 15.16(*b*) shows a texture-mapped image of the model. In this example, the texture comes from a real image of the dog taken from the same viewpoint at which the model is displayed. We really want to be able to display texture-mapped images from arbitrary viewpoints, as we did with rendering.

118 Definition. **Image-based rendering** is a technique that uses a set of real images of an object to produce an artificial image from an arbitrary viewpoint.

Image-based rendering can actually be accomplished without a geometric model of the object, by storing a very large number of images at different viewpoints and interpolating between them to produce arbitrary views. However, if we have a mesh model of the object and a small number of real images that cover the set of potential viewpoints, then the geometry plus the images can be used to produce very realistic renderings. The reconstruction system described in Chapter 13, which produced rough mesh models of objects from a set of range images and associated color images, can also render the objects from various user-selected viewpoints, using a technique called *view-based texturing*. Figure 15.17 shows the basic principle behind this approach. On the left, a false-color rendition of the object can be manipulated by the user with a mouse to rotate it to a desired viewpoint. In the center, the three closest views to that viewpoint have been retrieved. On the right, a texture-mapped image of the dog in the desired orientation is produced. To create each nonbackground pixel of the image, a ray is sent from the image pixel to the 3D model and then from the model to the appropriate pixel of each of the three closest views. The colors from these three pixels are blended to produce a value for the selected pixel in the created image. Rather than counting all three pixels equally, the blending algorithm takes into account how similar a stored view is to the required view, the direction of the ray from the object model to the

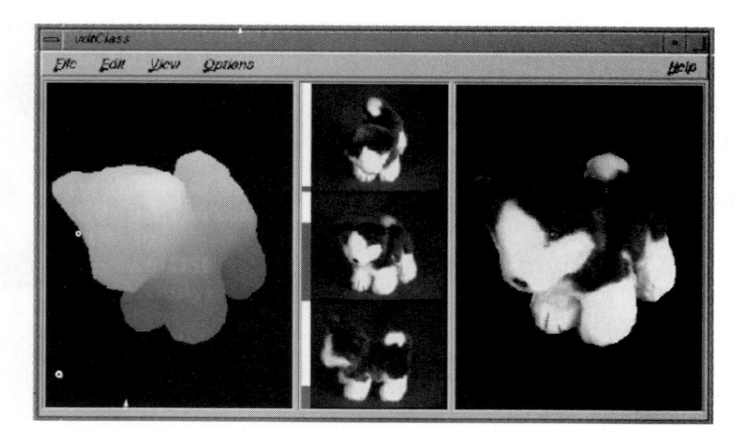

Figure 15.17 (left) Range data of a dog model; (center) three real color images from nearby viewpoints; and (right) the rendered image using a weighted combination of pixels from these views. (Courtesy of Kari Pulli.) See colorplate.

stored-view pixel, and the nearness of the stored-view pixel to the boundary of that view. It also uses a software variant of z-buffering for throwing out the contributions of pixels that are too far away to actually lie on the surface.

Full 3D mesh models of real objects take a long time to construct. View-based texturing does not actually need the full geometric model; instead it can work with only the original set of registered range and color images. Figure 15.18 illustrates this process for the dog model. Instead of a full mesh, a partial mesh is created for each of the sample views. These partial meshes can be used together with the color images of the object to produce renderings that are just as good as from a full mesh.

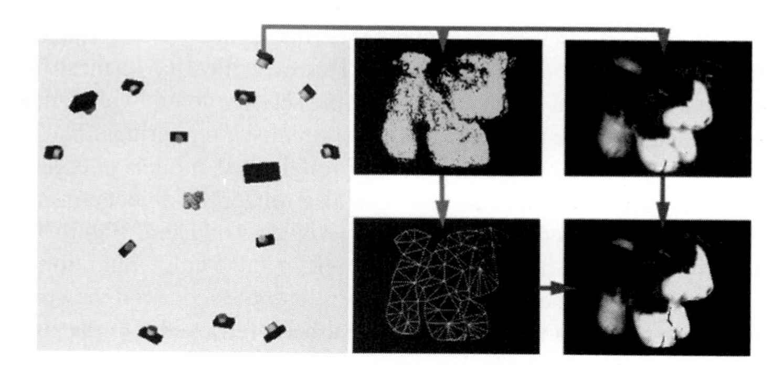

Figure 15.18 Registered range and color images from a small number of views of an object can be used to produce a high-quality rendered image, without ever constructing a full model of the object. (left) Potential viewpoints; (top center) range data from one of the viewpoints; (top right) color data from the same viewpoint; (bottom center) mesh constructed from the range data; and (bottom right) rendered image achieved by texture-mapping the color data onto the mesh. (Courtesy of Kari Pulli.) See colorplate.

Furthermore, there are some real objects that are not suitable for producing solid models. When an object has thin parts, such as the sail of a boat or the leaf of a plant, the mesh model would have to be of extremely high resolution to capture the topology. Since it is possible to capture range and color data from a number of views, view-based texturing can still be used to produce very realistic images of the object from arbitrary orientations. Figure 15.19 illustrates this process on a basket of flowers.

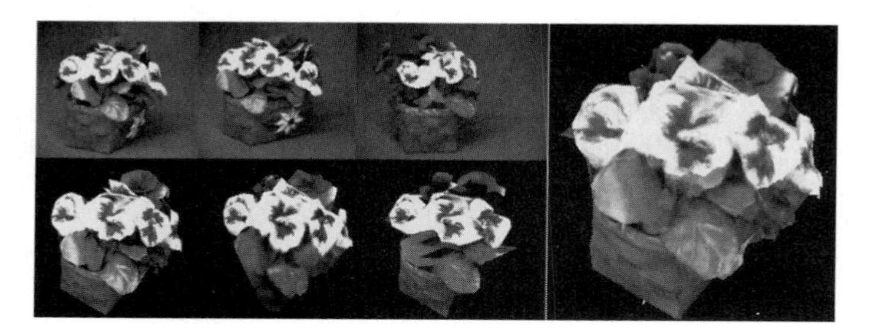

Figure 15.19 The same technique applied to an object for which construction of a full 3D model is nearly impossible, due to the thinness of parts of the object. (upper left) Three different color images of the object; (lower left) three images from a different, selected viewpoint constructed by mapping the pixels of the three original images onto the new viewpoint; and (right) final rendered image, which is a weighted combination of the three constructed images. (Courtesy of Kari Pulli.) See colorplate.

Exercise 15.5: Augmenting the image of a real cube.

Calibrate a camera using a method from Chapter 13 and a calibration *jig* which is a cube or small box (use the seven visible corners as control points). Design a 2D die face containing three dots to represent the number 3. For the top of the 3D cube, construct a mapping function g, as in Chapter 11, which maps a point $[x_m, y_m, z_m]$ from the top of the 3D cube to a 2D point $[x_t, y_t]$ within the square modeling the die face. This mapping should be linear and map the four corners of the top of the cube to the four corners of the die face. Generate and print the image containing the pixels from the real calibration scene everywhere except for the top of the cube, which should contain the pixels from the model of the die face. Repeat the above process using the same mapping function g but using an arbitrary square scanned image, perhaps of your own face.

Exercise 15.6: Synthesizing the image of a cube.

Use a real camera matrix obtained as in Exercise 15.5, or use one from Chapter 13. (a) Create a synthetic image of a cube posed somewhere in the FOV of the camera. (b) Create an image as in part (a), except that one of the faces of the cube should be the face of the die as in the previous problem. (*c) Texture map two of the faces with two different photos of human faces using mappings from Chapter 11.

15.9 HCI AND PSYCHOLOGICAL ISSUES

Clearly, there are many opportunities to enhance the quality and bandwidth of the interface between man and machine using devices described in this chapter. In VE, the primary goal is to provide a quality immersive experience as intended. Quality interaction using sight, touch, and force is needed for virtual surgery, for example. Individual differences among humans create difficulties in engineering VR systems; for example, the different sizes and shapes of the head complicates the design of HMDs and vision system differences complicates the control of displays based on stereo fusion. Also, different people have slightly different perceptions of objectively identical stimuli for color, roughness, and sound, etc.

VR systems can produce unintended effects. For example, motion sickness is usually unintended. Other unintended effects can be eyestrain, fatigue, or frustration; for example, all these can result from a stereo fusion subsystem that is not quite matched to the operator or to reality. Even worse, human operators of flight simulators have been known to suddenly become lost in their virtual environments! These issues provide many challenges to VR system designers.

15.10 REFERENCES

An overview of the state of the art in virtual reality is provided in the 1995 report edited by Durlach and Mavor. A broad set of papers reporting recent work on various aspects of VR is included in the 1995 collection edited by Barfield and Furness. Both books provide extensive background, including description of devices, definitions, excellent examples and figures and categorized references to the literature.

Virtual reality has diverse applications in medicine. Discussion of its use in planning surgery and in interactive segmentation of 3D imagery is given in a paper by Poston and Serra (1996). Our treatment of dextrous virtual work drew heavily from the ideas in that paper. Use of VR in rehabilitation and in therapy treating various phobias is described in a set of papers edited by Strickland (1997). Creating object models and defining their behavior and motion is a difficult problem of current research effort. For a description of the problems and some methods for their solution, see the papers by Green and Halliday (1996) and Deering (1996). The immersive power of VR systems can not only help humans, but it can also make them sick or irritated, especially if stereo and motion imagery is not very carefully presented: see the article by Viire (1997). The book by Stuart (1996) contains many useful tables summarizing the properties of human sensory capabilities and the input and output device characteristics needed for VR systems. The discussion of view-based texturing comes from the work of Pulli and others (1997). A recent collection that includes some of the work discussed here is *Mixed Reality* edited by Ohta and Tamura (1999).

1. Barfield, W., and T. Furness III, eds. 1995. *Virtual Environments and Advanced Interface Design.* Oxford University Press.
2. Biocca, F., and F. Levy. 1995. *Communication in the Age of Virtual Reality,* F. Biocca and M. R. Levy, eds. L. Erlbaum Associates, Hillsdale, NJ.

3. Deering, M. 1996. The Holosketch VR Sketching System, *Communications of the ACM,* v. 39(5):54–61.
4. Durlach, N., and A. Mavor, eds. 1995. *Virtual Reality: Scientific and Technological Challenges.* National Research Council, National Academy Press, Washington, D.C.
5. Green, M., and S. Halliday. 1996. A geometric modeling and animation system for virtual reality. *Communications of the ACM,* v. 39(5):46–53.
6. Ohta, Y., and H. Tamura, eds. 1999. *Mixed Reality: Merging Real and Virtual Worlds.* Ohmsha, Ltd., Tokyo, Japan; also distributed by Springer-Verlag, New York.
7. Poston, T., and L. Serra. 1996. Dextrous virtual work. *Communications of the ACM,* v. 39(5):37–45.
8. Pulli, K., M. Cohen, T. Duchamp, H. Hoppe, L. Shapiro, and W. Stuetzle. 1997. View-based rendering: visualizing real objects from scanned range and color data. *Proc. 8th Eurographics Workshop on Rendering* (June 1997).
9. Strickland, D., ed. 1997. Special issue on VR and health care. *Communications of the ACM,* v. 40(8).
10. Stuart, R. 1996. The Design of Virtual Environments. McGraw-Hill, New York.
11. Viire, E. 1997. Health and safety issues for VR. *Communications of the ACM,* v. 40(8):40–41.

16

Case Studies

This chapter describes two different commercial systems that use computer vision and pattern recognition techniques to solve real application problems. These application solutions let us see an entire system design that integrates different hardware and algorithms. Most—but not all—methods used have been treated in some previous section of this textbook. The first case studied is IBM's `VeggieVision` system for identification of produce at the supermarket checkout station. The second is an iris identification system for verification or recognition of a person's identity at an ATM or secure facility.

16.1 VEGGIE VISION: A SYSTEM FOR CHECKING OUT VEGETABLES

Use of barcodes has greatly reduced the amount of human labor required in selling products at the supermarket. Handling produce, however, continues to be labor intensive. Sometimes produce items, such as potatoes or apples, are packaged and barcoded in advance so that they can be handled in the same way as canned or boxed products. Many items, however, are loose so that the customer can choose individual pieces; for example, tomatoes or even green beans. Loose items may or may not be put in a plastic bag by the customer. In a typical store, such items are placed on a scale to determine the weight. The checker possibly has to identify the product and type its code into the register. This problem begs for an automated solution—why not have a camera in the scale to automatically identify the type of produce? Such a system would greatly simplify the job of the human checker and improve inventory control. A system called `VeggieVision` has in fact been developed at the IBM T. J. Watson Research Center. Laboratory testing has shown the system to be effective, and it is now under field test. An automated system has other advantages, such as more refined pricing by produce size or ripeness. The next sections give more details of the supermarket produce problem and the solution developed by IBM. We give special acknowledgement to Bolle, Connell, and others (1996) for providing documentation of `VeggieVision`. The reader can consult their publication cited in this chapter's Reference Section for more details than are given here.

16.1.1 Application Domain and Requirements

Roughly $m = 350$ different produce items are sold across the United States, but an individual market is likely to sell only about 150 items. Neither of these numbers present a difficult problem for automating produce identification. In order to be economical, an automated system must make an identification in approximately one second using computing equipment no more costly than what already is in current supermarket scanners and computers. It is desired that any new equipment be added into the same space as occupied by current store equipment and that no changes need to be made to the existing store environment.

The system must be adaptable to individual store environments because of several factors. First, not all stores carry the same produce items. Second, produce items change by season and even by day for the same store—consider, for example, a shipment of bananas that arrives somewhat green and gradually yellows. An effective system must be designed so that it can nicely adapt to such changes and have the capability of being extended to handle new items.

Finally, human operators must have an acceptable role in the operation of the overall system. This includes initial training to learn how to use the system, operation of the system in automatic mode, and making decisions when the automatic process is stymied for some reason. The overall system, including the machine and the human operator, must be much more efficient than the manual technique currently used in most stores. Figure 16.1 shows the

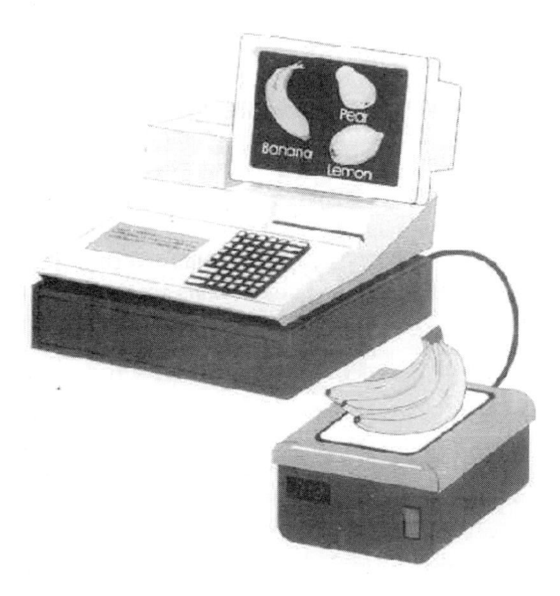

Figure 16.1 Designer's sketch of final supermarket system. (Likely produce items are displayed in case human interaction is needed.) (Courtesy of R. Bolle and J. Connell.)

desired overall system: A touch-sensitive display is included to show results to the checker and to allow the checker to make an identification in case the automatic system is unsure.

16.1.2 System Design

Hardware Components The scanning hardware had to fit within space similar to that already used for weighing produce and scanning barcodes. Moreover, it had to operate in various store environments without any special alterations. Figure 16.2 shows the scanner that was designed. Polarizing filters were used on both the light source and the camera: The direction of the camera filter is orthogonal to the direction of the filter used on the illumination in order to remove specular reflections from the produce. A digital signal processing chip (DSP) was chosen to economically perform image-processing operations within the one second cycle time target. The color camera and DSP act as a low rate input device for the cash register, which only needs an identifier and a weight for each item of produce placed on the scale.

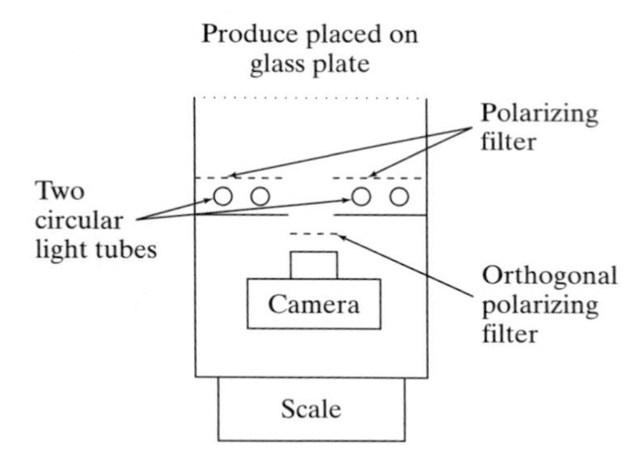

Figure 16.2 A vertical cross section of the scanner design shows color camera and polarizing light source within existing scale and barcode-reader.

Representation and Recognition Previous applications showed that color histograms promised to be effective features: Research and development confirmed this. Classical texture features did not perform well for this problem, so some problem-specific features were developed as described below. A simple shape feature was also used. Color, texture, shape, and size histograms for a produce image were combined to form a feature vector Q of dimension $d > 100$ to represent the unknown produce on the scale. Figure 16.3 shows only the color histograms for some apples (left) and some oranges (right).

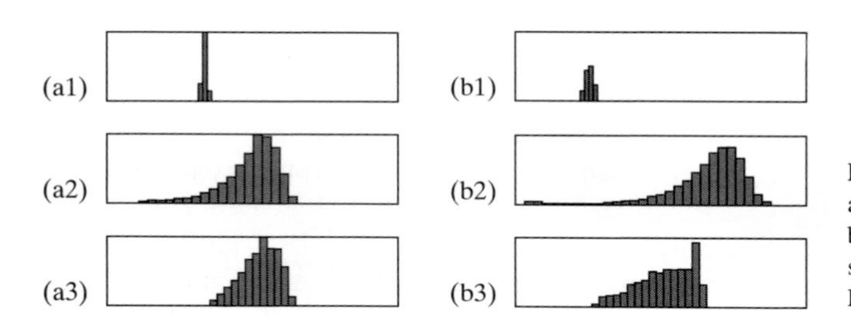

Figure 16.3 (left) Color histograms for apples and (right) oranges. From top to bottom the histograms are for hue, saturation, and intensity. (Courtesy of R. Bolle and J. Connell.)

Due to the need for the system to adapt to varying conditions, a nearest-neighbor classification scheme was chosen. Labeled samples of feature vectors from known produce items are stored in a simple memory array. With the maximum of $m = 350$ classes each with up to 10 samples, 3,500 samples can be stored. The DSP can easily compare a query feature vector Q to all 3,500 labeled samples within one second in order to find a set of the k nearest neighbors. There is no special data structure used to organize the training samples: This makes update simple. Each sample vector is stored with associated information so that the vector can be aged and possibly deleted from memory as the samples are used over time.

The distance between the query vector Q and the jth training sample of class L, ${}^LP_j$ is computed as in Chapter 8. Since all the individual features are counts from a histogram, the distance $d(Q, {}^LP_j)$ is the absolute value of the different counts Q and ${}^LP_j$.

$$d^j = d(Q, {}^LP_j) = \sum_{f \in F} w_f \, d(Q, {}^LP_{j,f}) \tag{16.1}$$

Thresholds control the identification and determine its certainty. There is a distance threshold t that defines whether or not Q is close enough to some sample or samples P_j: Let k_t be the number of neighbors close to Q selected from memory. The identification procedure is given in the next sections.

16.1.3 Identification Procedure

The overall algorithm for identifying the produce on the scale is given in Algorithm 16.1. Some of the steps are described in more detail in Section 16.1.4. Use of the nearest neighbors in the training samples to identify the produce is sketched in Figure 16.4.

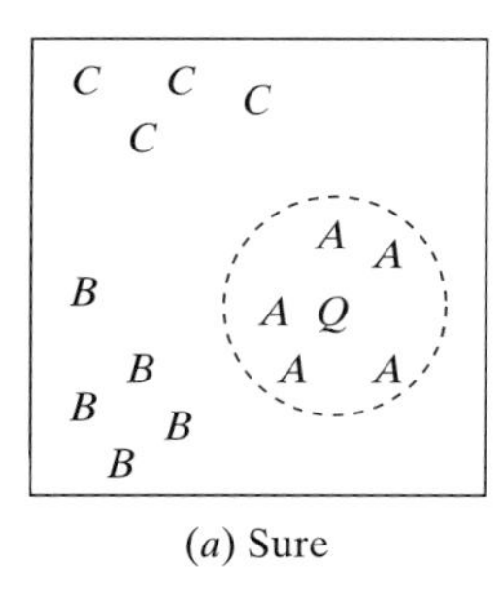

(*a*) Sure

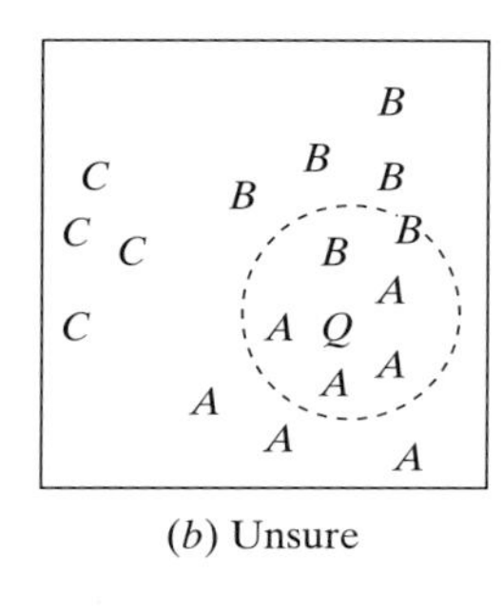

(*b*) Unsure

Figure 16.4 Sketch of concepts of decision-making in feature space: A 2D feature space is shown whereas the actual feature space is d-dimensional. (*a*) A *sure* identification results when all training samples within t of Q are from only one class. (*b*) The decision is unsure otherwise: Either the produce is scanned anew, or the system asks the operator to choose from close classes A and B.

16.1.4 More Details on the Process

Obtaining Images of Produce Images need to be taken with little control of the store environment. In particular, the camera within the scale will sense light from or reflecting from the ceiling. Two images are taken with the produce on the scale, one using the light source within the scale (lights-on) and one without this light source (lights-off). The three regions to be segmented are (1) the produce region is dark in the lights-off image and bright in the lights-on image; (2) the background region has similar brightness in both the lights-on and lights off images; and (3) if a plastic bag encloses the produce, it is not dark in

Identify the produce placed on the store checkout scale.

1. Upon operator signal, take lights-off image and lights-on image and extract foreground produce from background.
2. Make histogram of color features, texture features, shape features, and size features; concatenate them to form feature vector $\boldsymbol{Q}$.
3. Compare $\boldsymbol{Q}$ to each training sample ${}^{L}P_{j}$ in memory; discard all differences larger than t; sort the remaining samples into ascending order.
4. If the closest K neighbors all have the same label L, then label L is returned as the identity of $\boldsymbol{Q}$ and the identification is *sure*. In this case, the system can make the decision automatically.
5. If the decision is not sure for the first time: Ask the operator to reposition the produce and repeat the steps 1–4 above.
6. If the decision is not sure for the second time: Display up to N choices of labels from the sorted list for operator decision.
7. If appropriate, contribute $\boldsymbol{Q}$ to the set of training samples and possibly delete other training samples.

Algorithm 16.1 Overall flow of produce identification used in `VeggieVision`

the lights off image and not bright in the lights-on image. With this engineering, thresholds can be set so that the produce region is segmented out from the bag and background. To obtain quality color, polarized light is used within the scale to inhibit specular reflections, which will not be indicative of the produce surface. Moreover, the produce surfaces that are imaged do not receive uncontrolled illumination from outside the scale; thus, the sensed colors will be consistent even with variations of room lighting.

Computing Features Features are computed only for the produce region[s]. Features must be rotation invariant, but not size invariant. Histograms of four types of features are created and concatenated to form a single vector Q representing the unknown produce. The four features are color, texture, shape, and size.

The **color** of each pixel is transformed from RGB representation to (h, s, i) in HSI. Each value h, s and i is contributed to a histogram: Pixels where the intensity or saturation is low are not used due to numerical instability of the conversion. The three histograms are normalized with respect to the total area of the produce regions. Figure 16.3 shows the three separate color histograms for apples versus oranges.

A small set of **texture** features are computed at each pixel of only the green channel of the original color image. Center-surround masks of different sizes are used for computing texture features. Center-surround masks are just a center box region of positive weights and a surrounding background box region of negative weights. Computation is sped up by using a subsampled image. Both positive and negative responses to the masks are contributed to a histogram. The size of the central peak gives overall texture of the object; if the central peak is large, this means many low magnitude responses to the masks or very little texture. The spread of the histogram tells us about the level of texture contrast, such as sharp shadows between leaves versus subtle surface striping. Histogram asymmetry tells something about the size of the textons relative to the mask scale; for example, a skew toward positive differences suggests that the item has large leaves with narrow cracks between them versus something more *crinkly,* such as parsley.

A simple scheme is used to measure **shape.** The boundaries of the produce regions are smoothed and followed to compute curvature at each boundary pixel. Only the external boundary segments are used; the image frame and places where produce items touch each other are not used. The square of each curvature value is contributed to a histogram in order to yield better clustering. Spherical produce should result in a narrow peak corresponding to the radius; the actual location can differentiate between lemons from grapefruits, for example. Elongated produce, such as bananas and carrots, result in a broader set of values and a peak near zero. Leafy vegetables give broad distributions of curvature.

The fourth feature histogrammed is **size.** A *size* value is computed for each foreground pixel and not just for aggregates of pixels. Run-lengths are computed within the binary foreground mask in four directions (horizontal, vertical, and the two diagonals), creating four directional images. In each directional image, the directional size of a pixel is the overall length of the run of which it is a part. The size of a foreground pixel is then taken to be the minimum directional value at that pixel. Object size is thus defined without parametric models and without any segmentation beyond the original foreground versus background segmentation. A bunch of grapes segments into a *puffy cloud* foreground mask. Pixels on the outside bumps get small run-lengths while the interior pixels get much longer run-lengths. Thus the size histogram will have two peaks: One for the individual grapes, and one for the overall size of the bunch. A carrot will have a narrow peak in the size histogram at its characteristic width, which happens to be similar to the width of cherry tomatoes, and a peak around zero curvature in the shape histogram meaning that it is elongated, which tomatoes do not have.

Supervised Learning The use of nearest-neighbor classification provides for both bounded computation time and simple training and adaptation. Initially, the system can be trained by showing it several examples of the various produce items in the store and providing the class identification (inventory code). When the system is in use, the human operator can require that a new feature vector Q be added to the set of training samples. A sample can be deleted because it is redundant relative to the geometric structure or useage factor of the samples of its class. When training, a new sample that is correctly classified using existing samples is not memorized if it is within distance t_2 of the best match; otherwise, it is memorized. This allows multiple modes to form in feature space; for example, one for broccoli heads alone and one for broccoli with long stalks. When the limit of M samples has been reached for a class, the *least used* sample is erased. A usage count is incremented by $I+$ whenever a sample is the closest one and decremented by $I-$ when it is not.

Training `VeggieVision` from scratch in every store appears to be unnecessary. Experiments have shown that recognition performance will be depressed if training samples from a different store are used. However, the system can adapt, as described above. Human intervention will be high in the beginning as the system adjusts its training base to the produce actually leaving the store, but the overall human effort should be much less than beginning from scratch.

16.1.5 Performance

The developers have published the results of several experiments performed over a period of time. Details of the many variables studied can be found in the cited reference. In a recent study using 5,300 images over four different stores; the system was sure and correct 89 percent of the time; was either correct or had the correct category as the top choice for the operator 93 percent of the time; or was either correct or had the correct category listed somewhere in the first four operator choices 96 percent of the time. The operator may be asked once to reposition the produce. If `VeggieVision` is unsure in two tries, identification can be handled by the checker touching an icon on a display. It appears that the system would reduce labor considerably even if the checker would make all decisions using a touch-sensitive CRT.

Exercise 16.1

Assuming that bananas are rectangular, what should their shape histogram look like?

Exercise 16.2

Sketch and compare color, texture, and shape histograms expected for red apples versus yellow bananas.

Exercise 16.3

Suppose a customer combines three apples and two oranges in a plastic bag. Should our identification system be able to handle this? If so, how?

16.2 IDENTIFYING HUMANS VIA THE IRIS OF AN EYE

We now describe a system to identify persons by scanning the iris texture of an eye. The sensing hardware for an ATM environment is built by Sensar, which licenses from IriScan the software that performs feature extraction and matching. We give special acknowledgement to Gary Zhang of Sensar (1998) and John Daugman (1994, 1998) of Cambridge University for providing information and figures describing this application.

Identification of a person has always been an important problem for society. Correct identification must be established for commerical and legal transactions; for example, a person withdrawing cash from a bank account or changing an address of residence. Currently, this is often done by the person showing some document such as a driver's license or birth certificate to another person who controls in some way the action to be taken. In today's world many transactions are done via machines or computer networks: security and privacy are usually provided by the person (a) knowing a unique account number together with (b) a *password* or Personal Identification Number (*PIN*) associated with the account. These codes can be obtained, with or without permission, by other persons who may then do transactions thereby breaking the normal responsibility and control.

In addition to the very important applications in electronic commerce are those in police work. Fingerprints have been commonly used. Crime scenes are examined for fingerprints that might identify persons who were there. Fingerprints are also used in contexts where the person cooperates in the identification; for example, they are often used to identify workers in a secure environment. Fingerprints have been extensively studied and used for over one hundred years. Several electronic devices have been developed so that fingerprints from a cooperating person can be easily input to computer networks or other systems [see Jain and others (1999)]. Face recognition is also under intense development for identification and authentication (authorization) systems. These evolving systems have the capability of identifying people without their knowledge, such as in an airport, bank, or hotel. Thus they have obvious use in police work and in providing security, but also are problematic in terms of acceptability and privacy.

16.2.1 Requirements for Identification Systems

We consider systems performing one of two kinds of operations: (a) to identify a person, cooperating or not, from a large set of possibilities, and (b) to confirm that a cooperating person is indeed the one he or she claims to be. The latter case is often called authentication. Some of the requirements are not obvious, so we include a list of them. The system design will be dominated by the particular *biometric* used and how it is sensed and encoded for machine use. Three important biometrics are the appearance of the human (1) fingerprint, (2) face, and (3) iris of the eye. It will be argued below that the iris of the eye provides better information than the fingerprint or face.

1. The system must obtain information from a human with minimal inconvenience,
2. the biometric code must have little variance as it is obtained from the same person over time,
3. the biometric code obtained from one person must be significantly different from that of other persons (The set of persons to be discriminated will vary from one application to another.),
4. it must be very difficult to fool the system with fake data, such as an image printed on paper, and
5. the system must be cost effective relative to the particular application.

Before going on to describe the iris scanning system, it is instructive to make some comparisons among different biometrics regarding how they satisfy the above requirements. Besides those mentioned above, we add analysis of DNA as a biometric.

1. **Convenience in obtaining information:** fingerprint(fair); face(good); iris(good); DNA(poor). Cheap digital scanners exist for fingerprints but the user must carefully

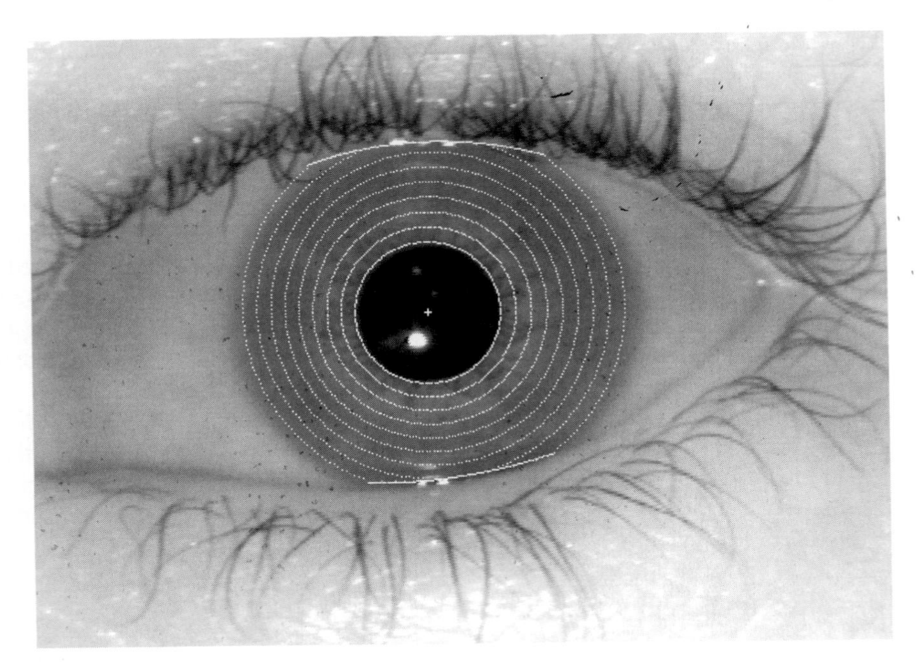

Figure 16.5 Narrow FOV image of the eye: Image processing has identified eight different bands of the iris from which texture features are extracted. (Contributed by John Daugman, Cambridge University. See Web pages www.cl.cam.ac.uk/~jgd1000/)

present a finger to them. A face can be effectively sensed by a cheap video camera with little inconvenience; more expensive optics and more control is needed to obtain a quality image of the iris. Obtaining DNA is, of course, an expensive off-line lab procedure, usually reserved only for important court cases.

2. **Low intraclass variance:** fingerprint(good); face(fair); iris(excellent); DNA (excellent). Note that strong deformations are added in taking fingerprints and that facial appearance can vary with pose, mood, hair, and age. The iris develops its texture before a child is born and changes very little over life, and a scanning system has been developed to produce consistent encodings of it.
3. **High interclass variance:** fingerprint(good); face(good); iris(excellent); DNA (excellent). While fingerprints provide excellent discrimination in the hands of experts, they are not quite as good when represented for automatic methods. Most people have doubles in terms of facial appearance—especially twins. Twins can, by themselves, create a 1 percent error rate! And, twins have the same DNA. Interestingly, twins do not have the same iris texture. In fact, the textures of the two eyes from the same person are just as uncorrelated as are the textures from eyes of different individuals.
4. **Difficult to fool:** fingerprint(good); face(good); iris(excellent); DNA(excellent). Several systems that use fingerprints or faces can be fooled by pictures or simple

models of a person's appearance. The pupil of the eye, which is inside the iris, undergoes size changes that can be tracked by a sensing system as a check against subterfuge.

5. **Cost effective:** fingerprint(fair); face(fair); iris(fair); DNA(poor). Fingerprints and faces can be sensed cheaply; however, methods of matching their representations can be expensive. Scanning the iris is expensive but matching is simple. Using DNA, of course, is expensive in time, human labor, and materials.

16.2.2 System Design

For concreteness, we describe the application of iris-scanning to identify users of an automatic teller machine (ATM). When the ATM user approaches the ATM, the iris of one eye is scanned and the *Sensar...Secure™ System* identifies the person in the customer records. The customer may then have access to the account or perhaps is asked to type a password for additional security. Variations in the application, such as opening a secure door, can be handled by varying the design parameters set below.

Careful scanner design and special optics are needed in order to obtain a high resolution image of a such a small object relative to the large 3D FOV in which it might appear. Moreover, 3D analysis is required in order to find the front person in case there is a queue waiting to use the machine. Once the front person's eye is located and scanned, patented software is used to obtain a $d = 2{,}048$-dimensional binary vector Q representing the gray-level texture of the iris. Matching this vector to a set of vectors representing some universe of persons is performed quite simply by computing minimum Hamming distance, which is just the number of bits in which two binary vectors differ.

Hardware Components A sketch of the *Sensar...Secure™* distributed processing architecture is given in Figure 16.6. The system is composed of four main units: (1) a general purpose computer that provides an interface with an application process and the sensing control and video processing units; (2) an optical platform with three cameras

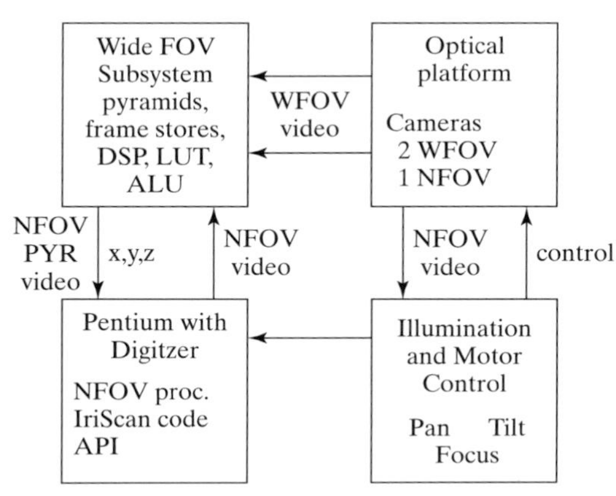

Figure 16.6 *Sensar...Secure™* distributed processing architecture.

that obtains both wide field of view (WFOV) images and near field of view images (NFOV); (3) the control unit for the optical platform; and (4) the video processing unit, which has special hardware for real-time processing of stereo video.

The two wide field of view cameras obtain a video stream that is used to locate the frontmost person in the field of view. The two video streams are passed to the signal processing unit, which performs the actual stereo processing using multiresolution pyramids. The x-y-z location of the designated eye of the person is passed to the main unit, which then uses that information to control the optical platform so that it can obtain the near field of view imagery of the eye. This cycle can be performed every half second so that a slowly moving person can be tracked. The near field of view video is processed by the main unit so that the specific eye region can be located and the iris code extracted. The overall process is given in Algorithm 16.2.

Note that the sensing hardware is more complex than what is used in many other systems and this limits the applications. The high cost is primarily a result of the requirement that sensing is passive as far as the user is concerned: The user can move freely in the work envelope and the system must locate her. This requirement necessitates the wide field of view sensing to find the subject followed by the near field of view sensing to obtain the needed image of the eye. The need for real-time stereo implies special hardware, hence the two resolution pyramids to speed up correlation of points in the two video streams.

Representation The ultimate representation of the eye and person is just a binary vector of dimension 2,048 (256 bytes of storage). A graphical representation of one such vector is given in Figure 16.7: 0 is printed black and 1 is printed white. Each bit of the code is determined by the sign of the result of correlating a specific Gabor filter with a specific neighborhood of the iris image. It is thus very important that the eye image be normalized for rotation before correlation is performed.

As shown in Figure 16.8, each bit of the iris code is determined by correlating a 2D Gabor wavelet with the image of the iris at a particular location (ρ_0, ϕ_0) on the iris and

Compute identity ID of closest person of scene.

1. Using wide FOV video and correlation-based stereo, locate closest head.
2. Identify location of face features using templates; then identify left [or right] eye at $[x, y, z]$.
3. Using $[x, y, z]$, near FOV monochrome camera obtains in-focus centered image I of eye.
4. 2,048 bit (256 byte) iris code $\boldsymbol{Q}$ obtained from image I of the eye using patented image processing software.
5. Iris code $\boldsymbol{Q}$ matched to those in database using XOR.
   If the two codes differ by fewer than K bits, then return person ID;
   else, return "reject".

Algorithm 16.2 Identification of a person enrolled in the database by iris scanning.

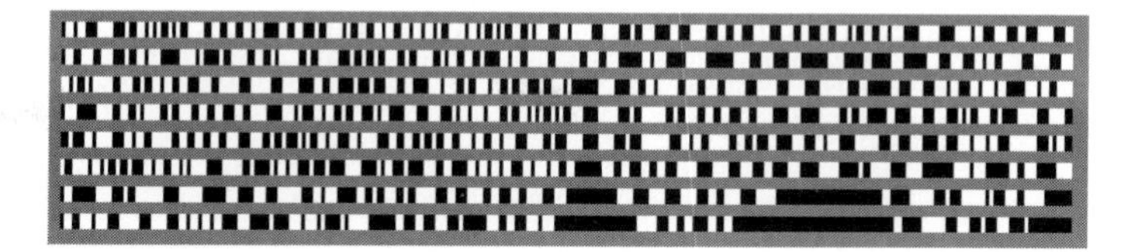

Figure 16.7 Graphical representation of 2,048 bit code representing the signs of the results of applying Gabor filters of different sizes to locations within the eight bands shown in Figure 16.5. (Contributed by John Daugman, Cambridge University.)

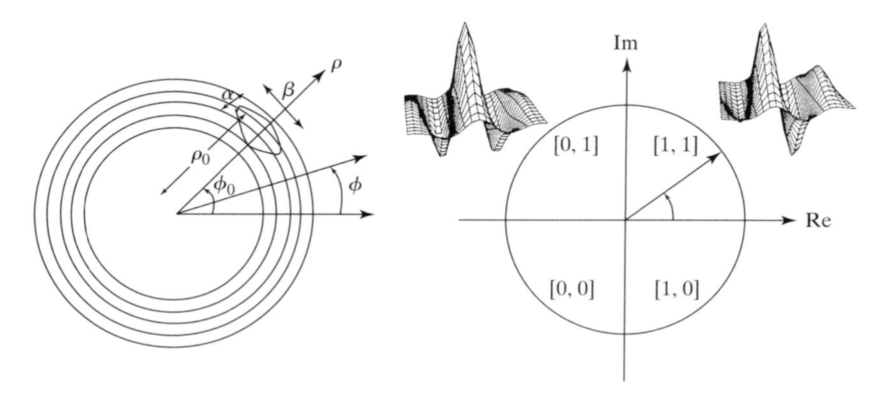

Figure 16.8 (left) Sketch of annular regions of the iris defined by a radial range $\rho_a \leq \rho \leq \rho_b$ and placement of a Gabor wavelet at location (ϕ_0, ρ_0) with spread parameters α and β; (right) shape of complex valued 2D Gabor wavelet. (Contributed by John Daugman, Cambridge University.)

with particular spread parameters α and β (demodulation). The cross section of the wavelet along the ρ direction is a Gaussian with spread α, while the cross section along direction ϕ is a Gaussian with spread β modulated by a sine wave. Each correlation of a wavelet with the picture function yields a complex number c as follows.

$$c = \int_\rho \int_\phi f(\rho, \phi)[\, e^{-j2\pi(\phi-\phi_0)} e^{-(\rho-\rho_0)^2/\alpha^2} e^{-(\phi-\phi_0)^2/\beta^2}]\rho\, d\rho\, d\phi \tag{16.2}$$

The complex correlation is converted to two bits of the iris code by testing its sign: *if* $(Re(c) \geq 0.0)$ *then* $b_{real} = 1$ *else* $b_{real} = 0$ and *if* $(Im(c) \geq 0.0)$ *then* $b_{img} = 1$ *else* $b_{img} = 0$. Clearly, any rotation of the image of the iris about the view direction will affect the location parameter ϕ_0. Any rotation will be small because the near field of view image can be obtained using information about the location of both eyes in the wide field of view image. During matching, the iris code is matched after undergoing slight rotations and the best match of these rotated codes to any code from the database of candidates is used. The ρ axis is defined in terms of the boundary of the pupil and the outer boundary of the iris: These two boundaries are assumed to be circular but not necessarily with the same center. They are found by integrating edge evidence in much the same way that the circular Hough transform does. The boundaries are defined by the two sets of parameters ρ, x_c, y_c that maximize the gradient magnitude around the circle

$$max_{(\rho, x_0, y_0)} \left| \frac{\partial}{\partial \rho} \oint \frac{f(x, y)}{2\pi\rho} ds \right| \tag{16.3}$$

Identification Procedure A sketch of the identification procedure is given in Algorithm 16.2. Given our previous discussion, all important details of the procedure have been covered. Comments on performance are given in the next section.

16.2.3 Performance

The time for the system to acquire and identify an iris scan varies with conditions and normally is within the range of one to five seconds. This should satisfy the requirements for an ATM system, but may not be fast enough to process persons moving through current airport security systems, for example. For the ATM application, the mechanical aspects of camera control are the limiting factors: Perhaps 90 percent of the time is spent on acquiring the image. After the image is passed to the algorithms, it takes $200 msec$ to locate the iris boundaries and to generate the IrisCode. Matches proceed at the rate of about 100,000 persons per second.

Most important is the probability of the system making an error in identification. From a theoretical model fitted to many test cases, Daugman (1998) made the following estimates of error rates. If 70 percent of the 2,048 bits must match in order to verify that the person is the one claimed, then the chances of accepting an imposter is about 1 chance in 6×10^9, while the chances of rejecting the true person is about 1 in 46,000. If the threshold is reduced to 66 percent, then the false accept and false reject rates are equal at about 1 chance in 1 million.

We conclude with a few words about the model used to estimate the above probabilities. The reader is referred to the Daugman paper (1998) for more detail. All possible pairs of iris scans from about 300 people were compared yielding the results shown in Figure 16.9. It was found that (a) when comparing different eyes, the distribution of Hamming distances (over 200,000 pairings) ranged between 0.4 and 0.6 of the bits; (b) a binomial distribution with $N = 266$ degrees of freedom and $p = 0.5 = q$ fit the observed distribution very well; and (c) perhaps surprisingly, the same type of distribution was observed in comparing the left eye and right eye scans from the same people, showing that scans from the two eyes of the same person are as uncorrelated as the scans from two different people. Figure 16.9 plots the distribution of Hamming distances between codes of different persons (right mode) alongside the distribution of Hamming distances between codes from the same person (left mode)—the crossover point is at probability 10^{-6}. The decision threshold need not be set at the crossover point. If a distance of at most 30 percent of the bits is tolerated, then the probabilty of accepting an incorrect match is one in six billion, which might be desirable when the cost of a false accept is much greater than the cost of a false reject.

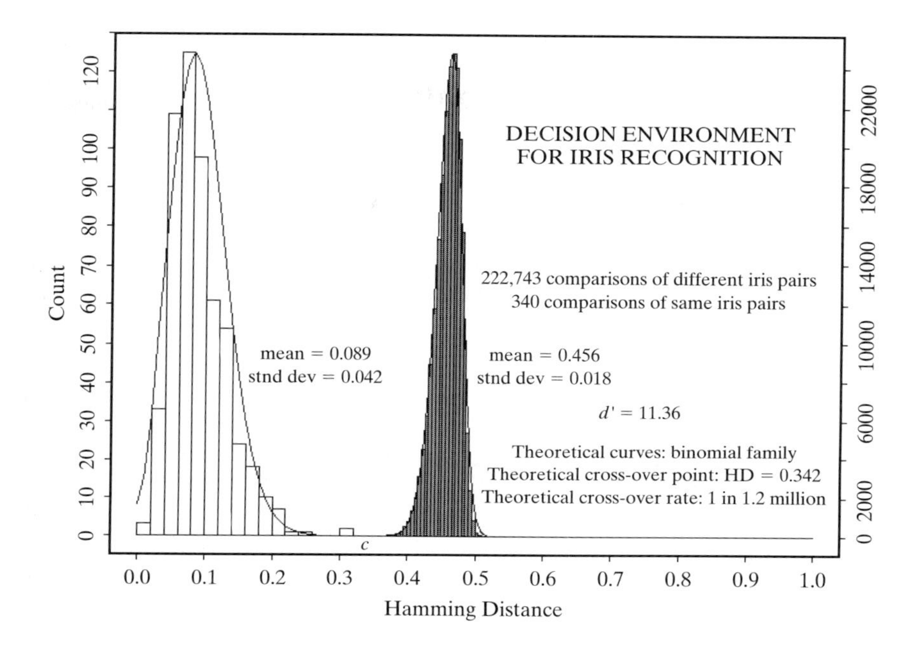

Figure 16.9 Distributions of Hamming distance for same eye (left mode) and different eyes (right mode). The crossover point is at 0.34 of the bits, where the probabilty of falsely dismissing a correct match equals the probability of falsely accepting an incorrect match: Both are about 10^{-6}. (Figure contributed by John Daugman of Cambridge University.)

16.3 REFERENCES

1. Bolle, R., J. Connell, N. Haas, R. Mohan, and G. Taubin. 1996. VeggieVision: a produce recognition system. *Proc. IEEE Workshop on Applications of Comput. Vision*.
2. Camus, T., U. M. Cahn von Seelen, G. G. Zhang, P. L. Venetianer, and M. Salganicoff. 1988. Sensar...Secure™ Iris Identification System, in *Proc. IEEE Workshop on Applications of Comput. Vision* (9–21 Oct. 1998), Princeton, NJ, 254–255.
3. Daugman, J. 1994. Biometric Personal Identification System Based on Iris Analysis. U.S. Patent No. 5,291,560 issued to John Daugman (March 1, 1994).
4. Daugman, J. 1998. Recognizing persons by their iris patterns. In *Biometrics: Personal Identification for a Networked Society,* A. Jain, R. Bolle, and S. Pankanti, eds. Kluwer Academic, Dordrecht, Netherlands and Norwell, MA.
5. DellaVecchia, M., T. Chmielewski, T. Camus, M. Salganicoff, and M. Negin. 1998. Methodology and apparatus for using the human iris as a robust biometric. *Ophthalmic Technologies VIII: SPIE Proc.* (Jan. 1998), 24–30.

Index

accidental alignments, 394
accumulator array, Hough
transform, 304–9
active contours models, 489–92
addition, image, 12
additive color systems, 192
affine calibration matrix, 431–37
affine mapping, 2D object recognition
via, 341–50
affine mapping functions, 329–39
rotation, 330–31
scaling, 329–30
translation, 332
affine method, 436–37
affine transformations, 338–41
3D, 413–21
affine warp, 334–35
aggregating: consistent neighboring edges into
curves, 301–3
motion trajectories, 321–24
albedo, definition, 204
algorithms, alignment, 497
back-propagation, 126
border, 295–97
boundary matching, 237–38
classical clustering, 281–82
classical connected components, using
union-find, 61–62, 65
classification, 101, 103–4
computing 3D surface coordinates using
calibrated camera and projector, 438
computing output image from input image,
139–41
conventions for defining, 555–56
converting RGB (red-blue-green) encoding
to HSI (hue-saturation-intensity)
encoding, 196
cylindrical warp of image region, 364–66
decision procedure, 99
deriving motion vectors for interesting
points, 260
detecting interesting image points, 258
discrete relaxation labeling, 357
finding straight line segments, 305–6, 308
flow of produce identification in Veggie
Vision, 552
geometric hashing offline
preprocessing, 349
geometric hashing online recognition, 350
graph-matching, 106
Greedy Exchange, 267, 268, 270
histogram equalization, 133
holecounting, 6
Hough transform for finding circles, 310
Hough transform for finding straight
lines, 306
identification by iris-scanning system, 558
image histogram, 84

interpretation tree search, 354
isodata clustering, 284
iterative K-means clustering, 282
iterative P3P solution, 443
labeling block edges via backtracking, 381
labeling block edges via discrete relaxation, 382
labeling edges of blocks via backtracking, 381
local-feature-focus, 343
O'Gorman and Clowes method for extracting straight lines, 305, 306, 308
perceptron learning, 122–23
photometric stereo with three light sources, 473
RAG (region adjacency graphs), 82
recognition-by-appearance using basis of principal components, 520
recursive labeling, 57–59
relational distance matching, 504
rigid transformation for aligning model triangle with congruent world triangle with, 421
row-by-row labeling, 59
Shi's clustering procedure, 289
single update stage for active contour, 491
space-carving, 464–67
tracking edges of binary edge image, 303
transformation from model features to image features using pose clustering, 344
union-find, 59
using color and motion to track ASL gestures, 322–24
watch-gear inspection, 68–71
alignment(s): accidental, 394
matching geometric models via, 496–504
smooth object, 501–4
3D-3D, 496–98
via transformation calculus, 419–21
2D-3D, 498–501. *See also* matching
ambient light, 207–8
definition, 208
analog images, definition, 29
angular field of view, 31
ANNs (artificial neural networks), 120–26
Mach band effect produced by, 153–55
appearance, 3D object recognition by, 516–23
application problems, 3–10
applications, of binary morphology, 68–71
AR (augmented reality), 530–32
architectural walkthrough, 529
arcs, detecting with Hough transform, 303–12
area, 73
arrays: CCD, 24–25, 26
PARENT, 59–61
pixel, 43–45
arrow junctions, 377
artificial neural networks (ANNs), 120–26
Mach band effect produced by, 153–55
artificial neurons (AN), 120–22
aspect graphs, 488–89
aspect, 488
assignment, definition, 351
auditory output, virtual reality (VR) systems, 539
augmented reality (AR), 530–32
definition, 535. *See also* virtual reality (VR) systems
autocorrelation, measuring texture by power spectrum and, 221–23
automatic thresholding, 85–89
axis (axes): best, 79–81
ellipse, lengths and orientations, 78–79
with least second moment, 81

B+-tree indexes, 245–47
background pixels, 51
backprojection, 200
back-propagation algorithm, 126
backtracking, labeling block edges via, 381
balloon models, 3D, 493–94
bandpass filtering, 181
basis, orthogonal, using, 160–62
basis images: computing, 519–21
for set of training images, 518–19
bay_above_bay, 112
bay_above_lake, 112
bay_num, 112
Bayesian classifier, definition, 115
Bayesian decision-making, 114–15
bays, 104, 112
best affine calibration matrix, 431–37
best axis, 79–81
binary decision trees, definition, 108
binary image(s), 24
analyzing, 51–91
closing, 65, 67
definition, 30
dilation, 65, 66–67
erosion, 65, 66–67
labeling connected components, 56–63
morphology, 63–73
opening, 65, 67

binary image(s) (*continued*)
run-coded, 37
translation, 66
binary morphology, 63–73
applications, 68–71
basic operations, 65–68
conditional dilation, 71–73
in medical imaging, 69
structuring elements, 63–65, 68–71
binary partition, 217
local, 217
bins/binning, 85, 308, 346
binsize, 85
blade, 372
definition, 374
blobs, 13; 504–6
block(s): labeling edges of via backtracking, 381
labeling edges of via discrete relaxation, 382
labeling of line drawings of, 377–83
blooming, 28
blur, relating resolution to, 406
Boolean features, 112
border algorithm, 295–97
boresighted multispectral sensors, 46
boundary(ies): coding, 292–93
cues from, 393–94
extraction, 295
illumination, 374
interpreting shape from, 391–92
matching, 237–38
in space-time, 321
bounding box, 76
box filter, definition, 136
box smoothing masks, 144
Burns line finder, 311–12

calibration: best affine calibration matrix, 431–37
of cameras, 431–37
of cameras, improved method of, 444–53
of cameras, example, 449–53
of projectors, 437
camera coordinate frame C, 44
camera effects: definitions, 272–73
ignoring, 274–76
camera model, 422–30
parameters, 436–37
camera pan, definition, 272
camera zoom, definition, 272
camera(s): calibration, 431–37
calibration example, 449–53
calibration, improved, 444–53
CCD (charge-coupled device), 22–24
computing 3D points using multiple, 428–30
data acquisition using, 461–63
extrinsic parameters, 445–49
human eye as, 26–27
image formation in, 24–26
intrinsic parameters, 445
posing for stereo configuration, 411–13
video, 26
Canny edge detector, 157–58
and linker, 297–301
case studies: identifying humans via iris of eye, 554–61
Veggie Vision, 548–54
category hierarchy, GRUFF, 515–16
Cauchy-Schwartz Inequality, 161, 162
CCDs (charge-coupled devices): arrays, 24–25, 26
cameras, 22–24, 24–26
variations, 28
centroid, 73
chain code, Freeman, 293
changes, detecting in video, 272–77
character recognition, 98–100
charge-coupled devices (CCDs): arrays, 24–25, 26
cameras, 22–24, 24–26
variations, 28
centroid, 73
child nodes, 107
chromatic distortion, 29
chrominance, 197
circle of confusion, 24
circles, finding with Hough transform, 309–10
circularity, 74
class mean, nearest, used in classification, 101–3
classes, definition, 94
classical connected components algorithm, using union-find, 61–62, 65
classification: algorithm, 101, 103–4
color used for, 198–99
common model for, 94–97
definition, 94
fuzzy, 124
nearest class mean used in, 101–3
nearest neighbors used in, 103–4. *See also* decision trees

classification system(s): building, 95–96
evaluating error rate of, 96
false alarms and false dismissals, 96–97
classifier(s), 95
definition, 94
implementing, 101–4
clearance primitive, 514
clipping, 28
closed form solutions for parameters, 314–15
closing of binary images, 65
definition, 67
clustering, 119
classical, algorithms, 282–82
isodata, 282–84
iterative K-means, 282
methods, 281
methods based on histograms, 284–86
Ohlander's recursive histogram-based technique, 285–86, 287
pose, 344–46
Shi's graph-partitioning technique, 286–89
CMY (cyan-magenta-yellow) subtractive color system, 193–94
code, Freeman chain, 293
coding, boundary, 292–93
collision, 245
color, 187–211
applications, 209
CMY (cyan-magenta-yellow) subtractive color system, 193–94
cube, 194
hexacone, 194, 195
histograms, 199–201, 231, 233
HSI (hue-saturation-intensity), 194–97
human perception, 209–10
images, 45–46
layout, 232–33
physics of, 188–91
pseudo, 210
RGB (red-green-blue) basis for, 191–93
segmentation, 201–2, 322–24
similarity measures, 231–44
triangle, 193, 195
used in Veggie Vision, 552
using for classification, 198–99
compression: data, 36
with JPEG (Joint Photographic Experts Group) format, 38–39
lossless, 36
lossy, 36
MPEG, for video, 261–62
with Motion JPEG (Joint Photographic Experts Group) format, 40
computer vision, definition, 1
conditional dilation: in binary morphology, 71–73
definition, 72
conditioning images, 128–86
confusion matrix, 106–7
definition, 106
connected components: algorithm, classical, using union-find, 61–62, 65
labeling, 56–63
labeling, using run-length encoding for, 62–63
consistent labeling, 351–53
definition, 351
constrained linear optimization, 456–57
constraints: epipolar, 402–3
hard, 490
integrating spatial, 472
ordering, 403
relational, symbolic matching and, 401
3D object recognition, 495–96
content-based image retrieval, 226–50
indexing for with multiple distance measures, 248
continuous relaxation labeling, 356–59
contours: active contour models, 489–92
detecting with Hough transform for lines and arcs, 303–12
identifying regions by, 295–312
identifying with Canny edge detector and linker, 297–301
internal contour energy, 491
intrinsic images, 371–77
of moving objects, 321
contrast, detecting, 141–43
contrast stretching, definition, 132
contributing points, 498
control points, 332–34
definition, 333
converting: RGB (red-blue-green) encoding to HSI (hue-saturation-intensity) encoding, 196
RGB (red-blue-green) to YUV, 197
convolution, 128
cross correlation and, 167–72
definition, 169
operation, 169–72
theorem, 182–83
co-occurrence matrices, 217–20
coordinate frames, 328–29, 413–15, 43–45

coordinate systems, 30–31, 413–15
raster-oriented, 30
coordinates, homogeneous, 329
corner(s), 377
detecting, 320–21
patterns, 4–6
correlation, 128
correspondence(s): cross-correlation, 400–1
epipolar constraint, 402–3
error versus coverage, 403
establishing, 400–3
ordering constraint, 403
pose from 2D-3D point , 455–56
symbolic matching and relational constraints, 401
in 3D-3D alignment, 496–98
counting: holes, 4–6
objects in an image, 54–56
coverage, error versus, 403
crease(s), 373, 377
definition, 374
cross correlation, 400–1
convolution and, 167–72
definition, 169
normalized, 170
crossbar inspection, 4–6
cubes: in octrees, 484–85
used in space-carving algorithm, 466–67
cues: boundaries and virtual lines, 393–94
depth from focus, 393
motion phenomena, 393
from non-accidental alignments, 394
shape from boundary, 391–92
shape from shading, 388
shape from texture, 388–91
3D in 2D images, 383–88
vanishing points, 392
curves: aggregating consistent neighboring edges into, 301–3
detecting with Hough transform, 303–12
segmenting via fitting, 317
cyan-magentag-yellow (CMY) subtractive color system, 193–94
cylinders: generalized-cylinder models, 483–84
cylindrical warp, of image region, 364–66
darkening with distance, 206–7
data: acquisition in 3D object reconstruction, 461–63
compression, 36
gloves, 534
multidimensional, decisions using, 117–19
range, 463–64, 465
databases: image, 226–30
image, queries, 228–30
organizing, 244–48
QBIC (Query by Image Content), 226–27
Decathlete game, 255–56
decision tree(s), 98, 107–14, 522
automatic construction of, 109
binary, definition, 108
nodes, 107
decision-making: Bayesian, 114–15
multidimensional data used for, 117–18
defect_cue, 69, 71
definition tree(s), GRUFF, 515
deformable models, physics-based and, 489–95
density, and direction of edges in analyzing texture, 215–17
depth: cues, 42
human perception of, 394
interpreting via focus, 393
3D cues in 2D images, 383–88
depth of field, 393
definition, 405
focus and, 404–6
depth perception, stereo, 397–403
derivative masks, 141–44
properties of, 143–44
detection: human edge, 153–55
LOG edge, Gaussian filtering and, 149–57
dextrous virtual work, 537–38
DFT (discrete Fourier transform), 179–81
difference operators for 2D images, 144–49
differencing 1D signals, 141–44
differencing masks, detecting edges using, 141–49
diffuse: definition, 204
reflection, 204–5
digital image(s), 3
definition, 29
formats, 35–40
picture functions and, 29–35
problems with, 27–29
dilation: of binary images, 65
of binary images, definition, 66–67
conditional, 71–73
conditional, definition, 72
dimensionality, high, 316
dimensions primitive, 514
direction and density of edges in analyzing texture, 215–17

discrete Fourier transform (DFT), 179–81
discrete relaxation: labeling, 354–56, 357
labeling block edges via, 382
discrimination, improving, 521–23
disparity, definition, 398
dissolve, definition, 272–73
distance: darkening with, 206–7
image distance measures, 230–44
measures, multiple, indexing for content-based image retrieval with, 248
pick-and-click, 234–35
relational, matching, 359–63
distortion: chromatic, 29
geometric, 27
radial, 366–67
distribution: normal, definition, 116
parametric models for, 116–17
probability, 114–15
document retrieval (DR), 97–98
DR (document retrieval), 97–98
dynamic thresholding, 89

edge(s), 377
aggregating consistent neighboring edges into curves, 301–3
block, labeling via backtracking, 381
block, labeling via discrete relaxation, 382
density and direction in analyzing texture, 215–17
detecting using differencing masks, 141–49
detecting with LOG filter, 151–53
human, detection of, 153–55
jump, 373
LOG, Gaussian filtering and detection of, 149–57
surface-edge-vertex models, 480–83
edge detector, Canny, 157–58
and linker, 297–301
edgeness per unit area, 216
eigenspace recognition by appearance, 522
8-neighbors, 52
elastic matching, 240
electromagnetic spectrum, 188
ellipse, 484
axes, lengths and orientations, 78–79
empirical error rate, definition, 96
empirical interpretation of error, 315
empirical reject rate, definition, 96
encapsulated postscript (EPS) format, 39
enclosure primitive, 514
encoding: octrees, 485–86
RGB (red-blue-green), conversion to HSI (hue-saturation-intensity), 196
run-length, using for connected components labeling, 62–63
YUV, 197
energy, minimizing, 491–92
enhancing images, 11–12, 128–86
definition, 130
entropy: computations, 110
of a set of events, definition, 109
epipolar: constraint, 402–3
geometry, 402–3
lines, definition, 402
plane, definition, 402
epipole, definition, 402
EPS (encapsulated postscript) format, 39
equalization, histogram, 132–34
erosion of binary images, 65
definition, 66–67
error(s): coverage versus, 403
definition, 316
empirical interpretation of, 315
false alarms and false dismissals, 96–97
rate, classification system, 96
statistical interpretation of, 315–16
estimation: pose, 453–60
pose estimation procedure, 439–44
Euclidean distance: definition, 100
scaled, definition, 103
even functions, 176
external corners, 4–6
external energy, 492
extracting non-iconic representations, 14
extractor, feature, 94
extremal axis length, 77
extremal points, 76–78
extrinsic camera parameters, 445–49
eye, as camera, 26–27. *See also* iris-scanning system
face(s), 377
finding, 240–41
identifying, 201–2
fade, definition, 272–73
false alarms, 96–97
false dismissals, 96–97
fast Fourier transform, 181–82
feature extraction, 498
feature extractor, 94
feature vector representation, 100
feedforward networks, multilayer, 123–26

field of view (FOV): angular, 31
definition, 31
file formats: GIF (Graphics Interchange Format), 38
JPEG (Joint Photographic Experts Group), 38–39
MPEG (Motion Picture Experts Group) for video, 39–40
TIFF (Tag Image File Format), 38. *See also* formats
file headers, 36
filtering: bandpass, 181, Gaussian, LOG edge detection and, 149–57
images, 128–86
LOG, Marr-Hildreth theory, 155–57
median, 137–41
filter(s): box, definition, 136
Gaussian, 136–37
LOG, detecting edges with, 151–53
masks as matched, 158–67
find procedure, 59–60
fishtank virtual reality, 539
fitting: constraints, 317
models to segments, 312–17
problems, 316–17
segmenting curves via, 317
flesh finding, 241–42
flight simulation, 529
FOC (focus of contraction), definition, 255
focus: depth of field and, 404–6
features, 341–42
interpreting depth from, 393
focus of contraction (FOC), 254
definition, 255
focus of expansion (FOE), 254, 393
definition, 255
FOE (focus of expansion), 254, 255
foreground pixels, 51
foreshortening, 42, 385–86
definition, 384
fork junctions, 377
formats: commonly used, 36–37
comparison of, 40
digital image, 35–40
EPS (encapsulated postscript), 39
GIF (Graphics Interchange Format), 38
JPEG (Joint Photographic Experts Group), 38–39
MPEG (Motion Picture Experts Group) for video, 39–40
PostScript, 39
TIFF (Tag Image File Format), 38
4-neighbors, 52
4-tuples, 508
Fourier analysis, 172
Fourier basis, 174–75
image processing operations using, 175
Fourier power spectrum, definition, 177
Fourier transform, 223
definition, 177
discrete, 179–81
fast, 181–82
FOV (field of view): angular, 31
definition, 31
frame buffer, 23–24
frame grabber, 23
frames of reference, 42–45
Freeman chain code, 293
Frei-Chen basis, 163–67
frequency, spatial, analysis of using sinusoids, 172–84
front image plane, 395
functional models, matching, 513–14
functional properties, GRUFF, 514–15
functions, odd and even, 176
fuzzy classification, 124

games, Decathlete, 255–56
Gamma correction, 131
gates, AND, OR, and NOT, 121, 125
Gaussian filter, 136–37
definition, 137
Gaussian filtering, LOG edge detection and, 149–57
Gaussian function, definition, 149
Gaussian noise, 136–37, 315
Gaussian smoothing, 156
masks, 144
Gaussians, useful properties of, 151
gear_body, 68, 71
generalized-cylinder models, 483–84
Generic Object Recognition Using Form and Function (GRUFF) system, 513–16, 517
geometric distortion, 27
geometric hashing, 346–50
geometric icons, 504–6
geometric models, matching via alignment, 496–504
geometry, used in Tsai calibration method, 446–47
geons, 504–6
GIF (Graphics Interchange Format), 38
gradient, texture, 385–87

Graphics Interchange Format (GIF), 38
graph-matching algorithms, 106
graph-partitioning clustering technique, Shi's, 286–89
graphs: aspect, 488–89
  region adjacency, 81–82
  region adjacency, definition, 81–82
gray-level mapping, 130–34
gray-scale image(s): definition, 30
  thresholding, 83–89
grayval/binsize, 85
Greedy Exchange algorithm, 267, 268, 270
grids, 437–39
group homogeneity, 86–88
GRUFF (Generic Object Recognition Using Form and Function) system, 513–16, 517
  category hierarchy, 515–16
  definition tree, 515
  functional properties, 514–15
  knowledge primitives, 514
  processing by, 517

Hamming distances, 560–61
haptic sense, definition, 540
hard constraints, 490
hash function, 244
hash indexes, 244–45
hash table(s), 244
  in relational indexing, 508
hashing, geometric, 346–50
HCI issues, in virtual reality (VR) systems, 546
head-mounted displays (HMDs), 530, 535–37
heuristics, for detection of zoom, 276
hexacone, 194, 195
hidden units, 124
high contrast, detecting, 141–43
high dimensionality, 316
highlight, definition, 206
histogram(s): clustering methods based on, 284–86
  color, 199–201, 231, 233
  comparing, 274, 276
  definition, 84
  equalization, 132–34
  mode seeking, 284–85
  Ohlander's recursive histogram based clustering technique, 285–86, 287
  shape, 236–37
  texture, 235
  using for threshold selection, 83–85
HMDs (head-mounted displays), 530, 535–37
hole_mask, 68, 70
hole_ring, 68, 69, 70
holes, counting, 4–6
homogeneous coordinates, definition, 329
Hough transform: algorithm, 306
  Burns line finder using principles of, 311
  for detecting lines and circular arcs, 303–12
  encoding gradient direction with, 318
  extensions, 310
  finding circles with, 309–10
  generalized, 310
HSI (hue-saturation-intensity), 194–97
  encoding, conversion from RGB (red-blue-green) encoding, 196
HSV (hue-saturation-value) system, 194
hue-saturation-intensity (HSI), 194–97
  encoding, conversion from RGB (red-blue-green) encoding, 196
human body, 3D models, 485
human edge detection, 153–55
human heart, modeling motion of, 494–95
human perception: color, 209
  depth, 394
  shading used in, 208–9
hyperplanes, 122

IBM, 226–27
identification: of humans via iris of eye, 554–61
  requirements for identification systems, 555–57
identifying regions: classical clustering algorithms, 281–82
  clustering methods, 281
  in image segmentation, 280–91
  region growing, 289–91
IDFT (inverse discrete Fourier transform), 180–81
illuminated objects, sensing, 189
illumination boundary, definition, 374
image addition, 12
image-based rendering, definition, 543
image collections, 227–28
image data, 36
image databases, 3–4, 226–30
  queries, 228–30
image distance measures, 230–44
image energy, 491–92

image enhancement: convolution and cross correlation, 167–72
definition, 130
detecting edges using differencing masks, 141–49
Gaussian filtering and log edge detection, 149–57
gray-level mapping, 130–32
histogram equalization, 132–34
image smoothing, 136–37
median filtering, 137–41
removal of small image regions, 134–35
image file formats: comparison of, 40
GIF (Graphics Interchange Format), 38
TIFF (Tag Image File Format) 38
image file header, 36
image flow: computing, 262–63
definition, 255
equation, 263–64
solving for by propagating constraints, 264–65
image formation, 24–26
image histograms, 83–85
image operations, 10–14
image plane, front, 395
image processing: 128–86
definition, 15–16
Fourier basis used for, 175
image quantization, spatial measurement and, 31–35
image registration, definition, 327
image representation, imaging and, 21–50
image restoration, definition, 130
image segmentation, 279–325
identifying regions, 280–91
image subtraction, 12, 253–54
image understanding, definition, 15–16
image warping, 12
imagery, real and synthetic, 542–45
image(s): acquisition of, 461–63
analog, definition, 29
basis, computing, 519–21
basis, for set of training images, 518–19
binary, analyzing, 51–91
binary, definition, 30
color, 45–46
computing features from, 13
computing output from input, 139–41
content-based, indexing for retrieval with multiple distance measures, 248
counting objects in, 54–56
digital, definition, 29
digital, formats, 35–40
digital, picture functions and, 29–35
digital, problems with; 27–29
enhancing, 11–12
filtering and enhancing, 128–86
gray-scale, definition, 30
gray-scale, thresholding, 83–89
improving, 129–30
intrinsic, 371–76
labeled, 292
labeled, definition, 30
masks applied to, 53–54
matching in 2D, 326–70
multiple, 12
multispectral, 45–46, 210
multispectral, definition, 30
perceiving 3D from 2D, 371–409
pseudo-colored, 30
range, 47–49
raw, 35
removing small regions from, 134–35
retrieving content-based, 226–50
run-coded binary, 37
smoothing, 136-37
thematic, 30, 210
tracking edges of binary edge image, 303
training, basis images for set of, 518–19
2D, 3D structure from, 42
2D, difference operators for, 144–49
2D, motion from sequences of, 251–78
3D cues in 2D images, 383–88
types of, 29–31. *See also* perspective imaging models
three-dimensional (3D) images
two-dimensional (2D) images
imaging: devices, 22–27
image representation and, 21–50. *See also* perspective imaging models
independent test data, definition, 96
indexes: B+−tree, 245–47
hash, 244–45
K-d tree, 247
R-tree, 247–48
spatial, 247–48
standard, 244–47
indexing: for content-based image retrieval with multiple distance measures, 248
relational, 363–64, 508, 510, 511. *See also* RIO object recognition system
input images, computing from output images, 139–41
inspection, crossbars, 4–6

integrated tracking, 271–72
integrating, spatial constraints, 472
intensity, 393
  mapping, 131
  values, 46. *See also* HSI (hue-saturation-intensity)
interactive segmentation of anatomical structure, 529
interest operators, 257–58
interesting points, 256–61
internal corners, 4–6
interposition, 42
  definition, 384
interpretation trees (IT), 352–54
  definition, 352
  line drawing, 380
intrinsic camera parameters, 445
intrinsic images, 371–76
  scene values, 375
invariant features, 14
inverse discrete Fourier transform (IDFT), 180–81
inverse perspective, 439
iris-scanning system, 554–61
  hardware components, 557–58
  performance, 560–61
  representation in, 558–60
  system design, 557–60
isodata clustering, 282–84
IT (interpretation trees), 352–54
  definition, 352
  line drawing, 380

Jacobian matrix, 441–42
Joint Photographic Experts Group (JPEG):
  format, 38–39
  Motion, 39–40
JPEG (Joint Photographic Experts Group):
  format, 38–39
  Motion, 39–40
jump edge, 373
  definition, 374
junction pixels, 301
junctions, 377
  types of, 377–78

K-d tree indexes, 247
K-means clustering, iterative, 282
keywords, 228–29
knowledge-based thresholding, 89
knowledge-directed thresholding, 285
knowledge primitives, GRUFF, 514

L-junctions, 377
label, definition, 351
LABEL field, 62–63
labeled image(s), 292
  definition, 30
labeling: block edges via discrete relaxation, 382
  connected components, 56–63
  connected components, using run-length encoding for, 62–63
  consistent, 351–53
  continuous relaxation, 356–59
  cubes, 466–67
  discrete relaxation, 354–56, 357
  edges of blocks via backtracking, 381
  line drawings of blocks, 377–83
  lines via relaxation, 381–83
  terms, 377
labeling algorithms: recursive, 57–59
  row-by-row, 59
labels function, 61
lake_num, 112
lakes, 112
Lambertian reflectance model, 469–71
Lambertian reflection, 204–5
laser light projectors, 438
Laws texture energy measures, 220–22, 224
layout, color, 232–33
leaf nodes, 107, 245–47
  quadtree, 294. *See also* nodes
learning: machine, 119
  supervised, 119
  supervised, on Veggie Vision, 553–54
  unsupervised, 119
least-squares: error criteria, definition, 313
  method, 312–14
  problem, defining, 431–36
lenses, 24. *See also* thin lens equation
LIDAR (light detection and range) devices, 47–48
lid_bottom_of_image, 112
lid_num, 112
lid_rightof_bay, 112
lids, 104, 112
light: ambient, 207–8
  ambient, definition, 208
  darkening with distance, 206–7
  diffuse reflection of, 204–5
  radiation from one source of, 203–4
  sensing, 21–22, 189
  specular reflection of, 205–6
  structured, 437–39

light: ambient (*continued*)
use of, 41–42
white, definition, 189
light detection and range (LIDAR) devices, 47–48
limb, 373
definition, 374
line drawings: interpretation tree for, 380
labeling drawings of blocks, 377–83
linear optimization, constrained, 456–57
linear transformations, scaling, 329–30
lines: Burns line finder, 311–12
detecting with Hough transform, 303–12
epipolar, 402
fitting, 312–14
labeling via relaxation, 381–83
straight, finding, 304–9
virtual, 393–94
linker, Canny edge detector and, 297–301
local binary partition, 217
local-feature-focus method, of object recognition, 341–44
location of model point and image point, 335–38
LOG edges, Gaussian filtering and detection of, 149–57
LOG filtering, Marr-Hildreth theory, 155–57
LOG filters, detecting edges with, 151–53
looming, 393
lossless compression, definition, 36
lossy compression, definition, 36
low-level features, detection of, 129–30
luminance, 197

Mach band effect, artificial neural network (ANN) used to produce, 153–55
machine learning, 119
machine vision, definition, 1
Magic Value, 37
magnetic resonance angiography (MRA), 47
magnetic resonance imaging (MRI), 6–7, 47, 210
mapping: affine, 2D object recognition via, 341–50
functions, affine, 329–39
gray-level, 130–34
polynomial, 367
texture, 542–45
mark, definition, 374
Marr-Hildreth theory, 155–57
mask(s), 134
applying to images, 53–54
box smoothing, 144
derivative, 141–44
differencing, detecting edges using, 141–49
Gaussian smoothing, 144, 151, 152
for implementation of LOG filter, 151, 152–53
as matched filters, 158–67
operations defined via, 167–68
origins, 54
Prewitt, 146, 148–49, 307
properties of derivative and smoothing, 143–44
Roberts, 146–47
Sobel, 146, 147
matching: boundary, 237–38
elastic, 240
functional models, 513–14
geometric models via alignment, 496–504
relational distance, 359–63
relational, 2D object recognition via, 350–64
relational models, 504–13
sketch, 238–40
symbolic, and relational constraints, 401
in 2D, 326–70
3D models and, 479–526. *See also* alignment
mathematical morphology, 63
matrix: best affine calibration, 431–37
co-occurrence, 217–20
perspective transformation, 423–26
MaxCol, 56
max-error criteria, definition, 313
maximum intensity projection (MIP), 47
MaxRow, 56
MDFs (most discriminating features), 522
mean radial distance, 75
measurement, spatial, image quantization and, 31–35
measure(s): color similarity, 231–33
distance, indexing for content-based image retrieval with multiple, 248
image distance, 230–44
object presence and relational similarity, 240–44
measuring: shape similarity, 235–40
texture, 215–23
texture similarity, 233–35

median: definition, 138
filtering, 137–41
MEFs (most expressive features), 521–22
memory, faster search of, 521–23
mesh: balloon models for 3D, 493–94
models, 472, 480, 481
regular, 480
triangular, 480
method of least squares, 312–14
microdensitometer, 45
MIP (maximum intensity projection), 47
mixed reality, 530
definition, 535
models: active contour, 489–92
balloon for 3D, 493–94
fitting to segments, 312–17
generalized-cylinder, 483–84
matching functional, 513–14
matching geometric via alignment, 496–504
mesh, 472, 480, 481
parametric, 116–17
perceptron, 120–23
perspective imaging, 395–97
physics-based and deformable, 489–95
relational, matching, 504–13
surface-edge-vertex, 480–83
3D, and matching, 479–526
3D relational, 504–6
true 3D versus view-class, 488–89
2D-3D alignment, 498–501
view-class relational, 506–13
wire-frame, 480. *See also* three dimensional (3D) models
moment. *See* second moment; second-order
morphology, binary image, 63–73
most discriminating features (MDFs), 522
most expressive features (MEFs), 521–22
motion: aggregating motion trajectories, 321–24
coherence, segmentation using, 321–24
computing paths of moving points, 265–72
modeling of human heart, 494–95
phenomena, 393
phenomena and applications, 251–53
structure perceived from, 472–75
from 2D image sequences, 251–78
in virtual reality (VR) systems, 540
motion field: definition, 254–55
point correspondences used to compute, 256–71
Motion Joint Photographic Experts Group (JPEG) format, 39–40
Motion JPEG (Joint Photographic Experts Group) format, 39–40
motion parallax, 386
definition, 387
Motion Picture Experts Group (MPEG): compression of video, 261–62
format for video, 39–40
motion vectors: computing, 254–65
deriving for interesting points, 260
MPEG (Motion Picture Experts Group): compression of video, 261–62
format for video, 39–40
MRA (magnetic resonance angiography), 47
MRI (magnetic resonance imaging), 6–7, 47, 210
multidimensional data, decisions using, 117–19
multilayer feedforward network, 123–26
multiple images, combining, 12
multispectral image(s), 45–46, 210
definition, 30

nearest-neighbor rule, 103
nearest neighbors, used in classification, 103–4
Necker Cube/Phenomena, 382
neighborhoods, pixels and, 51–52
neighbors: nearest, used in classification, 103–4
utility function, 57–58
neural nets, artificial, 119–26
neurons, 119–120
artificial, 120–22
nodes, 107, 245–47
octree, 484–85
quadtree, 294
noise, Gaussian, 136–37, 315
nominal resolution, definition, 31
non-accidental alignments, 394
non-iconic representations, extracting, 14
nonlinear optimization, 316
nonlinear warping, 364–68
nonmaximum suppression, 299
normalized dot product, 161–62
normalized RGB coordinates, 192–93
notation(s), 29–31
pixel values, 23

object coordinate frame O, 44
object counting, 54–56

object pose computation, 3D sensing and, 410–78
object presence, and relational similarity measures, 240–44
object recognition, 335–38
 2D, via affine mapping, 341–50
 2D, via relational matching, 350–64
 3D, classifying, 495–96
 3D, paradigms, 495–523
 of 3D objects by appearance, 516–23
 3D object recognition paradigms, 495–523
 by appearance, 516–23
 eigenspace recognition by appearance, 522
 Generic Object Recognition Using Form and Function (GRUFF) system, 513–16, 517
 geometric hashing method, 346–50
 local-feature-focus method, 341–44
 RIO system, 506–13
 TRIBORS system, 499–501. *See also* recognition
object reconstruction, 3D, 460–68
occlusion, 383–84
octree(s), 467, 484–86
odd functions, 176
offline preprocessing, 348, 349, 508
O'Gorman and Clowes algorithm, 305, 306, 308
Ohlander's recursive histogram-based clustering technique, 285–86, 287
one-dimensional (1D) signals, differencing, 141–44
online recognition, 348, 350
opening of binary images, 65–66
 definition, 67
operations, defining via masks, 167–69
operator(s): Canny, 157–58
 difference, 144–49
 interest, 257–58
 Prewitt, 146, 148–49
 Roberts cross, 146–47
 Sobel, 146, 147
optimization: constrained linear, 456–57
 nonlinear, 316
 and verification of pose, 460
ordering constraint, 403
organizing databases, 244–48
origins, mask, 54
orthogonal basis, using, 160–62
orthogonal transforms, definition, 331–32
orthographic projection(s), 426–28, 470
orthonormal transforms, definition, 331–32
Otsu method, automatic thresholding, 85–89
outliers, 315, 316
output images, computing from input images, 139–41
overlays, 292

page description language (PDL), 39
panning, 274
paradigms, 3D object recognition, 495–523
parallax, motion, 386, 387
parallel implementation 172
parallel list (PTLIST) array, 304, 305
parameters: camera model, 436–37
 closed form solutions for, 314–15
 extrinsic camera, 445–49
 intrinsic camera, 445
parametric models, for distribution, 116–17
PARENT arrays, 59–61
part, definition, 351
pattern recognition: concepts, 92–127
 problems, 92–93
PBM (Portable Bit Map), 37–38
PDL (page description language), 39
perception: human color, 209–10
 shading used in, 208–9
 of structure from motion, 472–75
perceptron model, 120–23
perimeter length, 74
perspective: imaging model, 395–97
 inverse, 439
 projections, 426–28
 transformation matrix, 423–26
perspective scaling, 385, 386
 definition, 384
Perspective 3 Point Problem (P3P), 439–44
PGM (Portable Gray Map), 37–38
Phong model of shading, 208
photography, model, 22
photometric stereo, 471–72
physics-based models, deformable and, 489–95
pick-and-click distance, 234–35
picture function(s): 2D, 175–79
 definition, 30
 digital images and, 29–35
pixel arrays, 43–45
pixel coordinate frame I, 43–44
pixel values, notations, 23
pixels: background, 51
 changing values of, 10–11
 definition, 3
 foreground, 51
 junction, 301

neighborhoods and, 51–52.
See also external corners; internal corners
planes, front image, 395
plates, 504–6
point correspondences, computing motion field with, 256–61
point operator, definition, 131
points: computing 3D points using multiple cameras, 428–30
contributing, 498
control, 332–34
location of, 335–38
pose from 2D-3D point correspondences, 455–56
representation of 2D, 328–29.
See also Perspective 3 Point Problem (P3P)
polygonal approximation, 293
polygons, Voronoi, 214–15
polynomial mappings, 367
Portable Bit Map (PBM), 37–38
Portable Gray Map (PGM), 37–38
pose: clustering, 344–46
definition, 344
estimation, 453–60
estimation procedure, 439–44
object pose computation and 3D sensing, 410–78
from 2D-3D point correspondences, 455–56
verification and optimization of, 460
in virtual reality (VR) systems, 539–40
PostScript, 39
power spectrum, 177–79
measuring texture by autocorrelation and, 221–23
precision: definition, 97
recall versus, 97–98
preprocessing, offline, 348, 349, 508
Prewitt, Judith, 146, 148
Prewitt masks, 307
Prewitt operator, 146
primary key, 244
primitives, knowledge, 514
prior_neighbors function, 61
probability distribution, 114–15
projection(s): orthographic, 426–28, 470
weak perspective, 426–28
projectors: calibration of, 437
laser light, 438
replacing camera with, 412–13
property tables, regions represented by, 294–95
proximity primitive, 514
pseudo color, 210
pseudo-colored images, 30
P3P (Perspective 3 Point Problem), 439–44
solution, 442–43
PTLIST (parallel list) array, 304, 305
pyramids, interpretation tree for line drawings of, 380

QBE (query-by-example), 229–30
QBIC (Query by Image Content) database, 226–27
quadtrees, 247–48
regions represented by, 294
quantization: effects, 29
image, spatial measurement and, 31–35
special quantization effects, 33
queries, image database, 228–30
query-by-example (QBE), 229–30
Query by Image Content (QBIC) database, 226–27
quicksort, modifying, 139
R-tree indexes, 247–48
radial distance: mean, 75
standard deviation of, 75
radial distortion, rectifying, 366–67
radiation, from one light source, 203–4
RAG (region adjacency graphs), 81–82
definition, 81–82
ramp, 33
range: data, 463–64, 465
images, 47–49
scanners, 47–49
raster order, 35
raster-oriented coordinate systems, 30
raw images, 35
ray tracing, 541–42
real image coordinate frame F, 44
real imagery, composing, 542–45
recall: definition, 98
precision versus, 97–98
receiver operating curve (ROC), 97
receptors, sensitivity of, 190–91
recognition: by alignment, 337
character, 98–100
eigenspace, by appearance, 522
online, 348, 350
structural pattern, 105
structural techniques, 104–6

recognition: by alignment (*continued*)
in Veggie Vision, 550–51. *See also* object recognition
recognition-by-alignment, definition, 337
reconstruction: 3D object, 460–68
surface, 464
recursive labeling algorithm, 57–59
red-green-blue (RGB): basis for color, 191–93
conversion to YUV, 197
encoding, conversion to HSI (hue-saturation-intensity) encoding, 196
reference frames, 42–45, 328–29
reflectance, Lambertian, 469–71
reflection(s), 338–39
diffuse, 204–5
specular, 205–6
specular, definition, 206
region adjacency graphs (RAG), 81–82
definition, 81–82
region properties, 73–81
regions, 377
boundary coding representing, 292–93
corners, 320–21
future, 297
growing, 289–91
identifying by contours, 295–312
identifying image segmentation, 280–91
labeled images representing, 292
labeled, finding borders of, 295–97
overlays representing, 292
past, 297
property tables representing, 294–95
quadtrees representing, 294
representing, 291–95
ribbons, 317–20
tracking existing boundaries of, 295–97
registration: image, 327
of views, 463–64
reject class, definition, 94
relation, definition, 351
relational constraints, symbolic matching and, 401
relational description, definition, 359
relational distance, matching, 359–63
relational indexing, 363–64, 508, 510, 511. *See also* RIO object recognition system
relational matching, 2D object recognition via, 350–64
relational models: matching, 504–13
view-class, 506–13
relational similarity measures, object presence and, 240–44
relative orientation primitive, 514
relaxation: continuous, 356–59
discrete, 354–56, 357
discrete, labeling block edges via, 382
labeling lines via, 381–83
removing: salt-and-pepper noise, 134–35
small components, 135–36
small regions from images, 134–35
rendering: 3D models, 540–42
definition, 540
image-based, 543
representation: of 3D models, 480–87
feature vector, 100
features used for, 98–100
in iris-scanning system, 558–60
mesh models, 472, 480, 481
surface-edge-vertex models, 480
in Veggie Vision, 550–51
resolution: definition, 31
nominal, definition, 31
relating to blur, 406
subpixel, definition, 31
resolving power, definition, 406
restoration, image, definition, 130
retrieval: of content-based images, 226–50
image, indexing for content-based with multiple distance measures, 248
problems, 3–4
RGB (red-blue-green): basis for color, 191–93
encoding, conversion to HSI (hue-saturation-intensity) encoding, 196
conversion to YUV, 197
ribbon(s), 317–20, 484
definition, 318
detecting straight, 319–20
rigid transformations, 331–32
RIO object recognition system, 506–13
features employed by, 507–8, 509
RMSE (root-mean-square error), definition, 313
Roberts basis, 162–63
Roberts masks, 146–47
robots, vision-guided, 9–10
ROC (receiver operating curve), 97
root-mean-square error (RMSE), definition, 313
rotation: arbitrary, 418–19

parameters for camera position, 445–49
2D, 330–31, 332–34
3D, 415–18
row-by-row labeling algorithm, 59
run-coded binary images, 37
run-length encoding, using for connected components labeling, 62–63
run-of-signs test, 315

salt-and-pepper noise, removing, 134–35
sampling_ring_spacer, 68, 71
sampling_ring_width, 68, 71
satellite images, 8–9
saturation. *See* HSI (hue-saturation-intensity)
scaled Euclidean distance, definition, 103
scaling: perspective, 385, 386
2D, 329–30, 332–34
3D, 415
scanners, range, 47–49
scattering, 27
scene change, definition, 272
SE (synthetic environment), 535. *See also* visual environment (VE)
searching, faster, 521–22
second moment: about axis, 80
axis with least, 81
second-order: column moment, 77
mixed moment, 77
row moment, 77
segmentation: color, 201–2, 322–24
image, 279–325
using motion coherence, 321–24
texture, 223–24
segmenting: curves via fitting, 317
video sequences, 273–74
segments: finding straight line, 304–9
models fitted to, 312–17
self-occluding surface, 373
sensing: illuminated objects, 189
light, 21–22
sensing devices, virtual reality (VR) systems, 539–40
sensor/transducer, 94
sensors, 45–49
LIDAR (light detection and range), 47–48
multispectral, 45–46
SFS (shape from shading), 388
shading, 187–211, 203–9
computing shape from, 468–72
human perception using, 208–9
interpreting shape from, 388
Phong model, 208
shadows, 393
shape(s): computing from shading, 468–72
histograms, 236–37
interpreting from boundaries, 391–92
interpreting from shading, 388
interpreting from texture, 388–91
similarity measures, 235–40
used in Veggie Vision, 553
shape-from-shading, definition, 469
shear, 338
Shi's graph-partitioning clustering technique, 286–89
shift theorem, 183–84
shot change, definition, 272
signal level, 516
signals: differencing 1D, 141–44
representing as combination of basis signals, 160–61
television, YIQ and YUV for, 197–98
similarity: color, 231–33
relational, 240–44
shape, 235–40
texture, 233–35
sinusoids, analysis of spatial frequency using, 172–84
size, used in Veggie Vision, 553
sketch matching, 238–40
slant, definition, 389
small components, removing, 135–36
small image regions, removing, 134–35
smooth object alignment, 501–4
smoothing: Gaussian, 156
image, 136–37
smoothing masks, 144, 167–69
properties of, 144
snakes, 489–92
Sobel masks, 146, 147
sonification, definition, 539
space-carving, 464–67
spatial constraints, integrating, 472
spatial frequency, analysis of using sinusoids, 172–84
spatial indexing, 247–48
spatial measurement, image quantization and, 31–35
spatial quantization effects, 33
spatial relationships, 242–44
spatio-temporal gradient magnitude, 321
specular reflection, 205–6
definition, 206
spin images, 498, 499, 500
stability primitive, 514

standard deviation, 102
of radial distance, 75
standard indexes, 244–47
statistical interpretation of error, 315–16
stereo: acquisition system, 461–63
configuration, 411–13
depth perception from, 397–403
displays, 399–400
photometric, 471–72
vision, establishing correspondences in, 400–3
stereoscopic display devices, 538–39
sticks, 504–6
stiffness, 494–95
still photos, JPEG (Joint Photographic Experts Group) format, 38–39
storing video sequences, 277
straight lines, finding segments of, 304–9
straight ribbons, detecting, 319–20
stretching, 130–31
contrast, definition, 132
strobe light, use of, 41–42
strongback, 492
structural pattern recognition, 105
structural techniques, recognition, 104–6
structured light, using, 42, 437–39
structure(s): corners, 320–21
identifying higher-level, 317–21
perceiving from motion, 472–75
ribbons, 317–20
3D, from 2D images, 42
union-find, 59–60
structuring elements, binary morphology, 63–65, 68–71
subpixel resolution, definition, 31
subtraction, image, 12, 253–54
superquadrics, 486–87
surface-edge-vertex models, 480–83
surface reconstruction, 464
surfaces, self-occluding, 373
surveillance, 253
symbolic matching, relational constraints and, 401
synthetic environment (SE), definition, 535. *See also* visual environment (VE)
synthetic imagery, composing, 542–45
system error, evaluating, 96

T-junctions, 377, 384
Tag Image File Format (TIFF), 38
teleoperation, 533–35
definition, 535
television signals, YIQ and YUV for, 197–98
temporal redundancy, 40
tests, run-of-signs, 315
tetrahedral elements, 494–95
texels, 213
texture described based on, 214–15
text applications, 8
texture, 212–25
description vector, 233–35
energy, 220–21, 224
histograms, 235
interpreting shape from, 388–91
measuring by autocorrelation and power spectrum, 221–23
measuring by binary partition, 217
measuring by co-occurrence matrices and features, 217–20
measuring by edge density and direction, 215–17
measuring by texture energy, 220–21, 224
quantitative measures of, 215–23
segmentation, 223–24
similarity measures, 233–35
statistical approach, 214
structural approach, 213
texel-based descriptions of, 214–15
used in Veggie Vision, 552–53
texture gradient, 42, 385–87
definition, 387
texture mapping, 542–45
definition, 542
texturing, view-based, 543–45
thematic images, 30, 210
theorems: convolution, 182–83
shift, 183
thin lens equation, 403–6
three-dimensional (3D) cues, in 2D images, 383–88
three-dimensional (3D) images: interpreting from boundaries, 391–92
interpreting from shading, 388
interpreting from texture, 388–91
interpreting from vanishing points, 392
labeling line drawings used to portray, 377–83
perceiving from 2D images, 371–409
three-dimensional (3D) models: alignment, 3D-3D, 496–98
alignment, 2D-3D, 498–501
balloon, 493–94
generalized-cylinder, 483–84

human body, 485
human heart, 494–95
matching and, 479–526
mesh, 472, 480, 481
octrees, 484–86
physics-based and deformable, 489–95
relational, 504–6
rendering, 540–42
representation methods, 480–87
superquadrics, 486–87
surface-edge-vertex, 480–83
true versus view-class models, 488–89. *See also* models
three-dimensional (3D) objects: recognition by appearance, 516–23
reconstruction, 460–68
RIO object recognition system, 506–13
three-dimensional (3D) points, computing using multiple cameras, 428–30
three-dimensional (3D) sensing, object pose computation and, 410–78
three-dimensional (3D) structure, from 2D images, 42
three-dimensional (3D)-3D alignment, 496–98
three dimensions (3D): affine transformations, 413–21
classifying 3D object recognition, 495–96
object recognition paradigms, 495–523
pose from 2D-3D point correspondences, 455–56
threshold: above, 83
below, 83
inside, 83
outside, 83
threshold values, 24
thresholding: automatic, 85–89
dynamic, 89
gray-scale images, 83–89
knowledge-based, 89
knowledge-directed, 285
TIF format. *See* TIFF
TIFF (Tag Image File Format), 38
tilt, definition, 389
tip_spacing, 68, 71
tracking, integrated, 271–72
training images, basis images for, 518–19
trajectories: aggregating motion trajectories, 321–24
computing, 265–72
trajectory of i, definition, 267
transformation(s): 2D, 327
3D affine, 338–41, 413–21
alignment via transformation calculus, 419–21
computing Tr = {RT}, 458–49
linear, 329–30
from model features to image features using local-feature-focus method, 343
perspective transformation matrix, 423–26
using pose clustering, 344
rigid, 331–32
transform(s): Fourier, 177, 179–81
orthogonal and orthormal, 331–32
translation: of binary images, definition, 66
parameters for camera position, 445–49
2D, 332–34
3D, 415
tree indexes: B+, 245–47
K-d, 247
R-, 247–48. *See also* triangle-tree
trees. *See* binary decision trees; decision trees; definition trees; interpretation trees; octrees; quadtrees; tree indexes; triangle-tree
triangle-tree, 248
triangles, aligning, 419–21
triangulation, 48
TRIBORS object recognition system, 499–501
trichromatic encoding, 191–93
triplets, 499–501
true 3D models, versus view-class models, 488–89
Tsai calibration method, 444–53
two-class problems, 96–97
two-dimensional (2D) images, 21
difference operators for, 144–49
motion from sequences of, 251–78
perceiving 3D images from, 371–409
3D cues in, 383–88
3D structure from, 42
types of, 29–31
two-dimensional (2D) models, 2D-3D alignment, 498–501
two-dimensional (2D) object recognition via relational matching, 350–64
two-dimensional (2D) picture functions, 175–79
two-dimensional (2D) transformation, definition, 327
two dimensions (2D): matching in, 326–70
pose from 2-D and 3D point correspondences, 455–56
registration of data, 326–28

union-find algorithms, 59
union-find structure, 59–60
union procedure, 59–60

vanishing point(s), 42, 392
VE (visual environment), 535, 538, 539
vector space, 160, 162
  of all signals, 158–60
  definitions, 159
vector(s): feature, 100
  motion, 254–65, 321–24
  motion, deriving for interesting points, 260
  texture description, 233–35
Veggie Vision, 548–54
  application domain and requirements, 549–50
  computing features, 552–53
  hardware components, 550
  identification procedure, 551
  obtaining images of produce, 551–52
  performance, 554
  representation and recognition, 550–51
  supervised learning on, 553–54
  system design, 550–51
verification: definition, 93
  and optimization of pose, 460
vertex, surface-edge-vertex models, 480–83
video: cameras, 26
  detecting significant changes in, 272–77
  MPEG (Motion Picture Experts Group) compression of, 261–62
  MPEG format for, 39–40
  segmenting sequences, 273–74
  storing sequences of, 277
view-based texturing, 543–45
view-class models: relational, 506–13
  versus true 3D models, 488–89
view classes, 488
virtual lines: cues from, 393–94
  definition, 394
virtual reality: definition, 535
  devices, 535–39
  fishtank, 539
virtual reality (VR) systems, 527–47
  applications, 529–30
  architectural walkthrough, 529
  augmented reality, 530–32
  dextrous virtual work, 537–38
  features, 528–29
  flight simulation, 529
  haptic sense and, 540
  HCI and psychological issues, 546
  head-mounted displays (HMDs), 530, 535–37
  interactive segmentation of anatomical structure, 529
  motion in, 540
  sensing devices, 539–40
  stereoscopic display devices, 538–39
  teleoperation, 533–35
  visual output, 539
vision, stereo, 397–403
visual environment (VE), 535, 538, 539
visual event, 489
visual output, virtual reality (VR) systems, 539
Voronoi polygons, 214–15
voxels, 485
VR (virtual reality) systems, 527–47

warp, affine, 334–35
warping: images, 12
  nonlinear, 364–68
wavelets, 182
weak perspective projections, 426–28
weights, 54
white light, definition, 189
wipe, definition, 272–73
wire-frame models, 480
within-group variance, 86–88
world coordinate frame W, 44
wrap-around, 28

X-ray devices, 46–47

YIQ, encoding for television signals, 197–98
YUV: conversion from RGB (red-blue-green) to, 197
  encoding for television signals, 197–98

z-buffer, 542
zero crossings, 143, 144, 154, 155
zoom, camera, definition, 272
zooming, 254–55, 274–75